Graduate Texts in Mathematics

Graduate Texts in Mathematics bridge the gap between passive study and creative understanding, offering graduate-level introductions to advanced topics in mathematics. The volumes are carefully written as teaching aids and highlight characteristic features of the theory. Although these books are frequently used as textbooks in graduate courses, they are also suitable for individual study.

Manfred Einsiedler · Thomas Ward

Unitary Representations and Unitary Duals

Manfred Einsiedler
Department of Mathematics
ETH
Zürich, Switzerland

Thomas Ward
Durham University
Durham, UK

ISSN 0072-5285 ISSN 2197-5612 (electronic)
Graduate Texts in Mathematics
ISBN 978-3-032-03898-2 ISBN 978-3-032-03899-9 (eBook)
https://doi.org/10.1007/978-3-032-03899-9

Mathematics Subject Classification: 43A65, 43A90, 22D10, 22D40, 22E46, 22D35

This Springer imprint is published by the registered company Springer Nature Switzerland AG
The registered company address is: Gewerbestrasse 11, 6330 Cham, Switzerland

Preface

The review by Kunze [59] of Mackey's book on the theory of unitary group representations [67] may help to justify the highly selective choice of material we present, as it begins:

> "It is probably impossible to write a comprehensive book on the theory of unitary representations. The subject, which logically begins in a modest way with complex representations of finite groups, proceeds to general compact groups, and goes on to treat a variety of noncompact groups, is simply too vast."

We make no attempt to be comprehensive and refer, for example, to the monographs of Fulton and Harris [32, Part I], Curtis and Reiner [20], or Kowalski [57] for the representation theory of finite groups, a large and important subject in itself. We also do not consider semi-simple Lie groups beyond the examples $\mathrm{SU}_2(\mathbb{R})$, $\mathrm{SO}_3(\mathbb{R})$, $\mathrm{SL}_2(\mathbb{R})$, and $\mathrm{SL}_3(\mathbb{R})$. For the theory of finite-dimensional representations of other semi-simple Lie groups, we refer to Fulton and Harris [32, Part II–IV], and for the theory of unitary representations of these groups we refer to Knapp [55], but suggest that the uninitiated start with this book.

Given that we do not treat these other topics, it is reasonable to ask what this book is about. It started as a project to understand the so-called property (τ) in number theory.(1) As this developed, and as we accumulated material either directly required or useful for that objective, two things became clear. First, to meaningfully reach property (τ) from our chosen starting point in a single volume of manageable size, we would have to maintain a narrow focus on that target. Second, along the way we joined a long line of mathematicians in discovering that the theory of unitary representations is rich and beautiful, and found that narrow focus to be restricting ourselves to a bread roll and water in a fine restaurant with an extensive menu. As a result, we abandoned *pro tem* the project of presenting a proof of property (τ), and instead settled on introducing the theory of unitary representations, in a way that suits our wider interests.

Some readers may be aware of our other projects [22, 24, 25, 26]. As a result it should not come as a surprise that those wider interests come from the role played by unitary representations in ergodic theory, dynamical systems, homogeneous dynamics, and number theory. In particular, this makes us favour non-compact groups, although we do cover the basic theory for representation of compact groups. We also discuss in detail the representation theory of abelian groups and Pontryagin duality. These topics are useful in ergodic theory as well as being crucial for the general theory of unitary representations. As we will see later, the connections between ergodic theory and unitary representations go in both directions. In fact for some groups with normal abelian subgroups all irreducible unitary representations can be obtained from ergodic constructions, and in some cases this link shows that it is impossible to classify all irreducible unitary representations.

However, as mentioned our original goal was to discuss property (τ), which relates to a concrete estimate for the decay of certain functions (matrix coefficients) on a group. It turns out that proving this estimate is significantly easier for the group $\mathrm{SL}_3(\mathbb{R})$ than it is for $\mathrm{SL}_2(\mathbb{R})$. Moreover, $\mathrm{SL}_3(\mathbb{R})$ has property (T) and the desired estimate is a general fact that does not require refined knowledge about the unitary representations in question. In contrast, for the group $\mathrm{SL}_2(\mathbb{R})$ the desired estimate is not a general phenomenon. In fact, proving the decay estimates for $\mathrm{SL}_2(\mathbb{R})$ and specific representations requires more knowledge. As a result we will discuss $\mathrm{SL}_3(\mathbb{R})$ (without classifying its irreducible unitary representations) before $\mathrm{SL}_2(\mathbb{R})$, its irreducible representations, and the (lack of) decay estimates.

Even though the topics above dominate the later chapters we also sought to develop the theory from the ground up, building mostly on functional analysis. However, we do not believe in developing an abstract theory without informative examples. Hence much of the effort throughout has gone into understanding various examples. We believe it is precisely this mixture of theory, examples, and further theory required to understand other aspects of the examples that makes the theory of unitary representations so compelling.

Prerequisites

We will assume throughout that the reader is familiar with both the content of and the more standard notation from linear algebra, real analysis, complex analysis in one variable, and functional analysis including the spectral theorem for self-adjoint operators. For convenience we use the monograph [25] as a source for functional analysis, but of course the required material can be found in many places.

Notation and Conventions

The general notational conventions we adopt are outlined in Section 1.1.2, with some more specific notation introduced throughout the text listed in an index of notation on page 554. We use $\mathbb{N} = \{1, 2, \dots\}$, $\mathbb{N}_0 = \mathbb{N} \cup \{0\}$, and $\mathbb{Z}$ to denote the natural numbers, non-negative integers, and integers; $\mathbb{Q}$, $\mathbb{R}$, $\mathbb{C}$ denote the rational numbers, real numbers and complex numbers; $\mathbb{T} = \mathbb{R}/\mathbb{Z}$ denotes the additive circle. We will use $\mathbb{S}^1$ with two different meanings, as the subset $\{(x, y) \in \mathbb{R}^2 \mid x^2 + y^2 = 1\}$ of $\mathbb{R}^2$ and as the compact multiplicative circle group $\{z \in \mathbb{C} \mid |z| = 1\}$. Throughout, compact and locally compact spaces are implicitly assumed to be Hausdorff.

Organisation

There are 255 exercises in the text, 161 of them with hints in an appendix. All of these contribute to the reader's understanding of the material. A small number are essential to the development (of the ideas in the section or of later theories). These 50 are denoted 'Essential Exercise' to highlight their significance. In addition to footnotes of immediate relevance a small number of end notes deal with tangential or historical remarks. These are marked in the text with numbers in parentheses and assembled at page 545.

Unlike some other mathematical theories, the theory of unitary representations has a particularly broad logical structure. In other words, the story we wish to explain is not built like a single tall tower, in which each chapter requires the reader to have mastered every aspect of the previous ones. Instead, there are some interesting theorems in the last four chapters that could be understood even if a large part of the text has been skipped. Because of this, and in light of the length of the text, we have tried to mark sections that could be skipped without affecting the majority of the later chapters with an asterisk. However, there is no clear algorithm to decide whether a certain section should be so marked. Hence, contrary to our intention and depending on the precise objectives of the reader or lecturer, a marked section might be crucial and a non-marked one might be less relevant. At a more modest level there are several places where, for example, we have detoured in order to check the measurability of some construction. The more confident reader is encouraged to skip this kind of argument as soon as it becomes familiar.

To further help the reader, we indicate the main dependencies between the various chapters in a *Leitfaden* on page ix, and list a few possible goals with an indication of what should be covered to reach them.

- To understand unitary representations of abelian groups, cover Chapters 1 and 2.
- To see examples of infinite-dimensional irreducible representations and the classification of the unitary duals of some metabelian groups cover Chapter 1, most of Chapter 2, and Chapter 5. In particular, this contains the classification of irreducible unitary representations of the Heisenberg group. The latter is also a crucial input to the discussion of the Weil representation in Section 8.8.
- To understand the Fell topology and a few interesting examples in Chapter 6, one should first cover Chapter 1 and most of Chapters 2 and 5.
- To understand the representation theory of compact groups, cover Chapter 1 and then move straight on to Chapters 3 and 4.
- To understand decay of matrix coefficients for $\mathrm{SL}_3(\mathbb{R})$, cover Chapter 1, some of Chapter 2 (focusing on the spectral theorem and diagonal spectral measures and neglecting duality theory), use Section 4.1 to gain familiarity with Lie algebras, read Section 5.1 (focusing only on the behaviour of diagonal spectral measures), and then move straight to Chapter 7.
- To understand the notion of temperedness for unitary representations of $\mathrm{SL}_2(\mathbb{R})$ cover Chapter 1, Sections 3.1 and 3.2, Section 4.1, Sections 6.1, 6.2,

and 6.3, Sections 7.1 and 7.3, and then pick the sections of interest in Chapter 8.

- To understand the unitary dual of $\mathrm{SL}_2(\mathbb{R})$ cover Chapter 1, unitary representations of the circle group (as, for example, a basic example of Sections 2.1 and 2.2), Sections 3.1 and 3.2, Section 4.1, Section 7.1, Sections 8.3, 8.4, and 8.5, and then move to Chapter 9.
- To understand the Fourier transform on the hyperbolic plane cover Chapter 1, the Fourier transform on $\mathbb{R}$ (as, for example, a very special case of Sections 2.1, 2.2, and 2.3), Section 4.1, Section 7.1, Sections 8.3 and 8.5, and then pick the sections of interest in Chapter 9.

Acknowledgements

We are grateful to Menny Aka, Valeria Ambrosio, Marc Burger, Segev Gonen Cohen, Emilio Corso, Alexander Furlong, Constantin Kogler, Elon Lindenstrauss, Manuel Lüthi, René Pfitscher, Roland Prohaska, Lior Silberman, Akshay Venkatesh, and others for comments, corrections, and suggestions. We also would like to thank the students of the courses in Fall 2019 and Spring 2022 at ETH in Zürich for their interest and for their useful feedback.

Manfred Einsiedler, Zürich
Thomas Ward, Durham
September 2, 2025

Leitfaden

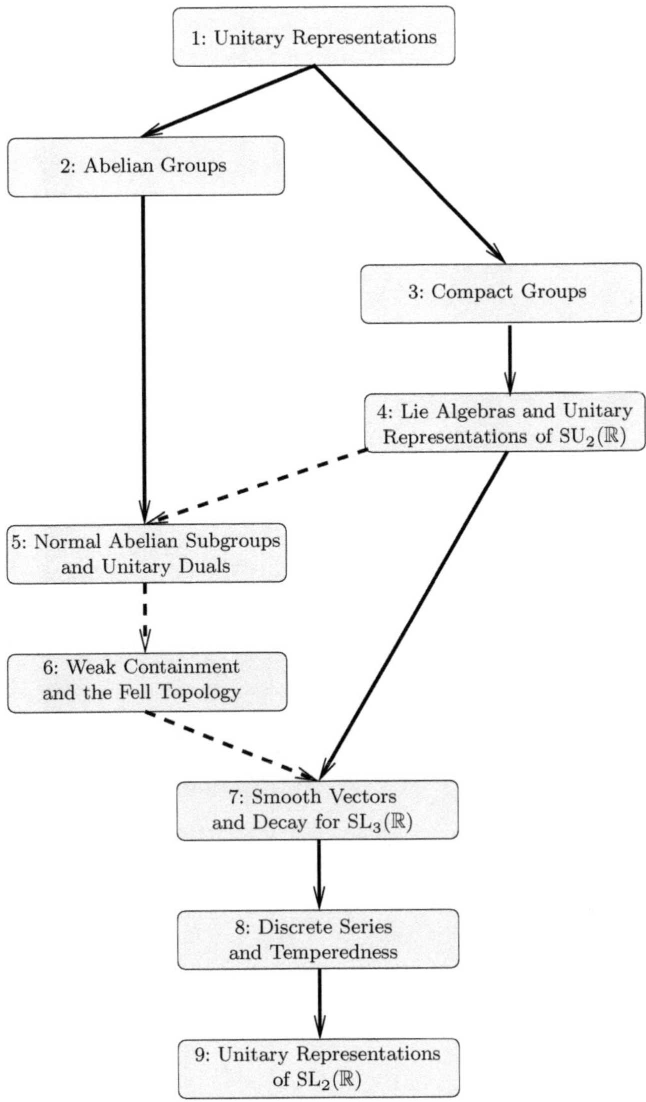

1: Unitary Representations
2: Abelian Groups
3: Compact Groups
4: Lie Algebras and Unitary Representations of $SU_2(\mathbb{R})$
5: Normal Abelian Subgroups and Unitary Duals
6: Weak Containment and the Fell Topology
7: Smooth Vectors and Decay for $SL_3(\mathbb{R})$
8: Discrete Series and Temperedness
9: Unitary Representations of $SL_2(\mathbb{R})$

Contents

Chapter 1
Unitary Representations

In this chapter we develop the initial theory of unitary representations, and discuss the principal examples and constructions.

1.1 Unitary Representations

1.1.1 Why Study Unitary Representations?

We start by offering several different answers to the question raised in the title of the subsection, in an order influenced by the interests of the authors—and with many important details ignored for the moment.

(1) One of the central objects of study in ergodic theory is a measure-preserving action $(g, x) \mapsto g \cdot x$ of a topological group G on a measure space (X, μ). One of the main tools used to study these systems is the associated linear action π of G on the space of functions $f \colon X \to \mathbb{C}$ defined by[†]

$$\big(\pi_g(f)\big)(x) = f(g^{-1} \cdot x).$$

The assumption that the action preserves the measure μ implies that

$$\|\pi_g(f)\|_2 = \|f\|_2$$

for all $f \in L^2_\mu(X)$, so that π defines a unitary representation of G on the Hilbert space $\mathcal{H}_\pi = L^2_\mu(X)$.

[†] The reader is aware that the symbol π has a more illustrious conventional meaning, as *quantitas in quam cum multiflicetur diameter, proveniet circumferencia* (the quantity which, when the diameter is multiplied by it, gives the circumference). We will have to use it in both meanings, often in close proximity. Both uses are so standard that we can only apologise, and hope the context provides clarity.

M. Einsiedler and T. Ward, *Unitary Representations and Unitary Duals*,
Graduate Texts in Mathematics 308, https://doi.org/10.1007/978-3-032-03899-9_1

(2) Modern number theory uses unitary representations heavily, as many problems in number theory have a naturally arising large group of symmetries. This group of symmetries may be intrinsic to the problem but not immediately evident. The unitary representations arising then often become associated to certain L-functions. The most basic examples of these are Dirichlet L-functions, which are associated to Dirichlet characters. At the same time, a unitary character (that is, a continuous homomorphism) $\chi\colon G \to \mathbb{S}^1$ on a topological group G gives rise to the most basic unitary representation of G on the Hilbert space $\mathcal{H}_\chi = \mathbb{C}$, defined by

$$v \stackrel{g}{\longmapsto} \chi(g)v$$

for all $g \in G$ and $v \in \mathbb{C}$.

(3) Given a problem in $\mathbb{R}^d$ with rotational symmetry, for example a partial differential equation involving the Laplace operator on a sphere or a ball, the unitary representation theory of the group $\mathrm{SO}_d(\mathbb{R})$ may help to reduce the problem to a potentially easier lower-dimensional sub-problem.

(4) In physics many different groups of symmetries appear naturally, sometimes in surprising places. For instance, the representation theory of Lie groups may be used to explain data concerning the hydrogen atom and subatomic particles.

(5) A beautiful setting that combines all of the topics above (and others) is the Quantum Unique Ergodicity (QUE) conjecture of Rudnick and Sarnak (we refer to the original paper [79] and a survey of Sarnak [82] for more on this). The uncertainty principle in quantum physics states that a quantum particle whose position is known very precisely must have a large uncertainty in its momentum, and *vice-versa*. The QUE conjecture, when specialised to the setting of a hyperbolic universe M of finite volume, goes much further. It states that if the energy of a quantum particle is known and very large, then the particle's position and direction of movement are not simply highly uncertain but in fact are nearly equidistributed in the unit tangent bundle of M (that is, nearly equidistributed in both position and direction). In mathematical terms, the conjecture concerns the eigenfunctions $\phi_1, \phi_2, \dots$ of the Laplace–Beltrami operator Δ on M and corresponding eigenvalues $\lambda_1, \lambda_2, \dots$ with $|\lambda_j| \to \infty$ as $j \to \infty$. If vol denotes the volume on M, and each ϕ_j is normalized to have $\|\phi_j\|_2 = 1$, then the conjecture states in particular that $|\phi_j|^2 \,\mathrm{dvol} \to \mathrm{dvol}$ in the weak* topology as $j \to \infty$. The only case of the conjecture known to hold (under mild additional assumptions) concerns arithmetically defined hyperbolic surfaces, in which case it is called the Arithmetic Quantum Unique Ergodicity (AQUE) conjecture. The arithmetic nature of the space here gives rise to additional symmetries, and in these cases the proof of equidistribution by Lindenstrauss [62] and Soundararajan [87] combines the theory of unitary representations for $\mathrm{SL}_2(\mathbb{R})$, number theory, and ergodic theory.

1.1.2 Standing Assumptions

We begin our formal discussion by recalling that a *topological group* is a group G that carries a topology with respect to which the maps $(g, h) \mapsto gh$ and $g \mapsto g^{-1}$ are continuous as maps $G \times G \to G$ and $G \to G$ respectively. A *locally compact* (*compact*, *connected*, and so on) group is a topological group for which the topological space is locally compact (compact, connected, and so on). We similarly extend algebraic properties to topological groups. Recall from the preface that, throughout, any compact or locally compact space is assumed to be Hausdorff.

Let us make the following simple observation in the setting of topological groups. If H, G are topological groups and $\Psi\colon H \to G$ is a homomorphism, then Ψ is continuous if and only if Ψ is continuous at the identity $e \in H$. In fact, if Ψ is continuous at the identity the composition $H \to H \to G \to G$ defined by

$$h \longmapsto hh_0^{-1} \longmapsto \Psi(hh_0^{-1}) \longmapsto \Psi(hh_0^{-1})\Psi(h_0)$$

is continuous at $h_0 \in H$ because translation is continuous. However, this map equals Ψ, and we obtain continuity of Ψ at any $h_0 \in H$, as claimed.

In pursuit of our interests in Lie-theoretic, dynamical, and number-theoretic applications of representation theory, we limit the generality to what is necessary as it simplifies some discussions and reduces the prerequisites. For instance, we will use the following assumptions and notation:

- G (sometimes H) denotes a locally compact, σ-compact, metric group equipped with a left-invariant Haar[†] measure denoted m_G (or m if there is only one group involved). As a cryptic reminder of the standing assumptions we often speak of 'the group G'. The identity of a multiplicative group will be denoted e, except in the case of explicitly presented matrix groups where the identity will be denoted I. The connected component containing the identity of a topological group G will be denoted G^o. For an abelian group G we will often use the conventional additive notation with identity denoted 0. Elements of the groups will be denoted by g, h and occasionally also by k, ℓ, or t. The function spaces $L^p_m(G)$ will be denoted $L^p(G)$ when the measure is clear, and we write $\int f \,\mathrm{d}m$ as shorthand for $\int_G f(g)\,\mathrm{d}m(g)$ when the domain of integration is clear.
- X (sometimes Y) denotes a locally compact, σ-compact, metric space often carrying a locally finite Borel measure μ (or ν). Measurability of sets or maps is always understood with respect to the Borel σ-algebra $\mathcal{B}_X$ on X. We will again write $\int f \,\mathrm{d}\mu$ when the domain of integration is clear from the context.
- For a continuous G-action on X, we denote the action of $g \in G$ by

$$X \ni x \longmapsto g{\boldsymbol\cdot}x \in X.$$

[†] We refer to [25, Sec. 10.1] for a discussion of Haar measures.

As usual 'action' means that $e \cdot x = x$ for all $x \in X$ and $(gh) \cdot x = g \cdot (h \cdot x)$ for all $g, h \in G$ and $x \in X$, and 'continuous' means that the map

$$G \times X \ni (g, x) \longmapsto g \cdot x \in X$$

is continuous with respect to the product topology on $G \times X$.
- $\mathcal{H}$ denotes a separable complex Hilbert space, with elements written u, v, or w (and occasionally a, b), and $\mathrm{B}(\mathcal{H})$ denotes the space of bounded linear operators $\mathcal{H} \to \mathcal{H}$. We will write $\langle \cdot, \cdot \rangle$ and $\| \cdot \|$ for the inner product and norm unless we want to emphasise the Hilbert space, in which case we write $\langle \cdot, \cdot \rangle_{\mathcal{H}}$ and $\| \cdot \|_{\mathcal{H}}$.
- We will also use j, k, ℓ, m, n as indices.

1.1.3 Definition of Unitary Representations

Definition 1.1 (Unitary representation). A *unitary representation* π of a topological group G on a complex Hilbert space $\mathcal{H}_\pi$ is a map $\pi \colon G \to \mathrm{B}(\mathcal{H}_\pi)$, usually written $\pi \colon g \mapsto \pi_g$, such that:

(1) π is a *representation*, meaning that $\pi_e = I$ is the identity and $\pi_{g_1 g_2} = \pi_{g_1} \pi_{g_2}$ for all $g_1, g_2 \in G$;
(2) π is *unitary*, meaning that $\pi_g^* = \pi_g^{-1} = \pi_{g^{-1}}$ for all $g \in G$, where π_g^* denotes the adjoint of π_g; and
(3) π is *continuous* with respect to the strong operator† topology on $\mathrm{B}(\mathcal{H}_\pi)$, meaning that for any $v \in \mathcal{H}_\pi$ the map $G \ni g \mapsto \pi_g v \in \mathcal{H}_\pi$ is continuous.

As in the definition above, we will decorate the Hilbert space with the letter denoting the representation as a subscript, so that it is clear that π is a unitary representation on $\mathcal{H}_\pi$, ρ is a unitary representation on $\mathcal{H}_\rho$, and so on. While we generally write π_g for $\pi(g)$ to reduce the number of parentheses needed, we will use $\pi(g)$ if the argument g is too complex to be placed in a subscript. To avoid trivialities we will always assume that $\mathcal{H}_\pi \neq \{0\}$.

One of the fundamental classes of unitary representations comes from the following notion. A *unitary character on G* is a continuous group homomorphism

$$\chi \colon G \longrightarrow \mathbb{S}^1 = \{z \in \mathbb{C} \mid |z| = 1\}.$$

This defines a unitary representation of G on $\mathcal{H}_\chi = \mathbb{C}$ by defining the action of $g \in G$ to be the map $v \mapsto \chi_g v = \chi(g) v$ for all $v \in \mathcal{H}_\chi$. We will also encounter more general characters. For instance,‡ a continuous group homomorphism from G to the multiplicative group $\mathbb{C}^\times = \{z \in \mathbb{C} \mid z \neq 0\}$ is called a *character* on G.

---

† The uniform (that is, operator) norm topology cannot be used in this context as many, or even most, natural examples would not satisfy this requirement (see Exercise 1.8).

‡ In Chapters 3, 4 and 9 other generalizations will be important.

As we will discuss in Section 2.1, unitary characters are important and exist in great abundance for abelian groups. However, for a non-abelian group G it may happen[†] that the only unitary character of G is the *trivial character* defined by $\mathbb{1}_G(g) = \mathbb{1}(g) = 1$ for all $g \in G$ associated to the *trivial representation* $\mathbb{1}_G$

$$\mathbb{C} \ni v \stackrel{g}{\longmapsto} \mathbb{1}_G(g)v = v$$

for all $g \in G$. More generally, we call a unitary representation π *trivial* if it satisfies $\pi_g = I$ for every $g \in G$.

We finish this section with a few more basic notions. For this let π be a unitary representation of the group G. If a subspace $\mathcal{V} < \mathcal{H}_\pi$ is π_g-invariant for each $g \in G$, then we will simply say that $\mathcal{V}$ is *π-invariant* (or just *invariant*). It is clear that if $\mathcal{V}$ is invariant, then its closure $\overline{\mathcal{V}}$ is also invariant (as π_g is a continuous operator for every $g \in G$). This is a rather simple notion, but is fundamental for many of the following discussions.

We also say a vector $v \in \mathcal{H}_\pi$ is *π-invariant* if $\pi_g v = v$ for all $g \in G$. We denote the subspace of invariant vectors by $\mathcal{H}_\pi^G$; it is the maximal subspace of $\mathcal{H}_\pi$ with the property that the restriction of π to $\mathcal{H}_\pi^G$ is trivial. It is automatically π-invariant and closed.

1.2 First Examples and Results

1.2.1 Continuous Actions and the Regular Representation

As indicated in Section 1.1.1, one important motivation for the study of unitary representations is the following set-up.

Definition 1.2. Suppose the group G acts continuously[‡] on the space X. We say that G *preserves a measure μ on X*, or equivalently that μ is an *invariant measure*, if $\mu(g^{-1}\boldsymbol{\cdot} B) = \mu(B)$ for all measurable $B \subseteq X$ and $g \in G$.

Proposition 1.3 (Action-associated representation). *Suppose that the group G acts continuously on X preserving a locally finite Borel measure μ on X. Then*

$$\big(\pi_g^X(f)\big)(x) = f(g^{-1}\boldsymbol{\cdot} x)$$

for $x \in X$, $f \in L^2_\mu(X)$, and $g \in G$, defines a unitary representation π^X of G on $\mathcal{H}_{\pi^X} = L^2_\mu(X)$.

This fundamental link between actions and representations arises in multiple settings and at many points in the history of these ideas. In ergodic theory this

---

[†] In particular, because the kernel of any character is a normal subgroup of G, and G might be a non-abelian simple group (see also Exercise A.1 as well as Exercises 1.82, 1.89, and 1.91).

[‡] As explained in Section 1.1.2, we assume implicitly that G and X are locally compact, σ-compact, and metric.

is sometimes called the Koopman [56] or von Neumann representation [73, 75] following their use of it in relation to spectral properties of dynamical systems in the early 1930s. In representation theory this is most closely associated to the important work of Mackey in the 1950s [64, 65, 66].

The reader is invited to work out the proof of Proposition 1.3 (see below for a generalization). An important special case of this is given in the following definition.

Definition 1.4 (Regular representation). The (left) regular representation $\lambda = \lambda^G$ of G is defined by

$$\lambda_{g_0}(f)(g) = f(g_0^{-1}g)$$

for all $f \in \mathcal{H}_\lambda = L^2_m(G)$ and $g_0, g \in G$.

Notice that the (left) regular representation is indeed a special case of the action-associated representation, by letting G act on itself by left multiplication. If G is unimodular (that is, if the left Haar measure is also right-invariant) then we can also use right multiplication to define the right regular representation similarly. However, in general right multiplication only preserves the 'measure class' of the left Haar measure but not the measure itself. Hence the right regular representation, and many other examples, motivate a generalization of Proposition 1.3.

For this, let us first recall that for a measurable map $T\colon X \to Y$ and a measure μ on X the formula

$$T_*\mu(B) = \mu(T^{-1}B),$$

for all measurable B, defines the *push-forward measure* $T_*\mu$ on Y. Using monotone convergence, the substitution rule

$$\int_Y f(y)\,\mathrm{d}T_*\mu(y) = \int_X f(T(x))\,\mathrm{d}\mu(x) \tag{1.1}$$

for any measurable $f \geqslant 0$ follows quickly from the definition. We note that if Y is equipped with a measure ν, then T is called *measure-preserving* if $T_*\mu = \nu$.

Moreover, let us recall that a measure ν on X is *absolutely continuous* with respect to another measure μ on X (also written as $\nu \ll \mu$) if $\mu(N) = 0$ implies $\nu(N) = 0$ for any measurable $N \subseteq X$. For σ-finite measures (and hence, in particular, for locally finite measures on the σ-compact X) this is equivalent to the existence of a measurable non-negative function on X, the *Radon–Nikodym derivative* $\frac{\mathrm{d}\nu}{\mathrm{d}\mu}$, with the property that

$$\int f \frac{\mathrm{d}\nu}{\mathrm{d}\mu}\,\mathrm{d}\mu = \int f\,\mathrm{d}\nu \tag{1.2}$$

for any measurable function $f \geqslant 0$ (see for instance [25, Prop. 3.29]). The Radon–Nikodym derivative $\frac{\mathrm{d}\nu}{\mathrm{d}\mu}$ is uniquely determined μ-almost surely by the property (1.2).

Definition 1.5. Two σ-finite measures μ and ν on the space X define the same *measure class* if $\nu \ll \mu$ and $\mu \ll \nu$. A measurable map $T\colon X \to X$ *preserves the measure class* of μ if $T_*\mu$ and μ define the same measure class.

A group action *preserves the measure class* of μ if the action of each element of the group preserves the measure class of μ. In this case μ is also called *quasi-invariant* for the action.

Suppose now that G acts continuously on X, preserving the measure class of a locally finite measure μ on X. We define

$$J(g,x) = \frac{\mathrm{d}\mu}{\mathrm{d}g_*^{-1}\mu}(x)$$

for $g \in G$ and almost every $x \in X$. Combining (1.1) and (1.2), we see that $J(g,\cdot)$ is almost surely characterized by the property

$$\int J(g,g^{-1}\boldsymbol{\cdot} y)f(g^{-1}\boldsymbol{\cdot} y)\,\mathrm{d}\mu(y) = \int J(g,x)f(x)\,\mathrm{d}g_*^{-1}\mu(x) = \int f\,\mathrm{d}\mu \tag{1.3}$$

for all non-negative measurable functions $f \geqslant 0$. It also suffices to consider only characteristic functions of the form $f = \mathbb{1}_B$ for measurable sets $B \subseteq X$. For $g_1, g_2 \in G$ and $f \geqslant 0$ we may apply the identity above to g_1 and the map $X \ni y \mapsto J(g_2, g_2^{-1}\boldsymbol{\cdot} y)f(g_2^{-1}\boldsymbol{\cdot} y)$ as well as to g_2 and f, which leads to

$$\begin{aligned}\int J(g_1,g_1^{-1}\boldsymbol{\cdot} z)J(g_2,g_2^{-1}g_1^{-1}\boldsymbol{\cdot} z)f(g_2^{-1}g_1^{-1}\boldsymbol{\cdot} z)\,\mathrm{d}\mu(z)&\\ = \int J(g_2,g_2^{-1}\boldsymbol{\cdot} y)f(g_2^{-1}\boldsymbol{\cdot} y)\,\mathrm{d}\mu(y) = \int f\,\mathrm{d}\mu.&\end{aligned}$$

Combining this identity with the characterization in (1.3) for $g = g_1g_2$ and setting $x = g_2^{-1}g_1^{-1}\boldsymbol{\cdot} z$, this implies that

$$J(g_1g_2,x) = J(g_1,g_2\boldsymbol{\cdot} x)J(g_2,x) \tag{1.4}$$

for μ-almost every $x \in X$. We may think of the cocycle equation (1.4) as a generalization of the chain rule for the Jacobian of a C^1-transformation: The Jacobian for g_1g_2 at $x \in X$ is the product of the Jacobians of g_1 at $g_2\boldsymbol{\cdot} x$ and the Jacobian of g_2 at x (see Exercise 1.11).

In the literature one can also find other definitions of cocycle equations. The choice is to some extent a matter of taste, much like choosing to use column vectors and left actions or row vectors and right actions. We decided to use the formulation in (1.4) and (1.5) for our discussions, emphasising the similarity to the chain rule. This influences many of the formulas to come.

Proposition 1.6 (Unitarily normalized representation). *Suppose that the group G acts continuously on X, and let μ be a locally finite Borel measure on X whose measure class is preserved by the G-action. Suppose also that the map $c\colon G \times X \to \mathbb{C}^\times$ is continuous, satisfies the* cocycle equation

$$c(g_1 g_2, x) = c(g_1, g_2{\bullet}x) c(g_2, x) \tag{1.5}$$

for $g_1, g_2 \in G$ and $x \in X$, and that its squared absolute value is equal almost surely to the Radon–Nikodym derivative

$$|c(g, \cdot)|^2 = J(g, \cdot) = \frac{\mathrm{d}\mu}{\mathrm{d}g_*^{-1}\mu} \tag{1.6}$$

for $g \in G$. Then the normalized *(and possibly* twisted*)* representation *π^c defined by*

$$\left(\pi_g^c f\right)(x) = c(g, g^{-1}{\bullet}x) f(g^{-1}{\bullet}x) \tag{1.7}$$

for $f \in \mathcal{H}_{\pi^c} = L^2_\mu(X)$, $g \in G$, and $x \in X$, is a unitary representation.

As the proof will show, building the square root

$$|c(g, x)| = J(g, x)^{\frac{1}{2}} = \Big(\frac{\mathrm{d}\mu}{\mathrm{d}g_*^{-1}\mu}(x)\Big)^{\frac{1}{2}}$$

of the Radon–Nikodym derivative into the definition (1.7) is necessary to obtain a unitary representation, and we will refer to this as the (*unitarily*) *normalized representation* associated to the action. Allowing the function c to be complex-valued gives us more flexibility in constructing unitary representations. We will refer to the case where c is indeed complex-valued as a *twisted normalized representation.*

Proof of Proposition 1.6. For simplicity, we write π for π^c. For a function f on X, $g \in G$, and $x \in X$, we define $\pi_g(f)(x) = c(g, g^{-1}{\bullet}x) f(g^{-1}{\bullet}x)$ as in the proposition. Now calculate

$$\begin{aligned}
\pi_{g_1}\left(\pi_{g_2}(f)\right)(x) &= c(g_1, g_1^{-1}{\bullet}x)\pi_{g_2}(f)(g_1^{-1}{\bullet}x) \\
&= c(g_1, g_1^{-1}{\bullet}x) c(g_2, g_2^{-1}g_1^{-1}{\bullet}x) f(g_2^{-1}g_1^{-1}{\bullet}x) \\
&= c(g_1 g_2, g_2^{-1}g_1^{-1}{\bullet}x) f((g_1 g_2)^{-1}{\bullet}x) = \pi_{g_1 g_2}(f)(x)
\end{aligned}$$

for $g_1, g_2 \in G$ and $x \in X$, by (1.5) at $g_2^{-1}g_1^{-1}{\bullet}x$. If $\|f\|_2 < \infty$ and $g \in G$, then we also have

$$\begin{aligned}
\|\pi_g(f)\|_2^2 &= \int_X |c(g, g^{-1}{\bullet}x) f(g^{-1}{\bullet}x)|^2 \,\mathrm{d}\mu(x) \\
&= \int_X |c(g, y) f(y)|^2 \,\mathrm{d}g_*^{-1}\mu(y) = \int_X |f(x)|^2 \,\mathrm{d}\mu(x)
\end{aligned}$$

$$\begin{aligned}\left(\pi_g^{\mathbb{S}^1,\xi} f\right)(v) &= c(g, g^{-1}\bullet v) f(g^{-1}\bullet v)\\ &= \left\| g\Big(\frac{1}{\|g^{-1}v\|} g^{-1}v\Big)\right\|^{1+\xi\mathrm{i}} f(g^{-1}\bullet v)\\ &= \|g^{-1}v\|^{-1-\xi\mathrm{i}} f(g^{-1}\bullet v).\end{aligned}$$

Throughout the text we will give a number of exercises, some of which will have hints in an appendix starting on page 525; those flagged as 'essential' exercises will be used in a crucial way in the ensuing discussion.

Exercise 1.8. Let λ be the regular representation of $\mathbb{R}$ on $L^2(\mathbb{R})$. Show that

$$\|\lambda_g - I\|_{\mathrm{op}} \geqslant \frac{1}{\sqrt{2}},$$

where $\|\cdot\|_{\mathrm{op}}$ denotes the operator norm (or, by working a bit harder, that $\|\lambda_g - I\|_{\mathrm{op}} = 2$) for all $g \in \mathbb{R}\smallsetminus\{0\}$.

Exercise 1.9. Suppose μ and ν are two σ-finite measures defining the same measure class on the space X. Show that $\frac{\mathrm{d}\mu}{\mathrm{d}\nu} = \left(\frac{\mathrm{d}\nu}{\mathrm{d}\mu}\right)^{-1}$ almost surely.

Exercise 1.10. Prove that the function in (1.8) satisfies the cocycle equation (1.5) for any $g_1, g_2 \in \mathrm{SL}_2(\mathbb{R})$ and $v \in \mathbb{S}^1$.

Exercise 1.11. (a) Let $d \in \mathbb{N}$, let $X \subseteq \mathbb{R}^d$ be open and non-empty, and let m_X denote the restriction of Lebesgue measure m to X. Suppose g acts continuously in the C^1-topology on X (that is, $x \mapsto g\bullet x$ is differentiable and the derivative depends continuously on $(g, x) \in G \times X$). Express

$$J(g, x) = \frac{\mathrm{d}m_X}{\mathrm{d}g_*^{-1}m_X}(x) = \left(\frac{\mathrm{d}g_*^{-1}m_X}{\mathrm{d}m_X}(x)\right)^{-1}$$

for $(g, x) \in G \times X$ in terms of a Jacobian and conclude that $c_\xi = J^{\frac{1}{2}+\mathrm{i}\xi}$ satisfies the assumptions in Proposition 1.6 for any $\xi \in \mathbb{R}$. Deduce that the term $J(g, g^{-1}\bullet x)^{\frac{1}{2}}$ (appearing implicitly in the definition (1.7) of the unitary representation) is the Jacobian of the action of g^{-1} at x.
(b) Generalize this to a d-dimensional smooth manifold X. For this, assume that μ is a *smooth measure* on X in the following sense: For any chart map $\Psi\colon O \to \Psi(O) \subseteq X$ for the manifold defined on an open subset $O \subseteq \mathbb{R}^d$, there exists some $D_O \in C(O)$ with $D_O > 0$ and $\mu|_{\Psi(O)} = \Psi_*\left(D_O \,\mathrm{d}m|_O\right)$.

Exercise 1.12. Let $c\colon G \times X \to \mathbb{C}^\times$ be a cocycle as in Proposition 1.6. Show that

$$c(g, g^{-1}\bullet x) = c(g^{-1}, x)^{-1}$$

for $g \in G$ and $x \in X$.

1.2.2 The Continuity Requirement

Proving the continuity requirement for a unitary representation is usually not difficult. To simplify this process in the many examples that follow, we extract an abstract principle from the proof of Proposition 1.6. Let us write $\langle S\rangle_{\mathbb{C}}$ for the complex linear hull of a subset S of a given complex vector space.

Lemma 1.13 (Criteria for continuity). *Let $\pi\colon G \to \mathrm{B}(\mathcal{H}_\pi)$ be a representation on a Hilbert space $\mathcal{H}_\pi$ taking unitary values, so that $\pi_{g_1 g_2} = \pi_{g_1}\pi_{g_2}$ and $\pi_g^* = \pi_g^{-1} = \pi_{g^{-1}}$ for all $g_1, g_2, g \in G$. To prove the continuity requirement in Definition 1.1(3) it suffices to find a subset $D \subseteq \mathcal{H}_\pi$ satisfying one of the following properties:*

(1) *The linear hull $\langle \pi_G D\rangle_{\mathbb{C}}$ is dense in $\mathcal{H}_\pi$ and $G \ni g \mapsto \pi_g(v)$ is continuous for every $v \in D$.*
(2) *The linear hull $\langle D\rangle_{\mathbb{C}}$ is dense and $G \ni g \mapsto \pi_g(v)$ is continuous at the identity $e \in G$ for every $v \in D$.*

PROOF. Suppose D satisfies the assumptions in (1). Continuity of the vector space operations and the group multiplication imply that the function

$$G \ni g \longmapsto \pi_g(\widetilde{v})$$

is continuous for every $\widetilde{v} \in \widetilde{D} = \langle \pi_G D\rangle_{\mathbb{C}}$. For a given $w \in \mathcal{H}_\pi$, we can choose a sequence (v_n) in $\widetilde{D}$ with $v_n \to w$ as $n \to \infty$. Now notice that by unitarity of π_g for $g \in G$ the functions

$$G \ni g \longmapsto \pi_g(v_n)$$

converge uniformly to $G \ni g \mapsto \pi_g(w)$ as $n \to \infty$. In fact, we have

$$\sup_{g\in G} \|\pi_g(v_n) - \pi_g(w)\| = \|v_n - w\| \longrightarrow 0$$

as $n \to \infty$. However, as uniform convergence of a sequence of continuous functions ($G \ni g \mapsto \pi_g v_n$ with $v_n \in \widetilde{D}$) implies that the limit ($g \mapsto \pi_g w$ for the given $w \in \mathcal{H}_\pi$) is continuous too, we deduce that the representation is continuous.

So suppose now that $G \ni g \mapsto \pi_g v$ is continuous at $e \in G$ for every $v \in D$ as in (2). We can again take linear combinations and apply uniform convergence as above to see that this extends to all elements of $\overline{\langle D\rangle_{\mathbb{C}}} = \mathcal{H}_\pi$. It remains to extend the continuity to all of G. So suppose that $w \in \mathcal{H}$ and $g_n \to g \in G$ as $n \to \infty$. Then

$$\pi_{g_n} w - \pi_g w = \pi_g(\pi_{g^{-1}g_n} w - w) \longrightarrow 0$$

since $g^{-1}g_n \to e$ as $n \to \infty$ and π_g is unitary (and hence continuous). □

Exercise 1.14. Show that the continuity requirement in Definition 1.1(3) can equivalently be expressed in the following way: π is *weakly continuous*, that is continuous with respect to the weak operator topology on $\mathrm{B}(\mathcal{H})$, meaning that for any $u, v \in \mathcal{H}$ the map

$$G \ni g \longmapsto \langle \pi_g u, v\rangle \in \mathbb{C}$$

is continuous.

1.2.3 An Aside on Haar Measures

We wish to discuss the right regular representation in greater detail. However, for this we have to recall a few more facts about Haar measures (see [25, Sec. 10.1] for the details).

One fundamental property is that the left Haar measure $m = m_G$ is unique up to a scalar multiple. If now $\theta\colon G \to G$ is an automorphism of G (as a topological group) then $\theta_* m$ is again a left Haar measure, since:

- $(\theta_* m)(K) = m\left(\theta^{-1}K\right) < \infty$ for every compact set $K \subseteq G$;
- $(\theta_* m)(O) = m\left(\theta^{-1}(O)\right) > 0$ for every non-empty open set $O \subseteq G$; and
- $(\theta_* m)(gB) = m\left(\theta^{-1}(gB)\right) = m\left(\theta^{-1}(g)\theta^{-1}B\right) = m\left(\theta^{-1}B\right) = \theta_* m(B)$ for every $g \in G$ and Borel measurable $B \subseteq G$.

Hence, by the uniqueness of Haar measure, we have $\theta_* m = \operatorname{mod}(\theta) m$ for some scalar $\operatorname{mod}(\theta) > 0$. In other words, the Radon–Nikodym derivative $\frac{\mathrm{d}\theta_* m}{\mathrm{d}m}$ is constant and equal to $\operatorname{mod}(\theta)$.

Applying this in the case of an inner automorphism $\theta_{g_0}\colon G \to G$ defined by $\theta_{g_0}(g) = g_0 g g_0^{-1}$, we obtain the *modular character*, a function

$$\Delta_G\colon G \longrightarrow \mathbb{R}_{>0}$$

defined by $\Delta_G(g_0) = \operatorname{mod}(\theta_{g_0})$ for all $g_0 \in G$. When the group is clear from the context we will usually write $\Delta = \Delta_G$. Also note that G is unimodular if and only if Δ is equal to the constant function $\mathbb{1}_G$.

Lemma 1.15 (Modular character). *The modular character is a continuous group homomorphism $\Delta = \Delta_G\colon G \to \mathbb{R}_{>0}$ satisfying*

$$\int f(gg_0^{-1})\,\mathrm{d}m(g) = \Delta(g_0)\int f\,\mathrm{d}m$$

or, equivalently,

$$\Delta(g_0)\int f(gg_0)\,\mathrm{d}m(g) = \int f\,\mathrm{d}m$$

for any non-negative measurable or integrable function f on G.

Proof. Let $g_0, g_1 \in G$. Then

$$\theta_{g_0g_1}(g) = g_0g_1g(g_0g_1)^{-1} = \theta_{g_0}\left(\theta_{g_1}(g)\right)$$

for all $g \in G$, and hence

$$\begin{aligned}\Delta(g_0g_1)m &= \left(\theta_{g_0g_1}\right)_* m = \left(\theta_{g_0}\right)_*\left(\theta_{g_1}\right)_* m \\ &= \left(\theta_{g_0}\right)_*\left(\Delta(g_1)m\right) = \Delta(g_0)\Delta(g_1)m.\end{aligned}$$

To prove continuity of Δ, fix some compact neighbourhood K of $e \in G$ and some positive ε. By regularity of m, there exists an open set $U \supseteq K$ with

$$m(U) < (1+\varepsilon)m(K).$$

By continuity of the group operations and compactness of K, there exists some symmetric neighbourhood $V = V^{-1}$ of $e \in G$ such that $VKV \subseteq U$. This now implies that for any $g_0 \in V$ we have

$$\Delta(g_0) = \frac{(\theta_{g_0})_* m(K)}{m(K)} = \frac{m(g_0^{-1} K g_0)}{m(K)} \leqslant \frac{m(U)}{m(K)} < 1 + \varepsilon,$$

and by symmetry of V we also have $\Delta(g_0) > (1+\varepsilon)^{-1}$. As $\varepsilon > 0$ was arbitrary, this gives continuity of Δ at $e \in G$, and hence continuity by the argument on page 3.

Now let f be a non-negative measurable or integrable function on G. Then

$$\begin{aligned}\int f(gg_0^{-1})\,\mathrm{d}m(g) &= \int f(g_0 g g_0^{-1})\,\mathrm{d}m(g) = \int f(\theta_{g_0} g)\,\mathrm{d}m(g)\\ &= \int f\,\mathrm{d}(\theta_{g_0})_* m = \Delta(g_0)\int f\,\mathrm{d}m,\end{aligned}$$

which implies the lemma. □

Sometimes we will need to use the right Haar measure on G, which is related to the left Haar measure via the following lemma.

Lemma 1.16 (Right Haar measure). *The right Haar measure $m_G^{(\mathrm{r})}$ on G (denoted by $m^{(\mathrm{r})}$ when the group G is clear from the context) can be defined by*

$$\mathrm{d}m_G^{(\mathrm{r})} = \mathrm{d}\imath_* m_G = \Delta^{-1}\,\mathrm{d}m_G,$$

where $\imath\colon G \to G$ is the inversion map sending g to g^{-1}. Using this definition we have, equivalently,

$$\int f\,\mathrm{d}m^{(\mathrm{r})} = \int f(g^{-1})\,\mathrm{d}m(g) = \int f(g)\Delta(g)^{-1}\,\mathrm{d}m(g)$$

and

$$\int f(g)\,\mathrm{d}m(g) = \int f(g^{-1})\Delta(g)^{-1}\,\mathrm{d}m(g) \tag{1.9}$$

for any measurable non-negative function f on G.

Proof. We begin by noting that $\imath_* m$ is a right Haar measure, since

$$(\imath_* m)(Bg) = m(\imath^{-1}(Bg)) = m(g^{-1}\imath^{-1}(B)) = \imath_* m(B)$$

for any measurable $B \subseteq G$ and $g \in G$. Furthermore, $\Delta^{-1}\,\mathrm{d}m$ defines another right Haar measure. Indeed, for a measurable non-negative function f we have

$$\int f(gg_0)\Delta(g)^{-1}\,\mathrm{d}m(g) = \Delta(g_0)\int f(gg_0)\Delta(gg_0)^{-1}\,\mathrm{d}m(g)$$
$$= \int f(g)\Delta(g)^{-1}\,\mathrm{d}m$$

by Lemma 1.15 applied to $f\Delta^{-1}$, which gives the claim.

By uniqueness of right Haar measure, it follows that the measure $\Delta^{-1}\,\mathrm{d}m$ must be a scalar multiple of the measure $\imath_* m$. That is, there exists some $\alpha > 0$ so that $\Delta^{-1}\,\mathrm{d}m = \alpha\imath_* m$. We claim that $\alpha = 1$. By continuity of Δ there exists a compact neighbourhood $V = V^{-1}$ of $e \in G$ such that $|\Delta(g)^{-1} - 1| < \varepsilon$ for all $g \in V$. Together with $\imath_* m(V) = m(V^{-1}) = m(V)$, this implies that

$$|1-\alpha|m(V) = \left|m(V) - \int_V \Delta^{-1}\,\mathrm{d}m\right| = \left|\int_V (1-\Delta^{-1})\,\mathrm{d}m\right| < \varepsilon m(V).$$

Dividing by $m(V)$ and recalling that $\varepsilon > 0$ was arbitrary, we see that $\alpha = 1$ as claimed.

Applying the above to the function $G \ni g \mapsto f(g^{-1})$ for a non-negative measurable f on G also gives the last formula in the lemma, and so concludes the proof. □

Example 1.17 (Affine group in one dimension). As an example of a non-unimodular group, we consider the group defined by

$$G = \left\{ \begin{pmatrix} a & b \\ & 1 \end{pmatrix} \;\middle|\; a > 0, b \in \mathbb{R} \right\}.$$

We note that G is called the *affine group in one dimension* or the '$ax+b$' group, because of the natural action of the elements of G on $x \in \mathbb{R}$ via

$$\begin{pmatrix} a & b \\ & 1 \end{pmatrix}\begin{pmatrix} x \\ 1 \end{pmatrix} = \begin{pmatrix} ax+b \\ 1 \end{pmatrix}.$$

We will study the unitary representations of G in detail in Chapter 5.

To describe the (left) Haar measure on G, we use the coordinates $a > 0$ and $b \in \mathbb{R}$ for the element

$$g_{a,b} = \begin{pmatrix} a & b \\ & 1 \end{pmatrix} \in G.$$

With these coordinates, the Haar measure m can be defined by the formula

$$\mathrm{d}m(g_{a,b}) = \frac{\mathrm{d}a}{a^2}\,\mathrm{d}b, \tag{1.10}$$

or, equivalently, by

$$\int_G f \, \mathrm{d}m = \int_0^\infty \int_{-\infty}^\infty f\left(\begin{pmatrix} a & b \\ & 1 \end{pmatrix}\right) \mathrm{d}b \, \frac{\mathrm{d}a}{a^2}$$

for any measurable function $f \geqslant 0$. To see this, let $g_{a_0,b_0} \in G$ and calculate

$$\begin{aligned}\int_G f(g_{a_0,b_0} g) \, \mathrm{d}m(g) &= \int_0^\infty \int_{-\infty}^\infty f\left(\begin{pmatrix} a_0 & b_0 \\ & 1 \end{pmatrix}\begin{pmatrix} a & b \\ & 1 \end{pmatrix}\right) \mathrm{d}b \, \frac{\mathrm{d}a}{a^2} \\ &= \int_0^\infty \int_{-\infty}^\infty f\left(\begin{pmatrix} a_0 a & a_0 b + b_0 \\ & 1 \end{pmatrix}\right) \underbrace{\mathrm{d}b}_{=\frac{\mathrm{d}\widetilde{b}}{a_0}} \underbrace{\frac{\mathrm{d}a}{a^2}}_{=\frac{a_0 \, \mathrm{d}\widetilde{a}}{(\widetilde{a})^2}} = \int_G f \, \mathrm{d}m\end{aligned}$$

by using the substitutions $\widetilde{b} = a_0 b + b_0$ and $\widetilde{a} = a_0 a$. As this holds for all measurable functions $f \geqslant 0$, it follows that (1.10) indeed defines the Haar measure on G.

Next we prove that

$$\Delta(g_{a_0,b_0}) = a_0^{-1} \tag{1.11}$$

for $g_{a_0,b_0} \in G$. For this, we calculate $g_{a_0,b_0}^{-1} = \begin{pmatrix} a_0^{-1} & -a_0^{-1} b_0 \\ & 1 \end{pmatrix}$ and consider the integral

$$\begin{aligned}\int_G f(g g_{a_0,b_0}^{-1}) \, \mathrm{d}m(g) &= \int_0^\infty \int_{-\infty}^\infty f\left(\begin{pmatrix} a & b \\ & 1 \end{pmatrix}\begin{pmatrix} a_0^{-1} & -a_0^{-1} b_0 \\ & 1 \end{pmatrix}\right) \mathrm{d}b \, \frac{\mathrm{d}a}{a^2} \\ &= \int_0^\infty \int_{-\infty}^\infty f\left(\begin{pmatrix} a a_0^{-1} & -a a_0^{-1} b_0 + b \\ & 1 \end{pmatrix}\right) \underbrace{\mathrm{d}b}_{=\mathrm{d}\widetilde{b}} \underbrace{\frac{\mathrm{d}a}{a^2}}_{=\frac{a_0 \, \mathrm{d}\widetilde{a}}{(\widetilde{a} a_0)^2}} = a_0^{-1} \int_G f \mathrm{d}m\end{aligned}$$

with the substitution $\widetilde{b} = -a a_0^{-1} b_0 + b$ and $\widetilde{a} = a a_0^{-1}$. Together with Lemma 1.15, this calculation shows (1.11).

Exercise 1.18. (a) Calculate the left and the right Haar measures of the group

$$G = \left\{ \begin{pmatrix} a & b \\ & 1 \end{pmatrix} \;\middle|\; a \in \mathbb{C}^\times, b \in \mathbb{C} \right\}$$

and its modular character.
(b) Repeat (a) for the group

$$G = \left\{ \begin{pmatrix} a_1 & x & z \\ & a_2 & y \\ & & a_3 \end{pmatrix} \;\middle|\; a_1, a_2, a_3 > 0 \text{ and } x, y, z \in \mathbb{R} \right\}.$$

Exercise 1.19. (a) Show that every compact group is unimodular.
(b) Show that if G has finite (left or right) Haar measure, then G is compact.

1.2.4 The Right-Regular Representation

We are now ready to discuss the right regular representation $\rho = \rho^G$. The action of $g_0 \in G$ on G by right multiplication is defined by

$$R_{g_0} : G \ni g \longmapsto g g_0^{-1} \in G,$$

and so the *normalized right regular representation* $\rho_{g_0} : L^2(G) \to L^2(G)$ is defined by

$$\rho_{g_0}(f)(g) = \Delta(g_0)^{\frac{1}{2}} f(g g_0). \tag{1.12}$$

Indeed, Lemma 1.15 and the uniqueness of the Radon–Nikodym derivative together show that $\Delta(g_0)$ is the (constant) Radon–Nikodym derivative $\frac{\mathrm{d}(R_{g_0})_* m}{\mathrm{d}m}$ for every $g_0 \in G$. Hence the cocycle

$$c(g,x) = \left(\frac{\mathrm{d}m}{\mathrm{d}(R_{g_0^{-1}})_* m}\right)^{\frac{1}{2}} = \left(\frac{\mathrm{d}(R_{g_0^{-1}})_* m}{\mathrm{d}m}\right)^{-\frac{1}{2}} = \left(\Delta(g_0^{-1})\right)^{-\frac{1}{2}} = \Delta(g_0)^{\frac{1}{2}}$$

satisfies the assumptions of Proposition 1.6 and leads to (1.12).

We now show that the right regular representation is—from an abstract point of view—not all that different from the left regular representation, even when $\Delta \neq \mathbb{1}_G$.

Definition 1.20 (Isomorphism and equivariance). Let π and ρ be unitary representations of G. A bounded operator $B \colon \mathcal{H}_\pi \to \mathcal{H}_\rho$ is *intertwining* (or *equivariant*) if $B \circ \pi_g = \rho_g \circ B$ for all $g \in G$. An intertwining unitary isomorphism $U \colon \mathcal{H}_\pi \to \mathcal{H}_\rho$ is called a *unitary isomorphism* between π and ρ. When such an isomorphism exists, we say that π and ρ are *isomorphic*.

As an example of an isomorphism between unitary representations, we show that the left regular and normalized right regular representations are isomorphic.

Using the preparations from Section 1.2.4, we define

$$U(f)(g) = \Delta(g)^{-\frac{1}{2}} f(g^{-1})$$

for $g \in G$ and a function $f \in L^2(G)$, and obtain from (1.9)

$$\|U(f)\|_2^2 = \int |f(g^{-1})|^2 \Delta(g)^{-1} \,\mathrm{d}m(g) = \int |f|^2 \,\mathrm{d}m = \|f\|_2^2$$

and

$$U^2(f)(g) = \Delta(g)^{-\frac{1}{2}} U(f)(g^{-1}) = \Delta(g)^{-\frac{1}{2}} \Delta(g^{-1})^{-\frac{1}{2}} f(g) = f(g).$$

Hence $U \colon L^2(G) \to L^2(G)$ is unitary with $U^2 = I$. Moreover, note that

$$U\left(\lambda_{g_0}(f)\right)(g) = \Delta(g)^{-\frac{1}{2}} \lambda_{g_0}(f)(g^{-1}) = \Delta(g)^{-\frac{1}{2}} f(g_0^{-1} g^{-1})$$

and by (1.12) also

$$\rho_{g_0}\left(U(f)\right)(g)=\Delta(g_0)^{\frac{1}{2}}U(f)(gg_0)=\Delta(g_0)^{\frac{1}{2}}\Delta(gg_0)^{-\frac{1}{2}}f(g_0^{-1}g^{-1})$$

for any $g,g_0\in G$ and $f\in L^2(G)$ or, equivalently, $U\circ\lambda_{g_0}=\rho_{g_0}\circ U$. Together, these show that U is an isomorphism between the left regular and normalized right regular representations (in both cases using the left Haar measure to define $L^2(G)$).

1.2.5 Containment, Invariant Subspaces, and Direct Sums

The requirement that U is unitary instead of just a bounded isomorphism in Definition 1.20 is not a severe restriction, due to the following lemma.

Lemma 1.21 (Obtaining intertwining isometries). *Suppose that π and ρ are two unitary representations of G and $B\colon\mathcal{H}_\pi\to\mathcal{H}_\rho$ is bounded, intertwining, and injective. Then there is a unitary isomorphism*

$$U\colon\mathcal{H}_\pi\longrightarrow\operatorname{Im}U=\overline{\operatorname{Im}B}$$

between π and the restriction of ρ to the ρ-invariant closed subspace $\operatorname{Im}U$. In particular, if B has dense image then π and ρ are isomorphic.

As we will see, this is a consequence of the polar decomposition of bounded operators, which we will essentially prove together with the lemma using the functional calculus for bounded self-adjoint operators.

PROOF OF LEMMA 1.21. Taking the adjoint of the equation defining equivariance gives

$$\pi_{g^{-1}}\circ B^*=B^*\circ\rho_{g^{-1}}$$

for all $g\in G$, since the representations are themselves unitary. In other words B^* intertwines ρ and π, which in turn shows that

$$B^*B\colon\mathcal{H}_\pi\to\mathcal{H}_\pi$$

is intertwining too, since

$$B^*B\pi_g=B^*\rho_gB=\pi_gB^*B \tag{1.13}$$

for all $g\in G$.

Next notice that $\langle B^*Bv,v\rangle=\langle Bv,Bv\rangle>0$ for all non-zero $v\in\mathcal{H}_\pi$, since B is assumed to be injective. Thus B^*B is also injective. Moreover, B^*B is self-adjoint.

Finally we claim that B^*B has dense image. Indeed if $v\in(\operatorname{Im}B^*B)^\perp$, then $\langle B^*Bv,w\rangle=\langle v,B^*Bw\rangle=0$ for all $w\in\mathcal{H}_\pi$. However, this implies that $B^*Bv=0$ and so $v=0$ since B^*B is injective. To summarize, we have shown that B^*B is an intertwining, injective, positive, self-adjoint operator with dense image.

We define the self-adjoint operator $A = \sqrt{B^*B}$ by using the functional calculus (see [25, Sec. 12.4]). Also recall from the definition of the continuous functional calculus that A can be obtained as a uniform limit of a sequence of operators $(p_n(B^*B))$, where p_n is a polynomial in one variable for each $n \geqslant 1$. It is easy to see (by the same argument as in (1.13)) that $p_n(B^*B)\colon \mathcal{H}_\pi \to \mathcal{H}_\pi$ is intertwining, which implies the same for A by the uniform convergence. We note that $A^2 = B^*B$ implies that A is injective with dense image, since B^*B is injective with dense image.

We now define U, initially as U_0 on the dense subspace $\operatorname{Im} A$ by

$$U_0(Av) = Bv \in \mathcal{H}_\rho,$$

and note that

$$\|U_0(Av)\|^2 = \langle Bv, Bv\rangle = \langle B^*Bv, v\rangle = \left\langle A^2v, v\right\rangle = \langle Av, Av\rangle = \|Av\|^2$$

for all $v \in \mathcal{H}_\pi$. Hence U_0 is an isometry on the dense subspace $\operatorname{Im} A$ and therefore has a unique extension to an isometry $U\colon \mathcal{H}_\pi \to \mathcal{H}_\rho$. Note that $\operatorname{Im} U_0 = \operatorname{Im} B$ and that $\operatorname{Im} U = \overline{\operatorname{Im} U_0}$ is a closed subspace of $\mathcal{H}_\rho$, since U is an isometry.

Finally, $v \in \mathcal{H}_\pi$ and $g \in G$ implies that

$$U_0\pi_g Av = U_0 A\pi_g v = B\pi_g v = \rho_g Bv = \rho_g U_0 Av,$$

or $U_0\pi_g = \rho_g U_0$. Using the density of $\operatorname{Im} A$, this gives $U\pi_g = \rho_g U$ for all elements $g \in G$, as claimed. □

Let us introduce the following convenient terminology for the conclusion of Lemma 1.21.

Definition 1.22 (Containment). A unitary representation π of G is said to be *contained in* another unitary representation ρ, written $\pi < \rho$, if there exists a ρ-invariant closed subspace $\mathcal{V}$ of $\mathcal{H}_\rho$ with the property that π is isomorphic to the unitary representation $\rho|_{\mathcal{V}}$ obtained by restricting ρ to $\mathcal{V}$.

We conclude the section with a few basic constructions which we will use often, without explicit reference, and whose proofs we leave as exercises.

Essential Exercise 1.23 (Invariant subspaces). Let π be a unitary representation of G and suppose $\mathcal{V} < \mathcal{H}_\pi$ is a closed π-invariant subspace. Then the orthogonal projection $P_{\mathcal{V}}\colon \mathcal{H}_\pi \to \mathcal{V} \subseteq \mathcal{H}_\pi$ is intertwining, the orthogonal complement $\mathcal{V}^\perp$ is also π-invariant, and π is isomorphic to the direct sum $\pi^{\mathcal{V}} \oplus \pi^{\mathcal{V}^\perp}$ of the representations obtained by restricting π to the invariant subspace $\mathcal{V}$, respectively $\mathcal{V}^\perp$ (see Exercise 1.24 below).

Essential Exercise 1.24 (Direct sums of unitary representations). Let S be a finite or countable index set, and let π_n be a unitary representation of G

on $\mathcal{H}_n$ for every $n \in S$. Show that $\pi_\oplus(g)(v_n)_{n\in S} = (\pi_n(g)v_n)_{n\in S}$ for $g \in G$ and $(v_n)_{n\in S} \in \mathcal{H}_\oplus$ with

$$\mathcal{H}_\oplus = \bigoplus_{n\in S} \mathcal{H}_n = \left\{ (w_n)_{n\in S} \;\middle|\; w_n \in \mathcal{H}_n, \left\|(w_n)_{n\in S}\right\|_\oplus^2 = \sum_{n\in S} \|w_n\|^2 < \infty \right\}$$

defines a unitary representation $\pi_\oplus = \bigoplus_{n\in S} \pi_n$, which we will refer to as the *direct sum representation.* For $S = \{1,2\}$ we will simply write $\pi_1 \oplus \pi_2$ for the direct sum representation.

Implicit in Exercise 1.24 is the construction of the *multiple direct sum* of a given Hilbert space, which we will also denote as follows. If $\mathcal{H}$ is a Hilbert space and $n \in \mathbb{N}$, then $\mathcal{H}^n$ is the *direct sum* (or, equivalently, the direct product) of n copies of $\mathcal{H}$, and

$$\mathcal{H}^\infty = \left\{ v = (v_1, v_2, \ldots) \;\middle|\; v_j \in \mathcal{H} \text{ for all } j \in \mathbb{N} \text{ and } \sum_{j\in\mathbb{N}} \|v_j\|^2 < \infty \right\}$$

is the *infinite direct sum* equipped with the inner product

$$\langle (v_j), (w_j) \rangle_{\mathcal{H}^\infty} = \sum_{j\in\mathbb{N}} \langle v_j, w_j \rangle_{\mathcal{H}}.$$

If π is a unitary representation of the group G we will refer to $\pi^n = \bigoplus_{k=1}^n \pi$ as the nth *direct sum unitary representation* on $\mathcal{H}_\pi^n$ for any $n \in \mathbb{N}$ and, if $n = \infty$, the *infinite direct sum unitary representation* on $\mathcal{H}_\pi^\infty$ as in Exercise 1.24.

1.3 Irreducible Representations and Schur's Lemma

Definition 1.25 (Irreducibility). A unitary representation π of G on a non-trivial Hilbert space $\mathcal{H}_\pi$ is *irreducible* if $\{0\}$ and $\mathcal{H}_\pi$ are the only closed π-invariant subspaces of $\mathcal{H}_\pi$.

It is clear that the unitary representation defined by a unitary character is always irreducible, since in that case $\mathcal{H}_\pi = \mathbb{C}$ has no non-trivial proper subspaces. We will, however, see many higher-dimensional irreducible unitary representations, both finite- and infinite-dimensional, in this volume. An irreducible representation is the analogue of an elementary or indivisible particle—it cannot be decomposed into constituent parts. Somewhat surprisingly, we will have many examples of irreducible representations that we will nonetheless analyze using special subspaces—after restricting attention to certain subgroups of G.

One of the goals for a given group might be to find all of its irreducible unitary representations, which makes the following terminology useful. We will achieve this classification goal for many groups, and will also see that it is more or less impossible for some other groups.

Definition 1.26 (Unitary dual). The *unitary dual* $\widehat{G}$ of a group is defined to be the collection of all equivalence classes $[\pi]$ (with respect to isomorphism) of irreducible unitary representations π of G.

As we have no interest in set-theoretic issues (strictly speaking, the collection of all irreducible unitary representations is not in general a set, but rather a class), we will be happy if we find a complete set of representatives of the equivalence classes of the unitary dual. Moreover, we will frequently write $\pi \in \widehat{G}$ in order to simplify the notation.

Using direct sums it is clear that one can build other, more general, representations from irreducible ones, but it is much less clear whether any unitary representation can be decomposed into irreducible ones.

Essential Exercise 1.27. Let π be a unitary representation on a Hilbert space $\mathcal{H}_\pi$ of finite dimension. Show that π is isomorphic to a finite direct sum of irreducible unitary representations that are contained in π.

For unitary representations on infinite dimensional Hilbert spaces the answer to the question of whether it can be decomposed into irreducible representations depends on the precise formulation of 'decomposed', and also on the type of group whose representations we are studying. We will return to this question in Sections 2.2, 2.7, and 3.3.

For now, we content ourselves with studying the properties of irreducible representations.

1.3.1 Twisting (Irreducible) Unitary Representations

The following result can be useful in constructing additional irreducible representations.

Lemma 1.28 (Tensor products with characters). *Let χ be a unitary character of G, and let π be a unitary representation of G. Then*

$$(\chi \otimes \pi)(g) = \chi_g \pi_g$$

for $g \in G$ defines a unitary representation $\chi \otimes \pi$ on $\mathcal{H}_{\chi\otimes\pi} = \mathcal{H}_\pi$. The representation $\chi \otimes \pi$ is irreducible if and only if π is irreducible.

PROOF. We note that $(\chi \otimes \pi)(g)$ is unitary for every $g \in G$, and that

$$(\chi\otimes\pi)(g_1g_2) = \chi_{g_1g_2}\pi_{g_1g_2} = \chi_{g_1}\pi_{g_1}\chi_{g_2}\pi_{g_2} = (\chi\otimes\pi)(g_1)(\chi\otimes\pi)(g_2)$$

for every $g_1, g_2 \in G$, since $\chi_{g_2} \in \mathbb{S}^1$ is a scalar. To see continuity of the representation, we let $v \in \mathcal{H}_\pi$ and suppose that the sequence (g_n) in G satisfies $g_n \to e$ as $n \to \infty$. Then we also have

$$\|(\chi \otimes \pi)(g_n)v - v\| \leqslant \|\chi_{g_n}\pi_{g_n}v - \chi_{g_n}v\| + |\chi_{g_n} - 1|\|v\| \longrightarrow 0$$

as $n \to \infty$, giving continuity by Lemma 1.13. Finally, note that a closed subspace $\mathcal{V} \subseteq \mathcal{H}_\pi$ is invariant under π_g for $g \in G$ if and only if it is invariant under $(\chi \otimes \pi)(g) = \chi_g \pi_g$ for $g \in G$, which implies the last statement in the lemma. □

1.3.2 Schur's Lemma

The most fundamental property of irreducible representations is traditionally called Schur's lemma. Despite the modest title, it and its immediate corollaries in this section will have widespread use in the theory to come, and will be cited more than thirty times in this volume.

Theorem 1.29 (Schur's lemma[(2)]). *Suppose that π is an irreducible unitary representation and ρ is a unitary representation of the group G.*

(a) *If $B\colon \mathcal{H}_\pi \to \mathcal{H}_\pi$ is bounded and intertwining, then $B = \alpha I$ for some $\alpha \in \mathbb{C}$.*
(b) *If $B\colon \mathcal{H}_\pi \to \mathcal{H}_\rho$ is bounded and intertwining, then $B^*B = \alpha I$ for some constant $\alpha \geqslant 0$. If in addition $B \neq 0$ (equivalently, if $\alpha > 0$), then a scalar multiple of B is an intertwining isometry, and so π is contained in ρ.*

Proof. Suppose first that $B\colon \mathcal{H}_\pi \to \mathcal{H}_\pi$ is bounded and intertwining satisfying $B = B^*$ but with $B \neq \alpha I$ for all $\alpha \in \mathbb{R}$. By the spectral theorem for self-adjoint operators (see [25, Th. 12.55]) this implies that the spectrum $\sigma(B)$ of the operator B consists of more than just one point. This in turn implies that there exist functions $f_j \in C(\sigma(B))$ with $f_1 f_2 = 0$ and with $\|f_j\|_\infty = 1$ for $j = 1, 2$. By the continuous functional calculus (see [25, Sec. 12.4.1]) there exist operators $f_j(B) \in \mathrm{B}(\mathcal{H}_\pi)$ with $\|f_j(B)\| = 1$ for $j = 1, 2$ and with $f_1(B)f_2(B) = 0$. Moreover, each $f_j(B)$ can be obtained as a a uniform limit of polynomials in B, so $f_j(B)$ is intertwining for $j = 1, 2$. It follows that $\ker f_1(B)$ is closed, π-invariant, non-trivial since

$$\{0\} \subsetneq \operatorname{Im} f_2(B) \subseteq \ker f_1(B),$$

and proper since $f_1(B) \neq 0$. As this contradicts our assumption of irreducibility of π we deduce that $B = \alpha I$ for some $\alpha \in \mathbb{R}$.

If $B\colon \mathcal{H}_\pi \to \mathcal{H}_\pi$ is bounded and intertwining, then B^*, $\frac{B+B^*}{2}$, and $\frac{B-B^*}{2\mathrm{i}}$ are all bounded and intertwining too. Applying the argument above to the two self-adjoint intertwining operators $\frac{B+B^*}{2}$, and $\frac{B-B^*}{2\mathrm{i}}$, we see that

$$B = \tfrac{B+B^*}{2} + \mathrm{i}\tfrac{B-B^*}{2\mathrm{i}} = \alpha I$$

for some $\alpha \in \mathbb{C}$ as claimed in (a).

To prove (b), again notice that $B^* \colon \mathcal{H}_\rho \to \mathcal{H}_\pi$ is intertwining, and hence

$$B^*B \colon \mathcal{H}_\pi \to \mathcal{H}_\pi$$

is also. By (a) this implies that $B^*B = \alpha I$ for some $\alpha \geqslant 0$ since B^*B is a positive operator. If $\alpha > 0$, then we can replace B by the intertwining isometry defined by $U = \alpha^{-\frac{1}{2}}B$, which gives an isomorphism between $\mathcal{H}_\pi$ and a closed subspace of $\mathcal{H}_\rho$. □

Exercise 1.30 (Converse to Schur's lemma). Let π be a unitary representation of G. Suppose that intertwining for $B \in \mathrm{B}(\mathcal{H}_\pi)$ implies that $B = \alpha I$ for some $\alpha \in \mathbb{C}$. Show that π is irreducible.

Exercise 1.31 (Goursat's lemma for unitary representations). Let π and ρ be irreducible unitary representations of G. Describe all closed invariant subspaces of $\mathcal{H}_\pi \oplus \mathcal{H}_\rho$. Show, in particular, that either there are precisely four closed invariant subspaces (including the trivial one), or there are infinitely many.

1.3.3 Irreducible Representations Leading to Characters

Schur's lemma is all we need for the first (rather abstract) classification result for irreducible unitary representations.

Corollary 1.32 (Characters on the centre). *Let π be an irreducible unitary representation of the group G. Then π_g is equal to $\chi_g I$ for any element g of the centre*

$$C_G = \{g \in G \mid gh = hg \text{ for all } h \in G\}$$

of G, where $\chi = \chi^\pi \colon C_G \to \mathbb{S}^1$ is a unitary character on the closed subgroup C_G. In particular, if G is abelian, then any irreducible unitary representation is one-dimensional and is induced by a unitary character on G.

Proof. For $g_0 \in C_G$ the operator $B = \pi_{g_0}$ is intertwining since

$$B \circ \pi_g = \pi_{g_0 g} = \pi_{g g_0} = \pi_g \circ B$$

for all $g \in G$. Schur's lemma (Theorem 1.29) implies that $\pi_{g_0} = B = \chi_{g_0} I$ for some scalar $\chi_{g_0} \in \mathbb{C}$. Since B is unitary, $\chi_{g_0} \in \mathbb{S}^1$. For $g_0, g_1 \in C_G$ we also have

$$(\chi_{g_0} I)\,(\chi_{g_1} I) = \pi_{g_0}\pi_{g_1} = \pi_{g_0 g_1} = \chi_{g_0 g_1} I,$$

which shows that $\chi \colon C_G \to \mathbb{S}^1$ is a homomorphism. Finally, if $v \in \mathcal{H}_\pi$ is a unit vector, then

$$\chi_{g_0} = \chi_{g_0}\|v\|^2 = \langle \pi_{g_0} v, v\rangle$$

shows that $g_0 \mapsto \chi_{g_0}$ is continuous on C_G.

If in addition G is abelian, then the continuous character χ constructed above is defined on all of G, and $\pi_g = \chi_g I$ for all $g \in G$. This implies that any subspace is π-invariant. By irreducibility of π, we deduce that $\dim \mathcal{H}_\pi = 1$. □

Even if G is not abelian, the corollary above and the following related definition are useful for classifying unitary representations.

Definition 1.33 (Central characters). Let π be a unitary representation of G such that the restriction of π to the centre is given by $\pi_g = \chi_g I$ for a unitary character χ on C_G and all $g \in C_G$. Then χ is called the *central character of* π.

Example 1.34. If G is abelian, H satisfies our standing assumptions, and τ is an irreducible representation of $G \times H$, then G belongs to the centre of $G \times H$ and we may apply Corollary 1.32. It follows that $\tau|_G = \chi I$ for a character χ on G. Let $\rho = \tau|_H$ and extend both χ and ρ trivially to the product $G \times H$. Then $\tau = \chi \otimes \rho$ as in Lemma 1.28 for the (necessarily irreducible) unitary representation $\rho = \tau|_H$ of H.

The following exercise highlights the differences in the use of characters in Lemma 1.28 and Corollary 1.32.

Exercise 1.35. (a) Show that $\mathrm{SL}_2(\mathbb{R})$ has no non-trivial characters, but has a non-trivial centre and a non-trivial central character.
(b) Show that the representations $\pi^{\mathbb{S}^1,\xi}$ of $\mathrm{SL}_2(\mathbb{R})$ on $L^2(\mathbb{S}^1)$ found in Example 1.7 are not irreducible for any $\xi \in \mathbb{R}$.

1.3.4 Unbounded Intertwining Operators

Occasionally we will need a generalization of Schur's lemma for densely defined closed intertwining operators (see [25, Ch. 13] for more background on such operators).

Definition 1.36 (Closed operators). Let $\mathcal{H}, \mathcal{H}'$ be Hilbert spaces with a subspace $D_T \subseteq \mathcal{H}$, and $T\colon D_T \to \mathcal{H}'$ a linear map. We say that T is a *densely defined* operator from $\mathcal{H}$ to $\mathcal{H}'$ if $\overline{D_T} = \mathcal{H}$, and is *closed* if

$$\mathrm{Graph}(T) = \{(v, Tv) \mid v \in D_T\} \subseteq \mathcal{H} \oplus \mathcal{H}'$$

is a closed subspace.

Definition 1.37 (Closed intertwining operators). Let π and ρ be unitary representations of G. A densely defined closed operator T from $\mathcal{H}_\pi$ to $\mathcal{H}_\rho$ is *intertwining* if D_T is π-invariant and $T \circ \pi_g = \rho_g \circ T$ on D_T for all $g \in G$.

Corollary 1.38 (Schur's lemma for closed operators). *Let π be an irreducible unitary representation of the group G and ρ another unitary representation of G. If T is a densely defined closed intertwining operator from $\mathcal{H}_\pi$ to $\mathcal{H}_\rho$, then T is bounded and T is either the zero operator or is a multiple of a unitary isomorphism to a closed subspace of $\mathcal{H}_\rho$. Moreover, if $\rho = \pi$ then $T = \alpha I$ for some $\alpha \in \mathbb{C}$.*

PROOF. We will prove that T is in fact defined everywhere and is bounded, which will allow us to apply Schur's lemma (Theorem 1.29) for bounded operators.

By equivariance of T, we have that $(u, Tu) \in \operatorname{Graph}(T)$ implies

$$(\pi_g u, \rho_g T u) \in \operatorname{Graph}(T)$$

for all $g \in G$. Since we also assumed that $\operatorname{Graph}(T)$ is closed, it follows that $\operatorname{Graph}(T)$ is a closed $\pi \oplus \rho$-invariant subspace of $\mathcal{H}_\pi \oplus \mathcal{H}_\rho$, which implies that its orthogonal projection operator

$$P_{\operatorname{Graph}(T)} \colon \mathcal{H}_\pi \oplus \mathcal{H}_\rho \longrightarrow \operatorname{Graph}(T) \subseteq \mathcal{H}_\pi \oplus \mathcal{H}_\rho$$

is intertwining. Moreover, the embedding operator $\imath_1 \colon \mathcal{H}_\pi \to \mathcal{H}_\pi \oplus \mathcal{H}_\rho$ and the projection $P_1 \colon \mathcal{H}_\pi \oplus \mathcal{H}_\rho \to \mathcal{H}_\pi$ to the first coordinate are also intertwining from π to $\pi \oplus \rho$, respectively from $\pi \oplus \rho$ to π.

Hence we can apply Schur's lemma to the bounded intertwining operator $P_1 P_{\operatorname{Graph}(T)} \imath_1 \colon \mathcal{H}_\pi \to D_T \subseteq \mathcal{H}_\pi$ defined by

$$P_1 P_{\operatorname{Graph}(T)} \imath_1 \colon v \longmapsto P_{\operatorname{Graph}(T)}(v, 0) = (w, Tw) \longmapsto w$$

(see Figure 1.2). It follows that $P_1 P_{\operatorname{Graph}(T)}(v, 0) = \alpha_0 v$ for some $\alpha_0 \in \mathbb{C}$ and

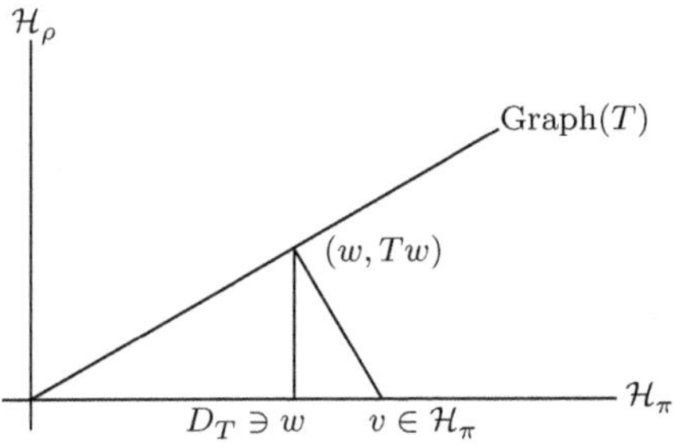

Fig. 1.2: The operator $P_1 P_{\operatorname{Graph}(T)}$.

all $v \in \mathcal{H}$. For a non-zero $v \in D_T$ (which must exist since T is densely defined) we have $(v, 0) \not\perp (v, Tv) \in \operatorname{Graph}(T)$ and so $P_{\operatorname{Graph}(T)}(v, 0) = (w, Tw) \neq 0$, which implies that $w = \alpha_0 v \neq 0$ and so $\alpha_0 \neq 0$. For a general $v \in \mathcal{H}_\pi$ this implies that $P_{\operatorname{Graph}(T)}(v, 0) = (\alpha_0 v, T(\alpha_0 v))$, hence also $\alpha_0 v \in D_T$ and so v lies in D_T. Moreover, $Tv = \alpha_0^{-1} T(\alpha_0 v) = \alpha_0^{-1} P_2 P_{\operatorname{Graph}(T)}(v, 0)$ for $v \in \mathcal{H}_\pi$ implies that T can be obtained from the orthogonal projection

$$P_2\colon \mathcal{H}_\pi \oplus \mathcal{H}_\rho \longrightarrow \mathcal{H}_\rho$$

onto the second coordinate, and the orthogonal projection $P_{\mathrm{Graph}(T)}$. Hence T is bounded and applying Schur's lemma (Theorem 1.29) gives the corollary. □

The following exercise gives a generalization of Lemma 1.21 to closed intertwining operators.

Exercise 1.39. Suppose π and ρ are unitary representations and T is a densely defined closed intertwining operator from $\mathcal{H}_\pi$ to $\mathcal{H}_\rho$. Show that $\pi|_{(\ker T)^\perp}$ is unitarily isomorphic to $\rho|_{\overline{\mathrm{Im}\, T}}$, and hence is contained in ρ.

1.3.5 Why the Hilbert Space is Always Complex*

†We want to show via a quick example how Schur's lemma in the formulation of Theorem 1.29(a) fails for irreducible representations on *real* Hilbert spaces. For this we define the multiplicative subgroup $G = \{\pm 1, \pm \mathbf{i}, \pm \mathbf{j}, \pm \mathbf{k}\}$ of the Hamiltonian quaternions

$$\mathbb{H} = \mathbb{R} + \mathbb{R}\mathbf{i} + \mathbb{R}\mathbf{j} + \mathbb{R}\mathbf{k},$$

which we will also consider as a real Hilbert space by declaring $1, \mathbf{i}, \mathbf{j}, \mathbf{k}$ an orthonormal basis. We define the representation π of G by left-multiplication on $\mathbb{H}$ and obtain that π is unitary and irreducible (since for any non-zero vector $v \in \mathbb{H}$ the vectors $v, \mathbf{i}v, \mathbf{j}v, \mathbf{k}v$ form a real basis of $\mathbb{H}$). Furthermore, for any $b \in \mathbb{H}$ the map $B\colon \mathbb{H} \to \mathbb{H}$ defined by $B\colon v \mapsto vb$ satisfies

$$\pi_g \circ B = B \circ \pi_g$$

for all $g \in G$, and hence Theorem 1.29(a) fails for this irreducible real representation.

In fact Schur's lemma over fields that are not algebraically closed involves in general the study of division algebras over that field. For our purposes and the applications we have in mind, this would only be an algebraic distraction, so we will always work with complex Hilbert spaces.

Exercise 1.40. Show that the representation theory of abelian groups also behaves differently over the reals, by finding a two-dimensional irreducible representation of $C_3 = \mathbb{Z}/3\mathbb{Z}$ over the reals.

---

† As explained on page vii, some sections (this being an example) are marked with a * if they are not crucial for the majority of the later chapters.

1.4 Finite Groups

Before we delve further into the functional-analytic aspects of the theory needed to study non-discrete locally compact groups, we wish to consider finite groups. For these many of the upcoming tools and the general theory are much simpler.

1.4.1 Modules Over the Group Ring

First of all it is helpful to extend a given unitary representation π of a finite group G to give $\mathcal{H}_\pi$ a module structure over a ring canonically associated to G. We will do this for a general group in Section 1.5.

In the case of a finite group this ring is given by the *group ring*

$$\mathbb{C}[G] = \left\{ \sum_{g\in G} c_g g \,\middle|\, c = (c_g) \in \mathbb{C}^G \right\}$$

consisting of formal complex linear combinations of group elements.† This becomes a ring on defining multiplication via the distributive law:

$$\begin{aligned}\left(\sum_{g\in G} c_g g\right)\left(\sum_{g\in G} d_g g\right) &= \sum_{g_1,g_2\in G} c_{g_1} d_{g_2} g_1 g_2\\ &= \sum_{g\in G}\left(\sum_{g_1 g_2 = g} c_{g_1} d_{g_2}\right) g\\ &= \sum_{g\in G}\left(\sum_{h\in G} c_h d_{h^{-1}g}\right) g\\ &= \sum_{g\in G}\left(\sum_{h\in G} c_{gh^{-1}} d_h\right) g.\end{aligned}$$

We note that the resulting multiplication is a form of *convolution*, and the general convolutions introduced in the next section will be fundamental for the whole theory.

Given a unitary representation π of the finite group G, we can now define the *convolution operators* on $\mathcal{H}_\pi$ by linear extension as

$$\pi_*\left(\sum_{g\in G} c_g g\right) : \mathcal{H}_\pi \ni v \longmapsto \sum_{g\in G} c_g \pi_g v \tag{1.14}$$

† Identifying $g \in G$ with the Dirac measure δ_g at g, we can identify $\mathbb{C}[G]$ with complex measures on G; even more concretely in this setting we can identify $\mathbb{C}[G]$ additively with $\mathbb{C}^G$ by viewing c as a vector.

for $\sum_{g \in G} c_g g \in \mathbb{C}[G]$.

Exercise 1.41. Let π be a unitary representation of a finite group G.
(a) Show that $\mathbb{C}[G]$ is a ring containing G in its group of units.
(b) Show that the convolution operator (1.14) gives $\mathcal{H}_\pi$ the structure of a module over $\mathbb{C}[G]$.

1.4.2 Decomposition Into Cyclic and Irreducible Subspaces

Let G be a finite group, π a unitary representation of G, and $v \in \mathcal{H}_\pi$. The smallest π-invariant subspace of $\mathcal{H}_\pi$ containing v is given by

$$\langle v \rangle_\pi = \pi_* \left(\mathbb{C}[G]\right) v.$$

It is called the *cyclic subspace generated by* v. If $\mathcal{H}_\pi = \langle v \rangle_\pi$, then $\mathcal{H}_\pi$ is called a *cyclic representation* and v is called a *generator*. In the case of a finite group, we have

$$\dim \langle v \rangle_\pi \leqslant \dim \mathbb{C}[G] = |G| < \infty,$$

which also implies that $\langle v \rangle_\pi$ is closed. In particular, every irreducible unitary representation of G is finite-dimensional.

It is quite straightforward to show that a general unitary representation π is a countable direct sum of cyclic subspaces (see Lemma 1.61). Applying Exercise 1.27 (see also its hint on page 526) to each of the finite-dimensional cyclic subspaces in the decomposition of π, we also see that π can be written as a countable direct sum of irreducible (and finite-dimensional) subspaces.

As a consequence of the construction above we also see that $\widehat{G}$ *separates points* in G in the following sense. If $g_1 \neq g_2$ in G then there exists $\pi \in \widehat{G}$ with $\pi_{g_1} \neq \pi_{g_2}$. Indeed, for the regular representation λ of G on $L^2(G) \cong \mathbb{C}[G]$ we have $\lambda_{g_1} \neq \lambda_{g_2}$ since, for example, $\lambda_g(\mathbb{1}_{\{e\}}) = \mathbb{1}_{\{g\}}$ for $g \in G$. However, by the argument above λ is a direct sum of irreducible representations and so there must exist $\pi \in \widehat{G}$ with $\pi_{g_1} \neq \pi_{g_2}$. We will prove this separation result in the general setting near the end of this chapter in Section 1.7.4.

As we will see in Section 3.3, the study of unitary representations of compact groups has many similarities to the case of finite groups. Another fundamental result for finite groups is the description of the left regular representation in terms of all elements of $\widehat{G}$ given by the Peter–Weyl theorem. This result also extends to compact groups, as discussed in Section 3.4.

Exercise 1.42. Let G be a finite group. Show that every cyclic representation (and so, in particular, every irreducible representation) is contained in $L^2(G)$.

1.4.3 Abelian Groups

For a finite abelian group G, Corollary 1.32 shows that $\widehat{G}$ consists only of characters. Recall also that G is isomorphic to a direct product of cyclic groups, which allows one to reduce the general case to the case of cyclic groups as in the next exercise. We note that Chapter 2 is devoted to obtaining a firm understanding of locally compact abelian groups, their unitary representations, and their characters.

Essential Exercise 1.43. For $n \geqslant 2$ denote the cyclic group with n elements by $\mathrm{C}_n = \mathbb{Z}/n\mathbb{Z}$, and let k_0 be a generator of C_n. Show that every (unitary or otherwise) representation π of C_n can be split into a sum of n invariant subspaces $\mathcal{V}_t$ for $t = 0, \ldots, n-1$ such that $\pi_{k_0}|_{\mathcal{V}_t}$ is multiplication by $\mathrm{e}^{2\pi\mathrm{i}t/n}$.

1.4.4 Dihedral Groups*

As a concrete example of a complete description of the unitary dual of a non-abelian group, we discuss here the dihedral groups.

The dihedral group D_n for $n \geqslant 2$ is defined as the group of all Euclidean symmetries of the regular n-gon in $\mathbb{R}^2$. For convenience, we identify $\mathbb{R}^2$ with $\mathbb{C}$, let $M = 0$ be the centre, and let $P_1 = 1$ be a vertex of the regular n-gon. One symmetry is the rotation k_0 through angle $\frac{2\pi}{n}$, defined by $k_0 \colon \mathbb{C} \ni z \mapsto \mathrm{e}^{\frac{2\pi\mathrm{i}}{n}} z$, which generates an abelian subgroup $\mathrm{C}_n < \mathrm{D}_n$ isomorphic to $\mathbb{Z}/n\mathbb{Z}$. However, there are also reflections, for example $r_0 \colon \mathbb{C} \ni z \mapsto \overline{z}$ belongs to D_n.

More formally, let C_n be the cyclic group of order n generated by $k_0 \in \mathrm{C}_n$ and define the dihedral group D_n as the semi-direct product $\mathbb{Z}/2\mathbb{Z} \ltimes \mathrm{C}_n$ generated by C_n and the involution $r_0 \in \mathbb{Z}/2\mathbb{Z}$ with $r_0 h r_0 = h^{-1}$ for $h \in \mathrm{C}_n$.

We will study irreducible unitary representations of D_n via their restriction to the abelian normal subgroup C_n, leading to two types of representation.

(Old) We say that an irreducible representation π of D_n is an *old representation* with respect to $\mathrm{C}_n \lhd \mathrm{D}_n$ if $\pi|_{\mathrm{C}_n}$ is trivial. It follows that these can also be thought of as unitary representations of $\mathrm{D}_n/\mathrm{C}_n \cong \mathbb{Z}/2\mathbb{Z}$. Thus there are two old unitary representations: the trivial representation $\mathbb{1}$ and the representation defined by the character sign, where

$$\mathrm{sign}(g) = \begin{cases} 1 & \text{if } g \in \mathrm{C}_n, \\ -1 & \text{if } g \in r_0\mathrm{C}_n. \end{cases}$$

(New) We say that an irreducible representation π of D_n is a *new representation* with respect to $\mathrm{C}_n \lhd \mathrm{D}_n$ if $\pi|_{\mathrm{C}_n}$ is non-trivial. As we will see $\mathcal{H}_\pi$ will be a direct sum of eigenspaces for π_{k_0}. Since π is assumed to be irreducible for D_n, these eigenspaces cannot also be invariant under r_0 unless there is only one.

Construction of New Representations

Let χ be a character of C_n that satisfies $\chi(C_n) \not\subseteq \mathbb{R}$, so that $\chi \neq \overline{\chi}$. From this (automatically unitary) character, we define a new two-dimensional representation using a very special case of the induced representation construction denoted in this case by $\pi = \operatorname{Ind}_{C_n}^{D_n}(\chi)$. We let $\mathcal{H}_\pi = \mathbb{C}^2$ and let C_n act on the first coordinate via the character χ and on the second via $\overline{\chi}$, so

$$\pi_g(\alpha_1, \alpha_2) = (\chi(g)\alpha_1, \overline{\chi}(g)\alpha_2)$$

for all $g \in C_n$ and $(\alpha_1, \alpha_2) \in \mathbb{C}^2$. For the element r_0 we define

$$\pi_{r_0}(\alpha_1, \alpha_2) = (\alpha_2, \alpha_1)$$

for all $(\alpha_1, \alpha_2) \in \mathbb{C}^2$. We note that $\pi|_{C_n}$ and $\pi|_{\langle r_0 \rangle}$ are unitary representations of C_n, resp. $\langle r_0 \rangle \cong \mathbb{Z}/2\mathbb{Z}$. To see that this extends to a representation of D_n, we verify that

$$\begin{aligned}
\pi_{r_0}\pi_h\pi_{r_0^{-1}}(\alpha_1, \alpha_2) &= \pi_{r_0}\pi_h(\alpha_2, \alpha_1) \\
&= \pi_{r_0}(\chi(h)\alpha_2, \overline{\chi(h)}\alpha_1) \\
&= (\overline{\chi(h)}\alpha_1, \chi(h)\alpha_2) = \pi_{h^{-1}}(\alpha_1, \alpha_2)
\end{aligned}$$

for $h \in C_n$ and $(\alpha_1, \alpha_2) \in \mathbb{C}^2$ and note that this matches the calculation $r_0 h r_0^{-1} = h^{-1}$ for $h \in C_n$.We now define $\pi(hr_0^j) = \pi(h)\pi(r_0^j)$ for $h \in C_n$ and $j \in \{0, 1\}$, which implies that

$$\begin{aligned}
\pi(h_1 r_0^{j_1})\pi(h_2 r_0^{j_2}) &= \pi(h_1)\pi(r_0)^{j_1}\pi(h_2)\pi(r_0^{j_2}) \\
&= \pi(h_1)\underbrace{\pi(r_0)^{j_1}\pi(h_2)\pi(r_0)^{-j_1}}_{=\pi(r_0^{j_1}h_2 r_0^{-j_1})}\pi(r_0)^{j_1+j_2} \\
&= \pi(h_1 r_0^{j_1} h_2 r_0^{-j_1})\pi(r_0^{j_1+j_2}) = \pi(h_1 r_0^{j_1} h_2 r_0^{j_2}),
\end{aligned}$$

for all $h_1, h_2 \in C_n$ and $j_1, j_2 \in \{0, 1\}$, as required.

To see that this is an irreducible representation, suppose that $\mathcal{V} \subseteq \mathcal{H}_\pi$ is non-trivial and π-invariant. Then it is also invariant under π_{k_0}, which is the diagonal matrix with eigenvalues $\chi(k_0) \neq \overline{\chi(k_0)}$. Since $\mathcal{V}$ is non-trivial and invariant under π_{k_0}, it follows that $\mathcal{V}$ must contain the eigenspace $\mathbb{C} \times \{0\}$ or the eigenspace $\{0\} \times \mathbb{C}$. As π_{r_0} switches these two subspaces, and $\mathcal{V}$ is invariant under π_{r_0}, we must have $\mathcal{V} = \mathbb{C}^2$, and so π is irreducible. We also note that the new unitary representation defined by χ is isomorphic to the unitary representation defined by $\overline{\chi}$.

Suppose now that χ is a character of C_n with $\chi(k_0) = -1$, which is only possible if n is even (as the multiplicative order of $\chi(k_0)$ must be a divisor of n). Then the construction above does not give an irreducible representation (as the diagonal $\{(\alpha, \alpha) \mid \alpha \in \mathbb{C}\}$ and its orthogonal complement are invariant under π).

In this case the subgroup $C_n^2 = \langle k_0^2 \rangle$ is a normal subgroup of D_n with

$$D_n/C_n^2 \cong \mathbb{Z}/2\mathbb{Z} \times \mathbb{Z}/2\mathbb{Z}$$

and we can extend χ in two ways to a unitary character of D_n. These define two more irreducible unitary representations (corresponding to the diagonal and its orthogonal complement).

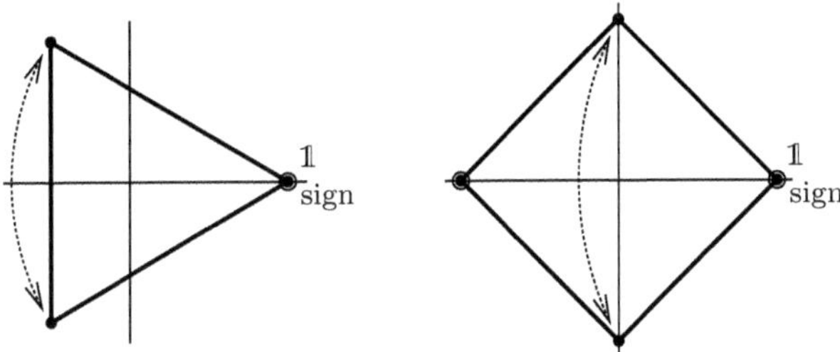

Fig. 1.3: The characters of C_n for $n = 3$ and $n = 4$ naturally form an equilateral triangle and square in $\mathbb{C}$ respectively. The real characters give rise to two irreducible unitary representations as indicated by the circles around the dots. The non-real characters arise in pairs, and each pair together gives rise to one two-dimensional unitary representation indicated by the dashed arrows.

Completeness

We now show that the constructions above give the complete list of irreducible unitary representations of D_n. So let π be an irreducible unitary representation of D_n. If $\pi|_{C_n}$ is trivial, then π is an old representation defined by a representation of $D_n/C_n \cong \mathbb{Z}/2\mathbb{Z}$, which is defined by a unitary character by Corollary 1.32 (either $\mathbb{1}$ or sign).

So suppose $\pi|_{C_n}$ is non-trivial. Then (by Exercise 1.43) $\mathcal{H}_\pi$ must have a non-trivial eigenspace $\mathcal{V}_\chi$ for a non-trivial character χ on C_n. For $v \in \mathcal{V}_\chi$ we have $\pi_h v = \chi(h)v$, and so

$$\pi_h \pi_{r_0} v = \pi_{r_0} \pi_{r_0} \pi_h \pi_{r_0} v = \pi_{r_0} \pi_{h^{-1}} v = \overline{\chi(h)} \pi_{r_0} v$$

for all $h \in C_n$. This shows that $\pi_{r_0}(\mathcal{V}_\chi) \subseteq \mathcal{V}_{\overline{\chi}}$. Using the same argument for $\overline{\chi}$ gives that $\pi_{r_0} \mathcal{V}_\chi = \mathcal{V}_{\overline{\chi}}$ is the eigenspace for eigenvalue $\overline{\chi}$.

If now $\chi = \overline{\chi}$, then n is even and $\mathcal{V}_\chi$ is invariant under $\langle C_n, r_0 \rangle = D_n$, and so $\mathcal{V}_\chi = \mathcal{H}_\pi$. Hence π is a representation induced by $D_n/C_n^2 \cong \mathbb{Z}/2\mathbb{Z} \times \mathbb{Z}/2\mathbb{Z}$ and so is defined by a character on this group by Corollary 1.32.

Finally, suppose that $\chi \neq \overline{\chi}$. Then we may choose some unit vector $v \in \mathcal{V}_\chi$ and see that $\mathcal{V} = \mathbb{C}v + \mathbb{C}\pi_{r_0} v$ is invariant under $\pi(C_n)$ and π_{r_0}, and hence must be all of $\mathcal{H}_\pi$. This implies that π is isomorphic to the new two-dimensional unitary representation constructed above.

1.4.5 Generalizations*

We will generalize the method used in Section 1.4.4 in an infinite setting in Chapter 5, building on the results in Chapter 2. For now we give a few more finite groups that can be treated similarly.

Exercise 1.44 (The affine group over $\mathbb{F}_p$). Let $p \in \mathbb{N}$ be a prime, and define the characteristic p analogue

$$G = \mathbb{F}_p^\times \ltimes \mathbb{F}_p = \left\{ \begin{pmatrix} a & b \\ 0 & 1 \end{pmatrix} \,\middle|\, a \in \mathbb{F}_p^\times, b \in \mathbb{F}_p \right\}$$

of the affine group. Show that $\widehat{G}$ consists of $(p-1)$ old representations defined by characters on $\mathbb{F}_p^\times \cong \mathbb{Z}/(p-1)\mathbb{Z}$ and one new irreducible representation on $\mathbb{C}^{p-1}$.

Exercise 1.45 (The symmetric group S_3). (a) Use the above to describe the unitary dual of the symmetric group $\mathrm{S}_3 \cong \mathrm{D}_3$ explicitly. Note that the alternating subgroup $\mathrm{A}_3 \lhd \mathrm{S}_3$ corresponds to the normal subgroup $\mathrm{C}_3 \lhd \mathrm{D}_3$ considered above.
(b) Let V be a complex vector space carrying a linear representation ρ of S_3. Define the *symmetric subspace* $V_{\mathrm{s}} = V^{\mathrm{S}_3}$ of invariant vectors, the *skew-symmetric subspace*

$$V_{\mathrm{ss}} = \{v \in V \mid \rho(\tau)v = \mathrm{sign}(\tau)v \text{ for all } \tau \in \mathrm{S}_3\},$$

and the *cycle-symmetric subspace*

$$V_{\mathrm{cs}} = \{v \in V \mid v + \rho(\sigma)v + \rho(\sigma^2)v = 0\},$$

where $\sigma = (231)$ is the cyclic permutation and generator of A_3. Show that these subspaces are ρ-invariant, and that

$$V = V_{\mathrm{s}} \oplus V_{\mathrm{ss}} \oplus V_{\mathrm{cs}}.$$

Find intertwining projection maps from V to V_{s}, to V_{ss}, and to V_{cs}. Furthermore, show that for any non-zero $v \in V_{\mathrm{s}} \cup V_{\mathrm{ss}}$ the restriction of ρ to $\langle \rho(\mathrm{S}_3)v \rangle$ is isomorphic to one of the old irreducible unitary representations of S_3 described in (a). What happens for $v \in V_{\mathrm{cs}}$?
(c) Apply part (b) to the vector space $V = \{f \colon \mathbb{R}^3 \to \mathbb{C}\}$ equipped with the representation induced by permutation of the coordinates in $\mathbb{R}^3$. Show that in this case each of the subspaces in part (b) is non-trivial.

For the group S_4 the above procedure can be repeated to a large extent. However, for $n \geqslant 5$ the normal subgroup $\mathrm{A}_n \lhd \mathrm{S}_n$ is simple, and so the procedure discussed here does not apply. We will not discuss the representations of S_n in detail, referring to Fulton and Harris [32, Ch. 4] or Kowalski [57] for the details. As mentioned before, many general results about the representation theory of finite groups will follow as a special case of the discussions in Chapter 3 concerning compact groups.

1.5 Convolution Operators

In this section we introduce two Banach algebras related to the group G which will allow us to create additional linear structures associated to unitary representations.

1.5.1 The Banach Algebra $L^1(G)$

The multiplication of $f_1, f_2 \in L^1(G)$ in the Banach algebra $L^1(G)$ is defined by the convolution

$$\begin{aligned} f_1 * f_2(g) &= \int f_1(h) f_2(h^{-1}g) \, \mathrm{d}m(h) \\ &= \int f_1(gk) f_2(k^{-1}) \, \mathrm{d}m(k) && (1.15) \\ &= \int f_1(g\ell^{-1}) f_2(\ell) \Delta(\ell)^{-1} \, \mathrm{d}m(\ell) && (1.16) \end{aligned}$$

for any $g \in G$, where we used the measure-preserving substitution $h = gk$ and in (1.16) the possibly non-measure-preserving substitution $\ell = \imath(k) = k^{-1}$ from Lemma 1.16. The defining Banach algebra inequality follows by using Fubini's theorem since

$$\begin{aligned} \|f_1 * f_2\|_1 &= \int |f_1 * f_2(g)| \, \mathrm{d}m(g) \\ &\leqslant \iint |f_1(h)| |f_2(h^{-1}g)| \, \mathrm{d}m(h) \, \mathrm{d}m(g) \\ &= \int |f_1(h)| \int |f_2(g)| \, \mathrm{d}m(g) \, \mathrm{d}m(h) = \|f_1\|_1 \|f_2\|_1 \end{aligned}$$

(which also proves that the integral defining $f_1 * f_2(g)$ exists for m-almost every $g \in G$). As G is assumed to be locally compact σ-compact metric, the Banach algebra $L^1(G)$ is separable (see [25, Prop. 3.91]).

On occasion we will look at *convolution powers*, that is, powers of functions $f \in L^1(G)$ with respect to convolution. We will use the notation f^{*n} defined inductively by $f^{*1} = f$ and $f^{*(n+1)} = f * f^{*n}$ for $n \in \mathbb{N}$.

In the absence of a convolution identity, an often adequate replacement is provided by the next lemma.

Proposition 1.46 (Approximate convolution identity). *Let (B_n) be a decreasing sequence of compact neighbourhoods of $e \in G$ that form a basis of the neighbourhood of e, and let (ψ_n) be a sequence in $L^1(G)$ with the property that* $\operatorname{supp} \psi_n \subseteq B_n$, $\psi_n \geqslant 0$, *and* $\int \psi_n \, \mathrm{d}m = 1$ *for all* $n \geqslant 1$. *Then we have*

$$\lim_{n \to \infty} \psi_n * f = \lim_{n \to \infty} f * \psi_n = f$$

for all $f \in L^1(G)$*, where the convergence is in* $L^1(G)$*.*

PROOF. By the density of $C_c(G) \subseteq L^1(G)$ (see [25, Prop. 2.51]) and the Banach algebra inequality, it is enough to consider $f \in C_c(G)$. Then, by continuity of f,

$$\left|\psi_n * f(g) - f(g)\right| \leqslant \int_{B_n} \psi_n(h) \left|f(h^{-1}g) - f(g)\right| \mathrm{d}m(h) \longrightarrow 0$$

as $n \to \infty$ for every $g \in G$. Clearly

$$|\psi_n * f(g)| \leqslant \|f\|_\infty$$

for all $n \geqslant 1$ and $\psi_n * f(g) = 0$ for $g \notin B_1 \operatorname{supp} f$ which gives $\psi_n * f \to f$ in $L^1(G)$ as $n \to \infty$ by dominated convergence.

For convolution on the right we use the third formula (1.16) in the definition of convolution in the form

$$f * \psi_n(g) = \int f(g\ell^{-1}) \Delta(\ell)^{-1} \psi_n(\ell) \, \mathrm{d}m(\ell)$$

and argue as before using continuity of $f \in C_c(G)$ and of Δ. In fact we again have

$$|f * \psi_n(g) - f(g)| \leqslant \int_{B_n} \left|f(g\ell^{-1})\Delta(\ell)^{-1} - f(g)\right| \psi_n(\ell) \, \mathrm{d}m(\ell) \longrightarrow 0$$

as $n \to \infty$ for every $g \in G$, $\|f * \psi_n\|_\infty \leqslant \|f\|_\infty \|\Delta^{-1}|_{B_1}\|_\infty$ and

$$\operatorname{supp}(f * \psi_n) \subseteq (\operatorname{supp} f) B_1$$

for all $n \geqslant 1$. Together with dominated convergence, this gives the lemma. □

The Banach algebra $L^1(G)$ is also a Banach $*$-algebra, where the involution or star operator $f \mapsto f^*$ is defined by

$$f^*(g) = \overline{f(g^{-1})} \Delta(g)^{-1}.$$

Indeed, it is clear that f^* is a conjugate-linear function of f,

$$\|f^*\|_1 = \int |f(g^{-1})| \Delta(g)^{-1} \, \mathrm{d}m(g) = \|f\|_1$$

by Lemma 1.16, and

$$(f^*)^* (g) = \overline{f^*(g^{-1})} \Delta(g)^{-1} = f(g) \Delta(g^{-1})^{-1} \Delta(g)^{-1} = f(g)$$

for all $g \in G$. Finally, for $f_1, f_2 \in L^1(G)$ we have

$$\begin{aligned} f_1^* * f_2^*(g) &= \int f_1^*(h) f_2^*(h^{-1}g) \, \mathrm{d}m(h) \\ &= \int \overline{f_1(h^{-1})} \Delta(h)^{-1} \overline{f_2(g^{-1}h)} \Delta(h^{-1}g)^{-1} \, \mathrm{d}m(h) \\ &= \overline{\int f_2(g^{-1}h) f_1(h^{-1}) \, \mathrm{d}m(h)} \Delta(g)^{-1} \\ &= \overline{f_2 * f_1(g^{-1})} \Delta(g)^{-1} = (f_2 * f_1)^* (g) \end{aligned}$$

by definition and the second formula (1.15) in the definition of convolution.

We conclude our discussion of $L^1(G)$ by noting that if G is abelian, then $L^1(G)$ is also commutative. Indeed, for $f_1, f_2 \in L^1(G)$ and $g \in G$

$$f_1 * f_2(g) = \int f_1(h) f_2(g-h) \, \mathrm{d}m(h) = \int f_2(k) f_1(g-k) \, \mathrm{d}m(k) = f_2 * f_1(g)$$

by the measure preserving substitution $k = g - h$ (formally by $\Delta \equiv 1$ and Lemma 1.16).

Exercise 1.47. Show that $L^1(G)$ has a unit if and only if G is discrete.

Exercise 1.48. Generalize Proposition 1.46 by relaxing the condition on ψ_n and assuming that $\psi_n \geqslant 0$, $\int \psi_n \, \mathrm{d}m = 1$, and $\int_{B_n} \psi_n \, \mathrm{d}m \to 1$ as $n \to \infty$ instead of $\operatorname{supp} \psi_n \subseteq B_n$ for all $n \in \mathbb{N}$.

Essential Exercise 1.49. Show, for $f_1 \in L^1(G)$ and $f_2 \in L^\infty(G)$ (and also for $f_1 \in C_c(G)$ and $f_2 \in C(G)$), that the convolution product $f_1 * f_2$ exists everywhere and defines a continuous function on G. Show that

$$\operatorname{supp}(f_1 * f_2) \subseteq (\operatorname{supp} f_1)(\operatorname{supp} f_2)$$

and conclude that $C_c(G) * C_c(G) \subseteq C_c(G)$.

1.5.2 The Banach Algebra $M(G)$

The Banach $*$-algebra $L^1(G)$ is actually a closed sub-algebra of the much larger Banach $*$-algebra $M(G)$, called the *measure algebra*, which we introduce now.

For this, recall that a complex-valued (Borel) measure ν on G can be defined† by $\mathrm{d}\nu = f_\nu \, \mathrm{d}\mu$, where μ is a σ-finite measure on G and $f_\nu \in L^1_\mu(G)$ (see [25, App. B.5]). We define $M(G)$ to be the space of all complex-valued measures $\mathrm{d}\nu = f_\nu \, \mathrm{d}\mu$ on G with the norm $\|\nu\| = \int |f_\nu| \, \mathrm{d}\mu$. By the Riesz representation theorem (see [25, Thm. 7.54]), $M(G)$ can be identified as a Banach space with the dual‡ $C_0(G)'$ so that $\|\nu\|$ is equal to the operator norm

† In effect, this is describing a measure by how it integrates functions. For convenience, we will talk about both ν and $\mathrm{d}\nu$ as a measure.

‡ The dual of a Banach space V is often denoted V^*, but we have chosen to write V' to avoid confusion with the $*$-operator in $L^1(G)$, the adjoint, and several convolution operators.

associated to the linear functional $C_0(G) \ni F \mapsto \int F \,\mathrm{d}\nu$. For $\nu \in M(G)$ and $\mathrm{d}\nu = f_\nu \,\mathrm{d}\mu$ as above, we define $\overline{\nu}$ by $\mathrm{d}\overline{\nu} = \overline{f_\nu} \,\mathrm{d}\mu$ or, equivalently, as the functional $C_0(G) \ni F \mapsto \int F\overline{f_\nu} \,\mathrm{d}\mu$.

Using the fact that G is a group, we can now equip $M(G)$ with a multiplication and an involution. The convolution product of $\nu_1, \nu_2 \in M(G)$ is defined by the functional sending $F \in C_0(G)$ to

$$\begin{aligned}\int F \,\mathrm{d}(\nu_1 * \nu_2) &= \iint F(g_1 g_2) \,\mathrm{d}\nu_1(g_1) \,\mathrm{d}\nu_2(g_2)\\ &= \iint F(g_1 g_2) f_{\nu_1}(g_1) f_{\nu_2}(g_2) \,\mathrm{d}\mu_1(g_1) \,\mathrm{d}\mu_2(g_2),\end{aligned} \tag{1.17}$$

where as above $\mathrm{d}\nu_1 = f_{\nu_1} \,\mathrm{d}\mu_1$, $\mathrm{d}\nu_2 = f_{\nu_2} \,\mathrm{d}\mu_2$, and the order of integration is irrelevant due to Fubini's theorem. We note that the definition immediately implies that the convolution of measures is commutative if the group G is abelian. The convolution is also associative since for $F \in C_0(G)$ and $\nu_1, \nu_2, \nu_3 \in M(G)$ we have

$$\begin{aligned}\int F \,\mathrm{d}(\nu_1 * (\nu_2 * \nu_3)) &= \iint F(g_1 h) \,\mathrm{d}(\nu_2 * \nu_3)(h) \,\mathrm{d}\nu_1(g_1)\\ &= \iiint F(g_1 g_2 g_3) \,\mathrm{d}\nu_1(g_1) \,\mathrm{d}\nu_2(g_2) \,\mathrm{d}\nu_3(g_3)\\ &= \int F \,\mathrm{d}((\nu_1 * \nu_2) * \nu_3).\end{aligned}$$

Here we used from the first to the second line the definition of $\nu_2 * \nu_3$ and the fact that $F(g_1 \cdot) \in C_0(G)$ for any fixed $g_1 \in G$.

The Banach algebra inequality $\|\nu_1 * \nu_2\| \leqslant \|\nu_1\| \|\nu_2\|$ for $\nu_1, \nu_2 \in M(G)$ follows from

$$\begin{aligned}\left|\int F \,\mathrm{d}(\nu_1 * \nu_2)\right| &\leqslant \iint |F(g_1 g_2)| |f_{\nu_1}(g_1)| |f_{\nu_2}(g_2)| \,\mathrm{d}\mu_1(g_1) \,\mathrm{d}\mu_2(g_2)\\ &\leqslant \|F\|_\infty \int |f_{\nu_1}(g_1)| \,\mathrm{d}\mu_1(g_1) \int |f_{\nu_2}(g_2)| \,\mathrm{d}\mu_2(g_2)\\ &= \|F\|_\infty \|\nu_1\| \|\nu_2\|,\end{aligned}$$

where $\mathrm{d}\nu_1 = f_{\nu_1} \,\mathrm{d}\mu_1$ and $\mathrm{d}\nu_2 = f_{\nu_2} \,\mathrm{d}\mu_2$ as above, and $F \in C_0(G)$. Notice that $M(G)$ is always a unital algebra, since the Dirac measure δ_e is a multiplicative unit in $M(G)$ (recall that the Dirac measure δ_g for a given $g \in G$ is defined by $\delta_g(B) = 1$ if $g \in B$ and $\delta_g(B) = 0$ otherwise).

We define the involution $\nu \mapsto \nu^*$ on $M(G)$ by

$$\int F \,\mathrm{d}\nu^* = \overline{\int \overline{F}(g^{-1}) \,\mathrm{d}\nu(g)} = \int F(g^{-1}) \,\mathrm{d}\overline{\nu}(g) = \int F(g^{-1}) \overline{f_\nu}(g) \,\mathrm{d}\mu(g) \tag{1.18}$$

for $F \in C_0(G)$. We also have $(\nu^*)^* = \nu$ since

$$\int F \,\mathrm{d}(\nu^*)^* = \overline{\int \overline{F}(g^{-1}) \,\mathrm{d}\nu^*(g)} = \int F\big((g^{-1})^{-1}\big) \,\mathrm{d}\nu(g) = \int F \,\mathrm{d}\nu$$

for $F \in C_0(G)$. The involution is again conjugate-linear, satisfies $\|\nu^*\| = \|\nu\|$ since

$$\left|\int F \,\mathrm{d}\nu^*\right| \leqslant \int |F(g^{-1})||f_\nu(g)| \,\mathrm{d}\mu(g) \leqslant \|F\|_\infty \|\nu\|,$$

and has $(\nu_1 * \nu_2)^* = \nu_2^* * \nu_1^*$ since

$$\begin{aligned}
\int F \,\mathrm{d}(\nu_1 * \nu_2)^* &= \overline{\int \overline{F}(g^{-1}) \,\mathrm{d}\nu_1 * \nu_2(g)} \\
&= \overline{\iint \overline{F}((g_1 g_2)^{-1}) \,\mathrm{d}\nu_1(g_1)\nu_2(g_2)} \\
&= \iint F(g_2^{-1} g_1^{-1}) \,\mathrm{d}\overline{\nu_1}(g_1) \,\mathrm{d}\overline{\nu_2}(g_2) = \int F \,\mathrm{d}(\nu_2^* * \nu_1^*)
\end{aligned}$$

for $F \in C_0(G)$.

We may identify $f \in L^1(G)$ with the finite complex-valued measure ν_f defined by $\mathrm{d}\nu_f = f \,\mathrm{d}m$, which shows that $L^1(G)$ is a closed subspace of the Banach space $M(G)$. It is also a sub-algebra, meaning that the convolutions of functions in $L^1(G)$ and of measures in $M(G)$ are compatible. Indeed, for any $F \in C_0(G)$ we have

$$\begin{aligned}
\int F \,\mathrm{d}\nu_{f_1 * f_2} &= \int F(g) f_1 * f_2(g) \,\mathrm{d}m \\
&= \iint F(g) f_1(h) f_2(h^{-1}g) \,\mathrm{d}m(h) \,\mathrm{d}m(g) \\
&= \iint F(hk) f_1(h) f_2(k) \,\mathrm{d}m(h) \,\mathrm{d}m(k) = \int F \,\mathrm{d}(\nu_{f_1} * \nu_{f_2})
\end{aligned}$$

by the substitution $k = h^{-1}g$ for a fixed h. In particular—something we failed to verify above—it follows that convolution in $L^1(G)$ is associative.

Moreover, we claim that the involution on $M(G)$, when restricted to $L^1(G)$, also gives the previously defined involution on $L^1(G)$. To see this, assume that $\mathrm{d}\nu_f = f \,\mathrm{d}m$ for some $f \in L^1(G)$. Then

$$\int F \,\mathrm{d}\nu_f^* = \int F(g^{-1}) \overline{f(g)} \,\mathrm{d}m = \int F(g) \overline{f(g^{-1})} \Delta(g)^{-1} \,\mathrm{d}m = \int F \,\mathrm{d}\nu_{f^*}$$

for any $F \in C_0(G)$ by Lemma 1.16.

We note that even though our definitions (1.17) of the convolution product and (1.18) of the involution only used $F \in C_0(G)$, these formulas hold more generally, and in particular for $F \in C_b(G)$. To see this, choose a sequence (χ_n) in $C_c(G)$ with $0 \leqslant \chi_n \leqslant 1$ for all $n \geqslant 1$ and with $\lim_{n\to\infty} \chi_n(g) = 1$ for all $g \in G$ (by using Urysohn's lemma and our standing assumptions on G). Then (1.17) and (1.18) hold for $F_n = F\chi_n$, and applying dominated convergence gives the same formulas for $F \in C_b(G)$.

Finally, notice that the convolution of the Dirac measure δ_g for $g \in G$ and the measure ν_f corresponding to some $f \in L^1(G)$ is given by

$$\delta_g * \nu_f = \nu_{\lambda_g f} \tag{1.19}$$

since

$$\begin{aligned}\int F \,\mathrm{d}(\delta_g * \nu_f) &= \iint F(hk)\,\mathrm{d}\delta_g(h)\,\mathrm{d}\nu_f(k) = \int F(gk)f(k)\,\mathrm{d}m(k)\\ &= \int F(h)f(g^{-1}h)\,\mathrm{d}m(h) = \int F\,\mathrm{d}\nu_{\lambda_g f}\end{aligned}$$

for any $F \in C_0(G)$.

In many ways $M(G)$ is too big for its own good. For example, it is not separable (unless G is discrete, in which case $M(G) = L^1(G)$), and the dual space of $M(G)$ is (depending on mathematical taste) monstrous. For this and other reasons we will mostly restrict ourselves to $L^1(G)$, whose dual space is simply $L^\infty(G)$.

Exercise 1.50. Show that $L^1(G) \subseteq M(G)$ is a two-sided ideal.

Exercise 1.51. Prove the analogue $\nu_f * \delta_g = \nu_{\rho_{g^{-1}} f}$ to (1.19) for all $f \in L^1(G)$ and $g \in G$, where $\rho_g f(h) = \Delta(g) f(hg)$ for $g, h \in G$ and $f \in L^1(G)$.

1.5.3 Convolution and Unitary Representations

Now let π be a unitary representation of G on the Hilbert space $\mathcal{H}_\pi$. As we now show, $\mathcal{H}_\pi$ is automatically a module over the Banach algebras $L^1(G)$ and $M(G)$. Indeed, for any $\nu \in M(G)$ and $u \in \mathcal{H}_\pi$ we can define the operator $\pi_*(\nu)$ by the weak integral

$$\pi_*(\nu)u = \int \pi_g u \,\mathrm{d}\nu(g),$$

which means we define $\pi_*(\nu)u$ by the Fréchet–Riesz representation theorem (see [25, Cor. 3.19] for example) and the formula

$$\langle \pi_*(\nu)u, w\rangle = \int \langle \pi_g u, w\rangle \,\mathrm{d}\nu(g) \tag{1.20}$$

for all $w \in \mathcal{H}_\pi$. Indeed, the right-hand side of (1.20) depends sesqui-linearly on the pair (u, w) (that is, linearly on u and conjugate-linearly on w) and satisfies

$$\left|\int \langle \pi_g u, w\rangle \,\mathrm{d}\nu\right| = \left|\int \langle \pi_g u, w\rangle f(g)\,\mathrm{d}\mu(g)\right|$$
$$\leqslant \int |\langle \pi_g u, w\rangle||f(g)|\,\mathrm{d}\mu(g) \leqslant \|u\|\|w\|\|\nu\| \tag{1.21}$$

where $\mathrm{d}\nu = f\,\mathrm{d}\mu$ for $f \in L^1_\mu(G)$, and μ is a σ-finite measure. This allows us to apply the Fréchet–Riesz representation theorem to define the vector $\pi_*(\nu)u$ as the unique vector satisfying (1.20). By (1.21) and the Cauchy–Schwarz inequality, we have $\|\pi_*(\nu)u\| \leqslant \|\nu\|\|u\|$. Hence the so-defined operator $\pi_*(\nu)$ also satisfies

$$\|\pi_*(\nu)\| \leqslant \|\nu\|,$$

which we will use frequently without explicit reference. We refer[†] to [25, Sec. 3.5.4] for more details, to Exercise 1.57 for a definition using a strong (or Riemann) integral, and to Exercise 1.58 for another equivalent definition in the case of an action-associated representation. We will call $\pi_*(\nu)$ the *convolution operator* associated to $\nu \in M(G)$.

Next we verify that $\pi_*(\nu_1^*) = \pi_*(\nu_1)^*$ and $\pi_*(\nu_1 * \nu_2) = \pi_*(\nu_1)\pi_*(\nu_2)$ for any $\nu_1, \nu_2 \in M(G)$. To see this, let $u, w \in \mathcal{H}_\pi$. Then

$$\begin{aligned}\langle \pi_*(\nu_1^*)u, w\rangle &= \int \langle \pi_g u, w\rangle \,\mathrm{d}\nu_1^*(g) = \int \langle \pi_{g^{-1}} u, w\rangle \,\mathrm{d}\overline{\nu_1}(g)\\ &= \int \langle u, \pi_g w\rangle \,\mathrm{d}\overline{\nu_1}(g) = \int \overline{\langle \pi_g w, u\rangle}\,\mathrm{d}\overline{\nu_1}(g)\\ &= \overline{\langle \pi_*(\nu_1)w, u\rangle} = \langle u, \pi_*(\nu_1)w\rangle\end{aligned}$$

shows the first claim. For the second, we have

$$\begin{aligned}\langle \pi_*(\nu_1 * \nu_2)u, w\rangle &= \int \langle \pi_g u, w\rangle \,\mathrm{d}\nu_1 * \nu_2(g)\\ &= \iint \langle \pi_{g_1}\pi_{g_2} u, w\rangle \,\mathrm{d}\nu_1(g_1)\,\mathrm{d}\nu_2(g_2)\\ &= \int \underbrace{\int \langle \pi_{g_2} u, \pi_{g_1}^* w\rangle \,\mathrm{d}\nu_2(g_2)}_{=\langle \pi_*(\nu_2)u, \pi_{g_1}^* w\rangle}\,\mathrm{d}\nu_1(g_1)\\ &= \int \langle \pi_{g_1}\pi_*(\nu_2)u, w\rangle \,\mathrm{d}\nu_1(g_1) = \langle \pi_*(\nu_1)\pi_*(\nu_2)u, w\rangle,\end{aligned}$$

as required.

As mentioned above, we often identify $f \in L^1(G)$ with the associated finite complex-valued measure $\mathrm{d}\nu_f = f\,\mathrm{d}m$, which by the above defines a bounded operator $\pi_*(f) = \pi_*(\nu_f)$ on $\mathcal{H}_\pi$ with $\|\pi_*(f)\| \leqslant \|f\|_1$.

[†] In [25] we denoted $\pi_*(\nu)u$ by $\nu \underset{\pi}{*} u$ to emphasise the similarity between the operator $\pi_*(\nu)$ and convolution. We have chosen here the notation $\pi_*(\nu)$ as it is closer to the standard notation $\pi(\nu)$ but still contains a reminder of the similarities to convolution.

Notice that the unitary representation π is automatically contained in the module structure of $\mathcal{H}_\pi$ with respect to $M(G)$ since for any Dirac measure δ_g we have[†]

$$\pi_*(\delta_g) = \int \pi_h \, \mathrm{d}\delta_g(h) = \pi_g.$$

The same is not true when we restrict the module structure to $L^1(G)$ (as we often will), but it nearly is.

Proposition 1.52 (Operators for an approximate identity). *Let π be a unitary representation of the group G, let (ψ_n) be an approximate identity as in Proposition* 1.46*, and let $g \in G$. Then*

$$\pi_*\left(\lambda_g \psi_n\right) = \pi_g \pi_*(\psi_n) \longrightarrow \pi_g$$

in the strong operator topology as $n \to \infty$.

Proof. By (1.19) we have $\lambda_g \psi_n = \delta_g * \psi_n$ if we identify any $f \in L^1(G)$ with its associated complex-valued measure $\mathrm{d}\nu_f = f \, \mathrm{d}m$. Therefore

$$\pi_*(\lambda_g \psi_n) = \pi_*(\delta_g * \psi_n) = \pi_*(\delta_g)\pi_*(\psi_n) = \pi_g \pi_*(\psi_n)$$

by the discussion above. Since π_g is continuous, it suffices to consider the case $g = e$.

Let $u \in \mathcal{H}_\pi$. By continuity of the representation, for every $\varepsilon > 0$ there exists some n_0 with the property that $g \in B_{n_0}$ implies $\|\pi_g u - u\| < \varepsilon$. Thus for $n \geqslant n_0$ and $w \in \mathcal{H}_\pi$ we have

$$\begin{aligned} |\langle \pi_*(\psi_n)u, w\rangle - \langle u, w\rangle| &= \left| \int_{B_n} \left(\langle \pi_g u, w\rangle - \langle u, w\rangle\right) \psi_n(g) \, \mathrm{d}m(g) \right| \\ &\leqslant \int_{B_n} \left|\langle \pi_g u - u, w\rangle\right| \psi_n(g) \, \mathrm{d}m(g) \leqslant \varepsilon \|w\|, \end{aligned}$$

where we used the fact that $\psi_n \geqslant 0$, $\int \psi_n \, \mathrm{d}m = \|\psi_n\|_1 = 1$ for all $n \geqslant 1$, and Cauchy–Schwarz. Since $w \in \mathcal{H}_\pi$ was arbitrary, this gives

$$\|\pi_*(\psi_n)u - u\| \leqslant \varepsilon$$

for $n \geqslant n_0$, as required. □

Corollary 1.53 (Invariance). *For a closed subspace $\mathcal{V} \subseteq \mathcal{H}_\pi$, the following are equivalent:*

(a) *$\mathcal{V}$ is invariant under π,*
(b) *$\mathcal{V}$ is invariant under $\pi_*(L^1(G))$, and*

[†] Strictly speaking, we should verify this identify by applying the operators to u in $\mathcal{H}_\pi$ and taking the inner product with w in $\mathcal{H}_\pi$ as in the definition of $\pi_*(\nu)$, but when these vectors are just distracting decorations we sometimes do not write them down.

(c) *$\mathcal{V}$ is invariant under $\pi_*(M(G))$.*

PROOF. Suppose that $\mathcal{V}$ is invariant under $\pi(G)$ as in (a). Then for $\mu \in M(G)$ and $v \in \mathcal{V}$ we have

$$\langle \pi_*(\mu)v, w\rangle = \int \langle \pi_g v, w\rangle \, \mathrm{d}\mu(g) = 0$$

for any $w \in \mathcal{V}^\perp$, so $\pi_*(\mu)v \in \mathcal{V}$ as required for (c). Clearly (c) implies (b), and finally (b) implies (a) by Proposition 1.52. □

Essential Exercise 1.54. Let π and ρ be unitary representations of G, and let $B\colon \mathcal{H}_\pi \to \mathcal{H}_\rho$ be a bounded intertwining operator. Show that

$$B \circ \pi_*(f) = \rho_*(f) \circ B$$

for all $f \in L^1(G)$, and $B \circ \pi_*(\nu) = \rho_*(\nu) \circ B$ for all $\nu \in M(G)$.

Exercise 1.55. Generalize Proposition 1.52 by allowing the more general approximate identities from Exercise 1.48.

Exercise 1.56. Suppose G is unimodular, and let $f \in L^1(G)$ have the property that for every $g_0 \in G$ we have $f(g_0 g) = f(g g_0)$ for almost every $g \in G$. Show that $\pi_*(f)\colon \mathcal{H}_\pi \to \mathcal{H}_\pi$ is intertwining for any unitary representation π of G.

The following two exercises may help the reader to familiarize herself with the notion of convolution operators.

Exercise 1.57. Let π be a unitary representation of G on $\mathcal{H}_\pi$.
(a) Let ν be a compactly supported complex-valued Borel measure on G and $K = \operatorname{supp}\nu$. For $v \in \mathcal{H}_\pi$ and a finite partition $\mathcal{P} = \{B_1, \dots, B_n\}$ of K and sample points $g_j \in B_j$ for $j = 1, \dots, n$, define the Riemann sum by

$$R_\nu(\mathcal{P}, (g_1, \dots, g_n)) = \sum_{j=1}^{n} \nu(B_j)\pi_{g_j} v.$$

Use uniform continuity of the map $K \ni g \mapsto \pi_g v \in \mathcal{H}_\pi$ to show that the Riemann sums converge in $\mathcal{H}_\pi$ as the diameter

$$\operatorname{diam}(\mathcal{P}) = \max_{j=1,\dots,n} \sup_{g_1, g_2 \in B_j} \mathsf{d}(g_1, g_2)$$

of the partition goes to zero. Show that the limit $R\text{-}\!\int_K \pi_g v \, \mathrm{d}\nu$ of the above Riemann sums satisfies

$$\left\| R\text{-}\!\int_K \pi_g v \, \mathrm{d}\nu \right\| \leqslant \|\nu\| \|v\|.$$

(b) Let $\nu \in M(G)$ and define $\nu_m = \nu|_{K_m}$ for $m \in \mathbb{N}$, where $K_1 \subseteq K_2 \subseteq \cdots$ is an increasing sequence of compact subsets of G with $G = \bigcup_{m=1}^\infty K_m$. Extend the above notion of a Riemann integral to an improper Riemann integral by showing that the limit

$$R\text{-}\!\int_G \pi_g v \, \mathrm{d}\nu = \lim_{m\to\infty} R\text{-}\!\int_{K_m} \pi_g v \, \mathrm{d}\nu_m$$

exists and is independent of the choice of the sequence (K_m).
(c) Show that $R\text{-}\!\int_G \pi_g v \, \mathrm{d}\nu = \pi_*(\nu)(v)$ for any $v \in \mathcal{H}_\pi$ and any $\nu \in M(G)$.

Exercise 1.58. Suppose that G acts continuously on X preserving a locally finite measure μ on X, giving rise to the action-associated representation π as in Proposition 1.3. Show that for any $f \in L^2_\mu(X)$ and $\nu \in M(G)$ the integral

$$\int_G f(g^{-1}\bullet x)\,\mathrm{d}\nu(g)$$

exists for μ-almost every $x \in X$ and equals $\pi_*(\nu)f \in L^2_\mu(X)$ almost surely.

Exercise 1.59. Let π be a unitary representation of G on $\mathcal{H}_\pi$, let ν be a complex-valued measure on G, $\varepsilon > 0$, $v, w \in \mathcal{H}_\pi$ with $\|\pi_g v - w\| < \varepsilon$ for all $g \in \operatorname{supp}\nu$. Show that

$$\|\pi_*(\nu)v - \nu(G)w\| < \varepsilon\|\nu\|.$$

Exercise 1.60. Let $f \in L^1(G)$ with $f > 0$ and $\int_G f\,\mathrm{d}m = 1$, and let π be a unitary representation of G.
(a) Show that $v \in \mathcal{H}_\pi$ is invariant if and only if $\pi_*(f)v = v$.
(b) Suppose in addition that $f^* = f$. Show that $\pi_*(f)^n v \to P_{\mathcal{H}_\pi^G} v$ as $n \to \infty$ for any vector $v \in \mathcal{H}_\pi$.

1.6 Cyclic Representations and Matrix Coefficients

In this section we introduce several fundamental notions for the study of unitary representations, all of which relate to the question of how to decompose a given unitary representation into simpler constituents.

1.6.1 Decomposition into Cyclic Subspace

As already mentioned in Section 1.4.2, it is natural to study the smallest invariant subspaces containing a particular element.

Definition 1.61 (Cyclic subspaces). Let π be a unitary representation of G and $v \in \mathcal{H}_\pi$. The smallest closed π-invariant subspace of $\mathcal{H}_\pi$ containing v is called the *cyclic subspace generated by* v and denoted by $\langle v\rangle_\pi$. We say that π is *cyclic* if there exists a vector $v \in \mathcal{H}_\pi$, called a *generator*, with $\langle v\rangle_\pi = \mathcal{H}_\pi$.

We note that due to Corollary 1.53 we may equivalently use invariance under the unitary representation or invariance under the associated module structures to define the cyclic subspace generated by a vector. More formally we have for a unitary representation π of G and a vector $v \in \mathcal{H}_\pi$ the identity

$$\langle v\rangle_\pi = \overline{\langle \pi(G)v\rangle_\mathbb{C}} = \overline{\pi_*(L^1(G))v} = \overline{\pi_*(M(G))v},$$

where $\langle\cdot\rangle_\mathbb{C}$ again denotes the linear hull over $\mathbb{C}$.

Lemma 1.62 (Decomposition into cyclic subspaces). *Let π be a unitary representation of G. Then there exists a finite or countable*[†] *collection of pairwise orthogonal cyclic subspaces $\langle v_n\rangle_\pi \subseteq \mathcal{H}_\pi$ such that $\mathcal{H}_\pi = \bigoplus_{n\geqslant 1}\langle v_n\rangle_\pi$.*

PROOF. Let $w_1, w_2, \ldots \in \mathcal{H}$ be an orthonormal basis. Let $v_1 = w_1$ and define $\mathcal{V}_1$ to be the space $\langle v_1\rangle_\pi$. Let v_2 be the orthogonal projection of w_2 onto $\mathcal{V}_1^\perp$. By invariance of $\mathcal{V}_1$ we have $\pi_g(v_2) \perp \mathcal{V}_1$ for all $g \in G$, which gives $\langle v_2\rangle_\pi \perp \mathcal{V}_1$. In particular, we see that $\mathcal{V}_2 = \mathcal{V}_1 \oplus \langle v_2\rangle_\pi$ is an orthogonal direct sum of two cyclic subspaces that contains both w_1 and w_2.

Next define v_3 as the orthogonal projection of w_3 onto $\mathcal{V}_2^\perp$ and obtain

$$w_3 \in \mathcal{V}_3 = \mathcal{V}_2 \oplus \langle v_3\rangle_\pi$$

as before. Repeating the argument inductively defines the cyclic subspaces. Since all basis vectors belong to the direct sum of these cyclic subspaces, the direct sum equals $\mathcal{H}_\pi$ and the lemma follows. □

Exercise 1.63. Show that $\dim \mathcal{H}_\pi^G \leqslant 1$ for any cyclic representation π of G.

1.6.2 Matrix Coefficients

The following definition is both fundamental for the theory to come, as well as of great interest in applications.

Definition 1.64 (Matrix coefficients). Let π be a unitary representation of G and $v, w \in \mathcal{H}_\pi$. The function $\varphi_v^\pi \in C_b(G) = C(G) \cap L^\infty(G)$ defined for $g \in G$ by

$$\varphi_v^\pi(g) = \langle \pi_g v, v\rangle$$

is a *diagonal (or principal) matrix coefficient for* v. More generally, the function $\varphi_{v,w}^\pi \in C_b(G)$ defined by

$$\varphi_{v,w}^\pi(g) = \langle \pi_g v, w\rangle$$

is a *(non-diagonal) matrix coefficient*[‡] *for* v, w. If the unitary representation is clear from the context we will also simply write $\varphi_v = \varphi_v^\pi$ and $\varphi_{v,w} = \varphi_{v,w}^\pi$ for the matrix coefficients of $v, w \in \mathcal{H}_\pi$.

We note that for the diagonal matrix coefficient of a vector $v \in \mathcal{H}_\pi$ we have $\|\varphi_v\|_\infty = \varphi_v(e) = \|v\|^2$ and, more generally, we have $\|\varphi_{v,w}\|_\infty \leqslant \|v\|\|w\|$ for $v, w \in \mathcal{H}_\pi$. The following observation, together with Lemma 1.62, shows that the study of matrix coefficients is in a sense equivalent to the study of unitary representations.

[†] Recall the standing assumption that $\mathcal{H}_\pi$ is separable.

[‡] If v, w belong to an orthonormal basis, then the function indeed represents an entry of a possibly infinite matrix describing the operator π_g.

Proposition 1.65 (Equal matrix coefficients). *Let π and ρ be two unitary representations of G. Suppose that the principal matrix coefficients defined by vectors $v \in \mathcal{H}_\pi$ and $w \in \mathcal{H}_\rho$ agree, meaning that $\langle \pi_g v, v\rangle = \langle \rho_g w, w\rangle$ for all g in G. Then π restricted to the cyclic subspace $\langle v\rangle_\pi$ is unitarily isomorphic to ρ restricted to the cyclic subspace $\langle w\rangle_\rho$ via a unitary isomorphism that sends v to w.*

PROOF. Fix $g_1, \dots, g_n \in G$ and $\alpha_1, \dots, \alpha_n \in \mathbb{C}$. Then the norm of the element $\sum_k \alpha_k \pi_{g_k}(v)$ can be expressed in terms of matrix coefficients, since

$$\begin{aligned}\Big\|\sum_k \alpha_k \pi_{g_k}(v)\Big\|^2 &= \Big\langle \sum_k \alpha_k \pi_{g_k}(v), \sum_\ell \alpha_\ell \pi_{g_\ell}(v)\Big\rangle \\ &= \sum_{k,\ell} \alpha_k \overline{\alpha_\ell} \Big\langle \pi_{g_\ell^{-1} g_k} v, v\Big\rangle = \sum_{k,\ell} \alpha_k \overline{\alpha_\ell} \varphi_v^\pi(g_\ell^{-1} g_k). \end{aligned} \tag{1.22}$$

In particular, equality of the matrix coefficients $\varphi_v^\pi = \varphi_w^\rho$ implies that

$$\Big\|\sum_k \alpha_k \pi_{g_k}(v)\Big\| = \Big\|\sum_k \alpha_k \rho_{g_k}(w)\Big\|, \tag{1.23}$$

and so $\sum_k \alpha_k \pi_{g_k}(v) = 0$ if and only if $\sum_k \alpha_k \rho_{g_k}(w) = 0$. Thus the map U defined by

$$U\colon \sum_k \alpha_k \pi_{g_k}(v) \longmapsto \sum_k \alpha_k \rho_{g_k}(w) \tag{1.24}$$

is well-defined, injective, and an isometry by (1.23). It follows that U extends to an isometry between $\langle v\rangle_\pi$ and $\langle w\rangle_\rho$. It is clear from the definition that $U(\pi_g(u)) = \rho_g(U(u))$ for all $g \in G$ whenever u is a finite sum as in (1.24). By continuity of π_g and ρ_g, this implies that U is intertwining and the proposition follows. □

We note that the matrix coefficients already featured implicitly in Section 1.5.3 in the definition of the convolution operators $\pi_*(\nu)$ and $\pi_*(f)$ for $\nu \in M(G)$ and $f \in L^1(G)$ since

$$\int \varphi_{u,w} \,\mathrm{d}\nu = \int \langle \pi_g u, w\rangle \,\mathrm{d}\nu(g) = \langle \pi_*(\nu)u, w\rangle$$
$$\int \varphi_{u,w} f \,\mathrm{d}m = \int \langle \pi_g u, w\rangle f(g) \,\mathrm{d}m(g) = \langle \pi_*(f)u, w\rangle$$

for all $u, w \in \mathcal{H}_\pi$. We will frequently use these formulas, often without explicit reference.

Finally, the map that sends $u, w \in \mathcal{H}_\pi$ to their matrix coefficient $\varphi_{u,w}$ has interesting equivariance properties for any unitary representation π of G, which will be important later. Indeed, for $u, w \in \mathcal{H}_\pi$ and $g, h \in G$ we have

$$\varphi_{\pi_g u, w}(h) = \langle \pi_h \pi_g u, w\rangle = \varphi_{u,w}(hg)$$

and

$$\varphi_{u,\pi_g w}(h) = \langle \pi_h u, \pi_g w \rangle = \langle \pi_g^{-1}\pi_h u, w \rangle = \varphi_{u,w}(g^{-1}h).$$

Exercise 1.66. Let

$$\pi_\oplus = \bigoplus_{n=1}^{\infty} \pi_n$$

be as in Exercise 1.24 and let $v = \sum_{n=1}^{\infty} v_n$ and $w = \sum_{n=1}^{\infty} w_n$ be elements of $\mathcal{H}_{\pi_\oplus}$ with $v_n, w_n \in \mathcal{H}_n = \mathcal{H}_{\pi_n}$ for $n \geqslant 1$. Show that

$$\varphi_{v,w}^{\pi_\oplus} = \sum_{n=1}^{\infty} \varphi_{v_n,w_n}^{\pi_n},$$

and that the series converges uniformly on G.

Exercise 1.67. Let π be a cyclic unitary representation of G with generator $w \in \mathcal{H}_\pi$. Show that the map $\mathcal{H}_\pi \ni v \mapsto \varphi_{v,w} \in C(G)$ is linear and injective.

Exercise 1.68. Let π and ρ be irreducible unitary representations of G. Let $v, w \in \mathcal{H}_\pi$ and $a, b \in \mathcal{H}_\rho$ be non-zero with $\|v\| = \|a\|$, $\|w\| = \|b\|$, and $\varphi_{v,w}^{\pi} = \varphi_{a,b}^{\rho}$. Show that π and ρ are isomorphic, and that there exists a unitary isomorphism $U \colon \mathcal{H}_\pi \to \mathcal{H}_\rho$ between π and ρ with $U(v) = a$ and $U(w) = b$.

Exercise 1.69. Let G be a compact group, and let π be an irreducible unitary representation of G. Show that $\pi < \lambda$.

1.6.3 Positive-Definite Functions and Continuous GNS

The following notion identifies a crucial property of matrix coefficients.

Definition 1.70 (Continuous positive-definite function). A continuous function ϕ on G is called *positive-definite* if

$$\sum_{k,\ell} \alpha_k \overline{\alpha_\ell} \phi(g_\ell^{-1} g_k) \geqslant 0$$

for any finite list $g_1, \dots, g_n \in G$ and scalars $\alpha_1, \dots, \alpha_n \in \mathbb{C}$.

Lemma 1.71 (Diagonal matrix coefficient). *A diagonal matrix coefficient of a unitary representation is a positive-definite function.*

PROOF. This follows at once from (1.22). □

We will adopt where convenient the notational convention that ϕ is used for an abstract positive-definite function, while φ_v is used for concrete matrix coefficients.

The converse of Lemma 1.71 is given by our introductory version of the Gelfand–Naimark–Segal (GNS) construction.[(3)]

Proposition 1.72 (Gelfand–Naimark–Segal: continuous version). *Let the function ϕ on G be continuous and positive-definite. Then there exists a cyclic representation π^ϕ and a generator $v_\phi \in \mathcal{H}_\phi = \mathcal{H}_{\pi^\phi}$ such that ϕ is the diagonal matrix coefficient of v_ϕ for π^ϕ. In particular, $\|\phi\|_\infty = \phi(e) = \|v_\phi\|^2$.*

For the proof, the following abstract principle will be helpful. Suppose $\mathcal{V}$ is a complex vector space and $\langle \cdot, \cdot \rangle : \mathcal{V} \times \mathcal{V} \to \mathbb{C}$ is sesqui-linear and positive semi-definite. Then $\langle \cdot, \cdot \rangle$ satisfies the *polarization identity*

$$\langle v, w \rangle = \frac{1}{4} \sum_{n=0}^{3} \mathrm{i}^n \langle v + \mathrm{i}^n w, v + \mathrm{i}^n w \rangle$$

for all $v, w \in \mathcal{V}$ by sesqui-linearity (and expanding the right hand side). Using the positive semi-definiteness and sesqui-linearity again, we also obtain the conjugate-symmetry

$$\begin{aligned} \overline{\langle v, w \rangle} &= \frac{1}{4} \sum_{n=0}^{3} \underbrace{\mathrm{i}^{-n}\mathrm{i}^{-n}\mathrm{i}^{n}}_{=\mathrm{i}^{-n}} \underbrace{\langle v + \mathrm{i}^n w, v + \mathrm{i}^n w \rangle}_{\in \mathbb{R}} \\ &= \frac{1}{4} \sum_{n=0}^{3} \mathrm{i}^{-n} \left\langle \mathrm{i}^{-n} v + w, \mathrm{i}^{-n} v + w \right\rangle = \langle w, v \rangle \end{aligned} \tag{1.25}$$

for all $v, w \in \mathcal{V}$. In particular, sesqui-linearity and positive semi-definiteness of $\langle \cdot, \cdot \rangle$ together imply that $\langle \cdot, \cdot \rangle$ is a semi-inner product on $\mathcal{V}$.

PROOF. For the first step of the conjuring trick that defines $\mathcal{H}_\phi$ and π^ϕ, we will use the group ring

$$\mathbb{C}[G] = \left\{ r = \sum_{k=1}^{n} \alpha_k g_k \;\middle|\; n \in \mathbb{N}, g_1, \dots, g_n \in G, \alpha_1, \dots, \alpha_n \in \mathbb{C} \right\}$$

(see also Section 1.4.1). For elements $\sum_k \alpha_k g_k$ and $\sum_\ell \beta_\ell h_\ell$ in $\mathbb{C}[G]$ we define the ϕ-product

$$\left\langle \sum_k \alpha_k g_k, \sum_\ell \beta_\ell h_\ell \right\rangle_\phi = \sum_{k,\ell} \alpha_k \overline{\beta_\ell} \phi(h_\ell^{-1} g_k).$$

It is clear that $\langle \cdot, \cdot \rangle_\phi$ depends linearly on the first argument and conjugate-linearly on the second. Moreover, the definition of positive-definiteness shows that

$$\left\langle \sum_k \alpha_k g_k, \sum_k \alpha_k g_k \right\rangle_\phi \geqslant 0$$

for any $\sum_k \alpha_k g_k \in \mathbb{C}[G]$.

Using the argument before the proof, we see that $\langle \cdot, \cdot \rangle_\phi$ is also conjugate-symmetric, and hence defines a semi-inner product on $\mathbb{C}[G]$. We define the associated semi-norm $\|r\|_\phi = \sqrt{\langle r, r \rangle_\phi}$ for $r \in \mathbb{C}[G]$, its kernel

$$W_0 = \{r \in \mathbb{C}[G] \mid \|r\|_\phi = 0\},$$

and define $\mathcal{H}_\phi$ to be the completion of $\mathbb{C}[G]/W_0$ with respect to the norm $\|\cdot\|_\phi$ induced on it. Note that $\mathcal{H}_\phi$ is a Hilbert space with respect to the inner product, again denoted by $\langle\cdot,\cdot\rangle_\phi$, induced by the ϕ-product.

For $g \in G$ we define the linear map $\pi_g \colon \mathbb{C}[G] \to \mathbb{C}[G]$ by

$$\pi_g \colon \sum_k \alpha_k g_k \longmapsto \sum_k \alpha_k g g_k,$$

which satisfies

$$\|\pi_g(r)\|_\phi^2 = \Big\langle \sum_k \alpha_k g g_k, \sum_k \alpha_k g g_k \Big\rangle_\phi = \sum_{k,\ell} \alpha_k \overline{\alpha_\ell} \phi(g_\ell^{-1} g_k) = \|r\|_\phi^2$$

for $r = \sum_k \alpha_k g_k \in \mathbb{C}[G]$ by definition of the ϕ-product. Thus $\pi_g(W_0) = W_0$ and π_g defines an isometry, again denoted π_g, from $\mathcal{H}_\phi$ to $\mathcal{H}_\phi$. Since $\pi_e = I$ and since we have

$$\pi_g(\pi_h(r)) = \pi_g\Big(\sum_k \alpha_k h g_k\Big) = \sum_k \alpha_k g h g_k = \pi_{gh}(\nu)$$

for $r = \sum_k \alpha_k g_k \in \mathbb{C}[G]$ and all $g, h \in G$, we see that $g \mapsto \pi_g$ is a homomorphism from G into the group of unitary operators on $\mathcal{H}_\phi$.

It remains to show the continuity requirement, that is to show that

$$G \ni g \longmapsto \pi_g v \in \mathcal{H}_\phi$$

is continuous for every $v \in \mathcal{H}_\phi$. For this, consider $v_\phi = e + W_0$ and suppose that $g_n \to g$ as $n \to \infty$. Then

$$\begin{aligned}
\|\pi_{g_n}(v_\phi) - \pi_g(v_\phi)\|_\phi^2 &= \langle g_n - g, g_n - g\rangle_\phi \\
&= \|g_n\|_\phi^2 - 2\Re\langle g_n, g\rangle + \|g\|_\phi^2 \\
&= \phi(e) - 2\Re\phi(g^{-1}g_n) + \phi(e) \longrightarrow 0
\end{aligned}$$

as $n \to \infty$ by continuity of ϕ. Using linear combinations and density this extends to continuity of the representation π—more formally set $D = \{v_\phi\}$, note that $\langle \pi_G(D)\rangle_{\mathbb{C}} = \langle \pi_g(e + W_0) \mid g \in G\rangle_{\mathbb{C}} = \mathbb{C}[G] + W_0$ is dense in $\mathcal{H}_\pi$, and apply Lemma 1.13(1).

By our standing assumption G is separable, which implies the same for $\pi_G(v_\phi)$ and so also for the cyclic representation $\langle v_\phi \rangle_\pi = \mathcal{H}_\phi$ with generator $v_\phi = e + W_0$. Finally, the principal matrix coefficient of v_ϕ is given by

$$\varphi_{v_\phi}(g) = \langle \pi_g(e + W_0), e + W_0\rangle_\phi = \langle g, e\rangle_\phi = \phi(g)$$

for all $g \in G$. □

Proposition 1.65, Lemma 1.71, and Proposition 1.72 together give a one-to-one correspondence between positive-definite functions ϕ and their *associated cyclic representations* π^ϕ. Knowing that every unitary representation can be decomposed into cyclic representations (by Lemma 1.62), this suggest that we should study the set of positive-definite functions more closely. Since the latter is contained in the linear space $C_b(G)$, this gives a common setting in which to discuss the relationship between different cyclic unitary representations (on otherwise unrelated Hilbert spaces).

1.6.4 Irreducibility and Extremality

Notice first that an irreducible representation is always cyclic since any non-zero vector can be used as a generator. The property of irreducibility of a cyclic representation has a very satisfying characterization in terms of the associated matrix coefficients.

Proposition 1.73 (Extremality). *Let*

$$\mathcal{P}(G) = \{\phi \in C_b(G) \mid \phi \textit{ is positive-definite}\}$$

and

$$\mathcal{P}^1(G) = \{\phi \in \mathcal{P}(G) \mid \phi(e) = 1\}.$$

Then $\mathcal{P}(G) = [0, \infty)\mathcal{P}^1(G)$ is a convex cone and $\mathcal{P}^1(G)$ is a convex subset of $C_b(G)$. An element ϕ of $\mathcal{P}^1(G)$ is an extremal point of $\mathcal{P}^1(G)$ if and only if the associated cyclic representation π^ϕ is irreducible.

We recall that a point p of a convex set P is called extremal if the relation $\alpha_1 p_1 + \alpha_2 p_2 = p$ for $\alpha_1, \alpha_2 \in (0, 1)$ with $\alpha_1 + \alpha_2 = 1$ and $p_1, p_2 \in P$ forces $p_1 = p_2 = p$.

Proof of Proposition 1.73. Clearly if $\alpha_1, \alpha_2 \geqslant 0$ and $\phi_1, \phi_2 \in \mathcal{P}(G)$ then the linear combination $\alpha_1\phi_1 + \alpha_2\phi_2$ also belongs to $\mathcal{P}(G)$, and so $\mathcal{P}(G)$ is a convex cone. If in addition $\alpha_1 + \alpha_2 = 1$ and $\phi_1, \phi_2 \in \mathcal{P}^1(G)$, then

$$\alpha_1\phi_1 + \alpha_2\phi_2 \in \mathcal{P}^1(G),$$

and hence $\mathcal{P}^1(G)$ is convex. Finally, if $\phi \in \mathcal{P}(G)\smallsetminus\{0\}$, then $\phi(e) = \|\phi\|_\infty > 0$ by the correspondence between positive-definite functions and matrix coefficients (Proposition 1.72). Hence $\phi_1 = \phi(e)^{-1}\phi \in \mathcal{P}^1(G)$ and as $\phi \in \mathcal{P}(G)\smallsetminus\{0\}$ was arbitrary it follows that $\mathcal{P}(G) = [0, \infty)\mathcal{P}^1(G)$.

It remains to prove the equivalence in the last part of the proposition.

Extremality implies irreducibility. Suppose first that $\phi \in \mathcal{P}^1(G)$ is an extremal point and suppose for the purpose of a contradiction that $\mathcal{H}_\phi$ is a sum $\mathcal{V}_1 \oplus \mathcal{V}_2$ for two non-trivial closed π^ϕ-invariant subspaces $\mathcal{V}_1, \mathcal{V}_2 \subseteq \mathcal{H}_\phi$ (see also the left-hand part of Figure 1.4). Let us write $\pi = \pi^\phi$ for brevity, let v_ϕ denote the generator of $\mathcal{H}_\phi$, and split v_ϕ into its components by writing

$$v_\phi = v_1 + v_2$$

with $v_1 \in \mathcal{V}_1$ and $v_2 \in \mathcal{V}_2$. Since $\langle \pi_G v_\phi \rangle_\mathbb{C}$ is dense in $\mathcal{H}_\phi$, it follows that $\langle \pi_G v_j \rangle_\mathbb{C}$ is dense in $\mathcal{V}_j$, so $\mathcal{V}_j$ is cyclic with generator v_j, for $j = 1, 2$. Since $\mathcal{V}_j$ is assumed to be non-trivial, this implies that $\|v_j\|^2 > 0$ for $j = 1, 2$. Thus

$$\begin{aligned}\phi(g) = \langle \pi_g(v_\phi), v_\phi \rangle &= \langle \pi_g v_1, v_1 \rangle + \langle \pi_g v_2, v_2 \rangle \\ &= \|v_1\|^2 \langle \pi_g \widetilde{v}_1, \widetilde{v}_1 \rangle + \|v_2\|^2 \langle \pi_g \widetilde{v}_2, \widetilde{v}_2 \rangle\end{aligned}$$

for $g \in G$, where we set $\widetilde{v}_j = \|v_j\|^{-1} v_j$ for $j = 1, 2$. Notice that

$$\phi_j(g) = \langle \pi_g \widetilde{v}_j, \widetilde{v}_j \rangle$$

for $j = 1, 2$ belong to $\mathcal{P}^1(G)$. Thus we have shown that

$$\phi = \|v_1\|^2 \phi_1 + \|v_2\|^2 \phi_2$$

(with $\|v_1\|^2 + \|v_2\|^2 = \|v_\phi\|^2 = 1$ and $\|v_1\|^2, \|v_2\|^2 > 0$) is expressing ϕ as a convex combination in terms of $\phi_1, \phi_2 \in \mathcal{P}^1(G)$. The assumed extremality therefore implies that $\phi = \phi_1 = \phi_2$. However, by Proposition 1.65, this implies that the cyclic representation $\mathcal{V}_1$ is unitarily isomorphic to $\mathcal{V}_2$, where the unitary isomorphism $U\colon \mathcal{V}_1 \to \mathcal{V}_2$ sends the generator $\widetilde{v}_1 \in \mathcal{V}_1$ to the generator $\widetilde{v}_2 \in \mathcal{V}_2$. Therefore the subspace

$$\mathcal{W} = \left\{ w + \tfrac{\|v_2\|}{\|v_1\|} U w \;\middle|\; w \in \mathcal{V}_1 \right\} \subseteq \mathcal{V}_1 \oplus \mathcal{V}_2 = \mathcal{H}_\phi$$

(the graph of $\frac{\|v_2\|}{\|v_1\|} U$) is closed and π-invariant. However,

$$v_\phi = v_1 + v_2 = v_1 + \frac{\|v_2\|}{\|v_1\|} U v_1 \in \mathcal{W}$$

now shows that the generator v_ϕ of $\mathcal{H}_\phi$ belongs to the π-invariant closed subspace $\mathcal{W}$. Since $\mathcal{W}$ is a proper subspace, this is a contradiction. It follows that π^ϕ must be irreducible.

Irreducibility implies extremality. Suppose now that π is an irreducible unitary representation and $\phi = \varphi_{v_\pi} \in \mathcal{P}^1(G)$ for some $v_\pi \in \mathcal{H}_\pi$. We need to show that ϕ is an extremal point in $\mathcal{P}^1(G)$.

Suppose therefore that $\phi = \alpha_1 \phi_1 + \alpha_2 \phi_2$ with $\alpha_1, \alpha_2 \in (0, 1)$, $\alpha_1 + \alpha_2 = 1$, and $\phi_1, \phi_2 \in \mathcal{P}^1(G)$. We wish to show that $\phi_1 = \phi_2$. Applying Proposition 1.72 to ϕ_j for $j = 1, 2$ gives cyclic representations π_j with generators $v_j \in \mathcal{H}_{\pi_j}$ of norm $\|v_j\| = \phi_j(e) = 1$ and principal matrix coefficients

$$\phi_j(g) = \varphi_{v_j}(g) = \langle \pi_j(g)(v_j), v_j \rangle$$

for $g \in G$. This shows that

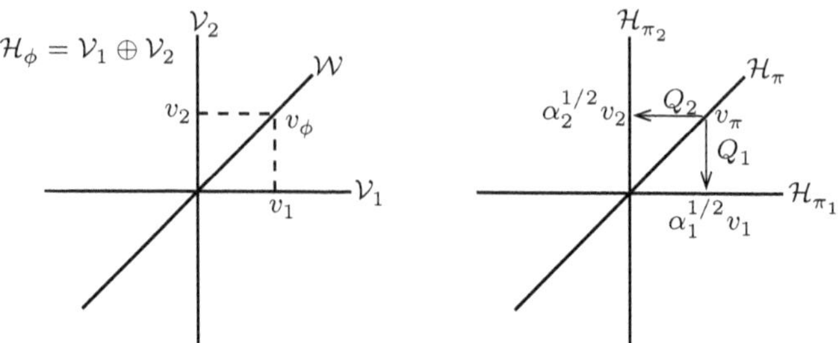

Fig. 1.4: In both directions of the proof, we relate the given representation with a direct sum representation as indicated in these figures. However, the logic of the argument is quite different in the two parts.

$$\alpha_1^{\frac{1}{2}} v_1 \oplus \alpha_2^{\frac{1}{2}} v_2 \in \mathcal{H}_{\pi_1} \oplus \mathcal{H}_{\pi_2}$$

has the principal matrix coefficient

$$\begin{aligned}\Big\langle (\pi_1 \oplus \pi_2)(g)\Big(\alpha_1^{\frac{1}{2}} v_1 \oplus \alpha_2^{\frac{1}{2}} v_2\Big), \alpha_1^{\frac{1}{2}} v_1 \oplus \alpha_2^{\frac{1}{2}} v_2\Big\rangle& \\ &= \alpha_1 \langle \pi_1(g)v_1, v_1\rangle + \alpha_2 \langle \pi_2(g)v_2, v_2\rangle \\ &= (\alpha_1\phi_1 + \alpha_2\phi_2)(g) = \phi(g)\end{aligned}$$

for $g \in G$. Using Proposition 1.65 we see that $\mathcal{H}_\pi$, its representation π, and its generator v_π can be identified with the closed subspace

$$\mathcal{H}_\pi = \Big\langle \alpha_1^{\frac{1}{2}} v_1 \oplus \alpha_2^{\frac{1}{2}} v_2 \Big\rangle_{\pi_1 \oplus \pi_2} \subseteq \mathcal{H}_{\pi_1} \oplus \mathcal{H}_{\pi_2},$$

the restriction of $\pi_1 \oplus \pi_2$ to this subspace, and the generator $v_\pi = \alpha_1^{\frac{1}{2}} v_1 \oplus \alpha_2^{\frac{1}{2}} v_2$.

To obtain the desired conclusion we consider the intertwining projection operators $P_j \colon \mathcal{H}_{\pi_1} \oplus \mathcal{H}_{\pi_2} \to \mathcal{H}_{\pi_j}$ and their restrictions $Q_j = P_j|_{\mathcal{H}_\pi}$ to the subspace $\mathcal{H}_\pi = \langle v_\pi \rangle_{\pi_1 \oplus \pi_2}$ (see also the right-hand part of Figure 1.4). Note that $Q_i(v_\pi) = \alpha_j^{\frac{1}{2}} v_j \neq 0$ for $j = 1, 2$. By Schur's lemma (Theorem 1.29) we obtain that $\alpha_j^{-\frac{1}{2}} Q_j \colon \mathcal{H}_\pi \to \mathcal{H}_{\pi_j}$ is an intertwining isometry from $\mathcal{H}_\pi$ into $\mathcal{H}_{\pi_j}$ that sends $v_\pi = \alpha_1^{\frac{1}{2}} v_1 \oplus \alpha_2^{\frac{1}{2}} v_2$ to the generator $v_j \in \mathcal{H}_{\pi_j}$. This shows that the matrix coefficient ϕ of v_π in $\mathcal{H}_\pi$ coincides with the matrix coefficient ϕ_j of $v_j \in \mathcal{H}_{\pi_j}$ for $j = 1, 2$. To summarize: we have shown that $\phi = \alpha_1\phi_1 + \alpha_2\phi_2$ with $\phi_1, \phi_2 \in \mathcal{P}^1(G)$, $\alpha_1, \alpha_2 \in (0, 1)$ and $\alpha_1 + \alpha_2 = 1$ implies $\phi = \phi_1 = \phi_2$, so $\phi = \varphi_{v_\pi}$ is an extremal point of $\mathcal{P}^1(G)$. □

1.7 GNS Construction and Corollaries

The Krein–Milman and Choquet theorems (see [25, Th. 8.80 & 8.90]) both explain ways in which elements of a convex set can be determined by (or even expressed in terms of) extreme points of the convex set. However, both of the theorems require the assumption that the convex set under consideration is compact with respect to some locally convex topology. As compactness is frequently obtained from the Banach–Alaoglu theorem concerning the closed unit ball of a dual Banach space (see [25, Th. 8.10]), we wish to view $\mathcal{P}^1(G)$ as a subset of a dual space. In order to achieve this objective, we have to (seemingly) relax the definition of positive-definite functions.

As we will see, the results of this section form the functional analytic foundations for many of the following chapters.

1.7.1 Measurable Positive-Definite Functions

Recall that $L^\infty(G) = L^1(G)'$, where $L^\infty(G)$ is equipped with the *essential supremum norm*

$$\|f\|_\infty = \inf\bigl\{\lambda > 0 \mid m(\{g \in G \mid |f(g)| > \lambda\}) = 0\bigr\}.$$

Definition 1.74 (Measurable positive-definite function). We call a function $\phi \in L^\infty(G)$ *measurable positive-definite* if

$$\iint f(g_1)\overline{f(g_2)}\phi(g_2^{-1}g_1)\,\mathrm{d}m(g_1)\,\mathrm{d}m(g_2) \geqslant 0$$

for any $f \in L^1(G)$.

Essential Exercise 1.75. Show that every continuous positive-definite function is also a measurable positive-definite function.

The converse to the exercise above is more surprising, and is provided by the full version of the GNS-construction.

Theorem 1.76 (Measurable Gelfand–Naimark–Segal construction). *Every measurable positive-definite function $\phi \in L^\infty(G)$ has a continuous representative, which in turn is a matrix coefficient of a generator $v_\phi \in \mathcal{H}_\pi$ for some cyclic unitary representation $\pi = \pi^\phi$.*

We note that after Theorem 1.76 is proved, we will not distinguish between the notions positive definite (as in Definition 1.70) and measurable positive definite (as in Definition 1.74). For the proof the following lemma will be useful.

Lemma 1.77 (The ϕ-product). *For $\phi \in L^\infty(G)$ we define the* ϕ-product *of $f_1, f_2 \in L^1(G)$ by*

$$\langle f_1, f_2 \rangle_\phi = \iint f_1(g_1)\overline{f_2(g_2)}\phi(g_2^{-1}g_1)\,\mathrm{d}m(g_1)\,\mathrm{d}m(g_2).$$

Then the ϕ-product exists for all $f_1, f_2 \in L^1(G)$ and satisfies the continuity bound

$$\left|\langle f_1, f_2 \rangle_\phi\right| \leqslant \|\phi\|_\infty \|f_1\|_1 \|f_2\|_1. \tag{1.26}$$

Moreover, it can also be expressed by the convolution formula

$$\langle f_1, f_2 \rangle_\phi = \int f_2^* * f_1 \phi \,\mathrm{d}m \tag{1.27}$$

for any $f_1, f_2 \in L^1(G)$. In particular, ϕ is measurable positive-definite if and only if

$$\langle f, f \rangle_\phi = \int f^* * f \phi \,\mathrm{d}m \geqslant 0 \tag{1.28}$$

for all $f \in L^1(G)$.

PROOF. Assume first that $\phi \in \mathscr{L}^\infty(G)$ (that is, measurably defined on all of G and bounded). Let $f_1, f_2 \in L^1(G)$. Using the measure-preserving substitution replacing g_1 by $h = g_2^{-1}g_1$ and the potentially non-measure preserving substitution $k = g_2^{-1}$ from Lemma 1.16 we obtain

$$\begin{aligned}\langle f_1, f_2 \rangle_\phi &= \iint f_1(g_1)\overline{f_2(g_2)}\phi(g_2^{-1}g_1)\,\mathrm{d}m(g_1)\,\mathrm{d}m(g_2)\\ &= \iint f_1(g_2h)\overline{f_2(g_2)}\phi(h)\,\mathrm{d}m(h)\,\mathrm{d}m(g_2)\\ &= \iint \overline{f_2(k^{-1})}\Delta(k)^{-1}f_1(k^{-1}h)\phi(h)\,\mathrm{d}m(k)\,\mathrm{d}m(h)\\ &= \int f_2^* * f_1(h)\phi(h)\,\mathrm{d}m(h),\end{aligned}$$

where as before $f_2^*(k) = \overline{f_2(k^{-1})}\Delta(k)^{-1}$ for $k \in G$. This implies the convolution formula (1.27) but also the continuity bound (1.26) by bounding $\phi(h)$ by its essential supremum and using the properties of the involution and convolution in $L^1(G)$. In particular, it follows that the ϕ-inner product only depends on the equivalence class $\phi \in L^\infty(G)$.

The last claim in the lemma follows by noting that the expression appearing in Definition 1.74 is $\langle f, f \rangle_\phi$ and combining this with (1.27). □

PROOF OF THEOREM 1.76. The proof is another conjuring trick and, in particular, is similar to the proof of the continuous version in Proposition 1.72. However, since ϕ is not defined pointwise we have to start the construction with $L^1(G)$ instead of $\mathbb{C}[G]$, which requires additional arguments. In fact, it may be helpful for the reader to compare the various steps below to the easier but to a large extent analogous steps in the proof of Proposition 1.72. We assume without loss of generality that $\phi \neq 0$.

DEFINING THE HILBERT SPACE $\mathcal{H}_\phi$. We begin by studying the ϕ-product in Lemma 1.77. Clearly $\langle \cdot, \cdot \rangle_\phi$ depends linearly on the first, and conjugate-linearly on the second, argument. By assumption ϕ is measurable positive-definite and so we also have $\langle f, f \rangle_\phi \geqslant 0$ for all $f \in L^1(G)$. By the argument just after Proposition 1.72, specifically (1.25), it follows that the ϕ-product is a semi-inner product on $L^1(G)$. We also define the associated semi-norm

$$\|f\|_\phi = \sqrt{\langle f, f \rangle_\phi}$$

for $f \in L^1(G)$ and the kernel $W_0 = \{f \in L^1(G) \mid \|f\|_\phi = 0\}$. The Hilbert space $\mathcal{H}_\phi$ is defined as the completion of $L^1(G)/W_0$ with respect to the norm induced on $L^1(G)/W_0$, which we again denote by $\|\cdot\|_\phi$. From (1.26) we also obtain another form of the continuity bound

$$\|f + W_0\|_\phi \leqslant \|\phi\|_\infty^{\frac{1}{2}} \|f\|_1 \tag{1.29}$$

for all $f \in L^1(G)$. In particular, the map from $L^1(G)$ to $\mathcal{H}_\phi$ is continuous. Moreover, $\mathcal{H}_\phi$ is separable since $L^1(G)$ is separable and its image $L^1(G)/W_0$ is dense in $\mathcal{H}_\phi$.

DEFINING THE REPRESENTATION π^ϕ. To define the unitary representation, notice that

$$\begin{aligned} \|\lambda_g f\|_\phi^2 &= \iint f(g^{-1}g_1)\overline{f(g^{-1}g_2)}\phi(g_2^{-1}g_1)\,\mathrm{d}m(g_1)\,\mathrm{d}m(g_2) \\ &= \iint f(h_1)\overline{f(h_2)}\phi(h_2^{-1}h_1)\,\mathrm{d}m(h_1)\,\mathrm{d}m(h_2) = \|f\|_\phi^2 \end{aligned}$$

for $f \in L^1(G)$ and $g \in G$, by definition of the ϕ-product and the measure preserving substitution $h_j = g^{-1}g_j$ for $j = 1, 2$. We define the unitary operator π_g as the unique extension of λ_g (which remains well-defined modulo W_0) to the completion $\mathcal{H}_\phi$.

Continuity of this unitary representation follows by applying Lemma 1.13 with $D = C_c(G)/W_0$. Indeed, for $f \in C_c(G)$, continuity of

$$G \ni g \longmapsto \lambda_g f \in L^1(G)$$

follows by the same argument as for the left regular representation on $L^2(G)$ (see also [25, Lem. 3.74]). Finally, we apply the fact that the map from $L^1(G)$ to $L^1(G)/W_0 \subseteq \mathcal{H}_\phi$ is continuous by (1.29) to conclude that π is indeed a unitary representation.

THE INNER PRODUCT WITH AN APPROXIMATE IDENTITY. In the following, (ψ_n) will be an approximate identity as in Proposition 1.46 (and will be needed as a substitute for the generator $v_\phi = e + W_0$ in the proof of Proposition 1.72). Since

$$L^1(G) \ni f \longmapsto f^* \in L^1(G)$$

is an isometric involution, we see that (ψ_n^*) is also an approximate identity and so the convolution formula of the ϕ-product in (1.27) and Proposition 1.46 imply the limit formula

$$\lim_{n\to\infty}\langle f,\psi_n\rangle_\phi = \lim_{n\to\infty}\int(\psi_n^* * f)\phi\,\mathrm{d}m = \int f\phi\,\mathrm{d}m \tag{1.30}$$

for any $f \in L^1(G)$.

DEFINING THE GENERATOR v_ϕ. In the proof of Proposition 1.72 we introduced the vector $v_\phi = e + W_0$ which finished the proof very quickly by realizing the continuous function as the principal matrix coefficient of v_ϕ. The analogue of this is the following key step. We claim that for the approximate identity (ψ_n) the limit

$$v_\phi = \lim_{n\to\infty}(\psi_n + W_0) \in \mathcal{H}_\phi \tag{1.31}$$

exists in the weak topology.

For this, notice first that by the continuity bound (1.29) we have

$$\|\psi_n + W_0\|_\phi \leqslant \|\phi\|_\infty^{\frac12}\|\psi_n\|_1 = \|\phi\|_\infty^{\frac12}$$

for all $n \in \mathbb{N}$. We combine this with (1.30) to see that the limit

$$L(w) = \lim_{n\to\infty}\langle w,\psi_n\rangle_\phi$$

exists for all $w \in \mathcal{H}_\phi$. Indeed, fixing $w \in \mathcal{H}_\phi$, let $\varepsilon > 0$, and choose some function $f \in L^1(G)$ so that $\|w - (f + W_0)\|_\phi < \varepsilon$. Now apply (1.30) to find some N so that $|\langle f,\psi_k\rangle - \langle f,\psi_\ell\rangle| < \varepsilon$ for $k,\ell \geqslant N$. Using Cauchy–Schwarz, this now gives

$$\begin{aligned}|\langle w,\psi_k\rangle_\phi - \langle w,\psi_\ell\rangle_\phi| &\leqslant |\langle w-f,\psi_k\rangle_\phi| + |\langle f,\psi_k\rangle_\phi - \langle f,\psi_\ell\rangle_\phi| + |\langle f-w,\psi_\ell\rangle_\phi|\\ &\leqslant \varepsilon\|\phi\|_\infty^{\frac12} + \varepsilon + \varepsilon\|\phi\|_\infty^{\frac12}\end{aligned}$$

for $k,\ell \geqslant N$. Therefore $\big(\langle w,\psi_n\rangle_\phi\big)_n$ is a Cauchy sequence in $\mathbb{C}$ and so the limit $L(w)$ exists for any $w \in \mathcal{H}_\phi$. Moreover,

$$|L(w)| = \Big|\lim_{n\to\infty}\langle w,\psi_n\rangle_\phi\Big| \leqslant \lim_{n\to\infty}\|w\|_\phi\|\psi_n\|_\phi \leqslant \|w\|_\phi\|\phi\|_\infty^{\frac12}.$$

It follows that $\mathcal{H}_\phi \ni w \mapsto L(w) \in \mathbb{C}$ defines a bounded linear functional on $\mathcal{H}_\phi$. By Fréchet–Riesz representation there exists some $v_\phi \in \mathcal{H}_\phi$ with $L(w) = \langle w, v_\phi\rangle$ for all $v \in \mathcal{H}_\phi$. Combined with the definition of L, this gives the claim (1.31).

CONVOLUTION APPLIED TO v_ϕ. Next we claim that the weak limit v_ϕ above satisfies

$$\pi_*(f)v_\phi = f + W_0 \tag{1.32}$$

for all $f \in L^1(G)$. Note that this implies in particular that

$$\pi_*(L^1(G))v_\phi = L^1(G) + W_0 \subseteq \mathcal{H}_\phi$$

is dense, and so that v_ϕ is a generator for $\mathcal{H}_\phi$.

To see (1.32) we first let $f_1, f_2 \in L^1(G)$. Using the definition of $\pi_*(f)$ and Fubini's theorem, we calculate

$$\begin{aligned}\left\langle \pi_*(f)(f_1 + W_0), f_2 + W_0 \right\rangle_\phi &= \int \left\langle \pi_g(f_1 + W_0), f_2 + W_0 \right\rangle_\phi f(g)\,\mathrm{d}m(g) \\ &= \iiint f(g) f_1(g^{-1}g_1)\,\mathrm{d}m(g)\overline{f_2(g_2)}\phi(g_2^{-1}g_1)\,\mathrm{d}m(g_1)\,\mathrm{d}m(g_2) \\ &= \langle f * f_1 + W_0, f_2 + W_0 \rangle_\phi\,.\end{aligned}$$

Setting $f_1 = \psi_n$ this implies

$$\begin{aligned}\langle \pi_*(f)v_\phi, f_2 + W_0 \rangle &= \langle v_\phi, \pi_*(f)^*(f_2 + W_0) \rangle_\phi \\ &= \lim_{n\to\infty} \langle \psi_n + W_0, \pi_*(f)^*(f_2 + W_0) \rangle_\phi \\ &= \lim_{n\to\infty} \langle \pi_*(f)(\psi_n + W_0), f_2 + W_0 \rangle_\phi \\ &= \lim_{n\to\infty} \langle f * \psi_n + W_0, f_2 + W_0 \rangle_\phi \\ &= \lim_{n\to\infty} \langle f + W_0, f_2 + W_0 \rangle_\phi\end{aligned}$$

as (ψ_n) is an approximate identity and the map from $L^1(G)$ to $\mathcal{H}_\phi$ is continuous. Together with the density of $L^1(G) + W_0$ in $\mathcal{H}_\phi$ this gives (1.32).

THE MATRIX COEFFICIENT OF v_ϕ. Having obtained the existence of the limit in (1.31) we write $\varphi_{v_\phi}(g)$ for the matrix coefficient of v_ϕ. To conclude the proof of the theorem we now prove that $\varphi_{v_\phi} = \phi$ almost surely by calculating $\left\langle \pi_*(f)v_\phi, v_\phi \right\rangle_\phi$ in two different ways.

Using first the definition of $\pi_*(f)$ we see that

$$\left\langle \pi_*(f)v_\phi, v_\phi \right\rangle_\phi = \int \left\langle \pi_g v_\phi, v_\phi \right\rangle_\phi f(g)\,\mathrm{d}m(g) = \int f\varphi_{v_\phi}\,\mathrm{d}m.$$

Combining (1.32) with the definition of v_ϕ and the limit formula (1.30), we also have

$$\left\langle \pi_*(f)v_\phi, v_\phi \right\rangle_\phi = \left\langle f + W_0, v_\phi \right\rangle_\phi = \lim_{n\to\infty} \langle f, \psi_n \rangle_\phi = \int f\phi\,\mathrm{d}m$$

for every $f \in L^1(G)$. Since $\phi \in L^\infty(G)$ is uniquely determined by its associated functional on $L^1(G)$, we see that $\varphi_{v_\phi} \in C_b(G)$ is a continuous representative of the equivalence class $\phi \in L^\infty(G)$. □

1.7.2 Approximation by Irreducible Representations

By combining the GNS construction with the Krein–Milman theorem, we can now show that all cyclic representations can be approximated by convex combinations of irreducible representations. In the following we will refer to the topology on $C(G)$ corresponding to uniform convergence within compact subsets of G as the compact-open topology.

Corollary 1.78 (Approximation by convex combination). *The set*

$$\mathcal{P}^{\leqslant 1}(G) = \{\phi \in \mathcal{P}(G) \mid \phi(e) \in [0,1]\}$$

is convex and weak compact when considered as a subset of $L^\infty(G)$. Every element $\phi \in \mathcal{P}^1(G)$ can be approximated by a finite convex combination*

$$\sum_{j=1}^{n} \alpha_j \phi_j$$

of extremal elements $\phi_1, \ldots, \phi_n \in \mathcal{P}^1(G)$ (corresponding to unit vectors in irreducible representations) with $\alpha_1, \ldots, \alpha_n \in [0,1]$ and $\alpha_1 + \cdots + \alpha_n = 1$.

In fact, ϕ can even be approximated in the compact-open topology. That is, for every compact subset $K \subseteq G$ and every $\varepsilon > 0$ we can find a finite convex combination as above such that

$$\Big\| \phi - \sum_{j=1}^{n} \alpha_j \phi_j \Big\|_{K,\infty} = \sup_{g \in K} \Big| \phi(g) - \sum_{j=1}^{n} \alpha_j \phi_j(g) \Big| < \varepsilon.$$

Proof. By the GNS construction (Theorem 1.76) and Exercise 1.75 there is no need to distinguish between a continuous or a measurable positive-definite function if we identify a measurable positive-definite function with its continuous representative. Using the reformulation in (1.28) of the measurable definition of positive-definiteness (Definition 1.74) it is clear that the subset $\mathcal{P}(G) \subseteq L^\infty(G)$ is weak* closed. Since $\|\phi\|_\infty = \phi(e)$ (see Proposition 1.72), we obtain from the Banach–Alaoglu theorem (see [25, Th. 8.10]) that

$$\mathcal{P}^{\leqslant 1}(G) = \mathcal{P}(G) \cap \overline{B_1^{L^\infty(G)}(0)}$$

is weak* compact.

We now identify the extremal points of $\mathcal{P}^{\leqslant 1}(G)$. Since we have $\phi(e) \geqslant 0$ and $\phi(e) = \|\phi\|_\infty$ for every $\phi \in \mathcal{P}(G)$, it is clear that $0 \in \mathcal{P}^{\leqslant 1}(G)$ is an extremal point of $\mathcal{P}^{\leqslant 1}(G)$. Moreover, any $\phi \in \mathcal{P}^{\leqslant 1}(G)$ with $\phi(e) \in (0,1)$ cannot be extremal since $\phi = (1-\phi(e))0 + \phi(e)\left(\phi(e)^{-1}\phi\right)$ with $0 \neq \phi(e)^{-1}\phi \in \mathcal{P}^{\leqslant 1}(G)$. Looking at the norms, we see that if $\phi \in \mathcal{P}^1(G)$ is equal to $\alpha_1\phi_1 + \alpha_2\phi_2$ for $\alpha_1, \alpha_2 > 0$, $\alpha_1 + \alpha_2 = 1$ and $\phi_1, \phi_2 \in \mathcal{P}^{\leqslant 1}(G)$ then in fact $\phi_1, \phi_2 \in \mathcal{P}^1(G)$ also. The definition of extremality now shows that $\phi \in \mathcal{P}^1(G)$ is extremal as an element of $\mathcal{P}^{\leqslant 1}(G)$ if and only if it is extremal as an element of $\mathcal{P}^1(G)$. Taken together, it follows that the extremal points of $\mathcal{P}^{\leqslant 1}(G)$ are 0 and the extremal points of $\mathcal{P}^1(G)$. By Proposition 1.73 each of the latter corresponds to a unit vector in an irreducible unitary representation.

By the Krein–Milman theorem (see [25, Th. 8.80] for example) the set $\mathcal{P}^{\leqslant 1}(G)$ is equal to the weak* closure of the convex hull of its extremal points. So assume that $\phi \in \mathcal{P}^1(G)$, let $\mathcal{N}_0$ be an open neighbourhood of 0 in the weak* topology, and let $\varepsilon \in (0,1)$. Since $\overline{B_{1-\varepsilon}^{L^\infty(G)}(0)}$ is weak* closed,

$$\mathcal{N} = \left(\phi + \tfrac{1}{2}\mathcal{N}_0\right) \smallsetminus \overline{B_{1-\varepsilon}^{L^\infty(G)}(0)}$$

is an open neighbourhood of ϕ in the weak* topology. Thus, by the Krein–Milman theorem, there exists a finite convex combination $\sum_{j=1}^n \alpha_j \phi_j$ of pairwise distinct extremal points that belongs to $\mathcal{N}$. If the extreme point $0 \in \mathcal{P}^{\leqslant 1}(G)$ does not appear in the sum, we have obtained the desired approximation of ϕ by a convex combination of extremal points of $\mathcal{P}^1(G)$.

So suppose now that 0, say as $\phi_n = 0$, appears in the convex combination. Then the corresponding coefficient satisfies $\alpha_n < \varepsilon$, as otherwise we would have $\sum_{j=1}^n \alpha_j \phi_j \in \overline{B_{1-\varepsilon}^{L^\infty(G)}(0)}$. This implies that the convex combination

$$\phi' = \tfrac{1}{(1-\alpha_n)} \sum_{j=1}^{n-1} \alpha_j \phi_j \in \mathcal{P}^1(G)$$

of extremal points of $\mathcal{P}^1(G)$ also approximates ϕ. To see this formally, we note that $\| \sum_{j=1}^{n-1} \alpha_j \phi_j \| = \sum_{j=1}^{n-1} \alpha_j = 1 - \alpha_n$, and obtain

$$\begin{aligned}\phi - \phi' &= \phi - \sum_{j=1}^{n} \alpha_j \phi_j + \left(1 - \tfrac{1}{1-\alpha_n}\right) \sum_{j=1}^{n-1} \alpha_j \phi_j \\ &\in \tfrac{1}{2}\mathcal{N}_0 + \left(\tfrac{\alpha_n}{1-\alpha_n}\right) \overline{B_{1-\alpha_n}^{L^\infty(G)}(0)} \subseteq \tfrac{1}{2}\mathcal{N}_0 + \varepsilon \overline{B_{1}^{L^\infty(G)}(0)} \subseteq \mathcal{N}_0\end{aligned}$$

if we choose ε sufficiently small. This proves the first half of the corollary.

The fact that the approximation is possible even with respect to the compact-open topology follows from the above together with the result in the next subsection. □

1.7.3 Upgrading to Uniform Convergence on Compact Subsets

Proposition 1.79 (Weak* and compact-open topology). *The compact-open topology on* $\overline{B_1^{L^\infty(G)}(0)}$ *is stronger than the weak* topology and, on the subset* $\mathcal{P}^1(G)$ *(thought of as a subset of* $C_b(G) \subseteq L^\infty(G)$*), the two topologies agree.*

We note that after this result the remaining proofs of the chapter are shorter and, arguably, easier.

PROOF OF PROPOSITION 1.79. We split the proof into two parts.

COMPACT-OPEN IS STRONGER. For the easier half of the proposition, let ϕ_0 be in $\overline{B_1^{L^\infty(G)}}(0)$ (for example, $\phi_0 \in \mathcal{P}^1(G)$), let f_0 be a function in $L^1(G)$, and let $\varepsilon > 0$, which together define the weak* neighbourhood

$$\mathcal{N}_{f_0;\varepsilon} = \left\{ \phi \in \overline{B_1^{L^\infty(G)}(0)} \;\middle|\; \left| \int (\phi - \phi_0) f_0 \,\mathrm{d}m \right| < \varepsilon \right\}$$

of ϕ_0 in $\overline{B_1^{L^\infty(G)}(0)}$. By the density of $C_c(G)$ in $L^1(G)$ there exists $f \in C_c(G)$ with

$$\|f - f_0\|_1 \leqslant \tfrac{1}{4}\varepsilon,$$

which implies that

$$\left\{ \phi \in \overline{B_1^{L^\infty(G)}(0)} \;\middle|\; \|\phi - \phi_0\|_{\operatorname{supp} f,\infty} < \tfrac{\varepsilon}{2(1+\|f\|_1)} \right\} \subseteq \mathcal{N}_{f;\frac{\varepsilon}{2}} \subseteq \mathcal{N}_{f_0;\varepsilon}.$$

For the latter inclusion, note that

$$\left| \int (\phi - \phi_0) f_0 \,\mathrm{d}m \right| \leqslant \underbrace{\left| \int \phi (f_0 - f) \,\mathrm{d}m \right|}_{\leqslant \frac{1}{4}\varepsilon} + \left| \int (\phi - \phi_0) f \,\mathrm{d}m \right| + \underbrace{\left| \int \phi_0 (f - f_0) \,\mathrm{d}m \right|}_{\leqslant \frac{1}{4}\varepsilon} < \varepsilon$$

for any $\phi \in \mathcal{N}_{f;\frac{\varepsilon}{2}} \subseteq \overline{B_1^{L^\infty(G)}(0)}$. Since finite intersections of neighbourhoods of the form $\mathcal{N}_{f_0;\varepsilon}$ form a neighbourhood basis for the weak* topology, this proves that the compact-open topology is stronger than the weak* topology on $\overline{B_1^{L^\infty(G)}(0)}$.

WEAK* IS STRONGER. For the (more surprising) converse, we let $\phi_0 \in \mathcal{P}^1(G)$ and apply the GNS construction (Theorem 1.76) to find a cyclic unitary representation π with generator $v \in \mathcal{H}_\pi$ so that $\|v\| = 1$ and $\phi_0 = \varphi_v^\pi$. Next we let $K \subseteq G$ be a compact subset and $\varepsilon \in (0,1]$. We will need a few preparatory definitions. Using the continuity property of the representation π we find a compact neighbourhood $U = U^{-1}$ of the identity $e \in G$ with

$$\|\pi_h v - v\| < \varepsilon \tag{1.33}$$

for all $h \in U$. We define $f_0 = \frac{1}{m(U)} \mathbb{1}_U$. By continuity of the left regular representation on $L^1(G)$ there exists an open neighbourhood $V = V^{-1} \subseteq U$ of $e \in G$ such that

$$\left\| \lambda_h f_0 - f_0 \right\|_1 < \varepsilon \tag{1.34}$$

for all $h \in V$. Applying compactness of K, we can choose a finite cover

$$\bigcup_{\ell=1}^{n} g_\ell V \supseteq K. \tag{1.35}$$

Finally, we define $f_\ell = \frac{1}{m(U)}\mathbb{1}_{g_\ell U}$ for $\ell = 1, \ldots, n$, and use these to define a weak* neighbourhood $\mathcal{N}_{f_0,f_1,\ldots,f_n;\varepsilon} = \mathcal{N}_{f_0;\varepsilon} \cap \ldots \cap \mathcal{N}_{f_n;\varepsilon}$ of ϕ_0. We will show that any $\phi \in \mathcal{N}_{f_0,f_1,\ldots,f_n;\varepsilon} \cap \mathcal{P}^1(G)$ is within $O(\varepsilon^{\frac{1}{2}})$ of $\phi_0 = \varphi_v^\pi$ on K.

So suppose that $\phi \in \mathcal{N}_{f_0,\ldots,f_n;\varepsilon} \cap \mathcal{P}^1(G)$ and use the GNS construction (Theorem 1.76) to find a unitary representation ρ and some $a \in \mathcal{H}_\rho$ with $\|a\| = 1$ and $\phi = \varphi_a^\rho$ so that

$$\left| \int_G (\varphi_a^\rho - \varphi_v^\pi)\, f_\ell \,\mathrm{d}m \right| < \varepsilon \tag{1.36}$$

for $\ell = 0, \ldots, n$. We define $a' = \rho_*(f_0)a \in \mathcal{H}_\rho$, so that $\|a'\| \leqslant \|f_0\|_1 \|a\| = 1$, and notice that

$$\begin{aligned} \langle a', a \rangle &= \int f_0(h) \langle \rho_h a, a \rangle \,\mathrm{d}m(h) = \int f_0 \varphi_a^\rho \,\mathrm{d}m \\ &= \int f_0 \varphi_v^\pi \,\mathrm{d}m + O(\varepsilon) && \text{(by (1.36))} \\ &= \frac{1}{m(U)} \int_U \langle \pi_h v, v \rangle \,\mathrm{d}m(h) + O(\varepsilon) = 1 + O(\varepsilon) \end{aligned}$$

by (1.33) and Cauchy–Schwarz applied to $\langle \pi_h v - v, v \rangle$ for $h \in U$. Hence we obtain $\Re \langle a', a \rangle > 1 - O(\varepsilon)$ and

$$\|a' - a\|^2 = \|a'\|^2 - 2\Re \langle a', a \rangle + \|a\|^2 \ll \varepsilon,$$

which gives

$$\|a' - a\| \ll \varepsilon^{\frac{1}{2}}. \tag{1.37}$$

For $h \in V$ we also obtain

$$\|\rho_h a' - a'\| = \|(\rho_h \rho_*(f_0) - \rho_*(f_0))a\| = \|\rho_*(\lambda_h f_0 - f_0)a\| = O(\varepsilon) \tag{1.38}$$

by (1.34).

We now use these uniform continuity estimates together with the cover in (1.35). In fact, for every $g \in K$ there exists $\ell \in \{1, \ldots, n\}$ with $g = g_\ell h$ for some $h \in V$, so that Cauchy–Schwarz gives

$$\begin{aligned} \varphi_a^\rho(g) = \langle \rho_g a, a \rangle &= \langle \rho_{g_\ell} \rho_h a', a \rangle + O(\varepsilon^{\frac{1}{2}}) && \text{(by (1.37))} \\ &= \langle \rho_{g_\ell} a', a \rangle + O(\varepsilon^{\frac{1}{2}}). && \text{(by (1.38))} \end{aligned}$$

Next recall that

$$\rho_{g_\ell} a' = \rho_{g_\ell} \rho_*(f_0) a = \rho_*(\lambda_{g_\ell} f_0) a = \rho_*(f_\ell) a,$$

which leads to

$$\varphi_a^\rho(g) = \langle \rho_*(f_\ell)a, a\rangle + O(\varepsilon^{\frac{1}{2}}) = \int f_\ell \varphi_a^\rho \,\mathrm{d}m + O(\varepsilon^{\frac{1}{2}}).$$

By assumption, we have

$$\int f_\ell \varphi_a^\rho \,\mathrm{d}m = \int f_\ell \varphi_v^\pi \,\mathrm{d}m + O(\varepsilon).$$

Combining this with the above, and using the fact that the above applies to any positive-definite function $\phi = \varphi_a^\rho$ in the neighbourhood $\mathcal{N}_{f_0,\dots,f_n;\varepsilon} \cap \mathcal{P}^1(G)$ of $\phi_0 = \varphi_v^\pi$ and so in particular to itself, we obtain

$$\phi(g) = \varphi_a^\rho(g) = \phi_0(g) + O(\varepsilon^{\frac{1}{2}})$$

for all $g \in K$. Since $K \subseteq G$ was an arbitrary compact subset and $\varepsilon > 0$ was arbitrary, this concludes the proof that the weak* topology is stronger. □

Exercise 1.80. For now we simply note that the weak* and compact-open topologies are different on $\mathcal{P}^{\leqslant 1}(G)$ in general. Analyze where the proof above fails if you try to use it to show that the two topologies agree on $\mathcal{P}^{\leqslant 1}(G)$.

1.7.4 Separation of Points by Irreducible Representations

Corollary 1.81 (Gelfand–Raikov). *Let g_1 and g_2 be distinct elements of the group G. Then there exists an irreducible unitary representation π of G with $\pi_{g_1} \neq \pi_{g_2}$.*

Proof. Since $g_0 = g_1^{-1}g_2 \in G \smallsetminus \{e\}$, there exists a neighbourhood U of e with $g_0 \notin U$. By continuity of the map $G \ni g \mapsto g^{-1}$ and by continuity of multiplication there exists a compact neighbourhood V of $e \in G$ with $VV^{-1} \subseteq U$. We define

$$f = m(V)^{-\frac{1}{2}} \mathbb{1}_V,$$

which satisfies $\varphi_f^\lambda(e) = \|f\|_2^2 = 1$ and

$$\varphi_f^\lambda(g_0) = \frac{1}{m(V)} \int \mathbb{1}_V(g_0^{-1}g)\mathbb{1}_V(g)\,\mathrm{d}m(g) = 0$$

since $g_0^{-1}g \in V$ and $g \in V$ would imply that $g_0 = g\big(g_0^{-1}g\big)^{-1} \in VV^{-1} \subseteq U$.

In other words, the above defines a principal matrix coefficient $\varphi_f^\lambda \in \mathcal{P}^1(G)$ that separates g_0 from the identity $e \in G$. By Corollary 1.78, there exist irreducible representations π_j, unit vectors $v_j \in \mathcal{H}_{\pi_j}$, and scalars $\alpha_j \in [0,1]$ for $j = 1, \dots, n$ with $\sum_{j=1}^n \alpha_j = 1$ and with

$$\left\| \varphi_f^\lambda - \sum_{j=1}^n \alpha_j \varphi_{v_j}^{\pi_j} \right\|_{\{g_0\},\infty} < \frac{1}{2}.$$

Thus

$$\Big|\sum_{j=1}^{n}\alpha_j\varphi_{v_j}^{\pi_j}(g_0)\Big| < \frac{1}{2}.$$

Therefore, there exists some j with

$$\varphi_{v_j}^{\pi_j}(g_0) = \langle(\pi_j)_{g_0}v_j, v_j\rangle \neq 1 = \|v_j\|^2.$$

This implies for $\pi = \pi_j$ that $\pi_{g_1}^{-1}\pi_{g_2} = \pi_{g_0} \neq I$ and the corollary follows. □

Exercise 1.82. Suppose that every irreducible unitary representation of the group G is one-dimensional (that is, defined by a unitary character of G). Show that this implies that G is abelian.

1.7.5 Applying Choquet's Theorem*

For most of our discussions the corollaries to the GNS construction above will be sufficient. Occasionally we will, however, require a refinement of Corollary 1.78 that relies on Choquet's theorem instead of the Krein–Milman theorem. For this argument, we let $\mathcal{E}^1(G)$ be the set of extremal points of $\mathcal{P}^1(G)$.

Corollary 1.83 (Integral convex combination). *The set $\mathcal{E}^1(G)$ is a measurable subset of the compact metric space $\mathcal{P}^{\leqslant 1}(G)$. For any $\phi_0 \in \mathcal{P}^1(G)$, there exists a probability measure μ on $\mathcal{E}^1(G)$ with the property that ϕ_0 can be obtained as the integral convex combination*

$$\phi_0(g) = \int_{\mathcal{E}^1(G)} \phi(g)\,\mathrm{d}\mu(\phi) \tag{1.39}$$

for all $g \in G$.

Notice that by Proposition 1.73 the elements of $\mathcal{E}^1(G)$ are precisely the principal matrix coefficients of unit vectors in irreducible unitary representations, so that the argument above gives another way to recover general positive-definite functions from irreducible unitary representations.

PROOF OF COROLLARY 1.83. Recall from Corollary 1.78 that the set

$$\mathcal{P}^{\leqslant 1}(G) \subseteq L^\infty(G)$$

of all positive-definite functions ϕ with $\phi(e) \in [0,1]$ is compact with respect to the weak* topology. Finally, recall from the proof of Corollary 1.78 that the extremal points $\mathcal{E}^1(G)$ of $\mathcal{P}^1(G)$ together with $0 \in \mathcal{P}^{\leqslant 1}(G)$ comprise precisely the extremal points

$$\mathcal{E}(G) = \mathcal{E}^1(G) \cup \{0\} \tag{1.40}$$

of $\mathcal{P}^{\leqslant 1}(G) = [0,1]\mathcal{P}^1(G)$. By Choquet's theorem [25, Sec. 8.6.2], the set $\mathcal{E}(G)$ is measurable and there exists a Borel probability measure μ on $\mathcal{P}^{\leqslant 1}(G)$

with $\mu(\mathcal{E}(G)) = 1$, so that ϕ_0 is represented as the barycentre† of μ, meaning that it is the weak integral

$$\phi_0 = \int_{\mathcal{E}(G)} \phi \, \mathrm{d}\mu(\phi) \tag{1.41}$$

defined using weak* continuous linear functionals on $L^\infty(G)$. In fact the weak* continuous linear functionals on $L^\infty(G)$ are precisely the evaluation maps

$$\mathrm{ev}_f \colon L^\infty(G) \ni \phi \longmapsto \int_G f\phi \, \mathrm{d}m$$

for $f \in L^1(G)$ (see [25, Lem. 8.13]) and so (1.41) is actually the claim that

$$\int_G f\phi_0 \, \mathrm{d}m = \int_{\mathcal{E}(G)} \int_G f\phi \, \mathrm{d}m \, \mathrm{d}\mu(\phi)$$

for all $f \in L^1(G)$. We now apply Fubini's theorem on the right-hand side, and obtain

$$\int_G f\phi_0 \, \mathrm{d}m = \int_G f(g) \int_{\mathcal{E}(G)} \phi(g) \, \mathrm{d}\mu(\phi) \, \mathrm{d}m(g) \tag{1.42}$$

for all $f \in L^1(G)$. Moreover, dominated convergence shows that

$$G \ni g \longmapsto \int_{\mathcal{E}(G)} \phi(g) \, \mathrm{d}\mu(\phi)$$

is continuous. Hence continuity of ϕ_0 and (1.42) for all $f \in L^1(G)$ imply that

$$\phi_0(g) = \int_{\mathcal{E}(G)} \phi(g) \, \mathrm{d}\mu(\phi).$$

Setting $g = e$, we see that $\mu(\{0\}) = 1$ and so we obtain (1.39). □

Essential Exercise 1.84. Explain why the map $G \times \mathcal{E}(G) \ni (g, \phi) \mapsto \phi(g)$ is measurable (which is needed in order to apply Fubini's theorem to obtain (1.42)), where we use the Borel σ-algebra of the weak* topology on $\mathcal{E}(G)$.

Exercise 1.85. Use Proposition 1.79 and Corollary 1.83 to give a different proof of Corollary 1.78.

Exercise 1.86. Let $\phi_0 = \varphi_v^\pi$ for a unitary representation π and a unit vector $v \in \mathcal{H}_\pi$. Let μ be the probability measure on $\mathcal{E}^1(G)$ in Corollary 1.83. Show that $\langle v \rangle_\pi$ contains an invariant vector if and only if $\mu(\{\mathbb{1}\}) > 0$, where $\mathbb{1} \colon G \ni g \mapsto 1 \in \mathbb{C}$ is the constant function on G.

† We refer to [25, Sec. 8.6.2] for more details on this notion.

1.8 First Semi-Simple Phenomena*

To start to see the stark contrast between abelian groups and semi-simple groups (with compact and solvable groups in between in many ways), we prove two fundamental results for $\mathrm{SL}_2(\mathbb{R})$. We restrict to this specific group solely in the interests of brevity, since the results hold more generally for any non-compact simple Lie group (and with some complications also for non-compact semi-simple Lie groups).

Theorem 1.87 (Mautner phenomenon for $\mathrm{SL}_2(\mathbb{R})$). *If π is a unitary representation of $\mathrm{SL}_2(\mathbb{R})$ and $v \in \mathcal{H}_\pi$ is invariant (meaning that $\pi_g v = v$) under g either for*

$$g = a_t = \begin{pmatrix} \mathrm{e}^t & \\ & \mathrm{e}^{-t} \end{pmatrix}$$

for some $t \neq 0$, or for

$$g = u_s = \begin{pmatrix} 1 & s \\ & 1 \end{pmatrix}$$

for some $s \neq 0$, then v is invariant under all of $\mathrm{SL}_2(\mathbb{R})$.

Proof. Suppose that $v \in \mathcal{H}_\pi$ is invariant under a_t for $t \neq 0$. Replacing a_t by a_t^{-1} if necessary, we may assume that $t > 0$. Then

$$a_t^{-n} u_s a_t^n = \begin{pmatrix} \mathrm{e}^{-nt} & \\ & \mathrm{e}^{nt} \end{pmatrix} \begin{pmatrix} 1 & s \\ & 1 \end{pmatrix} \begin{pmatrix} \mathrm{e}^{nt} & \\ & \mathrm{e}^{-nt} \end{pmatrix} = \begin{pmatrix} 1 & \mathrm{e}^{-2nt} s \\ & 1 \end{pmatrix}$$

so that

$$\|\pi_{u_s} v - v\| = \|\pi_{a_t^{-n}} (\pi_{u_s} \pi_{a_t^n} v - \pi_{a_t^n} v)\| = \|\pi_{a_t^{-n} u_s a_t^n} v - v\| \longrightarrow 0$$

as $n \to \infty$ by continuity of the representation. For $\begin{pmatrix} 1 & \\ s & 1 \end{pmatrix}$ we can use the same argument but taking $n \to -\infty$ instead. Together we deduce that v is invariant under

$$H = \left\langle \left\{ \begin{pmatrix} 1 & s \\ & 1 \end{pmatrix}, \begin{pmatrix} 1 & \\ s & 1 \end{pmatrix} \,\middle|\, s \in \mathbb{R} \right\} \right\rangle.$$

However, it is easy to see (cf. Section A.1) that $H = \mathrm{SL}_2(\mathbb{R})$, which proves the first case of the theorem.

Now suppose that $v \in \mathcal{H}_\pi$ is invariant under $u_s = \begin{pmatrix} 1 & s \\ & 1 \end{pmatrix}$ for $s \neq 0$. Let us define the shorthand $g_\varepsilon = \begin{pmatrix} 1 & \\ \varepsilon & 1 \end{pmatrix}$ for $\varepsilon \in \mathbb{R}$ for convenience. Then

$$u_s^m g_\varepsilon u_s^n = \begin{pmatrix} 1 & ms \\ & 1 \end{pmatrix} \begin{pmatrix} 1 & \\ \varepsilon & 1 \end{pmatrix} \begin{pmatrix} 1 & ns \\ & 1 \end{pmatrix} = \begin{pmatrix} 1 + \varepsilon ms & ms(1 + \varepsilon ns) + ns \\ \varepsilon & 1 + \varepsilon ns \end{pmatrix}.$$

Choosing $\varepsilon = \varepsilon(n) = -\frac{1}{2ns}$ and $m = -2n$, this simplifes to give

$$u_s^{-2n} g_{\varepsilon(n)} u_s^n = \begin{pmatrix} 2 & 0 \\ \varepsilon(n) & \frac{1}{2} \end{pmatrix}.$$

Using continuity of the representation again, we see that

$$\|\pi_{a_{\log 2}} v - v\| = \lim_{n\to\infty} \|\pi_{u_s^{-2n} g_{\varepsilon(n)} u_s^n} v - v\| = \lim_{n\to\infty} \|\pi_{g_{\varepsilon(n)}} v - v\| = 0.$$

It follows that v is invariant under $a_{\log 2}$, and the first argument now applies to show that v is invariant under $\mathrm{SL}_2(\mathbb{R})$. □

Essential Exercise 1.88. Extend Theorem 1.87 by proving that if π is a unitary representation of $\mathrm{SL}_3(\mathbb{R})$ and $v \in \mathcal{H}_\pi$ is invariant under

$$\left\{ \begin{pmatrix} 1 & 0 & x_1 \\ & 1 & x_2 \\ & & 1 \end{pmatrix} \;\middle|\; x_1, x_2 \in \mathbb{R} \right\} \cong \mathbb{R}^2$$

then v is invariant under $\mathrm{SL}_3(\mathbb{R})$.

Exercise 1.89 (No non-trivial finite-dimensional representations). Use the method used to prove Theorem 1.87 to show that if π is a finite-dimensional unitary representation of $\mathrm{SL}_2(\mathbb{R})$, then $\pi_g = I$ for all $g \in \mathrm{SL}_2(\mathbb{R})$.

Our second initial result about $\mathrm{SL}_2(\mathbb{R})$ gives a significant strengthening of the Mautner phenomenon. We say that π *has invariant vectors* if there exists a non-zero vector $v \in \mathcal{H}_\pi$ that is invariant.

Theorem 1.90 (Howe–Moore for $\mathrm{SL}_2(\mathbb{R})$). *Let π be a unitary representation of $\mathrm{SL}_2(\mathbb{R})$ that does not have invariant vectors. Then every matrix coefficient $\varphi_{v,w}(g) = \langle \pi_g v, w \rangle$ converges to zero as $g \to \infty$ in $\mathrm{SL}_2(\mathbb{R})$.*

PROOF. Let (g_n) be a sequence in $\mathrm{SL}_2(\mathbb{R})$ with $g_n \to \infty$ as $n \to \infty$.

DIAGONAL CASE. Assume first that $g_n = \begin{pmatrix} \mathrm{e}^{t_n} & \\ & \mathrm{e}^{-t_n} \end{pmatrix}$ for some sequence (t_n) with $t_n \to \infty$ as $n \to \infty$. Notice that $\pi_{g_n} v \in B_{\|v\|}^{\mathcal{H}_\pi}(0)$ belongs to a fixed weak* compact subset of $\mathcal{H}_\pi$ so that there exists a weak* convergent subsequence (see [25, Prop. 8.11]), which for simplicity we again denote $(\pi_{g_n} v)$. Thus

$$\widetilde{v} = \lim_{n\to\infty} \pi_{g_n} v$$

exists in the weak* topology. We claim that $\widetilde{v} = 0$, which will follow from Theorem 1.87 and our assumption on π if we show that $\widetilde{v}$ is invariant under the action of $u = \begin{pmatrix} 1 & s \\ & 1 \end{pmatrix}$ for all $s \in \mathbb{R}$. To prove the latter, let $w \in \mathcal{H}_\pi$ and calculate

$$\langle \pi_u \widetilde{v}, w \rangle = \langle \widetilde{v}, \pi_u^* w \rangle = \lim_{n\to\infty} \langle \pi_{g_n} v, \pi_u^* w \rangle = \lim_{n\to\infty} \left\langle \pi_{g_n} \pi_{g_n^{-1} u g_n} v, w \right\rangle.$$

Now notice that $\lim_{n\to\infty} \|\pi_{g_n^{-1}ug_n} v - v\| = 0$ by continuity of the representation since $g_n^{-1}ug_n = \begin{pmatrix} 1 & \mathrm{e}^{-2t_n}s \\ & 1 \end{pmatrix} \to I$ as $n \to \infty$. Hence

$$\begin{aligned}|\langle \pi_u \widetilde{v}, w\rangle - \langle \widetilde{v}, w\rangle| &= \lim_{n\to\infty} \left| \left\langle \pi_{g_n} \pi_{g_n^{-1}ug_n} v, w \right\rangle - \left\langle \pi_{g_n} v, w \right\rangle \right| \\ &\leqslant \lim_{n\to\infty} \|\pi_{g_n^{-1}ug_n} v - v\| \|w\| = 0,\end{aligned}$$

which shows that $\pi_u \widetilde{v} = \widetilde{v}$ since w was arbitrary. As discussed above, this implies that $\widetilde{v} = 0$. By weak* compactness of $\overline{B_{\|v\|}^{\mathcal{H}_\pi}(0)}$ this implies that $\lim_{n\to\infty} \pi_{g_n} v = 0$ without any need to pass to a subsequence. Indeed, if this were not the case then we would be able to find a weak* neighbourhood U of 0 and a subsequence that is entirely outside U. Choosing a weak* convergent subsequence and applying the argument above then gives a contradiction.

Extending to the general case. If $g_n \in \mathrm{SL}_2(\mathbb{R})$ for all $n \geqslant 1$ and we have that $g_n \to \infty$ as $n \to \infty$, then we can apply the Cartan decomposition (see Section A.2) to write $g_n = k_n a_n k_n'$ with $k_n, k_n' \in K = \mathrm{SO}_2(\mathbb{R})$ and

$$a_n = \begin{pmatrix} \mathrm{e}^{t_n} & \\ & \mathrm{e}^{-t_n} \end{pmatrix}$$

with $t_n \to \infty$ as $n \to \infty$. Using compactness of K in the same way as above, we may assume that $k_n \to k$ and $k_n' \to k'$ as $n \to \infty$. By continuity of the representation and the case for (a_n) proved above, we may then find for $\varepsilon > 0$ some n_0 such that $n \geqslant n_0$ implies the three estimates indicated in the braces underneath the expression

$$\langle \pi_{a_n} \underbrace{\pi_{k_n'} v - \pi_{k'} v}_{\|\cdot\| < \varepsilon}, \pi_{k_n^{-1}} w \rangle + \langle \pi_{a_n} \pi_{k'} v, \underbrace{\pi_{k_n^{-1}} w - \pi_{k^{-1}} w}_{\|\cdot\| < \varepsilon} \rangle + \underbrace{\langle \pi_{a_n} \pi_{k'} v, \pi_{k^{-1}} w \rangle}_{\|\cdot\| < \varepsilon}.$$

However, recalling that $g_n = k_n a_n k_n'$, this sum is actually equal to $\langle \pi_{g_n} v, w \rangle$. Hence the Cauchy–Schwarz inequality gives

$$\left| \langle \pi_{g_n} v, w \rangle \right| \leqslant (\|w\| + \|v\| + 1)\,\varepsilon$$

for all $n \geqslant n_0$, which proves the theorem. □

Exercise 1.91 (No non-trivial finite-dimensional representations). Prove the same statement as in Exercise 1.89, but this time as a consequence of Theorem 1.90.

One is often interested in effective versions of Theorem 1.90. That is, whether a concrete estimate on the rate of decay of the matrix coefficients could be given. As we will see in Chapter 7, this is relatively straightforward for some groups (with $\mathrm{SL}_3(\mathbb{R})$ being the prime example), but is more delicate for $\mathrm{SL}_2(\mathbb{R})$.

Exercise 1.92 (Frobenius). Let $p > 2$ be a prime and $\mathbb{F}_p = \mathbb{Z}/p\mathbb{Z}$ the field with p elements. Show that any non-trivial unitary representation of the finite group $\mathrm{SL}_2(\mathbb{F}_p)$ has dimension at least $\frac{p-1}{2}$.

1.9 Summary and Outlook

The results of this chapter form the basis of all the discussions that follow. Most notably, we will need:

- Schur's lemma and its corollaries,
- familiarity with convolution in $L^1(G)$ and the convolution operators for unitary representations,
- the notions of matrix coefficients, positive-definite functions and their equivalences, and
- the corollaries to the Gelfand–Naimark–Seagal construction.

We have not yet seen many examples of unitary duals, and this should be corrected as soon as possible. There are two general classes of groups that are reasonable candidates to discuss next. In fact the unitary representation theory of abelian groups in Chapter 2 and of compact groups in Chapter 3 are the easiest to understand, and we could continue with either of these.

Chapter 2
Abelian Groups

In this chapter we completely classify unitary representations of locally compact σ-compact metric abelian groups. The assumptions that the group is locally compact and abelian are critical for the material developed in this chapter, but as before we will also assume that the group G is σ-compact and metric. We will keep the topological assumptions on G implicit and refer to G as *the abelian group G*.

2.1 Pontryagin Dual

In Corollary 1.32 we saw that every irreducible representation of the abelian group G is one-dimensional and hence defines a (continuous unitary) character. In the context of abelian groups the following terminology is often used instead of the phrase 'unitary dual' as in Definition 1.26.

Definition 2.1 (Pontryagin dual of the abelian group). The *dual group* (or *Pontryagin dual*) of the abelian group G is defined (as an abstract group) by

$$\widehat{G} = \{\chi \colon G \longrightarrow \mathbb{S}^1 \mid \chi \text{ is a continuous character}\}$$

with the group operations being pointwise product and inverse.

The reader not familiar with this notion of duality may use the following exercise as a warmup for (the conclusions of which will be special cases of the theory developed in this chapter).

Exercise 2.2 (Generalization of Fourier series). Suppose that the abelian group G is compact, and normalize the Haar measure m to satisfy $m(G) = 1$. Assume furthermore that $\widehat{G}$ separates points† in G, meaning that for every $g_1 \neq g_2$ in G there exists some $\chi \in \widehat{G}$ with $\chi(g_1) \neq \chi(g_2)$.

† We note that this assumption is in fact always satisfied. This can be checked directly in many examples, and in general follows from Corollary 1.81 or the more specialized Theorem 2.17.

M. Einsiedler and T. Ward, *Unitary Representations and Unitary Duals*,
Graduate Texts in Mathematics 308, https://doi.org/10.1007/978-3-032-03899-9_2

(a) Show that $\widehat{G}$ forms a countable orthonormal basis of $L^2(G)$.
(b) Show, for any unitary representation π, that $\mathcal{H}_\pi$ is the orthogonal direct sum of the *eigenspaces* $\mathcal{H}_\pi^\chi$ *of weight* $\chi \in \widehat{G}$ defined by

$$\mathcal{H}_\pi^\chi = \{w \in \mathcal{H}_\pi \mid \pi_g w = \chi(g)w \text{ for all } g \in G\}.$$

For many concrete groups it is not at all difficult to describe the dual group explicitly.

Essential Exercise 2.3 (Basic examples). Prove the following isomorphisms of (abstract) groups: (a) $\widehat{\mathbb{Z}} \cong \mathbb{T}$; (b) $\widehat{\mathbb{T}} \cong \mathbb{Z}$; and (c) $\widehat{\mathbb{R}} \cong \mathbb{R}$.

It is easy to give concrete examples of non-compact groups where the description of a unitary representation is in general more complicated than it is in Exercise 2.2(b). Let λ denote the regular representation of $\mathbb{R}$ on $L^2(\mathbb{R})$ and let χ denote a unitary character of $\mathbb{R}$. If now $\lambda_g f = \chi(g)f$ for all $g \in \mathbb{R}$ and some $f \in L^2(\mathbb{R})$, then $|f|$ would be constant, and so $f = 0$. The same applies to any non-compact abelian group G and shows that the regular representation on $L^2(G)$ cannot be a direct sum of irreducible representations. Nonetheless, we will arrive at a complete understanding of all unitary representations of abelian groups using only characters in this chapter.

2.1.1 Characters as Algebra Homomorphisms

Given explicit descriptions of a group and its dual, many of the theorems in this chapter have alternative and often simpler proofs. However, to handle the general case seamlessly the following is an important tool. We recall from Section 1.5.1 that $L^1(G)$ is a commutative Banach algebra.

Proposition 2.4 (Algebra homomorphisms of $L^1(G)$). *For the abelian group G every character $\chi \in \widehat{G}$ induces a continuous non-trivial algebra homomorphism from $\mathcal{A} = L^1(G)$ to $\mathbb{C}$ via the formula*

$$\chi_{\mathcal{A}}(f) = \int f\chi \,\mathrm{d}m$$

for $f \in L^1(G)$. Moreover, every continuous non-trivial algebra homomorphism from $\mathcal{A}$ to $\mathbb{C}$ is of this form, for some uniquely determined $\chi \in \widehat{G}$.

PROOF. Given a unitary character $\chi \in \widehat{G}$ and $f_1, f_2 \in L^1(G)$ we have

$$\begin{aligned}
\int f_1 * f_2(g)\chi(g)\,\mathrm{d}m(g) &= \iint f_1(h)f_2(g-h)\chi(g)\,\mathrm{d}m(h)\,\mathrm{d}m(g) \\
&= \iint f_1(h)f_2(k)\chi(h+k)\,\mathrm{d}m(k)\,\mathrm{d}m(h) \\
&= \Bigl(\int f_1\chi\,\mathrm{d}m\Bigr)\Bigl(\int f_2\chi\,\mathrm{d}m\Bigr)
\end{aligned}$$

by Fubini's theorem, which shows that $\chi_{\mathcal{A}}$ is indeed an algebra homomorphism. By the identification of the dual of $L^1(G)$ with $L^\infty(G)$ (see [25, Prop. 7.34]), $\chi_{\mathcal{A}}$ is continuous and non-trivial since $\|\chi_{\mathcal{A}}\| = \|\chi\|_\infty = 1$.

The converse is more involved. Let $\chi_{\mathcal{A}}\colon \mathcal{A} = L^1(G) \to \mathbb{C}$ be a non-trivial continuous algebra homomorphism. Then there is an element $\chi \in L^\infty(G)$ with $\|\chi\|_\infty < \infty$ with

$$\chi_{\mathcal{A}}(f) = \int_G f\chi \, \mathrm{d}m$$

for all $f \in L^1(G)$. We have to show that χ can be chosen in $C_b(G)$ and with the property that $\chi(gh) = \chi(g)\chi(h)$ for all $g, h \in G$.

By assumption $\chi_{\mathcal{A}} \neq 0$, so there exists some $f_0 \in L^1(G)$ with $\chi_{\mathcal{A}}(f_0) \neq 0$. Also, by the assumption on $\chi_{\mathcal{A}}$ and Fubini's theorem

$$\begin{aligned}\chi_{\mathcal{A}}(f)\chi_{\mathcal{A}}(f_0) = \chi_{\mathcal{A}}(f * f_0) &= \int_G \int_G f(g) f_0(h-g) \, \mathrm{d}m(g)\chi(h) \, \mathrm{d}m(h) \\ &= \int_G f(g) \int_G \lambda_g(f_0)\chi \, \mathrm{d}m \, \mathrm{d}m(g)\end{aligned}$$

for all $f \in L^1(G)$. We now define

$$\chi'(g) = (\chi_{\mathcal{A}}(f_0))^{-1} \int_G \lambda_g(f_0)\chi \, \mathrm{d}m,$$

so that $\chi_{\mathcal{A}}(f) = \int_G f\chi' \, \mathrm{d}m$ for all $f \in L^1(G)$, and in particular $\chi' = \chi$ almost everywhere.

Notice that χ' is defined using f_0 and χ essentially by convolution. This implies that $\chi' \in C_b(G)$, since

$$\begin{aligned}|\chi'(g) - \chi'(g_0)| &= |\chi_{\mathcal{A}}(f_0)|^{-1} \left| \int_G (\lambda_g f_0 - \lambda_{g_0} f_0)\chi \, \mathrm{d}m \right| \\ &\leqslant |\chi_{\mathcal{A}}(f_0)|^{-1} \|\lambda_g f_0 - \lambda_{g_0} f_0\|_1 \|\chi\|_\infty \longrightarrow 0\end{aligned}$$

as $g \to g_0$, since $G \ni g \mapsto \lambda_g f_0 \in L^1(G)$ is continuous. Simplifying the notation, we suppose that $\chi = \chi' \in C_b(G)$.

Now choose a sequence (B_n) of decreasing open neighbourhoods of $0 \in G$ that form a basis of the neighbourhoods at 0. Then the sequence (ψ_n) defined by

$$\psi_n = \frac{1}{m(B_n)} \mathbb{1}_{B_n}$$

for all $n \geqslant 1$ forms an approximate identity (see Proposition 1.46).

Now let $g_1, g_2 \in G$ be arbitrary. Then $\lambda_{g_1}\psi_n = \frac{1}{m(B_n)}\mathbb{1}_{B_n+g_1}$ and so

$$\chi_{\mathcal{A}}\left(\lambda_{g_1}\psi_n\right) = \frac{1}{m(B_n)} \int_{B_n+g_1} \chi \, \mathrm{d}m \longrightarrow \chi(g_1)$$

as $n \to \infty$ by continuity of χ, and similarly $\chi_{\mathcal{A}}(\lambda_{g_2}\psi_n) \to \chi(g_2)$ as $n \to \infty$. Moreover, (1.19) shows (by identifying $f \in L^1(G)$ with $\nu_f \in M(G)$) that

$$\lambda_{g_1}\psi_n * \lambda_{g_2}\psi_n = \delta_{g_1} * \psi_n * \delta_{g_2} * \psi_n = \delta_{g_1+g_2} * \psi_n * \psi_n = \lambda_{g_1+g_2}(\psi_n * \psi_n),$$

and using $\psi_n * \psi_n$ (which also forms an approximate identity by Proposition 1.46) in the same way as ψ_n above, we obtain from this

$$\begin{aligned}\chi_{\mathcal{A}}\left(\lambda_{g_1}\psi_n\right)\chi_{\mathcal{A}}\left(\lambda_{g_2}\psi_n\right) &= \chi_{\mathcal{A}}\left(\lambda_{g_1}\psi_n * \lambda_{g_2}\psi_n\right)\\ &= \chi_{\mathcal{A}}\left(\lambda_{g_1+g_2}\left(\psi_n * \psi_n\right)\right) \longrightarrow \chi(g_1+g_2)\end{aligned}$$

as $n \to \infty$. Thus with the above we obtain

$$\chi(g_1+g_2) = \chi(g_1)\chi(g_2)$$

for $g_1, g_2 \in G$. In other words, $\chi\colon G \to \mathbb{C}$ is a continuous homomorphism to the multiplicative structure of $\mathbb{C}$. Since χ is bounded and non-zero, it follows that χ is non-zero everywhere, and that it takes values in $\mathbb{S}^1$. This shows that $\chi_{\mathcal{A}}$ is defined by $\chi \in \widehat{G}$. □

Using the correspondence in Proposition 2.4 as an identification, we can and will consider the weak* topology on $\widehat{G} \subseteq L^\infty(G)$. The following corollary defines the important notion of *Fourier transform*, and gives the fundamental properties of the Fourier transform. These will be used frequently, and often without an explicit reference to this corollary.

Corollary 2.5 (Topology, structure, and functions on $\widehat{G}$). *For the abelian group G the dual group $\widehat{G}$ is a locally compact σ-compact metric group in the weak* topology (equivalently the compact-open topology). The Fourier (back) transform*

$$\check{f}(\chi) = \int f\chi \,\mathrm{d}m \tag{2.1}$$

for $f \in L^1(G)$ satisfies $\check{f} \in C_0(\widehat{G})$, $\widecheck{f_1 * f_2} = \check{f}_1\check{f}_2$, $\widecheck{f^*} = \overline{\check{f}}$, and $\|\check{f}\|_\infty \leqslant \|f\|_1$ for all $f, f_1, f_2 \in L^1(G)$. Moreover, $\{\check{f} \mid f \in L^1(G)\} \subseteq C_0(\widehat{G})$ is a dense sub-algebra with respect to the supremum norm on $\widehat{G}$.

We note that (2.1) is often referred to as the *Fourier back transform* (which is the reason for the notation). However, we decided to use this formula as a starting point, as we will see in the next section that it has (with this sign convention) important and conveniently described relationships to unitary representations. Because of the duality results that we will see later, the designation of one transform as the 'back' one and the other, implicitly, as the 'forward' one is rather arbitrary, and refers to a choice of a sign.

Proof. Let $\sigma(L^1(G))$ denote the set of all non-trivial continuous algebra homomorphisms $L^1(G) \to \mathbb{C}$. By Proposition 2.4 these all have norm one. Hence

$$\Omega = \sigma\big(L^1(G)\big) \cup \{0\}$$

may be described as the set of all $\chi \in L^\infty(G)$ with $\|\chi\|_\infty \leqslant 1$ and with

$$\int f_1 * f_2 \chi \, \mathrm{d}m = \int f_1 \chi \, \mathrm{d}m \int f_2 \chi \, \mathrm{d}m$$

for all $f_1, f_2 \in L^1(G)$. This implies that Ω is weak* closed, and hence weak* compact by the Banach–Alaoglu theorem (see [25, Th. 8.10]). Since $L^1(G)$ is separable, the weak* topology is metrizable when restricted to the closed unit ball in $L^\infty(G)$ (see [25, Prop. 8.11]). This implies that $\sigma(L^1(G)) = \Omega \smallsetminus \{0\}$ is locally compact, σ-compact and metrizable. We identify $\sigma(L^1(G))$ with $\widehat{G}$ as in Proposition 2.4. Moreover, recall from Proposition 1.79 that the weak* topology on $\mathcal{P}^1(G) \supseteq \widehat{G}$ coincides with the compact-open topology, which proves that $\widehat{G}$ is locally compact, σ-compact and metric with respect to the compact-open topology. In this topology it is easy to see that the group operations are continuous.

Let $f \in L^1(G)$ and define the Fourier transform as in the corollary. Equivalently, we may think of $\check{f}$ as the evaluation map

$$\mathrm{ev}_f \colon \chi \longmapsto \chi(f) = \int f \chi \, \mathrm{d}m$$

from $L^1(G)' \cong L^\infty(G)$ to $\mathbb{C}$ restricted to $\sigma(L^1(G)) \cong \widehat{G}$. Hence by the definition of the weak* topology we see that $\check{f}$ is continuous on $\widehat{G}$. Moreover, by setting

$$\check{f}(0) = \int f 0 \, \mathrm{d}m = 0$$

we see that $\check{f}$ has a continuous extension to the compact space

$$\Omega = \sigma(L^1(G)) \cup \{0\},$$

with $\check{f}(0) = 0$. In other words, $0 \in \Omega$ plays the role of ∞ in the one-point compactificiation of $\widehat{G}$, and $\check{f} \in C_0(\widehat{G})$.

The property

$$\widecheck{f_1 * f_2}(\chi) = \chi(f_1 * f_2) = \chi(f_1)\chi(f_2) = \check{f}_1 \check{f}_2(\chi)$$

for $f_1, f_2 \in L^1(G)$ is precisely the algebra homomorphism property of the map $\chi \in \sigma(L^1(G))$. In particular, this implies that the image of the Fourier transform is a sub-algebra of

$$C_0(\widehat{G}) \cong \{F \in C(\Omega) \mid F(0) = 0\}.$$

For $f \in L^1(G)$ we defined f^* in Section 1.5.1, which in the additive notation becomes $f^*(g) = \overline{f(-g)}$ for all $g \in G$. Since

$$\widecheck{f^*}(\chi) = \int_G f^* \chi \, \mathrm{d}m = \int_G \overline{f}(g)\chi(-g) \, \mathrm{d}m = \overline{\widecheck{f}(\chi)}$$

for all $\chi \in \Omega = \widehat{G} \cup \{0\}$, we see that the image of $L^1(G)$ in $C_0(\widehat{G})$ is closed under conjugation. The inequality $\|\widecheck{f}\|_\infty \leqslant \|f\|_1$ follows, as $\chi \in \widehat{G}$ satisfies $\|\chi\|_\infty = 1$ which gives $|\widecheck{f}(\chi)| = |\int f\chi \, \mathrm{d}m| \leqslant \|f\|_1$.

Finally we note that the sub-algebra $\{\widehat{f} \mid f \in L^1(G)\} \subseteq C(\Omega)$ also separates points: If $\widecheck{f}(\chi_1) = \widecheck{f}(\chi_2)$ for some $\chi_1, \chi_2 \in \Omega$ and all $f \in L^1(G)$, then χ_1 and χ_2 define the same functional on $L^1(G)$ and so $\chi_1 = \chi_2$. Applying the Stone–Weierstrass theorem ([25, Thm. 2.40]) we see that the algebra

$$\mathcal{A} = \{\widecheck{f} \mid f \in L^1(G)\} + \mathbb{C}\mathbb{1}$$

is dense in $C(\Omega)$. This shows that for a given $F \in C_0(\widehat{G})$ and $\varepsilon > 0$ there exists some $f \in L^1(G)$ and $\alpha \in \mathbb{C}$ such that $\|\widecheck{f} + \alpha\mathbb{1} - F\|_\infty < \varepsilon$. Since $F(0) = 0$ and $\widecheck{f}(0) = 0$ we see that $|\alpha| < \varepsilon$, and hence $\|\widecheck{f} - F\|_\infty < 2\varepsilon$ as required. □

Essential Exercise 2.6 (Basic examples). Show that the isomorphisms in Exercise 2.3 are also isomorphisms of topological groups. Furthermore, show that for any $d \in \mathbb{N}$ we have the isomorphism $\widehat{\mathbb{R}^d} \cong \mathbb{R}^d$ as topological groups.

Exercise 2.7 (Continuity and bound). Let $\mathcal{A}$ be a commutative Banach algebra over $\mathbb{C}$. Show that any algebra homomorphism $\chi \colon \mathcal{A} \to \mathbb{C}$ is continuous and satisfies $\|\chi\| \leqslant 1$.

Exercise 2.8. Upgrade Exercise 1.80 by showing that for $G = \mathbb{R}$ the weak* topology and the compact-open topology have different neighbourhoods of 0 in $\mathcal{E}^{\leqslant 1}(\mathbb{R}) = \widehat{\mathbb{R}} \cup \{0\} \subseteq \mathcal{P}^{\leqslant 1}(\mathbb{R})$.

2.2 Spectral Theory, First Formulations

Due to the results (and exercises) of the last section it is natural to have a more symmetric notation for the group and its Pontryagin dual, so we also use additive notation for $\widehat{G}$. In fact, we will write $t, t_0, t_1, \ldots$ for the elements of the additive group $\widehat{G}$ and for $t \in \widehat{G}$ we write $\chi_t \colon G \to \mathbb{S}^1$ for the associated multiplicative character. Furthermore we define the dual pairing $\langle \cdot, \cdot \rangle \colon G \times \widehat{G} \to \mathbb{S}^1$ by

$$G \times \widehat{G} \ni (g, t) \longmapsto \langle g, t \rangle = \chi_t(g) \in \mathbb{S}^1.$$

In particular, in this notation we also write

$$\widecheck{f}(t) = \int_G f\chi_t \, \mathrm{d}m = \int_G f(g) \, \langle g, t \rangle \, \mathrm{d}m(g)$$

for the Fourier transform of $f \in L^1(G)$ at $t \in \widehat{G}$.

2.2.1 Multiplication Representations

We start the discussions of unitary representations of the abelian group G by discussing an important class.

Lemma 2.9 (Multiplication representations). *For a σ-finite measure μ on $X = \widehat{G} \times \mathbb{N}$ the multiplication representation*

$$M_g(w)(t,n) = \langle g,t\rangle\, w(t,n) \tag{2.2}$$

for $g \in G$, $w \in L^2_\mu(X)$ and $(t,n) \in X$ defines a unitary representation M of G on $L^2_\mu(X)$. Moreover, the convolution operator $M_*(f)$ associated to $f \in L^1(G)$ is given by the multiplication operator

$$\begin{aligned} M_*(f) = M_{\check{f}} \colon L^2_\mu(X) &\longrightarrow L^2_\mu(X) \\ w &\longmapsto \check{f} w. \end{aligned}$$

Proof. Since $\langle g,t\rangle \in \mathbb{S}^1$ for all $(g,t) \in G \times \widehat{G}$ the operator

$$M_g \colon L^2_\mu(X) \to L^2_\mu(X)$$

is unitary for any $g \in G$. Moreover, if (g_n) is a sequence in G with $g_n \to g_0$ as $n \to \infty$, then continuity of the character χ_t implies that $\langle g_n, t\rangle \to \langle g,t\rangle$ as $n \to \infty$ for all $t \in \widehat{G}$. Given some $w \in L^2_\mu(X)$, this then implies that

$$\big\|M_{g_n} w - M_{g_0} w\big\|_2^2 = \int_X |\,\langle g_n,t\rangle - \langle g_0,t\rangle\,|^2 |w(t)|^2 \,\mathrm{d}\mu(t) \longrightarrow 0$$

as $n \to \infty$ by dominated convergence. It follows that $G \ni g \mapsto M_g$ is indeed a unitary representation of G on $L^2_\mu(X)$.

To see the last part of the lemma let $f \in L^1(G)$ and $v, w \in L^2_\mu(X)$. We have

$$\begin{aligned} \langle M_*(f)v, w\rangle_{L^2_\mu(X)} &= \int_G f(g)\,\langle M_g v, w\rangle_{L^2_\mu(X)}\,\mathrm{d}m(g) \\ &= \int_G f(g) \int_X \langle g,t\rangle\, v(t,n)\overline{w(t,n)}\,\mathrm{d}\mu(t,n)\,\mathrm{d}m(g) \\ &= \int_X \check{f}(t) v(t,n)\overline{w(t,n)}\,\mathrm{d}\mu(t,n) = \left\langle \check{f} v, w\right\rangle_{L^2_\mu(X)} \end{aligned}$$

by definition of the convolution operator $M_*(f)$ using weak integration, Fubini's theorem, and the definition of the Fourier transform $\check{f}$. As $v, w \in L^2_\mu(X)$ were arbitrary the lemma follows. □

2.2.2 Bochner's Theorem

We now return to the discussion of general unitary representations of the abelian group G, and recall from Section 1.6 that for this we need to understand positive-definite functions.

Theorem 2.10 (Bochner's theorem). *Let ϕ be a positive-definite function on the abelian group G. Then there exists a uniquely determined finite measure μ on $\widehat{G}$ such that*

$$\phi(g) = \int_{\widehat{G}} \langle g, t \rangle \, \mathrm{d}\mu(t) \tag{2.3}$$

for all $g \in G$.

We will refer to the measure μ satisfying (2.3) as a *(diagonal) spectral measure.*

We note that Theorem 2.10 follows at once from Corollary 1.83. Indeed in our case of an abelian group $\mathcal{E}^1(G)$ is equal to the dual group $\widehat{G}$ as in Definition 2.1, and so (1.39) becomes (2.3). For the reader who wants to avoid the application of Choquet's theorem (as in the proof of Corollary 1.83) we give a second proof relying on different ideas.

Proof of Theorem 2.10. Suppose the finite measures μ_1, μ_2 both satisfy (2.3). For $f \in L^1(G)$ we may then use Fubini's theorem to see that

$$\begin{aligned}\int_G f(g)\phi(g) \, \mathrm{d}m(g) &= \int_G \int_{\widehat{G}} f(g)\langle g, t \rangle \, \mathrm{d}\mu_j(t) \, \mathrm{d}m(g) \\ &= \int_{\widehat{G}} \check{f}(t) \, \mathrm{d}\mu_j(t)\end{aligned}$$

for $j = 1, 2$. However, $\mathcal{A} = \{\check{f} \mid f \in L^1(G)\} \subseteq C_0(\widehat{G})$ is dense with respect to the supremum norm by Corollary 2.5. Therefore, we obtain

$$\int_{\widehat{G}} F \, \mathrm{d}\mu_1 = \int_{\widehat{G}} F \, \mathrm{d}\mu_2$$

for all $F \in C_0(\widehat{G})$ and the uniqueness of the measure in the Riesz representation theorem implies that $\mu_1 = \mu_2$.

We will now prove the existence of the measure μ by defining a linear functional Λ on $C_0(\widehat{G})$ and applying the Riesz representation theorem. By density of $\mathcal{A} \subseteq C_0(\widehat{G})$, it is sufficient to define $\Lambda(\check{f})$ for $f \in L^1(G)$ and show that

$$|\Lambda(\check{f})| \leqslant \phi(0)\|\check{f}\|_\infty. \tag{2.4}$$

To this end, define

$$\Lambda(\check{f}) = \int f\phi \, \mathrm{d}m \tag{2.5}$$

for $f \in L^1(G)$ and notice that (2.4), once established, will in particular imply that $\Lambda(\check{f})$ is indeed well-defined for any $\check{f} \in C_0(\widehat{G})$, (meaning that it does not depend on the choice of f, but only on its Fourier transform $\check{f}$).

For the proof of the bound (2.4) for the functional in (2.5) we recall from the GNS construction (Theorem 1.76) that $\phi = \varphi_{v_0}$ is a matrix coefficient of some $v_0 \in \mathcal{H}_\pi$ for some unitary representation π of G. With this we have

$$\int f\phi \,\mathrm{d}m = \int f\varphi_{v_0} \,\mathrm{d}m = \int f(g) \langle \pi_g v_0, v_0 \rangle \,\mathrm{d}m(g) = \langle \pi_*(f)v_0, v_0 \rangle$$

for all $f \in L^1(G)$. Therefore

$$\left| \int f\phi \,\mathrm{d}m \right| \leqslant \|\pi_*(f)\|_{\mathrm{op}} \|v_0\|^2 = \|\pi_*(f)\|_{\mathrm{op}} \phi(0)$$

and so we need to estimate $\|\pi_*(f)\|_{\mathrm{op}}$. For this, we recall from Section 1.5.3 that $\pi_*(f)^* = \pi_*(f^*)$. Also recall that $L^1(G)$ is an abelian Banach algebra (since G is assumed to be abelian), which implies that $\pi_*(f)$ is a normal operator, and so its operator norm is equal to its spectral radius

$$\|\pi_*(f)\|_{\mathrm{op}} = \lim_{n\to\infty} \left\| \pi_*(f)^{2^n} \right\|_{\mathrm{op}}^{2^{-n}} = \lim_{n\to\infty} \left\| \pi_*\left(f^{*2^n}\right) \right\|_{\mathrm{op}}^{2^{-n}}.$$

Combining this with the bound

$$\|\pi_*(f^{*2^n})\|_{\mathrm{op}} \leqslant \|f^{*2^n}\|_1$$

we obtain

$$\|\pi_*(f)\|_{\mathrm{op}} \leqslant \lim_{n\to\infty} \|f^{*2^n}\|_1^{2^{-n}}.$$

Here the limit exists because of the spectral radius formula in the commutative Banach algebra $L^1(G)$ (see [25, Cor. 11.29]) and equals

$$\max_{\chi \in \sigma(L^1(G)) \cup \{0\}} |\chi(f)| = \|\check{f}\|_\infty$$

by the identification $\sigma(L^1(G)) = \widehat{G}$ in Proposition 2.4 and the proof of Corollary 2.5. To summarize, we have shown that (2.5) defines a continuous linear functional Λ on $C_0(\widehat{G})$ satisfying (2.4).

Applying the Riesz representation theorem (see [25, Th. 7.54]) to Λ, we find a finite measure μ on $\widehat{G}$ such that

$$\int_G f\phi \,\mathrm{d}m = \Lambda(\check{f}) = \int_{\widehat{G}} \check{f} \,\mathrm{d}\mu \tag{2.6}$$

for all $f \in L^1(G)$ and $\|\mu\| = \|\Lambda\| \leqslant \phi(0)$. Here μ is potentially complex-valued, and we may write $\mathrm{d}\mu = F \,\mathrm{d}|\mu|$ for some (positive) finite Borel measure $|\mu|$ on $\widehat{G}$ and some measurable function F satisfying

$$\|F\|_{L^1_{|\mu|}} = \|\Lambda\| \leqslant \phi(0).$$

Now set $f = \psi_n$ in (2.6) for some approximate identity (ψ_n) and let $n \to \infty$ to obtain

$$\phi(0) = \lim_{n\to\infty} \int \psi_n \phi \, \mathrm{d}m = \lim_{n\to\infty} \int \widecheck{\psi_n} F \, \mathrm{d}|\mu| = \int F \, \mathrm{d}|\mu|$$

since the bound $\|\widecheck{\psi_n}\|_\infty \leqslant 1$ and the convergence $\widecheck{\psi_n}(t) \to 1$ as $n \to \infty$ for every $t \in \widehat{G}$ allows us to apply dominated convergence. However,

$$\int |F| \, \mathrm{d}|\mu| = \|F\|_{L^1_{|\mu|}} = \|\Lambda\| \leqslant \phi(0) = \int F \, \mathrm{d}|\mu|$$

implies that F is real-valued and non-negative almost everywhere with respect to $|\mu|$, so that $\mathrm{d}\mu = F \, \mathrm{d}|\mu|$ is in fact a positive finite Borel measure.

Finally, fix some $g \in G$, set $f = \lambda_g \psi_n$ in (2.6) and let $n \to \infty$ to obtain

$$\phi(g) = \lim_{n\to\infty} \int \psi_n(h)\phi(g+h) \, \mathrm{d}m(h) = \lim_{n\to\infty} \int (\lambda_g \psi_n) \, \phi \, \mathrm{d}m = \lim_{n\to\infty} \int \widecheck{\lambda_g \psi_n} \, \mathrm{d}\mu.$$

Since $\|\widecheck{\lambda_g \psi_n}\|_\infty \leqslant \|\lambda_g \psi_n\|_1 = 1$ and

$$\widecheck{\lambda_g \psi_n}(t) = \int (\lambda_g \psi_n) \chi_t \, \mathrm{d}m = \int \psi_n(h) \chi_t(g+h) \, \mathrm{d}m(h) \longrightarrow \chi_t(g) = \langle g, t \rangle$$

as $n \to \infty$ for any $t \in \widehat{G}$, we may again apply dominated convergence to obtain

$$\phi(g) = \int \langle g, t \rangle \, \mathrm{d}\mu(t)$$

as claimed. □

The following exercise may help to develop an intuitive understanding of spectral measures.

Exercise 2.11. Fix a sequence $(t_n)_{n\in\mathbb{N}}$ in $\widehat{G}$, and define a unitary representation π of G on $\mathcal{H}_\pi = \ell^2(\mathbb{N})$ by

$$(\pi_g v)_n = \langle g, t_n \rangle v_n$$

for all $n \in \mathbb{N}$, $g \in G$, and $v = (v_n) \in \mathcal{H}_\pi$. Calculate μ_v for $v \in \mathcal{H}_\pi$, and interpret it as giving weights to the eigenvalues in $\widehat{G}$.

2.2.3 The Spectral Theorems

Using Bochner's theorem (Theorem 2.10) it is now quite straightforward to completely describe cyclic as well as general unitary representations of abelian groups.

Corollary 2.12 (Spectral theorem for cyclic representations). *Let π be a unitary representation of the abelian group G, and let $v \in \mathcal{H}_\pi$. Applying Bochner's theorem to the positive-definite function φ_v defined by*

$$\varphi_v(g) = \langle \pi_g v, v \rangle$$

for $g \in G$, we obtain the spectral measure μ_v for v. Then the cyclic representation on

$$\langle v \rangle_\pi = \overline{\langle \pi_g v \mid g \in G \rangle_{\mathbb{C}}}$$

is unitarily isomorphic to the unitary multiplication representation M of G on $L^2_{\mu_v}(\widehat{G})$ defined by

$$M_g(w)(t) = \langle g, t \rangle\, w(t)$$

for $g \in G$, $w \in L^2_{\mu_v}(\widehat{G})$, and $t \in \widehat{G}$. Moreover, the intertwining isometry sends $v \in \mathcal{H}_\pi$ to $\mathbb{1} \in L^2_{\mu_v}(\widehat{G})$ and in particular,

$$\|v\|^2 = \varphi_v(0) = \mu_v(\widehat{G}).$$

Finally, π is cyclic if and only if there exists a finite μ on $\widehat{G}$ such that π is isomorphic to the multiplication representation M on $L^2_\mu(\widehat{G})$.

PROOF. We first consider the case of a multiplication representation defined by a finite measure μ on $\widehat{G}$. We will show that the representation is cyclic and has the vector $\mathbb{1} \in L^2_\mu(\widehat{G})$ as a generator. Indeed, for $f \in L^1(G)$ Lemma 2.9 shows that $M_*(f)\mathbb{1} = \check{f} \in \langle \mathbb{1} \rangle_M$. By density of the set of these functions in $C_0(\widehat{G})$ with respect to the supremum norm (see Corollary 2.5) and since μ is assumed to be a finite measure, we see that $C_0(\widehat{G}) \subseteq \langle \mathbb{1} \rangle_M$. By density of $C_c(\widehat{G}) \subseteq L^2_\mu(\widehat{G})$ we obtain $\langle \mathbb{1} \rangle_M = L^2_\mu(\widehat{G})$ as claimed.

Now consider a unitary representation π and a vector $v \in \mathcal{H}_\pi$. Then the function defined by $\varphi_v(g) = \langle \pi_g v, v \rangle$ is positive-definite. Applying Bochner's theorem (Theorem 2.10) we find a finite measure μ_v on $\widehat{G}$ such that

$$\varphi_v(g) = \int_{\widehat{G}} \langle g, t \rangle \,\mathrm{d}\mu_v(t).$$

Notice that the matrix coefficient of $\mathbb{1} \in L^2_\mu(\widehat{G})$ is given by

$$\langle M_g \mathbb{1}, \mathbb{1} \rangle_{L^2_\mu(\widehat{G})} = \int_{\widehat{G}} \langle g, t \rangle \,\mathrm{d}\mu_v(t) = \varphi_v(g) = \langle \pi_g v, v \rangle$$

for all $g \in G$. By Proposition 1.65 this shows that π restricted to the cyclic representation $\langle v \rangle_\pi$ generated by v is unitarily isomorphic to $\langle \mathbb{1} \rangle_M = L^2_\mu(\widehat{G})$, and that we may assume that v is sent to $\mathbb{1}$.

The last statement in the corollary also follows now. If π is cyclic then we have found the isomorphism. If π is isomorphic to the multiplication representation on $L^2_\mu(\widehat{G})$ for a finite measure μ, then it must be cyclic since we already showed that $L^2_\mu(\widehat{G})$ is cyclic (with generator $\mathbb{1}$). □

Using the cyclic case, we easily obtain a similar description of a general representation.

Corollary 2.13 (Spectral theorem). *Let π be a unitary representation of the abelian group G. Then there exists a finite measure μ on $X = \widehat{G} \times \mathbb{N}$ and a unitary isomorphism between π and the unitary multiplication representation M of G on $L^2_\mu(X)$.*

PROOF. Let π be a unitary representation of G. Applying Lemma 1.62 we can split

$$\mathcal{H}_\pi = \bigoplus_{n\geqslant 1} \langle v_n \rangle_\pi$$

into a direct sum of cyclic representations. Applying the cyclic case in Corollary 2.12, we find a sequence of finite measures (μ_n) such that

$$\mathcal{H}_\pi = \bigoplus_{n\geqslant 1} \langle v_n \rangle_\pi \cong \bigoplus_{n\geqslant 1} L^2_{\mu_n}(\widehat{G}).$$

We use these measures to define a σ-finite measure μ on $X = \widehat{G} \times \mathbb{N}$ by setting

$$\mu(B) = \sum_{n=1}^{\infty} \mu_n\bigl(\{t \in \widehat{G} \mid (t,n) \in B\}\bigr)$$

for any Borel set $B \subseteq X$. This measure satisfies

$$L^2_\mu(X) \cong \bigoplus_{n\geqslant 1} L^2_{\mu_n}(\widehat{G})$$

where $f \in L^2_\mu(X)$ corresponds to the sequence (f_n) defined by $f_n(t) = f(t,n)$ for $t \in \widehat{G}$ and $n \geqslant 1$. Combining the above isomorphisms shows that π is isomorphic to the multiplication representation M on $L^2_\mu(X)$.

Next we claim that the multiplication representation on $L^2_\mu(X)$ is unitarily isomorphic to a multiplication representation defined by a finite measure μ' on X. Indeed, since μ is σ-finite there exists some strictly positive measurable function F with the property that F^2 is integrable. Now define the finite measure μ' by $\mathrm{d}\mu' = F^2\,\mathrm{d}\mu$ and $U = M_{F^{-1}}$ so that for $w \in L^2_\mu(X)$ we have $U(w) = F^{-1}w$ and

$$\|Uw\|^2_{L^2_{\mu'}(X)} = \int F^{-2}|w|^2\,\mathrm{d}\mu' = \int F^{-2}|w|^2F^2\,\mathrm{d}\mu = \|w\|^2_{L^2_\mu(X)},$$

which implies the claim (and hence the corollary) since $M_{F^{-1}}$ commutes with M_g for any $g \in G$. □

Spectral measures as in Corollaries 2.12 and 2.13 carry complete information about containment of one representation in another. We leave the following special case as an exercise, and return to this question more generally in Sections 2.5.1 and 2.7, where we will also prove more refined versions of the spectral theorem.

Essential Exercise 2.14 (Containment for characters). Let π be a unitary representation of the abelian group G, let $t_0 \in \widehat{G}$, and denote the corresponding character by χ_{t_0}. Characterize, in terms of spectral measures μ_v for $v \in \mathcal{H}_\pi$, the property that χ_{t_0} is contained in π.

Exercise 2.15 (Cyclicity for σ-finite measures). Let μ be a σ-finite measure on $\widehat{G}$. Show that the unitary representation M of Lemma 2.9 is cyclic.

Exercise 2.16 (Example with infinite multiplicity). Give an example of a unitary representation of the abelian group G for which we really have to use the space $X = \widehat{G} \times \mathbb{N}$ in Corollary 2.13, and could not have used $\widehat{G} \times \{1, \ldots, n\}$ for some $n \in \mathbb{N}$.

2.3 Plancherel Formula

We show in this section that by applying the spectral theorem (Corollary 2.12 and Corollary 2.13) to the regular representation of G we obtain a general formulation of the Fourier transform. We will use this in the next section to establish a duality principle between the abelian group G and its Pontryagin dual $\widehat{G}$.

For $t_0 \in \widehat{G}$ we write M_{t_0} for the multiplication operator

$$M_{t_0}(v)(g) = \langle g, t_0 \rangle v(g)$$

for $v \in L^2(G)$ and $g \in G$. Moreover, we will write $\widehat{\lambda}$ for the regular representation of $\widehat{G}$ on functions f on $\widehat{G}$, so that

$$\widehat{\lambda}_{t_0}(f)(t) = f(t - t_0)$$

for all $t, t_0 \in \widehat{G}$. The maps M_{t_0} and $\widehat{\lambda}_{-t_0}$ are linked to each other, which will be the driving force for the following result.

Theorem 2.17 (Plancherel formula). *Given the abelian group G and a normalization of its Haar measure $m = m_G$ there exists a normalization of the Haar measure $m_{\widehat{G}}$ on $\widehat{G}$ and a unitary isomorphism*

$$U\colon L^2(G) \longrightarrow L^2(\widehat{G})$$

which extends the map $f \mapsto \check{f}$ for $f \in L^1(G) \cap L^2(G)$ to all of $L^2(G)$ and has the equivariance properties

$$U \circ \lambda_g = M_g \circ U,$$
$$U \circ M_t = \widehat{\lambda}_{-t} \circ U$$

for all $g \in G$ *and* $t \in \widehat{G}$*. Moreover, the inverse* $U^{-1}\colon L^2(\widehat{G}) \to L^2(G)$ *is the unique isometric extension of the map*

$$L^1(\widehat{G}) \cap L^2(\widehat{G}) \ni F \longmapsto \widehat{F} \in C_0(G) \cap L^2(G),$$

where

$$\widehat{F}(g) = \int_{\widehat{G}} F(t)\overline{\langle g,t\rangle}\,\mathrm{d}m_{\widehat{G}}(t)$$

for any $g \in G$*.*

We split the proof of the theorem into several steps. Together we will see that after some cosmetic changes the theorem and the existence of the properly normalized Haar measure on $\widehat{G}$ will follow from the spectral theorem in Corollary 2.12.

Lemma 2.18 (Convolution on $L^2(G)$). *For the regular representation* λ *of the abelian group* G *on the space* $L^2(G)$ *and functions* $f \in L^1(G)$*,* $v \in L^2(G)$*, we have that*

$$\lambda_*(f)v = f * v \in L^2(G)$$

can be calculated almost everywhere by the integral formula defining convolution. In particular, $L^1(G) * L^2(G) \subseteq L^2(G)$*.*

PROOF. Let $f \in L^1(G)$ and $v \in L^2(G)$. Fixing another $w \in L^2(G)$ we may use Fubini's theorem to see

$$\begin{aligned}
\langle \lambda_*(f)v, w\rangle &= \int f(h)\,\langle \lambda_h v, w\rangle\,\mathrm{d}m(h)\\
&= \iint f(h)v(g-h)\overline{w(g)}\,\mathrm{d}m(g)\,\mathrm{d}m(h)\\
&= \int f * v(g)\overline{w(g)}\,\mathrm{d}m(g) = \langle f * v, w\rangle.
\end{aligned}$$

This shows that the integral defining $f * v$ exists almost everywhere, defines a function in $L^2(G)$, and equals $\lambda_*(f)v$. (We also refer to Exercise 1.58 and the hint on page 527 for a different argument.) □

Lemma 2.19 (Generators of $L^2(G)$). *The regular representation* λ *of the abelian group* G *on* $L^2(G)$ *is cyclic. In fact, there exists some*

$$\psi \in \mathcal{V} = L^1(G) \cap L^2(G)$$

with $\widecheck{\psi} > 0$ *on all of* $\widehat{G}$ *and every such* ψ *is a generator for the regular representation.*

PROOF. We first claim that there exists some function $\psi \in \mathcal{V} = L^1(G) \cap L^2(G)$ such that $\check{\psi} > 0$ on all of $\widehat{G}$. For this we let $\psi_n \in \mathcal{V}$ be an approximate identity as in Proposition 1.46 so that $\widecheck{\psi_n}(t) \to 1$ as $n \to \infty$ for every $t \in \widehat{G}$ (by continuity of χ_t). Now choose $c_n > 0$ decaying sufficiently rapidly so that

$$\psi = \sum_{n=1}^{\infty} c_n \psi_n^* * \psi_n$$

converges both in $L^1(G)$ and in $L^2(G)$, defining an element of $\mathcal{V}$. Together with the convergence above and $\widecheck{\psi_n^* * \psi_n} = \left|\widecheck{\psi_n}\right|^2 \geqslant 0$, this implies that $\check{\psi} > 0$ as claimed.

For the claim in the lemma that any such $\psi \in \mathcal{V}$ is a generator, we apply the spectral theorem in the form of Corollary 2.13 to the unitary representation λ (which we do not know to be cyclic yet). Hence we obtain a finite measure μ on $X = \widehat{G} \times \mathbb{N}$ and a unitary isomorphism

$$U \colon L^2(G) \longrightarrow L^2_\mu(X)$$

such that $U \circ \lambda_*(\psi) = M_{\check{\psi}} \circ U$ by Lemma 2.9. To see that ψ is a generator suppose that $v \in L^2(G)$ satisfies $v \in \langle \psi \rangle_\lambda^\perp$ and let $f \in \mathcal{V}$. Together with the identity $\lambda_*(f)\psi = f * \psi$ from Lemma 2.18 and commutativity of convolution we obtain

$$0 = \langle \lambda_*(f)\psi, v \rangle = \langle f * \psi, v \rangle = \langle \psi * f, v \rangle = \langle \lambda_*(\psi) f, v \rangle = \left\langle \check{\psi} U(f), U(v) \right\rangle.$$

Now vary f in the dense subspace $\mathcal{V} \subseteq L^2(G)$ and use continuity of the multiplication operator $M_{\check{\psi}} \colon w \in L^2_\mu(X) \mapsto \check{\psi} w \in L^2_\mu(X)$ to obtain that $U(v)$ is orthogonal to $\check{\psi} L^2_\mu(X)$. However, since $\check{\psi} > 0$ the image $\check{\psi} L^2_\mu(X)$ of the multiplication operator $M_{\check{\psi}}$ is dense, which in turn implies that $U(v) = 0$ and hence also $v = 0$. This implies the remaining claim of the lemma, namely that $L^2(G)$ is cyclic and is generated by any $\psi \in \mathcal{V}$ with $\check{\psi} > 0$ on all of $\widehat{G}$. □

Lemma 2.20 (Flattening the measure). *If we apply the spectral theorem in the form of Corollary 2.12 to the regular representation λ of the abelian group G, it is possible to replace the original measure ν on $\widehat{G}$ by a σ-finite measure μ defining the same measure class as ν, such that the map $U \colon L^2(G) \to L^2_\mu(G)$ also satisfies*

$$U(f) = \check{f} \tag{2.7}$$

for all $f \in \mathcal{V} = L^1(G) \cap L^2(G)$.

We note that the measure arising in Corollary 2.12 is not at all canonical, since it depends on the chosen generator. Lemma 2.20 'corrects' this issue.

PROOF OF LEMMA 2.20. By Lemma 2.19 we can apply Corollary 2.12 and assume that the unitary isomorphism between the regular representation λ and the multiplication representation M has the form $U_0\colon L^2(G) \to L^2_\nu(\widehat{G})$ for a finite measure ν on $\widehat{G}$. Using Lemmas 2.9 and 2.18 we obtain

$$\check{\psi} U_0(f) = U_0(\psi * f) = U_0(f * \psi) = \check{f} U_0(\psi) \tag{2.8}$$

almost everywhere (with respect to ν) and for any $f \in \mathcal{V} = L^1(G) \cap L^2(G)$. In particular,

$$U_0(f) = \check{\psi}^{-1} U_0(\psi) \check{f}.$$

As $U_0(\mathcal{V}) \subseteq L^2_\nu(\widehat{G})$ is dense, it follows that $U_0(\psi) \neq 0$ almost everywhere. With this we define the complex-valued measurable and non-vanishing function

$$F = \check{\psi} U_0(\psi)^{-1}$$

on $\widehat{G}$, the σ-finite measure μ on $\widehat{G}$ by

$$\frac{\mathrm{d}\mu}{\mathrm{d}\nu} = |F|^{-2},$$

and the map $U = M_F \circ U_0\colon L^2(G) \to L^2_\mu(\widehat{G})$ (with inverse $U_0^{-1} \circ M_{F^{-1}}$). The latter satisfies

$$\begin{aligned}\|U(f)\|^2_{L^2_\mu(\widehat{G})} &= \int_{\widehat{G}} |F|^2 |U_0(f)|^2 \,\mathrm{d}\mu = \int_{\widehat{G}} |U_0(f)|^2 |F|^2 \frac{\mathrm{d}\mu}{\mathrm{d}\nu} \,\mathrm{d}\nu \\ &= \|U_0(f)\|^2_{L^2_\nu(\widehat{G})} = \|f\|_2\end{aligned}$$

for all $f \in L^2(G)$, and by (2.8) also $U(f) = \check{f}$ as in (2.7) for $f \in \mathcal{V}$. Since any two multiplication operators commute, the new unitary isomorphism still satisfies the conclusion of the spectral thereom. □

Lemma 2.21 (Translation invariance). *Let μ be a σ-finite measure on $\widehat{G}$ satisfying $\|\check{f}\|_{L^2_\mu(\widehat{G})} = \|f\|_{L^2(G)}$ for all $f \in \mathcal{V} = L^1(G) \cap L^2(G)$. Then $\mu = m_{\widehat{G}}$ is a Haar measure on $\widehat{G}$.*

PROOF. By Lemma 2.18 we have $\mathcal{V} * \mathcal{V} \subseteq \mathcal{V}$. Together with Corollary 2.5 we then see that $\check{\mathcal{V}} = U(\mathcal{V})$ is contained in $C_0(\widehat{G}) \cap L^2_\mu(\widehat{G})$ and is a sub-algebra of $C_0(\widehat{G})$. Also recall from Lemma 2.19 that there exists some $\psi \in \mathcal{V}$ with $\check{\psi} > 0$.

Next we claim that μ is locally finite. Using $\check{\psi} \in C_0(\widehat{G})$ we see that

$$O = \left\{t \in \widehat{G} \mid \check{\psi}^2(t) > \tfrac{1}{2}\check{\psi}^2(t_0)\right\}$$

is a neighbourhood of $t_0 \in \widehat{G}$. Together with $\check{\psi}^2 \in L^1_\mu(\widehat{G})$, which follows from the fact that $\check{\psi} = U(\psi) \in L^2_\mu(G)$, it follows that $\mu(O) < \infty$. Since $t_0 \in \widehat{G}$ was arbitrary, it follows that μ is locally finite as claimed.

For $f \in \mathcal{V}$ and $t, t_0 \in \widehat{G}$ we have

$$\widecheck{(\chi_{t_0} f)}(t) = \int_G (\chi_{t_0} f)\chi_t \, \mathrm{d}m = \check{f}(t_0 + t) = \widehat{\lambda}_{-t_0} \check{f}(t). \tag{2.9}$$

Notice that

$$L^2(G) \ni f \longmapsto \chi_{t_0} f \in L^2(G)$$

is unitary, which implies that

$$\check{\mathcal{V}} \ni \check{f} \longmapsto \widehat{\lambda}_{-t_0} \check{f} \tag{2.10}$$

preserves the norm and inner products for all $t_0 \in \widehat{G}$. This is already a translation invariance claim for μ, but restricted to the class of functions $(\check{\mathcal{V}})^2$.

To prove that μ is translation-invariant, we show that

$$\int_{\widehat{G}} F(t) \, \mathrm{d}\mu(t) = \int_{\widehat{G}} F(t_0 + t) \, \mathrm{d}\mu(t)$$

for any $F \in C_c(\widehat{G})$ and $t_0 \in \widehat{G}$, which in turn we show by approximating F with elements of $(\check{\mathcal{V}})^2$.

Indeed, by density of $\mathcal{V}$ in $L^1(G)$ and by Corollary 2.5 we see that $\check{\mathcal{V}}$ is dense in $C_0(\widehat{G})$ with respect to the supremum norm. Clearly this approximation may be of little use since $\mu(\widehat{G})$ might be infinite. To overcome this we recall that $\check{\psi}^2 \in L^1_\mu(\widehat{G})$ is positive on $\widehat{G}$. Given some $F \in C_c(\widehat{G})$ and $\varepsilon > 0$, we can apply the density claim and find a function $f \in \mathcal{V}$ such that

$$\left\| \check{f} - \check{\psi}^{-2} F \right\|_\infty < \varepsilon.$$

We multiply this by $\check{\psi}^2$ and obtain

$$\left| \check{\psi}^2 \check{f} - F \right| < \varepsilon \check{\psi}^2 \tag{2.11}$$

on all of $\widehat{G}$. Integrating this inequality we obtain

$$\left| \int \check{\psi}^2 \check{f} \, \mathrm{d}\mu - \int F \, \mathrm{d}\mu \right| < \varepsilon \int \check{\psi}^2 \, \mathrm{d}\mu.$$

As noted in (2.10), the integrals of the functions $\check{\psi}^2 \check{f}, \check{\psi}^2 \in (\check{\mathcal{V}})^2$ remain unchanged when these are shifted by t_0. Shifting the estimate in (2.11) by t_0 and integrating again, this gives

$$\left| \int F(t_0 + t) \, \mathrm{d}\mu(t) - \int F(t) \, \mathrm{d}\mu(t) \right| < 2\varepsilon \int \check{\psi}^2 \, \mathrm{d}\mu.$$

As $\varepsilon > 0$ was arbitrary, we deduce that

$$\int F(t_0 + t) \, \mathrm{d}\mu(t) = \int F(t) \, \mathrm{d}\mu(t)$$

for any $F \in C_c(\widehat{G})$. Since $\mu \neq 0$ we see that μ is a Haar measure on $\widehat{G}$. □

We now show that the combination of the above lemmas gives the Plancherel formula for G.

PROOF OF THEOREM 2.17. By Lemma 2.19 and Lemma 2.20 we may apply the spectral theorem in the form of Corollary 2.12 and assume that the unitary isomorphism

$$U \colon L^2(G) \to L^2_\mu(\widehat{G})$$

satisfies $U(f) = \check{f}$ for all $f \in \mathcal{V} = L^1(G) \cap L^2(G)$. Applying Lemma 2.21, we also see that the measure is a Haar measure $\mu = m_{\widehat{G}}$. The formula

$$U \circ \lambda_g = M_g \circ U$$

holds by Corollary 2.12. Finally,

$$U(M_{t_0} f) = \widehat{\lambda}_{-t_0}(U(f))$$

holds initially only for $f \in \mathcal{V}$ (see (2.9)), but by density this extends to all of $L^2(G)$.

It remains to prove the description of the inverse of U on $L^1(\widehat{G}) \cap L^2(\widehat{G})$. So let $F \in L^1(\widehat{G}) \cap L^2(\widehat{G})$ and let (ψ_n) with $\psi_n = \frac{1}{m(B_n)} \mathbb{1}_{B_n}$ for $n \in \mathbb{N}$ be again an approximate identity (as in Proposition 1.46) for a basis (B_n) of the neighbourhood of $0 \in G$ with $B_n = -B_n$ for all $n \in \mathbb{N}$. For $g \in G$ we have

$$\begin{aligned}
\langle U^{-1} F, \lambda_g \psi_n \rangle &= \left\langle F, \widecheck{\lambda_g \psi_n} \right\rangle \\
&= \left\langle F, M_g \widecheck{\psi_n} \right\rangle \\
&= \int F(t) \overline{\langle g, t \rangle \widecheck{\psi_n}(t)} \, \mathrm{d}t \longrightarrow \int F(t) \overline{\langle g, t \rangle} \, \mathrm{d}t = \widehat{F}(g)
\end{aligned}$$

as $n \to \infty$ by using the isomorphism U and by dominated convergence (since we have $|\widecheck{\psi_n}(t)| \leqslant \|\psi_n\|_1 = 1$ and $\widecheck{\psi_n}(t) \to 1$ as $n \to \infty$ for every $t \in \widehat{G}$).

To obtain the desired conclusion, we interpret $\langle U^{-1} F, \lambda_g \psi_n \rangle$ as a convolution. Indeed, we have

$$\begin{aligned}
\langle U^{-1} F, \lambda_g \psi_n \rangle &= \int U^{-1}(F)(h) \underbrace{\psi_n(h - g)}_{= \psi_n(g-h)} \, \mathrm{d}m(h) \\
&= \int \psi_n(k) U^{-1}(F)(g - k) \, \mathrm{d}m(k) \\
&= \psi_n * U^{-1}(F)(g) = \lambda_*(\psi_n)\big(U^{-1}(F)\big)(g)
\end{aligned}$$

for almost every $g \in G$ by using the substitution $k = g - h$ and by Lemma 2.18. Using Proposition 1.52 (for the regular representation), we see that

$$G \ni g \longmapsto \langle U^{-1}F, \lambda_g \psi_n \rangle$$

converges as the function $\lambda_*(\psi_n)(U^{-1}F)$ in $L^2(G)$ to $U^{-1}F$ as $n \to \infty$.

Even though the two notions of convergence above are different, we now obtain $U^{-1}F = \widehat{F}$ almost everywhere. Indeed, L^2-convergence implies the existence of a subsequence along which there is pointwise convergence, which gives the desired equality. □

2.4 Pontryagin Duality

We note that the Plancherel formula in Theorem 2.17 already expresses some symmetry between G and $\widehat{G}$ (apart from a choice of sign). Using this, we will now establish a complete duality between G and its dual group $\widehat{G}$. For this we let $\widehat{\widehat{G}}$ denote the *Pontryagin bi-dual* (that is, the Pontryagin dual of the Pontryagin dual $\widehat{G}$) of the abelian group G. Moreover, for any $g \in G$ we define a map on $\widehat{G}$ by $\imath(g)(t) = \langle g, t \rangle = \chi_t(g)$ for $t \in \widehat{G}$. The properties of the Pontryagin bi-dual, the maps $\imath(g)$ for $g \in G$, and the map $\imath$ are given in the following result.

Theorem 2.22 (Pontryagin duality). *For the abelian group G the canonical map $\imath \colon G \to \widehat{\widehat{G}}$ is an isomorphism of topological groups.*

Let us start with some observations about the dual pairing $\langle \cdot, \cdot \rangle$.

Lemma 2.23. (1) *The map $\langle \cdot, \cdot \rangle : G \times \widehat{G} \longrightarrow \mathbb{S}^1$ is continuous.*
(2) *$\imath(g) \in \widehat{\widehat{G}}$ for any $g \in G$.*
(3) *The canonical map $\imath \colon G \to \widehat{\widehat{G}}$ is injective and continuous.*

PROOF. For the proof of (1) suppose (g_n) in G converges to $g \in G$ and (t_n) in $\widehat{G}$ converges to $t \in \widehat{G}$ as $n \to \infty$. Then $K = \{g_n \mid n \in \mathbb{N}\} \cup \{g\}$ is a compact subset of G. By Corollary 2.5 the convergence $t_n \to t$ implies in particular that (χ_{t_n}) converges uniformly to χ_t on K as $n \to \infty$. This implies that $\langle g_n, t_n \rangle = \chi_{t_n}(g_n) \to \chi_t(g) = \langle g, t \rangle$ as $n \to \infty$.

Property (2) now follows quickly from (1) since $\imath(g)(t) = \langle g, t \rangle \in \mathbb{S}^1$ depends continuously on $t \in \widehat{G}$, and (by the definition of the group structure on $\widehat{G}$) defines a homomorphism from $\widehat{G}$ to $\mathbb{S}^1$.

It remains to prove (3). For $g_1, g_2 \in G$ and $t \in \widehat{G}$, we have

$$\imath(g_1 - g_2)(t) = \chi_t(g_1 - g_2) = \chi_t(g_1)\overline{\chi_t(g_2)} = \big(\imath(g_1)\imath(g_2)^{-1}\big)(t),$$

so $\imath \colon G \to \widehat{G}$ is a homomorphism. Let g be in $G \smallsetminus \{0\}$. By the Gelfond–Raikov theorem (Corollary 1.81) there exists some $t \in \widehat{G}$ with $\langle g, t \rangle \neq 1$. (Alternatively, we could also note that $\lambda_g \neq I$ and apply the equivariance formulas in Theorem 2.17 to obtain the same conclusion.) In other words, $\imath(g) \in \widehat{\widehat{G}} \smallsetminus \{0\}$ and we see that $\imath \colon G \to \widehat{\widehat{G}}$ is injective.

To prove continuity of $\imath$, suppose that (g_n) converges to $g \in G$ as $n \to \infty$ and let $K \subseteq \widehat{G}$ be a compact subset. Then

$$L = (\{g_n \mid n \in \mathbb{N}\} \cup \{g\}) \times K \subseteq G \times \widehat{G}$$

is compact, and $\langle \cdot, \cdot \rangle$ restricted to L is uniformly continuous. This in particular implies that for every $\varepsilon > 0$ there exists some $N \geqslant 1$ such that

$$| \langle g_n, t \rangle - \langle g, t \rangle | < \varepsilon$$

for all $n \geqslant N$ and $t \in K$. In other words, the functions $\imath(g_n) \colon \widehat{G} \to \mathbb{S}^1$ and $\imath(g) \colon \widehat{G} \to \mathbb{S}^1$ are uniformly ε-close on K. Since the compact set $K \subseteq \widehat{G}$ and $\varepsilon > 0$ were arbitrary, Corollary 2.5 implies that $\imath(g_n) \to \imath(g)$ as $n \to \infty$, and it follows that $\imath$ is continuous. □

For the proof of Pontryagin duality, we will use the following general result concerning homomorphisms of topological groups in the special case of the canonical homomorphism $\imath \colon G \to \widehat{\widehat{G}}$.

Lemma 2.24. *Let $\theta \colon G \to G'$ be a continuous injective group homomorphism from the abelian group G to the abelian group G' satisfying $\theta_*(m_G) = m_{G'}$. Then θ is an isomorphism of topological groups.*

PROOF. Let U be an open neighbourhood of $0 \in G$, and let $V = -V$ be a compact neighbourhood of 0 with $V + V \subseteq U$. Using $\theta_* m_G = m_{G'}$, we see that the characteristic function $f = \mathbb{1}_{\theta(V)}$ is a non-trivial element of the intersection $L^1(G') \cap L^2(G')$. Moreover,

$$\begin{aligned} f * f(g') &= \int_{G'} \mathbb{1}_{\theta(V)}(h) \underbrace{\mathbb{1}_{\theta(V)}(g' - h)}_{= \mathbb{1}_{\theta(V)}(h - g')} \, \mathrm{d}m_{G'}(h) \\ &= \langle \lambda_{g'} \mathbb{1}_{\theta(V)}, \mathbb{1}_{\theta(V)} \rangle_{L^2(G')} = m_{G'}\big((\theta(V) + g') \cap \theta(V)\big) \end{aligned}$$

for all $g' \in G'$. This realises the function $f * f$ as the diagonal matrix coefficient for $\mathbb{1}_{\theta(V)} \in L^2(G')$, and gives $f * f \in C_b(G')$ and $f * f(0) > 0$. Therefore

$$(f * f)^{-1}\big((0, \infty)\big) \subseteq \theta(V) - \theta(V) = \theta(V - V) \subseteq \theta(U)$$

is a neighbourhood of $0 \in G'$. As U was an arbitrary neighbourhood of $0 \in G$ and θ is a homomorphism, this shows that θ maps open sets in G to open sets in G' (that is, θ is an open map).

In particular, $\theta(G) \subseteq G'$ is an open subgroup. However, this implies that $\theta(G)$ is also closed, since the continuity of the group operation implies that its complement $G' \smallsetminus \theta(G) = \bigcup_{g' \in G' \smallsetminus \theta(G)} \big(g' + \theta(G)\big)$ is open. Finally, using $\theta_* m_G = m_{G'}$ in the form

$$m_{G'}\big(G' \smallsetminus \theta(G)\big) = m_G\big(\theta^{-1}(G' \smallsetminus \theta(G))\big) = m_G(\emptyset) = 0$$

shows, by the properties of Haar measure, that the open set $G' \setminus \theta(G)$ must be empty. In other words, the continuous injective homomorphism $\theta \colon G \to G'$ is open and surjective, so is a homeomorphism between G and G'. □

We will now use the Plancherel formula and its equivariance properties to prove Pontryagin duality, by 'interpreting m_G as a spectral measure' for the regular representation $\widehat{\lambda}$ of $\widehat{G}$ on $L^2(\widehat{G})$.

PROOF OF THEOREM 2.22. Let $\imath \colon G \to \widehat{\widehat{G}}$ be the canonical homomorphism studied in Lemma 2.23. We define the measure $\mu = \imath_*(m_G)$ on $\imath(G) \subseteq \widehat{\widehat{G}}$. Because of Lemma 2.24, our main goal is to show that μ is the Haar measure on $\widehat{\widehat{G}}$.

For this, we let $U \colon L^2(G) \to L^2(\widehat{G})$ be the unitary isomorphism from the Plancherel formula (Theorem 2.17 applied to G). In particular, we have that $U^{-1}(F) = \widehat{F}$ is given by the Fourier transform for

$$F \in \mathcal{V}_{\widehat{G}} = L^1(\widehat{G}) \cap L^2(\widehat{G}).$$

With this and the substitution $h = -g$ on G, we obtain for the Fourier transform $\check{F}$ on $\widehat{\widehat{G}}$ and the measure $\mu = \imath_*(m_G)$ on $\widehat{\widehat{G}}$ that

$$\begin{aligned} \|\check{F}\|^2_{L^2_\mu(\widehat{G})} &= \int \left|\check{F}(\imath(g))\right|^2 \mathrm{d}m_G(g) \\ &= \int \left| \int F(t) \underbrace{\langle t, \imath(g)\rangle}_{=\langle g,t\rangle} \mathrm{d}m_{\widehat{G}}(t) \right|^2 \mathrm{d}m_G(g) \\ &= \int \left| \int F(t)\langle -h, t\rangle \,\mathrm{d}m_{\widehat{G}}(t) \right|^2 \mathrm{d}m_G(h) \\ &= \int |\widehat{F}|^2 \,\mathrm{d}m_G = \|F\|^2_{L^2(\widehat{G})}. \end{aligned}$$

In other words, the measure μ on $\widehat{\widehat{G}}$ satisfies the assumptions of Lemma 2.21 when applied to $\widehat{G}$. Therefore $\mu = m_{\widehat{\widehat{G}}}$ is a Haar measure on $\widehat{\widehat{G}}$.

Recalling that $\mu = \imath_*(m_G)$ and the fact that $\imath \colon G \to \widehat{\widehat{G}}$ is an injective continuous group homomorphism, Lemma 2.24 shows that $\imath \colon G \to \widehat{\widehat{G}}$ is in fact an isomorphism of topological groups. □

The automatic reflexivity of the abelian group G in the Pontryagin duality theorem (Theorem 2.22) allows us to prove many duality statements that are reminiscent of finite dimensional linear algebra. In fact these duality statements go much further, as we will see in the following subsections.

Exercise 2.25. Show (without using Theorem 2.22) that $\imath \colon G \to \widehat{\widehat{G}}$ is proper. Indeed, show first that if (g_n) is a sequence in G with $g_n \to \infty$ as $n \to \infty$ then $\lambda_{g_n}(v) \to 0$ in the weak* topology as $n \to \infty$ for any $v \in L^2(G)$. Now try to push this statement to a claim about the regular representation $\widehat{\lambda}_{\imath(g_n)}$ on $\widehat{\widehat{G}}$ and the elements $\imath(g_n) \in \widehat{\widehat{G}}$.

Exercise 2.26. Lemma 2.24 was phrased for abelian groups to avoid changing notation in the middle of the chapter. Show that the assumption that the groups G and G' are abelian can be dropped.

2.4.1 First Duality Results, Sums and Products

Proposition 2.27 (Compactness and discreteness). *For an abelian locally compact metric group compactness and discreteness are dual to each other in the following sense. If G is compact, then $\widehat{G}$ is discrete. If G is discrete, then $\widehat{G}$ is compact.*

PROOF. If G is compact and $m_G(G) = 1$, then the constant function $\mathbb{1}$ belongs to $L^1(G)$. As any χ_t for $t \in \widehat{G}\smallsetminus\{0\}$ is orthogonal to $\mathbb{1}$, we see that

$$\check{\mathbb{1}} = \mathbb{1}_{\{0\}} \in C_0(\widehat{G})$$

by Corollary 2.5. Hence $\widehat{G}$ is discrete.

If G is discrete and m_G is the counting measure, then $\mathbb{1}_{\{0\}} \in L^1(G) \cap L^2(G)$ satisfies $\check{\mathbb{1}}_{\{0\}} = \mathbb{1} \in L^2(\widehat{G})$ by Theorem 2.17. Hence $m_{\widehat{G}}(\widehat{G}) < \infty$, which in turn implies that $\widehat{G}$ is compact (see Exercise 1.19(b) and its hint on page 525). □

Exercise 2.28. Give another proof that discreteness of G implies compactness of $\widehat{G}$ using the compact-open topology on $\widehat{G}$ directly.

It will be convenient to use the notation

$$N_{\widehat{G}}(K, \varepsilon) = \left\{t \in \widehat{G} \mid |\langle g, t\rangle - 1| < \varepsilon \text{ for all } g \in K\right\}$$

for the neighbourhood of $0 \in \widehat{G}$ defined by a compact subset $K \subseteq G$ and $\varepsilon > 0$ in the compact-open topology of $\widehat{G}$. In the following, $G_1, G_2, \ldots$ will always denote abelian groups that are as usual locally compact, σ-compact, and metric, and we will refer to them simply as abelian groups.

For discrete abelian groups $G_1, G_2, \ldots$ we define the *direct sum* by

$$\sum_{n=1}^{\infty} G_n = \left\{(g_n) \in \prod_{n\in\mathbb{N}} G_n \mid g_n = 0 \text{ for all sufficiently large } n \in \mathbb{N}\right\},$$

which we again endow with the discrete topology.

Proposition 2.29 (Products and sums). *The Pontryagin dual $\widehat{G_1 \times G_2}$ of the direct product $G_1 \times G_2$ of the abelian groups G_1 and G_2 is canonically isomorphic to $\widehat{G_1} \times \widehat{G_2}$. If the abelian groups G_n for $n \in \mathbb{N}$ are compact, then the Pontryagin dual of the direct product $\prod_{n=1}^{\infty} G_n$ is canonically isomorphic to the* *direct sum* $\sum_{n=1}^{\infty}\widehat{G_n}$. *Finally, if the abelian groups* G_n *for* $n \in \mathbb{N}$ *are discrete, then the Pontryagin dual of the direct sum* $\sum_{n=1}^{\infty} G_n$ *is canonically isomorphic to* $\prod_{n=1}^{\infty}\widehat{G_n}$.

We note that the claimed isomorphisms are indeed quite natural. For instance, in the first statement we can use $(t_1, t_2) \in \widehat{G_1} \times \widehat{G_2}$ to induce a character on $G_1 \times G_2$ by the formula

$$\chi_{(t_1,t_2)}(g_1, g_2) = \langle (g_1, g_2), (t_1, t_2)\rangle = \langle g_1, t_2\rangle \langle g_2, t_2\rangle \tag{2.12}$$

for all $(g_1, g_2) \in G_1 \times G_2$. The isomorphism in the second and third are of the same nature. Hence we may and will interpret the above claims as equalities written

$$\widehat{G_1 \times G_2} = \widehat{G_1} \times \widehat{G_2},$$

$$\widehat{\prod_{n=1}^{\infty} G_n} = \sum_{n=1}^{\infty}\widehat{G_n}$$

if all the G_n are compact, and

$$\widehat{\sum_{n=1}^{\infty} G_n} = \prod_{n=1}^{\infty}\widehat{G_n}$$

if all the G_n are discrete.

PROOF OF PROPOSITION 2.29. It is easy to see that the natural group operations make $G_1 \times G_2$ again into a locally compact, σ-compact, metric abelian group. If $(t_1, t_2) \in \widehat{G_1} \times \widehat{G_2}$ then (2.12) defines an element $\chi_{(t_1,t_2)}$ of $\widehat{G_1 \times G_2}$. If, on the other hand, $\chi \in \widehat{G_1 \times G_2}$ we may compose χ with the embedding of G_1 into $G_1 \times G_2$, which defines a continuous homomorphism $G_1 \to G_1 \times G_2 \to \mathbb{S}^1$ by

$$g_1 \longmapsto (g_1, 0) \longmapsto \chi(g_1, 0).$$

It follows that there is a uniquely determined $t_1 \in \widehat{G_1}$ with $\chi(g_1, 0) = \langle g_1, t_1\rangle$ for all $g_1 \in G_1$. By the same argument there is a uniquely determined $t_2 \in \widehat{G_2}$ with $\chi(0, g_2) = \langle g_2, t_2\rangle$ for all $g_2 \in \widehat{G_2}$. Since χ is a homomorphism, this gives

$$\chi(g_1, g_2) = \chi(g_1, 0)\chi(0, g_2) = \chi_{(t_1,t_2)}(g_1, g_2)$$

for all $(g_1, g_2) \in G_1 \times G_2$. It follows that (2.12) defines an isomorphism between $\widehat{G_1} \times \widehat{G_2}$ and $\widehat{G_1 \times G_2}$, which is easily seen to be an isomorphism of groups.

To see continuity of this isomorphism in both directions, it is sufficient to consider neighbourhoods of the identity. So suppose first that $K_1 \subseteq G_1$ and $K_2 \subseteq G_2$ are compact and $\varepsilon > 0$. If now $\chi_{(t_1,t_2)} \in N_{\widehat{G_1 \times G_2}}(K_1 \times K_2, \varepsilon)$ then by restriction to $G_1 \times \{0\}$ and $\{0\} \times G_2$ we obtain $t_1 \in N_{\widehat{G_1}}(K_1, \varepsilon)$ and $t_2 \in N_{\widehat{G_2}}(K_2, \varepsilon)$. For the converse, we note that if $K \subseteq G_1 \times G_2$ is compact, then $K \subseteq K_1 \times K_2$ for the compact projections K_1 and K_2 of K to G_1 and G_2. If now $t_1 \in N_{\widehat{G_1}}(K_1, \frac{\varepsilon}{2})$ and $t_2 \in N_{\widehat{G_2}}(K_2, \frac{\varepsilon}{2})$ then

$$\left|\chi_{(t_1,t_2)}(g_1, g_2) - 1\right| = \left|\langle g_1, t_1\rangle(\langle g_2, t_2\rangle - 1) + \langle g_1, t_1\rangle - 1\right| < \varepsilon$$

for all $(g_1, g_2) \in K$, and hence $\chi_{(t_1,t_2)} \in N_{\widehat{G_1 \times G_2}}(K, \varepsilon)$.

Suppose now that $G_1, G_2, \dots$ are all compact and $G = \prod_{n=1}^{\infty} G_n$ is equipped with the product topology. Any $\chi \in \widehat{G}$ restricts as before to any factor G_m for $m \in \mathbb{N}$, and we obtain that there exists a uniquely determined $t_m \in \widehat{G_m}$ with

$$\chi(0, \dots, 0, g_m, 0, \dots) = \langle g_m, t_m\rangle$$

for all $g_m \in G_m$, where $(0, \dots, 0, g_m, 0, \dots)$ denotes the element of $\prod_{n=1}^{\infty} G_n$ that has g_m as its mth entry, and otherwise only zeroes. By continuity of χ there exists a neighbourhood U of $0 \in G$ such that $|\chi(g) - 1| < 1$ for all $g \in U$. By definition of the product topology there exists some $N \in \mathbb{N}$ such that

$$\{0\}^N \times \prod_{n=N+1}^{\infty} G_n \subseteq U.$$

Therefore,

$$\chi\Big(\{0\}^N \times \prod_{n=N+1}^{\infty} G_n\Big)$$

is a subgroup of $\mathbb{S}^1$ contained in $\{z \in \mathbb{S}^1 \mid |z - 1| < 1\}$, which must therefore be trivial. Hence $t_m = 0$ for all $m > N$ and χ is given by

$$\chi((g_n)_n) = \prod_{n=1}^{N} \langle g_n, t_n\rangle \tag{2.13}$$

for all $(g_n)_n \in G$. Using the projections from G to its factors G_n for $n \in \mathbb{N}$ it follows that (2.13) defines an element of $\widehat{G}$ for any $(t_n)_n \in \prod_{n=1}^{N} \widehat{G_n}$ and N in $\mathbb{N}$. Since $\widehat{G}$ and $\sum_{n=1}^{\infty} \widehat{G_n}$ are both discrete by Proposition 2.27, this proves the second claim.

The third claim follows from the second by applying it to $\widehat{G_n}$ and using Pontryagin duality (Theorem 2.22). □

Exercise 2.30. Prove the last claim in Proposition 2.29 directly, without relying on Pontryagin duality.

2.4.2 Dual Homomorphisms

Let now G_1 and G_2 be locally compact, σ-compact, metric, abelian groups and let $\theta\colon G_1 \to G_2$ be a continuous group homomorphism. Then there exists a dual map $\widehat{\theta}\colon \widehat{G_2} \to \widehat{G_1}$ defined by

$$\langle g, \widehat{\theta}(t)\rangle = \langle \theta(g), t\rangle \tag{2.14}$$

for $g \in G_1$ and $t \in \widehat{G_2}$.

Lemma 2.31 (Dual homomorphisms). *With the assumptions above, equation* (2.14) *defines a continuous group homomorphism* $\widehat{\theta}\colon \widehat{G_2} \to \widehat{G_1}$*, called the* dual homomorphism *of* θ. *Under the canonical isomorphism between the bi-duals and the original groups in Theorem* 2.22 *the dual homomorphism of* $\widehat{\theta}$ *is given by* θ. *Moreover, if* $\theta'\colon G_2 \to G_3$ *is another continuous group homomorphism with values in an abelian group* G_3*, then* $\widehat{\theta' \circ \theta} = \widehat{\theta} \circ \widehat{\theta'}\colon \widehat{G_3} \to \widehat{G_1}$.

PROOF. Clearly the right hand side of (2.14) belongs to $\mathbb{S}^1$ for every $g \in G_1$ and $t \in \widehat{G_2}$. Fixing t and varying g we can use the fact that θ is a continuous group homomorphism to see that the map $G_1 \ni g \mapsto \langle \theta(g), t\rangle$ indeed defines an element $\widehat{\theta}(t) \in \widehat{G_1}$.

For $g \in G_1$ and $t_1, t_2 \in G_2$ we have

$$\begin{aligned}
\bigl\langle g, \widehat{\theta}(t_1 + t_2)\bigr\rangle &= \langle \theta(g), t_1 + t_2\rangle \\
&= \langle \theta(g), t_1\rangle\langle \theta(g), t_2\rangle \\
&= \bigl\langle g, \widehat{\theta}(t_1)\bigr\rangle\bigl\langle g, \widehat{\theta}(t_2)\bigr\rangle = \bigl\langle g, \widehat{\theta}(t_1) + \widehat{\theta}(t_2)\bigr\rangle,
\end{aligned}$$

which shows that $\widehat{\theta}\colon \widehat{G_2} \to \widehat{G_1}$ is a homomorphism.

As $\widehat{\theta}$ is a homomorphism, it suffices to prove continuity at the identity of $\widehat{G_2}$ to obtain continuity on all of $\widehat{G_2}$. So let $K \subseteq G_1$ be some compact subset and $\varepsilon > 0$ and use these to define the neighbourhood $N_{\widehat{G_1}}(K, \varepsilon)$. Then $\theta(K) \subseteq G_2$ is compact (since θ is continuous), so it defines a neighbourhood $N_{\widehat{G_2}}(\theta(K), \varepsilon)$. It is now easy to see that $t \in N_{\widehat{G_2}}(\theta(K), \varepsilon)$ and $g \in K$ implies that

$$\bigl|\bigl\langle g, \widehat{\theta}(t)\bigr\rangle - 1\bigr| = |\langle \theta(g), t\rangle - 1| < \varepsilon$$

and hence $\widehat{\theta}(t) \in N_{\widehat{G_1}}(K, \varepsilon)$, which gives continuity of $\widehat{\theta}$ at $0 \in \widehat{G_2}$, as required.

For $g \in G_1$ and $t \in \widehat{G_2}$ we have[†]

$$\langle \widehat{\widehat{\theta}}(g), t\rangle = \bigl\langle g, \widehat{\theta}(t)\bigr\rangle = \langle \theta(g), t\rangle,$$

which proves that $\widehat{\widehat{\theta}} = \theta$.

Finally for θ' and $t \in \widehat{G_3}$ as in the last part of the lemma we have

$$\langle g, \widehat{\theta' \circ \theta}(t)\rangle = \langle \theta'(\theta(g)), t\rangle = \langle \theta(g), \widehat{\theta'}(t)\rangle = \langle g, \widehat{\theta} \circ \widehat{\theta'}(t)\rangle$$

for all $g \in G_1$, and the lemma follows. □

[†] More formally, this is to be interpreted as $\langle t, \widehat{\widehat{\theta}}(\imath(g))\rangle = \langle \widehat{\theta}(t), \imath(g)\rangle = \langle g, \widehat{\theta}(t)\rangle$.

Recall from Exercise 2.6 that $\widehat{\mathbb{R}^d} \cong \mathbb{R}^d$ via the isomorphism defined by the formula $\langle g, t\rangle = \mathrm{e}^{2\pi \mathrm{i} g\cdot t}$ for $g, t \in \mathbb{R}^d$, where $g \cdot t = \sum_{j=1}^{d} g_j t_j$ is the standard inner product on $\mathbb{R}^d$ for $d \geqslant 1$. Suppose now $d, e \geqslant 1$ and that $A\colon \mathbb{R}^d \to \mathbb{R}^e$ is a linear map. Then the dual homomorphism $\widehat{A}$ is equal to the dual map A^{t} in the sense of linear algebra, and so is defined by the transpose of the matrix defining A if we use the standard basis on $\mathbb{R}^d$ and $\mathbb{R}^e$. In fact,

$$\left\langle g, \widehat{A}t\right\rangle = \langle Ag, t\rangle = \mathrm{e}^{2\pi\mathrm{i}(Ag\cdot t)} = \mathrm{e}^{2\pi\mathrm{i}(g\cdot A^{\mathrm{t}}t)} = \langle g, A^{\mathrm{t}}t\rangle$$

for all $g \in \mathbb{R}^d$ and $t \in \mathbb{R}^e$, which proves the claim.

We finish the subsection by stating another duality claim for homomorphisms, which we will prove at the end of the next subsection as a corollary of our discussion regarding quotients.

Corollary 2.32 (Injectivity and dense image). *Let $\theta\colon G_1 \to G_2$ be a continuous group homomorphism from the abelian group G_1 to the abelian group G_2. Then*

(1) *θ is injective if and only if $\widehat{\theta}$ has dense image; and*
(2) *θ has dense image if and only if $\widehat{\theta}$ is injective.*

2.4.3 Subgroups and Quotients

For a closed subgroup H of the abelian group G, we define

$$G/H = \{g + H \mid g \in G\}$$

as the quotient group, and equip G/H with the quotient topology. Recall that G/H is abelian, and by the more general Proposition C.3 it also follows that G/H in the quotient topology is a locally compact σ-compact metric abelian group.

The *annihilator* $H^{\perp}$ of a closed subgroup (or even a subset) H of the abelian group G is defined by

$$H^{\perp} = \left\{t \in \widehat{G} \mid \langle h, t\rangle = 1 \text{ for all } h \in H\right\}.$$

Proposition 2.33 (Duality of subgroups and quotients). *Let $H \subseteq G$ be a closed subgroup of the abelian group G. Then $H^{\perp} \subseteq \widehat{G}$ is a closed subgroup which, together with the quotient group, satisfies the following duality statements.*

(1) *$\widehat{G/H} \cong H^{\perp}$ via the canonical pairing defined by $\langle g + H, t\rangle = \langle g, t\rangle$ for t in $H^{\perp}$ and $g + H$ in G/H.*
(2) *$\widehat{H} \cong \widehat{G}/H^{\perp}$ via the canonical pairing defined by $\langle h, t + H^{\perp}\rangle = \langle h, t\rangle$ for h in H and $t + H^{\perp}$ in $\widehat{G}/H^{\perp}$.*
(3) *$\left(H^{\perp}\right)^{\perp} = H$, where we identify $\widehat{\widehat{G}}$ with G using Theorem 2.22.*

For simplicity of notation we will also write the first two claims in the form $\widehat{G/H} = H^\perp$ and $\widehat{H} = \widehat{G}/H^\perp$.

Proof of Proposition 2.33. By definition,

$$H^\perp = \bigcap_{h\in H} \ker\bigl(\widehat{G} \ni t \longmapsto \langle h,t\rangle\bigr),$$

and so by Lemma 2.23 we see that $H^\perp$ is a closed subgroup of $\widehat{G}$.

We note that any $t \in H^\perp$ induces a well-defined homomorphism

$$\chi\colon G/H \ni g+H \longmapsto \langle g,t\rangle \in \mathbb{S}^1,$$

which by definition of the quotient topology is also continuous. On the other hand, a character χ on G/H induces by composition a character

$$\chi\circ p\colon G \to \mathbb{S}^1$$

on G which must correspond to some $t \in H^\perp$. This gives the identification between $\widehat{G/H}$ and $H^\perp$, which is clearly also compatible with the group structures. It remains to show that this identification is a homeomorphism.

By Proposition C.3 compact subsets $K' \subseteq G/H$ are precisely sets of the form $K' = p(K)$ for a compact set $K \subseteq G$. This shows that the neighbourhood $N_{\widehat{G/H}}(K',\varepsilon)$ of $0 \in \widehat{G/H}$ corresponds to the neighbourhood

$$H^\perp \cap N_{\widehat{G}}(K,\varepsilon)$$

of $0 \in H^\perp$ for any $K' = p(K)$ and $\varepsilon > 0$, which completes the proof of (1).

Next we prove (3). Clearly we have $H \subseteq \bigl(H^\perp\bigr)^\perp$. Suppose that $g \in G\smallsetminus H$. Recall from Lemma 2.23(3) that $\imath\colon G/H \to \widehat{\widehat{G/H}}$ is injective which, by definition, means that for $g+H \neq 0+H$ in G/H there exists a character on G/H that maps $g+H$ to $z \neq 1$ in $\mathbb{S}^1$. By (1), this implies the existence of $t \in H^\perp$ with $\langle g,t\rangle = z \neq 1$. However, this implies that $g \notin \bigl(H^\perp\bigr)^\perp$, proving (3).

The isomorphism in (2) now follows from Pontryagin duality (Theorem 2.22) by applying (1) to the subgroup $H^\perp$ in $\widehat{G}$. □

Example 2.34 (Duality between projection and embedding). Let $H \subseteq G$ be a closed subgroup of the abelian group G. Using the first isomorphism in Proposition 2.33 the dual of the canonical map $p\colon G \to G/H$ is the canonical embedding map from $\widehat{G/H} = H^\perp$ to $\widehat{G}$ since

$$\langle g,\widehat{p}(t)\rangle = \langle p(g),t\rangle = \langle g+H,t\rangle = \langle g,t\rangle.$$

for $g \in G$ and $t \in H^\perp$. Similarly we may use the second isomorphism to conclude that the dual of the embedding $\imath\colon H \to G$ is given by the canonical projection from $\widehat{G}$ to $\widehat{H} = \widehat{G}/H^\perp$ as

$$\langle h, \widehat{\imath}(t)\rangle = \langle h, t\rangle = \langle h, t + H^{\perp}\rangle$$

for all $h \in H$ and $t \in \widehat{G}$.

Using Proposition 2.33 we can now prove the duality claim regarding injectivity and dense image for homomorphisms.

PROOF OF COROLLARY 2.32. Suppose that θ has dense image and $t \in \widehat{G_2}$ satisfies $\widehat{\theta}(t) = 0$. Then we have $1 = \langle g, \widehat{\theta}(t)\rangle = \langle \theta(g), t\rangle$ for all $g \in G_1$, or equivalently $\chi_t(\operatorname{Im}\theta) = 1$. Since $\overline{\operatorname{Im}\theta} = G_2$ and χ_t is continuous, this implies that $t = 0$ and hence that $\widehat{\theta}$ is injective.

Suppose now that $\overline{\operatorname{Im}\theta} \neq G_2$. Then by Proposition 2.33 there exists a non-trivial $t \in (\overline{\operatorname{Im}\theta})^{\perp} \subseteq \widehat{G_2}$. For this t and all $g \in G_1$ we then have

$$\langle g, \widehat{\theta}(t)\rangle = \langle \theta(g), t\rangle = 1,$$

which implies that $\widehat{\theta}(t) = 0$ and hence that $\widehat{\theta}$ is not injective.

This proves (2), which with Pontryagin duality and the identity $\widehat{\widehat{\theta}} = \theta$ in Lemma 2.31 also implies (1). □

Exercise 2.35 (Kernel and closure of image). Let $\theta\colon G_1 \to G_2$ be a homomorphism as in Corollary 2.32.

(a) Show that $\left(\overline{\operatorname{Im}\theta}\right)^{\perp} = \ker\widehat{\theta}$.

(b) Show that $\left(\ker\theta\right)^{\perp} = \overline{\operatorname{Im}\widehat{\theta}}$.

Exercise 2.36 (Connectedness and torsion). Let G be a compact metric abelian group. Show that G is connected if and only if $\widehat{G}$ has no torsion elements (that is, if and only if $t \in \widehat{G}$ with $nt = 0$ for a natural number $n \geqslant 1$ implies that $t = 0$).

Exercise 2.37. Let $A \subseteq G$ be an arbitrary subset of a topological group. Prove that $(A^{\perp})^{\perp}$ is the closure of the group generated by A.

2.4.4 Projective and Direct Limits*

This and the following section are rather special and can easily be skipped without affecting the developments in the following chapters.

We wish to discuss two more constructions that are once again dual to each other under Pontryagin duality. We start with the projective limit, which in a sense generalizes the direct product, and is defined as follows. Suppose that (G_n) is a sequence of abelian groups, and $\theta_n\colon G_{n+1} \to G_n$ is a continuous surjective homomorphism with compact kernel for every $n \in \mathbb{N}$. Then the *projective limit* of the system (G_n, θ_n) is defined to be the closed subgroup

$$G = \varprojlim(G_n, \theta_n) = \left\{ (g_n) \in \prod_{n=1}^{\infty} G_n \;\middle|\; \theta_n(g_{n+1}) = g_n \text{ for every } n \in \mathbb{N} \right\}$$

of the product $\prod_{n=1}^{\infty} G_n$ equipped with the product topology.

The second construction is the direct limit, which in a sense generalizes the direct sum. For this, suppose that (H_n) is a sequence of abelian groups and $\imath_n\colon H_n \to H_{n+1}$ is a homomorphism such that $\imath_n\colon H_n \to \imath_n(H_n)$ is an isomorphism of topological groups between H_n and the open subgroup $\imath(H_n)$ of H_{n+1} for every $n \in \mathbb{N}$. We use $\imath_n$ to identify H_n with the subgroup $\imath_n(H_n)$ and define the *direct limit* of the system $(H_n, \imath_n)$ as

$$H = \varinjlim(H_n, \imath_n) = \bigcup_{n=1}^{\infty} H_n. \tag{2.15}$$

This is a small cheat, as we use the identification to suppress the set-theoretic construction of the direct limit in the category of sets,[†] that is of a set H_∞ and maps $\phi_n\colon H_n \to H_\infty$ with $H_\infty = \bigcup_{n=1}^{\infty} \phi_n(H_n)$ and $\phi_{n+1} \circ \imath_n = \phi_n$ for all $n \in \mathbb{N}$. As we want to focus here on the algebraic and topological properties of $\varinjlim(H_n, \imath_n)$, this will simplify our discussion a little. In concrete examples this step may need to be treated more carefully (see Exercise 2.38).

Exercise 2.38. Let $q > 1$ be an integer, define $H_n = \mathbb{Z}$, and $\imath_n\colon H_n \to H_{n+1}$ by $\imath_n(k) = qk$ for all $k \in H_n = \mathbb{Z}$ and $n \in \mathbb{N}$. Describe the direct limit $\varinjlim(\mathbb{Z}, \times q)$.

Using the fact that (2.15) is an increasing union, the group operations on H are defined in the obvious way: Given $h_1, h_2 \in H$ there exists some $n \in \mathbb{N}$ with $h_1, h_2 \in H_n$ and hence $h_1 h_2, h_1^{-1}$ are defined in H by the using the group operations in H_n.

The topology on $\varinjlim(H_n, \imath_n)$ is defined by the property that each H_n is (homeomorphically embedded as) an open subset of $\varinjlim(H_n, \imath_n)$. More precisely, let the abelian groups H_n and embeddings $\imath_n\colon H_n \to H_{n+1}$ be as in the definition of the direct limit $\varinjlim(H_n, \imath_n)$. Due to the assumed properties of $\imath_n$ we have that H_n can be considered to be an open subgroup of H_{n+1}. This allows us to define the topology on H as in (2.15) as the inductive topology, in which a subset $O \subseteq H$ is open if and only if $O \cap H_n$ is open for all $n \in \mathbb{N}$. In particular, $H_n \subseteq H$ is open for every $n \in \mathbb{N}$.

Proposition 2.39 (Projective and direct limits). *Under the assumptions above, the projective limit $\varprojlim(G_n, \theta_n)$ and the direct limit $\varinjlim(H_n, \imath_n)$ are again locally compact σ-compact metric abelian groups.*

PROOF. We first discuss the topological group $\varprojlim(G_n, \theta_n)$. From continuity of $\theta_n\colon G_{n+1} \to G_n$ for $n \in \mathbb{N}$ and the definition of the product topology, it follows that $G = \varprojlim(G_n, \theta_n)$ is a closed subset of $\prod_{n=1}^{\infty} G_n$. As the maps are also homomorphisms, G is in fact a closed subgroup, and hence is a metric abelian topological group.

[†] For example, one can use $H_\infty = \{1\} \times H_1 \sqcup \bigsqcup_{n \geqslant 2} \{n\} \times H_n \smallsetminus \imath_{n-1}(H_{n-1}) \subseteq \mathbb{N} \times \bigcup_{n \in \mathbb{N}} H_n$. We leave the definition of the maps ϕ_n and the proof of the identity $\phi_{n+1} \circ \imath_n = \phi_n$ for $n \in \mathbb{N}$ to the reader.

Given $(g_n) \in G = \varprojlim(G_n, \theta_n)$ we can find a compact neighbourhood K_1 of $g_1 \in G_1$. We apply Proposition C.3 to find a compact subset $K_2' \subseteq G$ with $\theta_1(K_2') = K_1$. Therefore $K_2 = \theta_1^{-1}(K_1) = K_2' + \ker\theta_1$ is a compact neighbourhood of g_2 satisfying $\theta_2(g_2) = g_1$. Iterating this argument we find a sequence (K_n) of compact subsets $K_n \subseteq G_n$ so that $\theta_n^{-1}(K_n) = K_{n+1}$ is a neighbourhood of g_{n+1} for all $n \in \mathbb{N}$. Therefore we obtain the neighbourhood

$$\left(K \times \prod_{n=2}^{\infty} G_n\right) \cap G = \left(\prod_{n=1}^{\infty} K_n\right) \cap G$$

of $(g_n) \in G$. By Tychonoff's theorem, $\prod_{n=1}^{\infty} K_n$ is compact and so G is locally compact. Writing G_1 as a countable union of compact sets and applying the argument above once more, we also see that G is σ-compact.

Suppose now that the abelian groups H_n and embeddings $\imath_n\colon H_n \to H_{n+1}$ are as in the definition of the direct limit $\varinjlim(H_n, \imath_n)$. Due to the assumed properties of $\imath_n$ we have that H_n can be considered to be an open subgroup of H_{n+1}. This allows us to define H as in (2.15) equipped with the obvious operations and the inductive topology. Recall that $H_n \subseteq H$ is open for every $n \in \mathbb{N}$. Since each H_n is locally compact, σ-compact, and has second countable topology, the same is true for H, and in particular H is metric. If a sequence in H converges to some $h \in H_n$, then by openness of H_n in H, all but finitely many terms of the sequence must lie in H_n. From this it is easy to conclude that H is also a topological group. □

For the discussion of the Pontryagin dual of projective and direct limits the following two notions and their relation will be useful. We say that a continuous homomorphism $\theta\colon G \to G'$ is a *proper projection* if it is onto and has compact kernel (which by using Proposition C.3 implies properness of the map). Furthermore we say that $\imath\colon H \to H'$ is an *open embedding* if it is an isomorphism between H and an open subgroup $\imath(H)$ of H'.

Lemma 2.40 (Proper projections and open embeddings). *Let G, G' and H, H' be locally compact σ-compact metric abelian groups.*

(1) *If $\theta\colon G \to G'$ is a proper projection, then $\widehat{\theta}\colon \widehat{G'} \to \widehat{G}$ is an open embedding.*
(2) *If $\imath\colon H \to H'$ is an open embedding, then $\widehat{\imath}\colon \widehat{H'} \to \widehat{H}$ is a proper projection.*

PROOF. We suppose that θ is a proper projection. Then surjectivity of θ implies injectivity of $\widehat{\theta}$ by Corollary 2.32. Moreover, the last claim in Proposition C.3 shows that $\theta = \overline{\theta} \circ p$, where $p\colon G \to G/\ker\theta$ is the canonical projection map and $\overline{\theta}\colon G/\ker\theta \to G'$ is an isomorphism. Dually, we then have $\widehat{\theta} = \widehat{p} \circ \widehat{\overline{\theta}}$, where $\widehat{\overline{\theta}}$ is an isomorphism and

$$\widehat{p}\colon \widehat{G/\ker\theta} = (\ker\theta)^{\perp} \to \widehat{G}$$

is the canonical embedding by Example 2.34. Furthermore we have

$$\widehat{G}/(\ker\theta)^{\perp} = \widehat{\ker\theta}$$

by Proposition 2.33. By assumption $\ker\theta$ is compact, which implies that $\widehat{\ker\theta}$ is discrete by Proposition 2.27. Therefore, $(\ker\theta)^{\perp}$ is an open subgroup of $\widehat{G}$ and it follows that $\widehat{\theta}$ embeds $\widehat{G'}$ onto the open subgroup $\widehat{\theta}(\widehat{G'}) = (\ker\theta)^{\perp} \subseteq \widehat{G}$ as claimed.

We suppose now that $\imath\colon H \to H'$ is a continuous embedding with the property that $\imath(H) \subseteq H'$ is an open subgroup and $\imath\colon H \to \imath(H)$ is a group isomorphism. Identifying H with $\imath(H)$, the dual homomorphism to $\imath\colon H \to H'$ is the canonical projection $\widehat{\imath}\colon \widehat{H'} \to \widehat{H'}/H^{\perp}$ with kernel $H^{\perp} = \widehat{H'/H}$ by Example 2.34. Since H'/H is discrete, Proposition 2.27 shows that $\ker\widehat{\imath} = H^{\perp}$ is compact. □

With these preparations, we can now prove the duality between the two limit constructions.

Proposition 2.41 (Duality of limits). *Let the projective limit $\varprojlim(G_n, \theta_n)$ and the direct limit $\varinjlim(H_n, \imath_n)$ be as in Proposition 2.39. Then*

$$\widehat{\varprojlim(G_n, \theta_n)} = \varinjlim(\widehat{G_n}, \widehat{\theta_n}),$$

where we use the open embedding $\imath_n = \widehat{\theta_n}\colon \widehat{G_n} \to \widehat{G_{n+1}}$ for all $n \in \mathbb{N}$ to define the direct limit. Dually,

$$\widehat{\varinjlim(H_n, \imath_n)} = \varprojlim(\widehat{H_n}, \widehat{\imath_n}),$$

where we use the proper projection $\theta_n = \widehat{\imath_n}\colon \widehat{H_{n+1}} \to \widehat{H_n}$ for $n \in \mathbb{N}$ to define the projective limit.

PROOF. Lemma 2.40 shows that if $(H_n, \imath_n)$ satisfies the assumptions for the construction of

$$H = \varinjlim(H_n, \imath_n)$$

then $(\widehat{H_n}, \widehat{\imath_n})$ satisfies the assumptions for the construction of

$$G = \varprojlim(\widehat{H_n}, \widehat{\imath_n}). \tag{2.16}$$

Suppose now $h \in H$ and $(t_n) \in G$. Then there exists an $m \in \mathbb{N}$ with $h \in H_m$, and we define

$$\langle h, (t_n)\rangle_{\lim} = \langle h, t_m\rangle.$$

We wish to prove that this is the dual pairing between H and $\widehat{H} \cong G$. Note that $h \in H_m \subseteq H_{m+1}$ and

$$\langle h, t_{m+1}\rangle = \langle \imath_m(h), t_{m+1}\rangle = \langle h, \widehat{\imath}_m(t_{m+1})\rangle = \langle h, t_m\rangle$$

since $\widehat{\imath_n}(t_{n+1}) = t_n$ for $n \in \mathbb{N}$ by construction of the projective limit (2.16). Thus the expression $\langle h, (t_n)\rangle_{\lim}$ is independent of the choice of m, which easily implies that

$$H \ni h \longmapsto \langle h, (t_n) \rangle_{\lim} \in \mathbb{S}^1 \tag{2.17}$$

defines a multiplicative homomorphism. Using the fact that H_1 is an open subgroup of H and $t_1 \in \widehat{H_1}$ we also obtain continuity of the character defined by (2.17). In other words, we have a well-defined homomorphism $\Phi\colon G \to \widehat{H}$, which sends (t_n) to the character in (2.17). Moreover, if we have $\Phi((t_n)) = 0$ for some $(t_n) \in G$, then $\langle H_n, t_n \rangle = 1$ for all $n \in \mathbb{N}$, and so Φ is injective.

Now let χ be a character on H. Restricting χ to any of the open subgroups H_n for $n \in \mathbb{N}$, we obtain the character $\chi|_{H_n}$ on H_n. Hence there exists a uniquely determined $t_n \in \widehat{H_n}$ with

$$\chi(h) = \langle h, t_n \rangle$$

for all $h \in H_n$. For $h \in H_n$ we also have $h = \imath_n(h) \in H_{n+1}$, and so

$$\langle h, t_n \rangle = \chi(h) = \langle \imath_n(h), t_{n+1} \rangle = \langle h, \widehat{\imath_n}(t_{n+1}) \rangle$$

for all $h \in H_n$. This implies that $t_n = \widehat{\imath}_n(t_{n+1})$ for all $n \in \mathbb{N}$. Therefore (t_n) lies in G, and we have shown that $\Phi\colon G \to \widehat{H}$ is onto and so gives the desired identification.

To see that $\widehat{H}$ and G are also isomorphic as topological groups, we note that $H = \bigcup_{n=1}^{\infty} H_n$ is an open cover and hence any compact set $K \subseteq H$ belongs to some H_m. It follows that $N_{\widehat{H}}(K, \varepsilon)$ with $K \subseteq H_m$ corresponds under the isomorphism from $\widehat{H}$ to $G = \varprojlim(\widehat{H_n}, \widehat{\imath_n})$ to the set

$$\{(t_n) \in G \mid t_m \in N_{\widehat{H_m}}(K, \varepsilon)\}.$$

As the latter is an open subset of G (with respect to the restriction of the product topology), we conclude that the map from G to $\widehat{H}$ is continuous. Proposition C.3 now implies that the above isomorphism from G to $\widehat{H}$ as abstract groups is in fact an isomorphism for topological groups.

The dual statement concerning the Pontryagin dual of $\varprojlim(G_n, \theta_n)$ follows from Lemma 2.40, the above, and Pontryagin duality (Theorem 2.22). □

2.4.5 Local Fields*

By definition a *local field* $\mathbb{K}$ is a locally compact σ-compact non-discrete metric field, where all field operations are assumed to be continuous. Our main goal here is to prove the following self duality statement.

Proposition 2.42 (Self-duality). *A local field $\mathbb{K}$ is isomorphic as an additive group to its own Pontryagin dual $\widehat{\mathbb{K}}$. In fact, for any non-trivial character $\chi \in \widehat{\mathbb{K}}$ we can define an isomorphism of topological groups*

$$\mathbb{K} \ni t \longmapsto \chi_t \in \widehat{\mathbb{K}}$$

by $\chi_t\colon \mathbb{K} \ni a \mapsto \chi_t(a) = \chi(at)$.

A useful tool, both for our discussion and for the further study and classification of local fields, is the *induced absolute value* defined by

$$|a|_{\mathbb{K}} = \frac{m_{\mathbb{K}}(aM)}{m_{\mathbb{K}}(M)},$$

where $a \in \mathbb{K}$, $m_{\mathbb{K}}$ is a Haar measure for the group $(\mathbb{K}, +)$, and $M \subseteq \mathbb{K}$ is any Borel subset with positive finite measure.

Lemma 2.43 (Properties of the absolute value). *Let $\mathbb{K}$ be a local field. The absolute value $|\cdot|_{\mathbb{K}}\colon \mathbb{K} \to [0,\infty)$ is positive in the sense that $|a|_{\mathbb{K}} > 0$ for all $a \in \mathbb{K}\smallsetminus\{0\}$, well-defined, multiplicative in the sense that*

$$|a_1 a_1|_{\mathbb{K}} = |a_1|_{\mathbb{K}} |a_2|_{\mathbb{K}}$$

for all $a_1, a_2 \in \mathbb{K}$, continuous, and proper. Moreover, $a_n \to 0$ as $n \to \infty$ if and only if $|a_n|_{\mathbb{K}} \to 0$ as $n \to \infty$.

Proof. That $|a|_{\mathbb{K}}$ is well-defined (that is, independent of the choice of Haar measure $m_{\mathbb{K}}$ and the set M) follows as in our discussion of the modular character in Section 1.2.4. In fact, for any $a \in \mathbb{K}\smallsetminus\{0\}$ defining $\mu_a(B) = m_{\mathbb{K}}(aB)$ for Borel subsets $B \subseteq \mathbb{K}$ gives a Haar measure μ_a on $(\mathbb{K}, +)$ and hence, by uniqueness, μ_a must be a positive multiple of $m_{\mathbb{K}}$. In particular, $|a|_{\mathbb{K}} > 0$ for all $a \in \mathbb{K}\smallsetminus\{0\}$. For $a = 0$ we have $aM = \{0\}$ for any M as in the definition of $|a|_{\mathbb{K}}$, and the assumption that $\mathbb{K}$ is non-discrete implies that $|0|_{\mathbb{K}} = 0$.

To see multiplicativity, suppose without loss of generality that a_1, a_2 are elements of $\mathbb{K}\smallsetminus\{0\}$. Then

$$|a_1 a_2|_{\mathbb{K}} = \frac{m_{\mathbb{K}}(a_1 a_2 M)}{m_{\mathbb{K}}(M)} = \frac{m_{\mathbb{K}}(a_1 a_2 M)}{m_{\mathbb{K}}(a_2 M)} \frac{m_{\mathbb{K}}(a_2 M)}{m_{\mathbb{K}}(M)} = |a_1|_{\mathbb{K}} |a_2|_{\mathbb{K}}$$

for any $M \subseteq \mathbb{K}$ as in the definition of $|\cdot|_{\mathbb{K}}$. We will assume in the following that $M \subseteq \mathbb{K}$ is a compact neighbourhood of $0 \in \mathbb{K}$.

For the proof of continuity of $|\cdot|\colon \mathbb{K} \to [0,\infty)$ at 0, we will need the following topological claim for the local field $\mathbb{K}$. For any open neighbourhood U of $0 \in \mathbb{K}$ there exists some neighbourhood V of $0 \in \mathbb{K}$ such that $VM \subseteq U$. Indeed, for any $a \in M$ there exists (by continuity of multiplication) open neighbourhoods V_a of 0 and O_a of a such that $V_a O_a \subseteq U$. By compactness of M we can find a finite subcover

$$M \subseteq O_{a_1} \cup \cdots \cup O_{a_m}$$

and so it follows that $VM \subseteq U$ for

$$V = V_{a_1} \cap \cdots V_{a_m}.$$

To see continuity of $|\cdot|_{\mathbb{K}}$ at 0, we let $\varepsilon > 0$ and choose U to be an open neighbourhood of 0 with measure $m_{\mathbb{K}}(U) < m_{\mathbb{K}}(M)\varepsilon$. Let V be an open neighbourhood of 0 with $VM \subseteq U$ as above. For $a \in V$ we then obtain

$$|a|_{\mathbb{K}} = \frac{m_{\mathbb{K}}(aM)}{m_{\mathbb{K}}(M)} \leqslant \frac{m_{\mathbb{K}}(U)}{m_{\mathbb{K}}(M)} < \varepsilon.$$

As $\varepsilon > 0$ was arbitrary it follows that $|\cdot|_{\mathbb{K}}$ is continuous at $0 \in \mathbb{K}$.

To see continuity of $|\cdot|_{\mathbb{K}}$ at 1 we first prove an analogue of the above topological claim. By local compactness and local finiteness of $m_{\mathbb{K}}$ there exists an open set $O \supseteq M$ with finite Haar measure. We define

$$B_n = \{a \in O \mid \mathsf{d}(a, M) < \tfrac{1}{n}\}$$

for $a \in \mathbb{K}$. Since $O \supseteq B_1 \supseteq B_2 \supseteq \cdots$ is a decreasing sequence, O has finite Haar measure, and

$$M = \bigcap_{n=1}^{\infty} B_n,$$

we have $m_{\mathbb{K}}(M) = \lim_{n\to\infty} m_{\mathbb{K}}(B_n)$. Let $\varepsilon > 0$ and choose $n \geqslant 1$ such that

$$m_{\mathbb{K}}(B_n) < (1+\varepsilon) m_{\mathbb{K}}(M).$$

Just as in the proof of the above topological claim, we can use compactness of M, openness of B_n, and the inclusion $1 \cdot M \subseteq B_n$ to find an open neighbourhood V of $1 \in \mathbb{K}$ with $VM \subseteq B_n$. For $a \in V$ we then obtain

$$|a|_{\mathbb{K}} = \frac{m_{\mathbb{K}}(aM)}{m_{\mathbb{K}}(M)} \leqslant \frac{m_{\mathbb{K}}(B_n)}{m_{\mathbb{K}}(M)} < 1+\varepsilon.$$

Therefore

$$(1+\varepsilon)^{-1} < |a|_{\mathbb{K}} < 1+\varepsilon$$

for all $a \in V$ with $a^{-1} \in V$, since $|a^{-1}|_{\mathbb{K}} = |a|_{\mathbb{K}}^{-1}$ by the multiplicative property. As $\varepsilon > 0$ was arbitrary and $\{a \in V \mid a^{-1} \in V\}$ is a neighbourhood of 1, we obtain the continuity of $|\cdot|_{\mathbb{K}}$ at $1 \in \mathbb{K}$. Continuity at $a_0 \in \mathbb{K} \smallsetminus \{0\}$ now follows from the identity

$$|a|_{\mathbb{K}} = |a a_0^{-1} a_0|_{\mathbb{K}} = |a a_0^{-1}|_{\mathbb{K}} |a_0|_{\mathbb{K}}$$

for $a \in \mathbb{K}$.

It remains to prove that $|\cdot|_{\mathbb{K}}$ is proper, and that a sequence (a_n) in $\mathbb{K}$ converges to 0 if $(|a_n|_{\mathbb{K}})$ converges to 0 as $n \to \infty$. Let $M \subseteq \mathbb{K}$ be a compact neighbourhood of $0 \in \mathbb{K}$ as above, and suppose in addition that $1 \in M$. Let $U_0 \subseteq M$ be an open neighbourhood of 0 with Haar measure satisfying $m_{\mathbb{K}}(U_0) \leqslant \frac{1}{2} m_{\mathbb{K}}(M)$. Let V_0 be an open neighbourhood of $0 \in \mathbb{K}$ satisfying $V_0 M \subseteq U_0$ as in the topological claim above. For the following we choose and fix some $t \in V_0 \smallsetminus \{0\}$. Then $|t|_{\mathbb{K}} \leqslant \frac{1}{2}$ (by the argument proving continuity at 0 above) and so

$$|t^n|_{\mathbb{K}} = |t|_{\mathbb{K}}^n \longrightarrow 0 \tag{2.18}$$

as $n \to \infty$. Moreover,

$$t = t \cdot 1 \in tM \subseteq U_0 \subseteq M,$$

and by induction $t^n \in M$ for all $n \geqslant 1$. By compactness of M the sequence (t^n) has a convergent subsequence. However, because of the established positivity and continuity of the absolute value, we conclude from (2.18) that 0 is the only possible limit point of a convergent subsequence. By compactness, we obtain $t^n \to 0 \in \mathbb{K}$ as $n \to \infty$.

For $\ell \in \mathbb{N}$ we now define $P_\ell = t^{-\ell}(M \smallsetminus tM)$ and obtain

$$\mathbb{K} = M \cup \bigcup_{\ell=1}^{\infty} P_\ell.$$

Indeed, for any $a \in \mathbb{K} \smallsetminus M$ we have $t^n a \to 0$ as $n \to \infty$, so there exists a minimal $\ell \geqslant 1$ with $t^\ell a \in M$ (satisfying $t^\ell a \notin tM$), and hence $a \in P_\ell$.

We also define

$$c = \min\{|a|_{\mathbb{K}} \mid a \in \overline{M \smallsetminus tM}\},$$

and note that $c > 0$ since $\overline{M \smallsetminus tM} \subseteq M$ is compact and does not contain 0. It now follows that any $a = t^{-\ell} a_0 \in P_\ell$ with $a_0 \in M \smallsetminus tM$ and $\ell \in \mathbb{N}$ has absolute value

$$|a|_{\mathbb{K}} = |t^{-\ell} a_0|_{\mathbb{K}} = |t|_{\mathbb{K}}^{-\ell} |a_0|_{\mathbb{K}} \geqslant 2^\ell c.$$

As

$$M \cup \bigcup_{\ell=1}^{L} P_\ell = t^{-L} M$$

is compact for every $L \geqslant 1$ and $|a|_{\mathbb{K}} \geqslant 2^{L+1} c$ for all $a \in \mathbb{K} \smallsetminus (t^{-L} M)$, it follows that the absolute value is a proper function.

Now suppose that (a_n) is a sequence in $\mathbb{K}$ with $|a_n|_{\mathbb{K}} \to 0$ as $n \to \infty$. Then $a_n \in M$ for all sufficiently large n since $|a|_{\mathbb{K}} \geqslant 2c$ if $a \notin M$ by the above and we may again use compactness of M, positivity, and continuity of the absolute value to conclude that $a_n \to 0$ as $n \to \infty$. □

Proof of Proposition 2.42. Let $\chi \in \widehat{\mathbb{K}}$ be any non-trivial character. We define $\chi_t \colon \mathbb{K} \ni a \mapsto \chi_t(a) = \chi(at)$ for any $t \in \mathbb{K}$ as in the proposition. Since the map $\mathbb{K} \ni a \mapsto at \in \mathbb{K}$ is a continuous homomorphism of the additive group $\mathbb{K}$ it follows that $\chi_t \in \widehat{\mathbb{K}}$ for any $t \in \mathbb{K}$. For $t_1, t_2, a \in \mathbb{K}$ we also have

$$\chi_{t_1+t_2}(a) = \chi\bigl((t_1 + t_2)a\bigr) = \chi(t_1 a)\chi(t_2 a) = \chi_{t_1}(a)\chi_{t_2}(a),$$

and so $\Phi \colon \mathbb{K} \ni t \mapsto \chi_t \in \widehat{\mathbb{K}}$ is a homomorphism. By assumption, χ is a non-trivial character. So let $a_0 \in \mathbb{K}$ be chosen with $\chi(a_0) \neq 1$. If now $t \in \mathbb{K} \smallsetminus \{0\}$ then

$$\chi_t(\tfrac{1}{t} a_0) = \chi(t \tfrac{1}{t} a_0) = \chi(a_0) \neq 1,$$

which shows that χ_t is non-trivial and hence Φ is injective.

To see continuity of Φ, let $M \subseteq \mathbb{K}$ be compact and fix $\varepsilon > 0$. By continuity of χ,

$$U = \{a \in \mathbb{K} \mid |\chi(a) - 1| < \varepsilon\}$$

is an open neighbourhood of $0 \in \mathbb{K}$. By compactness of M there exists an open neighbourhood V of 0 such that $VM \subseteq U$ (just as in the topological claim in the proof of Lemma 2.43). Therefore $t \in V$ gives $|\chi_t(a) - 1| = |\chi(at) - 1| < \varepsilon$ for all $a \in M$. Equivalently, $t \in V$ implies that

$$\chi_t \in N_{\widehat{\mathbb{K}}}(M, \varepsilon),$$

which gives continuity of Φ.

Next we claim that Φ is proper. So let (t_n) be a sequence in $\mathbb{K}$ with the property that $t_n \to \infty$ as $n \to \infty$. By Lemma 2.43, we have

$$a_n = t_n^{-1} a_0 \longrightarrow 0$$

as $n \to \infty$. This shows that (χ_{t_n}) has no convergent subsequence with respect to the compact-open topology. In fact, uniform convergence of $(\chi_{t_{n_k}})$ on the compact set $\{0\} \cup \{a_n \mid n \in \mathbb{N}\}$ to a character χ' would imply that $\lim_{k\to\infty} \chi_{t_{n_k}}(a_{n_k}) = \chi'(0) = 1$, but we have $\chi_{t_n}(a_n) = \chi(a_0)$ for all $n \geqslant 1$, and $\chi(a_0) \neq 1$. This shows that $\chi_{t_n} \to \infty$ as $n \to \infty$, and hence that Φ is proper.

It follows that $\Phi(\mathbb{K}) \subseteq \widehat{\mathbb{K}}$ is a closed subgroup of $\widehat{\mathbb{K}}$. To identify the subgroup $\Phi(\mathbb{K})$ we suppose that $a \in \Phi(\mathbb{K})^{\perp} \subseteq \mathbb{K}$, so that $\langle a, \Phi(t)\rangle = \chi_t(a) = 1$ for all $t \in \mathbb{K}$. If $a \neq 0$ we can set $t = a^{-1}a_0$ and obtain the contradiction $\chi_t(a) = \chi(aa^{-1}a_0) = \chi(a_0) \neq 1$. Hence $\Phi(\mathbb{K})^{\perp} = \{0\}$ and Proposition 2.33 implies that $\Phi(\mathbb{K}) = \widehat{\mathbb{K}}$. By the last claim in Proposition C.3 we now obtain that $\Phi\colon \mathbb{K} \to \widehat{\mathbb{K}}$ is an isomorphism of topological groups, which concludes the proof. □

We note that the above applies to the local fields $\mathbb{R}$ and $\mathbb{C}$. In particular, Proposition 2.42 and Proposition 2.29 together give a complete (but much longer) proof of Exercise 2.6. Let us briefly describe a second class of local fields that are especially important in number theory.

Fix a prime $p \in \mathbb{N}$. The local field $\mathbb{Q}_p$ is defined as the completion of $\mathbb{Q}$ with respect to the so-called *p-adic norm* defined by

$$|a|_p = \begin{cases} 0 & \text{if } a = 0; \\ p^{-k} & \text{if } a = p^k \frac{m}{n} \in \mathbb{Q}^{\times} \text{ for } k \in \mathbb{Z} \text{ and } m, n \in \mathbb{Z} \smallsetminus p\mathbb{Z}. \end{cases}$$

The definition of $|\cdot|_p$ implies that $|ab|_p = |a|_p |b|_p$ for all $a, b \in \mathbb{Q}$. Similarly, if $a = p^k \frac{m}{n}$ and $b = p^{\ell} \frac{r}{s}$ with $k, \ell \in \mathbb{Z}$ and $m, n, r, s \in \mathbb{Z} \smallsetminus p\mathbb{Z}$, then

$$\begin{aligned} |a + b|_p &= \left| p^{\min(k,\ell)} \left(p^{k-\min(k,\ell)} \tfrac{m}{n} + p^{\ell-\min(k,\ell)} \tfrac{r}{s} \right) \right|_p \\ &\leqslant \max(p^{-k}, p^{-\ell}) = \max\bigl(|a|_p, |b|_p\bigr) \leqslant |a|_p + |b|_p. \end{aligned}$$

As in the study of norms on vector spaces, it now follows that $|a - b|_p$ defines a metric on $\mathbb{Q}$ which extends to the completion $\mathbb{Q}_p$ of $\mathbb{Q}$. Moreover, the p-adic norm extends from $\mathbb{Q}$ to a continuous function on $\mathbb{Q}_p$ defined by setting

$$|a|_p = \mathsf{d}_p(a, 0)$$

for all $a \in \mathbb{Q}_p$. From these properties it follows that multiplication and addition extend continuously from $\mathbb{Q}$ to $\mathbb{Q}_p$. In fact, by the above discussion we have

$$|a + b|_p \leqslant \max\bigl(|a|_p, |b|_p\bigr) \leqslant |a|_p + |b|_p \tag{2.19}$$

and

$$|ab|_p = |a|_p |b|_p \tag{2.20}$$

for all $a, b \in \mathbb{Q}$. This implies that sums and products of Cauchy sequences in $\mathbb{Q}$ are again Cauchy sequences in $\mathbb{Q}$. Hence we obtain the definition of addition and multiplication on $\mathbb{Q}_p$ so that (2.19) and (2.20) also hold for $a, b \in \mathbb{Q}_p$.

To see that $\mathbb{Q}_p$ is a topological field, we claim that any Cauchy sequence (a_n) in $\mathbb{Q}\smallsetminus\{0\}$ with limit $a \in \mathbb{Q}_p\smallsetminus\{0\}$ satisfies that (a_n^{-1}) is again a Cauchy sequence. With $\delta = \frac{1}{2}|a|_p$ it follows that $|a_n|_p \geqslant \delta$ for all $n \geqslant N$ and some $N \geqslant 1$. Therefore

$$|a_m^{-1} - a_n^{-1}|_p \leqslant |a_m^{-1} a_n^{-1}(a_n - a_m)|_p \leqslant \delta^{-2}|a_m - a_n|_p$$

for all $m, n \geqslant N$, which implies the claim since (a_n) is a Cauchy sequence. Now the limit $b = \lim_{n\to\infty} a_n^{-1} \in \mathbb{Q}_p$ satisfies

$$ab = \lim_{n\to\infty} a_n a_n^{-1} = 1.$$

Similarly, one also obtains that $\mathbb{Q}_p\smallsetminus\{0\} \ni a \mapsto a^{-1} \in \mathbb{Q}_p\smallsetminus\{0\}$ is continuous.

The subgroup $\mathbb{Z}_p$ of $\mathbb{Q}_p$ is defined as the closure of $\mathbb{Z}$ in $\mathbb{Q}_p$, and so $|a|_p \leqslant 1$ for all $a \in \mathbb{Z}_p$. We recall that if $n \in \mathbb{Z}\smallsetminus p\mathbb{Z}$, then there exists some ℓ such that $n\ell \equiv 1$ modulo p. In other words, there exists some $a \in \mathbb{Z}$ with the property that $n\ell = 1 - ap$, which implies that

$$(n\ell)^{-1} = \frac{1}{1 - ap} = \sum_{j=0}^{\infty} (ap)^j$$

is a series in $\mathbb{Z}$ that converges in $\mathbb{Z}_p \subseteq \mathbb{Q}_p$ since

$$\left|\frac{1}{1 - ap} - \sum_{j=0}^{J} (ap)^j\right|_p = \left|\frac{1 - (1 - ap)\sum_{j=0}^{J}(ap)^j}{1 - ap}\right|_p = \left|\frac{(ap)^{J+1}}{1 - ap}\right|_p \leqslant p^{-(J+1)}$$

for $J \geqslant 1$. This implies that for $k \geqslant 0$ and $m, n \in \mathbb{Z}\smallsetminus p\mathbb{Z}$ we have

$$p^k \frac{m}{n} = p^k m\ell(n\ell)^{-1} \in \mathbb{Z}_p.$$

Since $|a|_p \in \{0\} \cup p^{\mathbb{Z}}$ for all $a \in \mathbb{Q}$, we obtain that

$$\mathbb{Z}_p = \overline{B}_1(0) = \{a \in \mathbb{Q}_p \mid |a|_p \leqslant 1\} = \{a \in \mathbb{Q}_p \mid |a|_p < p\} = B_p(0)$$

is the closed unit ball in $\mathbb{Q}_p$ and is simultaneously an open ball. Using (2.20) and multiplication by powers of p we deduce that open and closed balls with centre 0 have the form

$$\begin{aligned} p^k\mathbb{Z}_p = \overline{B}_{p^{-k}}(0) &= \left\{a \in \mathbb{Q}_p \mid |a|_p \leqslant p^{-k}\right\} \\ &= B_{p^{1-k}}(0) = \left\{a \in \mathbb{Q}_p \mid |a|_p < p^{1-k}\right\} \end{aligned}$$

for some $k \in \mathbb{Z}$. By (2.19), these sets form additive subgroups of $\mathbb{Q}_p$. Moreover, (2.20) shows that $\mathbb{Z}_p$ is a subring and that the sets $p^k\mathbb{Z}_p \subseteq \mathbb{Z}_p$ are ideals for any $k \geqslant 0$.

For $k \geqslant 1$ division by p^k with remainder in $\mathbb{Z}$ implies that

$$\mathbb{Z} = \bigsqcup_{m=0}^{p^k-1} (m + p^k\mathbb{Z}),$$

where the sets $m + p^k\mathbb{Z}$ for $m \in \{0, \dots, p^k - 1\}$ on the right-hand side are the closed balls of radius p^{-k} around m with respect to $|\cdot|_p$, intersected with $\mathbb{Z}$. Taking the closure in $\mathbb{Q}_p$, we obtain

$$\mathbb{Z}_p = \bigsqcup_{m=0}^{p^k-1} (m + p^k\mathbb{Z}_p). \tag{2.21}$$

As these balls have distance at least p^{-k+1}, the union remains disjoint. This shows that $\mathbb{Z}_p$ is totally bounded as a metric space. In particular, as $\mathbb{Z}_p$ is complete by definition, we see that $\mathbb{Z}_p$ is compact and so $\mathbb{Q}_p$ is locally compact and σ-compact.

Exercise 2.44. Show that any element $a \in \mathbb{Z}_p$ may be written as

$$a = \sum_{k=0}^{\infty} a_k p^k$$

with $a_k \in \{0, 1, \dots, p-1\}$ for $k \geqslant 0$, and that any $a \in \mathbb{Q}_p$ may be written as

$$a = \sum_{k=\ell}^{\infty} a_k p^k$$

with $a_k \in \{0, 1, \dots, p-1\}$ for all $k \geqslant \ell$ and some $\ell \in \mathbb{Z}$.

Exercise 2.45. Use (2.21) to show that $\mathbb{Z}_p/p^k\mathbb{Z}_p \cong \mathbb{Z}/p^k\mathbb{Z} = \mathrm{C}_{p^k}$ for all $k \geqslant 1$ and deduce that we may identify $\mathbb{Z}_p$ with the projective limit $\varprojlim(\mathrm{C}_{p^k}, \theta_k)$ where

$$\begin{aligned} \theta_k \colon \mathrm{C}_{p^{k+1}} &\longrightarrow \mathrm{C}_{p^k} \\ m + p^{k+1}\mathbb{Z} &\longmapsto m + p^k\mathbb{Z} \end{aligned}$$

is the canonical projection map for all $k \geqslant 1$.

Exercise 2.46. Show that $\mathbb{Q}_p$ can be obtained as the inductive limit $\mathbb{Q}_p = \varinjlim(p^{-\ell}\mathbb{Z}_p, \imath_\ell)$, where

$$\begin{aligned} \imath_\ell \colon p^{-\ell}\mathbb{Z}_p &\longrightarrow p^{-\ell-1}\mathbb{Z}_p \\ a &\longmapsto a \end{aligned}$$

is the inclusion map for all $\ell \geqslant 0$.

Exercise 2.47. Find an explicit formula for a non-trivial character on $\mathbb{Q}_p$, and use this to exhibit an explicit isomorphism between $\mathbb{Q}_p$ and $\widehat{\mathbb{Q}_p}$.

Exercise 2.48. (a) Show that the absolute value $|\cdot|_{\mathbb{Q}_p}$ in Lemma 2.43 agrees with the p-adic norm $|\cdot|_p$ on $\mathbb{Q}_p$.
(b) Describe the Haar measure on $\mathbb{Q}_p$.

Exercise 2.49. (a) Show that for a finite set S of primes in $\mathbb{N}$ the ring $\mathbb{Z}[\frac{1}{p} \mid p \in S]$ gives rise by diagonal embedding to a discrete subgroup of $\mathbb{R} \times \prod_{p\in S} \mathbb{Q}_p$.
(b) Show that

$$\widehat{\mathbb{Z}[\tfrac{1}{p} \mid p \in S]} \cong \big(\mathbb{R} \times \textstyle\prod_{p\in S} \mathbb{Q}_p\big)/\mathbb{Z}[\tfrac{1}{p} \mid p \in S].$$

(c) Show that for any non-empty finite set S of primes there is an injective homomorphism

$$\mathbb{R} \hookrightarrow \widehat{\mathbb{Z}[\tfrac{1}{p} \mid p \in S]}$$

with dense image.[(4)]

The case of the characteristic p local field defined by the field of formal Laurent series

$$\mathbb{F}_p(\!(X)\!) = \Big\{\sum_{k=\ell}^{\infty} a_k X^k \mid \ell \in \mathbb{Z}, a_k \in \mathbb{F}_p = \mathbb{Z}/p\mathbb{Z} \text{ for all } k \geqslant \ell\Big\}$$

is quite similar to the above. The norm $|\cdot| \colon \mathbb{F}_p(\!(X)\!) \to \mathbb{R}_{\geqslant 0}$ is defined by

$$\left|\sum_{k=\ell}^{\infty} a_k X^k\right| = \begin{cases} 0 & \text{if } a_k = 0 \text{ for all } k \in \mathbb{Z}, \\ p^{-\ell} & \text{if } \ell \in \mathbb{Z} \text{ and } a_\ell \neq 0. \end{cases}$$

The ring

$$\mathbb{F}_p[\![X]\!] = \Big\{\sum_{k=0}^{\infty} a_k X^k \mid \ell \in \mathbb{Z}, a_k \in \mathbb{F}_p \text{ for all } k \geqslant 0\Big\}$$

is an open and compact neighbourhood of 0, and all other open and closed balls about 0 can be obtained by multiplication by X^ℓ for $\ell \in \mathbb{Z}$. We leave the details as an exercise.

Exercise 2.50. Show that $\mathbb{F}_p(\!(X)\!)$ is a locally compact σ-compact metric field if we declare $\mathbb{F}_p(\!(X)\!)$ to be the inductive limit over ℓ of $X^{-\ell}\mathbb{F}_p[\![X]\!]$ and use the inductive topology on the latter group.

We conclude this topic by noting that any local field $\mathbb{K}$ is isomorphic to exactly one of the following three types of field:

- (Archimedean) $\mathbb{K} = \mathbb{R}$ or $\mathbb{K} = \mathbb{C}$;
- (Non-Archimedean of zero characteristic) $\mathbb{K} = \mathbb{Q}_p$ or $\mathbb{K}$ is a finite field extension of $\mathbb{Q}_p$ for some prime $p \in \mathbb{N}$;
- (Non-Archimedean of positive characteristic) $\mathbb{K} = \mathbb{F}_p((X))$ or $\mathbb{K}$ is a finite field extension of $\mathbb{F}_p((X))$, for some prime $p \in \mathbb{N}$.

We refer to Weil [90, Sec. I.3] for this classification of local fields.

2.5 Spectral Measures

We already encountered spectral measures in Bochner's theorem (Theorem 2.10) and in the description of cyclic representations (Corollary 2.12). In particular, Theorem 2.10 shows the existence and uniqueness of diagonal spectral measures as defined below. Here we study them in greater detail and also consider more general non-diagonal matrix coefficients, whose existence and uniqueness will be shown in Proposition 2.53.

Definition 2.51 (Spectral measures). Let π be a unitary representation of the abelian group G. For any $v \in \mathcal{H}_\pi$ the *(principal* or *diagonal) spectral measure* μ_v is the finite measure on $\widehat{G}$ with the property that

$$\langle \pi_g v, v \rangle = \int_{\widehat{G}} \langle g, t \rangle \, \mathrm{d}\mu_v(t)$$

for all $g \in G$. For $v, w \in \mathcal{H}_\pi$ the *(non-diagonal) spectral measure* $\mu_{v,w}$ is a finite complex-valued measure on $\widehat{G}$ such that

$$\langle \pi_g v, w \rangle = \int \langle g, t \rangle \, \mathrm{d}\mu_{v,w}(t)$$

for all $g \in G$.

In most of our discussion we will consider only one unitary representation π and will write μ_v and $\mu_{v,w}$ for the spectral measures of $v, w \in \mathcal{H}_\pi$ as in the above definition. If we want to emphasise the unitary representation used in the definition of the spectral measure, we will write μ_v^π or $\mu_{v,w}^\pi$. The main properties of spectral measures are summarized in the next result.

Proposition 2.52 (Geometry of diagonal spectral measures). *Let π be a unitary representation of the abelian group G. The (diagonal) spectral measures μ_v for $v \in \mathcal{H}_\pi$ satisfy the following for all $v, w, v_1, v_2, \dots \in \mathcal{H}_\pi$.*

(1) *We have $\|\mu_v\| = \mu_v(\widehat{G}) = \|v\|^2$.*
(2) *If $w \in \langle v \rangle_\pi$, then $\mu_w \ll \mu_v$. Indeed, if $w \in \langle v \rangle_\pi$ corresponds under a unitary isomorphism as in the cyclic spectral theorem (Corollary 2.12) to $F \in L^2_{\mu_v}(\widehat{G})$, then $\mathrm{d}\mu_w = |F|^2 \, \mathrm{d}\mu_v$.*

(3) *If* $\langle v_1\rangle_\pi \perp \langle v_2\rangle_\pi$, *then* $\mu_{v_1+v_2} = \mu_{v_1} + \mu_{v_2}$.
(4) *If* $v = \sum_{k=1}^\infty v_k$ *is convergent with* $\langle v_k\rangle_\pi \perp \langle v_\ell\rangle_\pi$ *for all* $k \neq \ell \in \mathbb{N}$, *then* $\mu_v = \sum_{k=1}^\infty \mu_{v_k}$.
(5) *If* $\mu_{v_1} \perp \mu_{v_2}$, *then* $\langle v_1\rangle_\pi \perp \langle v_2\rangle_\pi$.
(6) *If* $w \in \langle v\rangle_\pi$ *and* $\mu_v \ll \mu_w$, *then* $\langle w\rangle_\pi = \langle v\rangle_\pi$.
(7) *If* $\mu_{v_1} \perp \mu_{v_2}$, *then we even have* $\langle v_1 + v_2\rangle_\pi = \langle v_1\rangle_\pi \oplus \langle v_2\rangle_\pi$.

PROOF. The equality in (1) already appeared in Corollary 2.12. Applying Corollary 2.12, we have that π restricted to $\langle v\rangle_\pi$ is unitarily isomorphic to $L^2_{\mu_v}(\widehat{G})$, where v corresponds to $\mathbb{1}$. Let $F \in L^2_{\mu_v}(\widehat{G})$ correspond to w. Then

$$\langle \pi_g w, w\rangle = \langle M_g F, F\rangle_{L^2(\widehat{G},\mu_v)} = \int \langle g,t\rangle |F(t)|^2 \,\mathrm{d}\mu_v(t)$$

for all $g \in G$, showing that $\mathrm{d}\mu_w = |F|^2 \,\mathrm{d}\mu_v$, and hence (2).

Suppose that $\langle v_1\rangle_\pi \perp \langle v_2\rangle_\pi$ as in (3). Then

$$\begin{aligned}\int \langle g,t\rangle \,\mathrm{d}\mu_{v_1+v_2}(t) &= \langle \pi_g(v_1+v_2), v_1+v_2\rangle = \langle \pi_g v_1, v_1\rangle + \langle \pi_g v_2, v_2\rangle \\ &= \int \langle g,t\rangle \,\mathrm{d}\mu_{v_1}(t) + \int \langle g,t\rangle \,\mathrm{d}\mu_{v_2}(t) \\ &= \int \langle g,t\rangle \,\mathrm{d}(\mu_{v_1} + \mu_{v_2})(t)\end{aligned}$$

for all $g \in G$, which proves (3) by uniqueness of spectral measures in Bochner's theorem (Theorem 2.10).

Suppose now that $v = \sum_{k=1}^\infty v_k$ as in (4) and define $u_N = \sum_{k=1}^N v_k$, respectively $w_N = \sum_{k=N+1}^\infty v_k$, for some $N \geqslant 1$, so that $v = u_N + w_N$. By (3), we have $\mu_v = \mu_{u_N} + \mu_{w_N}$, and by induction also $\mu_{u_N} = \sum_{k=1}^N \mu_{v_k}$. By (1) we have $\|\mu_{w_N}\| = \|w_N\|^2 \to 0$ as $N \to \infty$, which implies that

$$\mu_v = \lim_{N\to\infty} \left(\mu_{u_N} + \mu_{w_N}\right) = \sum_{k=1}^\infty \mu_{v_k}.$$

Now suppose that $\mu_{v_1} \perp \mu_{v_2}$ as in (5). We write v_2 as a sum $v_2 = w + \widetilde{w}$ with $w \in \langle v_1\rangle_\pi$ and $\widetilde{w} \in \langle v_1\rangle_\pi^\perp$. Since $\langle w\rangle_\pi \subseteq \langle v_1\rangle_\pi$ and $\langle \widetilde{w}\rangle_\pi \perp \langle v_1\rangle_\pi$, we obtain $\mu_{v_2} = \mu_w + \mu_{\widetilde{w}}$ from (3). By (2) we have $\mu_w \ll \mu_{v_1}$, which together with the assumption $\mu_{v_1} \perp \mu_{v_2}$ implies $\mu_w = 0$ and so also $\|w\|^2 = \mu_w(\widehat{G}) = 0$ by (1). Therefore $v_2 = \widetilde{w} \in \langle v_1\rangle_\pi^\perp$, and (5) follows.

For the proof of (6), we apply the cyclic spectral theorem (Corollary 2.12) to the subspace $\langle v\rangle_\pi$ and assume without loss of generality that $\pi = M$ and that $v = \mathbb{1} \in L^2(\widehat{G}, \mu)$ for a finite measure $\mu = \mu_v$. We also recall from Lemma 2.9 that $M_*(f) = M_{\widetilde{\overline{f}}}$ for all $f \in L^1(G)$. It follows that $\widetilde{L^1(G)}w \subseteq \langle w\rangle_M$ and, by density of $\widetilde{L^1(G)} \subseteq C_0(\widehat{G})$, also $C_0(\widehat{G})w \subseteq \langle w\rangle_M$ for all $w \in L^2_\mu(\widehat{G})$.

Suppose now that $w \in L^2_\mu(\widehat{G})$ satisfies $\mu \ll \mu_w$. By (2) we have $\mathrm{d}\mu_w = |w|^2 \,\mathrm{d}\mu$, which implies that $w(t) \neq 0$ for μ-almost every $t \in \widehat{G}$. However, this implies $C_0\widehat{G})w$ is dense in $L^2_\mu(\widehat{G})$ and we obtain $\langle w\rangle = L^2_\mu(\widehat{G})$ as claimed in (6).

It remains to prove (7). By (5) we have $\langle v_1\rangle_\pi \perp \langle v_2\rangle_\pi$. Let $w = v_1 + v_2$ and conclude from (3) that $\mu_w = \mu_{v_1} + \mu_{v_2}$. Since $\mu_{v_1} \perp \mu_{v_2}$ we can decompose $\widehat{G}$ into $\widehat{G} = B_1 \sqcup B_2$ for some measurable sets $B_1, B_2 \subseteq \widehat{G}$ with $\mu_{v_j}(\widehat{G}\smallsetminus B_j) = 0$ for $j = 1, 2$. Using the cyclic spectral theorem (Corollary 2.12), the function $\mathbb{1}_{B_1} \in L^2(\widehat{G}, \mu_w)$ corresponds to some

$$u_1 \in \langle v_1 + v_2\rangle_\pi \subseteq \langle v_1\rangle_\pi \oplus \langle v_2\rangle_\pi,$$

which by (2) has spectral measure

$$\mathrm{d}\mu_{u_1} = \mathbb{1}_{B_1} \,\mathrm{d}\mu_w = \mathrm{d}\mu_{v_1}.$$

In particular, $\mu_{u_1} \perp \mu_{v_2}$, which by (5) implies $\langle u_1\rangle_\pi \perp \langle v_2\rangle_\pi$ and so $u_1 \in \langle v_1\rangle_\pi$. Since $\mu_{u_1} = \mu_{v_1}$, property (6) implies $\langle v_1\rangle_\pi = \langle u_1\rangle_\pi \subseteq \langle v_1 + v_2\rangle_\pi$. This also implies $\langle v_2\rangle_\pi \subseteq \langle v_1 + v_2\rangle_\pi$, which concludes the proof. □

We now discuss the main properties of the non-diagonal spectral measures. These will be crucial in the next section.

Proposition 2.53 (Non-diagonal spectral measures). *For any unitary representation π of the abelian group G there exists a map*

$$\begin{aligned} \mathcal{H}_\pi \times \mathcal{H}_\pi &\longrightarrow \mathcal{M}^{\mathbb{C}}(\widehat{G}) \\ (v, w) &\longmapsto \mu_{v,w} \end{aligned}$$

sending a pair (v, w) to the spectral measure $\mu_{v,w}$ satisfying

$$\langle \pi_g v, w\rangle = \int \langle g, t\rangle \,\mathrm{d}\mu_{v,w}(t) \tag{2.22}$$

for all $g \in G$ (as in Definition 2.51) and the following additional properties.

(1) *$\mu_{v,w}$ depends linearly on $v \in \mathcal{H}_\pi$ and conjugate-linearly on $w \in \mathcal{H}_\pi$.*
(2) *$\mu_{w,v} = \overline{\mu_{v,w}}$ for all $v, w \in \mathcal{H}_\pi$.*
(3) *$\|\mu_{v,w}\| \leqslant \|v\|\|w\|$ for all $v, w \in \mathcal{H}_\pi$.*
(4) *$\langle \pi_*(f)v, w\rangle = \int \check{f} \,\mathrm{d}\mu_{v,w}$ for all $f \in L^1(G)$.*
(5) *If the unitary isomorphism in Corollary 2.12 sends $v, w \in \langle u\rangle_\pi$ to the functions $F_v, F_w \in L^2_{\mu_u}(\widehat{G})$ respectively, then $\mathrm{d}\mu_{v,w} = F_v \overline{F_w} \,\mathrm{d}\mu_u$.*
(6) *If $v = \sum_{k=1}^\infty v_k$, $w = \sum_{k=1}^\infty w_k$, and $v_k, w_k \in \langle u_k\rangle_\pi$, $\langle u_k\rangle_\pi \perp \langle u_\ell\rangle_\pi$ for all $k \neq \ell \in \mathbb{N}$ and some sequence (u_k) in $\mathcal{H}_\pi$, then we have*

$$\mu_{v,w} = \sum_{k=1}^\infty \mu_{v_k,w_k}.$$

Moreover, for a pair $(v,w) \in \mathcal{H}_\pi \times \mathcal{H}_\pi$ the property (2.22) *for all $g \in G$ (alternatively, property* (4) *for all $f \in L^1(G)$) uniquely determines the measure $\mu_{v,w}$.*

Let us highlight a particular case in the following exercise.

Essential Exercise 2.54. Let π be a unitary representation of the abelian group G. Assume the existence and properties of the non-diagonal spectral measures as in Proposition 2.53. Show for $v,w \in \mathcal{H}_\pi$ that $\mu_{v,w} = 0$ if and only if $\langle v\rangle_\pi \perp \langle w\rangle_\pi$.

PROOF OF PROPOSITION 2.53. Let $v,w \in \mathcal{H}$ and assume that a finite complex-valued measure $\mu_{v,w}$ satisfies (2.22) for all $g \in G$. For any $f \in L^1(G)$ we then have

$$\begin{aligned}\langle \pi_*(f)v, w\rangle &= \int_G f(g)\langle \pi_g v, w\rangle \,\mathrm{d}m(g)\\ &= \int_G f(g)\int_{\widehat{G}}\langle g,t\rangle \,\mathrm{d}\mu_{v,w}(t)\,\mathrm{d}m(g) = \int_{\widehat{G}} \check{f}(t)\,\mathrm{d}\mu_{v,w}(t)\end{aligned}$$

by definition of the convolution operator, property (2.22), Fubini's theorem, and the definition of the Fourier transform. This shows (4) and that $\int_{\widehat{G}} \check{f}\,\mathrm{d}\mu_{v,w}$ is uniquely determined by v,w for all $f \in L^1(G)$. Since $\widecheck{L^1(G)}$ is dense in $C_0(\widehat{G})$ by Corollary 2.5 the identification of finite signed measures on $\widehat{G}$ with linear functionals on $C_0(\widehat{G})$ in the Riesz representation theorem (see [25, Th. 7.54]) implies the uniqueness claim.

Let us next turn to the existence part of the argument. For this we first prove (5). So suppose that $v,w \in \langle u\rangle_\pi$ are sent to $F_v, F_w \in L^2_{\mu_u}(\widehat{G})$ under the unitary isomorphism in Corollary 2.12. Then

$$\big\langle \pi_g v, w\big\rangle = \big\langle M_g F_v, F_w\big\rangle = \int \langle g,t\rangle\, F_v(t)\overline{F_w(t)}\,\mathrm{d}\mu_u(t)$$

for all $g \in G$, which already implies (5). To obtain the existence we set $u = v$ and decompose $w = w_1 + w_2$ with $w_1 \in \langle v\rangle_\pi$ and $w_2 \in \langle v\rangle_\pi^\perp$, which implies

$$\big\langle \pi_g v, w\big\rangle = \big\langle \pi_g v, w_1 + w_2\big\rangle = \big\langle \pi_g v, w_1\big\rangle$$

for all $g \in G$, and so $\mu_{v,w} = \mu_{v,w_1}$ exists by the above discussion, proving (5). This also implies (3) since

$$\|\mu_{v,w}\| = \int |\mathbb{1}\overline{F_{w_1}}|\,\mathrm{d}\mu_v \leqslant \|\mathbb{1}\|_2\|F_{w_1}\|_2 = \|v\|\|w_1\| \leqslant \|v\|\|w\|$$

by the Cauchy–Schwarz inequality.

To prove linearity in $v \in \mathcal{H}_\pi$ for a fixed $w \in \mathcal{H}_\pi$ notice that

$$\begin{aligned}\left\langle\pi_g(\alpha_1v_1+\alpha_2v_2),w\right\rangle &= \alpha_1\left\langle\pi_gv_1,w\right\rangle+\alpha_2\left\langle\pi_gv_2,w\right\rangle\\ &= \alpha_1\int\langle g,t\rangle\,\mathrm{d}\mu_{v_1,w}(t)+\alpha_2\int\langle g,t\rangle\,\mathrm{d}\mu_{v_2,w}(t)\\ &= \int\langle g,t\rangle\,\mathrm{d}(\alpha_1\mu_{v_1,w}+\alpha_2\mu_{v_2,w})(t)\end{aligned}$$

for any $g\in G$, $v_1,v_2\in\mathcal{H}_\pi$ and $\alpha_1,\alpha_2\in\mathbb{C}$. By uniqueness of the spectral measures, linearity in the first argument follows. Conjugate-linearity with respect to the second argument follows in the same way, which gives (1).

To prove (2) we fix $v,w\in\mathcal{H}_\pi$. Then

$$\begin{aligned}\int\langle g,t\rangle\,\mathrm{d}\mu_{w,v}(t) &= \langle\pi_gw,v\rangle=\overline{\langle v,\pi_gw\rangle}=\overline{\langle\pi_{-g}v,w\rangle}\\ &= \overline{\int\langle -g,t\rangle\,\mathrm{d}\mu_{v,w}(t)}=\int\langle g,t\rangle\,\mathrm{d}\overline{\mu_{v,w}}(t)\end{aligned}$$

for all $g\in G$, which implies (2) by the uniqueness claim.

It remains to prove (6). Suppose first that $v=v_1+v_2$, $w=w_1+w_2$ with $v_k,w_k\in\langle u_k\rangle$ for $k=1,2$ and some $u_1,u_2\in\mathcal{H}_\pi$ with $\langle u_1\rangle_\pi\perp\langle u_2\rangle_\pi$. Then we have

$$\begin{aligned}\langle\pi_gv,w\rangle &= \langle\pi_gv_1,w_1\rangle+\langle\pi_gv_2,w_2\rangle\\ &= \int\langle g,t\rangle\,\mathrm{d}\mu_{v_1,w_1}(t)+\int\langle g,t\rangle\,\mathrm{d}\mu_{v_2,w_2}(t)\\ &= \int\langle g,t\rangle\,\mathrm{d}(\mu_{v_1,w_1}+\mu_{v_2,w_2})(t)\end{aligned}$$

for all $g\in G$, which implies $\mu_{v,w}=\mu_{v_1,w_1}+\mu_{v_2,w_2}$. This extends by induction to finite sums, and by (3) to the general case (see also the proof of Proposition 2.52(4)). □

The ease with which we were able to encode various properties of unitary representations into properties of diagonal and non-diagonal spectral measures indicates how natural spectral measures are. In the following sections, we will also see how powerful the use of spectral measures can be.

A second way to obtain the existence of the non-diagonal spectral measures is to use the *polarization identity* as outlined in the next exercise.

Exercise 2.55. Let π be a unitary representation of the abelian group G. Show that

$$\mu_{v,w}=\tfrac{1}{4}\left(\mu_{v+w}-\mu_{v-w}+\mathrm{i}\mu_{v+\mathrm{i}w}-\mathrm{i}\mu_{v-\mathrm{i}w}\right)$$

defines the non-diagonal spectral measure for every $v,w\in\mathcal{H}_\pi$.

Exercise 2.56. Let π be a unitary representation of the abelian group G. Analyze the circumstances under which $v,w\in\mathcal{H}_\pi$ satisfy $\|\mu_{v,w}\|=\|v\|\|w\|$.

2.5.1 Containment

In this section we use spectral measures to give a complete characterisation of containment for cyclic unitary representations (thus for now avoiding questions concerning multiplicity, which we postpone to Section 2.7). The following generalizes Exercise 2.14.

Proposition 2.57 (Containment). *Let π be a cyclic representation with generator $v \in \mathcal{H}_\pi$ and let ρ be a cyclic representation with generator $w \in \mathcal{H}_\rho$ of the abelian group G. Then $\pi < \rho$ if and only if $\mu_v^\pi \ll \mu_w^\rho$.*

PROOF. We assume first that $\pi < \rho$. Simplifying the notation we may suppose that $\mathcal{H}_\pi = \langle v \rangle_\rho \subseteq \mathcal{H}_\rho = \langle w \rangle_\rho$ with $\pi = \rho|_{\mathcal{H}_\pi}$. By Proposition 2.52(2) this implies $\mu_v \ll \mu_w$ as claimed.

So assume now that $\mu_v^\pi \ll \mu_w^\rho$ and let

$$F = \left(\frac{\mathrm{d}\mu_v^\pi}{\mathrm{d}\mu_w^\rho} \right)^{\frac{1}{2}}.$$

Using Corollary 2.12 we assume that $\mathcal{H}_\rho = L^2(\widehat{G}, \mu_w^\rho)$. Since μ_v^π is a finite measure, we have $F^2 \in L^1(\widehat{G}, \mu_w^\rho)$ and so $F \in \mathcal{H}_\rho = L^2(\widehat{G}, \mu_w^\rho)$. Hence we have

$$\langle M_g F, F \rangle = \int \langle g, t \rangle\, F(t)^2 \,\mathrm{d}\mu_w^\rho(t) = \int \langle g, t \rangle \,\mathrm{d}\mu_v^\pi(t) = \langle \pi_g v, v \rangle$$

for all $g \in G$ by Corollary 2.12. Applying Proposition 1.65 we obtain $\mathcal{H}_\pi < \mathcal{H}_\rho$ as desired. □

2.6 Functional Calculus

As we have used many times, we note that a unitary representation π of G gives rise to a module structure on $\mathcal{H}_\pi$ for the Banach algebra $L^1(G)$. Also notice that for the abelian group G we have already seen the Banach algebra homomorphism

$$\widehat{\cdot}\colon L^1(G) \longrightarrow C_0(\widehat{G}) \subseteq \mathscr{L}^\infty(\widehat{G}).$$

Using the non-diagonal spectral measures from the previous section, we extend here the scalars of the module $\mathcal{H}_\pi$ to scalars in $\mathscr{L}^\infty(\widehat{G})$.

Proposition 2.58 (Functional calculus for $\mathscr{L}^\infty(\widehat{G})$). *For any unitary representation π of the abelian group G we have a module structure on $\mathcal{H}_\pi$ for the algebra $\mathscr{L}^\infty(\widehat{G})$ that extends the module structure for $L^1(G)$. More formally, for any $F \in \mathscr{L}^\infty(\widehat{G})$ there exists a bounded operator $\pi_{\mathrm{FC}}(F)$ on $\mathcal{H}_\pi$ that depends linearly on F and satisfies*

(1) $\|\pi_{\mathrm{FC}}(F)\|_{\mathrm{op}} \leqslant \|F\|_\infty$, *and if a measurable subset* $B \subseteq \widehat{G}$ *has the property that* $\mu_v(\widehat{G} \smallsetminus B) = 0$ *for all* $v \in \mathcal{H}_\pi$, *then we also have*

$$\|\pi_{\mathrm{FC}}(F)\|_{\mathrm{op}} \leqslant \|F\|_{B,\infty} = \sup\{|F(t)| \mid t \in B\}.$$

(2) $\pi_{\mathrm{FC}}(F)^* = \pi_{\mathrm{FC}}(\overline{F})$,
(3) $\pi_{\mathrm{FC}}(\check{f}) = \pi_*(f)$ *for all* $f \in L^1(G)$, *and*
(4) $\pi_{\mathrm{FC}}(F_1)\pi_{\mathrm{FC}}(F_2) = \pi_{\mathrm{FC}}(F_1F_2)$ *for* $F_1, F_2 \in \mathscr{L}^\infty(\widehat{G})$.
(5) *If* ρ *is a unitary representation of* G *and* $B\colon \mathcal{H}_\pi \to \mathcal{H}_\rho$ *is a bounded intertwining operator, then* $B \circ \pi_{\mathrm{FC}}(F) = \rho_{\mathrm{FC}}(F) \circ B$ *for* $F \in \mathscr{L}^\infty(\widehat{G})$.
(6) *If* $\mathcal{H}_\pi \cong L^2_\mu(X)$ *for a finite measure* μ *on* $X = \widehat{G} \times \mathbb{N}$ *as in the spectral theorem (Corollary 2.13), then* $\pi_{\mathrm{FC}}(F)$ *corresponds under the isomorphism to the multiplication operator* M_F *on* $L^2_\mu(X)$.

Moreover, for $F \in \mathscr{L}^\infty(\widehat{G})$ *the operator* $\pi_{\mathrm{FC}}(F)$ *is uniquely characterized by the formula*

$$\langle \pi_{\mathrm{FC}}(F)v, w\rangle = \int_{\widehat{G}} F \,\mathrm{d}\mu_{v,w} \tag{2.23}$$

for all $v, w \in \mathcal{H}_\pi$.

Proof. Let $B \subseteq \widehat{G}$ be as in (1) (for example, $B = \widehat{G}$). By Proposition 2.53(5)–(6) or Exercise 2.55, the assumption on B also implies that

$$\mu_{v,w}(\widehat{G} \smallsetminus B) = 0$$

for all $v, w \in \mathcal{H}_\pi$. We are going to define $\pi_{\mathrm{FC}}(F)$ for $F \in \mathscr{L}^\infty(\widehat{G})$ using (2.23) and the Fréchet–Riesz representation theorem. In fact, by Proposition 2.53 the map

$$\mathcal{H}_\pi \times \mathcal{H}_\pi \ni (v, w) \longmapsto \int F \,\mathrm{d}\mu_{v,w} = \int_B F \,\mathrm{d}\mu_{v,w}$$

depends linearly on $v \in \mathcal{H}_\pi$, conjugate-linearly on w, and satisfies the estimate

$$\left| \int_B F \,\mathrm{d}\mu_{v,w} \right| \leqslant \|F\|_{B,\infty} \|\mu_{v,w}\| \leqslant \|F\|_{B,\infty} \|v\| \|w\|$$

by Proposition 2.53(3). This implies that

$$\mathcal{H}_\pi \ni w \longmapsto \overline{\int F \,\mathrm{d}\mu_{v,w}}$$

is a bounded linear functional, which therefore has the form $w \mapsto \langle w, v_F\rangle$ for some uniquely determined $v_F \in \mathcal{H}_\pi$ with $\|v_F\| \leqslant \|F\|_{B,\infty}\|v\|$. Equivalently, we have $\langle v_F, w\rangle = \int F \,\mathrm{d}\mu_{v,w}$. Using the linearity of $\mu_{v,w}$ in v, we also see that v_F depends linearly on $v \in \mathcal{H}_\pi$ and so uniquely defines a bounded operator

$$\mathcal{H}_\pi \ni v \longmapsto \pi_{\mathrm{FC}}(F)v = v_F \in \mathcal{H}_\pi$$

with $\|\pi_{\mathrm{FC}}(F)\|_{\mathrm{op}} \leqslant \|F\|_{B,\infty}$ as claimed in (1). Note that a function F in $\mathscr{L}^\infty(B)$ can be extended trivially (or in any measurable way) to all of $\widehat{G}$, and also gives rise to the operator $\pi_{\mathrm{FC}}(F)$ independent of the extension.

Clearly $\int F \,\mathrm{d}\mu_{v,w}$ depends, for a given $v, w \in \mathcal{H}_\pi$, linearly on F, which implies that $\pi_{\mathrm{FC}}(F)$ depends linearly on $F \in \mathscr{L}^\infty(\widehat{G})$.

Given $F \in \mathscr{L}^\infty(\widehat{G})$, we have

$$\begin{aligned}\langle \pi_{\mathrm{FC}}(F)^* v, w\rangle &= \langle v, \pi_{\mathrm{FC}}(F) w\rangle = \overline{\langle \pi_{\mathrm{FC}}(F) w, v\rangle}\\ &= \int \overline{F} \,\mathrm{d}\overline{\mu_{w,v}} = \int \overline{F} \,\mathrm{d}\mu_{v,w} = \langle \pi_{\mathrm{FC}}(\overline{F}) v, w\rangle\end{aligned}$$

by Proposition 2.53(2) for any $v, w \in \mathcal{H}_\pi$, which implies the conjugation formula $\pi_{\mathrm{FC}}(F)^* = \pi_{\mathrm{FC}}(\overline{F})$ as claimed in (2).

For $F = \check{f}$ with $f \in L^1(G)$ we have

$$\left\langle \pi_{\mathrm{FC}}(\check{f}) v, w\right\rangle = \int \check{f} \,\mathrm{d}\mu_{v,w} = \langle \pi_*(f) v, w\rangle$$

for all $v, w \in \mathcal{H}_\pi$, by definition of the spectral measure $\mu_{v,w}$ in Proposition 2.53. Therefore $\pi_{\mathrm{FC}}(\check{f}) = \pi_*(f)$ as claimed in (3).

Next we prove (6), and assume first that $\mathcal{H}_\pi = \langle u\rangle_\pi \cong L^2_\mu(\widehat{G})$ as in the cyclic spectral theorem (Corollary 2.12). By Proposition 2.53(5) for $v, w \in \langle u\rangle_\pi$ corresponding to $F_v, F_w \in L^2_\mu(\widehat{G})$ we have $\mathrm{d}\mu_{v,w} = F_v \overline{F_w} \,\mathrm{d}\mu_u$, and so

$$\langle \pi_{\mathrm{FC}}(F) v, w\rangle_{\mathcal{H}_\pi} = \int F F_v \overline{F_w} \,\mathrm{d}\mu_u = \langle M_F F_v, F_w\rangle_{L^2_\mu(\widehat{G})}$$

for all $v, w \in \langle u\rangle_\pi$ and all $F \in \mathscr{L}^\infty(\widehat{G})$. This proves the claim for cyclic representations. However, by linearity and continuity of all operators involved, this extends to the general case.

To prove (4), we now simply apply the spectral theorem in Corollary 2.13 and property (6) proved above. Under the unitary intertwining isomorphism

$$\mathcal{H}_\pi \cong L^2_\mu(X)$$

for $X = \widehat{G} \times \mathbb{N}$ we have that $\pi_{\mathrm{FC}}(F_1), \pi_{\mathrm{FC}}(F_2), \pi_{\mathrm{FC}}(F_1 F_2)$ correspond to the operators $M_{F_1}, M_{F_2}, M_{F_1 F_2}$ respectively, satisfying $M_{F_1} M_{F_2} = M_{F_1 F_2}$.

It remains to prove (5). So suppose that ρ is a unitary representation, and let $v \in \mathcal{H}_\pi$ and $w \in \mathcal{H}_\rho$. Then

$$\int \langle g, t\rangle \,\mathrm{d}\mu^\rho_{Bv,w}(t) = \langle \rho_g B v, w\rangle = \langle B \pi_g v, w\rangle = \langle \pi_g v, B^* w\rangle = \int \langle g, t\rangle \,\mathrm{d}\mu^\pi_{v,B^*w}(t)$$

for all $g \in G$, which implies that $\mu^\rho_{Bv,w} = \mu^\pi_{v,B^*w}$ by uniqueness of spectral measures. We note that $\mu^\rho_{Bv,w}$ is a spectral measure defined by ρ, and μ^π_{v,B^*w} is a spectral measure defined by π. For $F \in \mathscr{L}^\infty(\widehat{G})$ we then have

$$\begin{aligned}\langle B\pi_{\mathrm{FC}}(F)v, w\rangle &= \langle \pi_{\mathrm{FC}}(F)v, B^*w\rangle \\ &= \int F \,\mathrm{d}\mu^{\pi}_{v,B^*w} \\ &= \int F \,\mathrm{d}\mu^{\rho}_{Bv,w} \\ &= \langle \rho_{\mathrm{FC}}(F)Bv, w\rangle\,.\end{aligned}$$

As this holds for all $v \in \mathcal{H}_\pi$ and $w \in \mathcal{H}_\rho$, property (5), and hence the proposition, follow. □

The following exercises should help the reader to become more familiar with the concepts introduced here.

Essential Exercise 2.59 (Functional calculus and invariance). Let π be a unitary representation of the abelian group G. Show that a closed subspace $\mathcal{V}$ of $\mathcal{H}_\pi$ is invariant under π if and only if it is invariant under $\pi_{\mathrm{FC}}(\mathscr{L}^\infty(\widehat{G}))$.

Essential Exercise 2.60 (Functional calculus & spectral measures). Let π be a unitary representation of the abelian group G. Show that

$$\mathrm{d}\mu_{\pi_{\mathrm{FC}}(F)v} = |F|^2 \,\mathrm{d}\mu_v, \; \mathrm{d}\mu_{\pi_{\mathrm{FC}}(F)v,w} = F \,\mathrm{d}\mu_{v,w}$$

for $v, w \in \mathcal{H}_\pi$ and $F \in \mathscr{L}^\infty(\widehat{G})$.

Exercise 2.61. Let π be a unitary representation of the abelian group G. Show that

$$\pi_g \pi_{\mathrm{FC}}(\mathbb{1}_{\{t\}}) = \langle g, t\rangle \pi_{\mathrm{FC}}(\mathbb{1}_{\{t\}})$$

for all elements g in G and t in $\widehat{G}$.

2.6.1 Projection-Valued Measures

As a special case of the functional calculus above, we obtain for a unitary representation π of the abelian group G a *projection-valued measure* defined by

$$\mathcal{B}(\widehat{G}) \ni B \longmapsto \Pi_B = \pi_{\mathrm{FC}}(\mathbb{1}_B),$$

which gives for every Borel subset $B \subseteq \widehat{G}$ an orthogonal projection Π_B onto the subspace of $\mathcal{H}_\pi$ corresponding to '*the generalized sum of eigenspaces for all characters in B*' as we will explain now.

The projection-valued measure allows us to reconstruct the functional calculus and, in particular, to reconstruct the unitary representation. In fact if $F \in \mathscr{L}^\infty(\widehat{G})$ and $F_0 = \sum_{j=1}^n c_j \mathbb{1}_{B_j}$ is an approximation of F by a simple function, then

$$\Big\|\pi_{\rm FC}(F)-\sum_{j=1}^{n}c_j\varPi_{B_j}\Big\|_{\rm op}=\|\pi_{\rm FC}(F-F_0)\|_{\rm op}\leqslant\|F-F_0\|_\infty$$

shows that $\pi_{\rm FC}(F)$ can be approximated by the finite sums

$$\sum_{j=1}^{n}c_j\varPi_{B_j}=\pi_{\rm FC}(F_0).$$

We make the definition

$$\int_{\widehat{G}}F_0\,\mathrm{d}\varPi=\sum_{j=1}^{n}c_j\varPi_{B_j}.$$

With this interpretation we can now define

$$\int_{\widehat{G}}F\,\mathrm{d}\varPi=\pi_{\rm FC}(F)=\lim_{F_0:\|F_0-F\|_\infty\to 0}\pi_{\rm FC}(F_0),$$

where convergence holds in the uniform topology.

Also notice that for $g\in G$ we may define a function $F_g\in\mathscr{L}^\infty(\widehat{G})$ by setting $F_g(t)=\langle g,t\rangle$. Since

$$\big\langle\pi_{\rm FC}(F_g)v,w\big\rangle=\int_{\widehat{G}}\langle g,t\rangle\,\mathrm{d}\mu_{v,w}(t)=\big\langle\pi_g v,w\big\rangle$$

for all $v,w\in\mathcal{H}_\pi$, this has the property that

$$\pi_{\rm FC}(F_g)=\pi_g=\int_{\widehat{G}}\langle g,\cdot\rangle\,\mathrm{d}\varPi. \tag{2.24}$$

As mentioned above, this suggests that for any measurable $B\subseteq\widehat{G}$ we may think of $\operatorname{Im}\varPi_B=\varPi_B\mathcal{H}_\pi$ as the generalized sum of the eigenspaces for characters[†] χ_t with $t\in B$. The generalization concerns the integral and the fact that, strictly speaking, there may not be a single eigenvector, but only approximate eigenvectors. Similarly, $\pi_{\rm FC}(F)$ can now be interpreted as the operator that multiplies (generalized) eigenvectors in $\mathcal{H}_\pi$ for eigenvalue $t\in\widehat{G}$ by $F(t)$. Informally, we may also write (2.24) in the form

$$\pi_g v=\int_{\widehat{G}}\langle g,t\rangle\varPi_t v\,\mathrm{d}t$$

and think of $\varPi_t v$ as the projection of $v\in\mathcal{H}_\pi$ to the eigenspace in $\mathcal{H}_\pi$ corresponding to the character χ_t. However, as the latter may be trivial for all $t\in\widehat{G}$ this formally makes no sense, and this is the reason we prefer the notation of (2.24). This provides a useful viewpoint, but does not provide additional for-

[†] Whenever we have a collection of operators that commute with each other, an eigenvalue is really a function on the collection of operators, and here—in the context of unitary representations—a unitary character on the group.

mal properties. We refer to [25, Sec. 12.7] for a more thorough discussion of spectral-valued measures.

Exercise 2.62. If B_1 and B_2 are disjoint measurable sets, show that Π_{B_1} and Π_{B_2} are orthogonal projections.

Essential Exercise 2.63. Let π be a unitary representation of the abelian group G. Let $B_1, B_2, \ldots$ be a sequence of pairwise disjoint measurable subsets of $\widehat{G}$ and define $B = \bigsqcup_{n=1}^{\infty} B_n$. Show that

$$\Pi_B = \pi_{\mathrm{FC}}(\mathbb{1}_B) = \sum_{n=1}^{\infty} \pi_{\mathrm{FC}}(\mathbb{1}_{B_n}) = \sum_{n=1}^{\infty} \Pi_{B_n},$$

where the convergence holds in the strong operator topology. Find an example where the convergence of the series fails in the uniform topology.

2.7 Spectral Theory and Multiplicity

In this section we will revisit the spectral theorem for the abelian group G, and derive with some more work a version that presents all the information regarding multiplicity.

2.7.1 Maximal Spectral Type

Proposition 2.64 (Maximal spectral type). *Let π be a unitary representation of the abelian group G. Then there exists a vector $v_{\max} \in \mathcal{H}_\pi$ with spectral measure $\mu_{\max} = \mu_{v_{\max}}$ with the property that $\mu_v \ll \mu_{\max}$ for any $v \in \mathcal{H}_\pi$. The measure class of $\mu_{\max}$ is uniquely characterized by this property, and is called the* maximal spectral type. *Moreover, given some $v_0 \in \mathcal{H}_\pi$ the vector $v_{\max}$ can be chosen so that $v_0 \in \langle v_{\max} \rangle_\pi$.*

PROOF. We apply the spectral theorem in the form of Corollary 2.13, and find orthonormal vectors $v_1, v_2, \ldots$ of $\mathcal{H}_\pi$ such that

$$\mathcal{H}_\pi = \bigoplus_{n \geqslant 1} \langle v_n \rangle_\pi \cong \bigoplus_{n \geqslant 1} L^2_{\mu_n}(\widehat{G})$$

with $\mu_n = \mu_{v_n}$ for all $n \geqslant 1$. We now define $v_{\max} = \sum_{n \geqslant 1} \frac{1}{n} v_n$ and obtain from Proposition 2.52(4) that the spectral measure $\mu_{\max} = \mu_{v_{\max}}$ is given by

$$\mu_{\max} = \sum_{n \geqslant 1} \frac{1}{n^2} \mu_n.$$

For $u \in \mathcal{H}_\pi$ we then have $u = \sum_{n \geqslant 1} u_n$ with $u_n \in \langle v_n \rangle_\pi$ for $n \geqslant 1$, which also implies that $\mu_{u_n} \ll \mu_n \ll \mu_{\max}$ for all $n \geqslant 1$. It follows from Proposition 2.52(4) and the definition of absolute continuity that

$$\mu_u = \sum_{n \geqslant 1} \mu_{u_n} \ll \mu_{\max}.$$

As $u \in \mathcal{H}_\pi$ was arbitrary, it follows that $\mu_{\max}$ is indeed a maximal spectral measure for π.

We now prove the claimed uniqueness of the measure class of $\mu_{\max}$. Suppose that $v \in \mathcal{H}_\pi$ is another vector with the property that $\mu_u \ll \mu_v$ for all $u \in \mathcal{H}_\pi$. Setting $u = v_{\max}$ gives $\mu_{\max} = \mu_{v_{\max}} \ll \mu_v$. Setting $u = v$ in the discussion above gives $\mu_v \ll \mu_{\max}$, and hence $\mu_{\max}$ and μ_v lie in the same measure class.

Now let $v_0 \in \mathcal{H}_\pi$ be arbitrary, and write $\mathcal{V} = \langle v_0 \rangle_\pi^\perp$, so that we may write $\mathcal{H}_\pi = \langle v_0 \rangle_\pi \oplus \mathcal{V}$. We apply the above to find a vector $w \in \mathcal{V}$ of maximal spectral type for the representation $\pi|_{\mathcal{V}}$. Next apply the Lebesgue decomposition theorem (see [25, Sec. 3.1.3]) to μ_w and μ_{v_0} to write μ_w as a sum $\mu_w = \mu_{\text{abs}} + \mu_{\text{sing}}$, where $\mu_{\text{abs}} \ll \mu_{v_0}$ and $\mu_{\text{sing}} = \mu_w|_B \perp \mu_{v_0}$ for some measurable subset $B \subseteq \widehat{G}$. By the cyclic spectral theorem (Corollary 2.12) there exists some $w_{\text{sing}} \in \langle w \rangle_\pi \subseteq \mathcal{V}$ corresponding to $\mathbb{1}_B \in L^2_{\mu_w}(\widehat{G})$, which by Proposition 2.52(2) has spectral measure $\mathrm{d}\mu_{w_{\text{sing}}} = \mathbb{1}_B \,\mathrm{d}\mu_w = \mathrm{d}\mu_{\text{sing}} \perp \mu_{v_0}$. Equivalently, we may set $w_{\text{sing}} = \pi_{\text{FC}}(\mathbb{1}_B) w$ and apply Exercises 2.59 and 2.60. We define $v_{\max} = v_0 + w_{\text{sing}}$ and note that

$$\mu_{v_{\max}} = \mu_{v_0} + \mu_{\text{sing}}$$

by Proposition 2.52(3) and $\langle w \rangle_\pi \subseteq \mathcal{V} = \langle v_0 \rangle_\pi^\perp$. Proposition 2.52(7) also implies that $\langle v_{\max} \rangle_\pi = \langle v_0 \rangle_\pi \oplus \langle w_{\text{sing}} \rangle_\pi$, and in particular $\langle v_{\max} \rangle_\pi$ contains v_0. For any $u = u_0 + u_1 \in \mathcal{H}_\pi = \langle v_0 \rangle_\pi \oplus \mathcal{V}$ with $u_0 \in \langle v_0 \rangle_\pi$ and $u_1 \in \mathcal{V}$ we again have, by various parts of Proposition 2.52, that

$$\mu_u = \mu_{u_0} + \mu_{u_1} \ll \mu_{v_0} + \mu_w = \mu_{v_0} + \mu_{\text{abs}} + \mu_{\text{sing}} \ll \mu_{v_0} + \mu_{\text{sing}} = \mu_{v_{\max}}.$$

This concludes the proof. □

Corollary 2.65 (Spectral theorem with descending measures). *Let π be a unitary representation of the abelian group G. Then there exists a (possibly finite) sequence of vectors $u_1, u_2, \ldots$ in $\mathcal{H}_\pi$ such that $\mathcal{H}_\pi = \bigoplus_{n \geqslant 1} \langle u_n \rangle_\pi$ and $\mu_{\max} = \mu_{u_1} \gg \mu_{u_2} \gg \cdots$.*

Proof. Let $v_1, v_2, \ldots$ be a basis of $\mathcal{H}_\pi$. Applying Proposition 2.64 gives a vector $u_1 = v_{\max}$ whose spectral measure represents the maximal spectral type and whose cyclic representation contains v_1.

Now project the vector v_2 to $\langle u_1 \rangle_\pi^\perp$ and apply Proposition 2.64 above again to find a vector u_2 with the property that $v_2 \in \langle u_1 \rangle_\pi \oplus \langle u_2 \rangle_\pi$ and such that μ_{u_2} represents the maximal spectral type of $\pi|_{\langle u_1 \rangle_\pi^\perp}$.

Proceeding like this we can make sure that each of the basis vectors v_n in $\mathcal{H}_\pi$ belongs to $V_n = \langle u_1 \rangle_\pi \oplus \langle u_2 \rangle_\pi \oplus \cdots \oplus \langle u_n \rangle_\pi$ and that the maximal spectral type on the orthogonal complement $V_n^\perp$ is realized by u_{n+1} for $n \in \mathbb{N}$. This gives the corollary. □

Essential Exercise 2.66. Let π be a unitary representation of the abelian group G, and let $F \in \mathscr{L}^\infty(\widehat{G})$. Show that $\|\pi_{\mathrm{FC}}(F)\|_{\mathrm{op}}$ is equal to the essential supremum norm of F with respect to $\mu_{\max}$.

Weakening the notion of maximal spectral measure, we also obtain the following topological object related to a unitary representation. For this, note that the support of two measures in the same measure class agree, which makes the following independent of the choice of the maximal measure.

Definition 2.67 (Support of π). Let π be a unitary representation of the abelian group G. Then the support $\operatorname{supp}(\pi) \subseteq \widehat{G}$ of π is defined as the support of the maximal spectral measure from Proposition 2.64.

Exercise 2.68. Let π be a unitary representation of an abelian group. Show that any two vectors in $\mathcal{H}_\pi$ of maximal spectral type (that is, vectors whose spectral measures are maximal spectral measures) have isomorphic cyclic representations.

2.7.2 Spectral Multiplicity

Theorem 2.69 (Spectral theorem with multiplicities). *Let π be a unitary representation of the abelian group G. Then there exists a collection*

$$(\mu_1, \mu_2, \ldots, \mu_\infty) = (\mu_n)_{n \in \mathbb{N} \cup \{\infty\}}$$

of finite measures on $\widehat{G}$ that satisfy the following properties and so completely describe the unitary representation π.

(1) *For $m \neq n \in \mathbb{N} \cup \{\infty\}$ the measures $\mu_m \perp \mu_n$ are singular. In fact, if $\mu_{\max}$ is a maximal spectral measure as in Proposition 2.64 then there is a measurable partition $\{P_n \mid n \in \mathbb{N} \cup \{\infty\}\}$ of $\widehat{G}$ with the property that $\mu_n = \mu_{\max}|_{P_n}$ for all $n \in \mathbb{N} \cup \{\infty\}$.*
(2) *The representation π is unitarily isomorphic to the multiplication representation on $\bigoplus_{n \in \mathbb{N}} \big(L^2_{\mu_n}(P_n)\big)^n \oplus \big(L^2_{\mu_\infty}(P_\infty)\big)^\infty$, where $\big(L^2_{\mu_\infty}(P_\infty)\big)^\infty$ denotes the Hilbert space direct sum of countably many copies of $L^2_{\mu_\infty}(P_\infty)$.*
(3) *The isomorphism in (2) is not in general canonical. Nonetheless, the subspace $\mathcal{H}_\pi^{(n)}$ of $\mathcal{H}_\pi$ corresponding to $\big(L^2_{\mu_n}(P_n)\big)^n$ for $n \in \mathbb{N} \cup \{\infty\}$ can be defined using spectral measures alone.*
(4) *The measure class of μ_n for $n \in \mathbb{N} \cup \{\infty\}$ is uniquely determined by (1) and (2).*

PROOF OF (1) AND (2). Let $u_1, u_2, \dots$ be as in Corollary 2.65. We claim that it is possible to replace $u_2, u_3, \dots$ by another sequence $w_2, w_3, \dots$ without changing their respective cyclic subspaces so that $\frac{\mathrm{d}\mu_{w_n}}{\mathrm{d}\mu_{\max}} \in \{0,1\}$ almost everywhere with respect to $\mu_{\max} = \mu_{u_1}$. To see this, fix $n \geqslant 2$, let $F_n = \frac{\mathrm{d}\mu_{u_n}}{\mathrm{d}\mu_{\max}}$ and

$$B_n = \{t \in \widehat{G} \mid F_n(t) > 0\}.$$

Then the function

$$\widetilde{F_n}(t) = \begin{cases} F_n^{-\frac{1}{2}}(t) & t \in B_n, \\ 0 & t \in \widehat{G} \smallsetminus B_n \end{cases}$$

satisfies $\widetilde{F_n} \in L^2_{\mu_{u_n}}(\widehat{G})$ since

$$\int_{\widehat{G}} \widetilde{F_n}^2 \,\mathrm{d}\mu_{u_n} = \int_{B_n} F_n^{-1} \,\mathrm{d}\mu_{u_n} = \int_{B_n} F_n^{-1} F_n \,\mathrm{d}\mu_{\max} = \mu_{\max}(B_n) < \infty.$$

We see from Proposition 2.52(2) that the vector $w_n \in \langle u_n \rangle_\pi$ corresponding to $\widetilde{F_n} \in L^2_{\mu_{u_n}}(\widehat{G})$ has spectral measure

$$\mu_{w_n} = |\widetilde{F_n}|^2 \,\mathrm{d}\mu_{u_n} = |\widetilde{F_n}|^2 F_n \,\mathrm{d}\mu_{\max} = \mu_{\max}|_{B_n},$$

since $|\widetilde{F_n}|^2 F_n = \mathbb{1}_{B_n}$. Moreover, μ_{w_n} defines the same measure class as μ_{u_n} (since $\widetilde{F_n} > 0$ μ_{u_n}-almost everywhere). Hence by Proposition 2.52(6) we see that $\langle w_n \rangle_\pi = \langle u_n \rangle_\pi$. This proves the claim.

So we suppose now that the vectors $u_1, u_2, \dots$ are as in Corollary 2.65 with $\mu_{\max} = \mu_{u_1}$, and in addition that $\frac{\mathrm{d}\mu_{u_n}}{\mathrm{d}\mu_{\max}} = \mathbb{1}_{B_n}$ for some measurable set $B_n \subseteq \widehat{G}$ for all $n \geqslant 2$. Since $\mu_{\max} = \mu_{u_1} \gg \mu_{u_2} \gg \cdots$ and

$$\mu_{u_n}(\widehat{G} \smallsetminus B_n) = 0$$

for all $n \geqslant 2$, we have

$$0 = \mu_{u_m}(\widehat{G} \smallsetminus B_n) = \mu_{\max}(B_m \smallsetminus B_n)$$

by definition of absolute continuity for all $m \geqslant n$. We now replace B_2 by the set $B_1 \cap B_2$, B_3 by the set $B_1 \cap B_2 \cap B_3$ and so on, without changing their defining properties. Hence we may also assume that these subsets comprise a descending chain

$$\widehat{G} = B_1 \supseteq B_2 \supseteq \cdots$$

of measurable subsets with the property that $\mu_{u_n} = \mu_{\max}|_{B_n}$ for all $n \in \mathbb{N}$. We now define $\mu_n = \mu_{\max}|_{P_n}$ where $P_n = B_n \smallsetminus B_{n+1}$ for all $n \in \mathbb{N}$, and $\mu_\infty = \mu|_{P_\infty}$ where $P_\infty = \bigcap_{n \geqslant 1} B_n$. This gives the unitary intertwining isomorphism

$$L^2_{\mu_{u_n}}(\widehat{G}) = L^2_{\mu_{u_n}}(B_n) \cong \bigoplus_{k\in\{n,\dots,\infty\}} L^2_{\mu_k}(P_k),$$

where the projection map from the left to the factors on the right is simply multiplication by the characteristic function of P_k for $n \leqslant k \leqslant \infty$. Together with Corollary 2.65 this gives

$$\mathcal{H}_\pi \cong \bigoplus_{n\geqslant 1} L^2_{\mu_{u_n}}(\widehat{G}) \cong \bigoplus_{n\geqslant 1}\Big(\bigoplus_{k\in\{n,\dots,\infty\}} L^2_{\mu_k}(P_k)\Big) \cong \bigoplus_{n\in\mathbb{N}\cup\{\infty\}} \big(L^2_{\mu_n}(P_n)\big)^n.$$

The first isomorphism is intertwining by Corollary 2.65, the second is realized by multiplication operators and so commutes with the multiplication representation, and the final isomorphism is simply a permutation of the invariant factors. This proves (1) and (2). □

2.7.3 The Multiplicity Subspaces

Next we will prove that the multiplicity n subspace $\mathcal{H}_\pi^{(n)}$ in Theorem 2.69(3) is canonical for every $n \in \mathbb{N}\cup\{\infty\}$. The canonical description of $\mathcal{H}_\pi^{(n)}$ uses spectral measures, and will be easy to prove using our knowledge of them together with some measure-theoretic constructions.

For every $n \in \mathbb{N}$ the subspace $\mathcal{H}_\pi^{(\geqslant n)}$ is given by

$$\mathcal{H}_\pi^{(\geqslant n)} = \{v \in \mathcal{H}_\pi \mid v \text{ has spectral multiplicity at least } n\},$$

where we say that $v \in \mathcal{H}_\pi$ has *spectral multiplicity at least* n if there exist n vectors

$$v_1 = v, v_2, \dots, v_n \in \mathcal{H}_\pi$$

such that $\langle v_k\rangle_\pi \perp \langle v_\ell\rangle_\pi$ for $1 \leqslant k < \ell \leqslant n$ and $\mu^\pi_{v_k} = \mu^\pi_v$ for $k \leqslant n$. We also define

$$\mathcal{H}_\pi^{(n)} = \mathcal{H}_\pi^{(\geqslant n)} \cap \big(\mathcal{H}_\pi^{(\geqslant n+1)}\big)^\perp \tag{2.25}$$

as the relative orthogonal complement of $\mathcal{H}_\pi^{(\geqslant n+1)}$ within $\mathcal{H}_\pi^{(\geqslant n)}$. Similarly, for $n = \infty$ the *infinite multiplicity subspace* $\mathcal{H}_\pi^{(\infty)}$ is given by

$$\mathcal{H}_\pi^{(\infty)} = \bigcap_{n\in\mathbb{N}} \mathcal{H}_\pi^{(\geqslant n)}. \tag{2.26}$$

We note that the above definition of $\mathcal{H}_\pi^{(n)}$ for $n \in \mathbb{N}\cup\{\infty\}$ is canonical, since spectral measures are uniquely determined by a given vector and the unitary representation.

Before starting with the argument for part (3) of the theorem, we wish to set up some helpful further notation. It will be convenient to write $\mathcal{V}_\infty = \ell^2(\mathbb{N})$

and $\mathcal{V}_n = \mathbb{C}^n$ for $n \in \mathbb{N}$, and in either case $\|\cdot\|_{\mathcal{V}}$ for its standard Hilbert space norm and $\langle \cdot, \cdot \rangle_{\mathcal{V}}$ for its inner product. For any $n \in \mathbb{N} \cup \{\infty\}$ we may and will think of an element $v = (v_j)_j \in \left(L^2_{\mu_n}(P_n)\right)^n$ as a $\mathcal{V}_n$-valued function

$$P_n \ni t \longmapsto (v_j(t))_j \in \mathcal{V}_n$$

satisfying $\|v\|^2 = \sum_j \|v_j\|^2 = \int_{P_n} \|(v_j(t))_j\|^2_{\mathcal{V}} \, \mathrm{d}\mu_n(t)$ (by a simple form of Fubini's theorem). We denote the space of measurable, square-integrable, $\mathcal{V}_n$-valued functions by

$$L^2_{\mu_n}(P_n, \mathcal{V}_n) = \left\{ v \colon P_n \to \mathcal{V}_n \;\middle|\; \|v\|^2 = \int_{P_n} \|v(t)\|^2_{\mathcal{V}} \, \mathrm{d}\mu_n(t) < \infty \right\}.$$

Using this notation, parts (1) and (2) of the theorem, and the fact that the definition of $\mathcal{H}_\pi^{(n)}$ in (2.25) and (2.26) is canonical, we may and will assume that

$$\mathcal{H}_\pi = \bigoplus_{n \in \mathbb{N} \cup \{\infty\}} L^2_{\mu_n}(P_n, \mathcal{V}_n) \tag{2.27}$$

and that the representation is given by the multiplication representation. Using the identity $\mu_n = \mu_{\max}|_{P_n}$ for the measurable partition

$$\{P_n \mid n \in \mathbb{N} \cup \{\infty\}\}$$

of $\widehat{G}$, we may even identify an element $v \in \mathcal{H}_\pi$ with a function on $\widehat{G}$, which satisfies $v|_{P_n} \in L^2_{\mu_n}(P_n, \mathcal{V}_n)$ for all $n \in \mathbb{N} \cup \{\infty\}$ and for which

$$\widehat{G} \ni t \longmapsto \|v(t)\|_{\mathcal{V}}$$

is square-integrable with respect to $\mu_{\max}$. For $n \in \mathbb{N} \cup \{\infty\}$ and a measurable subset $B \subseteq P_n$ we will also write $L^2_{\mu_n}(B, \mathcal{V}_n)$ for the subspace of $L^2_{\mu_n}(P_n, \mathcal{V}_n)$ consisting of functions that vanish on $P_n \smallsetminus B$.

Using these identifications it remains to prove that $\mathcal{H}_\pi^{(n)} = L^2_{\mu_n}(P_n, \mathcal{V}_n)$ for $n \in \mathbb{N} \cup \{\infty\}$. For this we need the following measurable construction.

Lemma 2.70 (Measurable selection of orthonormal basis). *Let $n \in \mathbb{N}$ and $n \leqslant m \in \mathbb{N} \cup \{\infty\}$. Then there exists a measurable map sending a vector v_1 in $\{v \in \mathcal{V}_m \mid \|v\| = 1\}$ to a list $(v_1, v_2, \ldots, v_n) \in (\mathcal{V}_m)^n$ consisting of n orthonormal entries.*

PROOF. Given v_1 we may apply the Gram–Schmidt orthonormalization procedure to the list of vectors $v_1, e_1, e_2, \ldots$. As usual, we discard at each step the vector if its projection onto the orthogonal complement of the span of the previous vectors vanishes. This gives a definition of the required orthonormal basis in a measurable way. In fact, on each of the measurable subsets $\langle e_1 \rangle_{\mathbb{C}}$, $\langle e_1, e_2 \rangle_{\mathbb{C}} \smallsetminus \langle e_1 \rangle_{\mathbb{C}}, \ldots$ of $\mathcal{V}_n$ with the same discarding steps, the new vectors depend continuously on v_1. For instance, for the second vector v_2 we have $v_2 = e_2$ if $v_1 \in \langle e_1 \rangle_{\mathbb{C}}$ and

$$v_2 = \frac{1}{\|v_1 - \langle v_1, e_1 \rangle e_1\|} (v_1 - \langle v_1, e_1 \rangle e_1)$$

if $v_1 \notin \langle e_1 \rangle_{\mathbb{C}}$. □

Exercise 2.71. Give a more careful general definition of v_k for $k \in \{2, \dots, n\}$ in Lemma 2.70.

PROOF OF THEOREM 2.69(3). As mentioned above, the definition of $\mathcal{H}_\pi^{(n)}$ for $n \in \mathbb{N} \cup \{\infty\}$ given in (2.25)–(2.26) using spectral measures is canonical (as spectral measures are uniquely determined by a given vector and the unitary representation). This and parts (1) and (2) allow us to assume that π is the multiplication representation on the Hilbert space in (2.27). It remains to prove that

$$\mathcal{H}_\pi^{(\geqslant n)} = \bigoplus_{m \geqslant n} L^2_{\mu_m}(P_m, \mathcal{V}_m)$$

for $n \in \mathbb{N}$, where m is allowed to be ∞. This will give $\mathcal{H}_\pi^{(n)} = L^2_{\mu_n}(P_n, \mathcal{V}_n)$ for $n \in \mathbb{N}$ and $\mathcal{H}_\pi^{(\infty)} = L^2_{\mu_\infty}(P_\infty, \mathcal{V}_\infty)$.

For this, we note that the definition of spectral measures shows that

$$\mathrm{d}\mu_v^\pi(t) = \sum_{n \in \mathbb{N} \cup \{\infty\}} \|v(t)\|^2_{\mathcal{V}} \,\mathrm{d}\mu_n(t) = \|v(t)\|^2_{\mathcal{V}} \,\mathrm{d}\mu_{\max} \tag{2.28}$$

and, more generally,

$$\mathrm{d}\mu_{v,w}^\pi(t) = \sum_{n \in \mathbb{N} \cup \{\infty\}} \langle v(t), w(t) \rangle_{\mathcal{V}} \,\mathrm{d}\mu_n(t) = \langle v(t), w(t) \rangle_{\mathcal{V}} \,\mathrm{d}\mu_{\max} \tag{2.29}$$

for $t \in \widehat{G}$ and all $v, w \in \mathcal{H}_\pi$ (also see Proposition 2.53).

We suppose first that $v \in \bigoplus_{m \geqslant n} L^2_{\mu_m}(P_m, \mathcal{V}_m)$. By using (2.28) we see that the spectral measure is given by $F^2 \,\mathrm{d}\mu_{\max}$, where $F(t) = \|v(t)\|_{\mathcal{V}}$ for all t in $\bigcup_{m \geqslant n} P_m$ and $F(t) = 0$ otherwise. We also define

$$B = \{t \in \bigcup_{m \geqslant n} P_m \mid F(t) \neq 0\}.$$

For $t \in B$ we apply Lemma 2.70 to $\widetilde{v}_1(t) = F^{-1} v(t)$ to obtain the orthonormal vectors $\widetilde{v}_1(t), \widetilde{v}_2(t), \dots, \widetilde{v}_n(t)$. We extend these functions trivially to all of $\widehat{G}$. By measurability of the construction in Lemma 2.70, we now obtain the vectors

$$v_k = F \widetilde{v}_k \in \bigoplus_{m \geqslant n} L^2_{\mu_m}(P_m, \mathcal{V}_m)$$

for $k \in \{2, \dots, n\}$. We also define $v_1 = v$. This ensures that

$$\|v_k(t)\|_{\mathcal{V}} = F(t) = \|v(t)\|_{\mathcal{V}}$$

for all $k \geqslant 1$ and $t \in \widehat{G}$, which implies that these n vectors have the same spectral measures as v by (2.28). Moreover, we also have $\langle v_k(t), v_\ell(t)\rangle_{\mathcal{V}} = 0$ by construction for $k \neq \ell$ and all $t \in P_n$. Hence $\langle v_k\rangle_\pi \perp \langle v_\ell\rangle_\pi$ by (2.29) and Exercise 2.54 (see also the hint on page 530), which proves the requirement for v to have spectral multiplicity at least n.

Suppose now that $v \in \mathcal{H}_\pi^{(\geqslant n)} \subseteq \bigoplus_{m\in\mathbb{N}\cup\{\infty\}} L^2_{\mu_m}(P_m, \mathcal{V}_m)$ has the property that there exist n vectors $v_1 = v, v_2, \dots, v_n$ with orthogonal cyclic representations and spectral measures equal to μ_v. As in the previous step, we can define $B = \{t \in \widehat{G} \mid \|v(t)\|_{\mathcal{V}} \neq 0\}$. We note that $\langle v_k(t), v_\ell(t)\rangle_{\mathcal{V}} = 0$ for almost every $t \in \widehat{G}$ (with respect to $\mu_{\max}$) for $1 \leqslant k < \ell \leqslant n$. Indeed, Exercise 2.54 shows that we have $\mu^\pi_{v_k,v_\ell} = 0$, which, together with (2.29), gives $\langle v_k(t), v_\ell(t)\rangle_{\mathcal{V}} = 0$ for almost every $t \in \widehat{G}$. This shows that the n vectors $v_1(t), v_2(t), \dots, v_n(t)$ are, for almost every $t \in B$, orthogonal. For $t \in P_m$ and $m < n$ the vectors $v_k(t)$ take values in $\mathcal{V}_m = \mathbb{C}^m$, which shows that $\mu_{\max}(B \cap P_m) = 0$. Hence we have the opposite inclusion

$$\mathcal{H}_\pi^{(\geqslant n)} \subseteq \bigoplus_{m\geqslant n} L^2_{\mu_m}(P_m, \mathcal{V}_m).$$

Together with the first part of the proof, this gives equality, and so that

$$\mathcal{H}_\pi^{(n)} = L^2_{\mu_n}(P_n, \mathcal{V}_n)$$

for $n \in \mathbb{N} \cup \{\infty\}$. □

Proof of (4) in Theorem 2.69. By the above, we know that $\mathcal{H}_\pi^{(n)} \subseteq \mathcal{H}_\pi$ is a canonical invariant subspace that is isomorphic to the multiplication representation on $\left(L^2_{\mu_n}(P_n)\right)^n$. However, by the properties of spectral measures in Proposition 2.64 (see also Proposition 2.52(2) and (4)) this implies that μ_n is a maximal spectral measure of $\pi|_{\mathcal{H}_\pi^{(n)}}$, which in particular implies that its measure class is uniquely and canonically determined by π. □

We note that the phrase *countable Lebesgue spectrum* is used, for example, in dynamical systems to refer to the case of a measure-preserving invertible transformation (corresponding to the case $G = \mathbb{Z}$ in Proposition 1.3) where $\mu_{\max} = \mu_\infty$ represents the measure class of the Lebesgue measure on $\widehat{\mathbb{Z}} = \mathbb{T}$. Other possibilities are described using the terminology *pure point* (or *discrete*) *spectrum*, *continuous spectrum* or *mixed spectrum*, but these do not reflect any multiplicity information. We refer to Cornfeld, Fomin and Sinaĭ [17] for more on how spectral methods are used in the study of measure-preserving dynamical systems.

2.8 The Centralizer of the Representation

At times it is useful to understand the centralizer of a unitary representation. Let π be a unitary representation of the abelian group G, and $\mu_{\max}$ a maximal spectral measure associated to π as in Proposition 2.64. Applying Theorem 2.69, we find a partition $\{P_n \mid n \in \mathbb{N} \cup \{\infty\}\}$ and measures $\mu_n = \mu_{\max}|_{P_n}$ so that

$$\mathcal{H}_\pi \cong \bigoplus_{n \in \mathbb{N} \cup \{\infty\}} L^2_{\mu_n}(P_n, \mathcal{V}_n), \tag{2.30}$$

where we again write $\mathcal{V}_n = \mathbb{C}^n$ for $n \in \mathbb{N}$, $\mathcal{V}_\infty = \ell^2(\mathbb{N})$, $L^2_{\mu_n}(P_n, \mathcal{V}_n)$ for the space of square-integrable $\mathcal{V}_n$-valued functions on P_n for $n \in \mathbb{N} \cup \{\infty\}$, and the representation on the right is the multiplication representation M.

Proposition 2.72 (The centralizer of π). *With the above assumptions the centralizer*

$$C(\pi) = \{B \in \mathrm{B}(\mathcal{H}_\pi) \mid B \text{ is intertwining}\}$$

of π corresponds under the isomorphism (2.30) *to the set of all operators*

$$T = \bigoplus_{n \in \mathbb{N} \cup \{\infty\}} T_n$$

with a pointwise definition in the following sense:

(1) *For $n \in \mathbb{N}$ the operator T_n can be identified with an n-by-n matrix having entries in the ring $L^\infty_{\mu_n}(P_n)$. More formally, $T_n \in \mathrm{Mat}_{n,n}(L^\infty_{\mu_n}(P_n))$ maps $v \in L^2_{\mu_n}(P_n, \mathcal{V}_n)$ to $T_n v \in L^2_{\mu_n}(P_n, \mathcal{V}_n)$ defined by*

$$(T_n v)(t) = T_n(t) v(t) \in \mathcal{V}_n = \mathbb{C}^n$$

for $t \in P_n$.

(2) *The operator T_∞ can be defined by a measurable map $T_\infty \colon P_\infty \to \mathrm{B}(\mathcal{V}_\infty)$ such that*

$$(T_\infty v)(t) = T_\infty(t) v(t)$$

for $t \in P_\infty$ and $v \in L^2_{\mu_\infty}(P_\infty, \mathcal{V}_\infty)$.

(3) *We have* $\|T\| = \sup_{n \in \mathbb{N} \cup \{\infty\}} \operatorname{ess\,sup}_{P_n} \|T_n(\cdot)\|_{\mathrm{op}} < \infty$.

We note that measurability of the map $T_\infty \colon P_\infty \to \mathrm{B}(\mathcal{V}_\infty)$ is defined by measurability of the inner products

$$P_\infty \ni t \longmapsto \langle T_\infty(t) v, w \rangle_{\mathcal{V}}$$

for all pairs $v, w \in \mathcal{V}_\infty$ (also see Exercise 2.73).

Exercise 2.73. Show that the Borel σ-algebra $\mathcal{B}_\tau$ generated by open sets on $\mathrm{B}(\mathcal{V}_\infty)$ with respect to the strong operator topology τ and the Borel σ-algebra $\mathcal{B}_{\tau_w}$ generated by open sets on $\mathrm{B}(\mathcal{V}_\infty)$ with respect to the weak operator topology τ_w coincide. Also show that the above notion of measurability agrees with measurability of the map $t \mapsto T_\infty(t)$ with respect to the σ-algebra $\mathcal{B}_\tau = \mathcal{B}_{\tau_w}$.

PROOF OF PROPOSITION 2.72. To simplify notation, we assume that the isomorphism in (2.30) is an identity. Suppose that T is in the centralizer of π.

LINEARITY OVER $\mathscr{L}^\infty(\widehat{G})$. For $F \in \mathscr{L}^\infty(\widehat{G})$ and

$$v \in \mathcal{H}_\pi = \bigoplus_{n \in \mathbb{N} \cup \{\infty\}} L^2_{\mu_n}(P_n, \mathcal{V}_n)$$

we have, by the properties of the measurable functional calculus in Proposition 2.58(5) and (6), that T also commutes with $\pi_{\mathrm{FC}}(F) = M_F$, so that

$$T(Fv) = T(M_F v) = M_F T(v) = FT(v). \tag{2.31}$$

Put in algebraic terms, we see that T is linear with respect to the $\mathscr{L}^\infty(\widehat{G})$-module structure of $\mathcal{H}_\pi$. Below we will upgrade this to the statement that T itself has a pointwise definition, as in the proposition.

Applying this for the characteristic functions of the elements of the partition $\{P_n \mid n \in \mathbb{N} \cup \{\infty\}\}$ we obtain

$$T(L^2_{\mu_n}(P_n, \mathcal{V}_n)) = T(\mathbb{1}_{P_n} \mathcal{H}_\pi) = \mathbb{1}_{P_n} T(\mathcal{H}_\pi) \subseteq L^2_{\mu_n}(P_n, \mathcal{V}_n)$$

for $n \in \mathbb{N} \cup \{\infty\}$. We will describe the restriction of T to $L^2_{\mu_n}(P_n, \mathcal{V}_n)$ for every $n \in \mathbb{N} \cup \{\infty\}$.

MATRIX COEFFICIENTS LIE IN $L^\infty_{\mu_n}(P_n)$. We fix $n \in \mathbb{N} \cup \{\infty\}$ and two vectors $v, w \in \mathcal{V}_n$ that will also be considered as constant functions on P_n belonging to $L^2_{\mu_n}(P_n, \mathcal{V}_n)$. We will now describe the inner product $\langle (Tv)(t), w \rangle_{\mathcal{V}}$ as a function of $t \in P_n$. In fact, we claim that

$$\Big(P_n \ni t \longmapsto F_{v,w}(t) = \langle (Tv)(t), w \rangle_{\mathcal{V}}\Big) \in L^\infty_{\mu_n}(P_n)$$

and, moreover,

$$\|F_{v,w}\|_\infty \leqslant \|T\|_{\mathrm{op}} \|v\|_{\mathcal{V}} \|w\|_{\mathcal{V}}. \tag{2.32}$$

For this, we define

$$B = \{t \in P_n \mid |F_{v,w}(t)| > \|T\|_{\mathrm{op}} \|v\|_{\mathcal{V}} \|w\|_{\mathcal{V}}\}, \tag{2.33}$$

set $f = \mathbb{1}_B$ and $f_2 = \arg(F_{v,w}) \mathbb{1}_B$. Then $f_1, f_2 \in L^\infty_{\mu_n}(P_n) \cap L^2_{\mu_n}(P_n)$ with $\|f_1\|_2 = \|f_2\|_2 = \mu(B)^{\frac{1}{2}}$. Together with the $\mathscr{L}^\infty(\widehat{G})$-linearity in (2.31), we deduce that the integral in

$$\begin{aligned}\langle Tf_1v, f_2w\rangle_{L^2_{\mu_n}(P_n,\mathcal{V}_n)} &= \langle f_1Tv, f_2w\rangle_{L^2_{\mu_n}(P_n,\mathcal{V}_n)} \\ &= \int_B f_1(t)\overline{f_2(t)}\,\underbrace{\langle (Tv)(t), w\rangle_{\mathcal{V}}}_{=F_{v,w}}\,\mathrm{d}\mu_n(t) \\ &= \int_B |F_{v,w}|\,\mathrm{d}\mu_n\end{aligned}$$

exists and is bounded by

$$\begin{aligned}\|T\|_{\mathrm{op}}\|f_1v\|_2\|f_2w\|_2 &= \|T\|_{\mathrm{op}}\|f_1\|_2\|v\|_{\mathcal{V}}\|f_2\|_2\|w\|_{\mathcal{V}} \\ &= \|T\|_{\mathrm{op}}\|v\|_{\mathcal{V}}\|w\|_{\mathcal{V}}\mu(B).\end{aligned}$$

Combining the last two facts and recalling the definition of B in (2.33), this implies that $\mu(B) = 0$, and (2.32) follows.

THE CASE OF $n \in \mathbb{N}$. We now fix some $n \in \mathbb{N}$ and define

$$T_n(t) = \big((Te_1)(t), (Te_2)(t), \dots, (Te_n)(t)\big) \in \mathrm{Mat}_{n,n}(\mathbb{C})$$

for $t \in P_n$ or, keeping $t \in P_n$ implicit,

$$T_n = (Te_1, Te_2, \dots, Te_n),$$

where we again write e_k for the kth basis vector of $\mathcal{V}_n$ considered as a constant function in $L^2_{\mu_n}(P_n, \mathcal{V}_n)$. Writing Te_k as a column vector belonging to $L^2_{\mu_n}(P_n, \mathcal{V}_n)$, we obtain from the above that $T_n \in \mathrm{Mat}_{n,n}\big(L^\infty_{\mu_n}(P_n)\big)$. Indeed, for $j, k \in \{1, \dots, n\}$ the (j, k)th matrix entry is given by

$$T_{n,j,k} = \langle Te_k, e_j\rangle_{\mathcal{V}} \in L^\infty_{\mu_n}(P_n).$$

It follows that we can use T_n to define a bounded operator on $L^2_{\mu_n}(P_n, \mathcal{V}_n)$ by

$$L^2_{\mu_n}(P_n, \mathcal{V}_n) \ni \begin{pmatrix} v_1 \\ \vdots \\ v_n \end{pmatrix} \longmapsto T_n \begin{pmatrix} v_1 \\ \vdots \\ v_n \end{pmatrix} = \begin{pmatrix} T_{n,1,1}v_1 + \cdots + T_{n,1,n}v_n \\ \vdots \\ T_{n,n,1}v_1 + \cdots + T_{n,n,n}v_n \end{pmatrix}.$$

To see that the operator T restricted to $L^2_{\mu_n}(P_n, \mathcal{V}_n)$ agrees with the operator defined by T_n, we suppose that $v = \sum_k v_k e_k \in L^2_{\mu_n}(P_n, \mathcal{V}_n)$. If we assume in addition that $v_k \in L^\infty_{\mu_n}(P_n)$ for $k = 1, \dots, n$, then we can use (2.31) to obtain

$$Tv = \sum_k v_k T(e_k) = \sum_{j,k} v_k \underbrace{\langle T(e_k), e_j\rangle_{\mathcal{V}}}_{T_{n,j,k}} e_j = T_n v.$$

Since $L^\infty_{\mu_n}(P_n)$ is a dense subspace of $L^2_{\mu_n}(P_n)$, we deduce that T restricted to $L^2_{\mu_n}(P_n, \mathcal{V}_n)$ coincides with the operator defined above by the matrix T_n.

THE CASE $n = \infty$. Next we study the case $n = \infty$. Just as in the finite-dimensional case, we will use T to define operators $T_\infty(t) \in \mathrm{B}(\mathcal{V}_\infty)$ for almost every $t \in P_\infty$, use these to define an operator T_∞ on $L^2_{\mu_\infty}(P_\infty, \mathcal{V}_\infty)$, and show that T restricted to $L^2_{\mu_\infty}(P_\infty, \mathcal{V}_\infty)$ is equal to T_∞. For this, we let

$$\mathcal{W} = \{w_1, w_2, \ldots\}$$

be a dense subset of $\mathcal{V}_\infty$ that is also a subspace over the field $\mathbb{Q}[\mathrm{i}]$. We may also suppose $\mathcal{W}$ contains the standard orthonormal basis $e_1, e_2, \ldots$ of $\mathcal{V}_\infty$. Fixing indices $j, k \in \mathbb{N}$, considering $w_j, w_k \in L^2_{\mu_\infty}(P_\infty, \mathcal{V}_\infty)$ again as constant functions on P_∞, and using (2.32) we see that $\langle (Tw_j)(t), w_k\rangle_{\mathcal{V}} \in L^\infty_{\mu_\infty}(P_\infty)$ is bounded by $\|T\|_{\mathrm{op}}\|w_j\|_{\mathcal{V}}\|w_k\|_{\mathcal{V}}$ for almost every $t \in \mathcal{P}_\infty$. Moreover, we also have that $\langle (Tw_j)(t), w_k\rangle_{\mathcal{V}}$ depends sesqui-linearly on the vectors w_j, w_k as elements of the vector space $\mathcal{W}$ over $\mathbb{Q}[\mathrm{i}]$, almost surely. Collecting these countably many null sets, applying density of $\mathcal{W}$ in $\mathcal{V}_\infty$, and applying the Riesz representation theorem on $\mathcal{V}_\infty$, it follows that for almost every $t \in P_\infty$ we have

$$\langle (Tw_j)(t), w_k\rangle_{\mathcal{V}} = \langle T_\infty(t)w_j, w_k\rangle_{\mathcal{V}}$$

for some bounded operator $T_\infty(t) \in \mathrm{B}(\mathcal{V}_\infty)$ satisfying $\|T_\infty(t)\|_{\mathrm{op}} \leqslant \|T\|_{\mathrm{op}}$. Measurability of $P_\infty \ni t \mapsto T_\infty(t)$ (that is, of the map $t \mapsto \langle T_\infty(t)v, w\rangle_{\mathcal{V}}$ for all $v, w \in \mathcal{V}_\infty$) follows once again from density of $\mathcal{W}$.

Similarly to the finite-dimensional case we can use $T_\infty(t)$ for $t \in P_\infty$ to define a bounded operator T_∞ on $L^2_{\mu_\infty}(P_\infty, \mathcal{V}_\infty)$ by sending $v \in L^2_{\mu_\infty}(P_\infty, \mathcal{V}_\infty)$ to the function $T_\infty v$ defined by $(T_\infty v)(t) = T_\infty(t)v(t)$ for $t \in P_\infty$. To see measurability, let $v = \sum_k v_k e_k$ so that

$$(T_\infty v)(t) = \sum_{j,k} \langle T_\infty(t)e_k, e_j\rangle_{\mathcal{V}} v_k(t) e_j$$

for $t \in P_\infty$. Moreover,

$$\|T_\infty v\|^2_{L^2_{\mu_\infty}(P_\infty,\mathcal{V}_\infty)} = \int_{P_\infty} \underbrace{\|(T_\infty v)(t)\|^2_{\mathcal{V}}}_{\leqslant \|T\|^2_{\mathrm{op}}\|v(t)\|_2^2} \,\mathrm{d}\mu_\infty(t) \leqslant \|T\|^2_{\mathrm{op}}\|v\|^2_{L^2_{\mu_\infty}(P_\infty,\mathcal{V}_\infty)}.$$

This shows that T_∞ gives a well-defined operator with norm $\|T_\infty\|_{\mathrm{op}} \leqslant \|T\|_{\mathrm{op}}$.

It remains to see that the operator T_∞ coincides with T when it is restricted to $L^2_{\mu_\infty}(P_\infty, \mathcal{V}_\infty)$. The construction of T_∞ implies that

$$\big\langle (Te_k)(t), e_j\big\rangle_{\mathcal{V}} = \big\langle T_\infty(t)e_k, e_j\big\rangle_{\mathcal{V}}$$

for almost every $t \in P_\infty$ and all $j, k \in \mathbb{N}$. This implies that

$$(Te_k)(t) = T_\infty(t)e_k$$

for all $k \in \mathbb{N}$. For $N \in \mathbb{N}$ and $f_1, \ldots, f_N \in L^\infty_{\mu_\infty}(P_\infty)$ we now have by (2.31) that

$$T\Big(\sum_{k\leqslant N} f_k e_k\Big) = \sum_{k\leqslant N} f_k T e_k = \sum_{k\leqslant N} f_k T_\infty e_k = T_\infty\Big(\sum_{k\leqslant N} f_k e_k\Big).$$

Varying $N \in \mathbb{N}$ and the functions in $L^\infty_{\mu_\infty}(P_\infty)$, this gives a dense subset of $L^2_{\mu_\infty}(P_\infty, \mathcal{V}_\infty)$. It follows that T_∞ is indeed the restriction of the operator T to $L^2_{\mu_\infty}(P_\infty, \mathcal{V}_\infty)$.

THE OPERATOR NORMS. We note that we could also have used the above argument for $n \in \mathbb{N}$, so that we also have $\|T_n(t)\|_{\mathrm{op}} \leqslant \|T\|_{\mathrm{op}}$ for almost every $t \in P_n$ and $n \in \mathbb{N}$. In particular, we have

$$\sup_{n\in\mathbb{N}\cup\{\infty\}} \operatorname*{ess\,sup}_{t\in P_n} \|T_n(t)\|_{\mathrm{op}} \leqslant \|T\|_{\mathrm{op}}.$$

To see the opposite inequality, let S denote the supremum on the left and let $v, w \in \mathcal{H}_\pi$. Then

$$\begin{aligned}|\langle Tv, w\rangle| &\leqslant \sum_{n\in\mathbb{N}\cup\{\infty\}} \int_{P_n} |\langle T_n(t)v(t), w(t)\rangle_{\mathcal{V}}| \,\mathrm{d}\mu_{\max}(t)\\ &\leqslant \int_{\widehat{G}} S\|v(t)\|_{\mathcal{V}}\|w(t)\|_{\mathcal{V}} \,\mathrm{d}\mu_{\max}(t) \leqslant S\|v\|\,\|w\|\end{aligned}$$

by the description of $\mathcal{H}_\pi$, the pointwise definition of T obtained above, the Cauchy–Schwarz inequality on $\mathcal{V}_n$ for $n \in \mathbb{N}\cup\{\infty\}$, the definition of S, and finally the Cauchy–Schwarz inequality on $L^2_{\mu_{\max}}(\widehat{G})$. However, this implies that $\|T\|_{\mathrm{op}} \leqslant S$.

THE CONVERSE. The converse statement that any $T = \bigoplus_{n\in\mathbb{N}\cup\{\infty\}} T_n$ as in the proposition defines a bounded operator we leave as an exercise (see Exercise 2.74). As the unitary representation corresponds under the isomorphism to the multiplication representation, it follows that a so-defined operator T belongs to the centralizer of π. This gives the proposition. □

Exercise 2.74. Show that any operator $T = \bigoplus_{n\in\mathbb{N}\cup\{\infty\}} T_n$ as in Proposition 2.72 defines a bounded intertwining operator by the following steps.

(a) Let $n \in \mathbb{N}$, let (X, μ) be a finite measure space and let T lie in $\mathrm{Mat}_{n,n}\big(L^\infty_\mu(X)\big)$. Show that T induces a bounded operator on $L^2_\mu\big(X, \mathcal{V}_n\big)$ satisfying

$$(Tv)(t) = T(t)v(t)$$

for $v \in L^2_\mu\big(X, \mathcal{V}_n\big)$ and almost every $t \in X$ (where we use matrix multiplication on the right), and that $\|T\|_{\mathrm{op}} = \operatorname{ess\,sup}_{t\in X} \|T(t)\|_{\mathrm{op}}$.

(b) Let $n = \infty$ and let (X, μ) be a finite measure space. Show that a measurable map

$$T\colon X \ni t \longmapsto \mathrm{B}(\mathcal{V}_\infty)$$

with $\operatorname{ess\,sup}_{t\in X} \|T(t)\|_{\mathrm{op}} < \infty$ induces a bounded operator T on $L^2_\mu\big(X, \mathcal{V}_\infty\big)$ satisfying

$$(Tv)(t) = T(t)v(t)$$

for $v \in L^2_\mu\big(X, \mathcal{V}_\infty\big)$ and almost every $t \in \widehat{G}$, and that T satisfies $\|T\|_{\mathrm{op}} = \operatorname{ess\,sup}_{t\in X} \|T(t)\|_{\mathrm{op}}$.

(c) Let μ be a finite measure on $\widehat{G}$, and let $\{P_n \mid n \in \mathbb{N} \cup \{\infty\}\}$ be a countable measurable partition of $\widehat{G}$. Suppose T_n is defined as in (a) or (b), using $\mu_n = \mu|_{P_n}$ for $n \in \mathbb{N} \cup \{\infty\}$. Suppose, moreover, that

$$\sup_{n \in \mathbb{N} \cup \{\infty\}} \operatorname{ess\,sup}_{t \in P_n} \|T(t)\|_{\mathrm{op}} < \infty.$$

Show that this implies that $T = \bigoplus_{n \in \mathbb{N} \cup \{\infty\}} T_n$ is a bounded operator on

$$\bigoplus_{n \in \mathbb{N} \cup \{\infty\}} L^2_{\mu_n}(P_n, \mathcal{V}_n)$$

that is intertwining for the canonical multiplication representation of G.

2.9 Summary and Outlook

The rather complete understanding of unitary representations of abelian groups in terms of:

- spectral measures, which completely encode matrix coefficients;
- the measurable functional calculus, which allows us to isolate parts of the spectrum at will;
- the spectral theorem with complete multiplicity data; and
- their centralizer

obtained in this chapter is rewarding and important in itself. However, it will also be the key for understanding the unitary dual and unitary representations of other groups. Most notably, this applies to groups with normal abelian subgroups as discussed in Chapter 5. Moreover, the abelian theory (together with Section 5.1) will also be important for understanding certain aspects of unitary representations of semi-simple groups like $\mathrm{SL}_3(\mathbb{R})$, as we will see in Chapter 7.

For a Banach algebra $\mathcal{A}$ the Gelfand transform (see [25, Sec. 11.3.3]) gives a canonical correspondence between closed subsets of the Gelfand dual $\sigma(\mathcal{A})$ (the set of characters on $\mathcal{A}$) and closed ideals in $\mathcal{A}$ as follows. If $\mathcal{I} \subseteq \mathcal{A}$ is a closed ideal then the hull

$$\nu(\mathcal{I}) = \{h \in \sigma(\mathcal{A}) \mid \widehat{f}(h) = 0 \text{ for all } f \in \mathcal{I}\}$$

is a closed subset of $\sigma(\mathcal{A})$. If $N \subseteq \sigma(\mathcal{A})$ is a closed subset then the kernel

$$\imath(N) = \{f \in \mathcal{A} \mid \widehat{f} = 0 \text{ on } N\}$$

is a closed ideal in $\mathcal{A}$. It is clear that $\imath(\nu(\mathcal{I})) \supseteq \mathcal{I}$ and $\nu(\imath(N)) \supseteq N$, and (for example) if $\mathcal{A}$ is a commutative C^* algebra then the maps ν and $\imath$ are continuous bijections inverse to each other. One of the most influential results in studying harmonic analysis on locally compact abelian groups is a theorem due to Wiener:(5) If $\mathcal{I}$ is a closed ideal in the Banach algebra $L^1(G)$ and $\nu(\mathcal{I}) = \emptyset$, then $\mathcal{I} = L^1(G)$. If G is discrete then $L^1(G)$ has a unit, so any proper ideal is contained in a maximal ideal and these are the kernels of characters on $L^1(G)$, which can be used to show Wiener's theorem in this setting. The importance of Wiener's theorem lies in the non-compact and non-discrete case, where it is important for results in spectral synthesis and abstract Tauberian theorems. We refer to Katznelson [49] for more on this in the setting of $\mathbb{R}$ and Folland [30] for locally compact abelian groups.

Pontryagin duality and the Plancherel formula are useful tools for constructing and understanding concrete abelian groups. Moreover, local fields as introduced in Section 2.4.5 are of fundamental importance to modern number theory.(6)

The reader may continue with Chapter 3 or 5, returning to Chapters 3 and 4 when needed.

Chapter 3
Compact Groups

In this chapter we will obtain fundamental results concerning all unitary representations of compact groups. We will assume implicitly that the group considered is metric as well and simply refer to it as *the compact group* G. As we will see, the description of general unitary representations in terms of irreducible representations is easier from the analytic point of view than in the abelian case, due to complete reducibility of unitary representations into irreducible representations. However, from an algebraic point of view the discussions here are harder, since the irreducible representations can be of any finite dimension instead of just one-dimensional. For convenience, we will assume throughout the chapter that the Haar measure m on G is normalized to satisfy $m(G) = 1$. We recall that any compact group is unimodular (see Exercise 1.19(a) and its hint on page 526).

We begin the chapter, however, by discussing two fundamental constructions for unitary representations that are particularly important for compact groups.

3.1 The Contragredient Representation

For the first construction, recall from the Fréchet–Riesz representation theorem that any Hilbert space $\mathcal{H}$ is isometrically isomorphic to its dual $\mathcal{H}'$. However, this canonical isomorphism sending $w \in \mathcal{H}$ to the map $w' \colon v \mapsto \langle v, w \rangle$ in $\mathcal{H}'$ is conjugate-linear over $\mathbb{C}$, and so in particular is not an isomorphism of complex Hilbert spaces. Transporting a unitary representation π via this conjugate-linear isomorphism to the dual $\mathcal{H}'_\pi$ may therefore create a representation not unitarily isomorphic to π.

Definition 3.1 (Contragredient). For a Hilbert space $\mathcal{H}$ the dual $\mathcal{H}'$ may be endowed with the inner product

$$\langle v', w' \rangle_{\mathcal{H}'} = \overline{\langle v, w \rangle_{\mathcal{H}}} = \langle w, v \rangle_{\mathcal{H}} \tag{3.1}$$

M. Einsiedler and T. Ward, *Unitary Representations and Unitary Duals*,
Graduate Texts in Mathematics 308, https://doi.org/10.1007/978-3-032-03899-9_3

for any $v, w \in \mathcal{H}$ and (by the Cauchy–Schwarz inequality) this inner product induces the operator norm on $\mathcal{H}'_\pi$. For a unitary representation π of G on $\mathcal{H}_\pi$ the *contragredient representation* $\overline{\pi}$ is defined as the unitary representation induced by the dual operators on $\mathcal{H}_{\overline{\pi}} = \mathcal{H}'_\pi$. That is,

$$\overline{\pi}_g \colon \mathcal{H}'_\pi \ni v' \longmapsto \overline{\pi}_g(v') = v' \circ \pi_g^{-1} \in \mathcal{H}'_\pi$$

for $g \in G$.

Let us make a few comments that should help the reader to become familiar with this construction, which may be confusing at first sight.

First, it is easy to see that $\mathcal{H}'_\pi$ is again a (complex) Hilbert space. For example, by (3.1),

$$\langle v', w' \rangle_{\mathcal{H}'} = \langle w, v \rangle_{\mathcal{H}}$$

depends linearly on $v' \in \mathcal{H}'_\pi$ and conjugate-linearly on $w' \in \mathcal{H}'_\pi$ by conjugate-linearity of the isomorphism between $\mathcal{H}'_\pi$ and $\mathcal{H}_\pi$.

We verify that $\overline{\pi}$ is, as claimed, a unitary representation. For the representation $\overline{\pi}$ we note that for $g \in G$ and $v' \in \mathcal{H}'$ the linear functional $\overline{\pi}_g(v') = v' \circ \pi_g^{-1}$ sends any $w \in \mathcal{H}$ to

$$v'(\pi_g^{-1}(w)) = \langle \pi_g^{-1} w, v \rangle_{\mathcal{H}_\pi} = \langle w, \pi_g v \rangle_{\mathcal{H}_\pi} = (\pi_g v)'(w),$$

which shows that $\overline{\pi}_g(v') = (\pi_g v)'$. In other words the diagram

$$\begin{array}{ccc}
\mathcal{H}_\pi \ni v & \xmapsto{\pi_g} & \pi_g v \in \mathcal{H}_\pi \\
\big\downarrow & & \big\downarrow \\
\mathcal{H}'_\pi \ni v' & \xmapsto{\overline{\pi}_g} & \overline{\pi}_g v' \in \mathcal{H}'_\pi
\end{array}$$

commutes. From this, we see that

$$\begin{aligned}
\overline{\pi}_g(\alpha_1 v'_1 + \alpha_2 v'_2) &= \overline{\pi}_g((\overline{\alpha_1} v_1 + \overline{\alpha_2} v_2)') \\
&= (\pi_g(\overline{\alpha_1} v_1 + \overline{\alpha_2} v_2))' \\
&= (\overline{\alpha_1} \pi_g v_1 + \overline{\alpha_2} \pi_g v_2)' = \alpha_1 \overline{\pi}_g v'_1 + \alpha_2 \overline{\pi}_g v'_2
\end{aligned}$$

for any $\alpha_1, \alpha_2 \in \mathbb{C}$ and $v'_1, v'_2 \in \mathcal{H}'_\pi$. Similarly, we have

$$\|\overline{\pi}_g v'\|_{\mathcal{H}'_\pi} = \|(\pi_g v)'\|_{\mathcal{H}'_\pi} = \|\pi_g v\|_{\mathcal{H}_\pi} = \|v\|_{\mathcal{H}_\pi} = \|v'\|_{\mathcal{H}'_\pi}$$

for all $v \in \mathcal{H}$ and $g \in G$, which shows that π_g is unitary for all $g \in G$. The homomorphism property and the continuity requirement follow in the same way from the respective properties of π.

Next we calculate the matrix coefficient

$$\varphi^{\overline{\pi}}_{v',w'}(g) = \langle \overline{\pi}_g v', w' \rangle_{\mathcal{H}'_\pi} = \langle (\pi_g v)', w' \rangle_{\mathcal{H}'_\pi}$$
$$= \langle w, \pi_g v \rangle_{\mathcal{H}_\pi} = \overline{\langle \pi_g v, w \rangle_{\mathcal{H}_\pi}} = \overline{\varphi^{\pi}_{v,w}(g)} \qquad (3.2)$$

of $v', w' \in \mathcal{H}'_\pi$ at $g \in G$, and obtain the conjugate of the matrix coefficient of $v, w \in \mathcal{H}_\pi$. In particular, if the operator π_g is described by a matrix with respect to some orthonormal basis of $\mathcal{H}$, then $\overline{\pi}_g$ is described by the complex conjugate of the same matrix with respect to the dual basis of $\mathcal{H}'$.

In the special case of a one-dimensional unitary representation defined by a unitary character χ on G, this shows that the contragredient is the unitary representation defined by the complex conjugate character $\overline{\chi}$ on G. If χ does not have purely real values then χ is not isomorphic to $\overline{\chi}$ as a unitary representation.

We remark that $\overline{\pi}$ is irreducible if and only if π is, since there is a conjugate-linear intertwining isomorphism $v \mapsto v'$ between π and $\overline{\pi}$.

Exercise 3.2. (a) Show that the contragredient of the left regular representation is isomorphic to the left regular representation.
(b) Give a sufficient criterion for a general unitary representation π to be isomorphic to its contragredient $\overline{\pi}$, that in particular applies to (a).

Exercise 3.3. Let G be abelian as in Chapter 2, and let π be a unitary representation of G. Apply the spectral theorem (Corollary 2.13, or the more refined Theorem 2.69) to π. Describe the contragredient of π in terms of the data arising in the spectral theorem.

3.2 The Tensor Product Representation

The second construction we wish to present is the definition of the tensor product of two representations, which will include the definition in Lemma 1.28 as a very special case.

3.2.1 Basic Construction and Properties

In the following, we let $\mathcal{V}$ and $\mathcal{W}$ denote two separable Hilbert spaces.

Proposition 3.4 (Tensor product). *There exists a Hilbert space $\mathcal{V} \otimes \mathcal{W}$ together with a bilinear map $\mathcal{V} \times \mathcal{W} \ni (v, w) \mapsto v \otimes w \in \mathcal{V} \otimes \mathcal{W}$ such that for all $v, v_1, v_2 \in \mathcal{V}$ and $w, w_1, w_2 \in \mathcal{W}$ we have*

(1) $\langle v_1 \otimes w_1, v_2 \otimes w_2 \rangle_{\mathcal{V}\otimes\mathcal{W}} = \langle v_1, v_2 \rangle_{\mathcal{V}} \langle w_1, w_2 \rangle_{\mathcal{W}}$,
(2) $\|v \otimes w\|_{\mathcal{V}\otimes\mathcal{W}} = \|v\|_{\mathcal{V}} \|w\|_{\mathcal{W}}$, *and*
(3) $\{v \otimes w \mid v \in \mathcal{V}, w \in \mathcal{W}\}$ *spans a dense subspace of* $\mathcal{V} \otimes \mathcal{W}$.

Moreover, the subspace spanned by $\{v \otimes w \mid v \in \mathcal{V}, w \in \mathcal{W}\}$ is isomorphic to the tensor product in the sense of linear algebra $\mathcal{V} \otimes_{\mathrm{la}} \mathcal{W}$, the map

$$\mathcal{V} \times \mathcal{W} \ni (v, w) \longmapsto v \otimes w$$

is continuous, and the tensor product $\mathcal{V} \otimes \mathcal{W}$ is separable.

Recall that the tensor product in the sense of linear algebra, which we have written as $\mathcal{V} \otimes_{\mathrm{la}} \mathcal{W}$, of the complex vector spaces $\mathcal{V}$ and $\mathcal{W}$ is defined as the universal vector space together with the bilinear map

$$\mathcal{V} \times \mathcal{W} \ni (v, w) \longmapsto v \otimes w \in \mathcal{V} \otimes_{\mathrm{la}} \mathcal{W}$$

with the property that for any other bilinear map

$$B\colon \mathcal{V} \times \mathcal{W} \ni (v, w) \longmapsto B(v, w) \in \mathcal{Z}$$

with values in another complex vector space $\mathcal{Z}$ there is a unique linear map

$$L\colon \mathcal{V} \otimes_{\mathrm{la}} \mathcal{W} \longrightarrow \mathcal{Z}$$

such that $B(v, w) = L(v \otimes w)$ for all $(v, w) \in \mathcal{V} \times \mathcal{W}$. Moreover, we recall that $\mathcal{V} \otimes_{\mathrm{la}} \mathcal{W}$ can be constructed by taking the formal linear hull of all pure tensors $v \otimes w$ and forming the quotient by the subspace generated by the necessary relations to enforce bilinearity of the map

$$\mathcal{V} \times \mathcal{W} \ni (v, w) \longmapsto v \otimes w \in \mathcal{V} \otimes_{\mathrm{la}} \mathcal{W};$$

we refer to Hungerford [42, Sec. IV.5] for the details and the general setting of tensor products of modules. We will use this construction to give a canonical definition of the inner product on $\mathcal{V} \otimes_{\mathrm{la}} \mathcal{W}$, and define $\mathcal{V} \otimes \mathcal{W}$ as its completion. After the following abstract and coordinate-free construction, we will discuss a more concrete viewpoint in Corollary 3.7.

Proof of Proposition 3.4. We want to define $\mathcal{V} \otimes \mathcal{W}$ as the completion of the tensor product in the sense of linear algebra $\mathcal{V} \otimes_{\mathrm{la}} \mathcal{W}$ with respect to some inner product, which we now construct.

Fix some $v_1 \in \mathcal{V}$ and $w_1 \in \mathcal{W}$. Then the map

$$\mathcal{V} \times \mathcal{W} \ni (v, w) \longmapsto \langle v, v_1 \rangle \langle w, w_1 \rangle \in \mathbb{C},$$

is bilinear, and so by the universal property of the tensor product there exists a unique linear functional $L_{(v_1,w_1)}\colon \mathcal{V} \otimes_{\mathrm{la}} \mathcal{W} \ni t \mapsto L_{(v_1,w_1)}(t) \in \mathbb{C}$ with

$$L_{(v_1,w_1)}(v \otimes w) = \langle v, v_1 \rangle \langle w, w_1 \rangle \tag{3.3}$$

for all $v \in \mathcal{V}$ and $w \in \mathcal{W}$.

We now take the conjugate and note that the expression $\overline{L_{(v_1,w_1)}(t)}$ depends bilinearly on (v_1, w_1) for any fixed $t \in \mathcal{V} \otimes_{\mathrm{la}} \mathcal{W}$. Indeed, for a pure tensor $t = v \otimes w$ this follows from (3.3), and the general case follows by taking sums of pure tensors. In other words, $\mathcal{V} \times \mathcal{W} \ni (v_1, w_1) \mapsto \overline{L_{(v_1,w_1)}}$ is bilinear and has values in the complex vector space

$$\mathcal{Z} = \{S\colon \mathcal{V} \otimes_{\mathrm{la}} \mathcal{W} \to \mathbb{C} \mid S \text{ is conjugate-linear}\}.$$

By the universal property of the tensor product, it follows that there exists a linear map $I_p\colon \mathcal{V} \otimes_{\mathrm{la}} \mathcal{W} \to \mathcal{Z}$ with $I_p(v_1 \otimes w_1) = \overline{L_{(v_1,w_1)}}$ and so, by (3.3),

$$I_p(v_1 \otimes w_1)(v_2 \otimes w_2) = \overline{L_{(v_1,w_1)}(v_2, w_2)} = \langle v_1, v_2\rangle\langle w_1, w_2\rangle$$

for all $v_1, v_2 \in \mathcal{V}$ and $w_1, w_2 \in \mathcal{W}$.

We claim that

$$\langle t_1, t_2\rangle_{\otimes} = I_p(t_1)(t_2)$$

defines an inner product for tensors $t_1, t_2 \in \mathcal{V} \otimes_{\mathrm{la}} \mathcal{W}$, which will allow us to define $\mathcal{V} \otimes \mathcal{W}$ as its completion. From the construction (and in particular, from the definition of $\mathcal{Z}$), it is clear that $I_p(t_1)(t_2)$ depends linearly on $t_1 \in \mathcal{V} \otimes_{\mathrm{la}} \mathcal{W}$ and conjugate-linearly on $t_2 \in \mathcal{V} \otimes \mathcal{W}$. For the conjugate symmetry we note that for pure tensors $v_1 \otimes w_1, v_2 \otimes w_2 \in \mathcal{V} \otimes_{\mathrm{la}} \mathcal{W}$ we have

$$\overline{I_p(v_1 \otimes w_1)(v_2 \otimes w_2)} = \langle v_2, v_1\rangle \langle w_2, w_1\rangle = I_p(v_2 \otimes w_2)(v_1 \otimes w_1).$$

By sesqui-linearity, this extends to the identity

$$\overline{\langle t_1, t_2\rangle_{\otimes}} = \overline{I_p(t_1)(t_2)} = I_p(t_2)(t_1) = \langle t_2, t_1\rangle_{\otimes}$$

for all tensors $t_1, t_2 \in \mathcal{V} \otimes_{\mathrm{la}} \mathcal{W} = \langle v \otimes w \mid v \in \mathcal{V}, w \in \mathcal{W}\rangle_{\mathbb{C}}$.

For definiteness, let $t \in \mathcal{V} \otimes_{\mathrm{la}} \mathcal{W}$ be non-zero and write t as a sum

$$t = \sum_{\ell} v_\ell \otimes w_\ell$$

for some $v_1, \ldots, v_L \in \mathcal{V}$, and $w_1, \ldots, w_L \in \mathcal{W}$. Now choose an orthonormal basis $(e_1, \ldots, e_J)$ of $\langle v_\ell \mid \ell = 1, \ldots, L\rangle_{\mathbb{C}} \subseteq \mathcal{V}$ and an orthonormal basis $(f_1, \ldots, f_K)$ of $\langle w_\ell \mid \ell = 1, \ldots, L\rangle_{\mathbb{C}} \subseteq \mathcal{W}$ for some $J, K \in \mathbb{N}$. Expressing each v_ℓ (resp. w_ℓ) in terms of this orthonormal basis, putting these into the expression for t, and expanding bilinearly, we obtain

$$t = \sum_{j,k} t_{j,k} e_j \otimes f_k$$

with $t_{j,k} \in \mathbb{C}$ for $1 \leqslant j \leqslant J$ and $1 \leqslant k \leqslant K$. If all of these coefficients vanish, then clearly $t = 0$. If not, we use sesqui-linearity and the construction of $\langle \cdot, \cdot\rangle_{\otimes}$ to obtain

$$\begin{aligned}\langle t, t\rangle_{\otimes} &= \sum_{\substack{j_1,k_1\\ j_2,k_2}} t_{j_1,k_1}\overline{t_{j_2,k_2}} \left\langle e_{j_1} \otimes f_{k_1}, e_{j_2} \otimes f_{k_2}\right\rangle_{\otimes} \\ &= \sum_{\substack{j_1,k_1\\ j_2,k_2}} t_{j_1,k_1}\overline{t_{j_2,k_2}} \left\langle e_{j_1}, e_{j_2}\right\rangle \left\langle f_{k_1}, f_{k_2}\right\rangle = \sum_{j,k} |t_{j,k}|^2 > 0.\end{aligned}$$

Having shown that $\langle\cdot,\cdot\rangle_\otimes$ is an inner product on $\mathcal{V}\otimes_{\text{la}}\mathcal{W}$, we define $\mathcal{V}\otimes\mathcal{W}$ as the completion of $\mathcal{V}\otimes_{\text{la}}\mathcal{W}$ with respect to the norm $\|\cdot\|_\otimes$ induced by $\langle\cdot,\cdot\rangle_\otimes$. This shows the existence.

To see that the properties of $\mathcal{V}\otimes\mathcal{W}$ imply that $\langle v\otimes w\mid v\in\mathcal{V},w\in\mathcal{W}\rangle_\mathbb{C}$ is isomorphic to $\mathcal{V}\otimes_{\text{la}}\mathcal{W}$, note that the universal property of $\mathcal{V}\otimes_{\text{la}}\mathcal{W}$ applied to the bilinear map $\otimes\colon\mathcal{V}\times\mathcal{W}\to\mathcal{V}\otimes\mathcal{W}$ provides a linear map

$$L\colon\mathcal{V}\otimes_{\text{la}}\mathcal{W}\to\langle v\otimes w\mid v\in\mathcal{V},w\in\mathcal{W}\rangle_\mathbb{C}$$

with $L(v\otimes w)=v\otimes w$, which is an isometric bijection when $\mathcal{V}\otimes_{\text{la}}\mathcal{W}$ is given the inner product $\langle\cdot,\cdot\rangle_\otimes$ constructed above. (This also implies uniqueness; see Exercise 3.5.)

It remains to show continuity of the tensor product map, since this will then imply that $\{v\otimes w\mid v\in\mathcal{V},w\in\mathcal{W}\}$ is separable, and hence its closed linear hull $\mathcal{V}\otimes\mathcal{W}$ is also separable. So suppose that $v_n\to v$ in $\mathcal{V}$ and $w_n\to w$ in $\mathcal{W}$ as $n\to\infty$. Then

$$\begin{aligned}\|v_n\otimes w_n-v\otimes w\|_\otimes&\leqslant\|v_n\otimes w_n-v\otimes w_n\|_\otimes+\|v\otimes w_n-v\otimes w\|_\otimes\\&=\|v_n-v\|\|w_n\|+\|v\|\|w_n-w\|\longrightarrow0\end{aligned}$$

as $n\to\infty$, as required. □

Exercise 3.5. Show that properties (1) to (3) in Proposition 3.4 uniquely determine $\mathcal{V}\otimes\mathcal{W}$ up to isomorphism.

Exercise 3.6. (a) Let $v_1,v_2\in\mathcal{V}$ be linearly independent and $w_1,w_2\in\mathcal{W}$ be linearly independent. Show that an element of the form $v_1\otimes w_1+v_2\otimes w_2$ is not of the form $v\otimes w$ for any $v\in\mathcal{V}$ and $w\in\mathcal{W}$.
(b) Show that $\mathcal{V}\otimes\mathcal{W}'$ is canonically isomorphic to the space $\operatorname{HS}(\mathcal{W},\mathcal{V})$ of Hilbert–Schmidt operators from $\mathcal{W}$ to $\mathcal{V}$ (see [25, Ex. 6.53]).
(c) Show that if both $\mathcal{V}$ and $\mathcal{W}$ are infinite dimensional, then $\mathcal{V}\otimes_{\text{la}}\mathcal{W}$ is a proper subspace of $\mathcal{V}\otimes\mathcal{W}$.

Corollary 3.7 (Tensor products of L^2-spaces). *Let μ and ν be locally finite measures on X respectively on Y. Then the tensor product*

$$L^2_\mu(X)\otimes L^2_\nu(Y)$$

is canonically isomorphic to $L^2_{\mu\times\nu}(X\times Y)$. Under this isomorphism the tensor product $f_X\otimes f_Y$ of $f_X\in L^2_\mu(X)$ and $f_Y\in L^2_\nu(Y)$ corresponds to the function $f_X\otimes f_Y(x,y)=f_X(x)f_Y(y)$ for $(x,y)\in X\times Y$. Moreover, for closed subspaces $\mathcal{V}\subseteq L^2_\mu(X)$ and $\mathcal{W}\subseteq L^2_\nu(Y)$ the tensor product $\mathcal{V}\otimes\mathcal{W}$ is canonically isomorphic to the closed subspace of $L^2_{\mu\times\nu}(X\times Y)$ generated by elements $f_X\otimes f_Y$ for $f_X\in\mathcal{V}$ and $f_Y\in\mathcal{W}$.

In particular, given bases $(e_j\mid j\in J)$ and $(f_k\mid k\in K)$ of separable Hilbert spaces $\mathcal{V}$ and $\mathcal{W}$ respectively, with $J,K\subseteq\mathbb{N}$, we obtain an isomorphism

$$\mathcal{V}\otimes\mathcal{W}\cong\ell^2(J)\otimes\ell^2(K)=\ell^2(J\times K).$$

In other words, the family $(e_j \otimes f_k \mid j \in J, k \in K)$ *forms an orthonormal basis of* $\mathcal{V} \otimes \mathcal{W}$.

Proof. Let $B\colon L^2_\mu(X) \times L^2_\nu(Y) \to L^2_{\mu\times\nu}(X \times Y)$ be defined by

$$B(f_X, f_Y)(x, y) = f_X(x) f_Y(y)$$

for $f_X \in L^2_\mu(X)$, $f_Y \in L^2_\nu(Y)$ and $(x, y) \in X \times Y$. By Fubini's theorem we have $B(f_X, f_Y) \in L^2_{\mu\times\nu}(X \times Y)$ and $\|B(f_X, f_Y)\|_2 = \|f_X\|_2 \|f_Y\|_2$. Since B is bilinear, it induces a linear map

$$\imath\colon L^2_\mu(X) \otimes_{\mathrm{la}} L^2_\nu(Y) \longrightarrow L^2_{\mu\times\nu}(X \times Y)$$

with $\imath(f_X \otimes f_Y) = B(f_X, f_Y)$ for all $f_X \in L^2_\mu(X)$ and $f_Y \in L^2_\nu(Y)$. Moreover, by Proposition 3.4 and Fubini's theorem again, we have

$$\begin{aligned}
\left\langle f_X \otimes f_Y, \widetilde{f_X} \otimes \widetilde{f_Y} \right\rangle_\otimes &= \left\langle f_X, \widetilde{f_X} \right\rangle \left\langle f_Y, \widetilde{f_Y} \right\rangle \\
&= \int_X f_X \overline{\widetilde{f_X}} \,\mathrm{d}\mu \int_Y f_Y \overline{\widetilde{f_Y}} \,\mathrm{d}\nu \\
&= \int_{X\times Y} f_X(x) f_Y(y) \overline{\widetilde{f_X}(x) \widetilde{f_Y}(y)} \,\mathrm{d}\mu(x) \,\mathrm{d}\nu(y) \\
&= \left\langle \imath(f_X \otimes f_Y), \imath(\widetilde{f_X} \otimes \widetilde{f_Y}) \right\rangle_{L^2(X\times Y)}
\end{aligned}$$

for all $f_X, \widetilde{f_X} \in L^2_\mu(X)$ and $f_Y, \widetilde{f_Y} \in L^2_\nu(Y)$. For any tensor

$$t \in L^2_\mu(X) \otimes_{\mathrm{la}} L^2_\nu(Y)$$

we write $t = \sum_\ell v_\ell \otimes w_\ell$ and use the above to obtain

$$\begin{aligned}
\langle t, t \rangle_\otimes &= \sum_{\ell_1, \ell_2} \langle v_{\ell_1} \otimes w_{\ell_1}, v_{\ell_2} \otimes w_{\ell_2} \rangle_\otimes \\
&= \sum_{\ell_1, \ell_2} \langle \imath(v_{\ell_1} \otimes w_{\ell_1}), \imath(v_{\ell_2} \otimes w_{\ell_2}) \rangle_{L^2_{\mu\times\nu}(X\times Y)} = \langle \imath(t), \imath(t) \rangle_{L^2_{\mu\times\nu}(X\times Y)}.
\end{aligned}$$

Thus $\imath$ is an isometry and it extends from $L^2_\mu(X) \otimes_{\mathrm{la}} L^2_\nu(Y)$ to its completion $L^2_\mu(X) \otimes L^2_\nu(Y)$. Since the image of the unique extension is complete, and contains all characteristic functions of the form $\mathbb{1}_{B_X \times B_Y}$ for finite measure sets $B_X \subseteq X$ and $B_Y \subseteq Y$, it follows that $\imath$ is an isomorphism.

The argument above also applies to closed subspaces $\mathcal{V}$ and $\mathcal{W}$ of $L^2_\mu(X)$ and $L^2_\nu(Y)$ respectively, and defines an isomorphism between $\mathcal{V} \otimes \mathcal{W}$ and a closed subspace of $L^2_{\mu\times\nu}(X \times Y)$.

Recalling that $\ell^2(J)$ for a finite or countably infinite index set J is equal to $L^2(J)$ with respect to the counting measure, the final claim follows from the above. □

With the exception of not being canonical, the final claim in Corollary 3.7 gives a convenient way of thinking about the tensor product of Hilbert spaces. Fix a basis $(e_j \mid j \in J)$ of $\mathcal{V}$ and a basis $(f_k \mid k \in K)$ of $\mathcal{W}$ for some finite or countable index sets J and K. Then let us write $e_j \otimes f_k$ for the basis vector in $\ell^2(J \times K)$ corresponding to the index $(j,k) \in J \times K$. This gives the identification

$$\mathcal{V} \otimes \mathcal{W} \cong \ell^2(J) \otimes \ell^2(K) = \ell^2(J \times K),$$

and for $v = \sum_j \alpha_j e_j$ and $w = \sum_k \beta_k f_k$ the tensor product map

$$v \otimes w = \sum_{j,k} \alpha_j \beta_k e_j \otimes f_k.$$

Exercise 3.8. Show that for convergent series $v = \sum_{m=1}^{\infty} v_m$ in $\mathcal{V}$ and $w = \sum_{n=1}^{\infty} w_n$ in $\mathcal{W}$, we have that $v \otimes w = \sum_{m=1}^{\infty} \sum_{n=1}^{\infty} v_m \otimes w_n$ converges also. Moreover, absolute convergence of the series for v and w implies absolute convergence of the series for $v \otimes w$.

Corollary 3.9 (Tensor operators). *Let $A\colon \mathcal{V} \to \mathcal{V}$ and $B\colon \mathcal{W} \to \mathcal{W}$ be bounded operators. Then there exists a uniquely determined bounded operator*

$$A \otimes B\colon \mathcal{V} \otimes \mathcal{W} \longrightarrow \mathcal{V} \otimes \mathcal{W}$$

with

$$A \otimes B(v \otimes w) = Av \otimes Bw \tag{3.4}$$

for $v \in \mathcal{V}$ and $w \in \mathcal{W}$. Moreover,

- $\|A \otimes B\|_{\mathrm{op}} = \|A\|_{\mathrm{op}} \|B\|_{\mathrm{op}}$;
- $(A \otimes B)^* = A^* \otimes B^*$;
- *if A and B are self-adjoint, then $A \otimes B$ is self-adjoint;*
- *if A and B are unitary, then $A \otimes B$ is unitary; and*
- $(A \otimes B)(A' \otimes B') = (AA') \otimes (BB')$ *for bounded operators $A'\colon \mathcal{V} \to \mathcal{V}$ and $B'\colon \mathcal{W} \to \mathcal{W}$.*

PROOF. Let (f_k) be a basis of $\mathcal{W}$, where we implicitly let k run over all $k \in \mathbb{N}$ with $k \leqslant \dim \mathcal{W}$. We define

$$A \otimes I\colon \mathcal{V} \otimes \mathcal{W} \longrightarrow \mathcal{V} \otimes \mathcal{W}$$

by using the isomorphism[†]

$$\mathcal{V}^{\dim \mathcal{W}} \ni (v_k)_k \longmapsto \sum_k v_k \otimes f_k \in \bigoplus_k \mathcal{V} \otimes f_k = \mathcal{V} \otimes \mathcal{W} \tag{3.5}$$

and applying A in each component. More formally, we use the isomorphism in (3.5) and its inverse to make the definition

[†] To see that this is indeed an isomorphism, choose a basis of $\mathcal{V}$ and use the last statement of Corollary 3.7.

$$(A \otimes I)\Big(\sum_k v_k \otimes f_k\Big) = \sum_k (Av_k) \otimes f_k$$

for any sequence $(v_k) \in \mathcal{V}^{\dim \mathcal{W}}$ with $\sum_k \|v_k\|^2 < \infty$. We claim that this indeed defines a bounded operator satisfying $\|A \otimes I\|_{\text{op}} = \|A\|_{\text{op}}$. In fact, for any

$$t = \sum_k v_k \otimes f_k \in \mathcal{V} \otimes \mathcal{W}$$

we have

$$\begin{aligned} \|(A \otimes I)t\|_{\otimes}^2 &= \Big\|\sum_k (Av_k) \otimes f_k\Big\|_{\otimes}^2 \\ &= \sum_k \|(Av_k) \otimes f_k\|_{\otimes}^2 \\ &= \sum_k \|Av_k\|^2 \leqslant \|A\|_{\text{op}}^2 \sum_k \|v_k\|^2 = \|A\|_{\text{op}}^2 \|t\|^2, \end{aligned}$$

which gives $\|A \otimes I\|_{\text{op}} \leqslant \|A\|_{\text{op}}$. Together with

$$\sup_{v \in \mathcal{V}, \|v\| \leqslant 1} \|(A \otimes I)(v \otimes f_k)\|_{\otimes} = \sup_{v \in \mathcal{V}, \|v\| \leqslant 1} \|Av\| = \|A\|_{\text{op}}$$

for any k, we in fact obtain $\|A \otimes I\|_{\text{op}} = \|A\|_{\text{op}}$.

By continuity of $A \otimes I$ and bilinearity and continuity of the tensor product, we also have

$$A \otimes I(v \otimes w) = \sum_k \beta_k \underbrace{(A \otimes I)(v \otimes f_k)}_{(Av \otimes f_k)} = Av \otimes w \tag{3.6}$$

for $v \in \mathcal{V}$ and $w = \sum_k \beta_k f_k \in \mathcal{W}$.

Given another bounded operator $B \colon \mathcal{W} \to \mathcal{W}$ we define $I \otimes B$ similarly, also satisfying $\|I \otimes B\| = \|B\|$ and the relation

$$(I \otimes B)(v \otimes w) = v \otimes Bw$$

for all $v \in \mathcal{V}$ and $w \in \mathcal{W}$. Using (3.6) and the latter identity we obtain

$$(A \otimes I)(I \otimes B) = (I \otimes B)(A \otimes I),$$

which suggests the definition

$$A \otimes B = (A \otimes I)(I \otimes B),$$

satisfying (3.4). The remaining properties we leave as an exercise. □

Essential Exercise 3.10. Prove the remaining properties of $A \otimes B$ claimed in Corollary 3.9.

Proposition 3.11 (Outer tensor product). *Given unitary representations π of G and ρ of H, there exists a unitary representation $\pi \otimes \rho$ of $G \times H$, which we call the* outer *or* Kronecker tensor product representation, *on the tensor product $\mathcal{H}_\pi \otimes \mathcal{H}_\rho$ with the property that*

$$\left\langle (\pi \otimes \rho)_{(g,h)}(v_1 \otimes w_1), v_2 \otimes w_2 \right\rangle = \left\langle \pi_g v_1, v_2 \right\rangle \left\langle \rho_h w_1, w_2 \right\rangle \tag{3.7}$$

for all $g \in G$, $h \in H$, $v_1, v_2 \in \mathcal{H}_\pi$, and $w_1, w_2 \in \mathcal{H}_\rho$.

PROOF. We define $(\pi \otimes \rho)_{(g,h)} = \pi_g \otimes \rho_h$ for $g \in G$ and $h \in H$, which satisfies (3.7) due to the definition of $\pi_g \otimes \rho_h$ in (3.4) and the definition of the tensor product in Proposition 3.4. Moreover, Corollary 3.9 also implies that $(\pi \otimes \rho)_{(g,h)}$ is unitary for all $(g,h) \in G \times H$ and that $\pi \otimes \rho$ is a homomorphism. It remains to prove continuity, where we will apply Lemma 1.13 for $D = \{v \otimes w \mid v \in \mathcal{V}, w \in \mathcal{W}\}$. So assume $v \in \mathcal{V}$ and $w \in \mathcal{W}$. Then $G \ni g \mapsto \pi_g v \in \mathcal{V}$ and $H \ni h \mapsto \rho_h w \in \mathcal{W}$ are continuous by assumption, and

$$G \times H \ni (g,h) \longmapsto (\pi_g \otimes \rho_h)(v \otimes w) = \pi_g v \otimes \rho_h w$$

is therefore continuous by Proposition 3.4. This shows the continuity of the representation. □

Exercise 3.12 (Tensor product of cyclic representations). Let π be a cyclic representation of G with generator $v_0 \in \mathcal{H}_\pi$ and let ρ be a cyclic representation of H with generator $w_0 \in \mathcal{H}_\rho$. Show that $v_0 \otimes w_0 \in \mathcal{H}_\pi \otimes \mathcal{H}_\rho$ is a generator for $\pi \otimes \rho$.

In the case of $G = H$ we define the *inner tensor representation* $\pi \otimes \rho$ of G by

$$(\pi \otimes \rho)_g = \pi_g \otimes \rho_g$$

for all $g \in G$. We will always make clear whether we are dealing with an inner or an outer tensor product.

Exercise 3.13. Let $K \subseteq G$ be a compact subset. Show that the linear hull of all diagonal and non-diagonal matrix coefficients of all irreducible unitary representations of G restricted to K spans a dense subspace of $C(K)$.

Exercise 3.14. Let G and H be abelian groups as in Chapter 2. Let π be a unitary representation of G, and let ρ be a unitary representation of H.
(a) Describe the spectral measure of $v \otimes w$ for $v \in \mathcal{H}_\pi$ and $w \in \mathcal{H}_\rho$ with respect to the outer tensor product representation $\pi \otimes \rho$ of $G \times H$ on $\mathcal{H}_\pi \otimes \mathcal{H}_\rho$ in terms of the spectral measures of v and w.
(b) Assume that $G = H$ and repeat the above for the inner tensor product representation.

This concludes the material required for the discussion of unitary representations of compact groups.

3.2.2 Irreducibility of Outer Tensor Products

Proposition 3.15 (Irreducibility of tensor products). *Given irreducible unitary representations π of G and ρ of H, the outer tensor product representation $\pi \otimes \rho$ of $G \times H$ on $\mathcal{H}_\pi \otimes \mathcal{H}_\rho$ is again irreducible.*

The following lemma, which is based on Schur's lemma, will be essential for the proof.

Lemma 3.16. *Let π be an irreducible unitary representation of G and $\mathcal{W}$ a Hilbert space. Suppose that $T \in \mathrm{B}(\mathcal{H}_\pi \otimes \mathcal{W})$ is intertwining for the representation $\pi \otimes I$ of G defined by $(\pi \otimes I)_g = \pi_g \otimes I$ for $g \in G$. Then $T = I \otimes B$ for some $B \in \mathrm{B}(\mathcal{W})$.*

Proof. We fix a basis (f_k) of $\mathcal{W}$, and will again use the unitary isomorphism

$$\mathcal{H}_\pi \otimes \mathcal{W} = \bigoplus_k \mathcal{H}_\pi \otimes f_k \cong \mathcal{H}_\pi^{\dim \mathcal{W}}. \tag{3.8}$$

By the properties of tensor operators in Corollary 3.9, this isomorphism is also intertwining between $\pi \otimes I$ and $\pi^{\dim \mathcal{W}}$ on $\mathcal{H}_\pi^{\dim \mathcal{W}}$. For $k \in \mathbb{N}$ with $k \leqslant \dim \mathcal{W}$, we write

$$\imath_k \colon \mathcal{H}_\pi \ni v \longmapsto v \otimes f_k \in \mathcal{H}_\pi \otimes \mathcal{W}$$

for the intertwining embedding into the kth subspace $\mathcal{H}_\pi \otimes f_k \subseteq \mathcal{H}_\pi \otimes \mathcal{W}$, and

$$P_k \colon \mathcal{H}_\pi \otimes \mathcal{W} \ni t = \sum_\ell v_\ell \otimes f_\ell \longmapsto v_k$$

for the intertwining projection.

For $T \in \mathrm{B}(\mathcal{H}_\pi \otimes \mathcal{W})$ as in the lemma and indices $k, \ell \leqslant \dim \mathcal{W}$ we then have an intertwining operator

$$T_{k,\ell} = P_k \circ T \circ \imath_\ell \colon \mathcal{H}_\pi \longrightarrow \mathcal{H}_\pi,$$

which by Schur's lemma (Theorem 1.29) must equal $b_{k,\ell} I$ for some $b_{k,\ell} \in \mathbb{C}$, where I denotes the identity on $\mathcal{H}_\pi$. For a unit vector $v \in \mathcal{H}_\pi$ and the basis vector f_ℓ, this shows that

$$T(v \otimes f_\ell) = \sum_k \underbrace{P_k\big(T(v \otimes f_\ell)\big)}_{=T_{k,\ell}(v)} \otimes f_k \tag{3.9}$$

$$= \sum_k v \otimes (b_{k,\ell} f_k)$$

$$= v \otimes \sum_k b_{k,\ell} f_k. \tag{3.10}$$

In fact the sum in (3.9) over $k \in \mathbb{N}$ converges due to the unitary isomorphism in (3.8), and (3.10) follows since

$$\mathcal{W} \ni w \longmapsto v \otimes w \in v \otimes \mathcal{W} \tag{3.11}$$

is an isometry.

Said differently, the above shows that $T(v \otimes \mathcal{W}) \subseteq v \otimes \mathcal{W}$ for any unit vector $v \in \mathcal{H}_\pi$. Using the isometry in (3.11) and its inverse, it follows that there exists a bounded operator $B_v \in \mathrm{B}(\mathcal{W})$ with

$$T(v \otimes w) = v \otimes B_v w$$

for all $w \in \mathcal{W}$.

Let $e_1, e_2, \dots \in \mathcal{H}_\pi$ be an orthonormal basis, let $j, k \leqslant \dim \mathcal{H}_\pi$ with $j \neq k$ be two indices, and define $v = \frac{1}{\sqrt{2}}(e_j + e_k)$. Then for $w \in \mathcal{W}$ we have

$$\begin{aligned} e_j \otimes B_{e_j} w + e_k \otimes B_{e_k} w &= T(e_j \otimes w + e_k \otimes w) \\ &= T\Big(\underbrace{\tfrac{1}{\sqrt{2}}(e_j + e_k)}_{v} \otimes \sqrt{2} w\Big) \\ &= \tfrac{1}{\sqrt{2}}(e_j + e_k) \otimes B_v(\sqrt{2} w) \\ &= e_j \otimes B_v(w) + e_k \otimes B_v(w), \end{aligned}$$

which implies that $B = B_{e_j} = B_{e_k}$ is independent of the basis vector. Using the fact that the linear hull of the subspaces $e_j \otimes \mathcal{W}$ for $j \leqslant \dim \mathcal{H}_\pi$ is dense and T is bounded, we obtain $T = I \otimes B$. □

With this lemma, we are ready to prove the proposition.

PROOF OF PROPOSITION 3.15. Suppose that $T \in \mathrm{B}(\mathcal{H}_\pi \otimes \mathcal{H}_\rho)$ is intertwining for the unitary representation $\pi \otimes \rho$ of $G \times H$. Restricting the unitary representation to G, we may apply Lemma 3.16 and obtain $T = I \otimes B$ for a bounded operator B on $\mathcal{H}_\rho$. Similarly, we obtain $T = A \otimes I$ for a bounded operator A on $\mathcal{H}_\pi$. We claim that this implies $T = sI$ for some $s \in \mathbb{C}$. Applying this to the intertwining projection operator $A = P_\mathcal{V}$ corresponding to a closed $\pi \otimes \rho$-invariant subspace $\mathcal{V} \subseteq \mathcal{H}_\pi \otimes \mathcal{H}_\rho$, it follows that either $\mathcal{V} = \{0\}$ or $\mathcal{V} = \mathcal{H}_\pi \otimes \mathcal{H}_\rho$ and so $\pi \otimes \rho$ is irreducible.

To see the claim, suppose first that $v \in \mathcal{H}_\pi$ and $w \in \mathcal{H}_\rho$ are unit vectors and extend these to orthonormal bases (e_j) of $\mathcal{H}_\pi$ with $e_1 = v$ and (f_k) of $\mathcal{H}_\rho$ with $f_1 = w$. By the last claim in Corollary 3.7 we know that the vectors $e_j \otimes f_k$ for $(j,k) \in J \times K$ form an orthonormal basis of $\mathcal{H}_\pi \otimes \mathcal{H}_\rho$. Let $(j,k) \neq (1,1)$. If $j \neq 1$ then we use $T = I \otimes B$ to obtain

$$\langle T(v \otimes w), e_j \otimes f_k \rangle = \langle v \otimes Bw, e_j \otimes f_k \rangle = \underbrace{\langle v, e_j \rangle}_{=0} \langle Bw, f_k \rangle = 0.$$

Similarly, if $k \neq 1$ then we use $T = A \otimes I$ to obtain that $T(v \otimes w)$ is orthogonal to $e_j \otimes f_k$. As this holds for any $(j,k) \neq (1,1)$ it follows that

$$T(v \otimes w) = s_{v,w} v \otimes w$$

for some $s_{v,w} \in \mathbb{C}$ depending on the unit vectors $v \in \mathcal{H}_\pi$ and $w \in \mathcal{H}_\rho$.

Let us now fix an orthonormal basis $e_1, e_2, \dots$ of $\mathcal{H}_\pi$ and an orthonormal basis $f_1, f_2, \dots$ of $\mathcal{H}_\rho$. By the above, there exists for every pair (j,k) of indices some $s_{j,k} \in \mathbb{C}$ with $T(e_j \otimes f_k) = s_{j,k} e_j \otimes f_k$. Using also $v = \frac{1}{\sqrt{2}}(e_j + e_{j'})$ for a second index $j' \leqslant \dim \mathcal{H}_\pi$, we obtain

$$s_{j,k} e_j \otimes f_k + s_{j',k} e_{j'} \otimes f_k = \sqrt{2} T\Big(\underbrace{\tfrac{1}{\sqrt{2}}(e_j + e_{j'})}_{v} \otimes f_k\Big) = s_{v,e_k}(e_j + e_{j'}) \otimes f_k$$

which implies that $s_{j,k} = s_{j',k} = s_k$ is independent of the first index. The argument for the second index is similar. Hence $T(e_j \otimes f_k) = s e_j \otimes f_k$ for a fixed $s \in \mathbb{C}$ and any pair (j,k) of indices. Since the vectors $e_j \otimes f_k$ for all pairs (j,k) form a basis, it follows that $T = sI$, as claimed. □

The converse to Proposition 3.15 does not hold for all pairs of groups, but does hold if at least one of the two groups is somewhat reasonable.

Proposition 3.17 (Partial characterization for products). *Let τ be an irreducible unitary representation of $G \times H$. Suppose that there exists a unitary representation $\pi \in \widehat{G}$ and $K \in \mathbb{N} \cup \{\infty\}$ such that for the restriction of τ to G we have $\mathcal{H}_\tau \cong \mathcal{H}_\pi^K$. Then there exists an irreducible representation $\rho \in \widehat{H}$ so that τ is isomorphic to the outer tensor product representation $\pi \otimes \rho$.*

PROOF. Let $\mathcal{W}$ be a K-dimensional Hilbert space so that

$$\mathcal{H}_\tau \cong \mathcal{H}_\pi^K \cong \mathcal{H}_\pi \otimes \mathcal{W}$$

as in (3.8). We may therefore assume that $\mathcal{H}_\tau = \mathcal{H}_\pi \otimes \mathcal{W}$, and that τ restricted to G is equal to $\pi \otimes I$. Note that τ_h for $h \in H$ is intertwining for $\tau|_G = \pi \otimes I$. By Lemma 3.16 this implies for $h \in H$ that $\tau_h = I \otimes \rho_h$ for an operator ρ_h in $\mathrm{B}(\mathcal{W})$. Since $\tau|_H$ is a unitary representation and $\mathcal{W} \cong v \otimes \mathcal{W}$ for any unit vector $v \in \mathcal{H}_\pi$, it follows that ρ is a unitary representation of H on $\mathcal{H}_\rho = \mathcal{W}$. Thus τ is the outer tensor product $\pi \otimes \rho$ on $\mathcal{H}_\tau = \mathcal{H}_\pi \otimes \mathcal{H}_\rho$. Irreducibility of τ now also implies that ρ must be irreducible. □

To summarize, Proposition 3.15 shows that the outer tensor product of irreducible unitary representations of G and H give rise to irreducible unitary representations of $G \times H$. Conversely, by Proposition 3.17 irreducible unitary representations of $G \times H$ arise in this way if the restriction to G is easily described with *one* irreducible representation of G. It may feel like the latter should always hold, since different irreducible representations should give rise to invariant subspaces, but this line of argument only works if it is somehow possible to separate different elements of $\widehat{G}$ by, for example, intertwining operators. We will see many examples where this can be done.

3.3 Structure of Unitary Representations

For the remainder of the chapter, we consider a metrizable compact group G. As mentioned in the introduction to the chapter, compactness of G has important consequences for its unitary representations.

Theorem 3.18 (Finite dimension). *Let π be an irreducible unitary representation of the compact group G. Then $\mathcal{H}_\pi$ has finite dimension.*

For non-compact groups this can fail quite badly. Indeed, Exercise 1.89 shows that for the group $\mathrm{SL}_2(\mathbb{R})$ there are no non-trivial finite-dimensional unitary representations.

Theorem 3.19 (Decomposability[(7)]). *Let ρ be a unitary representation of the compact group G. Then $\mathcal{H}_\rho$ is a direct sum of mutually orthogonal irreducible subspaces. More precisely, for every $[\pi] \in \widehat{G}$ define the linear hull $\mathcal{H}_\rho^{[\pi]}$ of all subspaces $\mathcal{V} \subseteq \mathcal{H}_\rho$ such that $\rho|_\mathcal{V} \cong \pi$. Then these subspaces are closed, mutually orthogonal,*

$$\mathcal{H}_\rho = \bigoplus_{[\pi]\in\widehat{G}} \mathcal{H}_\rho^{[\pi]},$$

and ρ restricted to $\mathcal{H}_\rho^{[\pi]}$ is isomorphic to $\mathrm{mult}(\pi, \rho) \in \mathbb{N}_0 \cup \{\infty\}$ *many copies of π. Here the* multiplicity $\mathrm{mult}(\pi, \rho)$ *is uniquely determined by ρ, but the isomorphism between $\mathcal{H}_\rho^{[\pi]}$ and $\mathcal{H}_\pi^{\mathrm{mult}(\pi,\rho)}$ is not.*

We have already seen in the paragraph after Exercise 2.3 (see also Theorem 2.17 and Proposition 2.57) that the regular representation of $\mathbb{R}$ on $L^2(\mathbb{R})$ does not even contain one irreducible subspace, also showing how Theorem 3.19 is a special property of compact groups.

To lighten the notation, we will sometimes write π as an abbreviation for $[\pi]$, as for example in the notation $\mathrm{mult}(\pi, \rho)$ for $[\pi] \in \widehat{G}$ and an arbitrary unitary representation ρ of G.

3.3.1 Equivariant Maps

For the proof of Theorems 3.18 and 3.19 we will need to construct intertwining maps using the next two lemmas.

Lemma 3.20 (Existence of intertwining operator). *Let π and ρ be unitary representations of the compact group G, and let $A \in \mathrm{B}(\mathcal{H}_\pi, \mathcal{H}_\rho)$ be a bounded operator. Then*

$$\widetilde{A}v = \int_G \rho_g A \pi_{g^{-1}} v \,\mathrm{d}m(g) \tag{3.12}$$

for $v \in \mathcal{H}_\pi$ defines another bounded operator from $\mathcal{H}_\pi$ to $\mathcal{H}_\rho$ that is intertwining.

PROOF. The integral in (3.12) is to be understood weakly, as in our definition of convolution operators. More precisely, for $v \in \mathcal{H}_\pi$ we define $\widetilde{A}v \in \mathcal{H}_\rho$ using the Fréchet–Riesz representation theorem and the formula

$$\langle \widetilde{A}v, w\rangle = \int_G \langle \rho_g A \pi_{g^{-1}} v, w\rangle \, dm(g)$$

for every $w \in \mathcal{H}_\rho$, where the function $g \mapsto \langle \rho_g A \pi_{g^{-1}} v, w\rangle = \langle A\pi_{g^{-1}} v, \rho_{g^{-1}} w\rangle$ is continuous (by continuity of the unitary representation and the inner product) and satisfies $|\langle \rho_g A \pi_{g^{-1}} v, w\rangle| \leqslant \|A\|\|v\|\|w\|$. Hence $\widetilde{A}$ is a well-defined linear operator with $\|\widetilde{A}\| \leqslant \|A\|$ (see also the argument in Section 1.5.3).

The intertwining property for $\widetilde{A}$ now follows quickly from the properties of Haar measure. Indeed,

$$\widetilde{A}\pi_{g_0} v = \int_G \rho_g A \pi_{g^{-1} g_0} v \, dm(g) = \int_G \rho_{g_0} \rho_k A \pi_{k^{-1}} v \, dm(k) = \rho_{g_0} \widetilde{A} v$$

for $g_0 \in G$ by the substitution $k = g_0^{-1} g$. □

Lemma 3.21 (Existence of compact intertwining operator). *Let π be a unitary representation of the compact group G. For a unit vector $u \in \mathcal{H}_\pi$ we define the operator T by*

$$Tv = \int_G \langle v, \pi_g u\rangle \pi_g u \, dm(g)$$

for all $v \in \mathcal{H}_\pi$ Then T is positive, self-adjoint, intertwining, non-trivial, and compact. Moreover, $u \in (\ker T)^\perp$.

PROOF. The intertwining property for T follows from Lemma 3.20, since

$$\pi_g A \pi_{g^{-1}} v = \langle \pi_{g^{-1}} v, u\rangle \pi_g u = \langle v, \pi_g u\rangle \pi_g v$$

for $Av = \langle v, u\rangle u$ and $g \in G$, for all $v \in \mathcal{H}_\pi$. Hence $\widetilde{A} = T$.

To prove positivity, let $v \in \mathcal{H}_\pi$ and calculate

$$\langle Tv, v\rangle = \int_G \underbrace{\langle v, \pi_g u\rangle \langle \pi_g u, v\rangle}_{=|\langle v, \pi_g u\rangle|^2} \, dm(g) \geqslant 0.$$

Positivity also implies that T is self-adjoint, which can also be checked directly.

If now $v \in \ker T$, then $\langle Tv, v\rangle = 0$ and so by the argument above and the continuity of the map $g \mapsto \langle v, \pi_g u\rangle$ it follows that $\langle v, \pi_g u\rangle = 0$ for all $g \in G$. In particular, $u \perp \ker T$ and so $T \neq 0$.

It remains to prove that T is compact, which we will do by approximating T uniformly by operators with finite-dimensional range. For this, note first that $g \mapsto \pi_g u$ is uniformly continuous on G by compactness. Hence, given some $\varepsilon > 0$ there exists a finite measurable partition $\mathcal{P} = \{B_1, \dots, B_n\}$ of G with $\max\{\operatorname{diam} B_j \mid j = 1, \dots, n\}$ small enough to ensure that $g, g_j \in B_j$ implies that $\|\pi_g u - \pi_{g_j} u\| < \varepsilon$. It follows that

$$\begin{aligned}&\|\langle v, \pi_g u\rangle \pi_g u - \langle v, \pi_{g_j} u\rangle \pi_{g_j} u\| \\ &\qquad \leqslant |\langle v, \pi_g u\rangle| \|\pi_g u - \pi_{g_j} u\| + |\langle v, \pi_g u - \pi_{g_j} u\rangle| \|\pi_{g_j} u\| \leqslant 2\varepsilon \|v\|\end{aligned}$$

for all $v \in \mathcal{H}_\pi$, since u is a unit vector. Using this partition and the sample points $g_j \in B_j$ for $j = 1, \dots, n$ we define the Riemann sum approximation

$$T_{\mathcal{P}} \colon \mathcal{H}_\pi \ni v \longmapsto T_{\mathcal{P}} v = \sum_{j=1}^{n} \langle v, \pi_{g_j} u\rangle m(B_j) \pi_{g_j} u$$

with values in $\langle \pi_{g_1} u, \dots, \pi_{g_n} u\rangle$. This defines a bounded operator on $\mathcal{H}_\pi$ with

$$\begin{aligned}\|T - T_{\mathcal{P}}\| &= \sup_{\|v\| \leqslant 1} \|Tv - T_{\mathcal{P}} v\| \\ &= \sup_{\|v\| \leqslant 1} \left\| \sum_{j=1}^{n} \int_{B_j} \left(\langle v, \pi_g u\rangle \pi_g u - \langle v, \pi_{g_j} u\rangle \pi_{g_j} u \right) \mathrm{d}m(g) \right\| \leqslant 2\varepsilon.\end{aligned}$$

As $T_{\mathcal{P}}$ has finite-dimensional range and $\varepsilon > 0$ was arbitrary, the lemma follows from [25, Lemma 6.7]. □

3.3.2 Proof of Theorems

PROOF OF THEOREM 3.18. Suppose that π is an irreducible unitary representation and that $u \in \mathcal{H}_\pi$ is a unit vector. Applying Lemma 3.21 we find a non-trivial compact intertwining operator $T \in \mathrm{B}(\mathcal{H}_\pi)$. By Schur's lemma (Theorem 1.29), irreducibility implies that $T = \lambda I$ for some $\lambda \in \mathbb{C}$. However, as $T \neq 0$ is a compact operator, we see that $\dim \mathcal{H}_\pi < \infty$ (see, for example, the discussion after [25, Def. 6.2]). □

PROOF OF THEOREM 3.19: DECOMPOSABILITY. Let ρ be a unitary representation of the compact group G and let $u \in \mathcal{H}_\rho$ be a unit vector. Apply Lemma 3.21 again to obtain an intertwining compact operator T so that $u \in (\ker T)^\perp$. Since T is compact and self-adjoint, there exists a (possibly finite) sequence (λ_n) of non-zero eigenvalues of T such that

$$\mathcal{H}_\rho = \bigoplus_{n \geqslant 1} \mathcal{W}_n \oplus \ker T,$$

where $\mathcal{W}_n = \{v \in \mathcal{H}_\rho \mid Tv = \lambda_n v\}$ is the finite-dimensional eigenspace of T with eigenvalue λ_n (see [25, Th. 6.27]). As T is intertwining it follows that each subspace $\mathcal{W}_n$ is ρ-invariant. Indeed $g \in G$ and $v \in \mathcal{W}_n$ implies that

$$T\rho_g v = \rho_g T v = \lambda_n \rho_g v,$$

and so $\rho_g v \in \mathcal{W}_n$. Using induction on the dimension and the fact that every invariant subspace has an invariant complement (see Exercise 1.24) it follows that every finite-dimensional representation and, in particular, each of the subspaces $\mathcal{W}_n$ is a finite direct sum of irreducible subspaces. It follows that $\bigoplus_{n\geqslant 1} \mathcal{W}_n$ is at most a countable direct sum of irreducible subspaces. Recall that $u \in \bigoplus_{n\geqslant 1} \mathcal{W}_n$.

To deduce the theorem from this, let $v_1, v_2, \dots$ be an orthonormal basis of $\mathcal{H}_\rho$. Apply the argument above, first to $u = v_1$ to obtain a direct sum $\mathcal{W}$ of irreducible subspaces that contains v_1. If $v_2 \notin \mathcal{W}$ project this vector to $\mathcal{W}^\perp$, and define the operator T as above but for the restriction of π to $\mathcal{W}^\perp$ and the normalized projection of v_2 to $\mathcal{W}^\perp$. This produces a further collection of mutually orthogonal irreducible subspaces in $\mathcal{W}^\perp$ such that v_1 and v_2 belong to the direct sum $\widetilde{W}$ of $\mathcal{W}$ and this new irreducible subspace.

Repeating the argument inductively produces mutually orthogonal irreducible subspaces whose direct sum contains all the basis vectors. □

To complete the proof of Theorem 3.19 it remains to discuss the subspaces $\mathcal{H}_\rho^{[\pi]}$ and the multiplicity $\mathrm{mult}(\pi, \rho)$ for all $[\pi] \in \widehat{G}$. For this, let us first summarize what we have obtained thus far in a convenient notation. Given the representation ρ we have found (finite-dimensional) irreducible subspaces $\mathcal{V}_1, \mathcal{V}_2, \dots < \mathcal{H}_\rho$ such that $\mathcal{H}_\rho = \bigoplus_{n\geqslant 1} \mathcal{V}_n$. For any $[\pi] \in \widehat{G}$ we denote those subspaces $\mathcal{V}_n$ with $\rho|_{\mathcal{V}_n} \cong \pi$ by $\mathcal{V}_\ell^\pi$ for $\ell = 1, \dots, m([\pi]) = m(\pi)$, where $m(\pi) \in \mathbb{N} \cup \{\infty\}$ is the total number of such subspaces. In this notation, we have shown that

$$\mathcal{H}_\rho = \bigoplus_{[\pi]\in\widehat{G}} \underbrace{\bigoplus_{\ell=1}^{m(\pi)} \mathcal{V}_\ell^\pi}_{=\mathcal{W}^\pi} = \bigoplus_{[\pi]\in\widehat{G}} \mathcal{W}^\pi. \tag{3.13}$$

The following lemma will be useful for the second part of the proof of Theorem 3.19, but is also of independent interest (see also Exercise 3.23).

Lemma 3.22 (Bound on multiplicity). *If ρ is a cyclic representation of the compact group G and*

$$\mathcal{H}_\rho = \bigoplus_{[\pi]\in\widehat{G}} \bigoplus_{\ell=1}^{m(\pi)} \mathcal{V}_\ell^\pi,$$

then $m(\pi) \leqslant \dim \mathcal{H}_\pi$ for every $[\pi] \in \widehat{G}$.

PROOF. Fix some $[\pi] \in \widehat{G}$. Projecting the generator to $\bigoplus_{\ell=1}^{m(\pi)} \mathcal{V}_\ell^\pi$ it follows that the latter is also cyclic, and we may suppose that

$$\mathcal{H}_\rho = \bigoplus_{\ell=1}^{m(\pi)} \mathcal{V}_\ell^\pi.$$

Let $d = \dim \mathcal{H}_\pi$ and suppose that $m = m(\pi) > d$. Projecting the generator to the orthogonal sum of the first $d+1$ subspaces, we may assume that m is $d+1$ and that $\mathcal{H}_\rho = \mathcal{H}_\pi^{d+1}$ is cyclic with generator $v = (v_1, \ldots, v_{d+1})$. Permuting the indices if necessary, it follows that $v_{d+1} = \alpha_1 v_1 + \cdots + \alpha_d v_d$ for some choice of $\alpha_1, \ldots, \alpha_d \in \mathbb{C}$. Now define the proper subspace

$$\mathcal{W} = \{(w_1, \ldots, w_d, \alpha_1 w_1 + \cdots + \alpha_d w_d) \mid w_1, \ldots, w_d \in \mathcal{H}_\pi\} < \mathcal{H}_\pi^{d+1}$$

and notice that it is invariant under $\rho = \bigoplus_{\ell=1}^{d+1} \pi$ and contains the vector v that generates $\mathcal{H}_\pi^{d+1}$. This contradiction proves the lemma. □

PROOF OF THEOREM 3.19 CONTINUED: MULTIPLICITIES. Let ρ be a unitary representation of G. As already done in the theorem, we define for every irreducible representation π the subspace $\mathcal{H}_\rho^{[\pi]}$ as the linear hull over those subspaces $\mathcal{V} \subseteq \mathcal{H}_\rho$ such that $\rho|_\mathcal{V} \cong \pi$. Furthermore, let $\mathcal{V}_\ell^\pi$ for $\ell = 1, \ldots, m(\pi)$ and $\mathcal{W}^\pi = \bigoplus_{\ell=1}^{m(\pi)} \mathcal{V}_\ell^\pi$ for all $\pi \in \widehat{G}$ be as in (3.13) so that the space $\mathcal{H}_\rho$ is the orthogonal direct sum of the subspaces $\mathcal{W}^\pi$ for $\pi \in \widehat{G}$. We wish to show that $\mathcal{H}_\rho^{[\pi_0]} = \mathcal{W}^{\pi_0}$ for every $[\pi_0] \in \widehat{G}$.

Suppose for a moment that π_1 and π_2 are inequivalent irreducible representations of G and $\mathcal{V}_1, \mathcal{V}_2 \subseteq \mathcal{H}_\rho$ are invariant subspaces such that $\rho|_{\mathcal{V}_j}$ is isomorphic to π_j for $j = 1, 2$. Using invariance of $\mathcal{V}_2$ we see first that the orthogonal projection $P\colon \mathcal{H}_\rho \to \mathcal{V}_2$ is intertwining. Composing the projection with the intertwining isometries $\mathcal{H}_{\pi_1} \to \mathcal{V}_1$ and $\mathcal{V}_2 \to \mathcal{H}_{\pi_2}$ gives now an intertwining map $\mathcal{H}_{\pi_1} \to \mathcal{H}_{\pi_2}$. By Schur's lemma (Theorem 1.29) it follows that $P|_{\mathcal{V}_1} = 0$ and so $\mathcal{V}_1 \perp \mathcal{V}_2$.

We fix some $[\pi_0] \in \widehat{G}$ and return to our discussion of the subspaces $\mathcal{W}^{\pi_0}$ and $\mathcal{H}_\rho^{[\pi_0]}$. Let $\mathcal{V}_0 \subseteq \mathcal{H}_\rho$ be an irreducible subspace with $\rho|_{\mathcal{V}_0} \cong \pi_0$, let π be an irreducible representation with $\pi \neq [\pi_0]$, and let $\mathcal{V}_\ell^\pi \subseteq \mathcal{W}^\pi$ be one of the subspaces in the definition of $\mathcal{W}^\pi$. Then the argument above shows $\mathcal{V}_0 \perp \mathcal{V}_\ell^\pi$. Varying the subspaces we obtain $\mathcal{H}_\rho^{[\pi_0]} \perp \mathcal{W}^\pi$ and varying $[\pi] \in \widehat{G} \smallsetminus \{[\pi_0]\}$, we obtain $\mathcal{H}_\rho^{[\pi_0]} \subseteq \mathcal{W}^{\pi_0}$, since $\mathcal{H}_\rho$ is the orthogonal direct sum $\bigoplus_{[\pi] \in \widehat{G}} \mathcal{W}^\pi$.

For the converse inclusion, let $v \in \mathcal{W}^{\pi_0} \smallsetminus \{0\}$ and apply the first part of the theorem to the cyclic representation $\langle v \rangle_\rho \subseteq \mathcal{W}^{\pi_0}$ generated by v. This shows that $\langle v \rangle_\rho = \bigoplus_{k=1}^K \mathcal{V}_k'$ for irreducible subspaces $\mathcal{V}_k' \subseteq \langle v \rangle_\rho$ and $K \in \mathbb{N} \cup \{\infty\}$. By the above we have $\mathcal{H}_\rho^{[\pi]} \subseteq \mathcal{W}^\pi$ for $[\pi] \in \widehat{G} \smallsetminus \{[\pi_0]\}$, which implies that $\rho|_{\mathcal{V}_k'} \cong \pi_0$ for all k. By Lemma 3.22 we now see that $K \leqslant \dim \mathcal{H}_\pi$ and hence that $\langle v \rangle_\rho$ is a finite sum of irreducible subspaces, which implies that $v \in \mathcal{H}_\rho^{[\pi_0]}$. As $v \in \mathcal{W}^\pi \smallsetminus \{0\}$ was arbitrary, we see that $\mathcal{W}^{\pi_0} = \mathcal{H}_\rho^{[\pi_0]}$, and in particular that $\mathcal{H}_\rho^{[\pi_0]}$ is closed.

As $\mathcal{H}_\rho^{[\pi_0]}$ is canonically defined, it follows that

$$\dim \mathcal{H}_\rho^{[\pi_0]} = m(\pi_0) \dim \mathcal{H}_{\pi_0}$$

is independent of the choices that were made to arrive at the splitting of $\mathcal{H}$ into irreducible subspaces $\mathcal{V}_1, \mathcal{V}_2, \ldots$. As $\dim \mathcal{H}_{\pi_0} < \infty$ by Theorem 3.18, this implies that

$$\mathrm{mult}(\pi_0, \rho) = m(\pi_0) = \frac{\dim \mathcal{H}_\rho^{[\pi_0]}}{\dim \mathcal{H}_{\pi_0}}$$

is well-defined.

Finally, note that the isomorphism U_π between $\mathcal{H}_\rho^\pi$ and $\mathcal{H}_\pi^{\mathrm{mult}(\pi,\rho)}$ can be changed in many ways. For instance, U_π can be composed with a permutation of the subspaces if $\mathrm{mult}(\pi, \rho) > 1$ or multiplied by a scalar of absolute value one whenever $\mathcal{H}_\rho^\pi \neq \{0\}$. □

Exercise 3.23. (a) Let π be a finite-dimensional irreducible unitary representation of G. Show that the unitary representation on $\mathcal{H}_\pi^{\dim \mathcal{H}_\pi}$ is cyclic.
(b) Let $I \subseteq \mathbb{N}$ and $[\pi_j] \in \widehat{G}$ for $j \in I$ such that $j \neq k$ in I implies that $\pi_j \not\sim \pi_k$ (equivalently, $[\pi_j] \neq [\pi_k]$). Show that $\bigoplus_{j\in I} \mathcal{H}_{\pi_j}^{\dim \mathcal{H}_{\pi_j}}$ is cyclic.

Exercise 3.24. Let $G = \mathbb{Z}/2\mathbb{Z} \ltimes \mathbb{T}$, with group operation defined by

$$(a_1, x_1) \cdot (a_2, x_2) = (a_1 + a_2, x_1 + (-1)^{a_1} x_2).$$

Describe the unitary dual $\widehat{G}$.

3.3.3 Containment

The description of unitary representations in Theorem 3.19 allows us to describe containment quite clearly.

Corollary 3.25 (Characterization of containment). *Let ρ_1 and ρ_2 be unitary representations of the compact group G. Then $\rho_1 < \rho_2$ if and only if we have* $\mathrm{mult}(\pi, \rho_1) \leqslant \mathrm{mult}(\pi, \rho_2)$ *for every* $[\pi] \in \widehat{G}$.

PROOF. Suppose that $\rho_1 < \rho_2$. Then we may assume that $\mathcal{H}_{\rho_1} \subseteq \mathcal{H}_{\rho_2}$ and that $\rho_1 = \rho_2|_{\mathcal{H}_{\rho_1}}$. Applying the decomposition theorem (Theorem 3.19) to $\mathcal{H}_{\rho_1}$ and $\mathcal{H}_{\rho_1}^\perp \subseteq \mathcal{H}_{\rho_2}$, it follows that

$$\mathrm{mult}(\pi, \rho_1) \leqslant \mathrm{mult}(\pi, \rho_1) + \mathrm{mult}(\pi, \rho_2|_{\mathcal{H}_{\rho_1}^\perp}) = \mathrm{mult}(\pi, \rho_2)$$

for every $[\pi] \in \widehat{G}$.

For the converse, we suppose that $\mathrm{mult}(\pi, \rho_1) \leqslant \mathrm{mult}(\pi, \rho_2)$ for every π in $\widehat{G}$, and obtain from the decomposition theorem for ρ_1 and ρ_2 that

$$\mathcal{H}_{\rho_1} \cong \bigoplus_{[\pi]\in\widehat{G}} \mathcal{H}_\pi^{\mathrm{mult}(\pi,\rho_1)} \subseteq \bigoplus_{[\pi]\in\widehat{G}} \mathcal{H}_\pi^{\mathrm{mult}(\pi,\rho_2)} \cong \mathcal{H}_{\rho_2}$$

as required. □

3.3.4 Irreducible Representations of Products

In this section we let G denote a compact metric group and H a locally compact, σ-compact metric group. We wish to relate the irreducible representations in $\widehat{G}$ and $\widehat{H}$ to the irreducible representations of $G \times H$, giving the converse to Proposition 3.11.

Proposition 3.26 (The unitary dual of compact products). *Let G be a compact metric group and H a locally compact, σ-compact metric group. Then the irreducible representations of the direct product $G \times H$ are precisely of the form $\pi \otimes \rho$ for $[\pi] \in \widehat{G}$ and $[\rho] \in \widehat{H}$.*

PROOF. The irreducibility of $\pi \otimes \rho$ for $[\pi] \in \widehat{G}$ and $[\rho] \in \widehat{H}$ holds more generally, and was established in Proposition 3.15.

For the converse, we suppose that τ is an irreducible unitary representation of $G \times H$. We restrict τ to the compact group G and apply Theorem 3.19. It follows that

$$\mathcal{H}_\tau = \bigoplus_{[\pi] \in \widehat{G}} \mathcal{H}_\tau^{[\pi]},$$

where $\mathcal{H}_\tau^{[\pi]}$ is the linear hull of all irreducible subspaces $\mathcal{V} < \mathcal{H}_\tau$ isomorphic to $\mathcal{H}_\pi$. Since the direct factors of $G \times H$ commute, it follows for $h \in H$ and a closed subspace $\mathcal{V} < \mathcal{H}_\tau$ that $\tau_h \mathcal{V} < \mathcal{H}_\tau$ is isomorphic to $\mathcal{V}$ with respect to the restriction of τ to G. Therefore $\tau_h \mathcal{H}_\tau^{[\pi]} < \mathcal{H}_\tau^{[\pi]}$ for $[\pi] \in \widehat{G}$ and $h \in H$. By irreducibility of $\mathcal{H}_\tau$ we must have $\mathcal{H}_\tau = \mathcal{H}_\tau^{[\pi]}$ for some $[\pi] \in \widehat{G}$. Let

$$K = \operatorname{mult}(\pi, \tau|_G) \in \mathbb{N} \cup \{\infty\}$$

so we may assume that $\mathcal{H}_\tau \cong \mathcal{H}_\pi^K$ by Theorem 3.19. By applying Proposition 3.17 it follows that there exists an irreducible unitary representation $\rho \in \widehat{H}$ on a K-dimensional Hilbert space so that $\tau = \pi \otimes \rho$. □

3.3.5 The Unitary Assumption

In all of our discussions up to this point (and in many of the following ones) we always start by assuming that the representation in question is unitary. In the context of compact groups this assumption can be weakened significantly.

Proposition 3.27 (Averaging the inner product). *Let $\mathcal{H}$ be a Hilbert space and suppose that π is a continuous representation of the compact group G on $\mathcal{H}$*

(so properties (1) *and* (3) *of Definition* 1.1 *hold, but we do not assume* (2)*). Then*

$$\langle v, w \rangle_\pi = \int \langle \pi_g v, \pi_g w \rangle_{\mathcal{H}} \, \mathrm{d}m(g)$$

for $v, w \in \mathcal{H}$ defines an inner product on $\mathcal{H}$ with the property that the induced norm is equivalent to the norm induced by $\langle \cdot, \cdot \rangle_{\mathcal{H}}$ and π is a unitary representation with respect to $\langle \cdot, \cdot \rangle_\pi$. In particular, this applies to every continuous finite-dimensional representation.

PROOF. Using the continuity of $G \ni g \mapsto \pi_g v$ for $v \in \mathcal{H}$, it is easy to see that $\langle \cdot, \cdot \rangle_\pi$ is well-defined and satisfies the axioms of an inner product. Moreover, also by compactness of G and the continuity requirement of the representation we have that $\{\pi_g v \mid g \in G\} \subseteq \mathcal{H}$ is bounded for every $v \in \mathcal{H}$. As $\mathcal{H}$ is a Hilbert space, the Banach–Steinhaus theorem on uniform boundedness (see [25, Th. 4.1]) applies and shows that $M = \sup\{\|\pi_g\| \mid g \in G\}$ is finite. Notice that $\|\pi_g v\|_{\mathcal{H}} \leqslant M\|v\|_{\mathcal{H}}$, which may be applied to g^{-1} and its action on the vector $w = \pi_g v$ to see that

$$\frac{1}{M}\|v\|_{\mathcal{H}} \leqslant \|\pi_g v\|_{\mathcal{H}} \leqslant M\|v\|_{\mathcal{H}}$$

for all $v \in \mathcal{H}$ and $g \in G$.

Let $\|\cdot\|_\pi$ denote the norm induced by $\langle \cdot, \cdot \rangle_\pi$. By integration, we now obtain

$$\frac{1}{M^2}\|v\|_{\mathcal{H}}^2 \leqslant \|v\|_\pi^2 = \int \|\pi_g v\|_{\mathcal{H}}^2 \, \mathrm{d}m(g) \leqslant M^2\|v\|_{\mathcal{H}}^2$$

for all v, which shows that $\|\cdot\|_{\mathcal{H}}$ and $\|\cdot\|_\pi$ are equivalent norms.

Finally, note that

$$\|\pi_g v\|_\pi^2 = \int \|\pi_{hg} v\|_{\mathcal{H}}^2 \, \mathrm{d}m(h) = \int \|\pi_k v\|_{\mathcal{H}}^2 \, \mathrm{d}m(k) = \|v\|_\pi^2$$

for all $v \in \mathcal{H}$, showing that π_g is unitary for all $g \in G$ with respect to the inner product $\langle \cdot, \cdot \rangle_\pi$ on $\mathcal{H}_\pi$. As $\|\cdot\|_{\mathcal{H}}$ and $\|\cdot\|_\pi$ are equivalent norms, it follows that $\mathcal{H}_\pi = \mathcal{H}$ is still a Hilbert space when equipped with the latter norm and that π still satisfies the continuity requirement, and so gives a unitary representation. □

3.4 The Regular Representation*

Having obtained the complete description of any unitary representation of the compact group G, we now move on to the regular representation λ. Here we will, in particular, also obtain complete knowledge of the multiplicities $\mathrm{mult}(\pi, \lambda)$ of a given irreducible unitary representation π of G. As a first step towards that goal, we study the matrix coefficients of unitary representations (which are continuous functions on G, and in particular are elements of $L^2(G)$ by compactness).

3.4.1 Schur Orthogonality

Definition 3.28. For a unitary representation π of the compact group G we define $\mathcal{M}_\pi = \langle \varphi_{u,v} \mid u, v \in \mathcal{H}_\pi \rangle$ to be the linear hull of all matrix coefficient for vectors in $\mathcal{H}_\pi$.

It is clear that $\mathcal{M}_\pi$ only depends on π up to unitary equivalence and so we may also define $\mathcal{M}_{[\pi]} = \mathcal{M}_\pi$ for any $[\pi] \in \widehat{G}$.

Proposition 3.29 (Dimension bound). *Let π be a unitary representation of the compact group G. Then $\mathcal{M}_\pi$ is invariant under the left- and right regular representations, is a two-sided ideal in $L^1(G)$ with respect to convolution, and satisfies*

$$\dim \mathcal{M}_\pi \leqslant (\dim \mathcal{H}_\pi)^2.$$

PROOF. To see the invariance under the regular representation let $u, v \in \mathcal{H}_\pi$ and $g, h \in G$. Then

$$\lambda_g(\varphi_{u,v})(h) = \varphi_{u,v}(g^{-1}h) = \langle \pi_{g^{-1}h} u, v \rangle = \langle \pi_h u, \pi_g v \rangle = \varphi_{u,\pi_g v}(h)$$

and

$$\rho_g(\varphi_{u,v})(h) = \varphi_{u,v}(hg) = \langle \pi_{hg} u, v \rangle = \langle \pi_h \pi_g u, v \rangle = \varphi_{\pi_g u, v}(h)$$

show that $\lambda_g \mathcal{M}_\pi \subseteq \mathcal{M}_\pi$ and $\rho_g \mathcal{M}_\pi \subseteq \mathcal{M}_\pi$. For the claim regarding convolution let $f \in L^1(G)$ to see that

$$\begin{aligned} f * \varphi_{u,v}(g) &= \int f(h) \varphi_{u,v}(h^{-1}g) \, \mathrm{d}m(h) \\ &= \int f(h) \langle \pi_{h^{-1}g} u, v \rangle \, \mathrm{d}m(h) \\ &= \int f(h) \langle \pi_g u, \pi_h v \rangle \, \mathrm{d}m(h) \\ &= \langle \pi_g u, \pi_*(\overline{f}) v \rangle = \varphi_{u, \pi_*(\overline{f}) v}(g) \end{aligned}$$

by definition of the convolution operator $\pi_*(\overline{f})$.

Similarly, we define $\widetilde{f}(k) = f(k^{-1})$ for $k \in G$ and obtain

$$\begin{aligned} \varphi_{u,v} * f(g) &= \int \varphi_{u,v}(h) f(h^{-1}g) \, \mathrm{d}m(h) \\ &= \int \langle \pi_g \pi_k u, v \rangle \, \widetilde{f}(k) \, \mathrm{d}m(k) \\ &= \Big\langle \pi_g \pi_*(\widetilde{f}) u, v \Big\rangle = \varphi_{\pi_*(\widetilde{f}) u, v}(g). \end{aligned}$$

It follows that $L^1(G) * \mathcal{M}_\pi \cup \mathcal{M}_\pi * L^1(G) \subseteq \mathcal{M}_\pi$.

Finally, if $w_1, w_2, \ldots, w_d \in \mathcal{H}_\pi$ is an orthonormal basis, then sesqui-linearity shows that $\mathcal{M}_\pi$ is spanned by φ_{w_i,w_j} for $i, j = 1, \ldots, d$. However, if $\mathcal{H}_\pi$ is infinite-dimensional, then there is nothing to prove. □

Theorem 3.30 (Schur orthogonality relations). *Let π and ρ be irreducible unitary representations of the compact group G and consider the subspaces $\mathcal{M}_\pi$ and $\mathcal{M}_\rho$ inside $L^2(G)$. If π and ρ are not unitarily equivalent, then $\mathcal{M}_\pi \perp \mathcal{M}_\rho$. Moreover, if $\{w_i \mid i = 1, \ldots, d_\pi\}$ is an orthonormal basis of $\mathcal{H}_\pi$ then*

$$\{\sqrt{d_\pi}\pi_{i,j} \mid i, j = 1, \ldots, d_\pi\}$$

is an orthonormal basis of $\mathcal{M}_\pi$, where $\pi_{i,j} = \varphi^\pi_{w_i,w_j}$ for $i, j = 1, \ldots, d_\pi$ and d_π is $\dim \mathcal{H}_\pi$. In particular, $\dim \mathcal{M}_\pi = d_\pi^2$.

Proof. We will apply Lemma 3.20 for various choices of A. We fix some vector $u_0 \in \mathcal{H}_\pi$, some $v_0 \in \mathcal{H}_\rho$ and consider the map A defined by

$$Au = \langle u, u_0 \rangle v_0$$

for $u \in \mathcal{H}_\pi$, which gives rise to

$$\widetilde{A}u = \int \langle \pi_{g^{-1}} u, u_0 \rangle \rho_g v_0 \, dm(g).$$

By definition we have

$$\left\langle \widetilde{A}u, v \right\rangle = \int \underbrace{\langle u, \pi_g u_0 \rangle}_{\overline{\varphi^\pi_{u_0,u}}} \underbrace{\langle \rho_g v_0, v \rangle}_{\varphi^\rho_{v_0,v}} \, dm(g) = \langle \varphi^\rho_{v_0,v}, \varphi^\pi_{u_0,u} \rangle$$

for all $u \in \mathcal{H}_\pi$ and $v \in \mathcal{H}_\rho$, where we write $\varphi^\pi_{u_0,u}$ for the matrix coefficient of $u_0, u \in \mathcal{H}_\pi$ defined by π and $\varphi^\rho_{v_0,v}$ for the matrix coefficient of $v_0, v \in \mathcal{H}_\rho$ defined by ρ.

Suppose now that π and ρ are not unitarily equivalent. Then $\widetilde{A} = 0$ by Schur's lemma (Theorem 1.29) and so the argument above shows that

$$\varphi^\pi_{u_0,u} \perp \varphi^\rho_{v_0,v}$$

for all $u_0, u \in \mathcal{H}_\pi$ and $v_0, v \in \mathcal{H}_\rho$, or equivalently that $\mathcal{M}_\pi \perp \mathcal{M}_\rho$.

Suppose now that $w_1, \ldots, w_{d_\pi}$ is an orthonormal basis of $\mathcal{H}_\pi$, let $\rho = \pi$ in the above discussion, and set $u_0 = w_k$ and $v_0 = w_j$ so that the linear map A has trace $\operatorname{tr}(A) = \delta_{k,j}$. Since $d_\pi = \dim \mathcal{H}_\pi < \infty$ we can obtain $\widetilde{A}$ also via the (matrix valued, Riemann-) integral

$$\widetilde{A} = \int \pi_g A \pi_g^{-1} \, dm(g).$$

Taking the trace we see that

$$\operatorname{tr} \widetilde{A} = \int \operatorname{tr}(\pi_g A \pi_g^{-1}) \, \mathrm{d}m(g) = \operatorname{tr}(A) = \delta_{k,j}.$$

By Schur's lemma (Theorem 1.29), $\widetilde{A} = \lambda I$ on $\mathcal{H}_\pi$ for some $\lambda \in \mathbb{C}$, which gives $\operatorname{tr} \widetilde{A} = \lambda d_\pi = \delta_{k,j}$. Hence we see that

$$\widetilde{A} = \tfrac{1}{d_\pi} \delta_{k,j} I.$$

We now apply $\widetilde{A}$ (which we defined using the basis vectors $u_0 = w_k$, respectively $v_0 = w_j$) to $u = w_\ell$ and take the inner product with $v = w_i$ to obtain

$$\tfrac{1}{d_\pi} \delta_{k,j} \delta_{\ell,i} = \left\langle \tfrac{1}{d_\pi} \delta_{k,j} w_\ell, w_i \right\rangle = \left\langle \widetilde{A} u, v \right\rangle = \left\langle \varphi^\pi_{v_0,v}, \varphi^\pi_{u_0,u} \right\rangle = \left\langle \pi_{i,j}, \pi_{k,\ell} \right\rangle,$$

where $\pi_{i,j} = \varphi^\pi_{w_i,w_j}$ and $\pi_{k,\ell} = \varphi^\pi_{w_k,w_\ell}$. Multiplying by d_π gives

$$\left\langle \sqrt{d_\pi} \pi_{i,j}, \sqrt{d_\pi} \pi_{k,\ell} \right\rangle = \delta_{k,i} \delta_{\ell,j}.$$

As in the proof of Proposition 3.29 sesqui-linearity of the matrix coefficients shows that $\mathcal{M}_\pi$ is generated by the vectors $\pi_{i,j}$ for $i, j = 1, \ldots, d_\pi$. Since these are orthogonal, it follows that $\dim \mathcal{M}_\pi = d_\pi^2$ and the theorem follows. □

3.4.2 A Dense Algebra

Recall that in the study of compact abelian groups a trigonometric polynomial is a finite linear combination of characters of the group, and that these form a dense sub-algebra of the space of continuous functions on the group with respect to the supremum norm. In the context of compact groups the following is the appropriate generalization, which as we will see will have the same properties.

Definition 3.31 (Matrix coefficient algebra). We define the *matrix coefficient algebra* of the compact group G to be the linear hull

$$\mathcal{M}(G) = \langle \mathcal{M}_\pi \mid [\pi] \in \widehat{G} \rangle.$$

The purpose of this section is to develop the main properties of $\mathcal{M}(G)$ as summarized in the following theorem.

Theorem 3.32 (Dense algebra). *For the compact group G the subspace*

$$\mathcal{M}(G) \subseteq C(G)$$

is a dense sub-algebra with respect to pointwise multiplication.

Our first step is to justify the nomenclature 'algebra' for $\mathcal{M}(G)$.

Lemma 3.33 (Algebra). *For the compact group G we have*

$$\mathcal{M}(G) = \langle \mathcal{M}_\rho \mid \rho \text{ is a unitary representation of } G \text{ with } \dim \mathcal{H}_\rho < \infty \rangle,$$

and that $\mathcal{M}(G)$ is a sub-algebra of $C(G)$, where $\mathcal{M}_\rho$ is defined as in Definition 3.28 for any unitary representation ρ of G.

PROOF. Let us write $\mathcal{F}$ for the linear span of all matrix coefficients of all finite-dimensional unitary representations of G. We first prove that $\mathcal{M}(G) = \mathcal{F}$ as claimed in the lemma. By Theorem 3.18 we have $\mathcal{M}_\pi \subseteq \mathcal{F}$ for all $\pi \in \widehat{G}$ and so $\mathcal{M}(G) \subseteq \mathcal{F}$. Suppose now that ρ is a finite-dimensional unitary representation of G and let $\mathcal{H}_\rho = \bigoplus_{n=1}^N \mathcal{V}_n$ be the decomposition of $\mathcal{H}_\rho$ into irreducible representations. Given $u = \sum_{n=1}^N u_n$ and $v = \sum_{n=1}^N v_n$ with $u_n, v_n \in \mathcal{V}_n$ for $n = 1, \ldots, N$ we have

$$\varphi^\rho_{u,v} = \sum_{n=1}^N \varphi^\rho_{u_n,v_n} \in \langle \mathcal{M}_{\rho|_{\mathcal{V}_n}} \mid n = 1, \ldots, N \rangle \subseteq \mathcal{M}(G).$$

Since $u, v \in \mathcal{H}_\rho$ and ρ were arbitrary as in the definition of $\mathcal{F}$ we obtain the opposite inclusion $\mathcal{F} \subseteq \mathcal{M}(G)$.

Suppose now that $\varphi^{\rho_1}_{u_1,v_1}$ and $\varphi^{\rho_2}_{u_2,v_2}$ are matrix coefficients for the finite-dimensional unitary representations ρ_1 and ρ_2 respectively, as in the definition of $\mathcal{F}$. Recalling from Section 3.1 the construction of the inner tensor product representation $\rho_1 \otimes \rho_2$ and especially Proposition 3.11 (which in the case at hand is easier as we are currently only dealing with finite-dimensional representations) we see that the product

$$\varphi^{\rho_1}_{u_1,v_1}(g)\varphi^{\rho_2}_{u_2,v_2}(g) = \langle (\rho_1 \otimes \rho_2)_g u_1 \otimes u_2, v_1 \otimes v_2 \rangle = \varphi^{\rho_1 \otimes \rho_2}_{u_1 \otimes u_2, v_1 \otimes v_2}(g)$$

is again a matrix coefficient for a finite-dimensional representation $\rho_1 \otimes \rho_2$ on $\mathcal{H}_{\rho_1} \otimes \mathcal{H}_{\rho_2}$ for the vectors $u_1 \otimes u_2$ and $v_1 \otimes v_2$. This implies that $\mathcal{M}(G) = \mathcal{F}$ is a sub-algebra of $C(G)$. □

PROOF OF THEOREM 3.32. We are going to apply the Stone–Weierstrass theorem to $\mathcal{M}(G)$. By Lemma 3.33 we know that $\mathcal{M}(G)$ is a sub-algebra of $C(G)$. Using the trivial representation we also see that the constant function $\mathbb{1}$ belongs to $\mathcal{M}(G)$. Moreover, given a matrix coefficient $\varphi^\rho_{u,v}$ of a finite-dimensional unitary representation ρ and vectors $u, v \in \mathcal{H}_\rho$, the formula (3.2) shows that $\overline{\varphi^\rho_{u,v}}$ is also a matrix coefficient of a finite-dimensional unitary representation, namely the contragredient $\overline{\rho}$. Therefore, $\mathcal{M}(G)$ is closed under conjugation.

Finally, recall from the Gelfand–Raikov theorem (Corollary 1.81) that for any $g_1 \neq g_2$ in G there exists a unitary representation $\pi \in \widehat{G}$ with the property that $\pi_{g_1} \neq \pi_{g_2}$. Hence there exist $u, v \in \mathcal{H}_\pi$ with $\varphi^\pi_{u,v}(g_1) \neq \varphi^\pi_{u,v}(g_2)$. It follows that $\mathcal{M}(G)$ also separates points. By the Stone–Weierstrass theorem, this shows that $\mathcal{M}(G)$ is dense in $C(G)$. □

3.4.3 The Peter–Weyl Theorem

The material of Section 3.3 and our preparations in Sections 3.4.1 and 3.4.2 leads naturally to the following complete description of the regular representation of a compact group.(8)

Theorem 3.34 (Peter–Weyl). *The regular representation of the compact group G is isomorphic to the direct sum of irreducible representations where each $\pi \in \widehat{G}$ appears with multiplicity $d_\pi = \dim \mathcal{H}_\pi$. More precisely, we have the decomposition*

$$L^2(G) = \bigoplus_{[\pi]\in\widehat{G}} \mathcal{M}_\pi,$$

and the subspace $\mathcal{M}_\pi$ is invariant for every $[\pi] \in \widehat{G}$. In fact the right regular representation ρ restricted to $\mathcal{M}_\pi$ is isomorphic to d_π copies of π, and the left regular representation λ restricted to $\mathcal{M}_\pi$ is isomorphic to d_π copies of the contragredient $\overline{\pi}$ of π.

As we will see, the proof consists of reviewing what we have obtained so far.

PROOF OF THEOREM 3.34. By Theorem 3.30 we know that $\mathcal{M}_{\pi_1} \perp \mathcal{M}_{\pi_2}$ if $\pi_1, \pi_2 \in \widehat{G}$ have $\pi_1 \not\simeq \pi_2$. Therefore the direct sum $\bigoplus_{[\pi]\in\widehat{G}} \mathcal{M}_\pi$ is an orthogonal decomposition of a subspace of $L^2(G)$. Moreover, by Theorem 3.32 the linear hull $\mathcal{M} = \langle \mathcal{M}_\pi \mid [\pi] \in \widehat{G} \rangle$ is dense in $C(G)$ and so also in $L^2(G)$ in the L^2 norm, which implies that $L^2(G) = \bigoplus_{[\pi]\in\widehat{G}} \mathcal{M}_\pi$.

Let us first consider the right regular representation ρ. Fix some $\pi \in \widehat{G}$, some orthonormal basis $w_1, \ldots, w_{d_\pi}$ of $\mathcal{H}_\pi$, where $d_\pi = \dim \mathcal{H}_\pi$, and finally some index $\ell \in \{1, \ldots, d_\pi\}$. Then the map U_ℓ defined by $U_\ell v = \sqrt{d_\pi}\varphi_{v,w_\ell}$ for $v \in \mathcal{H}_\pi$ satisfies

$$\rho_g(U_\ell v)(h) = \sqrt{d_\pi}\varphi_{v,w_\ell}(hg) = \sqrt{d_\pi}\langle \pi_h \pi_g v, w_\ell \rangle = U_\ell(\pi_g v)(h)$$

for all $g, h \in G$ and has $\|U_\ell w_k\| = \|\sqrt{d_\pi}\pi_{k,\ell}\| = 1$ for any $k \in \{1, \ldots, d_\pi\}$ by Schur's orthogonality relations in Theorem 3.30. By Schur's lemma (Theorem 1.29)

$$U_\ell \colon \mathcal{H}_\pi \longrightarrow \operatorname{Im} U_\ell \subseteq \mathcal{M}_\pi$$

is an isometric isomorphism. By Schur's orthogonality relations we also have

$$\operatorname{Im} U_{\ell_1} \perp \operatorname{Im} U_{\ell_2}$$

if $\ell_1 \neq \ell_2$ in $\{1, \ldots, d_\pi\}$ and hence the right regular representation restricted to $\mathcal{M}_\pi$ is isomorphic to d_π copies of π.

For the left regular representation we again fix some $k \in \{1, \ldots, d_\pi\}$ and define B_k on $\mathcal{H}_\pi$ by $B_k v = \sqrt{d_\pi}\varphi_{w_k,v}$ satisfying

$$\lambda_g(B_k v)(h) = \sqrt{d_\pi}\varphi_{w_k,v}(g^{-1}h) = \sqrt{d_\pi}\langle \pi_h w_k, \pi_g v \rangle = B_k(\pi_g v)(h)$$

for all $g, h \in G$. However, the map B_k is conjugate-linear, which we can correct by using the same notation as in Section 3.1 and considering instead the linear map $B'_k\colon \mathcal{H}'_\pi \ni v' \mapsto B_k v = \sqrt{d_\pi}\varphi_{w_k,v}$. Now we have

$$\lambda_g B'_k v' = \lambda_g B_k v = B_k \pi_g v = B'_k \overline{\pi_g} v'$$

for all $g \in G$ and $v' \in \mathcal{H}'_\pi$, which shows that B'_k is an isomorphism between the contragredient representation $\overline{\pi}$ and the restriction of the left regular representation to $\operatorname{Im} B'_k$. As in the case of the right regular representation the multiplicity is d_π. □

Exercise 3.35. Show that $\widehat{G}$ is countable.

Exercise 3.36. Consider the left-right representation γ of $G \times G$ on $L^2(G)$ defined by

$$\gamma_{(g_1,g_2)}(f)(h) = f(g_1^{-1} h g_2)$$

for all $f \in \mathcal{H}_\gamma = L^2(G)$ and $g_1, g_2, h \in G$. Show that

$$L^2(G) = \bigoplus_{[\pi]\in\widehat{G}} \mathcal{M}_\pi$$

is precisely the decomposition into inequivalent irreducible representations for the unitary representation γ of $G \times G$ in $\mathcal{H}_\gamma = L^2(G)$.

Exercise 3.37. Let π be a cyclic unitary representation of a compact group. Show that π is contained in the regular representation, that is $\pi < \lambda$.

Exercise 3.38. What additional consequences about $(d_\pi \mid [\pi] \in \widehat{G})$ can be derived from the Peter–Weyl theorem (Theorem 3.34) if G is a finite group? (See also Exercise 3.52.)

3.5 The Space of Conjugacy Classes*

Recall that for two elements g_1, g_2 of a group G we say that g_1 is *conjugate* to g_2 if there exists some $k \in G$ with $g_2 = k g_1 k^{-1}$, equivalently if g_1 and g_2 are on the same G-orbit under the action of G on itself by inner automorphisms. The equivalence classes (or G-orbits) are called the *conjugacy classes* and we write $G^\sharp$ for the space of conjugacy classes. We also write

$$\begin{aligned} p\colon G &\longrightarrow G^\sharp \\ g &\longmapsto [g] = \{kgk^{-1} \mid k \in G\} \end{aligned}$$

for the canonical projection onto the space of conjugacy classes.

If G is a topological group we also use the map p to define the quotient topology on $G^\sharp$. In general, this map and the resulting topological space may not be well-behaved. For example, $G^\sharp$ may not be Hausdorff.

Exercise 3.39. Show that the images under p of $\begin{pmatrix} 1 & 0 \\ 0 & 1 \end{pmatrix}$ and $\begin{pmatrix} 1 & 1 \\ 0 & 1 \end{pmatrix}$ are distinct but do not have disjoint neighbourhoods in the space $\mathrm{SL}_2(\mathbb{R})^\sharp$. Show that p is not a closed map, specifically that the image of $\left\{ \begin{pmatrix} a & \\ & 1/a \end{pmatrix} \,\middle|\, a > 0 \right\}$ is not closed.

However, if G is compact the situation is much better[(9)] and we can use the results of this chapter to give an introduction to harmonic analysis on $G^\sharp$. Moreover, as we will see this will also lead to a better understanding of the unitary representations of G.

3.5.1 The Topological Space

Lemma 3.40 (Topology on $G^\sharp$). *For the compact metric group G the quotient topology on $G^\sharp$ is compact and metrizable. In fact the metric $\mathsf{d}_\sharp$ on $G^\sharp$ defined by*

$$\mathsf{d}_\sharp([g_1],[g_2]) = \inf_{\substack{h_1 \in [g_1], \\ h_2 \in [g_2]}} \mathsf{d}(h_1,h_2)$$

for $[g_1],[g_2] \in G^\sharp$ is compatible with the quotient topology.

PROOF. Since G is compact, the equivalence classes $[g]$ for $g \in G$ are (as images under the continuous map $G \ni k \mapsto kgk^{-1}$) also compact. It follows that the distance between $[g_1]$ and $[g_2]$ is actually a minimum and so is positive unless $[g_1] = [g_2]$. From this it is easy to see that $\mathsf{d}_\sharp$ defines a metric on $G^\sharp$, and it remains to show that the topology induced by $\mathsf{d}_\sharp$ coincides with the quotient topology.

Given $g \in G$ and $\varepsilon > 0$ it is also clear that

$$\bigcup_{\substack{k_1 \in G \\ k_2 \in G}} k_1 B_\varepsilon(k_2 g k_2^{-1}) k_1^{-1} = \{h \in G \mid \mathsf{d}_\sharp([h],[g]) < \varepsilon\} = p^{-1} B_\varepsilon^{G^\sharp}([g]) \subseteq G$$

is open. It follows that every metric open set is open in the quotient topology. For the converse, suppose that $O^\sharp \subseteq G^\sharp$ is an open neighbourhood of $[g] \in G^\sharp$ in the quotient topology. Hence $O = p^{-1}(O^\sharp)$ is an open set containing $[g]$ and is invariant under conjugation by elements of G. It follows that $[g]$ and $G \smallsetminus O$ are disjoint compact sets so that their distance

$$\varepsilon = \min_{h \in [g], k \in G \smallsetminus O} \mathsf{d}(h,k) = d_\sharp([g], G \smallsetminus O)$$

is positive. Suppose now that

$$\mathsf{d}_\sharp([h],[g]) = \mathsf{d}(k_1 h k_1^{-1}, k_2 g k_2^{-1}) < \varepsilon$$

for some $h, k_1, k_2 \in G$. Since $k_2 g k_2^{-1} \in [g]$ it follows from the definition of ε that $k_1 h k_1^{-1} \in O$ or equivalently $[h] \in O^\sharp$. This shows that $B_\varepsilon([g]) \subseteq O^\sharp$, which implies that $O^\sharp$ is also a neighbourhood of $[g]$ with respect to the metric $\mathsf{d}_\sharp$. Together with the above we obtain that $\mathsf{d}_\sharp$ induces the quotient topology.

Since $G^\sharp = p(G)$ is a continuous image of the compact group G, $G^\sharp$ is also compact and the lemma follows. □

3.5.2 The Centre of the Convolution Algebra

Many groups (and, in particular, many semi-simple groups) have very small or trivial centre; see, for example, the discussion of $\mathrm{SU}_2(\mathbb{R})$ in the next chapter. This and Schur's lemma (Theorem 1.29) make the following result important. For this and the following, we endow $G^\sharp$ with the push-forward of the Haar measure m on G, which we again denote by m. Moreover, we identify a function on G that is invariant under conjugation by all $g \in G$ with a function on $G^\sharp$ and obtain in this way the inclusions $C(G^\sharp) \subseteq C(G)$, $L^1(G^\sharp) \subseteq L^1(G)$, and $L^2(G^\sharp) \subseteq L^2(G)$.

Proposition 3.41 (The centre of the convolution algebra). *The centre of $L^1(G)$ for the compact group G is given by $L^1(G^\sharp)$. That is, $f_c \in L^1(G)$ satisfies $f_c * \psi = \psi * f_c$ for all $\psi \in L^1(G)$ if and only if $f_c \in L^1(G^\sharp)$.*

PROOF. We suppose first that $f_c \in L^1(G^\sharp)$, so that $f_c(ghg^{-1}) = f_c(h)$ for all $g, h \in G$. For $\psi \in L^1(G)$ we then have

$$f_c * \psi(g) = \int f_c(g\ell^{-1})\psi(\ell)\,\mathrm{d}m(\ell) = \int f_c(\ell^{-1}g)\psi(\ell)\,\mathrm{d}m(\ell) = \psi * f_c(g)$$

for almost every $g \in G$ by the definition of convolution in (1.16), as required.

For the converse, we suppose now that $f_c \in L^1(G)$ satisfies $f_c * \psi = \psi * f_c$ for all $\psi \in L^1(G)$. We wish to apply this to $\psi = \lambda_g \psi_n$ for some $g \in G$ and an approximate identity (ψ_n) as in Proposition 1.46. In fact, using the compactness of G we see that conjugation is uniformly continuous and so we may choose a decreasing sequence (B_n) of neighbourhoods of $e \in G$ that are invariant under conjugation and satisfy $\bigcap_{n\geqslant 1} B_n = \{e\}$. We set

$$\psi_n = \frac{1}{m(B_n)}\mathbb{1}_{B_n} \in L^1(G^\sharp)$$

for all $n \geqslant 1$. Also recall (1.19) and Exercise 1.51, which we can express as saying that $\delta_g * f = \lambda_g f$ and $f * \delta_g = \rho_{g^{-1}} f$ for all $f \in L^1(G)$ and $g \in G$. It follows that

$$\big(\delta_g * \psi_n\big)(h) = \psi_n(g^{-1}h) = \psi_n(hg^{-1}) = \big(\psi_n * \delta_g\big)(h)$$

for $f = \psi_n$ and $g, h \in G$. By the assumption on f_c and associativity of convolution in $M(G)$, this gives

$$(f_c * \delta_g) * \psi_n = f_c * \underbrace{(\delta_g * \psi_n)}_{=\psi} = \underbrace{(\psi_n * \delta_g)}_{=\psi} * f_c = \psi_n * (\delta_g * f_c).$$

Letting $n \to \infty$ and applying Proposition 1.46, we obtain $f_c * \delta_g = \delta_g * f_c$ and so

$$f_c(hg^{-1}) = \rho_{g^{-1}}(f_c)(h) = \lambda_g(f_c)(h) = f_c(g^{-1}h)$$

for every $g \in G$ and almost every $h \in G$, or equivalently

$$f_c(ghg^{-1}) = f_c(h) \tag{3.14}$$

for every $g \in G$ and almost every $h \in G$.

Strictly speaking, the null set excluded in this statement may depend on the element $g \in G$, but one can replace f_c by an equivalent function so that (3.14) then holds for all $g, h \in G$. We refer to [24, Prop. 8.3] or Exercise 3.42. This proves the converse. □

Exercise 3.42. Show that (3.14) for every $g \in G$ and almost every $h \in G$ implies that there exists a null set $N \subseteq G$ such that $h \in G \smallsetminus N$ implies (3.14) for almost every $g \in G$, and conclude that

$$f_c(h) = \int_G f_c(ghg^{-1})\, \mathrm{d}m(g)$$

for almost every $h \in G$.

Exercise 3.43. Generalize Proposition 3.41 to the measure algebra by showing that the centre of $M(G)$ is given by

$$M(G^\sharp) = \{\mu \in M(G) \mid \mu \text{ is invariant under conjugation}\}.$$

For unitary representations of G we obtain the following consequence.

Proposition 3.44 (Unitary representations and the centre of $L^1(G)$). *Let π be a unitary representation of the compact group G. Then the convolution operators $\pi_*(f_c)$ for $f_c \in L^1(G^\sharp)$ are intertwining. If $\pi \in \widehat{G}$, then $\pi_*(f_c)$ is a multiple of the identity on $\mathcal{H}_\pi$.*

PROOF. Let π be a unitary representation of the compact group G, $g \in G$, and $f_c \in L^1(G^\sharp)$. Then

$$\begin{aligned}
\langle \pi_*(f_c)\pi_g u, v\rangle &= \int f_c(h)\langle \pi_{hg}u, v\rangle\, \mathrm{d}m(h) \\
&= \int f_c(kg^{-1})\langle \pi_k u, v\rangle\, \mathrm{d}m(k) \\
&= \int f_c(g^{-1}k)\langle \pi_k u, v\rangle\, \mathrm{d}m(k) \\
&= \int f_c(\ell)\langle \pi_{g\ell}u, v\rangle\, \mathrm{d}m(\ell) = \langle \pi_g \pi_*(f_c)u, v\rangle
\end{aligned}$$

for all $u, v \in \mathcal{H}_\pi$, which implies that

$$\pi_*(f_c)\pi_g = \pi_g \pi_*(f_c)$$

for all $g \in G$ as required. For $\pi \in \widehat{G}$ the final conclusion follows from Schur's lemma (Theorem 1.29). □

The following 'projection' from $L^1(G)$ to $L^1(G^\sharp)$ will be useful.

Lemma 3.45 (Averaging projection). *Given the compact group G we let*

$$A\colon L^1(G) \to L^1(G^\sharp)$$

be the averaging projection *defined by*

$$A(f)(g) = \int_G f(hgh^{-1})\,\mathrm{d}m(h)$$

for $g \in G$ and $f \in L^1(G)$. Then $A(f) = f$ for all $f \in L^1(G^\sharp)$, and $A(f)$ lies in $C(G^\sharp)$ for all f in $C(G)$. Moreover, for a finite-dimensional representation π of G we also have

$$\pi_*(A(f)) = \int_G \pi_h \pi_*(f) \pi_h^{-1}\,\mathrm{d}m(h)$$

for all $f \in L^1(G)$.

Proof. Let $f \in L^1(G)$. Since $g \mapsto f(kgk^{-1})$ is integrable with equal norms for all $k \in G$ and $m(G) = 1$, Fubini's theorem implies that the integral defining $A(f)(g)$ exists for almost every $g \in G$ and that $\|A(f)\|_1 \leqslant \|f\|_1$. By right-invariance of m, we also have $A(f) \in L^1(G^\sharp)$. If $f \in L^1(G^\sharp)$ we have

$$f(kgk^{-1}) = f(g)$$

for $g, k \in G$, and so $A(f) = f$ as $m(G) = 1$. The continuity of $A(f)$ for f in $C(G)$ follows from dominated convergence.

Suppose now that π is a finite-dimensional representation of G and u, v are in $\mathcal{H}_\pi$. Then

$$\begin{aligned}
\langle \pi_*(A(f))u, v\rangle &= \iint f(hgh^{-1})\langle \pi_g u, v\rangle\,\mathrm{d}m(h)\,\mathrm{d}m(g)\\
&= \iint f(k)\langle \pi_{h^{-1}kh} u, v\rangle\,\mathrm{d}m(k)\,\mathrm{d}m(h)\\
&= \iint f(k)\langle \pi_\ell \pi_k \pi_\ell^{-1} u, v\rangle\,\mathrm{d}m(k)\,\mathrm{d}m(\ell)\\
&= \int \langle \pi_\ell \pi_*(f)\pi_\ell^{-1} u, v\rangle\,\mathrm{d}m(\ell) = \left\langle \int \pi_\ell \pi_*(f)\pi_\ell^{-1}\,\mathrm{d}m(\ell) u, v\right\rangle,
\end{aligned}$$

which gives the lemma as $u, v \in \mathcal{H}_\pi$ were arbitrary. □

3.5.3 Characters

In the context of non-abelian groups the following notion of characters is important. We note however that these are not multiplicative characters (except for some exceptional cases) as we have considered in Chapter 1, and particularly in Chapter 2.

Definition 3.46 (Characters). Given a finite-dimensional representation ρ of a group G the *character of the representation* ρ is the complex-valued function χ_ρ defined by $\chi_\rho(g) = \operatorname{tr}(\rho_g)$ for $g \in G$.

We note that for a finite-dimensional representation ρ of a group G and elements $g, k \in G$ we have

$$\chi_\rho(kgk^{-1}) = \operatorname{tr}(\rho_k \rho_g \rho_k^{-1}) = \operatorname{tr}(\rho_g) = \chi_\rho(g),$$

which shows that the character of a representation can be considered as a function on $G^\sharp$. Hence, if ρ is a continuous finite-dimensional representation of a topological group, then χ_ρ is a continuous function on $G^\sharp$. Moreover, the characters of two isomorphic finite-dimensional representations are equal. In particular, if G is compact, then for every $[\pi] \in \widehat{G}$ we have a well-defined continuous character $\chi_{[\pi]} = \chi_\pi \in C(G^\sharp)$. Moreover, for $[\pi] \in \widehat{G}$ we have

$$\chi_{[\pi]} = \pi_{1,1} + \pi_{2,2} + \cdots + \pi_{d_\pi, d_\pi} \tag{3.15}$$

by definition, where as before $\pi_{k,k} = \varphi_{w_k, w_k}$ for $k = 1, \ldots, d_\pi$ and an orthonormal basis $w_1, \ldots, w_{d_\pi}$ of $\mathcal{H}_\pi$.

The characters appear quickly when the averaging projection is applied.

Proposition 3.47 (Averaging). *Let π be an irreducible representation of the compact group G, and let $w_1, \ldots, w_{d_\pi} \in \mathcal{H}_\pi$ be an orthonormal basis of $\mathcal{H}_\pi$. Then we have*

$$A(\pi_{m,n}) = \tfrac{1}{d_\pi} \delta_{m,n} \chi_{[\pi]}$$

for all $m, n \in \{1, \ldots, d_\pi\}$.

Proof. Let $m, n \in \{1, \ldots, d_\pi\}$. By the Schur orthogonality relations (Theorem 3.30) we have for all $k, \ell \in \{1, \ldots, d_\pi\}$ that

$$\begin{aligned} \langle \pi_*(d_\pi \overline{\pi_{m,n}}) w_k, w_\ell \rangle_{\mathcal{H}_\pi} &= \int d_\pi \overline{\pi_{m,n}(g)} \underbrace{\langle \pi_g w_k, w_\ell \rangle_{\mathcal{H}_\pi}}_{\pi_{k,\ell}(g)} \, dm(g) \\ &= d_\pi \langle \pi_{k,\ell}, \pi_{m,n} \rangle_{L^2(G)} = \delta_{km} \delta_{\ell n}. \end{aligned} \tag{3.16}$$

In other words, $\pi_*(d_\pi \overline{\pi_{m,n}})$ is the elementary linear map that sends w_m to w_n and all other basis vectors to zero. In particular, the map

$$\mathcal{M}_\pi \ni \phi \longmapsto \pi_*(d_\pi \overline{\phi}) \in \mathrm{B}(\mathcal{H}_\pi) \tag{3.17}$$

is a conjugate-linear isomorphism.

By Proposition 3.29, the subspace $\mathcal{M}_\pi \subseteq L^1(G)$ is invariant under the left and right regular representations, which implies that

$$G \ni g \longmapsto \left(\lambda_{h^{-1}}\rho_{h^{-1}}\pi_{m,n}(g) = \pi_{m,n}(hgh^{-1})\right)$$

belongs to $\mathcal{M}_\pi$ for all $h \in G$. Taking the integral over $h \in G$ as in the definition of the averaging projection of Lemma 3.45 and applying the Schur orthogonality relations (Theorem 3.30) we see that

$$F = A(\pi_{m,n}) = \sum_{k,\ell} \beta_{k,\ell}\pi_{k,\ell} \in C(G^\sharp)$$

for some matrix $(\beta_{k,\ell}) \in \operatorname{Mat}_{d_\pi,d_\pi}(\mathbb{C})$. Since $\pi_*\left(d_\pi\overline{F}\right)$ is intertwining by Proposition 3.44, it follows by Schur's lemma (Theorem 1.29) that

$$\pi_*(d_\pi\overline{F}) = \alpha I.$$

Together with the properties of the map in (3.17), this gives $\overline{\beta_{k,\ell}} = \alpha\delta_{k,\ell}$ for all $k, \ell \in \{1, \dots, d_\pi\}$, and so $A(\pi_{m,n}) = \overline{\alpha}\chi_{[\pi]}$ by (3.15). To calculate α we take the trace in $\mathcal{H}_\pi$ and obtain

$$d_\pi\alpha = \operatorname{tr}(\alpha I) = \operatorname{tr}\left(\pi_*(d_\pi\overline{F})\right) = \operatorname{tr}\left(\pi_*(d_\pi\overline{\pi_{m,n}})\right) = \delta_{m,n}$$

by Lemma 3.45 and the above description of the linear map $\pi_*(d_\pi\overline{\pi_{m,n}})$ in (3.16). This gives the lemma. □

The characters also have special properties for convolution.

Corollary 3.48 (Convolution of characters). *Let π, ρ be irreducible unitary representations of the compact group G. Then*

$$\chi_{[\pi]} * \chi_{[\pi]} = \tfrac{1}{d_\pi}\chi_{[\pi]}$$

and, if $[\pi] \neq [\rho]$ are inequivalent, then

$$\chi_{[\pi]} * \chi_{[\rho]} = 0.$$

PROOF. Suppose first that $[\pi] \neq [\rho] \in \widehat{G}$ are inequivalent. By Proposition 3.29, both $\mathcal{M}_{[\pi]}$ and $\mathcal{M}_{[\rho]}$ are two-sided ideals, which implies that

$$\chi_{[\pi]} * \chi_{[\rho]} \in \mathcal{M}_{[\pi]} \cap \mathcal{M}_{[\rho]}$$

for $\chi_{[\pi]} \in \mathcal{M}_{[\pi]}$ and $\chi_{[\rho]} \in \mathcal{M}_{[\rho]}$. However, by the Schur orthogonality relation (Theorem 3.30) we have $\mathcal{M}_{[\pi]} \perp \mathcal{M}_{[\rho]}$ and hence $\chi_{[\pi]} * \chi_{[\rho]} = 0$.

For the former claim we let $[\pi] \in \widehat{G}$ and note that

$$f = \chi_{[\pi]} * \chi_{[\pi]} \in \mathcal{M}_{[\pi]}$$

satisfies

$$\pi_*(d_\pi \overline{f}) = \pi_*\big(d_\pi \overline{\chi_{[\pi]}} * \overline{\chi_{[\pi]}}\big) = \tfrac{1}{d_\pi}\pi_*\big(d_\pi \overline{\chi_{[\pi]}}\big)\pi_*\big(d_\pi \overline{\chi_{[\pi]}}\big).$$

However, in the proof of Proposition 3.47 we saw that (3.17) is a conjugate-linear isomorphism sending $\chi_{[\pi]}$ to the identity. This gives $\pi_*(d_\pi \overline{f}) = \frac{1}{d_\pi} I$ and hence $f = \frac{1}{d_\pi}\chi_{[\pi]}$, which concludes the proof. □

Exercise 3.49. Let $[\pi] \in \widehat{G}$ be an irreducible representation of the compact group G. Let $\pi_{m,n}$ for $m, n \in \{1, \dots, d_\pi\}$ be as in the Schur orthogonality relations (Theorem 3.30). Calculate $\pi_{m,n} * \pi_{k,\ell}$ for all $m, n, k, \ell \in \{1, \dots, d_\pi\}$, and deduce the formula in Corollary 3.48 for $\chi_{[\pi]} * \chi_{[\pi]}$ from this.

3.5.4 Dense Algebra

Corollary 3.50 (Dense algebra). *For the compact group G the linear span*

$$\mathcal{M}(G^\sharp) = \langle \chi_\pi \,|\, [\pi] \in \widehat{G} \rangle = \langle \chi_\rho \,|\, \rho \text{ a finite-dimensional representation of } G \rangle$$

is a dense sub-algebra of $C(G^\sharp)$ with respect to pointwise multiplication.

PROOF. The equivalence of the two descriptions of $\mathcal{M}(G^\sharp)$ is a consequence of the fact that every irreducible representation is finite-dimensional by Theorem 3.18, and since a finite-dimensional representation ρ has $\mathcal{H}_\rho = \bigoplus_{j=1}^n \mathcal{H}_{\pi_j}$ for some irreducible representations $\pi_1, \dots, \pi_n$ so that the character decomposes as

$$\chi_\rho = \sum_{j=1}^n \chi_{\pi_j}.$$

Suppose now that $f \in C(G^\sharp)$ and $\varepsilon > 0$. Then by Theorem 3.32 there exist finitely many irreducible representations $\pi_1, \dots, \pi_n$ and vectors $v_j, w_j \in \mathcal{H}_{\pi_j}$ for $j = 1, \dots, n$ so that

$$\Big\| f - \sum_{j=1}^n \alpha_j \varphi_{v_j, w_j}^{\pi_j} \Big\|_\infty < \varepsilon$$

for some $\alpha_1, \dots, \alpha_n \in \mathbb{C}$.

Applying the averaging projection $A\colon C(G) \to C(G^\sharp)$ from Lemma 3.45 we obtain the corollary. Indeed,

$$A\big(\varphi_{v_j, w_j}^{\pi_j}\big) = \beta_j \chi_{[\pi_j]}$$

for some $\beta_j \in \mathbb{C}$ and $j = 1, \dots, n$. Hence

$$\left\| f - \sum_{j=1}^{n} \alpha_j \beta_j \chi_{[\pi_j]} \right\|_\infty < \varepsilon,$$

which gives the corollary, as $\varepsilon > 0$ was arbitrary.

To see that $\mathcal{M}(G^\sharp)$ is a sub-algebra we note that, for finite-dimensional representations ρ_1, ρ_2 of G, we have

$$\chi_{\rho_1 \otimes \rho_2}(g) = \operatorname{tr}\bigl(\rho_1(g) \otimes \rho_2(g)\bigr) = \operatorname{tr}\bigl(\rho_1(g)\bigr) \operatorname{tr}\bigl(\rho_2(g)\bigr) = \chi_{\rho_1}(g)\chi_{\rho_2}(g)$$

for all $g \in G$. □

3.5.5 Characters as an Orthonormal Basis

Corollary 3.51 (Characters form an orthonormal basis). *The characters $\chi_{[\pi]}$ for $[\pi] \in \widehat{G}$ form an orthonormal basis of $L^2(G^\sharp)$.*

PROOF. For inequivalent $[\pi] \neq [\pi'] \in \widehat{G}$ we have $\mathcal{M}_\pi \perp \mathcal{M}_{\pi'}$ by the Schur orthogonality relation (Theorem 3.30). With $\chi_\pi \in \mathcal{M}_\pi$ and $\chi_{\pi'} \in \mathcal{M}_{\pi'}$ this implies that $\langle \chi_\pi, \chi_{\pi'} \rangle = 0$ (where it does not matter whether we take the inner product in $L^2(G)$ or in $L^2(G^\sharp)$). For a given $[\pi] \in \widehat{G}$ we have

$$\chi_\pi = \pi_{1,1} + \pi_{2,2} + \cdots + \pi_{d_\pi, d_\pi},$$

where as before $\pi_{m,n} = \varphi_{w_m, w_n}$ for an orthonormal basis $w_1, \ldots, w_{d_\pi}$ of $\mathcal{H}_\pi$. Since $\|\pi_{m,m}\| = \frac{1}{\sqrt{d_\pi}}$ it follows that $\|\chi_\pi\| = 1$. By Corollary 3.50 we know that the collection $\chi_{[\pi]}$ for $[\pi] \in \widehat{G}$ forms an orthonormal basis of $L^2(G^\sharp)$. □

Exercise 3.52. What additional consequences about $(d_\pi \mid [\pi] \in \widehat{G})$ can be derived from Corollary 3.51 if G is finite? (See also Exercise 3.38.)

3.6 Summary and Outlook

As we have seen in this chapter, unitary representations of compact groups have many special properties, since

- all unitary representations are (at most countable) direct sums of irreducible finite-dimensional representations;
- the unitary dual is countable and so maybe should be considered 'discrete'; and
- the regular representation contains all irreducible representations with prescribed multiplicity.

In the next chapter we will discuss the group $\mathrm{SU}_2(\mathbb{R})$ as a particularly important example for the material of this chapter, and in particular will describe its conjugacy classes and its characters. This will lead us to the topic of unitary representations of (simple) Lie groups.

Chapter 4
Lie Algebras and Unitary Representations of $\mathrm{SU}_2(\mathbb{R})$

To continue the discussion of unitary representations, we need to introduce Lie algebras and their (for now) finite-dimensional representations. In particular this will allow us to describe the representation theory of the compact group $\mathrm{SU}_2(\mathbb{R})$, which represents an important example. For this we will sometimes use the following notational conventions in addition to the standing assumptions and notations of Section 1.1.

- If G is a real (or complex) Lie group (which is not assumed to be connected), then we write $\mathfrak{g} = \mathrm{Lie}\, G$ for its real (or complex) Lie algebra and write $\mathbf{a}, \mathbf{b}, \mathbf{c}, \mathbf{d}, \mathbf{e}, \mathbf{f} \in \mathfrak{g}$ for its elements. Recall that there is a smooth map $\exp\colon \mathfrak{g} \to G$ with a local inverse $\log\colon B_\delta^G(I) \to \mathfrak{g}$ defined on some neighbourhood $B_\delta^G(I)$ of the identity $I \in G$ with $\delta > 0$.
- We will use the letters s, t to denote real numbers.

4.1 Finite-Dimensional Representation Theory of $\mathrm{SL}_2(\mathbb{R})$

For the classification of simple (and semi-simple) Lie groups and their finite-dimensional representations the most important Lie group to understand first is $\mathrm{SL}_2(\mathbb{R})$. As we will see later (in Chapters 8 and 9), this remains true for the theory of unitary representations. As a warm-up for this discussion as well as because of its independent interest, we study in this chapter the twin sibling $\mathrm{SU}_2(\mathbb{R})$ of $\mathrm{SL}_2(\mathbb{R})$. In fact, as we will explain shortly these two groups are strongly related: $\mathrm{SU}_2(\mathbb{R})$ is the *compact real form*, and $\mathrm{SL}_2(\mathbb{R})$ is the *split real form*, of the complex Lie group $\mathrm{SL}_2(\mathbb{C})$, and they have identical descriptions of their finite-dimensional representations (which can be made unitary for $\mathrm{SU}_2(\mathbb{R})$ but are not unitary except in trivial cases for $\mathrm{SL}_2(\mathbb{R})$; see Exercise 1.89 and 1.91). As before, we will always study representations on complex vector spaces.

M. Einsiedler and T. Ward, *Unitary Representations and Unitary Duals*,
Graduate Texts in Mathematics 308, https://doi.org/10.1007/978-3-032-03899-9_4

4.1.1 A Quick Reminder on Closed Linear Groups

We wish to recall the most basic concepts for Lie group, with a focus on closed linear groups and refer to Knapp [54, Sec. 0] for more details. In fact, many facts concerning the connection between Lie groups and their Lie algebras are easy to verify for closed linear Lie groups $G < \mathrm{GL}_n(\mathbb{R})$ using only the properties of the exponential map

$$\begin{aligned}\exp\colon \mathrm{Mat}_{n,n}(\mathbb{R}) &\longrightarrow \mathrm{GL}_n(\mathbb{R})\\ m &\longmapsto \exp(m) = \sum_{k=0}^{\infty} \frac{1}{k!} m^k.\end{aligned}$$

One such basic property is that the derivative of exp is the identity map on $\mathrm{End}(\mathbb{R}^n) = \mathrm{Mat}_{n,n}(\mathbb{R})$. Here $\mathrm{Mat}_{n,n}(\mathbb{R})$ is also referred to as the Lie algebra $\mathfrak{gl}_n(\mathbb{R})$ of $\mathrm{GL}_n(\mathbb{R})$, and thought of as the tangent space of $\mathrm{GL}_n(\mathbb{R})$ at the identity I. Using this, we can calculate, for example, the *adjoint representation*, denoted Ad, of $\mathrm{GL}_n(\mathbb{R})$ on its Lie algebra $\mathfrak{gl}_n(\mathbb{R})$. By definition we need to take the derivative of conjugation at the identity for any $g \in \mathrm{GL}_n(\mathbb{R})$ to obtain

$$\mathrm{Ad}_g(\mathbf{m}) = \frac{\mathrm{d}}{\mathrm{d}t}\Big|_{t=0} \big(g \exp(t\mathbf{m}) g^{-1}\big) = g \frac{\mathrm{d}}{\mathrm{d}t}\Big|_{t=0} \big(\exp(t\mathbf{m})\big) g^{-1} = g\mathbf{m}g^{-1}$$

for all $\mathbf{m} \in \mathfrak{gl}_n(\mathbb{R})$. Here the last three expressions involve the usual product of matrices. Moreover, the *adjoint representation*, denoted ad, of $\mathfrak{gl}_n(\mathbb{R})$ on $\mathfrak{gl}_n(\mathbb{R})$ is defined as the derivative of the map

$$\mathrm{Ad}\colon \mathrm{GL}_n(\mathbb{R}) \ni g \longmapsto \mathrm{Ad}_g \in \mathrm{GL}\big(\mathfrak{gl}_n(\mathbb{R})\big)$$

at I. This derivative is a map from $\mathfrak{gl}_n(\mathbb{R})$ into the Lie algebra $\mathrm{End}\big(\mathfrak{gl}_n(\mathbb{R})\big)$ of the group $\mathrm{GL}\big(\mathfrak{gl}_n(\mathbb{R})\big)$. For $\mathbf{a} \in \mathfrak{gl}_n(\mathbb{R})$ the derivative $\mathrm{ad}_{\mathbf{a}}$ will map $\mathbf{m} \in \mathfrak{gl}_n(\mathbb{R})$ to

$$\mathrm{ad}_{\mathbf{a}}(\mathbf{m}) = \frac{\mathrm{d}}{\mathrm{d}t}\Big|_{t=0} \big(\exp(t\mathbf{a})\mathbf{m}\exp(-t\mathbf{a})\big) = \mathbf{am} - \mathbf{ma}, \tag{4.1}$$

where we used the chain and product rules.

We also call the map

$$\begin{aligned}[\cdot,\cdot]\colon \mathfrak{gl}_n(\mathbb{R}) \times \mathfrak{gl}_n(\mathbb{R}) &\longrightarrow \mathfrak{gl}_n(\mathbb{R})\\ (\mathbf{a},\mathbf{m}) &\longmapsto [\mathbf{a},\mathbf{m}] = \mathbf{am} - \mathbf{ma}\end{aligned}$$

the *Lie bracket*. This is a bilinear map satisfying

$$[\mathbf{a},\mathbf{b}] = -[\mathbf{b},\mathbf{a}]$$

for all $\mathbf{a},\mathbf{b} \in \mathfrak{gl}_n(\mathbb{R})$. Moreover, the *Jacobi identity*

$$[\mathbf{a},[\mathbf{b},\mathbf{c}]]+[\mathbf{b},[\mathbf{c},\mathbf{a}]]+[\mathbf{c},[\mathbf{a},\mathbf{b}]]=0$$

can be checked by a direct calculation. Equivalently, we have

$$[[\mathbf{a},\mathbf{b}],\mathbf{c}]=[\mathbf{a},[\mathbf{b},\mathbf{c}]]-[\mathbf{b},[\mathbf{a},\mathbf{c}]]$$

for all $\mathbf{a},\mathbf{b},\mathbf{c}\in\mathfrak{gl}_n(\mathbb{R})$, which can also be written as

$$\mathrm{ad}_{[\mathbf{a},\mathbf{b}]}=[\mathrm{ad}_{\mathbf{a}},\mathrm{ad}_{\mathbf{b}}]=\mathrm{ad}_{\mathbf{a}}\circ\mathrm{ad}_{\mathbf{b}}-\mathrm{ad}_{\mathbf{b}}\circ\mathrm{ad}_{\mathbf{a}}$$

for all $\mathbf{a},\mathbf{b}\in\mathfrak{gl}_n(\mathbb{R})$, where $\mathrm{ad}_{\bullet}$ is defined in (4.1).

We note that these facts generalize without too much effort to closed linear subgroups G of $\mathrm{GL}_n(\mathbb{R})$, and that these cases will suffice for all of our discussions.

Essential Exercise 4.1. (a) Show that any continuous homomorphism

$$g_{\bullet}:\mathbb{R}\ni t\longmapsto g_t\in\mathrm{GL}_n(\mathbb{R})$$

is differentiable.
(b) Show that the map

$$\mathfrak{gl}_n(\mathbb{R})\ni\mathbf{m}\longmapsto\big(\mathbb{R}\ni t\mapsto g_t=\exp(t\mathbf{m})\big)$$

is a bijection between elements of the Lie algebra $\mathfrak{gl}_n(\mathbb{R})$ and differentiable one-parameter subgroups as in (a).

4.1.2 The Lie Groups $\mathbf{SU_2}(\mathbb{R})$ and $\mathbf{SL_2}(\mathbb{R})$

We recall the definition of the real Lie group

$$\mathrm{SL}_2(\mathbb{R})=\left\{g=\begin{pmatrix}a&b\\c&d\end{pmatrix}\in\mathrm{Mat}_{22}(\mathbb{R})\mid\det g=ad-bc=1\right\},$$

called the *special linear group in* 2 *dimensions*, and its Lie algebra

$$\mathfrak{sl}_2(\mathbb{R})=\{\mathbf{m}\in\mathrm{Mat}_{22}(\mathbb{R})\mid\operatorname{tr}\mathbf{m}=0\},$$

which contains all elements of the form

$$\mathbf{m}=\begin{pmatrix}a&b\\c&-a\end{pmatrix}$$

for $a,b,c\in\mathbb{R}$. As in any Lie group, the Lie algebra has the property that the image $\exp(\mathfrak{sl}_2(\mathbb{R}))$ contains an open neighbourhood of the identity $I\in\mathrm{SL}_2(\mathbb{R})$ and so can be used to describe $\mathrm{SL}_2(\mathbb{R})$ locally.

We will frequently use the basis of $\mathfrak{sl}_2(\mathbb{R})$ given by

$$\mathbf{a} = \begin{pmatrix} 1 & 0 \\ 0 & -1 \end{pmatrix}, \mathbf{e} = \begin{pmatrix} 0 & 1 \\ 0 & 0 \end{pmatrix}, \mathbf{f} = \begin{pmatrix} 0 & 0 \\ 1 & 0 \end{pmatrix}, \tag{4.2}$$

which we will refer to as the $\mathfrak{sl}_2$-*triple*. These three elements satisfy the following relations

$$\left.\begin{aligned} [\mathbf{a}, \mathbf{e}] &= \begin{pmatrix} 1 & 0 \\ 0 & -1 \end{pmatrix}\begin{pmatrix} 0 & 1 \\ 0 & 0 \end{pmatrix} - \begin{pmatrix} 0 & 1 \\ 0 & 0 \end{pmatrix}\begin{pmatrix} 1 & 0 \\ 0 & -1 \end{pmatrix} = 2\mathbf{e} \\ [\mathbf{a}, \mathbf{f}] &= \begin{pmatrix} 1 & 0 \\ 0 & -1 \end{pmatrix}\begin{pmatrix} 0 & 0 \\ 1 & 0 \end{pmatrix} - \begin{pmatrix} 0 & 0 \\ 1 & 0 \end{pmatrix}\begin{pmatrix} 1 & 0 \\ 0 & -1 \end{pmatrix} = -2\mathbf{f} \\ [\mathbf{e}, \mathbf{f}] &= \begin{pmatrix} 0 & 1 \\ 0 & 0 \end{pmatrix}\begin{pmatrix} 0 & 0 \\ 1 & 0 \end{pmatrix} - \begin{pmatrix} 0 & 0 \\ 1 & 0 \end{pmatrix}\begin{pmatrix} 0 & 1 \\ 0 & 0 \end{pmatrix} = \mathbf{a}. \end{aligned}\right\} \tag{4.3}$$

In other words, with respect to the map $\mathrm{ad}_{\mathbf{a}} = [\mathbf{a}, \cdot]\colon \mathfrak{sl}_2(\mathbb{R}) \to \mathfrak{sl}_2(\mathbb{R})$, $\mathbf{e}$ is an eigenvector with eigenvalue 2, $\mathbf{f}$ is an eigenvector with eigenvalue -2, $\mathbf{a}$ is an eigenvector with eigenvalue 0, and $\mathbf{e}$ and $\mathbf{f}$ together generate $\mathbf{a}$. As a consequence, the number 2 will be quite prevalent in the representation theory of $\mathrm{SL}_2(\mathbb{R})$. Much of what we wish to discuss here will use these simple relations. We also note that we have been using versions of the relations (4.3) within the group $\mathrm{SL}_2(\mathbb{R})$ already in Section 1.8.

We will also use the complex Lie group

$$\mathrm{SL}_2(\mathbb{C}) = \left\{ g = \begin{pmatrix} a & b \\ c & d \end{pmatrix} \in \mathrm{Mat}_{22}(\mathbb{C}) \mid \det g = ad - bc = 1 \right\},$$

and its complex Lie algebra

$$\mathfrak{sl}_2(\mathbb{C}) = \{\mathbf{m} \in \mathrm{Mat}_{22}(\mathbb{C}) \mid \mathrm{tr}\,\mathbf{m} = 0\},$$

which also has the Lie algebra elements $\mathbf{a}$, $\mathbf{e}$, and $\mathbf{f}$ as a basis over $\mathbb{C}$. A more formal way of stating this is to say that there is an isomorphism

$$\mathfrak{sl}_2(\mathbb{C}) \cong \mathfrak{sl}_2(\mathbb{R}) \otimes_{\mathbb{R}} \mathbb{C}, \tag{4.4}$$

where $\mathfrak{sl}_2(\mathbb{R}) \otimes_{\mathbb{R}} \mathbb{C}$ is the complex Lie algebra obtained by bilinearly extending the Lie bracket map from $\mathbb{R}$ to $\mathbb{C}$. Because of the isomorphism in (4.4) we also say that $\mathfrak{sl}_2(\mathbb{R})$ is a *real form* or, more specifically, the *split real form* of $\mathfrak{sl}_2$. We now turn our attention to the only other real form (up to isomorphism).

The *special unitary group*

$$\mathrm{SU}_2(\mathbb{R}) = \{g \in \mathrm{Mat}_{22}(\mathbb{C}) \mid g^* g = I, \det g = 1\},$$

is a real Lie group with Lie algebra

$$\mathfrak{su}_2(\mathbb{R}) = \{\mathbf{m} \in \mathrm{Mat}_{22}(\mathbb{C}) \mid \mathbf{m}^* + \mathbf{m} = 0, \mathrm{tr}\,\mathbf{m} = 0\}.$$

We note that these are not a complex Lie group and Lie algebra since the adjoint operation is adjoint-linear over $\mathbb{C}$. The elements of $\mathfrak{su}_2(\mathbb{R})$ have the form

$$\mathbf{m} = \begin{pmatrix} a\mathrm{i} & b\mathrm{i} - c \\ b\mathrm{i} + c & -a\mathrm{i} \end{pmatrix}$$

for $a, b, c \in \mathbb{R}$ and we may use the basis

$$\mathbf{b}_1 = \begin{pmatrix} \mathrm{i} & 0 \\ 0 & -\mathrm{i} \end{pmatrix}, \quad \mathbf{b}_2 = \begin{pmatrix} 0 & \mathrm{i} \\ \mathrm{i} & 0 \end{pmatrix}, \quad \mathbf{b}_3 = \begin{pmatrix} 0 & -1 \\ 1 & 0 \end{pmatrix} \tag{4.5}$$

with the relations

$$\left.\begin{aligned} [\mathbf{b}_1, \mathbf{b}_2] &= \begin{pmatrix} \mathrm{i} & 0 \\ 0 & -\mathrm{i} \end{pmatrix}\begin{pmatrix} 0 & \mathrm{i} \\ \mathrm{i} & 0 \end{pmatrix} - \begin{pmatrix} 0 & \mathrm{i} \\ \mathrm{i} & 0 \end{pmatrix}\begin{pmatrix} \mathrm{i} & 0 \\ 0 & -\mathrm{i} \end{pmatrix} = 2\mathbf{b}_3 \\ [\mathbf{b}_2, \mathbf{b}_3] &= \begin{pmatrix} 0 & \mathrm{i} \\ \mathrm{i} & 0 \end{pmatrix}\begin{pmatrix} 0 & -1 \\ 1 & 0 \end{pmatrix} - \begin{pmatrix} 0 & -1 \\ 1 & 0 \end{pmatrix}\begin{pmatrix} 0 & \mathrm{i} \\ \mathrm{i} & 0 \end{pmatrix} = 2\mathbf{b}_1 \\ [\mathbf{b}_3, \mathbf{b}_1] &= \begin{pmatrix} 0 & -1 \\ 1 & 0 \end{pmatrix}\begin{pmatrix} \mathrm{i} & 0 \\ 0 & -\mathrm{i} \end{pmatrix} - \begin{pmatrix} \mathrm{i} & 0 \\ 0 & -\mathrm{i} \end{pmatrix}\begin{pmatrix} 0 & -1 \\ 1 & 0 \end{pmatrix} = 2\mathbf{b}_2. \end{aligned}\right\} \tag{4.6}$$

Since the elements in (4.5) are also a basis of $\mathfrak{sl}_2(\mathbb{C})$ over $\mathbb{C}$, we obtain once again an isomorphism

$$\mathfrak{sl}_2(\mathbb{C}) \cong \mathfrak{su}_2(\mathbb{R}) \otimes_{\mathbb{R}} \mathbb{C}.$$

Once more we say that $\mathfrak{su}_2(\mathbb{R})$ is a real form of $\mathfrak{sl}_2(\mathbb{C})$, which is called the *compact real form* since $\mathrm{SU}_2(\mathbb{R})$ is compact.

To better understand $\mathfrak{su}_2(\mathbb{R})$ and $\mathrm{SU}_2(\mathbb{R})$, we may also present these in equivalent ways as in the next lemma.

Lemma 4.2 (Two isomorphisms). (a) *The vector space $\mathbb{R}^3$ equipped with the cross product is a Lie algebra isomorphic to $\mathfrak{su}_2(\mathbb{R})$.*
(b) *The sphere $\mathbb{S}^3 \subseteq \mathbb{H}$ inside the four-dimensional Hamiltonian quaternions*

$$\mathbb{H} = \mathbb{R} + \mathbb{R}\mathbf{i} + \mathbb{R}\mathbf{j} + \mathbb{R}\mathbf{k}$$

forms a real Lie group with respect to multiplication so that $\mathbb{S}^3 \cong \mathrm{SU}_2(\mathbb{R})$ as real Lie groups.

Proof of Lemma 4.2(a). The cross product map

$$\begin{pmatrix} x \\ y \\ z \end{pmatrix} \times \begin{pmatrix} \alpha \\ \beta \\ \gamma \end{pmatrix} = \begin{pmatrix} y\gamma - z\beta \\ z\alpha - x\gamma \\ x\beta - y\alpha \end{pmatrix}$$

for

$$\begin{pmatrix} x \\ y \\ z \end{pmatrix}, \begin{pmatrix} \alpha \\ \beta \\ \gamma \end{pmatrix} \in \mathbb{R}^3$$

is bilinear and antisymmetric, just as a Lie bracket is. Also note that we have the relations $e_1 \times e_2 = e_3$, $e_2 \times e_3 = e_1$, and $e_3 \times e_1 = e_2$. We define a linear map $\varphi\colon \mathbb{R}^3 \to \mathfrak{su}_2(\mathbb{R})$ by

$$\varphi(e_1) = \tfrac{1}{2}\mathbf{b}_1,$$
$$\varphi(e_2) = \tfrac{1}{2}\mathbf{b}_2,$$
$$\varphi(e_3) = \tfrac{1}{2}\mathbf{b}_3.$$

Then φ is a linear isomorphism which satisfies $\varphi(a \times b) = [\varphi(a), \varphi(b)]$. In fact this follows first for a and b being any two basis vectors in positive order by dividing the relations in (4.6) by 4. By antisymmetry of the cross product and the Lie bracket this extends to any two basis vectors. Finally, by bi-linearity of the cross product and the Lie bracket this extends to all $a, b \in \mathbb{R}^3$. As $\mathfrak{su}_2(\mathbb{R})$ is a Lie algebra, the same therefore holds for $\mathbb{R}^3$ with the cross product, proving (a). □

PROOF OF LEMMA 4.2(b). We recall first that the Hamiltonian quaternions are defined by

$$\mathbb{H} = \mathbb{R} + \mathbb{R}\mathbf{i} + \mathbb{R}\mathbf{j} + \mathbb{R}\mathbf{k}$$

where $\mathbf{i}$, $\mathbf{j}$, $\mathbf{k}$ are formal symbols satisfying the four relations†

$$\mathbf{i}^2 = \mathbf{j}^2 = \mathbf{k}^2 = -1,$$
$$\mathbf{ij} = -\mathbf{ji} = \mathbf{k},$$
$$\mathbf{jk} = -\mathbf{kj} = \mathbf{i}, \text{ and}$$
$$\mathbf{ki} = -\mathbf{ik} = \mathbf{j}.$$

We may identify $\mathbb{H}$ with

$$\left\{\begin{pmatrix} z & -\overline{w} \\ w & \overline{z} \end{pmatrix} \mid z, w \in \mathbb{C}\right\} \subseteq \mathrm{Mat}_{22}(\mathbb{C}).$$

In fact, a direct calculation shows that the matrices $\mathbf{b}_1$, $\mathbf{b}_2$, $\mathbf{b}_3$ in (4.5) satisfy the relations between $\mathbf{i}, \mathbf{j}, \mathbf{k}$ above, so that

$$\varphi\colon \mathbb{H} \ni x + a\mathbf{i} + b\mathbf{j} + c\mathbf{k} \longmapsto xI + a\mathbf{b}_1 + b\mathbf{b}_2 + c\mathbf{b}_3 = \begin{pmatrix} x + a\mathrm{i} & b\mathrm{i} - c \\ b\mathrm{i} + c & x - a\mathrm{i} \end{pmatrix} \in \mathrm{Mat}_{2,2}(\mathbb{C})$$

is an algebra isomorphism to a sub-algebra of $\mathrm{Mat}_{2,2}(\mathbb{C})$.

The norm operator defined by

$$N(x + a\mathbf{i} + b\mathbf{j} + c\mathbf{k}) = x^2 + a^2 + b^2 + c^2 = \det \varphi(x + a\mathbf{i} + b\mathbf{j} + c\mathbf{k})$$

for $x, a, b, c \in \mathbb{R}$ satisfies $N(\mathbf{gh}) = N(\mathbf{g})N(\mathbf{h})$ for all $\mathbf{g}, \mathbf{h} \in \mathbb{H}$. It follows that

$$\mathbb{S}^3 = \{\mathbf{g} \in \mathbb{H} \mid N(\mathbf{g}) = 1\}$$

is a Lie group under multiplication, which is mapped under φ to

† These relations are easy to reconstruct because of their symmetry under cyclic permutations.

$$\varphi(\mathbb{S}^3) = \left\{ g = \begin{pmatrix} z & -\overline{w} \\ w & \overline{z} \end{pmatrix} \mid z, w \in \mathbb{C}, \det g = |z|^2 + |w|^2 = 1 \right\}.$$

Note that

$$g = \begin{pmatrix} z & -\overline{w} \\ w & \overline{z} \end{pmatrix} \in \varphi(\mathbb{S}^3)$$

satisfies

$$g^* g = \begin{pmatrix} \overline{z} & \overline{w} \\ -w & z \end{pmatrix} \begin{pmatrix} z & -\overline{w} \\ w & \overline{z} \end{pmatrix} = \begin{pmatrix} |z|^2 + |w|^2 & 0 \\ 0 & |z|^2 + |w|^2 \end{pmatrix} = I$$

and so $\varphi(\mathbb{S}^3) \subseteq \mathrm{SU}_2(\mathbb{R})$. Conversely, if

$$g = \begin{pmatrix} z & z_1 \\ w & w_1 \end{pmatrix} \in \mathrm{SU}_2(\mathbb{R}),$$

then $\left\| \begin{pmatrix} z \\ w \end{pmatrix} \right\| = 1$ and

$$0 = \left\langle \begin{pmatrix} z \\ w \end{pmatrix}, \begin{pmatrix} z_1 \\ w_1 \end{pmatrix} \right\rangle = \left\langle \begin{pmatrix} z \\ w \end{pmatrix}, \begin{pmatrix} -\overline{w} \\ \overline{z} \end{pmatrix} \right\rangle$$

implies that $\begin{pmatrix} z_1 \\ w_1 \end{pmatrix} = \alpha \begin{pmatrix} -\overline{w} \\ \overline{z} \end{pmatrix}$ for some $\alpha \in \mathbb{C}$. Taking the determinant this gives

$$1 = \det g = \det \begin{pmatrix} z & -\alpha\overline{w} \\ w & \alpha\overline{z} \end{pmatrix} = \alpha|z|^2 + \alpha|w|^2 = \alpha.$$

This proves that $\varphi(\mathbb{S}^3) = \mathrm{SU}_2(\mathbb{R})$, and hence part (b) of the lemma. □

The following lemma says more about the above Lie groups from a topological point of view.

Lemma 4.3 (Topological properties). (a) *The groups* $\mathrm{SL}_2(\mathbb{R})$, $\mathrm{SU}_2(\mathbb{R})$, *and* $\mathrm{SL}_2(\mathbb{C})$ *are connected.*
(b) *The Lie groups* $\mathrm{SU}_2(\mathbb{R})$ *and* $\mathrm{SL}_2(\mathbb{C})$ *are simply connected.*
(c) *The universal cover* $\widetilde{\mathrm{SL}_2(\mathbb{R})}$ *of* $\mathrm{SL}_2(\mathbb{R})$ *is a* $\mathbb{Z}$*-cover of* $\mathrm{SL}_2(\mathbb{R})$.

Proof. For $\mathrm{SU}_2(\mathbb{R})$ both (a) and (b) follow easily from Lemma 4.2(b) since

$$\mathrm{SU}_2(\mathbb{R}) \cong \mathbb{S}^3$$

is connected and simply connected.

The facts concerning $\mathrm{SL}_2(\mathbb{R})$ and $\mathrm{SL}_2(\mathbb{C})$ follow from the Iwasawa decomposition of these groups, as we will now show.

For $\mathrm{SL}_2(\mathbb{R})$, let

$$A = \left\{ \begin{pmatrix} t & \\ & t^{-1} \end{pmatrix} \,\middle|\, t > 0 \right\},$$

$$U = \left\{ \begin{pmatrix} 1 & x \\ & 1 \end{pmatrix} \middle| \, x \in \mathbb{R} \right\}$$

and $K = \mathrm{SO}_2(\mathbb{R})$. Then every element $g \in \mathrm{SL}_2(\mathbb{R})$ has a unique decomposition $g = kau$ with $k \in K$, $a \in A$, and $u \in U$ by the Iwasawa decomposition (see Exercise 4.4). This gives a homeomorphism $\mathrm{SL}_2(\mathbb{R}) \cong KAU \cong \mathbb{S}^1 \times \mathbb{R}^2$ which proves the claims in (a) and (c) for $\mathrm{SL}_2(\mathbb{R})$.

For $\mathrm{SL}_2(\mathbb{C})$ we have A as before,

$$U = \left\{ \begin{pmatrix} 1 & \alpha \\ & 1 \end{pmatrix} \middle| \, \alpha \in \mathbb{C} \right\},$$

and $K = \mathrm{SU}_2(\mathbb{R}) \cong \mathbb{S}^3$, which gives a homeomorphism

$$\mathrm{SL}_2(\mathbb{C}) = KAU \cong \mathbb{S}^3 \times \mathbb{R}^3,$$

and hence the remaining claims in the lemma. □

Exercise 4.4 (Euclidean meaning of Iwasawa decomposition). Give a proof of the Iwasawa decomposition for $\mathrm{SL}_2(\mathbb{R})$ (or of $\mathrm{SL}_d(\mathbb{R})$ for $d \geqslant 2$) and $\mathrm{SL}_2(\mathbb{C})$ using the Gram–Schmidt orthonormalization procedure.

4.1.3 A Quick Review of Lie Algebra Representations

Let us start by recalling that for a representation

$$\rho\colon G \ni g \longmapsto \rho_g \in \mathrm{GL}(W)$$

on a finite-dimensional vector space W of a (real or complex) Lie group G we can take the derivative of ρ at the identity to obtain the map

$$\mathrm{D}\rho\colon \mathfrak{g} \longrightarrow \mathfrak{gl}(W) = \mathrm{Hom}(W).$$

We note that the exponential map satisfies, together with the derivative $\mathrm{D}\rho$, the categorical property

$$\rho \circ \exp = \exp \circ \mathrm{D}\rho \tag{4.7}$$

(see Exercise 4.5).

Moreover, $\mathrm{D}\rho$ defines a representation of the Lie algebra $\mathfrak{g}$ of G on the vector space W, or, equivalently, $\mathrm{D}\rho$ is a Lie algebra homomorphism satisfying

$$\mathrm{D}\rho([\mathbf{b}, \mathbf{c}]) = [\mathrm{D}\rho(\mathbf{b}), \mathrm{D}\rho(\mathbf{c})] \tag{4.8}$$

for all $\mathbf{b}, \mathbf{c} \in \mathfrak{g}$. To see (4.8) we fix $g \in G$ and $\mathbf{c} \in \mathfrak{g}$. Then, by definition, the adjoint representation applied to g and $\mathbf{c}$ is given by

$$\mathrm{Ad}_g(\mathbf{c}) = \frac{\mathrm{d}}{\mathrm{d}t}\bigg|_{t=0} g \exp(t\mathbf{c}) g^{-1}.$$

For the element ρ_g and the Lie algebra element $\mathrm{D}\rho(\mathbf{c})$ for some $\mathbf{c} \in \mathfrak{g}$, this gives

$$\begin{aligned}
\mathrm{Ad}_{\rho_g}(\mathrm{D}\rho(\mathbf{c})) &= \frac{\mathrm{d}}{\mathrm{d}t}\Big|_{t=0} \rho_g \exp(\mathrm{D}\rho(t\mathbf{c}))\rho_g^{-1} \\
&= \frac{\mathrm{d}}{\mathrm{d}t}\Big|_{t=0} \rho_g \rho(\exp(t\mathbf{c}))\rho_g^{-1} && \text{(by (4.7))} \\
&= \frac{\mathrm{d}}{\mathrm{d}t}\Big|_{t=0} \rho\big(g\exp(t\mathbf{c})g^{-1}\big) = \mathrm{D}\rho\,\mathrm{Ad}_g(\mathbf{c}), && (4.9)
\end{aligned}$$

where we also used the homomorphism property of ρ and the chain rule for differentiation. If now $\mathbf{b}, \mathbf{c} \in \mathfrak{g}$ then we can set $g = \exp(t\mathbf{b})$ and use the defining property

$$[\mathbf{b}, \mathbf{c}] = \mathrm{ad}_{\mathbf{b}}(\mathbf{c}) = \frac{\mathrm{d}}{\mathrm{d}t}\Big|_{t=0} \mathrm{Ad}_{\exp(t\mathbf{b})}(\mathbf{c})$$

in the same way. In fact we have

$$\begin{aligned}
[\mathrm{D}\rho(\mathbf{b}), \mathrm{D}\rho(\mathbf{c})] &= \frac{\mathrm{d}}{\mathrm{d}t}\Big|_{t=0} \mathrm{Ad}_{\exp(\mathrm{D}\rho(t\mathbf{b}))}\big(\mathrm{D}\rho(\mathbf{c})\big) \\
&= \frac{\mathrm{d}}{\mathrm{d}t}\Big|_{t=0} \mathrm{Ad}_{\rho(\exp(t\mathbf{b}))}\big(\mathrm{D}\rho(\mathbf{c})\big) && \text{(by (4.7))} \\
&= \frac{\mathrm{d}}{\mathrm{d}t}\Big|_{t=0} \mathrm{D}\rho\,\mathrm{Ad}_{\exp(t\mathbf{b})}(\mathbf{c}) = \mathrm{D}\rho[\mathbf{b}, \mathbf{c}]. && \text{(by (4.9))}
\end{aligned}$$

Hence the study of finite-dimensional representations of G leads to the study of finite-dimensional representations of $\mathfrak{g}$. Below we will also denote the representation of $\mathfrak{g}$ induced from a representation ρ of G by ρ.

Let W be a finite-dimensional vector space carrying a representation ρ of a Lie group G, and let $V \subseteq W$ be a subspace. If V is invariant under $\rho(G)$, then by taking derivatives we see that V is also invariant under $\rho(\mathfrak{g})$. If G is connected, the reverse also holds. This follows since $\rho(\mathbf{c})V \subseteq V$ for $\mathbf{c} \in \mathfrak{g}$ implies by (4.7) that

$$\rho\big(\exp(\mathbf{c})\big)V = \exp\big(\rho(\mathbf{c})\big)V \subseteq V$$

for all $\mathbf{c} \in \mathfrak{g}$, which then extends to $\rho_g V \subseteq V$ for all $g \in G^o = G$ (since the subgroup generated by $\exp(\mathfrak{g})$ is open).

In particular, for the connected groups $\mathrm{SL}_2(\mathbb{R})$, $\mathrm{SL}_2(\mathbb{C})$, and $\mathrm{SU}_2(\mathbb{R})$ the notions of irreducibility for finite-dimensional representations of the Lie group or of the Lie algebra coincide.

As discussed above, any finite-dimensional representation of G gives rise to a representation of its Lie algebra $\mathfrak{g}$. However, the converse to this is a bit harder to prove and only states that every finite-dimensional representation of $\mathfrak{g}$ gives rise to a finite-dimensional representation of the universal cover $\widetilde{G}$ of G. In the case of $G = \mathrm{SL}_2(\mathbb{C})$ and $G = \mathrm{SU}_2(\mathbb{R})$ this and Lemma 4.3 explains the correspondence between the irreducible representations of G and its Lie algebra in Theorem 4.6 below. However, for $G = \mathrm{SL}_2(\mathbb{R})$ this correspondence is a special feature.

Essential Exercise 4.5. Let G be a Lie group with Lie algebra $\mathfrak{g} = \mathrm{Lie}(G)$, and let $\rho\colon G \to \mathrm{GL}(W)$ be a finite-dimensional representation of G.
(a) Show that ρ is differentiable (in doing so, use the fact that $\exp\colon \mathfrak{g} \to G$ is smooth and has a local inverse near the identity).
(b) Show the categorical property (4.7) of the exponential map.

4.1.4 Irreducible Representations of the Lie Algebra

We now specialize to the case $\mathfrak{g} = \mathfrak{sl}_2(\mathbb{C})$ and classify its irreducible representations in the next theorem, which is crucial for the representation theory of all semi-simple Lie groups and for the classification of semi-simple Lie groups.

Theorem 4.6 (Irreducible representations of $\mathfrak{sl}_2$). *The irreducible finite-dimensional representations of $\mathfrak{sl}_2(\mathbb{C})$ (respectively of $\mathfrak{sl}_2(\mathbb{R})$ or $\mathfrak{su}_2(\mathbb{R})$) are in a natural one-to-one correspondence with the elements of $\mathbb{N}_0$. In fact for every n in $\mathbb{N}_0$ the representation of $\mathrm{SL}_2(\mathbb{C})$ on the* symmetric tensor product

$$\mathrm{Sym}^n(\mathbb{C}^2) = \Big\{\sum_{k=0}^{n} \alpha_k e_1^{\odot k} \odot e_2^{\odot(n-k)} \mid \alpha_0, \ldots, \alpha_n \in \mathbb{C}\Big\}$$

gives rise to an irreducible representation of $\mathfrak{sl}_2(\mathbb{C})$ of dimension $(n+1)$. By restriction, we also obtain irreducible representations of $\mathrm{SL}_2(\mathbb{R})$, or $\mathrm{SU}_2(\mathbb{R})$, and of the Lie algebras $\mathfrak{sl}_2(\mathbb{R})$ and $\mathfrak{su}_2(\mathbb{R})$. Any irreducible finite-dimensional representation of $\mathfrak{sl}_2(\mathbb{C})$, of $\mathfrak{sl}_2(\mathbb{R})$, or of $\mathfrak{su}_2(\mathbb{R})$ is isomorphic to one of these.

Before we begin the proof of the theorem we first wish to describe the representations on $\mathrm{Sym}^n(\mathbb{C}^2)$ in more detail.

Given a finite-dimensional vector space W and a representation ρ of a group G on it, one can define the symmetric tensor product

$$\mathrm{Sym}^n(W) = \langle w_1 \odot w_2 \odot \cdots \odot w_n \mid w_1, \ldots, w_n \in W\rangle$$

as the linear hull of all formal commuting products of n vectors in W with the product map

$$W^n \ni (w_1, \ldots, w_n) \longmapsto w_1 \odot \cdots \odot w_n \in \mathrm{Sym}^n(W)$$

being multi-linear. In fact, using the n-fold tensor product $\bigotimes_{j=1}^n W$, we define

$$\mathrm{Sym}^n(W) = \bigotimes_{j=1}^{n} W / \langle w_1 \otimes \cdots \otimes w_n - w_{\sigma(1)} \otimes \cdots \otimes w_{\sigma(n)} \mid \sigma \in \mathrm{S}_n\rangle,$$

where S_n again denotes the symmetric group of $\{1,\ldots,n\}$. For $n=0$ we define $\mathrm{Sym}^0(W)=\mathbb{C}$, and refer to Hungerford [42] for more details. Moreover, any linear map $A\in\mathrm{Hom}(W,W)$ can be used to induce a linear map

$$\mathrm{Sym}^n(A)\in\mathrm{Hom}\big(\mathrm{Sym}^n(W),\mathrm{Sym}^n(W)\big)$$

with

$$\mathrm{Sym}^n(A)(w_1\odot\cdots\odot w_n)=(Aw_1)\odot\cdots\odot(Aw_n)$$

for all $w_1,\ldots,w_n\in W$.

Returning to the setting of Theorem 4.6, we note that $\mathbb{C}^2$ carries the standard representation of $\mathrm{SL}_2(\mathbb{R})$, $\mathrm{SL}_2(\mathbb{C})$, or $\mathrm{SU}_2(\mathbb{R})$ defined by

$$\mathbb{C}^2\ni v\longmapsto gv\in\mathbb{C}^2.$$

This defines the representation on $\mathrm{Sym}^n(\mathbb{C}^2)$ for all $n\in\mathbb{N}$, where $n=1$ corresponds to the standard representation. In the special case $n=0$ we use the trivial representation on $\mathrm{Sym}^n(\mathbb{C}^2)=\mathbb{C}$.

If we instead let W be the vector space of linear maps on $\mathbb{C}^2$, then $\mathrm{Sym}^n(W)$ becomes the space $\mathrm{Pol}_n(\mathbb{C}^2)$ of homogeneous polynomials of degree n on $\mathbb{C}^2$. The isomorphism $\mathrm{Sym}^n(\mathbb{C}^2)\cong\mathrm{Sym}^n(W)$ would also follows from the proof of the theorem below, but let us indicate briefly where it comes from. In fact, we will show that the standard representation on $\mathbb{C}^2$ and the representation ρ on W are isomorphic. Let

$$g=\begin{pmatrix}a&b\\c&d\end{pmatrix}\in\mathrm{SL}_2(\mathbb{C})$$

and then send the linear map f defined by $f(X_1,X_2)=\alpha X_1+\beta X_2$ to

$$\rho_g(f)=f\circ g^{-1}\colon\begin{pmatrix}X_1\\X_2\end{pmatrix}\longmapsto g^{-1}\begin{pmatrix}X_1\\X_2\end{pmatrix}\longmapsto(\alpha,\beta)g^{-1}\begin{pmatrix}X_1\\X_2\end{pmatrix}.$$

In terms of the basis X_1,X_2 (dual to the standard basis e_1,e_2 of $\mathbb{C}^2$), this corresponds to the map

$$\begin{aligned}\rho_g=(g^{-1})^{\mathrm{t}}&=\begin{pmatrix}d&-b\\-c&a\end{pmatrix}^{\mathrm{t}}\\&=\begin{pmatrix}d&-c\\-b&a\end{pmatrix}=\begin{pmatrix}&-1\\1&\end{pmatrix}\begin{pmatrix}a&b\\c&d\end{pmatrix}\begin{pmatrix}&1\\-1&\end{pmatrix},\end{aligned}$$

or $\rho_g=kgk^{-1}$ for all $g=\begin{pmatrix}a&b\\c&d\end{pmatrix}\in\mathrm{SL}_2(\mathbb{C})$, where $k=\begin{pmatrix}&-1\\1&\end{pmatrix}$.

This shows that the standard representation of $\mathrm{SL}_2(\mathbb{C})$ on $\mathbb{C}^2$ and its dual ρ on W are isomorphic, which is a special property of $\mathrm{SL}_2(\mathbb{C})$ due to the special rule for calculating g^{-1}. This isomorphism can be used to find the isomorphism between $\mathrm{Sym}^n(\mathbb{C}^2)$ and $\mathrm{Sym}^n(W)=\mathrm{Pol}_n(\mathbb{C}^2)$.

PROOF OF IRREDUCIBILITY. Let us now work with the representation on the space $\mathrm{Sym}^n(\mathbb{C}^2)$ of $G = \mathrm{SL}_2(\mathbb{C})$ or $G = \mathrm{SL}_2(\mathbb{R})$. The trivial representation corresponding to $n = 0$ is clearly irreducible.

We now assume that $n \geqslant 1$. For $\mathbf{a} = \begin{pmatrix} 1 & \\ & -1 \end{pmatrix}$ and $t \in \mathbb{R}$ we have

$$\exp(t\mathbf{a}) = \begin{pmatrix} \mathrm{e}^t & \\ & \mathrm{e}^{-t} \end{pmatrix}$$

and $\exp(t\mathbf{a})e_1 = \mathrm{e}^t e_1$ and $\exp(t\mathbf{a})e_2 = \mathrm{e}^{-t} e_2$ for the representation on $\mathbb{C}^2$ and hence

$$\begin{aligned} \mathrm{Sym}^n\big(\exp(t\mathbf{a})\big)\big(e_1^{\odot k} \odot e_2^{\odot(n-k)}\big) &= \big(\mathrm{e}^t e_1\big)^{\odot k} \odot \big(\mathrm{e}^{-t} e_2\big)^{\odot(n-k)} \\ &= \mathrm{e}^{kt-(n-k)t} e_1^{\odot k} \odot e_2^{\odot(n-k)} \end{aligned}$$

for $k = 0, \ldots, n$. This shows that $\exp(t\mathbf{a})$ acts diagonally on $\mathrm{Sym}^n(\mathbb{C}^2)$ with eigenvalues $\mathrm{e}^{-nt}, \mathrm{e}^{(-n+2)t}, \ldots, \mathrm{e}^{(n-2)t}, \mathrm{e}^{nt}$. Taking the derivative, we also obtain that the Lie algebra element $\mathbf{a}$ acts diagonally via

$$\mathrm{Sym}^n(\mathbf{a}) = \frac{\mathrm{d}}{\mathrm{d}t}\Big|_{t=0} \mathrm{Sym}^n\big(\rho(\exp(t\mathbf{a}))\big),$$

with eigenvalues $-n, -n+2, \ldots, n-2, n$.

For $\mathbf{e} = \begin{pmatrix} 0 & 1 \\ & 0 \end{pmatrix}$ and $t \in \mathbb{R}$ we have $\exp(t\mathbf{e}) = \begin{pmatrix} 1 & t \\ & 1 \end{pmatrix}$ and $\exp(t\mathbf{e})e_1 = e_1$ and $\exp(t\mathbf{e})e_2 = te_1 + e_2$ for the representation on $\mathbb{C}^2$ and hence

$$\begin{aligned} \mathrm{Sym}^n\big(\exp(t\mathbf{e})\big)\big(e_1^{\odot k} \odot e_2^{\odot(n-k)}\big) &= e_1^{\odot k}(te_1 + e_2)^{\odot(n-k)} \\ &= e_1^{\odot k}\Big(e_2^{\odot(n-k)} + te_1\binom{n-k}{1}e_2^{\odot(n-k-1)} + \cdots\Big), \end{aligned}$$

where the dots indicate the terms of order two and higher with respect to the variable t. Taking the derivative at $t = 0$, this gives

$$\mathrm{Sym}^n(\mathbf{e})(e_1^{\odot k} \odot e_2^{\odot(n-k)}) = (n-k)e_1^{\odot(k+1)} \odot e_2^{\odot(n-k-1)} \tag{4.10}$$

for $k = 0, \ldots, n$, which for $k = n$ should simply be read as $\mathrm{Sym}^n(\mathbf{e})(e_1^{\odot n}) = 0$. Similarly,

$$\mathrm{Sym}^n(\mathbf{f})(e_1^{\odot k} \odot e_2^{\odot(n-k)}) = ke_1^{\odot(k-1)} \odot e_2^{\odot(n-k+1)} \tag{4.11}$$

for all $k = 0, \ldots, n$, which for $k = 0$ should be read as $\mathrm{Sym}^n(\mathbf{f})(e_2^{\odot n}) = 0$.

Suppose now that $V \subseteq \mathrm{Sym}^n(\mathbb{C}^2)$ is non-trivial and invariant under Sym^n. Since $\mathrm{Sym}^n(\mathbf{a})$ is diagonal with $n+1$ different eigenvalues, each with multiplicity one, it follows that

$$e_1^{\odot k} \odot e_2^{\odot(n-k)} \in V$$

for some $k \in \{0, 1, \ldots, n\}$. However, using (4.10) and (4.11) we also see that

$$e_1^{\odot(k+1)} \odot e_2^{\odot(n-k-1)} \in V$$

if $k < n$, and

$$e_1^{\odot(k-1)} e_2^{\odot(n-k+1)} \in V$$

if $k > 0$. Iterating this shows that $V = \mathrm{Sym}^n(\mathbb{C}^2)$ contains all basis vectors, and irreducibility of our representation on $\mathrm{Sym}^n(\mathbb{C}^2)$ follows. □

As already visible in the proof above the eigenvectors of $\rho(\mathbf{a})$ and their eigenvalues play an important role in the theory. Hence they deserve a special name: the eigenvalues of $\rho(\mathbf{a})$ are called *weights* and the corresponding eigenvectors are called *weight vectors*. Moreover, the numbers 2 and -2 are called the *roots* and the vectors $\mathbf{e}, \mathbf{f} \in \mathfrak{sl}_2(\mathbb{C})$ the *root vectors*.

PROOF OF COMPLETENESS. We now will show that for $G = \mathrm{SL}_2(\mathbb{C})$ and for $G = \mathrm{SL}_2(\mathbb{R})$ the list of irreducible finite-dimensional representations above is complete. For this we let W carry an arbitrary finite-dimensional representation ρ of the Lie algebra $\mathfrak{g}$ of G. Later we will assume that ρ is irreducible, but the initial part of the construction is more general. Since W is finite-dimensional, $\rho(\mathbf{a}) \in \mathrm{Hom}(W)$ must have at least one weight, that is an eigenvalue of $\rho(\mathbf{a})$. Let us assume that $\lambda_0 \in \mathbb{C}$ is a weight with the property that $\Re\lambda_0$ is maximal in the set $\{\Re\lambda \mid \lambda \text{ is a weight}\}$. Let $w_0 \in W$ be a weight vector for weight λ_0. We will be using $\rho(\mathbf{f})$ and the following more general claim to find more weight vectors with weights $\lambda_0 - 2$, $\lambda_0 - 4$,... which will lead to a complete classification of W.

FUNDAMENTAL CALCULATION. If $v \in W$ is a weight vector for weight λ, then we claim that $\rho(\mathbf{e})v$ is either a weight vector for weight $\lambda + 2$ or is equal to zero. Similarly, $\rho(\mathbf{f})v$ is either a weight vector for weight $\lambda - 2$ or is equal to zero.

The proof of the claim is rather simple. Indeed, using the defining property (4.8) of a Lie algebra homomorphism and the defining properties of the $\mathfrak{sl}_2$-triple in (4.3), we have

$$\begin{aligned}
\rho(\mathbf{a})\rho(\mathbf{e})v &= \big(\rho(\mathbf{a})\rho(\mathbf{e}) - \rho(\mathbf{e})\rho(\mathbf{a})\big)v + \rho(\mathbf{e})\rho(\mathbf{a})v \\
&= \big[\rho(\mathbf{a}), \rho(\mathbf{e})\big]v + \rho(\mathbf{e})\lambda v \\
&= \rho\big([\mathbf{a}, \mathbf{e}]\big)v + \lambda\rho(\mathbf{e})v \\
&= \rho(2\mathbf{e})v + \lambda\rho(\mathbf{e})v = (\lambda + 2)\rho(\mathbf{e})v.
\end{aligned}$$

Similarly, we obtain

$$\begin{aligned}
\rho(\mathbf{a})\rho(\mathbf{f})v &= [\rho(\mathbf{a}), \rho(\mathbf{f})]v + \rho(\mathbf{f})\rho(\mathbf{a})v \\
&= \rho\big([\mathbf{a}, \mathbf{f}]\big)v + \lambda\rho(\mathbf{f})v = (\lambda - 2)\rho(\mathbf{f})v.
\end{aligned}$$

CONSTRUCTION OF EIGENVECTORS. Let $w_0 \in W$ be a weight vector with maximal real part of its weight λ_0 as above. By the fundamental calculation, $\rho(\mathbf{e})w_0$ would have weight $\lambda_0 + 2$, which by maximality of $\Re\lambda$ implies that

$$\rho(\mathbf{e})w_0 = 0. \tag{4.12}$$

On the other hand we can define

$$w_1 = \rho(\mathbf{f})w_0,\ w_2 = \rho(\mathbf{f})w_1,\ \ldots \tag{4.13}$$

and obtain weight vectors for weights $\lambda_0 - 2, \lambda_0 - 4, \ldots$. As eigenvectors for different eigenvalues are always linearly independent and $\dim W < \infty$ it follows that there must exist some $n \geqslant 0$ with $w_n = \rho(\mathbf{f})^n w_0 \neq 0$ but

$$\rho(\mathbf{f})w_n = 0. \tag{4.14}$$

AN INVARIANT SUBSPACE. We let $V = \langle w_0, \ldots, w_n \rangle \subseteq W$ and claim that V is invariant under ρ. Since V is generated by eigenvectors for $\rho(\mathbf{a})$, V is clearly invariant under $\rho(\mathbf{a})$. Moreover, by the construction in (4.13) and the property (4.14) we have $\rho(\mathbf{f})w_k \in V$ for all $k = 0, \ldots, n$ and hence $\rho(\mathbf{f})V \subseteq V$. It remains to study $\rho(\mathbf{e})$, where we will prove by induction that

$$\rho(\mathbf{e})w_k \begin{cases} = 0 & \text{for } k = 0; \\ \in \mathbb{C}w_{k-1} & \text{for } k > 0 \end{cases} \tag{4.15}$$

for $k = 0, \ldots, n$. Indeed, we know this for $k = 0$ by (4.12). If now (4.15) is already known for some $k \in \{0, \ldots, n-1\}$, then by construction

$$w_{k+1} = \rho(\mathbf{f})w_k$$

and

$$\begin{aligned} \rho(\mathbf{e})w_{k+1} &= \rho(\mathbf{e})\rho(\mathbf{f})w_k \\ &= [\rho(\mathbf{e}), \rho(\mathbf{f})]w_k + \rho(\mathbf{f})\rho(\mathbf{e})w_k \\ &= \rho(\mathbf{a})w_k + \rho(\mathbf{f})\rho(\mathbf{e})w_k \in \mathbb{C}w_k + \rho(\mathbf{f})\mathbb{C}w_{k-1} = \mathbb{C}w_k, \end{aligned}$$

by the inductive assumption. This shows the inductive step.

ASSUMING IRREDUCIBILITY. Suppose now in addition that ρ is irreducible. Since $V \subseteq W$ is non-trivial and invariant under ρ, it follows that $V = W$ has the basis $w_0, \ldots, w_n$ consisting of weight vectors for $\rho(\mathbf{a})$ for the weights

$$\lambda_0, \lambda_0 - 2, \ldots, \lambda_0 - 2n.$$

In particular, $\rho(\mathbf{a})$ is diagonalizable with these eigenvalues, and the trace of $\rho(\mathbf{a})$ is

$$\sum_{k=0}^{n} (\lambda_0 - 2k) = (n+1)\lambda_0 - 2\sum_{k=0}^{n} k = (n+1)\lambda_0 - (n+1)n.$$

Since $\rho(\mathbf{a}) = \rho([\mathbf{e}, \mathbf{f}]) = [\rho(\mathbf{e}), \rho(\mathbf{f})]$, we also know that $\operatorname{tr}\rho(\mathbf{a}) = 0$, and deduce that $\lambda = n \in \mathbb{N}_0$.

AN ISOMORPHISM. We now combine the arguments above with our discussion of $\mathrm{Sym}^n(\mathbb{C}^2)$ by constructing the graph of an isomorphism within

$$\widetilde{W} = \mathrm{Sym}^n(\mathbb{C}^2) \oplus W.$$

In fact the vector $v_0 = (e_1^{\odot n}, w_0) \in \widetilde{W}$ is a weight vector of weight n satisfying $\mathrm{Sym}^n(\mathbf{e}) \oplus \rho(\mathbf{e})v_0 = 0$ just as in the case of our original vector $w_0 \in W$. Applying the same argument as before, we produce weight vectors $v_0, v_1, \ldots, v_n$ in $\widetilde{W}$ such that $V = \langle v_0, v_1, \ldots, v_n \rangle$ is invariant. Moreover,

$$\dim V = n + 1 = \dim \mathrm{Sym}^n(\mathbb{C}^2) = \dim W$$

and the projections of V onto $\mathrm{Sym}^n(\mathbb{C}^2)$ respectively onto W are surjective, which follows from (4.11) for $\mathrm{Sym}^n(\mathbb{C}^2)$ and since $w_0, w_1, \ldots, w_n$ is a basis of W. This shows that $V = \mathrm{Graph}(\Phi)$ for some linear isomorphism $\Phi\colon \mathrm{Sym}^n(\mathbb{C}^2) \to W$ and invariance of V implies that ϕ is an isomorphism of the representations.

□

The maximal weight as it appeared in the proof above is called the *highest weight* and the corresponding weight vectors are called *highest weight vectors*.

Exercise 4.7. Use the arguments from the proof of Theorem 4.6 to show that for any finite-dimensional representation of $\mathfrak{sl}_2(\mathbb{C})$ a highest weight vector for highest weight λ_0 always generates an irreducible subrepresentation of dimension $\lambda_0 + 1$.

We refer to Fulton and Harris [32] for an accessible treatment of the theory of highest weight vectors for more general semi-simple groups. We will see similar mechanisms for creating more eigenvectors out of an initial eigenvector also for unitary representations later.

To summarize, we have proved Theorem 4.6 in the two cases $\mathfrak{g} = \mathfrak{sl}_2(\mathbb{C})$ and $\mathfrak{g} = \mathfrak{sl}_2(\mathbb{R})$, where we did not see any difference in the arguments as the $\mathfrak{sl}_2$-triple belonged to $\mathfrak{sl}_2(\mathbb{R})$ and all subspaces of the representation space are assumed to be complex subspaces. As we will now show, the extension to $\mathfrak{su}_2(\mathbb{R})$ does not require much apart from the right insight.

Proposition 4.8 (Complexification). *Any finite-dimensional (and, as always, complex) representation*

$$\rho\colon \mathfrak{su}_2(\mathbb{R}) \longrightarrow \mathrm{Hom}(W)$$

(or $\rho\colon \mathfrak{sl}_2(\mathbb{R}) \to \mathrm{Hom}(W)$) can be extended in a unique way to a $\mathbb{C}$-linear representation $\rho_{\mathbb{C}}\colon \mathfrak{sl}_2(\mathbb{C}) \to \mathrm{Hom}(W)$. A subspace $V \subseteq W$ is invariant under $\rho(\mathfrak{su}_2(\mathbb{R}))$ (respectively $\rho(\mathfrak{sl}_2(\mathbb{R}))$) if and only if it is invariant under $\rho_{\mathbb{C}}(\mathfrak{sl}_2(\mathbb{C}))$. In particular, they have the same list of irreducible finite-dimensional representations.

PROOF. Since $\mathrm{Hom}(W)$ is a complex vector space, $\mathfrak{su}_2(\mathbb{R})$ is a real vector space, and $\mathfrak{sl}_2(\mathbb{C}) = \mathfrak{su}_2(\mathbb{R}) \otimes_{\mathbb{R}} \mathbb{C}$ by the discussion in Section 4.1.2, every $\mathbb{R}$-linear map $\rho\colon \mathfrak{su}_2(\mathbb{R}) \to \mathrm{Hom}(W)$ has a uniquely defined $\mathbb{C}$-linear extension $\rho_{\mathbb{C}}\colon \mathfrak{sl}_2(\mathbb{C}) \to \mathrm{Hom}(W)$ satisfying

$$\rho_{\mathbb{C}}(\mathbf{b}_1 + \mathrm{i}\mathbf{b}_2) = \rho(\mathbf{b}_1) + \mathrm{i}\rho(\mathbf{b}_2)$$

for all $\mathbf{b}_1, \mathbf{b}_2 \in \mathfrak{su}_2(\mathbb{R})$. Moreover, if

$$\rho([\mathbf{b}_1, \mathbf{b}_2]) = [\rho(\mathbf{b}_1), \rho(\mathbf{b}_2)]$$

for all $\mathbf{b}_1, \mathbf{b}_2 \in \mathfrak{su}_2(\mathbb{R})$ then this property extends by bilinearity of $[\cdot, \cdot]$ and linearity of $\rho_{\mathbb{C}}$ from $\mathfrak{su}_2(\mathbb{R})$ to all $\mathbf{b}_1, \mathbf{b}_2 \in \mathfrak{sl}_2(\mathbb{C})$.

The remaining claims of the proposition follow from this. □

We have shown that $\mathfrak{sl}_2(\mathbb{C})$, $\mathfrak{sl}_2(\mathbb{R})$, and $\mathfrak{su}_2(\mathbb{R})$ as well as $\mathrm{SL}_2(\mathbb{C})$, $\mathrm{SL}_2(\mathbb{R})$, and $\mathrm{SU}_2(\mathbb{R})$ have the same list of irreducible finite-dimensional representations, which concludes the proof of Theorem 4.6. In this discussion, the Lie algebra $\mathfrak{su}_2(\mathbb{R})$ and its Lie group $\mathrm{SU}_2(\mathbb{R})$ were the odd ones out as they required extra effort (because $\mathfrak{su}_2(\mathbb{R})$ does not contain an $\mathfrak{sl}_2$-triple). However, in the next section the special properties of $\mathrm{SU}_2(\mathbb{R})$ are used to prove an important property of $\mathrm{SL}_2(\mathbb{C})$ and $\mathrm{SL}_2(\mathbb{R})$.

Exercise 4.9. Show that $\mathfrak{so}_n(\mathbb{R})$ and $\mathfrak{so}_{n-k,k}(\mathbb{R})$ with $k \in \{1, \ldots, n-2\}$ are real forms of $\mathfrak{so}_n(\mathbb{C})$, and conclude that Proposition 4.8 holds in the same way for these Lie algebras.

4.1.5 The Weyl Unitary Trick

Proposition 4.10 (Semi-simplicity of representations of $\mathfrak{su}_2(\mathbb{R})$). *Assume that W is a finite-dimensional representation of $\mathfrak{su}_2(\mathbb{R})$. If a complex subspace V of W is invariant under $\mathfrak{su}_2(\mathbb{R})$, then there exists an invariant complementary subspace V' so that $W = V \oplus V'$.*

PROOF. As explained in Section 4.1.3, W is also a representation space for the group $\mathrm{SU}_2(\mathbb{R})$ and V is invariant under $\mathrm{SU}_2(\mathbb{R})$. Fix some inner product on W and apply Proposition 3.27. Hence we may assume that W is a unitary representation of $\mathrm{SU}_2(\mathbb{R})$ and we may define $V' = V^\perp$ with respect to this inner product. □

Theorem 4.11 (Finite-dimensional representations of $\mathfrak{sl}_2$). *For finite-dimensional representations of $\mathfrak{sl}_2(\mathbb{C})$, $\mathfrak{sl}_2(\mathbb{R})$, and $\mathfrak{su}_2(\mathbb{R})$ we have the following properties.*

(a) *(Semi-simplicity) Any invariant subspace has an invariant complement.*
(b) *(Description) The representation is a finite direct sum of irreducible representations as described in Theorem 4.6.*

PROOF. For $\mathfrak{su}_2(\mathbb{R})$ part (a) is precisely the statement in Proposition 4.10. Part (b) follows from this by induction on the dimension. For $\mathfrak{sl}_2(\mathbb{C})$ and $\mathfrak{sl}_2(\mathbb{R})$ we combine Proposition 4.8 with the above. □

The argument above using compactness of $\mathrm{SU}_2(\mathbb{R}) \subseteq \mathrm{SL}_2(\mathbb{C})$ can in fact be used for all semi-simple real and complex Lie groups since there always exists a compact form[(10)] that can take the role of $\mathrm{SU}_2(\mathbb{R})$.

Exercise 4.12 (Clebsch–Gordan). For $m, n \in \mathbb{N}_0$ decompose the inner tensor product $\pi_m \otimes \pi_n$ into irreducible representations, where π_m denotes the irreducible representation $\mathrm{Sym}^n(\mathbb{C}^2)$ of $\mathrm{SU}_2(\mathbb{R})$.

4.2 Harmonic Analysis of $\mathrm{SU}_2(\mathbb{R})$ and Quotients*

4.2.1 Peter–Weyl Theorem for $\mathrm{SU}_2(\mathbb{R})$

†We now start the in-depth discussion of harmonic analysis on $\mathrm{SU}_2(\mathbb{R})$ and related spaces. Schur orthogonality and the Peter–Weyl theorem (Theorems 3.30 and 3.34) give a complete description of $L^2(G)$ for a compact group G, assuming a complete description of $\widehat{G}$. In Theorem 4.6 we have obtained the description of all irreducible finite-dimensional representations of $\mathrm{SU}_2(\mathbb{R})$. Combining these two (and calculating the inner product on $\mathrm{Sym}^n(\mathbb{C}^2)$) gives the following result. For this, we will be using the coordinate system $(z, w) \in \mathbb{C}^2$ with $|z|^2 + |w|^2 = 1$ for the elements

$$g = \begin{pmatrix} z & -\overline{w} \\ w & \overline{z} \end{pmatrix} \in \mathrm{SU}_2(\mathbb{R}). \tag{4.16}$$

Corollary 4.13 (Peter–Weyl for $\mathrm{SU}_2(\mathbb{R})$). *For $n \in \mathbb{N}_0$ and parameters k and ℓ in $\{0, \dots, n\}$ we define $\pi_{k,\ell}^{(n)}(g)$ for g as in (4.16) by*

$$\sqrt{k!(n-k)!\ell!(n-\ell)!} \sum_{\substack{i \in \{0,\dots,n-k\} \\ j \in \{0,\dots,k\} \\ i+j=\ell}} \frac{(-1)^{k-j}}{i!j!(n-k-i)!(k-j)!} z^{n-k-i} w^i \overline{z}^j \overline{w}^{k-j}.$$

Then the functions

$$g \longmapsto \sqrt{n+1}\,\pi_{k,\ell}^{(n)}(g)$$

for $k, \ell \in \{0, \dots, n\}$ are the normalized matrix coefficients associated to the irreducible representation on the $(n+1)$-dimensional space $\mathrm{Sym}^n(\mathbb{C}^2)$ for n in $\mathbb{N}_0$ (and using a convenient choice of orthonormal basis of $\mathrm{Sym}^n(\mathbb{C}^2)$). By also varying n in $\mathbb{N}_0$ we obtain an orthonormal basis of $L^2(\mathrm{SU}_2(\mathbb{R}))$.

PROOF. As already hinted at before the corollary, we have done all the work required for the corollary apart from determining the inner product on $\mathrm{Sym}^n(\mathbb{C}^2)$ for $n \in \mathbb{N}_0$. We note that the inner product is uniquely determined up to a positive scalar by the irreducibility of the representations and Schur's lemma (Theorem 1.29).

The case $n = 0$ corresponds to the trivial representation, and setting in addition k and ℓ to be 0 gives $\sqrt{1}\pi_{0,0}^{(0)} = 1$. The case $n = 1$ corresponds to the

† This section discusses the first non-trivial and quite important example of a compact simple group. In order to be completely explicit, the discussion is quite heavy with concrete formulas. As indicated in the title, these are not important for most of the subsequent discussions.

standard representation on $\mathbb{C}^2$ and the standard inner product on $\mathbb{C}^2$ makes the action unitary (by definition of $\mathrm{SU}_2(\mathbb{R})$). In this case we use the standard basis

$$w_0 = e_1 = \begin{pmatrix} 1 \\ 0 \end{pmatrix}, \quad w_1 = e_2 = \begin{pmatrix} 0 \\ 1 \end{pmatrix}$$

and obtain $\pi_{0,0}^{(1)}(g) = z$, $\pi_{0,1}^{(1)}(g) = w$, $\pi_{1,0}^{(1)}(g) = -\overline{w}$, $\pi_{1,1}^{(1)}(g) = \overline{z}$ for g as in (4.16). Setting $n = 1$ and using $k, \ell \in \{0, 1\}$ in the formula in the corollary, we obtain the same four functions on $\mathrm{SU}_2(\mathbb{R})$. By Theorem 3.30, the functions $\sqrt{2}\pi_{k,\ell}^{(1)}$ are then orthonormal for $k, \ell \in \{0, 1\}$.

So suppose now that $n \geqslant 2$. Then the vectors

$$\widetilde{w_k} = e_1^{\odot(n-k)} \odot e_2^{\odot k}$$

are eigenvectors for the elements

$$\begin{pmatrix} z & \\ & \overline{z} \end{pmatrix} \in \mathrm{SU}_2(\mathbb{R})$$

for all $z \in \mathbb{S}^1$ for eigenvalues $z^{n-k}\overline{z}^k = z^{n-2k}$ for $k = 0, \ldots, n$. As the eigenvalues are distinct, they must be pairwise orthogonal with respect to the desired inner product $\langle \cdot, \cdot \rangle$ on $\mathrm{Sym}^n(\mathbb{C}^2)$. We claim that the inner product can be chosen so that the vectors

$$w_k = \binom{n}{k}^{\frac{1}{2}} e_1^{\odot(n-k)} \odot e_2^{\odot k} \tag{4.17}$$

are an orthonormal basis of $\mathrm{Sym}^n(\mathbb{C}^2)$. This could be checked directly (for example, by showing that the representation of $\mathfrak{su}_2(\mathbb{R})$ only takes on anti-Hermitian matrices with respect to that basis). However, we will give a more conceptual argument for this.

We consider $\bigotimes_{j=1}^n(\mathbb{C}^2)$ and apply Proposition 3.11 (inductively extended to n factors) to define the unitary inner tensor product representation ρ of $\mathrm{SU}_2(\mathbb{R})$ on $\bigotimes_{j=1}^n(\mathbb{C}^2)$ so that

$$\rho_g(u_1 \otimes u_2 \otimes \cdots \otimes u_n) = (gu_1) \otimes (gu_2) \otimes \cdots \otimes (gu_n) \tag{4.18}$$

for all $u_1, \ldots, u_n \in \mathbb{C}^2$. Next note that there is a canonical intertwining map

$$\mathrm{Com}\colon \bigotimes_{j=1}^n(\mathbb{C}^2) \longrightarrow \mathrm{Sym}^n(\mathbb{C}^2)$$

that sends any tensor product to its commutative counterpart, that is,

$$\bigotimes_{j=1}^n(\mathbb{C}^2) \ni u_1 \otimes u_2 \otimes \cdots \otimes u_n \longmapsto u_1 \odot u_2 \odot \cdots \odot u_n \in \mathrm{Sym}^n(\mathbb{C}^2)$$

for all $u_1, \ldots, u_n \in \mathbb{C}^2$. In fact, as was already mentioned, $\mathrm{Sym}^n(\mathbb{C}^2)$ is defined as the quotient of $\bigotimes_{j=1}^n(\mathbb{C}^2)$ by the subspace generated by

$$u_1 \otimes u_2 \otimes \cdots \otimes u_n - u_{\sigma(1)} \otimes u_{\sigma(2)} \otimes \cdots \otimes u_{\sigma(n)}$$

for all $u_1, u_2, \ldots, u_n \in \mathbb{C}^2$ and permutations $\sigma \in \mathrm{S}_n$.

Also note that the permutation group S_n acts unitarily on $\bigotimes_{j=1}^n(\mathbb{C}^2)$ by setting

$$\lambda_\sigma(u_1 \otimes u_2 \otimes \cdots \otimes u_n) = u_{\sigma(1)} \otimes u_{\sigma(2)} \otimes \cdots \otimes u_{\sigma(n)} \tag{4.19}$$

for all $\sigma \in \mathrm{S}_n$ and $u_1, \ldots, u_n \in \mathbb{C}^2$ and then linearly extending this action. Moreover, using (4.18) and (4.19) it is easy to see that $\lambda_\sigma \rho_g = \rho_g \lambda_\sigma$ for all $\sigma \in \mathrm{S}_n$ and $g \in \mathrm{SU}_2(\mathbb{R})$. Therefore

$$V = \left\{ v \in \bigotimes_{j=1}^n(\mathbb{C}^2) \mid \lambda_\sigma(v) = v \text{ for all } \sigma \in \mathrm{S}_n \right\}$$

is invariant under ρ, and is non-trivial since $e_1 \otimes e_1 \otimes \cdots \otimes e_1 \in V$. Recall that $\bigotimes_{j=1}^n(\mathbb{C}^2)$ has

$$e_{j_1} \otimes e_{j_2} \otimes \cdots \otimes e_{j_n} \tag{4.20}$$

for $j_1, \ldots, j_n \in \{1, 2\}$ as an orthonormal basis, and note that λ_σ for $\sigma \in \mathrm{S}_n$ maps any such basis vector to another such basis vector. It follows from this that V is generated by the vectors

$$v_k = \sum_{\substack{j_1,\ldots,j_n \in \{1,2\} \\ n-k \text{ times } 1, \\ k \text{ times } 2}} e_{j_1} \otimes e_{j_2} \otimes \cdots \otimes e_{j_n} \tag{4.21}$$

for $k = 0, \ldots, n$. In fact, if one of the basis vectors $e_{j_1} \otimes e_{j_2} \otimes \cdots \otimes e_{j_n}$ with

$$|\{\ell \mid j_\ell = 1\}| = n - k$$

appears with a coefficient c in the expansion of some $v \in V$ with respect to the basis in (4.20), then all other basis vectors appearing in the sum (4.21) are images of the original basis vector under some $\sigma \in \mathrm{S}_n$. Hence these all appear with the same coefficient c in the vector v. Using this argument for all $k = 0, \ldots, n$ we deduce that v is a linear combination of $v_0, v_1, \ldots, v_n$.

Now note that the vector v_k in (4.21) has $\binom{n}{k}$ summands which are mutually orthogonal unit vectors in $\bigotimes_{j=1}^n(\mathbb{C}^2)$. Hence the vectors

$$\widetilde{v}_k = \binom{n}{k}^{-\frac{1}{2}} v_k$$

for $k = 0, \ldots, n$ comprise an orthonormal basis of V. Applying the intertwining map $\mathrm{Com}|_V : V \to \mathrm{Sym}^n(\mathbb{C}^2)$ we see that

$$\mathrm{Com}\,\widetilde{v}_k = \binom{n}{k}^{-\frac{1}{2}} \mathrm{Com}\, v_k = \binom{n}{k}^{-\frac{1}{2}} \binom{n}{k} e_1^{\odot(n-k)} \odot e_2^{\odot k} = w_k$$

for $k = 0, \ldots, n$. Since the $\widetilde{v}_k$ for $k = 0, \ldots, n$ form an orthonormal basis, this proves the claim that the basis in (4.17) is an orthonormal basis with respect to an inner product on $\mathrm{Sym}^n(\mathbb{C}^2)$ that makes ρ a unitary representation $\pi^{(n)}$ of $\mathrm{SU}_2(\mathbb{R})$.

We now calculate the matrix coefficients $\pi^{(n)}_{k,\ell} = \varphi_{w_k,w_\ell}$ for the basis vectors w_k, w_ℓ and $k, \ell = 0, \ldots, n$. Using the notation (4.16) once again, we have $ge_1 = ze_1 + we_2$ and $ge_2 = -\overline{w}e_1 + \overline{z}e_2$ and hence

$$\begin{aligned}
\pi^{(n)}_g w_k &= \binom{n}{k}^{\frac{1}{2}} (ze_1 + we_2)^{\odot(n-k)} \odot (-\overline{w}e_1 + \overline{z}e_2)^{\odot k} \\
&= \binom{n}{k}^{\frac{1}{2}} \left(\sum_{i=0}^{n-k} \binom{n-k}{i} z^{n-k-i} w^i e_1^{\odot(n-k-i)} \odot e_2^{\odot i} \right) \\
&\qquad \odot \left(\sum_{j=0}^{k} \binom{k}{j} (-1)^{k-j} \overline{w}^{k-j} \overline{z}^j e_1^{\odot(k-j)} \odot e_2^{\odot j} \right) \\
&= \binom{n}{k}^{\frac{1}{2}} \sum_{i=0}^{n-k} \sum_{j=0}^{k} \binom{n-k}{i} \binom{k}{j} (-1)^{k-j} z^{n-k-i} w^i \overline{z}^j \overline{w}^{k-j} e_1^{\odot(n-i-j)} \odot e_2^{\odot(i+j)}.
\end{aligned}$$

Taking the inner product with

$$w_\ell = \binom{n}{\ell}^{\frac{1}{2}} e_1^{\odot(n-\ell)} e_2^{\odot \ell}$$

selects only those terms in the sum above with $i + j = \ell$. In fact, multiplying and dividing $\pi^{(n)}_g w_k$ by the square root of $\binom{n}{\ell}$ and using $\|w_\ell\| = 1$, we obtain

$$\begin{aligned}
\pi^{(n)}_{k,\ell} &= \langle \pi^{(n)}_g w_k, w_\ell \rangle \\
&= \binom{n}{k}^{\frac{1}{2}} \binom{n}{\ell}^{-\frac{1}{2}} \sum_{\substack{i \in \{0,\ldots,n-k\} \\ j \in \{0,\ldots,k\} \\ i+j=\ell}} \binom{n-k}{i} \binom{k}{j} (-1)^{k-j} z^{n-k-i} w^i \overline{z}^j \overline{w}^{k-j} \\
&= \sqrt{k!(n-k)!\ell!(n-\ell)!} \sum_{\substack{i \in \{0,\ldots,n-k\} \\ j \in \{0,\ldots,k\} \\ i+j=\ell}} \frac{(-1)^{k-j}}{i!j!(n-k-i)!(k-j)!} z^{n-k-i} w^i \overline{z}^j \overline{w}^{k-j}.
\end{aligned}$$

Together with Theorem 3.34, this concludes the proof. □

4.2.2 Peter–Weyl Theorem for $SO_3(\mathbb{R})$

We briefly explain in this section how Corollary 4.13 also gives rise to a Peter–Weyl theorem for the group $SO_3(\mathbb{R})$. We will not, however, give the orthonormal basis explicitly in terms of the coordinates of $SO_3(\mathbb{R})$.

The connection between $SU_2(\mathbb{R})$ and $SO_3(\mathbb{R})$ is given by the following lemma. We recall that an isogeny between two semi-simple Lie groups is a finite-to-one surjection.

Lemma 4.14 (Isogeny for $SU_2(\mathbb{R})$). *We have*

$$SO_3(\mathbb{R}) \cong SU_2(\mathbb{R})/C,$$

where $C = \{\pm I\}$ is the centre of $SU_2(\mathbb{R})$.

Proof. We recall from Lemma 4.2 that $SU_2(\mathbb{R}) \cong \mathbb{S}^3 \subseteq \mathbb{H}$, and so in particular $SU_2(\mathbb{R})$ is simply connected (see also Lemma 4.3). We claim now that $SO_3(\mathbb{R})$ is also a connected three-dimensional Lie group. To see that it is connected, we let $g \in SO_3(\mathbb{R})$. Then g has a real eigenvector v with eigenvalue 1. In fact, if all eigenvalues are real, then $g = I$ or the eigenvalues must equal -1, -1, and 1. If there is a non-real complex eigenvalue λ then the eigenvalues must be λ, $\overline{\lambda}$, and 1. Hence in either case $g = I$ or g can be viewed as a rotation about some axis in $\mathbb{R}^3$. It follows that g belongs to a one-parameter subgroup and hence to the connected component of $I \in SO_3(\mathbb{R})$. As $g \in SO_3(\mathbb{R})$ was arbitrary, we deduce that $SO_3(\mathbb{R})$ is connected.

The Lie algebra of $SO_3(\mathbb{R}) = \{g \in SL_3(\mathbb{R}) \mid g^t g = I\}$ is given by

$$\mathfrak{so}_3(\mathbb{R}) = \{\mathbf{m} \in \mathfrak{sl}_3(\mathbb{R}) \mid \mathbf{m}^t + \mathbf{m} = 0\},$$

and so consists of all matrices of the form

$$\mathbf{m} = \begin{pmatrix} 0 & -\alpha & -\beta \\ \alpha & 0 & -\gamma \\ \beta & \gamma & 0 \end{pmatrix}$$

for $\alpha, \beta, \gamma \in \mathbb{R}$. It follows that $SO_3(\mathbb{R})$ is a three-dimensional connected Lie group as claimed.

To define the homomorphism $SU_2(\mathbb{R}) \to SO_3(\mathbb{R})$, we identify the vector

$$\begin{pmatrix} a \\ b \\ c \end{pmatrix} \in \mathbb{R}^3$$

with the Lie algebra element

$$\mathbf{m} = \begin{pmatrix} a\mathrm{i} & b\mathrm{i} - c \\ b\mathrm{i} + c & -a\mathrm{i} \end{pmatrix} \in \mathfrak{su}_2(\mathbb{R}), \tag{4.22}$$

and define

$$\rho_g(\mathbf{m}) = g\mathbf{m}g^{-1} = g\mathbf{m}g^*$$

for all $g \in \mathrm{SU}_2(\mathbb{R})$ and $\mathbf{m} \in \mathfrak{su}_2(\mathbb{R})$. Since

$$\mathfrak{su}_2(\mathbb{R}) = \{\mathbf{m} \in \mathfrak{gl}_2(\mathbb{C}) \mid \mathbf{m}^* = -\mathbf{m}, \operatorname{tr}\mathbf{m} = 0\},$$

it follows that $\rho_g(\mathfrak{su}_2(\mathbb{R})) \subseteq \mathfrak{su}_2(\mathbb{R})$. Moreover, for $\mathbf{m}$ as in (4.22), we have

$$\det \mathbf{m} = a^2 - (b\mathrm{i} - c)(b\mathrm{i} + c) = a^2 + b^2 + c^2$$

and $\det\big(\rho_g \mathbf{m}\big) = \det \mathbf{m}$ for all $g \in \mathrm{SU}_2(\mathbb{R})$. This shows that the adjoint representation ρ defines a homomorphism $\rho\colon \mathrm{SU}_2(\mathbb{R}) \to \mathrm{SO}_3(\mathbb{R})$.

Suppose now that $\rho_g = I$, so $g\mathbf{m}g^{-1} = \mathbf{m}$ for all $\mathbf{m} \in \mathfrak{su}_2(\mathbb{R})$. Since

$$\mathfrak{su}_2(\mathbb{R}) \otimes_{\mathbb{R}} \mathbb{C} = \mathfrak{sl}_2(\mathbb{C})$$

we see that $g\mathbf{m} = \mathbf{m}g$ for all $\mathbf{m} \in \mathfrak{sl}_2(\mathbb{C})$, which implies $g \in \mathbb{C}I$ by a direct calculation.

To summarize, if $g \in \ker \rho \subseteq \mathrm{SU}_2(\mathbb{R})$ then $g = I$ or $g = -I$. The converse of this statement is clear. The lemma follows from this: Since both $\mathrm{SU}_2(\mathbb{R})$ and $\mathrm{SO}_3(\mathbb{R})$ are three-dimensional and ρ has finite kernel, $\rho\big(\mathrm{SU}_2(\mathbb{R})\big)$ is also three-dimensional, which together with connectedness of $\mathrm{SO}_3(\mathbb{R})$ implies that

$$\rho\big(\mathrm{SU}_2(\mathbb{R})\big) = \mathrm{SO}_3(\mathbb{R}).$$

□

Corollary 4.15 (Peter–Weyl for $\mathrm{SO}_3(\mathbb{R})$). *The irreducible representation $\pi^{(n)}$ on $\mathrm{Sym}^n(\mathbb{C}^2)$ gives rise to a unitary representation of*

$$\mathrm{SO}_3(\mathbb{R}) \cong \mathrm{SU}_2(\mathbb{R})/C$$

if and only if n is even. In particular, the normalized matrix coefficients

$$\sqrt{n+1}\pi^{(n)}_{k,\ell}$$

for $k, \ell \in \{0, \dots, n\}$ and $n \in 2\mathbb{N}_0$ define an orthonormal basis of $L^2(\mathrm{SO}_3(\mathbb{R}))$.

PROOF. If π is an irreducible unitary representation of $\mathrm{SO}_3(\mathbb{R})$, then the isomorphism $\mathrm{SO}_3(\mathbb{R}) \cong \mathrm{SU}_2(\mathbb{R})/C$ can be used to consider π also as an irreducible unitary representation of $\mathrm{SU}_2(\mathbb{R})$, which we again denote by π. By Theorem 4.6, π is isomorphic to the representation $\pi^{(n)}$ on $\mathrm{Sym}^n(\mathbb{C}^2)$ for some $n \in \mathbb{N}_0$. By construction, we have $\pi^{(n)}(-I) = (-I)^n = I$, which implies that n is even.

On the other hand, if $n \in 2\mathbb{N}$ then $\pi^{(n)}(-I) = I$, which shows that $\pi^{(n)}$ descends to a unitary representation of $\mathrm{SO}_3(\mathbb{R})$.

The final claim follows from the Peter–Weyl theorem (Theorem 3.34) as in the proof of Corollary 4.13. □

Exercise 4.16. (a) Describe the unitary dial of $\mathrm{SU}_2(\mathbb{R}) \times \mathrm{SU}_2(\mathbb{R})$.
(b) Show (or recall) that $\mathrm{SO}_4(\mathbb{R})$ is isomorphic to a quotient of $\mathrm{SU}_2(\mathbb{R}) \times \mathrm{SU}_2(\mathbb{R})$ by a central subgroup of order two. Describe the unitary dual of $\mathrm{SO}_4(\mathbb{R})$.

4.2.3 The Unitary Representation on $L^2(\mathbb{S}^2)$

By the discussion in the previous section we know that $\mathrm{SU}_2(\mathbb{R})$ acts naturally on the unit sphere $\mathbb{S}^2 \subseteq \mathbb{R}^3$. We equip $\mathbb{S}^2$ with the natural surface area measure and obtain a measure-preserving action of $\mathrm{SU}_2(\mathbb{R})$ on $\mathbb{S}^2$. This in turn gives rise to a unitary representation of $\mathrm{SU}_2(\mathbb{R})$ on $L^2(\mathbb{S}^2)$ as in Proposition 1.3. We wish to use the description of irreducible representations of $\mathrm{SU}_2(\mathbb{R})$ in Theorem 4.6 to describe how $L^2(\mathbb{S}^2)$ splits into irreducible components.

Corollary 4.17 (Decomposition of $L^2(\mathbb{S}^2)$). *Using the unitary representation of $\mathrm{SU}_2(\mathbb{R})$ on $L^2(\mathbb{S}^2)$ as described above, we have*

$$L^2(\mathbb{S}^2) = \bigoplus_{n \in 2\mathbb{N}} \mathrm{Sym}^n(\mathbb{C}^2).$$

In other words, every irreducible unitary representation of $\mathrm{SU}_2(\mathbb{R})$ of even highest weight (equivalently, every irreducible representation of $\mathrm{SO}_3(\mathbb{R})$) appears in $L^2(\mathbb{S}^2)$ with multiplicity one.

PROOF. We will combine the description of $L^2(\mathrm{SU}_2(\mathbb{R}))$ in Corollary 4.13 with the isomorphism

$$\mathbb{S}^2 \cong \mathrm{SU}_2(\mathbb{R})/T \tag{4.23}$$

where

$$T = \left\{ \begin{pmatrix} \alpha & \\ & \overline{\alpha} \end{pmatrix} \;\middle|\; \alpha \in \mathbb{S}^1 \right\}$$

is the diagonal subgroup in $\mathrm{SU}_2(\mathbb{R})$.

For the proof of (4.23) we recall from Section 4.2.2 that $\mathrm{SU}_2(\mathbb{R})$ factors onto $\mathrm{SO}_3(\mathbb{R})$ and acts on $\mathfrak{su}_2(\mathbb{R}) \cong \mathbb{R}^3$ while preserving the quadratic form det. Let

$$v_0 = \begin{pmatrix} \mathrm{i} & \\ & -\mathrm{i} \end{pmatrix}.$$

For $g \in \mathrm{SU}_2(\mathbb{R})$ a quick calculation shows that $g v_0 g^{-1} = v_0$ if and only if $g \in T$. Moreover, as $\mathrm{SO}_3(\mathbb{R})$ acts transitively on $\mathbb{S}^2$ and $\mathrm{SU}_2(\mathbb{R})$ factors onto $\mathrm{SO}_3(\mathbb{R})$ by this action, we obtain $\mathbb{S}^2 = \rho(\mathrm{SU}_2(\mathbb{R}))v_0$, and (4.23) follows. More precisely, $gT \in \mathrm{SU}_2(\mathbb{R})/T$ corresponds to $g v_0 g^{-1} \in \mathbb{S}^2$ under the isomorphism and the action of $\mathrm{SU}_2(\mathbb{R})$ corresponds to left multiplication on $\mathrm{SU}_2(\mathbb{R})$.

Also recall that the Haar measure on a homogeneous space is unique up to positive proportionality, which implies that the surface area measure described above the corollary agrees up to a positive multiple with the push-forward of the Haar measure on $\mathrm{SU}_2(\mathbb{R})$ onto $\mathrm{SU}_2(\mathbb{R})/T$ (see also Exercise 4.18).

To summarize, we may consider $L^2(\mathbb{S}^2)$ as the subspace $\mathcal{V}$ of $L^2(\mathrm{SU}_2(\mathbb{R}))$ consisting of all functions on $\mathrm{SU}_2(\mathbb{R})$ that are right-invariant under T. In other words,

$$L^2(\mathbb{S}^2) \cong \mathcal{V} = \{f \in L^2(\mathrm{SU}_2(\mathbb{R})) \mid f \text{ has weight } 0 \text{ for } T\}.$$

Here we say that $f \in L^2(\mathrm{SU}_2(\mathbb{R}))$ has weight $m \in \mathbb{Z}$ for T if $f(gt) = \alpha^m f(g)$ for all diagonal elements

$$t = \begin{pmatrix} \alpha & \\ & \overline{\alpha} \end{pmatrix} \in T$$

and almost every $g \in \mathrm{SU}_2(\mathbb{R})$.

We again use the notation

$$g = \begin{pmatrix} z & -\overline{w} \\ w & \overline{z} \end{pmatrix}$$

for elements of $\mathrm{SU}_2(\mathbb{R})$, and the orthonormal basis comprising the functions

$$\sqrt{n+1}\pi_{k,\ell}^{(n)}$$

for $n \in \mathbb{N}_0$ and $k, \ell \in \{0, \ldots, n\}$ of $L^2(\mathrm{SU}_2(\mathbb{R}))$ from Corollary 4.13. Using the concrete formula for $\pi_{k,\ell}^{(n)}$ and the notation for g and t above, we see that

$$gt = \begin{pmatrix} \alpha z & -\overline{\alpha w} \\ \alpha w & \overline{\alpha z} \end{pmatrix}$$

and

$$\begin{aligned} \pi_{k,\ell}^{(n)}(gt) &= (k!(n-k)!\ell!(n-\ell)!)^{\frac{1}{2}} \\ &\quad \times \sum_{\substack{i \in \{0,\ldots,n-k\} \\ j \in \{0,\ldots,k\} \\ i+j=\ell}} \frac{(-1)^{k-j}}{i!j!(n-k-i)!(k-j)!} (\alpha z)^{n-k-i} (\alpha w)^i (\overline{\alpha z})^j (\overline{\alpha w})^{k-j} \\ &= \alpha^{n-2k} \pi_{k,\ell}^{(n)}(g). \end{aligned}$$

In other words, the orthonormal basis consists of eigenvectors for the right regular representation restricted to $T \cong \mathbb{S}^1$. Hence we can obtain an orthonormal basis of $\mathcal{V}$ by using only those normalized matrix coefficients $\sqrt{n+1}\pi_{k,\ell}^{(n)}$ with weight $n - 2k = 0$. We have shown that $\mathcal{V}$ has the orthonormal basis consisting of the functions $\sqrt{2k+1}\pi_{k,\ell}^{(2k)}$ with $k \in \mathbb{N}_0$ and $\ell \in \{0, \ldots, 2k\}$, and the left regular representation of $\mathrm{SU}_2(\mathbb{R})$ on

$$\mathcal{V}_{2k} = \left\langle \pi_{k,\ell}^{(2k)} \mid \ell \in \{0, \ldots, 2k\} \right\rangle$$

is isomorphic to $\mathrm{Sym}^{2k}(\mathbb{C}^2)$ by the argument used in the last part of the proof of the Peter–Weyl theorem (Theorem 3.34). This gives the corollary. □

Exercise 4.18. Show that the map corresponding to (4.23) sends Haar measure on $\mathrm{SU}_2(\mathbb{R})$ to a multiple of the surface area measure on $\mathbb{S}^2$.

4.2.4 Conjugacy Classes and Characters of $\mathrm{SU}_2(\mathbb{R})$

We wish to finish the discussion of the compact group $\mathrm{SU}_2(\mathbb{R})$ by describing $\mathrm{SU}_2(\mathbb{R})^\sharp$ and calculating the characters of $\mathrm{SU}_2(\mathbb{R})$.

Proposition 4.19 (Sato–Tate measure). *The trace* $\mathrm{tr}\colon \mathrm{SU}_2(\mathbb{R}) \to [-2,2]$ *descends to a homeomorphism*

$$\mathrm{tr}\colon \mathrm{SU}_2(\mathbb{R})^\sharp \longrightarrow [-2,2].$$

The push-forward of the Haar measure is given by

$$\mathrm{tr}_* m_{\mathrm{SU}_2(\mathbb{R})}\big((a,b)\big) = \frac{1}{\pi}\int_a^b \sqrt{1-\tfrac{t^2}{4}}\,\mathrm{d}t,$$

and is called the Sato–Tate measure.

Proof. Recall that $\mathrm{tr}(AB) = \mathrm{tr}(BA)$ for $A, B \in \mathrm{Mat}_{d,d}(\mathbb{C})$. In particular, the map $\mathrm{tr}\colon \mathrm{SU}_2(\mathbb{R})^\sharp \to \mathbb{R}$ defined by $\mathrm{tr}([g]) = \mathrm{tr}(g)$ is a well-defined continuous map. As the eigenvalues of any $g \in \mathrm{SU}_2(\mathbb{R})$ are of the form $\alpha, \overline{\alpha}$ for some $\alpha \in \mathbb{S}^1$, we see that $\mathrm{tr}(g)$ lies in $[-2,2]$. Varying $\alpha \in \mathbb{S}^1$ and using

$$t = \begin{pmatrix} \alpha & \\ & \overline{\alpha} \end{pmatrix} \in \mathrm{SU}_2(\mathbb{R}),$$

we also see that $\mathrm{tr}\colon \mathrm{SU}_2(\mathbb{R})^\sharp \to [-2,2]$ is surjective.

We claim that the trace map is also injective on $\mathrm{SU}_2(\mathbb{R})^\sharp$. So suppose that $\mathrm{tr}([g_1]) = \mathrm{tr}([g_2])$ for some $[g_1],[g_2] \in \mathrm{SU}_2(\mathbb{R})^\sharp$. Then the characteristic polynomials of g_1 and g_2 agree (since these are determined for 2×2 matrices by the trace and determinant). It follows that the eigenvalues α, $\overline{\alpha}$ of g_1 and g_2 are equal. Since $g_1 \in \mathrm{SU}_2(\mathbb{R})$ there exists an orthonormal basis $u_1, u_2 \in \mathbb{C}^2$ consisting of eigenvectors for α, resp. $\overline{\alpha}$. Multiplying u_2 by a scalar of absolute value one if necessary, we may assume $\det h = 1$ where $h = (u_1, u_2)$. It follows that $h \in \mathrm{SU}_2(\mathbb{R})$ and

$$g_1 = h \begin{pmatrix} \alpha & \\ & \overline{\alpha} \end{pmatrix} h^{-1}.$$

In other words,

$$[g_1] = \left[\begin{pmatrix} \alpha & \\ & \overline{\alpha} \end{pmatrix}\right],$$

and by symmetry between g_1, g_2 also $[g_1] = [g_2]$, as required.

As both $\mathrm{SU}_2(\mathbb{R})^\sharp$ and $[-2,2]$ are compact and tr is continuous, it follows that $\mathrm{tr}\colon \mathrm{SU}_2(\mathbb{R})^\sharp \to [-2,2]$ is a homeomorphism.

It remains to prove the explicit description of the image of Haar measure. For this, we again identify $\mathrm{SU}_2(\mathbb{R})$ with $\mathbb{S}^3 \subseteq \mathbb{H} \cong \mathbb{R}^4$. With this the Haar measure can be defined using the four-dimensional Lebesgue measure $m_{\mathbb{R}^4}$. In fact for $B \subseteq \mathbb{S}^3$ we define

$$m_{\mathbb{S}^3}(B) = m_{\mathbb{R}^4}\big(\{rv \mid r \in [0,1], v \in B\}\big) \tag{4.24}$$

and, since $\mathbb{S}^3$ acts linearly as a unimodular transformation on $\mathbb{R}^4$, it follows that $m_{\mathbb{S}^3}$ defines a measure on $\mathbb{S}^3$ with total measure $m_{\mathbb{S}^3}(\mathbb{S}^3) = m_{\mathbb{R}^4}\big(B_1^{\mathbb{R}^4}\big)$. This turns the description of $\mathrm{tr}_* \, m_{\mathbb{S}^3}$ into an exercise in multi-dimensional calculus.

In fact we will use four-dimensional spherical coordinates defined by

$$S\colon \begin{pmatrix} r \\ \theta \\ \phi \\ \psi \end{pmatrix} \longmapsto \begin{pmatrix} r\cos\theta \\ r\sin\theta\cos\phi \\ r\sin\theta\sin\phi\cos\psi \\ r\sin\theta\sin\phi\sin\psi \end{pmatrix}$$

with total derivative

$$\begin{pmatrix} \cos\theta & -r\sin\theta & 0 & 0 \\ \sin\theta\cos\phi & r\cos\theta\cos\phi & -r\sin\theta\sin\phi & 0 \\ \sin\theta\sin\phi\cos\psi & r\cos\theta\sin\phi\cos\psi & r\sin\theta\cos\phi\cos\psi & -r\sin\theta\sin\phi\sin\psi \\ \sin\theta\sin\phi\sin\psi & r\cos\theta\sin\phi\sin\psi & r\sin\theta\cos\phi\sin\psi & r\sin\theta\sin\phi\cos\psi \end{pmatrix}$$

and Jacobian determinant

$$r^3\sin^2\theta\sin\phi\det\begin{pmatrix} \cos\theta & -\sin\theta & 0 & 0 \\ \sin\theta\cos\phi & \cos\theta\cos\phi & -\sin\phi & 0 \\ \sin\theta\sin\phi\cos\psi & \cos\theta\sin\phi\cos\psi & \cos\phi\cos\psi & -\sin\psi \\ \sin\theta\sin\phi\sin\psi & \cos\theta\sin\phi\sin\psi & \cos\phi\sin\psi & \cos\psi \end{pmatrix}.$$

Expanding the remaining determinant along the first row, we see that it is given by

$$\Delta\cos^2\theta + \Delta\sin^2\theta = \Delta$$

where

$$\begin{aligned} \Delta &= \det\begin{pmatrix} \cos\phi & -\sin\phi & 0 \\ \sin\phi\cos\psi & \cos\phi\cos\psi & -\sin\psi \\ \sin\phi\sin\psi & \cos\phi\sin\psi & \cos\psi \end{pmatrix} \\ &= \cos^2\phi\det\begin{pmatrix} \cos\psi & -\sin\psi \\ \sin\psi & \cos\psi \end{pmatrix} + \sin^2\phi\det\begin{pmatrix} \cos\psi & -\sin\psi \\ \sin\psi & \cos\psi \end{pmatrix} = 1. \end{aligned}$$

A convenient domain for the spherical coordinates is

$$U = (0,\infty) \times (0,\pi) \times (0,\pi) \times (0,2\pi),$$

and spherical coordinates define a diffeomorphism S from U to a full measure open set $V \subseteq \mathbb{R}^4$.

Now let $a < b$ be in $[-2, 2]$, and define $\theta_a = \arccos \frac{a}{2}$ and $\theta_b = \arccos \frac{b}{2}$ so that

$$\mathrm{tr}^{-1}(a, b) = \left\{ \begin{pmatrix} x_1 \\ x_2 \\ x_3 \\ x_4 \end{pmatrix} \in \mathbb{S}^3 \;\middle|\; \frac{a}{2} < x_1 < \frac{b}{2} \right\}$$
$$= \left\{ \begin{pmatrix} \cos\theta \\ \sin\theta\cos\phi \\ \sin\theta\sin\phi\cos\psi \\ \sin\theta\sin\phi\sin\psi \end{pmatrix} \;\middle|\; \theta \in (\theta_b, \theta_a), \phi \in (0, \pi), \psi \in (0, 2\pi) \right\}.$$

Using the description of the Haar measure $m_{\mathbb{S}^3}$ in (4.24), this leads to

$$\mathrm{tr}_* \, m_{\mathbb{S}^3}\big((a, b)\big) = \int_0^1 r^3 \, \mathrm{d}r \int_{\theta_b}^{\theta_a} \sin^2\theta \, \mathrm{d}\theta \int_0^{\pi} \sin\phi \, \mathrm{d}\phi \int_0^{2\pi} \mathrm{d}\psi$$
$$= \tfrac{1}{4} \cdot 2 \cdot 2\pi \int_{\theta_b}^{\theta_a} \sin^2\theta \, \mathrm{d}\theta.$$

Instead of calculating the latter integral, we wish to rewrite it as an integral over $t \in [a, b]$ using $t = 2\cos\theta$ and $\mathrm{d}t = -2\sin\theta \, \mathrm{d}\theta$. This gives

$$\mathrm{tr}_* \, m_{\mathbb{S}^3}\big((a, b)\big) = \frac{\pi}{2} \int_a^b \sqrt{1 - \tfrac{t^2}{4}} \, \mathrm{d}t.$$

Normalizing the measure to be a probability gives the proposition. $\square$

Using the identification between $\mathrm{SU}_2(\mathbb{R})^\sharp$ and the interval $[-2, 2]$ we now describe the characters of $\mathrm{SU}_2(\mathbb{R})$. By Theorem 4.6 the irreducible representations of $\mathrm{SU}_2(\mathbb{R})$ are given by the nth symmetric tensor products $\mathrm{Sym}^n(\mathbb{C}^2)$ of the standard representation for all $n \in \mathbb{N}_0$. We define $t = z + \overline{z} \in [-2, 2]$ with $z \in \mathbb{S}^1$. Then the eigenvalues for the group element

$$\begin{pmatrix} z & \\ & \overline{z} \end{pmatrix} \in \mathrm{SU}_2(\mathbb{R})$$

and the basis vector $e_1^{\odot(n-k)} \odot e_2^{\odot k}$ are given by $z^{n-k}\overline{z}^k = z^{n-2k}$ for $n \in \mathbb{N}_0$ and $k \in \{0, \dots, n\}$. Hence the character χ associated to $\mathrm{Sym}^n(\mathbb{C}^2)$ is given by

$$\chi_n(t) = \chi_n\left(\begin{pmatrix} z & \\ & \overline{z} \end{pmatrix}\right) = \sum_{k=0}^{n} z^{n-2k}$$

for every $n \in \mathbb{N}_0$. Using the variable $t \in [-2, 2] \cong \mathrm{SU}_2(\mathbb{R})^\sharp$, the first few characters are given by

$$\begin{aligned}\chi_0(t) &= 1,\\ \chi_1(t) &= t,\\ \chi_2(t) &= z^2+1+z^{-2} = (z+\overline{z})^2-1 = t^2-1,\end{aligned}$$

and

$$\chi_3(t) = z^3+z+z^{-1}+z^{-3} = (z+\overline{z})^3-2(z+\overline{z}) = t^3-2t.$$

We conclude by linking the general character to certain classical polynomials. Using the notation $z = \mathrm{e}^{\mathrm{i}\theta} \in \mathbb{S}^1$ with $\theta \in [0, 2\pi)$, the eigenvalues of $\pi^{(n)}$ are given by

$$z^{n-2k} = \mathrm{e}^{\mathrm{i}(n-2k)\theta}$$

for $k = 0, \dots, n$, and so

$$\begin{aligned}\chi_n\left(\begin{pmatrix} z & \\ & \overline{z}\end{pmatrix}\right) &= \mathrm{e}^{\mathrm{i}n\theta} + \mathrm{e}^{\mathrm{i}(n-2)\theta} + \dots + \mathrm{e}^{-\mathrm{i}n\theta}\\ &= \mathrm{e}^{-\mathrm{i}n\theta}\frac{\mathrm{e}^{\mathrm{i}2(n+1)\theta}-1}{\mathrm{e}^{\mathrm{i}2\theta}-1}\\ &= \frac{\mathrm{e}^{\mathrm{i}(n+1)\theta}-\mathrm{e}^{-\mathrm{i}(n+1)\theta}}{\mathrm{e}^{\mathrm{i}\theta}-\mathrm{e}^{-\mathrm{i}\theta}} = \frac{\sin\big((n+1)\theta\big)}{\sin\theta}\end{aligned}$$

by the geometric series summation formula. Expressing this in terms of

$$t = \mathrm{tr}\begin{pmatrix} z & \\ & \overline{z}\end{pmatrix} = z+\overline{z} = 2\cos\theta$$

gives a formula for the character $\chi_n(t)$. Using the variable $T = \frac{t}{2} = \cos\theta$ instead, this would give rise to the *Chebyshev polynomials of the second kind.*

Exercise 4.20. Describe $\mathrm{SO}_3(\mathbb{R})^\sharp$ and the characters of $\mathrm{SO}_3(\mathbb{R})$.

Exercise 4.21. Describe $\big(\mathrm{SU}_2(\mathbb{R})\times\mathrm{SU}_2(\mathbb{R})\big)^\sharp$ and $\big(\mathrm{SO}_4(\mathbb{R})\big)^\sharp$.

4.3 Summary and Outlook

The main purpose of this chapter was to provide a concrete example of the unitary dual of a compact group, which was done using $\mathrm{SU}_2(\mathbb{R})$ and $\mathrm{SO}_3(\mathbb{R})$.

To achieve this, we actually classified all irreducible finite-dimensional representations of the complex Lie group $\mathrm{SL}_2(\mathbb{C})$, respectively of its Lie algebra $\mathfrak{sl}_2(\mathbb{C})$ by differentiation. As we will see in Chapter 7, generalizing this derivative representation to unitary representations is more delicate than it is in finite dimensions.

Chapter 5
Normal Abelian Subgroups and Unitary Duals

We discuss in this chapter the unitary duals of some solvable non-compact non-abelian groups G. In some cases we will find a complete description of the unitary dual $\widehat{G}$. This will give us a chance to see how the results of Chapter 2 can be of importance for other groups, to see concrete instances of infinite-dimensional irreducible unitary representations, and to see that the Fell topology on $\widehat{G}$ (to be introduced in Section 6.4) can have exotic properties. Moreover, we will use here a new type of construction that lifts a unitary representation from a subgroup of G to a unitary representation of G. This construction will also appear in other forms in the following chapters. The first cases considered here should help the reader to become familiar with this 'induced representation' construction.

However, we will also see that for some solvable groups G a complete description of the unitary dual is not a reasonable goal. As a result our focus will be on giving examples of both behaviours instead of a complete abstract discussion. Consequently, the only section of this chapter that will be important for later core developments (the above mentioned insights arising from examples notwithstanding) is Section 5.1.

5.1 Normal Abelian Subgroups

The results of Chapter 2 will be important in this chapter and also in later discussions concerning a non-abelian group G. In fact, if G has a closed normal abelian subgroup $A \lhd G$ then we will often be able to obtain useful information about a unitary representation π of G by restricting π to A, applying the results of Chapter 2, and using information about how the action of G on $A \lhd G$ (or rather, on its dual group $\widehat{A}$) and the spectral theory of $\pi|_A$ interact. Here $\pi|_A$ denotes the unitary representation of A on $\mathcal{H}_\pi$ obtained by restricting the homomorphism $\pi\colon G \to \mathrm{B}(\mathcal{H}_\pi)$ to A. We have seen this idea already in Section 1.4.4. In fact we will see applications of this approach even for simple groups G, where

M. Einsiedler and T. Ward, *Unitary Representations and Unitary Duals*,
Graduate Texts in Mathematics 308, https://doi.org/10.1007/978-3-032-03899-9_5

one first restricts to subgroups of G that themselves have a normal abelian subgroup.

5.1.1 Equivariance Properties

So suppose that G is a locally compact, σ-compact, metric group, and that A is a closed normal abelian subgroup of G. In this case any $g \in G$ induces an automorphism $\theta_g \colon A \to A$ defined by $\theta_g(a) = gag^{-1}$ for $a \in A$. Using Lemma 2.31 for the abelian group A we obtain for each $g \in G$ an automorphism $\widehat{\theta}_g \colon \widehat{A} \to \widehat{A}$. We recall that $\widehat{\theta}_{g_1 g_2} = \widehat{\theta_{g_1}\theta_{g_2}} = \widehat{\theta}_{g_2}\widehat{\theta}_{g_1}$ for $g_1, g_2 \in G$. Taking inverses, we see that

$$G \times \widehat{A} \ni (g,t) \longmapsto g\bullet t = \widehat{\theta}_g^{-1}(t) \in \widehat{A} \tag{5.1}$$

defines an action of G on $\widehat{A}$. This action, the following equivariance properties of spectral measures, and the functional calculus will be fundamental for the whole chapter.

For a unitary representation π of G, we will simply write μ_v (or $\mu_{v,w}$) for the spectral measure $\mu_v^{\pi|_A}$ on $\widehat{A}$ obtained from Corollary 2.12 for $\pi|_A$ and $v \in \mathcal{H}_\pi$ (resp. $\mu_{v,w}^{\pi|_A}$ obtained from Proposition 2.53 for $v, w \in \mathcal{H}_\pi$).

Proposition 5.1 (Spectral measures for normal subgroups). *Let A be a closed normal abelian subgroup of G and let π be a unitary representation of G. Then for $v, w \in \mathcal{H}_\pi$ and $g \in G$ we have*

$$\mu_{\pi_g v} = g_* \mu_v$$

and

$$\mu_{\pi_g v, \pi_g w} = g_* \mu_{v,w},$$

where on the right-hand sides we use the push-forward measures under the map $\widehat{A} \ni t \mapsto g\bullet t \in \widehat{A}$.

PROOF. Let $g \in G$ and $a \in A$. Then

$$\begin{aligned}
\int_{\widehat{A}} \langle a,t\rangle \,\mathrm{d}\mu_{\pi_g v,\pi_g w}(t) &= \langle \pi_a \pi_g v, \pi_g w\rangle = \langle \pi_{g^{-1}ag} v, w\rangle \\
&= \int_{\widehat{A}} \langle \theta_{g^{-1}}(a), t\rangle \,\mathrm{d}\mu_{v,w}(t) \\
&= \int_{\widehat{A}} \left\langle a, \widehat{\theta}_g^{-1} t\right\rangle \mathrm{d}\mu_{v,w}(t) = \int_{\widehat{A}} \langle a, s\rangle \,\mathrm{d}\big(\widehat{\theta}_g^{-1}\big)_* \mu_{v,w}(s)
\end{aligned}$$

for all $v, w \in \mathcal{H}_\pi$. Also recall the abbreviation $g\bullet t = \widehat{\theta}_g^{-1}(t)$ introduced in (5.1). Hence this proves the proposition by uniqueness of the non-diagonal spectral measure in Proposition 2.53. For $v = w \in \mathcal{H}_\pi$ we also obtain the case of the principal matrix coefficient in Corollary 2.12. □

We will now study the functional calculus for $\pi|_A$ in Proposition 2.58, which gives the Hilbert space $\mathcal{H}_\pi$ for a unitary representation π of G a module structure over $\mathscr{L}^\infty(\widehat{A})$ by letting $F \in \mathscr{L}^\infty(\widehat{A})$ act via $\pi_{\mathrm{FC}}(F) = (\pi|_A)_{\mathrm{FC}}(F)$.

Corollary 5.2 (Functional calculus for normal subgroups). *Let $A \lhd G$ be a closed normal abelian subgroup and let π be a unitary representation of G. Then*

$$\pi_g \pi_{\mathrm{FC}}(F)\pi_g^{-1} = \pi_{\mathrm{FC}}(F \circ \widehat{\theta}_g) = \pi_{\mathrm{FC}}(F \circ g^{-1})$$

for all $F \in \mathscr{L}^\infty(\widehat{A})$ and $g \in G$, where we again use the action in (5.1) in the last expression.

PROOF. For $v, w \in \mathcal{H}_\pi$ we have by the definition of the functional calculus for $\pi|_A$ in Proposition 2.58, Proposition 5.1, the abbreviation in (5.1), and the definition again that

$$\begin{aligned}\Big\langle \pi_{\mathrm{FC}}(F \circ \widehat{\theta}_g)v, w\Big\rangle &= \int_{\widehat{A}} F \circ \widehat{\theta}_g \,\mathrm{d}\mu_{v,w} = \int_{\widehat{A}} F \,\mathrm{d}(\widehat{\theta}_g)_*\mu_{v,w} \\ &= \int_{\widehat{A}} F \,\mathrm{d}\mu_{\pi_g^{-1}v, \pi_g^{-1}w} = \big\langle \pi_{\mathrm{FC}}(F)\pi_g^{-1}v, \pi_g^{-1}w\big\rangle \\ &= \big\langle \pi_g\pi_{\mathrm{FC}}(F)\pi_g^{-1}v, w\big\rangle\end{aligned}$$

for all $v, w \in \mathcal{H}_\pi$, which proves the corollary. □

It will also be convenient to use the projection-valued measures

$$\varPi_B = \pi_{\mathrm{FC}}(\mathbb{1}_B)$$

for measurable subsets $B \subseteq \widehat{A}$ as introduced in Section 2.6.1. Recall that these are orthogonal projections satisfying

$$\varPi_{B_1}\varPi_{B_2} = \varPi_{B_2}\varPi_{B_1} = \varPi_{B_1 \cap B_2}$$

for all measurable $B_1, B_2 \subseteq \widehat{A}$, and by Exercise 2.63 (see also the hint on page 531) we also have

$$\varPi_{\bigsqcup_{n=1}^\infty B_n} = \sum_{n=1}^\infty \varPi_{B_n},$$

with the sum converging in the strong operator topology whenever $B_1, B_2, \ldots$ are measurable mutually disjoint subsets of $\widehat{A}$.

Applying Corollary 5.2 to $F = \mathbb{1}_B$ for a measurable subset $B \subseteq \widehat{A}$ also gives the conjugacy formula

$$\varPi_{g \bullet B} = \pi_g \varPi_B \pi_g^{-1} \tag{5.2}$$

since

$$\varPi_{g\bullet B} = \pi_{\mathrm{FC}}\big(\mathbb{1}_{\widehat{\theta}_g^{-1}B}\big) = \pi_{\mathrm{FC}}\big(\mathbb{1}_B \circ \widehat{\theta}_g\big) = \pi_g\pi_{\mathrm{FC}}(\mathbb{1}_B)\pi_g^{-1} = \pi_g\varPi_B\pi_g^{-1}$$

by definition and Corollary 5.2. We will use this identity frequently. It may help to remember that Π_B is the projection to the subspace 'corresponding to generalized eigenvalues in B'. Hence (5.2) gives a natural description for how 'generalized eigenvalues' for $\pi|_A$ behave when π_g is applied (see Exercise 5.3).

Exercise 5.3. Let $A \lhd G$ be a closed normal abelian subgroup and let π be a unitary representation of G. For $v \in \mathcal{H}_\pi$ and $t \in \widehat{A}$ show that v is an eigenvector for $\pi|_A$ and t if and only if $\pi_g v$ is an eigenvector for $\pi|_A$ and $g \cdot t$. Deduce (5.2) for $B = \{t\}$ from this.

5.1.2 Analyzing the Action on the Dual

Next we wish to draw some first connections between unitary representations of G and dynamical, or rather ergodic-theoretical, properties of G acting on $\widehat{A}$. However, let us start with a basic observation.

Lemma 5.4 (Continuity). *Let $A \lhd G$ be a closed normal abelian subgroup. Then the action of G on $\widehat{A}$ defined by* (5.1) *is continuous.*

Proof. Let (g_n) be a sequence in G with $g_n \to g \in G$ as $n \to \infty$ and let (t_n) be a sequence in $\widehat{A}$ with $t_n \to t \in \widehat{A}$ as $n \to \infty$. Applying Lemma 2.23(1), the map

$$\varphi\colon G \times A \times \widehat{A} \ni (g, a, t) \longmapsto \langle g^{-1} a g, t\rangle = \langle \theta_g^{-1}(a), t\rangle \in \mathbb{S}^1$$

is continuous. Let $K \subseteq A$ be a compact subset. Then

$$M = \{g, g_n \mid n \in \mathbb{N}\} \times K \times \{t, t_n \mid n \in \mathbb{N}\} \subseteq G \times A \times \widehat{A}$$

is compact, and φ is uniformly continuous on M. Hence for every $\varepsilon > 0$ there exists an $N \in \mathbb{N}$ such that

$$|\langle a, g_n \cdot t_n\rangle - \langle a, g \cdot t\rangle| = \left|\langle \theta_{g_n}^{-1}(a), t_n\rangle - \langle \theta_g^{-1}(a), t\rangle\right| < \varepsilon$$

for all $n \geqslant N$ and $a \in K$. As this holds for any compact subset $K \subseteq A$ and $\varepsilon > 0$, we obtain the desired convergence $g_n \cdot t_n \to g \cdot t$ (in the compact-open topology of $\widehat{A}$ from Corollary 2.5) as $n \to \infty$. □

We now use this action to relate the maximal spectral measure obtained in Proposition 2.64 for the restriction $\pi|_A$ to the notion of quasi-invariance in Definition 1.5.

Lemma 5.5 (Quasi-invariance). *Let $A \lhd G$ be a closed normal abelian subgroup and let π be a unitary representation of G. Then the maximal measure is quasi-invariant under the action of G on $\widehat{A}$. That is, if $\mu_{\max}$ is a maximal measure on $\widehat{A}$ then for any $g \in G$ the push-forward measure $g_* \mu_{\max}$ under the action of g on $\widehat{A}$ lies in the same measure class as $\mu_{\max}$.*

PROOF. Let $v_{\max} \in \mathcal{H}_\pi$ be a vector of maximal spectral type, $\mu_{\max} = \mu_{v_{\max}}$, and $g \in G$. Then $\pi_g v_{\max} \in \mathcal{H}_\pi$ and Proposition 5.1 implies that

$$g_* \mu_{\max} = \mu_{\pi_g v_{\max}} \ll \mu_{\max},$$

since $\mu_{\max}$ is a maximal spectral measure (see Proposition 2.64). Applying this to g^{-1} gives $g_*^{-1}\mu_{\max} \ll \mu_{\max}$, hence $\mu_{\max} \ll g_*\mu_{\max}$, and the lemma follows. □

For irreducible unitary representations, more can be said.

Definition 5.6. Let the group G act continuously on the space X, and let μ be a quasi-invariant σ-finite measure on X. The measure μ is said to be *ergodic* if for any measurable set $B \subseteq X$ with $\mu(g \cdot B \triangle B) = 0$ for all $g \in G$ we either have $\mu(B) = 0$ or $\mu(X \setminus B) = 0$.

Proposition 5.7 (Ergodicity). *Let $A \lhd G$ be a closed normal abelian subgroup and let π be an irreducible unitary representation of G. Then the maximal measure $\mu_{\max}$ is ergodic for the action of G on $\widehat{A}$. Moreover, there exists one $n \in \mathbb{N} \cup \{\infty\}$ such that $\pi|_A$ is isomorphic to the multiplication representation on $\big(L^2_{\mu_{\max}}(\widehat{A})\big)^n$.*

PROOF. To better understand the functional calculus, we apply the spectral theorem to $\pi|_A$ in the form of Theorem 2.69. This shows that $\mathcal{H}_\pi$ is isomorphic to a direct sum of $L^2_{\mu_{\max}}(\widehat{G})$ and other L^2 spaces defined by subsets of $\widehat{G}$ and the measure $\mu_{\max}$. The operator $\pi_{\mathrm{FC}}(F)$ then simply takes the form of the multiplication operator M_F for any $F \in \mathscr{L}^\infty(\widehat{G})$ and remains well-defined for $F \in L^\infty_{\mu_{\max}}(\widehat{G})$. Moreover, $\Pi_B = \pi_{\mathrm{FC}}(\mathbb{1}_B)$ is the projection operator $M_{\mathbb{1}_B}$ for any measurable $B \subseteq \widehat{A}$.

Let $B \subseteq \widehat{A}$ be measurable with $\mu_{\max}(g \cdot B \triangle B) = 0$ for all $g \in G$ and define $\mathcal{V} = \Pi_B(\mathcal{H}_\pi)$. For $g \in G$ we now have

$$\pi_g \mathcal{V} = \pi_g \Pi_B(\mathcal{H}_\pi) = \pi_g \Pi_B \pi_g^{-1}(\mathcal{H}_\pi) = \Pi_{g \cdot B}(\mathcal{H}_\pi) = \Pi_B(\mathcal{H}_\pi) = \mathcal{V}$$

by the conjugacy formula (5.2) and since $\mathbb{1}_{g \cdot B} = \mathbb{1}_B$ $\mu_{\max}$-almost everywhere. This shows that $\mathcal{V}$ is a π-invariant closed subspace of $\mathcal{H}_\pi$. By irreducibility of π it follows that either $\mathcal{V} = \{0\}$ or $\mathcal{V} = \mathcal{H}_\pi$. Suppose first that

$$\mathcal{V} = \Pi_B(\mathcal{H}_\pi) = \{0\}.$$

Since the operator Π_B corresponds to the multiplication operator $M_{\mathbb{1}_B}$ this means that $\mu_{\max}(B) = 0$. The case $\mathcal{V} = \mathcal{H}_\pi$ similarly gives

$$\Pi_{\widehat{G} \setminus B}(\mathcal{H}_\pi) = \mathcal{V}^\perp = \{0\}$$

and so $\mu_{\max}(\widehat{G} \setminus B) = 0$. As this holds for all B as above, $\mu_{\max}$ is ergodic.

To see the last claim in the proposition, we first recall a different aspect of Theorem 2.69, which in particular defines a canonical decomposition

$$\mathcal{H}_\pi = \bigoplus_{k\in\mathbb{N}\cup\{\infty\}} \mathcal{H}_\pi^{(k)}$$

into closed subspaces $\mathcal{H}_\pi^{(k)} < \mathcal{H}_\pi$. In fact for every $n \in \mathbb{N}$

$$\mathcal{H}_\pi^{(\geqslant n)} = \bigoplus_{\substack{k\in\mathbb{N}\cup\{\infty\},\\ k\geqslant n}} \mathcal{H}_\pi^{(k)}$$

is precisely the (closed) subspace of all vectors with spectral multiplicity at least n with respect to $\pi|_A$ (see Section 2.7.3). Fix some $g \in G$ and apply π_g to a vector $v \in \mathcal{H}_\pi^{(\geqslant n)}$ as well as the vectors $v_2, v_3, \ldots, v_n$ that witness the fact that v has spectral multiplicity at least n. It follows from Proposition 5.1 that $\pi_g v$ and $\pi_g v_2, \ldots, \pi_g v_n$ again have identical spectral measures and perpendicular cyclic spaces. That is, $\pi_g v$ also has spectral multiplicity at least n. Hence $\mathcal{H}_\pi^{\geqslant n}$ is an invariant closed subspace. As π is irreducible, it follows that $\mathcal{H}_\pi^{\geqslant n}$ is equal to $\{0\}$ or $\mathcal{H}_\pi$ for every $n \in \mathbb{N}$. Therefore $\mathcal{H}_\pi = \mathcal{H}_\pi^{(n)}$ for some $n \in \mathbb{N} \cup \{\infty\}$, which implies the proposition by Theorem 2.69. □

5.2 Some Metabelian Groups

†We assume now that G is a locally compact, σ-compact, metric group, that A is a closed normal abelian subgroup of G, that $S < G$ is a closed abelian subgroup with $A \cap S = \{e\}$, and that

$$A \times S \ni (a, s) \longmapsto as \in G$$

is a homeomorphism of topological spaces. In other words, G is the semi-direct product $S \ltimes A$ (which may also be written $A \rtimes S$) of its subgroups A and S. We will refer to A as the normal abelian subgroup, and to S as the *complementary subgroup*. We note that under these assumptions G is called *metabelian*. We also recall that Section 5.1.2 already shows a connection between unitary representations and quasi-invariant measures.

5.2.1 A Construction Using Invariant Measures

In order to make our life a bit easier, we assume in the following that μ is an S-invariant σ-finite measure on $\widehat{A}$ with respect to the action introduced in (5.1).

† Strictly speaking, the material in this section is not needed for the following more general sections. Nonetheless we suggest that the reader work through most of the section, as it will provide motivation for some of the technical assumptions to come, as well as concrete examples of the general setup.

In order to have the standing assumption of Section 1.1.2 satisfied, we let X be a measurable subset of $\widehat{A}$ with $\mu(\widehat{A} \smallsetminus X) = 0$, and assume that one of the following two conditions holds:

- $X = S{\bullet}t_0$ is a single S-orbit for some $t_0 \in \widehat{A}$, and we will equip X with the quotient topology induced by the orbit map $S \ni s \mapsto s{\bullet}t_0 \in X$. This makes X into a locally compact, σ-compact, metric space, and the Borel σ-algebra of X agrees with the Borel σ-algebra of X viewed as a subset of $\widehat{A}$. We assume that μ is locally finite on X, which then makes it into the Haar measure on
$$X = S{\bullet}t_0 \cong S/\operatorname{Stab}_S(t_0).$$
- X has the property that the induced topology of X as a subset of $\widehat{A}$ is locally compact, σ-compact, and μ is locally finite when considered as a measure on X.

We will see that these assumptions on $G \cong S \ltimes A$, $X \subseteq \widehat{A}$, and the measure μ are a good compromise. They will allow us to give a framework for the construction of (irreducible) unitary representations that is easy to work with. Moreover, in some cases we will obtain from this construction all irreducible unitary representations of G.

Lemma 5.8 (Representations arising from invariant measures). *Every S-invariant σ-finite measure μ on $\widehat{A}$ as above gives rise to a unitary representation π^μ of G on $\mathcal{H}_\mu = L^2_\mu(X)$ by letting the elements of the normal abelian subgroup A act as multiplication operators and elements of the complementary subgroup S act via action-associated operators. More formally, we define*
$$\begin{aligned}
\big(\pi^\mu_a f\big)(t) &= \langle a,t\rangle f(t),\\
\big(\pi^\mu_s f\big)(t) &= f(s^{-1}{\bullet}t), \text{ and}\\
\big(\pi^\mu_{as} f\big)(t) &= \big(\pi^\mu_a\pi^\mu_s f\big)(t) = \langle a,t\rangle f(s^{-1}{\bullet}t)
\end{aligned}$$
for all $a \in A$, $s \in S$, $f \in L^2(\widehat{A},\mu)$, and $t \in \widehat{A}$.

PROOF. Lemma 2.9 shows that the multiplication representation $A \ni a \mapsto \pi^\mu_a$ defines a unitary representation of A. Also note that $S \ni s \mapsto \pi^\mu_s$ defines a unitary representation of S by Proposition 1.3.

For $a \in A$, $s \in S$ and $f \in L^2_\mu(X)$, we define $\pi^\mu_{as} f = \pi^\mu_a \pi^\mu_s f$. For $t \in \widehat{A}$ this means that
$$\pi^\mu_a\big(\pi^\mu_s f\big)(t) = \langle a,t\rangle\big(\pi^\mu_s f\big)(t) = \langle a,t\rangle f\big(s^{-1}{\bullet}t\big).$$
To see that π^μ is indeed a representation of $G \cong S \ltimes A$, we use the above for a and s^{-1}, recall the definition of the action of S on $\widehat{A}$ in (5.1), and calculate

$$\begin{aligned}\left(\pi_s^\mu \pi_a^\mu \pi_{s^{-1}}^\mu f\right)(t) &= \left(\pi_a^\mu \pi_{s^{-1}}^\mu f\right)\left(s^{-1}\raisebox{0.2ex}{.}t\right)\\ &= \left\langle a, s^{-1}\raisebox{0.2ex}{.}t\right\rangle f\left(ss^{-1}\raisebox{0.2ex}{.}t\right)\\ &= \langle \theta_s a, t\rangle f(t)\\ &= \left\langle sas^{-1}, t\right\rangle f(t) = \left(\pi_{sas^{-1}}^\mu f\right)(t).\end{aligned}$$

For $a_1, a_2 \in A$ and $s_1, s_2 \in S$ this now implies that

$$\pi_{a_1 s_1}^\mu \pi_{a_2 s_2}^\mu = \pi_{a_1}^\mu \underbrace{\pi_{s_1}^\mu \pi_{a_2}^\mu \pi_{s_1^{-1}}^\mu}_{=\pi_{s_1 a_2 s_1^{-1}}^\mu} \pi_{s_1}^\mu \pi_{s_2}^\mu = \pi_{a_1 s_1 a_2 s_1^{-1}}^\mu \pi_{s_1 s_2}^\mu = \pi_{a_1 s_1 a_2 s_2}^\mu$$

by definition, as required.

To see continuity of the representation π^μ of G, let $f \in L^2_\mu(X)$ and suppose that (a_n) and (s_n) are sequences in A and S, with $a_n \to a$ and $s_n \to s$ as $n \to \infty$. Then

$$\|\pi_{a_n s_n}^\mu f - \pi_{as}^\mu f\| \leqslant \|\pi_{a_n}^\mu\left(\pi_{s_n}^\mu f - \pi_s^\mu f\right)\| + \|\pi_{a_n}^\mu \pi_s^\mu f - \pi_a^\mu \pi_s^\mu f\| \longrightarrow 0$$

as $n \to \infty$, as required. □

Essential Exercise 5.9. Let $G = S \ltimes A$ as above and $t_0 \in \widehat{A}$. Show that the Haar measure on $S/\operatorname{Stab}_S(t_0) \cong S\raisebox{0.2ex}{.}t_0 \subseteq \widehat{A}$ is S-ergodic (as in Definition 5.6).

Lemma 5.10 (Ergodicity implies irreducibility). *If μ is an S-invariant ergodic locally finite measure on $X \subseteq \widehat{A}$ as above, then the unitary representation π^μ as in Lemma 5.8 is irreducible.*

PROOF. Suppose that $\mathcal{V} \subseteq \mathcal{H}_\mu = L^2_\mu(X)$ is a non-trivial π^μ-invariant closed subspace. Let $v_{\max}$ be a vector in $\mathcal{V}$ with maximal spectral type as in Proposition 2.64 for π^μ restricted to A and restricted to $\mathcal{V}$. We note that the spectral measure $\mu_{\max}$ of $v_{\max}$ is given by $|v_{\max}|^2 \,\mathrm{d}\mu$. Let $B = \{t \in \widehat{H} \mid v_{\max}(t) \neq 0\}$. We claim that B is (up to sets of measure zero) invariant under S. Indeed, for any $s \in S$

$$\pi_s^\mu v_{\max} = v_{\max} \circ \widehat{\theta}_s \in \mathcal{V}$$

has an absolutely continuous spectral measure $|v_{\max}|^2 \circ \widehat{\theta}_s \,\mathrm{d}\mu$ with respect to the spectral measure $\mathrm{d}\mu_{\max} = |v_{\max}|^2 \,\mathrm{d}\mu$. This implies that almost every $t \in \widehat{\theta}_s^{-1} B$ belongs to B. Using $\pi_{s^{-1}}^\mu v_{\max} \in \mathcal{V}$ in the same way, we see that $\mu(B \triangle \widehat{\theta}_s^{-1} B) = 0$ as claimed.

By ergodicity of μ, this implies that $\mu(B) = 0$ or $\mu(\widehat{A} \smallsetminus B) = 0$. As $\mathcal{V}$ is a non-trivial subspace we have $v_{\max} \neq 0$, and so $\mu(\widehat{A} \smallsetminus B) = 0$. Applying the measurable functional calculus for $\pi^\mu|_A$, we see that $\mathscr{L}^\infty(\widehat{A}) v_{\max} \subseteq \mathcal{V}$. However, since $v_{\max}(t) \neq 0$ for μ-almost every $t \in X$ the subspace

$$\mathscr{L}^\infty(\widehat{A}) v_{\max} \subseteq L^2_\mu(X)$$

is dense. Hence $\mathcal{V} = \mathcal{H}_\mu$ is irreducible, and the lemma follows. □

Lemma 5.11 (No isomorphisms). *Let μ_1 and μ_2 be two S-invariant and ergodic measures on $\widehat{A}$ as above that are not multiples of each other. Then π^{μ_1} and π^{μ_2} are not isomorphic.*

We note that in many of the concrete cases of the above framework that we will consider, the proof of Lemma 5.11 will be significantly easier than in the general case considered here.

PROOF OF LEMMA 5.11. We claim that $\mu_1 \perp \mu_2$. Note that the spectral measure of $v \in \mathcal{H}_{\mu_1} = L^2(\widehat{A}, \mu_1)$ for $\pi^{\mu_1}|_A$ is given by $|v|^2 \,\mathrm{d}\mu_1$, which applies similarly to elements of $\mathcal{H}_{\mu_2}$. Therefore by using the claim the spectral measures for elements of $\mathcal{H}_{\mu_1}$ and $\mathcal{H}_{\mu_2}$ are mutually singular. As spectral measures are canonically defined, the lemma follows from the claim.

For the proof of the claim we recall that, by assumption, both measures are σ-finite. Hence we can apply the Lebesgue decomposition theorem to μ_1 and μ_2, which allows us to write $\mu_2 = \mu_{\mathrm{abs}} + \mu_{\mathrm{sing}}$ for two uniquely determined measures

$$\mu_{\mathrm{abs}} = \mu_2|_{B_{\mathrm{abs}}} \ll \mu_1,$$
$$\mu_{\mathrm{sing}} = \mu_2|_{B_{\mathrm{sing}}} \perp \mu_1,$$

and a measurable partition $\widehat{A} = B_{\mathrm{abs}} \sqcup B_{\mathrm{sing}}$. Applying the push-forward by some $s \in S$, invariance of μ_1 and μ_2, and the uniqueness of the Lebesgue decomposition, we see that μ_{abs} and μ_{sing} are also invariant under S.

Next we define $F = \frac{\mathrm{d}\mu_{\mathrm{abs}}}{\mathrm{d}\mu_1}$, let $B \subset \widehat{A}$, and let $s \in S$. Then

$$\begin{aligned}\int_B F \,\mathrm{d}\mu_1 = \mu_{\mathrm{abs}}(B) = \mu_{\mathrm{abs}}(s^{-1}B) = \int_{s^{-1}B} F \,\mathrm{d}\mu_1 \\ = \int_{s^{-1}B} (F \circ s^{-1}) \circ s \,\mathrm{d}\mu_1 = \int_B F \circ s^{-1} \,\mathrm{d}\mu_1.\end{aligned}$$

As $B \subseteq \widehat{A}$ was arbitrary, this shows that $F = F \circ s^{-1}$ almost everywhere with respect to μ_1. As $s \in S$ was abitrary, ergodicity of μ_1 now implies that F must be constant μ_1-almost everywhere (see Exercise 5.12). Therefore μ_{abs} is a multiple of μ_1. Repeating the argument with $\mu_{\mathrm{abs}} \ll \mu_2$ we also see that μ_{abs} is a multiple of μ_2. However, this contradicts our assumption unless $\mu_{\mathrm{abs}} = 0$ and $\mu_1 \perp \mu_2$ as claimed. □

To summarize, we have shown (under some technical assumptions) that a σ-finite S-invariant and ergodic measure on $\widehat{A}$ gives rise to an irreducible unitary representation of $G = S \ltimes A$, and that these unitary representations are non-isomorphic for any two non-proportional ergodic measures. Depending on G there might be other irreducible representations (as we will see, for example, in Section 5.2.6).

Essential Exercise 5.12. Suppose that S acts on X and μ is an ergodic σ-finite measure on X. Let $F\colon X \to \mathbb{R}$ be a measurable function with the property that for all $s \in S$ we have $F(x) = F(s \cdot x)$ for almost every $x \in X$. Show that F is constant almost surely.

Exercise 5.13. Let $G = S \ltimes A$ and let μ_1, μ_2 be two S-invariant σ-finite measures on $\widehat{A}$ as above. Show that if $\mu_1 \ll \mu_2$ then $\pi^{\mu_1} < \pi^{\mu_2}$.

5.2.2 The Isometry Group of the Plane

We define the orientation-preserving isometry group $G = \mathrm{SO}_2(\mathbb{R}) \ltimes \mathbb{R}^2$ of the plane as the semi-direct product of $\mathrm{SO}_2(\mathbb{R})$ and $\mathbb{R}^2$. More concretely, we define

$$G = \left\{ \begin{pmatrix} k & a \\ 0 & 1 \end{pmatrix} \;\middle|\; k \in \mathrm{SO}_2(\mathbb{R}), a \in \mathbb{R}^2 \right\} \leqslant \mathrm{SL}_3(\mathbb{R})$$

and will consider the subgroups

$$K = \left\{ \begin{pmatrix} k_\phi & 0 \\ & 1 \end{pmatrix} \;\middle|\; k_\phi = \begin{pmatrix} \cos\phi & -\sin\phi \\ \sin\phi & \cos\phi \end{pmatrix} \text{ with } \phi \in \mathbb{R} \right\} \cong \mathrm{SO}_2(\mathbb{R})$$

and

$$A = \left\{ \begin{pmatrix} I & a \\ & 1 \end{pmatrix} \;\middle|\; a = \begin{pmatrix} a_1 \\ a_2 \end{pmatrix} \in \mathbb{R}^2 \right\} \cong \mathbb{R}^2,$$

satisfying $G = AK$, $A \lhd G$, and $K \cap A = \{I\}$. We will identify the complementary subgroup K with $\mathrm{SO}_2(\mathbb{R})$ and the normal abelian subgroup A with $\mathbb{R}^2$. We note that conjugation of $a \in A$ by $k \in K$ simply corresponds to the application of $k \in \mathrm{SO}_2(\mathbb{R})$ to $a \in \mathbb{R}^2$. By the discussion in Section 2.4.2, the homomorphism dual to the conjugation map θ_k is therefore given by $\widehat{\theta_k} = k^{\mathrm{t}} = k^{-1}$. Adapting the notation of Section 5.1 and (5.1), we have $k \cdot t = kt$ for $k \in \mathrm{SO}_2(\mathbb{R})$ and $t \in \widehat{A} \cong \mathbb{R}^2$.

The Irreducible Representations

We will roughly classify the unitary representations depending on whether A acts trivially or not.

(Old) Every irreducible unitary representation of

$$K \cong G/A$$

gives rise to a unitary representation of G, which we will refer to as an *old representation.* Since $G/A \cong \mathrm{SO}_2(\mathbb{R}) \cong \mathbb{T}$ we see that for every $n \in \mathbb{Z}$ there is an associated old irreducible unitary representation defined by the character

$$\chi_n\colon G \ni \begin{pmatrix} \cos\phi & -\sin\phi & a_1 \\ \sin\phi & \cos\phi & a_2 \\ & & 1 \end{pmatrix} \longmapsto \mathrm{e}^{\mathrm{i}n\phi}.$$

(New) We will refer to an irreducible unitary representation π of G for which $\pi|_A$ is a non-trivial representation of A as a *new representation.* As we have seen in Corollary 1.81, such representations must exist.

We now define for every $r > 0$ a new representation π^r on $\mathcal{H}_r = L^2(\mathbb{R}^2, \mu_r)$, where μ_r is the normalized arc length measure on the circle

$$r\mathbb{S}^1 = \{t \in \mathbb{R}^2 \mid \|t\| = r\}$$

of radius r. We will think of $r\mathbb{S}^1$ as a subset of the dual group $\widehat{A} \cong \widehat{\mathbb{R}^2} \cong \mathbb{R}^2$. In fact μ_r is a K-invariant and ergodic probability measure on

$$X = r\mathbb{S}^1 = K_\bullet \begin{pmatrix} r \\ 0 \end{pmatrix}.$$

As discussed in Section 5.2.1, we can now define $\pi^r|_A$ simply by the multiplication representation

$$\bigl(\pi^r_a f\bigr)(t) = \langle a, t\rangle f(t)$$

for $a \in A$, $f \in L^2(\widehat{A}, \mu_r)$, and $t \in \widehat{A}$. On the subgroup K we define

$$\bigl(\pi^r_k f\bigr)(t) = f\bigl(\widehat{\theta}_k t\bigr) = f(k^{-1}t)$$

for $k \in K$, $f \in L^2(\widehat{A}, \mu_r)$, and $t \in \widehat{A}$. Lemmas 5.8 and 5.10 and Exercise 5.9 show that this gives rise to an irreducible unitary representation π^r on the space $\mathcal{H}_r = L^2(\mathbb{R}^2, \mu_r)$.

We note that the conclusion of Lemma 5.11 is quite easy to see in the case at hand. For $r > 0$ and $f \in \mathcal{H}_r$ the spectral measure of f with respect to $\pi^r|_A$ has support in $r\mathbb{S}^1$. This already implies that for $r, s \in (0, \infty)$ with $r \neq s$ the representations π^r and π^s cannot be isomorphic.

The reader may benefit from analysing alongside our discussion a different but slightly easier semi-direct product. Hence we fix for the following exercises an integer $d \geqslant 2$, and define

$$K_d = \left\{ \begin{pmatrix} \cos\phi & -\sin\phi & 0 \\ \sin\phi & \cos\phi & 0 \\ & & 1 \end{pmatrix} \in K \;\middle|\; \phi \in \tfrac{2\pi}{d}\mathbb{Z} \right\} \cong \mathbb{Z}/d\mathbb{Z}$$

and $G_d = K_d A < G$. As before, we will distinguish between old representations (which correspond to elements of $\widehat{K_d} \cong \mathbb{Z}/d\mathbb{Z}$) and new representations.

Exercise 5.14. (a) Construct for every $t_0 \in \widehat{A}\smallsetminus\{0\}$ a new representation π of G_d on $\mathbb{C}^d$ such that for some $v \in \mathbb{C}^d$ we have $\pi_a v = \langle a, t_0\rangle v$ for all $a \in A$.
(b) Show that the new representation defined in (a) is indeed a unitary representation of G_d. Also show directly that the representation is irreducible.

(c) Characterize the pairs $t_0, t_1 \in \widehat{A} \smallsetminus \{0\}$ with the property that the new unitary representations from (a) associated to t_0 and t_1 are isomorphic.

Exercise 5.15. Prove the analogue of Proposition 5.16 for the group G_d.

Classification of the Unitary Dual

Proposition 5.16 (Description of $\widehat{G}$). *Let $G = K \ltimes A = \mathrm{SO}_2(\mathbb{R}) \ltimes \mathbb{R}^2$ be the isometry group of the plane as above. Then the set of old representations χ_n for $n \in \mathbb{Z}$ and of new irreducible representations π^r for $r \in (0, \infty)$ constructed above comprise the complete set of irreducible representations.*

We note that this proposition also follows quite easily from the main technical result of this chapter, namely Theorem 5.55, but believe it is worthwhile to prove the case at hand directly as a warm-up.

PROOF OF PROPOSITION 5.16. Let π be an irreducible representation of G. If $\pi|_A$ is trivial, then π induces a representation of $G/A \cong K = \mathrm{SO}_2(\mathbb{R}) \cong \mathbb{T}$ and so is an old representation defined by some weight $n \in \mathbb{Z} = \widehat{\mathbb{T}}$.

So let us assume that $\pi|_A$ is non-trivial. Let $\mu_{\max} = \mu_{v_{\max}}$ with $v_{\max} \in \mathcal{H}_\pi$ be a maximal spectral measure as in Proposition 2.64 for $\pi|_A$, and let

$$X = \operatorname{supp} \mu_{\max} \subseteq \widehat{A}$$

be its support. We claim that $X = r\mathbb{S}^1$ for some $r > 0$.

For the proof of the claim we first show that $X \subseteq \widehat{A}$ is invariant under the compact subgroup $K = \mathrm{SO}_2(\mathbb{R})$. Indeed, applying π_k for some $k \in K$ to $v_{\max}$ we obtain that $\mu_{\pi_k v_{\max}} = k_* \mu_{v_{\max}}$ defines the same measure class as $\mu_{v_{\max}}$ by Lemma 5.5. This implies that $kX = X$ for all $k \in K$. To prove the claim, we define the function $R \colon \widehat{A} \cong \mathbb{R}^2 \ni t \mapsto \|t\|$ and note that $R(kt) = R(t)$ for all $k \in K$ and $t \in \widehat{A}$. Together with Corollary 5.2, this shows that $B = \pi_{\mathrm{FC}}(R)$ is intertwining for π restricted to K. Since B is intertwining for π restricted to A by construction of the measurable functional calculus in Proposition 2.58, B is intertwining for π. By irreducibility of π and Schur's lemma (Theorem 1.29), it follows that $B = \pi_{\mathrm{FC}}(R) = rI$ for some constant r. By the spectral theorem (Corollary 2.65) and the measurable functional calculus (Proposition 2.58(6)), this amounts to the claim that

$$X = \operatorname{supp} \mu_{\max} = r\mathbb{S}^1.$$

Moreover, we must have $r > 0$ since $\pi|_A$ is assumed to be non-trivial.

Having found the right candidate for r, we still need to prove that π is isomorphic to π^r. For this we first notice that there exists some weight $n \in \mathbb{Z}$ and some eigenvector $v \in \mathcal{H}_\pi$ for $\pi|_K$ of weight n (since $K \cong \mathbb{T}$ is compact and abelian with $\widehat{K} \cong \mathbb{Z}$).

We define the measurable map $\mathrm{D} \in \mathscr{L}^\infty(\widehat{A})$ by taking the direction of the argument; that is,

$$\widehat{A} \ni t \longmapsto \mathrm{D}(t) = \begin{cases} 1 & \text{if } t = 0, \\ \mathrm{e}^{\mathrm{i}\phi} & \text{if } t = \|t\| k_\phi \begin{pmatrix} 1 \\ 0 \end{pmatrix} \text{ for some } k_\phi \in K. \end{cases}$$

We will use D together with the functional calculus in Proposition 2.58 for the restriction $\pi|_A$. By Corollary 5.2, we have

$$\pi_{\mathrm{FC}}(\mathrm{D} \circ k_\psi^{\mathrm{t}}) = \pi_{k_\psi} \pi_{\mathrm{FC}}(\mathrm{D}) \pi_{k_\psi}^{-1}$$

for $k_\psi \in K$, where

$$\mathrm{D} \circ k_\psi^{\mathrm{t}}(t) = \mathrm{D}\left(\|t\| k_{\phi-\psi} \begin{pmatrix} 1 \\ 0 \end{pmatrix} \right) = \mathrm{e}^{\mathrm{i}(\phi-\psi)} = \mathrm{e}^{-\mathrm{i}\psi} \mathrm{D}(t)$$

for all

$$t = \|t\| k_\phi \begin{pmatrix} 1 \\ 0 \end{pmatrix} \in \widehat{A} \smallsetminus \{0\}. \tag{5.3}$$

This now implies that $\pi_{\mathrm{FC}}(\mathrm{D})v$ is an eigenvector for $\pi|_K$ of weight $(n-1)$. Indeed,

$$\pi_{k_\psi} \pi_{\mathrm{FC}}(\mathrm{D}) v = \pi_{k_\psi} \pi_{\mathrm{FC}}(\mathrm{D}) \pi_{k_\psi}^{-1} \pi_{k_\psi} v = \pi_{\mathrm{FC}}\big(\mathrm{e}^{-\mathrm{i}\psi} \mathrm{D}\big) \mathrm{e}^{\mathrm{i}n\psi} v = \mathrm{e}^{\mathrm{i}(n-1)\psi} \pi_{\mathrm{FC}}(\mathrm{D}) v$$

for all $k_\psi \in K$. Similarly, $\pi_{\mathrm{FC}}(\overline{\mathrm{D}})v$ is an eigenvector of weight $(n+1)$. In other words, $\pi_{\mathrm{FC}}(\mathrm{D})$ lowers, and $\pi_{\mathrm{FC}}(\overline{\mathrm{D}})$ raises, the K-weight of any eigenvector. We also note that $\pi_{\mathrm{FC}}(\mathrm{D})$ and $\pi_{\mathrm{FC}}(\overline{\mathrm{D}}) = \pi_{\mathrm{FC}}(\mathrm{D})^*$ are unitary, since $|\,\mathrm{D}(t)| = 1$ for all $t \in \widehat{A}$. In particular, we have

$$\|v\| = \|\pi_{\mathrm{FC}}(\mathrm{D}) v\| = \|\pi_{\mathrm{FC}}(\overline{\mathrm{D}}) v\|.$$

Hence we may and will assume that $v = v_0 \in \mathcal{H}_\pi$ is an eigenvector for $\pi|_K$ of weight 0 with $\|v\| = 1$. For every $m \in \mathbb{Z}$ we also define $v_m = \pi_{\mathrm{FC}}(\mathrm{D}^m) v_0$ and notice that this implies that $\|v_m\| = 1$, $\pi_{\mathrm{FC}}(\mathrm{D}^n) v_m = v_{m+n}$, v_m has weight $-m$, and hence $\langle v_m, v_n \rangle = \delta_{m,n}$ for all $m, n \in \mathbb{Z}$. We define the closed subspace

$$\mathcal{V} = \langle v_m \mid m \in \mathbb{Z} \rangle \subseteq \mathcal{H}_\pi$$

and

$$U \colon \mathcal{V} \ni \sum_{m \in \mathbb{Z}} a_m v_m \longmapsto \sum_{m \in \mathbb{Z}} a_m \mathrm{D}^m \in \mathcal{H}_r = L^2(r\mathbb{S}^1, \mu_r).$$

We claim that $\mathcal{V} = \mathcal{H}_\pi$ and that U is an intertwining isomorphism from $\mathcal{H}_\pi$ to $\mathcal{H}_r$, which will imply the proposition.

Since μ_r is the normalized arc length measure on $r\mathbb{S}^1$, it follows that the functions D^m for $m \in \mathbb{Z}$ make up an orthonormal basis of $\mathcal{H}_r$. In particular, U

is a unitary isomorphism between $\mathcal{V}$ and $\mathcal{H}_r$. Moreover,

$$\begin{aligned}\pi^r_{k_\psi}\,\mathrm{D}^m(t) &= \pi^r_{k_\psi}\,\mathrm{D}^m\left(\|t\|k_\phi\begin{pmatrix}1\\0\end{pmatrix}\right)\\ &= \mathrm{D}^m\left(k_\psi^{-1}\|t\|k_\phi\begin{pmatrix}1\\0\end{pmatrix}\right) = \mathrm{e}^{\mathrm{i}m(\phi-\psi)} = \mathrm{e}^{-\mathrm{i}m\psi}\,\mathrm{D}^m(t)\end{aligned}$$

for all t as in (5.3) shows that D^m is also an eigenvector for $\pi^r|_K$ of weight $-m$. Therefore

$$U\colon \mathcal{V} \longrightarrow \mathcal{H}_r$$

is already intertwining for $\pi|_K$ as well as for the operators $\pi_{\mathrm{FC}}(\mathrm{D}^n)$ on $\mathcal{H}_\pi$, respectively $M_{\mathrm{D}^n} = \pi^r_{\mathrm{FC}}(\mathrm{D}^n)$ on $\mathcal{H}_r$ for all $n \in \mathbb{Z}$.

For the proof that $\mathcal{V}$ is invariant under $\pi|_A$ and that U is also intertwining for $\pi|_A$ (statements which are likely to both be unclear at this point) we will again use the functional calculus for $\pi|_A$. In fact, using the fact that $\pi|_A$ and $\pi^r|_A$ have their spectral measures supported on the same circle $r\mathbb{S}^1$, we will show that π_a for $a \in A$ can be expressed in terms of the operators $\pi_{\mathrm{FC}}(\mathrm{D}^n)$ for $n \in \mathbb{Z}$ which we used to define $\mathcal{V}$.

More precisely, on $r\mathbb{S}^1$ we have the identities

$$\begin{aligned}t_1 &= \tfrac{r}{2}\big(\mathrm{D}(t) + \mathrm{D}^{-1}(t)\big),\\ t_2 &= \tfrac{r}{2}\big(\mathrm{D}(t) - \mathrm{D}^{-1}(t)\big),\\ \mathrm{e}^{2\pi\mathrm{i}a_1t_1} &= \sum_{n=0}^{\infty}\frac{1}{n!}\Big(2\pi\mathrm{i}a_1\tfrac{r}{2}\big(\mathrm{D}(t) + \mathrm{D}^{-1}(t)\big)\Big)^n,\\ \mathrm{e}^{2\pi\mathrm{i}a_2t_2} &= \sum_{n=0}^{\infty}\frac{1}{n!}\Big(2\pi\mathrm{i}a_2\tfrac{r}{2}\big(\mathrm{D}(t) - \mathrm{D}^{-1}(t)\big)\Big)^n,\end{aligned}$$

where the two sums converge uniformly on $r\mathbb{S}^1$ for every

$$a = \begin{pmatrix}a_1\\a_2\end{pmatrix} \in A.$$

By the properties of the functional calculus in Proposition 2.58, this shows that we can express π_a for any $a \in A$ in terms of a, the number r, and $\pi_{\mathrm{FC}}(\mathrm{D}^n)$ for $n \in \mathbb{Z}$. Since $\mathcal{V}$ is invariant under the latter and U is intertwining for the latter, it follows that $\mathcal{V}$ is invariant under $\pi|_A$ and U is intertwining for $\pi|_A$. As $\mathcal{V}$ is closed and invariant, and π is assumed to be irreducible, it follows that $\mathcal{V} = \mathcal{H}_\pi$. This concludes the proof. □

One may wonder why we had to work quite so hard to obtain the isomorphism above, and in particular to obtain spectral multiplicity one for the restriction to the normal abelian subgroup $A \lhd G$. The following exercise shows that the latter is not automatic for cyclic representations.

Exercise 5.17 (Infinite spectral multiplicity of cyclic representations). Let G be the group $\mathrm{SO}_2(\mathbb{R}) \ltimes \mathbb{R}^2$ be as above and $r > 0$. Show that $(\pi^r)^\infty$ is cyclic.

Exercise 5.18. Let $G = \mathrm{SO}_2(\mathbb{R}) \ltimes \mathbb{R}^2$.
(a) Calculate the contragredient representation of all elements of $\widehat{G}$.
(b) Let $\pi, \rho \in \widehat{G}$ be irreducible unitary representations. Describe when the inner tensor product representation $\pi \otimes \rho$ is again irreducible.

The Full Isometry Group of the Plane

We will now discuss, through a series of exercises, the unitary dual $\widehat{G}$ of the full isometry group

$$G = \mathrm{O}_2(\mathbb{R}) \ltimes \mathbb{R}^2 = \left\{ \begin{pmatrix} k & a \\ & 1 \end{pmatrix} \;\middle|\; k \in \mathrm{O}_2(\mathbb{R}), a \in \mathbb{R}^2 \right\}.$$

Exercise 5.19 (Old representations). Show that the unitary dual of $\mathrm{O}_2(\mathbb{R})$ is given by

$$\widehat{\mathrm{O}_2(\mathbb{R})} = \{\mathbb{1}, \det\} \cup \{\delta^n \mid n \in \mathbb{N}\},$$

where $\mathbb{1}$ is the trivial representation of $\mathrm{O}_2(\mathbb{R})$, det is the unitary character defined by the determinant map, and δ^n for each $n \in \mathbb{N}$ is an irreducible unitary representation of $\mathrm{O}_2(\mathbb{R})$ on $\mathbb{C}^2$ so that $\delta^n|_{\mathrm{SO}_2(\mathbb{R})}$ has eigenspaces for weights $n, -n$.

Exercise 5.20 (New representations). Let $r > 0$. Show that the new representation π^r of $\mathrm{SO}_2(\mathbb{R})$ extends, by the same formulas, to a new representation $\pi^{r,+}$ of $\mathrm{O}_2(\mathbb{R})$ on $\mathcal{H}_r$. Show that $\pi^{r,-} = \det \otimes \pi^{r,+}$ defined by

$$G \ni g \longmapsto \pi^{r,-}(g) = \det(g)\pi^{r,+}(g)$$

defines an irreducible new representation $\pi^{r,-}$ that is not unitarily equivalent to $\pi^{r,+}$.

Exercise 5.21 (Completeness). Prove that

$$\widehat{\mathrm{O}_2(\mathbb{R}) \ltimes \mathbb{R}^2} = \widehat{\mathrm{O}_2(\mathbb{R})} \sqcup \{\pi^{r,+}, \pi^{r,-} \mid r \in (0, \infty)\}.$$

5.2.3 The Affine Group in One Dimension

As our next metabelian group, we wish to study the affine group in one dimension. In fact we have a choice of either studying the connected group

$$G_{>0} = \left\{ \begin{pmatrix} s & a \\ 0 & 1 \end{pmatrix} \;\middle|\; s \in \mathbb{R}_{>0}, a \in \mathbb{R} \right\} \cong \mathbb{R}_{>0} \ltimes \mathbb{R}$$

or the full affine group

$$G_\times = \left\{ \begin{pmatrix} s & a \\ 0 & 1 \end{pmatrix} \;\middle|\; s \in \mathbb{R}^\times, a \in \mathbb{R} \right\} \cong \mathbb{R}^\times \ltimes \mathbb{R}.$$

We will consider $G = G_{>0}$ and leave $G_\times$ as an exercise. We will use the letter s for elements of $\mathbb{R}^\times$, and the letter a for elements of $\mathbb{R}$. For convenience, we will use the abbreviations

$$g_s = \begin{pmatrix} s & 0 \\ & 1 \end{pmatrix} \text{ and } h_a = \begin{pmatrix} 1 & a \\ & 1 \end{pmatrix},$$

and define the abelian complementary subgroup $S = \{g_s \mid s \in \mathbb{R}_{>0}\}$ and the normal abelian subgroup $A = \{h_a \mid a \in \mathbb{R}\}$ so that $G = SA = AS$ and every element

$$g = \begin{pmatrix} s & a \\ & 1 \end{pmatrix} \in G$$

can be written uniquely as the product $g = h_a g_s$ with $h_a \in A$ and $g_s \in S$.

The Irreducible Representations

We will use the terminology from Section 5.2.2.

(Old) Every irreducible unitary representation of $\mathbb{R}_{>0} \cong S \cong G/A$ gives rise to an irreducible unitary representation of $G = G_{>0}$ by letting the normal subgroup A act trivially. Since $\mathbb{R}_{>0} \ni s \mapsto \log s \in \mathbb{R}$ is an isomorphism, we see that every such old representation is given by a unitary character

$$\chi_\alpha \colon G \ni h_a g_s \longmapsto \log s \in \mathbb{R} \longmapsto \mathrm{e}^{2\pi\mathrm{i}\alpha \log s} = s^{2\pi\mathrm{i}\alpha} \in \mathbb{S}^1$$

for some (uniquely determined) $\alpha \in \mathbb{R}$.

(New) We now define two new representations π^+ and π^- for which the normal subgroup $\mathbb{R} \cong A \lhd G$ acts non-trivially.

To define π^+, we will use the measure μ_+ on $\widehat{A}$ defined by

$$\mathrm{d}\mu_+ = \mathbb{1}_{(0,\infty)} \frac{\mathrm{d}t}{t},$$

where we use the isomorphism $\widehat{A} \cong \mathbb{R}$, the variable t to denote elements of $\widehat{A}$, and $\mathrm{d}t$ to denote the Lebesgue measure on $\widehat{A}$. Along with μ_+ we define the Hilbert space

$$\mathcal{H}_+ = L^2(\widehat{A}, \mu_+).$$

We note that the action of $g_s \in S$ on $\widehat{A}$ defined by

$$\widehat{A} \ni t \longmapsto g_s \cdot t = s^{-1} t \in \widehat{A}$$

preserves μ_+, since for any measurable $B \subseteq \mathbb{R}_{>0}$ we have

$$\mu_+(s^{-1}B) = \int_0^\infty \mathbb{1}_{s^{-1}B}(t) \frac{\mathrm{d}t}{t} = \int_0^\infty \mathbb{1}_B(st) \frac{\mathrm{d}t}{t} = \int_0^\infty \mathbb{1}_B(t') \frac{\mathrm{d}t'}{t'} = \mu_+(B)$$

via the substitution $t' = st$. As in Section 5.2.1, we now define the unitary representation on A by using the multiplication representation, and on S we use the action-associated representation. Putting these definitions together, we define once more

$$\pi_g^+ f(t) = \pi_{h_a g_s}^+ f(t) = \mathrm{e}^{2\pi \mathrm{i} a t} f(st)$$

for $g = h_a g_s = \begin{pmatrix} s & a \\ & 1 \end{pmatrix} \in G$, $f \in \mathcal{H}_+$, and $t \in \widehat{A}$.

Similarly, we define π^- on $\mathcal{H}_- = L^2(\widehat{\mathbb{R}}, \mu_-)$ using the measure μ_- defined by $\mathrm{d}\mu_- = \mathbb{1}_{(-\infty,0)} \frac{\mathrm{d}t}{|t|}$ on $\widehat{H} \cong \mathbb{R}$.

Lemmas 5.8 and 5.10 and Exercise 5.9 now show that π^+ and π^- are both irreducible unitary representations of G, and Lemma 5.11 shows that these are not isomorphic to each other.

Classification of the Unitary Dual

We will show the following result as a corollary of the main technical result (Theorem 5.55) of this chapter (see Section 5.6.2).

Proposition 5.22 (Description of $\widehat{G}$). *Let*

$$G = G_{>0} = \left\{ \begin{pmatrix} s & a \\ & 1 \end{pmatrix} \,\middle|\, s \in \mathbb{R}_{>0}, a \in \mathbb{R} \right\}$$

be as above. Then the set of old representations defined by characters on S and the two new representations π^+ and π^- comprise the complete set of irreducible representations of G.

5.2.4 The Heisenberg Group

We recall that the 3-dimensional Heisenberg group G is defined by

$$G = \left\{ \begin{pmatrix} 1 & x & z \\ 0 & 1 & y \\ 0 & 0 & 1 \end{pmatrix} \,\middle|\, x, y, z \in \mathbb{R} \right\}$$

and that the multiplication is given by

$$\begin{pmatrix} 1 & a & c \\ & 1 & b \\ & & 1 \end{pmatrix} \begin{pmatrix} 1 & x & z \\ & 1 & y \\ & & 1 \end{pmatrix} = \begin{pmatrix} 1 & a+x & c+z+ay \\ & 1 & b+y \\ & & 1 \end{pmatrix}$$

for all

$$\begin{pmatrix} 1 & a & c \\ & 1 & b \\ & & 1 \end{pmatrix}, \begin{pmatrix} 1 & x & z \\ & 1 & y \\ & & 1 \end{pmatrix} \in G.$$

More generally, for any $d \geqslant 1$ the $(2d+1)$-dimensional Heisenberg group is defined as the set $G = \mathbb{R}^{2d+1}$ whose elements we denote by

$$(a, b, c), (x, y, z) \in \mathbb{R}^d \times \mathbb{R}^d \times \mathbb{R},$$

equipped with the multiplication

$$(a, b, c) \cdot (x, y, z) = (a + x, b + y, c + z + \langle a, y \rangle).$$

This coincides with matrix multiplication if (x, y, z) is identified with the matrix

$$\begin{pmatrix} 1 & x_1 & x_2 & \cdots & x_n & z \\ & 1 & 0 & \cdots & 0 & y_1 \\ & & \ddots & \ddots & \vdots & \vdots \\ & & & 1 & 0 & y_{n-1} \\ & & & & 1 & y_n \\ & & & & & 1 \end{pmatrix} \in \mathrm{SL}_{n+2}(\mathbb{R}).$$

We note that the centre of G is given by the one-dimensional subgroup

$$C(G) = \{(0, 0, z) \mid z \in \mathbb{R}\}.$$

The Irreducible Representations

(Old) We say that a unitary representation is an old representation of G if the centre $C(G)$ acts trivially. Since $G/C(G) \cong \mathbb{R}^{2d}$ is abelian, it follows that any old irreducible representation of G is given by a character of

$$G/C(G) \cong \mathbb{R}^{2d}.$$

(New) We say that a unitary representation is a new representation of G if the centre $C(G)$ acts non-trivially. To construct the irreducible new representations of G, we define the normal subgroup

$$A = \{(0, y, z) \mid y \in \mathbb{R}^d, z \in \mathbb{R}\} \lhd G$$

corresponding to the 'last column' of the matrix group, and the 'complementary group'

$$S = \{(x, 0, 0) \mid x \in \mathbb{R}^d\} < G$$

so that once again $G = AS = SA \cong S \ltimes A$.

To apply the results of Section 5.1 we calculate the effect of the automorphism $\theta_x = \theta_{(x,0,0)}$ of A for all $x \in \mathbb{R}^d$. Indeed, for $(0, y, z) \in A$ we have

$$\begin{aligned}\theta_x\big((0,y,z)\big) &= (x,0,0)(0,y,z)(-x,0,0)\\ &= (x,y,z+\langle x,y\rangle)(-x,0,0)\\ &= (0,y,z+\langle x,y\rangle).\end{aligned}$$

Using the standard basis of $A \cong \mathbb{R}^{d+1}$ with the centre corresponding to the last coordinate, the automorphism θ_x has the matrix representation

$$\begin{pmatrix} I_d & 0 \\ x^{\mathrm{t}} & 1 \end{pmatrix},$$

so the dual automorphism $\widehat{\theta}_x$ on $\widehat{A} \cong \mathbb{R}^{d+1}$ has the matrix representation

$$\begin{pmatrix} I_d & x \\ 0 & 1 \end{pmatrix}.$$

Now let $\xi \in \mathbb{R}^\times$ and define

$$t_0 = \begin{pmatrix} 0 \\ \xi \end{pmatrix} \in \mathbb{R}^{d+1} \cong \widehat{A}.$$

Then the S-orbit of t_0 is given by

$$S \cdot t_0 = \left\{ \begin{pmatrix} I_d & -x \\ 0 & 1 \end{pmatrix} \begin{pmatrix} 0 \\ \xi \end{pmatrix} \;\middle|\; x \in \mathbb{R}^d \right\} = \left\{ \begin{pmatrix} t \\ \xi \end{pmatrix} \;\middle|\; t \in \mathbb{R}^d \right\}. \tag{5.4}$$

Using the Lebesgue measure on this affine subspace gives rise via Lemmas 5.8 and 5.10 to a new irreducible representation π^ξ of the Heisenberg group.

Classification of the Unitary Dual

The following result will once again follow from Theorem 5.55 (see Section 5.6.3).

Theorem 5.23 (Stone–von Neumann[11]). *Let $d \geqslant 1$. The unitary dual $\widehat{G}$ of the $(2d+1)$-dimensional Heisenberg group consists of the old representations given by unitary characters on $G/C(G) \cong \mathbb{R}^{2d}$ and the infinite-dimensional new irreducible representations π^ξ for $\xi \in \mathbb{R}^\times$. The representation π^ξ for $\xi \in \mathbb{R}^\times$ can also be defined by*

$$\big(\pi^\xi_{(x,y,z)} f\big)(t) = \mathrm{e}^{2\pi\mathrm{i}(\langle y,t\rangle + z\xi)} f(t + \xi x) \tag{5.5}$$

for $(x, y, z) \in G$, $f \in \mathcal{H}_{\pi^\xi} = L^2(\mathbb{R}^d)$, and $t \in \mathbb{R}^d$.

5.2.5 An Impossibly Complicated Dual

We now wish to apply the results of Section 5.2.1 to a concrete solvable group to see that sometimes it is more or less impossible to classify all irreducible unitary representations for a given group. For these negative results we will rely on some more background in ergodic theory (see [24], for example). We define the normal abelian subgroup to be

$$A = \left\{ \begin{pmatrix} 1 & a \\ 0 & 1 \end{pmatrix} \in \mathrm{Mat}_{2,2}(R) \;\middle|\; a \in R \right\}$$

where $R = \mathbb{F}_2[T, T^{-1}]$, and the complementary subgroup to be the diagonal subgroup $S \cong \mathbb{Z}$ in

$$G = \mathbb{Z} \ltimes R = \left\{ \begin{pmatrix} T^n & a \\ 0 & 1 \end{pmatrix} \in \mathrm{Mat}_{2,2}(R) \;\middle|\; n \in \mathbb{Z}, a \in R \right\}$$

to simplify some discussions. This group can also be described using the wreath product as $\mathbb{F}_2 \wr \mathbb{Z}$, and is called the *lamplighter group.*† With some modifications, the discussion also holds for

$$G_2 = \mathbb{Z} \ltimes \mathbb{Z}[\tfrac{1}{2}] = \left\{ \begin{pmatrix} 2^n & a \\ 0 & 1 \end{pmatrix} \in \mathrm{Mat}_{2,2}(\mathbb{Z}[\tfrac{1}{2}]) \;\middle|\; n \in \mathbb{Z}, a \in \mathbb{Z}[\tfrac{1}{2}] \right\}$$

and

$$G_M = \mathbb{Z} \ltimes \mathbb{Z}^2 = \left\{ \begin{pmatrix} M^n & a \\ 0 & 1 \end{pmatrix} \in \mathrm{GL}_3(\mathbb{Z}) \;\middle|\; n \in \mathbb{Z}, a \in \mathbb{Z}^2 \right\}$$

for some fixed hyperbolic $M \in \mathrm{GL}_2(\mathbb{Z})$, and for many other groups.

For

$$A \cong R = \mathbb{F}_2[T, T^{-1}] \cong \bigoplus_{n \in \mathbb{Z}} \mathbb{F}_2$$

the Pontryagin dual is given by $\widehat{A} \cong \mathbb{F}_2^{\mathbb{Z}}$ (see Proposition 2.29). In other words, in the language of ergodic theory $\widehat{A}$ is the full shift on 2 symbols. The automorphism θ of $A \cong R$ defined by $\theta(a) = Ta$ for $a \in R$ can be written as

$$\theta\big((c_n)_{n \in \mathbb{Z}}\big) = (c_{n-1})_{n \in \mathbb{Z}}$$

if we identify the polynomial $a = \sum_{n \in \mathbb{Z}} c_n T^n \in A$ with the (finitely supported) sequence $(c_n) \in \bigoplus_{n \in \mathbb{Z}} \mathbb{F}_2$ of its coefficients. The dual automorphism is therefore

† The name of this group comes from the following image. Envision an infinite street with lamps at each integer coordinate, with a lamplighter moving along the street. In this picture elements of G can be interpreted as instructions to the lamplighter to move along the street $\mathbb{Z}$ by distance one (corresponding to $\begin{pmatrix} T & 0 \\ & 1 \end{pmatrix}$) and to light or extinguish the lamp at that position (corresponding to $\begin{pmatrix} 1 & 1 \\ & 1 \end{pmatrix}$).

given by the left shift, so

$$\widehat{\theta}\big((t_n)_{n\in\mathbb{Z}}\big) = (t_{n+1})_{n\in\mathbb{Z}}$$

for $(t_n)_{n\in\mathbb{Z}} \in \widehat{A} \cong \mathbb{F}_2^{\mathbb{Z}}$.

The space of ergodic measures for the full shift is ridiculously large, and such measures cannot be meaningfully classified. We merely indicate a few ways to find invariant ergodic probability measures, and ask the reader to take on faith the fact that these constructions are very far from being exhaustive:

- (Periodic points and orbits) The full shift has many periodic points which may be obtained by taking any finite block of 0s and 1s and concatenating it infinitely often, and each such point gives rise to a finitely supported ergodic invariant measure simply by averaging along its orbit. Precisely, if $x \in \mathbb{F}_2^{\mathbb{Z}}$ has $\widehat{\theta}^n x = x$ for some $n \geqslant 1$ then $\frac{1}{n}\sum_{k=0}^{n-1}\delta_{\widehat{\theta}^k x}$ is such a measure, where δ_x denotes the point mass at x. Moreover, if the orbit of $x \in \mathbb{F}_2^{\mathbb{Z}}$ is infinite, then we can also use the σ-finite measure $\mu = \sum_{k\in\mathbb{Z}}\delta_{\widehat{\theta}^k x}$ for the construction of the irreducible unitary representation.
- (Bernoulli measures) For any $p \in (0,1)$ we can define the product measure using probabilities p and $1-p$ for the symbols $0 \in \mathbb{F}_2$ and $1 \in \mathbb{F}_2$ respectively. The resulting shift-invariant ergodic measure is called a *Bernoulli measure*.
- (Markov and Parry measures) For any finite set S of finite words (or blocks) in the alphabet $\mathbb{F}_2 = \{0,1\}$, one can define a closed $\widehat{\theta}$-invariant subset $X_S \subseteq \mathbb{F}_2^{\mathbb{Z}}$ of all sequences that do not contain any word from S. Under mild conditions on S one can find a natural ergodic invariant probability measure on X_S, called the *Parry measure*, which is in some sense the most uniformly distributed measure on X_S. Varying the set S gives another collection of ergodic invariant probability measures, a countable collection in this case. Allowing more general constructions (using different transition probabilities) gives uncountably many different ergodic invariant probability measures called *Markov measures*.
- (Coding a measure-preserving system) Given any ergodic measure-preserving system $T\colon X \to X$ on a probability space $(X, \mathcal{B}, \nu)$ we can define, for any measurable set $B \subseteq X$, a map

$$\Phi_B\colon X \to \mathbb{F}_2^{\mathbb{Z}}$$

  by sending $x \in X$ to the sequence $\big(\mathbb{1}_B(T^n x)\big)_{n\in\mathbb{Z}}$. Then $\mu_{T,B} = (\Phi_B)_*\nu$ is an ergodic invariant probability measure on $\mathbb{F}_2^{\mathbb{Z}}$. For this rather general construction it is very hard to see when precisely one gets different ergodic measures (though this is certainly going to happen often). So let us specialize this setup in the following construction.
- (Torus rotations) Let $d \geqslant 1$, $X = \mathbb{T}^d$ and $T = R_\alpha\colon x \mapsto x + \alpha$ be a rotation defined by some $\alpha = (\alpha_1, \dots, \alpha_d) \in \mathbb{R}^d$. Assuming that $1, \alpha_1, \dots, \alpha_d$ are linearly independent over $\mathbb{Q}$, this defines an ergodic measure-preserving system with respect to Lebesgue measure m. For any measurable set B in $\mathbb{T}^d$ we

now obtain a map Φ_B and an ergodic invariant probability measure $\mu_{\alpha,B}$ on $\mathbb{F}_2^{\mathbb{Z}}$. As T has 'very few factors', we can find a variety of ergodic measures by varying α and B. Using the spectral theory of the action-associated operators or entropy theory (see [22, Sec. 1.3]), we also see that these measures are all different from the Bernoulli, Parry, or Markov measures above, and they cannot coincide with any periodic point measure unless $m(B) \in \{0,1\}$. Using $X = \mathbb{T}^2$ and any 'picture with a frame' $B \subseteq \mathbb{T}^2$ as in Figure 5.1 and Exercise 5.24, we can construct a new irreducible unitary representation of the lamplighter group in such a way that different pictures (when considered modulo the Lebesgue measure) give rise to different irreducible unitary representations.

The selection of examples above is influenced by the mathematical interests of the authors. Many other examples and constructions of ergodic measures or irreducible unitary representations are possible (see also Section 5.2.6).

Fig. 5.1: For the purpose of Exercise 5.24, we say that a subset B of $\mathbb{T}^2$ 'has a frame' if there exists some $\varepsilon > 0$ with the property that $[0,1] \times [-\varepsilon,\varepsilon] + \mathbb{Z}^2 \subseteq B$ and $[-\varepsilon,\varepsilon] \times [0,1] + \mathbb{Z}^2 \subseteq B$ but the sets $(\varepsilon, 1-\varepsilon) \times \big((\varepsilon, 2\varepsilon) \cup (1-2\varepsilon, 1-\varepsilon)\big) + \mathbb{Z}^2$ and $\big((\varepsilon, 2\varepsilon) \cup (1-2\varepsilon, 1-\varepsilon)\big) \times (\varepsilon, 1-\varepsilon) + \mathbb{Z}^2$ are disjoint from B. These pictures define irreducible unitary representations of the lamplighter group that have no better description than that given by Exercise 5.24 and the pictures themselves.

In the following exercise, we invite the reader to prove the 'constructions of irreducible unitary representations via pictures' mentioned above.

Essential Exercise 5.24. Let $X = \mathbb{T}^2$, $\nu = m_{\mathbb{T}^2}$ be the Haar measure, and define $T\colon \mathbb{T}^2 \to \mathbb{T}^2$ by $T\colon \mathbb{T}^2 \ni x \mapsto x + \alpha \in \mathbb{T}^2$ for some $\alpha = (\alpha_1, \alpha_2) \in \mathbb{R}^2$. We suppose throughout that $1, \alpha_1, \alpha_2$ are linearly independent over $\mathbb{Q}$.
(a) Prove that $m_{\mathbb{T}^2}$ is an ergodic invariant measure for T. Show that any orbit $\{x + n\alpha \mid n \in \mathbb{Z}\}$ is dense in $\mathbb{T}^2$.
(b) Now suppose that $B \subseteq \mathbb{T}^2$ is an arbitrary measurable subset with a 'frame' as illustrated in Figure 5.1. Use B to define a factor map $\Phi_B\colon \mathbb{T}^2 \to \mathbb{F}_2^{\mathbb{Z}}$ sending a point $x \in \mathbb{T}^2$ to the sequence $\big(\mathbb{1}_B(x + n\alpha)\big)_{n\in\mathbb{Z}}$. Show that Φ_B is injective.
(c) Show that for two measurable subsets $B_1, B_2 \subseteq \mathbb{T}^2$ with frames, we have $\mu_{T,B_1} = \mu_{T,B_2}$ if and only if $m(B_1 \triangle B_2) = 0$.

(d) Now conclude, together with the results of this section, that any black and white picture on $[0,1]^2$ with a frame gives rise to an irreducible unitary representation of the lamplighter group uniquely associated to that picture.

5.2.6 The Discrete Heisenberg Group

The (3-dimensional) discrete Heisenberg group is defined by

$$G = \left\{ \begin{pmatrix} 1 & x & z \\ 0 & 1 & y \\ 0 & 0 & 1 \end{pmatrix} \,\middle|\, x, y, z \in \mathbb{Z} \right\}.$$

We will show that despite G being nilpotent, and hence as close as possible to being abelian without being abelian, its unitary dual is once again quite wild and has no simple description.

As in Section 5.2.4, we will again use the notation $(x, y, z) \in \mathbb{Z}^3$ for the matrices in G and the subgroups

$$\begin{aligned} C(G) &= \{(0,0,z) \mid z \in \mathbb{Z}\}, \\ A &= \{(0,y,z) \mid y, z \in \mathbb{Z}\}, \end{aligned}$$

and

$$S = \{(x,0,0) \mid x \in \mathbb{Z}\}.$$

Let us begin with the (by this stage obvious) candidates for irreducible unitary representations:

- old representations defined by characters on $G/C(G) \cong \mathbb{Z}^2$,
- new representations arising from S-orbits on $\widehat{A} \cong \mathbb{T}^2$, and
- new representations arising from S-invariant and ergodic probability measures on $\mathbb{T}^2$.

We now explain the sense in which the second type is itself 'unreasonable' and how to modify the third type to also give rise to an 'unreasonably large collection' of irreducible unitary representations.

By the same calculation as in Section 5.2.4, the dual automorphisms corresponding to elements of $S \cong \mathbb{Z}$ on $\mathbb{T}^2$ are defined by powers of the matrix

$$\widehat{\theta}_{(1,0,0)} = \begin{pmatrix} 1 & 1 \\ 0 & 1 \end{pmatrix} : \mathbb{T}^2 \ni \begin{pmatrix} x \\ \xi \end{pmatrix} \longmapsto \begin{pmatrix} x + \xi \\ \xi \end{pmatrix} \in \mathbb{T}^2.$$

Fixing some $\xi \in \mathbb{T} \smallsetminus \{0\}$, noting that $X = \mathbb{T} \times \{\xi\} \cong \mathbb{T}$ is invariant, and recalling that the S-action defined in (5.1) uses $\widehat{\theta}_{(1,0,0)}^{-1}$, we therefore need to study the rotation map

$$R_\xi \colon \mathbb{T} \ni x \longmapsto x - \xi \in \mathbb{T}.$$

For $\xi \in \mathbb{Q}/\mathbb{Z}$ every orbit under this map is finite. However, for an irrational ξ the orbit of every $x_0 \in \mathbb{T}$ is infinite, and given by $x_0 + \mathbb{Z}\xi \subseteq \mathbb{T}$. Hence each such orbit gives rise to an irreducible unitary representation.

What is unreasonable about this? After taking into account possible isomorphisms of irreducible representations, the irreducible unitary representations arising from ξ and orbits of S on $\widehat{A}$ are in one-to-one correspondence with elements of the quotient $\mathbb{T}/\mathbb{Z}\xi$. Since $\mathbb{Z}\xi$ is a dense subgroup of $\mathbb{T}$ for every irrational ξ, the quotient $\mathbb{T}/\mathbb{Z}\xi$ is 'unreasonable' because (for example) any cross-section (that is, a set containing a unique choice of representative x_0 for each coset $x_0+\mathbb{Z}\xi \subseteq \mathbb{T}$, and hence for its associated irreducible unitary representation of G) is necessarily non-measurable.

We now move on to the third type in the list above, namely irreducible unitary representations arising from S-invariant and ergodic probability measures. Unlike the discussion of the solvable discrete group in Section 5.2.5, the S-invariant and ergodic probability measures on $\widehat{A} = \mathbb{T}^2$ are not at all complicated.† If ξ is rational, then every orbit on $\mathbb{T} \times \{\xi\}$ is finite and the normalized sum of Dirac measures on the orbits are precisely all S-invariant and ergodic probability measures on $\mathbb{T} \times \{\xi\} \subseteq \mathbb{T}^2$. If, on the other hand, ξ is irrational, then the Lebesgue measure μ on $\mathbb{T} \times \{\xi\}$ is the only A-invariant and ergodic probability measure on $\mathbb{T} \times \{\xi\} \subseteq \mathbb{T}^2 \cong \widehat{A}$. Using this measure we may define a unitary representation π^ξ of G on $L^2(\mathbb{T})$. However, we have only scratched the surface of the possible constructions of irreducible unitary representations.

In fact for any $\alpha \in \mathbb{T}$ we may define the character

$$\chi^\alpha \colon G \ni (x, y, z) \longmapsto \mathrm{e}^{2\pi \mathrm{i} x\alpha}$$

and consider the irreducible unitary representation $\chi^\alpha \otimes \pi^\xi$ of G.

Exercise 5.25. Show that for $\alpha_1, \alpha_2 \in \mathbb{T}$ the unitary representations $\chi^{\alpha_1} \otimes \pi^\xi$ and $\chi^{\alpha_2} \otimes \pi^\xi$ are isomorphic if and only if $\alpha_2 - \alpha_1 \in \mathbb{Z}\xi$.

Notice that the exercise above shows that $\widehat{G}$ contains 'another copy of the unreasonable quotient' $\mathbb{T}/\mathbb{Z}\xi$ for every irrational $\xi \in \mathbb{T}$ that arises not from orbits but from twisting the representation π^ξ by characters. More general constructions are possible (using cocycles as in the next section).

5.2.7 The Mautner Group

The examples considered so far may give the impression that only discrete groups have unreasonably complicated unitary duals, while connected groups have better behaviour. To see that this is not the case, we consider here the connected group $G = G_{\mathrm{Mautner}}$ known as the *Mautner group.*

† The set of quasi-invariant measures (see Lemma 5.5), on the other hand, is more complicated. Nonetheless, even by restricting to invariant probability measures, we can construct many irreducible representations of G.

To define the Mautner group, we first recall the isometry group

$$G_2 = \mathrm{SO}_2(\mathbb{R}) \ltimes \mathbb{R}^2 \cong \mathbb{T} \ltimes \mathbb{R}^2$$

of the plane considered in Section 5.2.2, and define

$$G_4 = G_2 \times G_2 \cong \mathbb{T}^2 \ltimes (\mathbb{R}^2)^2.$$

We now let $S < \mathbb{T}^2$ be an immersed subgroup isomorphic to $\mathbb{R}$ and define the Mautner group by $G_{\mathrm{Mautner}} = S \ltimes (\mathbb{R}^2)^2$.

By this stage we expect that the description of $\widehat{G}$ (or our inability to give a description of it) has to do with the action of S on $\widehat{A}$ for $A = (\mathbb{R}^2)^2$. For the vector

$$t_0 = \left(\begin{pmatrix} 1 \\ 0 \end{pmatrix}, \begin{pmatrix} 1 \\ 0 \end{pmatrix} \right) \in (\mathbb{R}^2)^2 \cong \widehat{A}$$

the $\mathbb{T}^2$-orbit is free and given by $X = \mathbb{T}^2 \cdot t_0 = \mathbb{S}^1 \times \mathbb{S}^1$. The restriction of the S-action to X is isomorphic to the translation action of the dense subgroup S on $\mathbb{T}^2$. In particular, we can:

- use every coset in $\mathbb{T}^2/S$ (equivalently, every S-orbit in X) to obtain an irreducible unitary representation of G,
- use the Haar measure on $\mathbb{T}^2$ pushed down to an S-invariant and ergodic probability measure on X to obtain another irreducible unitary representation π^μ of G, or
- twist π^μ by characters of S.

In particular, the first and (due to a generalization of Exercise 5.25, also the) third type again gives rise to an unreasonable set of irreducible unitary representations.

5.3 A Plancherel-type Theorem*

We wish to show here a Plancherel type theorem for the metabelian groups considered in the previous section.[(12)] Once again we will not try to give the most general result of this type, but instead try to give a convenient framework able to handle the isometry group in Section 5.2.2, the affine group in Section 5.2.3, and the Heisenberg groups in Section 5.2.4. We will only need the irreducible representations discussed so far but not the completeness result promised for these groups.

We continue to assume that $G = S \ltimes A$ is the semi-direct product of a closed normal abelian subgroup $A \lhd G$ and a closed abelian subgroup $S < G$. We assume in addition the following structure for the action of $S < G$ on $\widehat{A}$ introduced in Section 5.1:

- There exists a measurable cross-section $T_0 \subseteq \widehat{A}$ such that

$$S \times T_0 \ni (s, t_0) \longmapsto s \cdot t_0 \in S \cdot T_0 \subseteq \widehat{A}$$

is a measurable isomorphism, and $m_{\widehat{A}}(\widehat{A} \setminus S \cdot T_0) = 0$.

Exercise 5.26. Describe a choice of T_0 for the isometry group in Section 5.2.2, the affine group in Section 5.2.3, and the Heisenberg group in Section 5.2.4.

For every $t_0 \in T_0$ we use the isomorphism

$$S \ni s \longmapsto t = s \cdot t_0 \in S \cdot T_0 \subseteq \widehat{A}$$

to push the Haar measure m_S to an S-invariant ergodic measure μ_{t_0} on $S \cdot t_0$, which then defines an irreducible unitary representation of G by Lemmas 5.8 and 5.10. For the following it will be more convenient to instead define this representation π^{t_0} on $L^2(S)$ by the formulas

$$\begin{aligned}(\pi_a^{t_0} f)(s) &= \langle a, s \cdot t_0 \rangle f(s),\\ (\pi_{s_0}^{t_0} f)(s) &= f(s_0^{-1} s)\end{aligned}$$

for all $a \in A$, $s, s_0 \in S$, and $f \in L^2(S)$.

Exercise 5.27. Show that $\pi^{t_0} \cong \pi^{\mu_{t_0}}$.

Proposition 5.28 (Plancherel). *Let the semi-direct product $G = S \ltimes A$ and the cross-section $T_0 \subseteq \widehat{A}$ be as above. We let $X = T_0 \times S$ and write $t_0(\cdot)$ for the projection $X \to T_0$ onto the first coordinate. Then there exists a σ-finite measure ν on X and an intertwining unitary isomorphism between the regular representation λ of G on $L^2(G)$ and the unitary representation π^X defined by*

$$\begin{aligned}(\pi_a^X f)(s, x) &= \langle a, s \cdot t_0(x) \rangle f(s, x),\\ (\pi_{s_0}^X f)(s, x) &= f(s_0^{-1} s, x)\end{aligned}$$

for all $a \in A$, $s, s_0 \in S$, $x \in X$, and $f \in L^2(S \times X, m_S \times \nu)$.

We note that the description above should be thought of as an integral decomposition

$$\lambda^G \cong \int_X \pi^{t_0(x)} \, \mathrm{d}\nu(x)$$

of the regular representation λ^G into irreducible representations π^{t_0} for t_0 in T_0, where ν plays the role of the spectral measure and the second factor S in X creates multiplicity in the decomposition.

Proof of Proposition 5.28. Let m_G be the left Haar measure on G. Using the coordinate system $sa \in G$ for $s \in S$ and $a \in A$, we may normalize the Haar measures so that $m_G \cong m_S \times m_A$ in these coordinates (see Exercise 5.29).

Using the fact that m_G is a product measure and A is abelian, we can apply the Plancherel formula (Theorem 2.17) as follows: For a given $f \in L^2(G)$ and m_S-almost every $s \in S$ and $m_{\widehat{A}}$-almost every $t \in \widehat{A}$ we may define the partial transform $f \mapsto \widetilde{f}$ by

$$\widetilde{f}(s,t) = \int f(sa)\langle a,t\rangle \,\mathrm{d}m_A(a), \tag{5.6}$$

and this satisfies

$$\|\widetilde{f}\|_{L^2(S\times\widehat{A},m_S\times m_{\widehat{A}})} = \|f\|_{L^2(G)}$$

(see Exercise 5.30). Recalling our assumptions that $m_{\widehat{A}}(\widehat{A}\smallsetminus S\boldsymbol{\cdot} T_0) = 0$ and that

$$S\times T_0 \ni (s,t_0) \longmapsto s\boldsymbol{\cdot} t_0 \in S\boldsymbol{\cdot} T_0$$

is a bijection, we restrict in the following to elements $t\in\widehat{A}$ of the form $t = s\boldsymbol{\cdot} t_0$ with $s\in S$ and $t_0\in T_0$. We define $U(f)$ for $f\in L^2(G)$ by

$$U(f)(s_1,t_0,s_2) = \widetilde{f}(s_1s_2^{-1}, s_2\boldsymbol{\cdot} t_0) \tag{5.7}$$

for $s_1,s_2\in S$ and $t_0\in T_0$. For $s_0\in S$ we then have

$$\begin{aligned} U(\lambda^G_{s_0}f)(s_1,t_0,s_2) &= \widetilde{\lambda^G_{s_0}f}(s_1s_2^{-1}, s_2\boldsymbol{\cdot} t_0)\\ &= \int f(s_0^{-1}s_1s_2^{-1}a)\langle a, s_2\boldsymbol{\cdot} t_0\rangle \,\mathrm{d}m_A(a) = U(f)(s_0^{-1}s_1,t_0,s_2) \end{aligned}$$

and for $a_0\in A$ we have

$$\begin{aligned} U(\lambda^G_{a_0}f)(s_1,t_0,s_2) &= \widetilde{(\lambda^G_{a_0}f)}(s_1s_2^{-1}, s_2\boldsymbol{\cdot} t_0)\\ &= \int f(a_0^{-1}s_1s_2^{-1}a)\langle a, s_2\boldsymbol{\cdot} t_0\rangle \,\mathrm{d}m_A(a)\\ &= \int f(s_1s_2^{-1}\underbrace{\theta_{s_1^{-1}s_2}(a_0^{-1})a}_{=a'})\langle a, s_2\boldsymbol{\cdot} t_0\rangle \,\mathrm{d}m_A(a)\\ &= \int f(s_1s_2^{-1}a')\langle \theta_{s_1^{-1}s_2}(a_0)a', s_2\boldsymbol{\cdot} t_0\rangle \,\mathrm{d}m_A(a')\\ &= \langle a_0, s_1\boldsymbol{\cdot} t_0\rangle \int f(s_1s_2^{-1}a)\langle a, s_2\boldsymbol{\cdot} t_0\rangle \,\mathrm{d}m_A(a)\\ &= \langle a_0, s_1\boldsymbol{\cdot} t_0\rangle U(f)(s_1,t_0,s_2). \end{aligned}$$

We now let $X = T_0\times S$, define ν to be the pull-back of the Haar measure $m_{\widehat{A}}$ to $T_0\times S$, and obtain

$$\begin{aligned} \big\|U(f)\big\|^2_{L^2(S\times X,m_S\times\nu)} &= \int \big|\widetilde{f}(s_1s_2^{-1}, s_2\boldsymbol{\cdot} t_0)\big|^2 \,\mathrm{d}m_S(s_1)\,\mathrm{d}\nu(t_0,s_2)\\ &= \int \big|\widetilde{f}(s, s_2\boldsymbol{\cdot} t_0)\big|^2 \,\mathrm{d}m_S(s)\,\mathrm{d}\nu(t_0,s_2)\\ &= \big\|\widetilde{f}\big\|^2_{L^2(S\times\widehat{A},m_S\times m_{\widehat{A}})} = \|f\|^2_{L^2(G)} \end{aligned}$$

for any $f \in L^2(G)$. Therefore $U\colon L^2(G) \to L^2(S \times X, m_S \times \nu)$ is an intertwining unitary isomorphism. □

Exercise 5.29. Verify that $m_G = m_S \times m_A$ defines a left Haar measure on $G = S \ltimes A$.

Exercise 5.30. Verify that (5.6) and (5.7) define unitary isomorphisms.

5.4 Measurable Cocycles

To understand the phenomena introduced in Section 5.2 in reasonable generality we need to develop some more tools.

5.4.1 The Radon–Nikodym Cocycle

We start by discussing the Jacobian from page 7 in greater detail.

Lemma 5.31 (The derivative cocycle). *Suppose G acts continuously on X, and let μ be a σ-finite measure whose measure class is preserved by the G-action. Then there exists a measurable real-valued function $J = J_\mu$ defined on a measurable subset of $G \times X$ with the following properties:*

- *$J(g, x)$ is defined for $g \in G$ for μ-almost every $x \in X$;*
- *for $g \in G$ we have $J(g, x) = \frac{\mathrm{d}\mu}{\mathrm{d}g_*^{-1}\mu}(x)$ for almost every $x \in X$;*
- *moreover J satisfies the cocycle equation $J(g_1g_2, x) = J(g_1, g_2{\bullet}x)J(g_2, x)$ for $g_1, g_2 \in G$ and almost every $x \in X$.*

For another σ-finite measure ν on X with the same measure class as μ we have $J_\mu(g, x) = F(g{\bullet}x)J_\nu(g, x)F(x)^{-1}$ for $g \in G$ and almost every $x \in X$, where $F = \frac{\mathrm{d}\nu}{\mathrm{d}\mu}$.

Proof. We first additionally suppose that μ is a probability measure. As X is σ-compact and metric, the Borel σ-algebra $\mathcal{B}_X$ is countably generated. Hence there exists a sequence of finite partitions $(\mathcal{P}_n)$ with $\mathcal{P}_n \subseteq \sigma(\mathcal{P}_{n+1})$ for $n \in \mathbb{N}$ and with

$$\mathcal{B}_X = \bigvee_{n=1}^{\infty} \sigma(\mathcal{P}_n).$$

Using $\mathcal{P}_n$ we define

$$f_n(g, x) = \begin{cases} 0 & \text{if } x \in P \in \mathcal{P}_n \text{ and } \mu(P) = 0; \\ \frac{\mu(g{\bullet}P)}{\mu(P)} & \text{if } x \in P \in \mathcal{P}_n \text{ and } \mu(P) \neq 0. \end{cases}$$

Notice that measurability of $P \in \mathcal{P}_n$ implies measurability of the pre-image

$$B_P = \{(g, x) \mid g^{-1}\boldsymbol{\cdot}x \in P\} \subseteq G \times X,$$

which implies measurability of the map

$$G \ni g \longmapsto \mu(g\boldsymbol{\cdot}P) = \int_X \mathbb{1}_{B_P}(g, x)\,\mathrm{d}\mu(x)$$

by Fubini's theorem. As $\mathcal{P}_n$ is measurable, we deduce that f_n is measurable as a function of $(g, x) \in G \times X$ for all $n \in \mathbb{N}$. Moreover, for fixed $g \in G$ and $P \in \mathcal{P}_n$ with $\mu(P) \neq 0$ we have

$$\int_P f_n(g, x)\,\mathrm{d}\mu(x) = \mu(g\boldsymbol{\cdot}P) = \int_P \frac{\mathrm{d}g_*^{-1}\mu}{\mathrm{d}\mu}\,\mathrm{d}\mu.$$

In other words, we have that

$$f_n(g, \cdot) = E\Big(\frac{\mathrm{d}g_*^{-1}\mu}{\mathrm{d}\mu} \mid \sigma(\mathcal{P}_n)\Big)$$

is a conditional expectation. The increasing martingale theorem (we refer to [24, Th. 5.5]) now implies for any fixed $g \in G$ that

$$\lim_{n\to\infty} f_n(g, x) = \frac{\mathrm{d}g_*^{-1}\mu}{\mathrm{d}\mu}(x)$$

for μ-almost every $x \in X$. It follows that

$$f\colon G \times X \ni (g, x) \longmapsto f(g, x) = \liminf_{n\to\infty} f_n(g, x) \in [0, \infty]$$

is measurable and satisfies

$$f(g, x) = \frac{\mathrm{d}g_*^{-1}\mu}{\mathrm{d}\mu}(x) \in [0, \infty) \tag{5.8}$$

for every $g \in G$ and, depending on g, for μ-almost every $x \in X$.

Fix some element $g \in G$ and set

$$N = \{x \mid f(g, x) = 0\}$$

to see that

$$0 = \int_N f(g, x)\,\mathrm{d}\mu(x) = \int_N \mathrm{d}g_*^{-1}\mu = \mu(g\boldsymbol{\cdot}N).$$

Since the G-action preserves the measure class of μ, it follows that $\mu(N) = 0$. We now define

$$J(g, x) = f(g, x)^{-1} \in (0, \infty)$$

whenever $f(g, x) \notin \{0, \infty\}$. For $g \in G$ and a measurable set $B \subseteq X$ we now have

$$\int_B J(g,x)\,\mathrm{d}g_*^{-1}\mu(x) = \int_B \underbrace{J(g,x)\frac{\mathrm{d}g_*^{-1}\mu}{\mathrm{d}\mu}}_{=1\text{ almost surely}}\,\mathrm{d}\mu(x) = \mu(B)$$

by (5.8). In other words,

$$J(g,x) = \frac{\mathrm{d}\mu}{\mathrm{d}g_*^{-1}\mu}(x)$$

for $g \in G$ and μ-almost every $x \in X$. Using the discussion from (1.3)–(1.4) on page 7, it follows that J satisfies all three properties in the lemma.

If now μ is only assumed to be σ-finite then we can choose a measurable function $F\colon X \to (0,\infty)$ with $\int F\,\mathrm{d}\mu = 1$, define the probability measure ν by $\mathrm{d}\nu = F\,\mathrm{d}\mu$, and apply the argument above to find a function J_ν as in the lemma applied to ν. Therefore it suffices to prove the last statement in the lemma.

So let μ and ν be two σ-finite quasi-invariant measures defining the same measure class, suppose that J_ν as in the lemma exists, and let $F = \frac{\mathrm{d}\nu}{\mathrm{d}\mu}$. We claim that

$$J_\mu(g,x) = F(g{\cdot}x)J_\nu(g,x)F(x)^{-1} = \frac{\mathrm{d}\mu}{\mathrm{d}g_*^{-1}\mu}(x) \tag{5.9}$$

for $g \in G$ and μ-almost every $x \in X$. For $g \in G$ and a measurable set $B \subseteq X$ we have

$$\begin{aligned}\int_B F(g{\cdot}x)J_\nu(g,x)F(x)^{-1}\,\mathrm{d}g_*^{-1}\mu(x) &= \int_{g\cdot B} J_\nu(g,g^{-1}{\cdot}y)F(g^{-1}{\cdot}y)^{-1}\underbrace{F(y)\,\mathrm{d}\mu(y)}_{=\mathrm{d}\nu(y)}\\ &= \int_B \underbrace{J_\nu(g,x)}_{=\frac{\mathrm{d}\nu}{\mathrm{d}g_*^{-1}\nu}(x)}\,F(x)^{-1}\,\mathrm{d}g_*^{-1}\nu(x)\\ &= \int_B F(x)^{-1}\,\mathrm{d}\nu(x) = \mu(B)\end{aligned}$$

by using (1.1) twice for the substitution $x = g^{-1}{\cdot}y$. As this holds for any measurable $B \subseteq X$, we obtain (5.9). □

5.4.2 Measurable Cocycles Define Representations

Our next goal is to generalize Proposition 1.6. For this let us write $\mathscr{U}(\mathcal{H})$ for the group of unitary operators on a Hilbert space $\mathcal{H}$. We will always equip $\mathscr{U}(\mathcal{H})$ with the strong operator topology.

Definition 5.32 (Measurable cocycle). Suppose G acts continuously on the space X, μ is a σ-finite measure whose measure class is preserved by the G-action, and $\mathscr{U}$ is a group. Let C be a function defined on a measurable subset of $G \times X$ with values in $\mathscr{U}$. We say C is a *cocycle* (for the G-action on X with values in $\mathscr{U}$) if $C(g,x) \in \mathscr{U}$ is defined for every $g \in G$ for μ-almost every $x \in X$ and C satisfies the *cocycle equation*

$$C(g_1 g_2, x) = C(g_1, g_2 \cdot x) C(g_2, x) \tag{5.10}$$

for $g_1, g_2 \in G$ and, depending on g_1, g_2, for μ-almost every $x \in X$. The cocycle is called *strict* if it is defined on all of $G \times X$ and satisfies (5.10) for $g_1, g_2 \in G$ and all $x \in X$. If $\mathscr{U}$ is equipped with a topology we say C is *measurable* if the pre-image of every open (or Borel) subset in $\mathscr{U}$ is measurable in $G \times X$.

We note that Lemma 5.31 defines a measurable cocycle with values in $\mathbb{R}_{>0}$. We will see in the next section how exact cocycles can arise naturally.

Proposition 5.33 (Cocycles giving representations). *Suppose that G acts continuously on X, that μ is a σ-finite measure on X whose measure class is preserved by the G-action, and that C is a measurable cocycle for the G-action with values in $\mathscr{U}(\mathcal{H})$ for a separable Hilbert space $\mathcal{H}$. Then*

$$\big(\pi_g^{\mu,C} f\big)(x) = J(g, g^{-1} \cdot x)^{\frac{1}{2}} C(g, g^{-1} \cdot x) f(g^{-1} \cdot x)$$

for $f \in L^2_\mu(X, \mathcal{H})$, $g \in G$, and $x \in X$ defines a unitary representation of G.

We note that the first paragraph of the proof of Proposition 1.6 applies (with minor modifications) if we define the measurable cocycle

$$c_{\mathrm{new}}(g,x) = J(g,x)^{\frac{1}{2}} C(g,x) \in U = \mathbb{R}_{>0} \mathscr{U}(\mathcal{H})$$

for $g \in G$ and $x \in X$. This shows that π_g is unitary for $g \in G$ and that

$$\pi^{\mu,C} \colon G \ni g \longmapsto \pi_g^{\mu,C} \in \mathscr{U}(L^2_\mu(X, \mathcal{H}))$$

is a homomorphism.

Essential Exercise 5.34. Show that $\pi_g^{\mu,C} \in \mathscr{U}(\mathcal{H}_C)$ is unitary for every $g \in G$ and that $G \ni g \mapsto \pi_g^{\mu,C}$ is a homomorphism.

Hence it remains to show continuity of the representation. We will do this by first showing that $\pi^{\mu,C}$ is 'measurable' in an appropriate sense and then showing that a measurable homomorphism must be continuous.

Lemma 5.35 (Unitary group). *For a separable Hilbert space $\mathcal{H}$ the group of unitary operators $\mathscr{U}(\mathcal{H})$ equipped with the strong operator topology is a separable metrizable topological group.*

PROOF. By definition, any $v \in \mathcal{H}$ and $\varepsilon > 0$ defines a neighbourhood

$$\mathscr{N}_{v,\varepsilon}(U_0) = \{U \in \mathscr{U}(\mathcal{H}) \mid \|Uv - U_0 v\| < \varepsilon\}$$

of $U_0 \in \mathscr{U}(\mathcal{H})$ in the strong operator topology. Moreover, these neighbourhoods form a sub-basis of the neighbourhoods of U_0.

Now fix $U_0 \in \mathscr{U}(\mathcal{H})$, $v \in \mathcal{H}$, $\varepsilon > 0$, and define $w = U_0^{-1}v$. For U in $\mathscr{N}_{w,\varepsilon}(U_0)$ we then have

$$\|U^{-1}v - U_0^{-1}v\| = \|U^{-1}U_0 w - w\| = \|U_0 w - Uw\| < \varepsilon,$$

which implies that $U^{-1} \in \mathscr{N}_{v,\varepsilon}(U_0^{-1})$. Therefore $\mathscr{U}(\mathcal{H}) \ni U \mapsto U^{-1} \in \mathscr{U}(\mathcal{H})$ is continuous with respect to the strong operator topology.

Similarly, fix $U_0, U_0' \in \mathscr{U}(\mathcal{H})$, $v \in \mathcal{H}$, $\varepsilon > 0$, and define $w = U_0'v$. For U in $\mathscr{N}_{w,\varepsilon/2}(U_0)$ and $U' \in \mathscr{N}_{v,\varepsilon/2}(U_0')$ we then have

$$\begin{aligned}\|UU'v - U_0U_0'v\| &\leqslant \|UU'v - UU_0'v\| + \|UU_0'v - U_0U_0'v\| \\ &= \|U'v - U_0'v\| + \|Uw - U_0w\| < \varepsilon.\end{aligned}$$

It follows that $\mathscr{U}(\mathcal{H}) \times \mathscr{U}(\mathcal{H}) \ni (U, U') \mapsto UU' \in \mathscr{U}(\mathcal{H})$ is continuous. To summarize, we have shown that $\mathscr{U}(\mathcal{H})$ is a topological group when equipped with the strong operator topology.

If $\mathcal{H}$ is finite-dimensional then the group $\mathscr{U}(\mathcal{H})$ of unitary operators is isomorphic to the Lie group $\mathrm{U}(\dim \mathcal{H})$ and the strong operator topology corresponds to the standard topology on $\mathrm{U}(\dim \mathcal{H})$, which implies the remaining claims in the lemma. So we assume from now on that $\mathcal{H}$ is infinite-dimensional and let $e_1, e_2, \ldots$ be an orthonormal basis. For $U_1, U_2 \in \mathscr{U}(\mathcal{H})$ we define

$$\mathsf{d}(U_1, U_2) = \sum_{n=1}^{\infty} \frac{1}{2^n} \|U_1 e_n - U_2 e_n\|.$$

It is a basic exercise in topology to check that $\mathsf{d}\colon \mathscr{U}(\mathcal{H}) \times \mathscr{U}(\mathcal{H}) \to [0, \infty)$ is a metric on $\mathscr{U}(\mathcal{H})$. We claim that the strong operator topology on $\mathscr{U}(\mathcal{H})$ agrees with the topology induced by d.

To prove the claim we fix some $U_0 \in \mathscr{U}(\mathcal{H})$, $v \in \mathcal{H}$, and $\varepsilon > 0$. As $(e_n)_{n\geqslant 1}$ is an orthonormal basis, there exist some $N \in \mathbb{N}$ and coefficients $c_1, \ldots, c_N \in \mathbb{C}$ with

$$\left\| v - \sum_{n=1}^{N} c_n e_n \right\| < \frac{\varepsilon}{4}.$$

Define $M = 1 + \sum_{n=1}^{N} |c_n|$ and suppose $U \in B_\delta(U_0)$ for $\delta = 2^{-N} M^{-1} \frac{\varepsilon}{2}$. This implies that

$$\|Ue_n - U_0 e_n\| \leqslant M^{-1}\frac{\varepsilon}{2}$$

for $n = 1, \ldots, N$. Multiplying by $|c_n|$ and summing over n also gives

$$\left\| U\left(\sum_{n=1}^{N} c_n e_n\right) - U_0\left(\sum_{n=1}^{N} c_n e_n\right) \right\| \leqslant \frac{\varepsilon}{2}$$

and so $\|Uv - U_0v\| < \varepsilon$. In other words we have $B_\delta(U_0) \subseteq \mathscr{N}_{v,\varepsilon}(U_0)$. This implies that the topology induced by d is stronger than the strong operator topology.

For the converse we again let $U_0 \in \mathscr{U}(\mathcal{H})$ and fix $\varepsilon > 0$. Let $N \in \mathbb{N}$ be chosen so that $\sum_{n=N+1}^{\infty} \frac{2}{2^n} = 2^{-N+1} < \frac{\varepsilon}{2}$. For any

$$U \in \mathscr{N}_{e_1,\varepsilon/2}(U_0) \cap \cdots \cap \mathscr{N}_{e_N,\varepsilon/2}(U_0)$$

this gives

$$\mathsf{d}(U, U_0) < \sum_{n=1}^{N} \frac{1}{2^n} \underbrace{\|Ue_n - U_0e_n\|}_{<\varepsilon/2} + \frac{\varepsilon}{2} < \varepsilon.$$

Hence $\mathscr{N}_{e_1,\varepsilon/2}(U_0) \cap \cdots \cap \mathscr{N}_{e_N,\varepsilon/2}(U_0) \subseteq B_\varepsilon(U_0)$. It follows that d induces the strong operator topology.

To see that $\mathscr{U}(\mathcal{H})$ is separable, we first note that the unit sphere

$$\mathbb{S}(\mathcal{H}) = \{v \in \mathcal{H} \mid \|v\| = 1\}$$

is separable and let $D = \{v_1, v_2, \ldots\} \in \mathbb{S}(\mathcal{H})$ be a dense set. For every $N \in \mathbb{N}$ we may therefore find for $Ue_1, \ldots, Ue_N \in \mathbb{S}(\mathcal{H})$ vectors $v_{k_1}, \ldots, v_{k_N} \in D$ with

$$\|Ue_n - v_{k_n}\| < \frac{1}{N}. \tag{5.11}$$

We choose for every $N \in \mathbb{N}$ and every choice $v_{k_1}, \ldots, v_{k_N} \in D$ that appear in this manner some $U = U(N, v_{k_1}, \ldots, v_{k_N})$ satisfying (5.11). This defines a countable subset $\mathscr{D}$ of $\mathscr{U}(\mathcal{H})$. We claim that $\mathscr{D}$ is dense in $\mathscr{U}(\mathcal{H})$. To see this fix $U_0 \in \mathscr{U}(\mathcal{H})$ and $\varepsilon > 0$. Let $N \in \mathbb{N}$ be chosen with $\frac{2}{N} < \frac{\varepsilon}{2}$. Applying U_0 to $e_1, \ldots, e_N$ we find as above vectors $v_{k_1}, \ldots, v_{k_N} \in D$ satisfying (5.11) for U_0. This shows for $U(N, v_{k_1}, \ldots, v_{k_N})$ that

$$\|U_0e_n - U(N, v_{k_1}, \ldots, v_{k_N})e_n\| < \frac{2}{N}$$

for $n = 1, \ldots, N$. Multiplying by 2^{-n} and summing over n gives

$$\mathsf{d}\left(U_0, U(N, v_{k_1}, \ldots, v_{k_N})\right) \leqslant \frac{2}{N} + 2 \cdot 2^{-N} < \varepsilon.$$

As $\varepsilon > 0$ and $U_0 \in \mathscr{U}(\mathcal{H})$ were arbitrary, the set $\mathscr{D}$ is dense in $\mathscr{U}(\mathcal{H})$ and hence $\mathscr{U}(\mathcal{H})$ is separable. □

Lemma 5.36 (Measurable maps). *Let $\mathcal{H}$ be a Hilbert space and let F be a map from X to $\mathscr{U}(\mathcal{H})$. Then the following are equivalent:*

(i) *F is measurable (that is, the pre-image under F of any open set in the strong operator topology is measurable).*
(ii) *The map $X \ni x \mapsto \langle F(x)v, w\rangle$ is measurable for any $v, w \in \mathcal{H}$.*

(iii) *The map* $X \ni x \mapsto \langle F(x)e_k, e_\ell \rangle$ *is measurable for* e_k, e_ℓ *belonging to a fixed orthonormal basis of* $\mathcal{H}$.

PROOF. For simplicity of notation we assume that $\mathcal{H}$ is infinite-dimensional throughout the proof. Assume (i) and note that $\mathscr{U}(\mathcal{H}) \ni U \mapsto Uv \in \mathcal{H}$ is continuous by definition of the strong operator topology for any $v \in \mathcal{H}$. Taking the inner product with $w \in \mathcal{H}$ realizes the map $x \mapsto \langle F(x)v, w \rangle$ as the composition of a measurable and a continuous map. It follows that (i) implies (ii), which implies (iii) trivially.

Assume now that (iii) holds, and let d be the metric defined in the proof of Lemma 5.35. For any $U_0 \in \mathcal{H}$ we then have

$$\mathsf{d}(F(x), U_0) = \sum_{n=1}^{\infty} \frac{1}{2^n} \|F(x)e_n - U_0 e_n\|.$$

Moreover, for $k \in \mathbb{N}$ we may express $U_0 e_k = \sum_{\ell=1}^{\infty} a_\ell e_\ell$ in terms of the basis. This gives

$$\|F(x)e_k - U_0 e_k\|^2 = 2 - 2\Re \langle F(x)e_k, U_0 e_k \rangle = 2 - 2\Re \sum_{\ell=1}^{\infty} a_\ell \langle F(x)e_k, e_\ell \rangle.$$

Using our assumption in (iii) we see that the map

$$X \ni x \longmapsto \|F(x)e_k - U_0 e_k\|$$

is measurable, which also implies that $X \ni x \mapsto \mathsf{d}(F(x), U_0)$ is measurable for any $U_0 \in \mathscr{U}(\mathcal{H})$. It follows that the pre-image of any open ball in $\mathscr{U}(\mathcal{H})$ is measurable in X. As $\mathscr{U}(\mathcal{H})$ is also separable, any open set in $\mathscr{U}(\mathcal{H})$ can be written as a countable union of open balls, which implies that F is measurable as in (i). □

Lemma 5.37 (Measurability). *The assumptions in Proposition* 5.33 *imply that the map*

$$\pi^{\mu,C} \colon G \ni g \longmapsto \pi_g^{\mu,C} \in \mathscr{U}(L^2_\mu(X, \mathcal{H}))$$

is measurable.

PROOF. By assumption the cocycle is defined on a measurable set $D \subseteq G \times X$ with $\mu(\{x \in X \mid (g,x) \notin D\}) = 0$ for $g \in G$, has values in $\mathscr{U}(\mathcal{H})$, and depends measurably on $(g,x) \in D$. We may extend the cocycle C to all of $G \times X$ by setting $C(g,x) = I$ for all $(g,x) \in G \times X \smallsetminus D$, which does not affect the definition of $\pi_g^{\mu,C}$ for $g \in G$. Let $e_1, e_2, \ldots$ be an orthonormal basis of $\mathcal{H}$, so that

$$G \times X \ni (g,x) \longmapsto \langle C(g,x)e_k, e_\ell \rangle$$

is measurable for all $k, \ell \in \mathbb{N}$ by assumption and Lemma 5.36(iii).

To prove that $\pi_g^{\mu,C}$ is measurable, we fix $f_1, f_2 \in L^2_\mu(X)$ and $k, \ell \in \mathbb{N}$. By definition of $\pi^{\mu,C}$ we have

$$\langle \pi_g^{\mu,C} f_1 e_k, f_2 e_\ell \rangle = \int_X J(g, g^{-1}\cdot x)^{\frac{1}{2}} f_1(g^{-1}\cdot x) f_2(x) \langle C(g, g^{-1}\cdot x) e_k, e_\ell \rangle_{\mathcal{H}} \mathrm{d}\mu(x). \tag{5.12}$$

Note that $J(g, g^{-1}\cdot x)$ and the inner products in $\mathcal{H}$ on the right-hand side depend measurably on $(g, x) \in G \times X$. It follows that the integrand in (5.12) also depends measurably on $(g, x) \in G \times X$. Moreover, since

$$x \longmapsto J(g, g^{-1}\cdot x)^{\frac{1}{2}} f_1(g^{-1}\cdot x), f_2 \in L^2_\mu(X)$$

we see that the integral in (5.12) is finite for every $g \in G$. It follows from Fubini's theorem that $G \ni g \mapsto \langle \pi_g^{\mu,C} f_1 e_k, f_2 e_\ell \rangle$ is measurable.

For the case of general $f_1, f_2 \in L^2_\mu(X, \mathcal{H})$ we note that

$$f_j(x) = \sum_{k=1}^{\infty} \langle f_j(x), e_k \rangle e_k$$

for μ-almost every $x \in X$ and $j = 1, 2$. Using sesqui-linearity we can reduce this case to the above and deduce that $G \ni g \mapsto \langle \pi_g^{\mu,C} f_1, f_2 \rangle$ is measurable. Together with Lemma 5.36(ii) this now implies the lemma. □

To conclude, we need the following miraculous upgrade.(13)

Lemma 5.38 (Continuity upgrade). *Let G be a locally compact σ-compact metric group, let $\mathscr{U}$ be a separable metric group, and let $F\colon G \to \mathscr{U}$ be a homomorphism. If F is measurable then it is continuous.*

PROOF. Let $V \subseteq \mathscr{U}$ be a neighbourhood of the identity in $\mathscr{U}$. Let V_0 be an open neighbourhood such that $V_0 V_0^{-1} \subseteq V$. Let $D \subseteq \mathscr{U}$ be a countable dense subset so that $\mathscr{U} = \bigcup_{u \in D} V_0 u$ is a countable open cover of $\mathscr{U}$. As F is measurable, the set $F^{-1}(V_0 u) \subseteq G$ is measurable for any $u \in D$. Since D is countable, there must exist some $u \in D$ for which $m_G(F^{-1}(V_0 u)) > 0$. We also write $B = F^{-1}(V_0 u)$.

This in turn implies that $N = \{g \in G \mid m_G(B \cap (g^{-1}B)) > 0\}$ is a non-empty neighbourhood of the identity $e \in G$. Indeed we may replace B by a measurable subset B' with $m_G(B') \in (0, \infty)$ if necessary. Then continuity of the left regular representation implies that $G \ni g \mapsto m_G(B' \cap g^{-1}B')$ is continuous, which implies that N is a neighbourhood of $e \in G$.

Let $g \in N$ and $h_1, h_2 \in B$ so that $h_1 = g^{-1}h_2$ or, equivalently, $g = h_2 h_1^{-1}$. Applying F, using the fact that F is a homomorphism, and the definitions of B and of V_0 we now obtain

$$F(g) = F(h_2)F(h_1)^{-1} \in (V_0 u)(V_0 u)^{-1} = V_0 V_0^{-1} \subseteq V.$$

In other words we have found a neighbourhood N of $e \in G$ with $F(N) \subseteq V$. As V was an arbitrary neighbourhood of the identity in $\mathscr{U}$ this proves the continuity of F at $e \in G$. Using the assumption that F is a homomorphism between topological groups then gives the lemma. □

We note that this concludes the proof of Proposition 5.33 as follows. The operator $\pi_g^{\mu,C}$ is unitary for $g \in G$ and $G \ni g \mapsto \pi_g^{\mu,C}$ is a homomorphism (see page 8), $G \ni g \mapsto \pi_g^{\mu,C} \in \mathscr{U}(L^2_\mu(X,\mathcal{H}))$ is measurable (with respect to the strong operator topology, by Lemma 5.37), which implies continuity by Lemma 5.38.

5.5 The Unitarily Normalized Induced Representation

Given a closed subgroup H of the group G and a unitary representation τ of H we wish to construct a related 'induced unitary representation' $\operatorname{Ind}_H^G \tau$ of G. We have already seen several examples—indeed we used special cases of this construction already in Section 1.4.4 and, more recently, in Section 5.2. The general definition is somewhat technical at first, and for this reason we try to motivate it further by discussing two special cases.

5.5.1 Induction for a Discrete Ambient Group

We suppose first that G is a discrete group and $H < G$ a subgroup. We may also choose a set $S \subseteq G$ of coset representatives so that $|gH \cap S| = 1$ for all $g \in G$. Suppose now that τ is a representation of H on a complex vector space V_H. We define the new vector space

$$V_G = \{f \colon G \to V_H \mid f(gh) = \tau_h^{-1} f(g) \text{ for } g \in G \text{ and } h \in H\}.$$

Note that the defining property for elements $f \in V_G$ means that $f(g)$ determines $f|_{gH}$. This allows an alternative description of V_G as in the following exercise.

Essential Exercise 5.39 (Induced representation). (a) Show that

$$\pi_g(f)(g_0) = f(g^{-1}g_0)$$

for $g, g_0 \in G$ and $f \in V_G$ defines a representation of G on V_G.
(b) Show that $V_G \ni f \mapsto f|_S \in V_H^S$ is a linear isomorphism.
(c) Show that V_G contains an H-invariant subspace on which the restriction to H is isomorphic to τ.

Assume now in addition that $V_H = \mathcal{H}_\tau$ is a Hilbert space and that τ is in fact a unitary representation of H. In this case we would like to modify the definition above to ensure that the new representation $\pi = \operatorname{Ind}_H^G \tau$ again defines a unitary representation on a Hilbert space. The key is to introduce a 'transverse square summability condition' to obtain

$$\mathcal{H}_\pi = \left\{ f\colon G \to \mathcal{H}_\tau \,\middle|\, f(gh) = \tau_h^{-1} f(g) \text{ for } g \in G, h \in H \text{ and } \sum_{gH \in G/H} \|f(g)\|^2 < \infty \right\}.$$

For this, note that $f \in \mathcal{H}_\pi$ implies $\|f(g)\|^2 = \|f(gh)\|^2$ for any $g \in G$ and $h \in H$, which allows us to think of $\|f(g)\|^2$ as a function of $gH \in G/H$.

Essential Exercise 5.40. Show that $\mathcal{H}_\pi \ni f \mapsto f|_S \in \bigoplus_{s \in S} \mathcal{H}_\tau$ is a linear isomorphism and deduce that the norm $\|\cdot\|_\pi\colon \mathcal{H}_\pi \to [0, \infty)$ defined by

$$\|f\|_\pi^2 = \sum_{gH \in G/H} \|f(g)\|^2 = \sum_{s \in S} \|f(s)\|^2$$

for $f \in \mathcal{H}_\pi$ makes $\mathcal{H}_\pi$ into a Hilbert space.
(b) Show that $G \ni g \mapsto \pi_g$ as in Exercise 5.39 restricted to $\mathcal{H}_\pi$ defines a unitary representation of G.
(c) Show that $\mathcal{H}_\pi$ contains a closed H-invariant subspace on which the restriction to H is isomorphic to $\mathcal{H}_\tau$.

5.5.2 Unitary Induction from a Discrete Subgroup

A second class of examples that is simpler than the general case but somewhat closer to it is obtained by assuming that H is a discrete subgroup in a non-discrete unimodular group G. In this case there exists a measurable fundamental domain $S \subseteq G$ satisfying $|gH \cap S| = 1$ for all $g \in G$ (which will follow, for example, from our subsequent more general discussions in Section 5.5.4).

Once again we assume that τ is a unitary representation of H and define a new representation $\pi = \operatorname{Ind}_H^G \tau$ of G by considering functions with the following properties:

(i) $f\colon G \to \mathcal{H}_\tau$ is measurable, with
(ii) $f(gh) = \tau_h^{-1} f(g)$ for $g \in G$ and $h \in H$, and
(iii) $\|f\|_\pi < \infty$, where

$$\|f\|_\pi^2 = \int_S \|f(g)\|_{\mathcal{H}_\tau}^2 \,\mathrm{d}m_G(g).$$

The Hilbert space $\mathcal{H}_\pi$ is then defined as the vector space of functions satisfying (i)–(iii) modulo the subspace of those functions f with $\|f\|_\pi = 0$. The unitary representation π is again defined by $\pi_g(f)(g_0) = f(g^{-1}g_0)$ for $g, g_0 \in G$ and $f \in \mathcal{H}_\pi$. The reader may work through the following exercise before we consider the general case.

Essential Exercise 5.41. (a) Show that $\|\cdot\|_\pi$ does not depend on the choice of the fundamental domain S.
(b) Show that $\mathcal{H}_\pi$ is a Hilbert space.
(c) Show that π is a unitary representation.

5.5.3 The General Definition

We return to the general case and suppose now that G is again a locally compact, σ-compact, metric group and $H < G$ is a closed subgroup.

Note that we wish to define a Hilbert space consisting of functions on G satisfying some square integrability condition on G/H. Hence we will have a need for a good choice of a measure on G/H. We note that there are many natural and important examples where G/H does not carry a σ-finite G-invariant measure.

Lemma 5.42 (G-invariant measure class). *Let $H < G$ be a closed subgroup of the group G. There exists a quasi-invariant finite measure ν on G/H. Any other quasi-invariant σ-finite measure on G/H defines the same measure class as ν.*

PROOF. For the existence, let $\delta\colon G \to (0,\infty)$ be integrable and define ν on G/H to be the push-forward of the measure defined by $\delta \,\mathrm{d}m_G$. For a measurable set $B \subseteq G/H$ we then have

$$\nu(B) = \int_G \mathbb{1}_B(g_0H)\delta(g_0)\,\mathrm{d}m_G(g_0)$$

by definition. If $\nu(B) > 0$ then $\mathbb{1}_B(g_0H) > 0$ for g_0 belonging to a set of positive measure for m_G. This implies for any fixed $g \in G$ that $\mathbb{1}_B(gg_0H) > 0$ on a set of positive measure for m_G, which together with $\delta > 0$ implies $\nu(gB) > 0$. Hence ν is a quasi-invariant finite measure on G/H.

Let us now assume that ν is any σ-finite quasi-invariant measure on G/H. We claim that null sets for ν in G/H can be characterized using the Haar measure m_G on G. Let $B \subseteq G/H$ be a measurable set. By quasi-invariance we have that $\nu(B) > 0$ if and only if $\nu(gB) > 0$ for all $g \in G$. To prove the claim we will integrate the latter with respect to m_G, which will allow us to obtain a characterization of $\nu(B) > 0$ using m_G only.

For this we define the measurable set

$$D = \{(g, g_0H) \in G \times G/H \mid g^{-1}g_0H \in B\}.$$

Fubini's theorem now shows that $\nu(B) > 0$ holds if and only if the integral

$$\begin{aligned}\int_G \nu(gB)\,\mathrm{d}m_G(g) &= \int_G \underbrace{\int_{G/H} \mathbb{1}_D(g, g_0H)\,\mathrm{d}\nu(g_0H)}_{=\nu(gB)}\,\mathrm{d}m_G(g)\\ &= \int_{G/H} m_G\bigl(\{g \mid g^{-1}g_0H \in B\}\bigr)\,\mathrm{d}\nu(g_0H)\end{aligned}$$

is positive. Moreover,

$$\begin{aligned} m_G(\{g \mid g^{-1}g_0 H \in B\}) &= m_G(g_0^{-1}\{g \mid g^{-1}g_0 H \in B\}) \\ &= m_G(\{k \mid k^{-1}H \in B\}) \end{aligned}$$

for any $g_0 \in G$ by left-invariance of m_G and by setting $k = g_0^{-1}g$. It follows that $\nu(B) > 0$ is equivalent to

$$m_G(\{k \in G \mid k^{-1}H \in B\}) > 0.$$

Hence we can characterize ν-null sets in G/H by the Haar measure m_G on G. This implies that the measure class of quasi-invariant σ-finite measures on G/H is uniquely determined. □

Definition 5.43 (Cross-section). We say that $S \subseteq G$ is a *cross-section* for $H < G$ if

(i) S is measurable,
(ii) $|S \cap gH| = 1$ for all $g \in G$, and
(iii) $S \ni s \mapsto sH \in G/H$ is a Borel isomorphism.

Lemma 5.44 (Cross-sections exist[(14)]). *Let $H < G$ be a closed subgroup of the group G. Then there exists a cross-section $S \subseteq G$ for H.*

We postpone the proof of this lemma to Section 5.5.4 as it is a distraction from our main objective, but note that it is quite straightforward in many concrete cases. Indeed, if G is a Lie group one can use a linear complement V to the Lie algebra of H inside the Lie algebra of G to define a 'local cross-section' which can be used countably often to patch together a global cross-section as in Definition 5.43.

We will build the general definition of the induced representation on our discussion in Section 5.4. For this we still need to find a cocycle for an action associated to G and H.

Clearly G acts on G/H. By the defining properties of a cross-section we also have $G/H \cong S$, which leads to a measurable action of G on S. To see this, note that an element $s \in S$ corresponds under the isomorphism to sH, which is mapped under the action of $g \in G$ to gsH. As S is a cross-section we have

$$(gsH) \cap S = \{g{\centerdot}s\}$$

for an element $g{\centerdot}s \in S$, which defines the image $g{\centerdot}s \in S$ of $s \in S$ under the action of $g \in G$. As G is acting on G/H and we have simply transported this action to S using the Borel isomorphism $G/H \cong S$, it follows that

$$G \times S \ni (g, s) \longmapsto g{\centerdot}s \in S$$

indeed defines an action. Moreover,

$$G \times S \ni (g, s) \longmapsto gsH \in G/H \longmapsto g{\centerdot}s \in S$$

is, being a composition of a continuous and a measurable map, also measurable.

Lemma 5.45 (Exact cocycles). *Let $H < G$ be a closed subgroup of the group G and let $S \subseteq G$ be a cross-section. For $g \in G$ and $s \in S$ we define $h_S(g,s) = (g\bullet s)^{-1}gs$ so that*

$$gs = \underbrace{(g\bullet s)}_{\in S}\underbrace{h_S(g,s)}_{\in H}.$$

Then $h_S\colon G\times S \to H$ defines an exact cocycle for the G-action on S with values in H. If τ is a unitary representation of H, then

$$\begin{aligned} C_{S,\tau}\colon G\times S &\longrightarrow \mathscr{U}(\mathcal{H}_\tau) \\ (g,s) &\longmapsto \tau(h_S(g,s)) \end{aligned}$$

defines an exact cocycle for the G-action on S with values in $\mathscr{U}(\mathcal{H}_\tau)$.

Proof. The discussion before the lemma defines the measurable action of G on S. We define $h_S(g,s) = (g\bullet s)^{-1}gs$ for $(g,s) \in G\times S$, which implies that $h_S\colon G\times S\to H$ is measurable.

Now fix $s\in S$ and $g_1,g_2\in G$. Then by using the definition of $h_S(g_1,s)$ and $h_S(g_2,g_1\bullet s)$ we see that

$$g_2g_1s = g_2\underbrace{(g_1\bullet s)}_{\in S}\underbrace{h_S(g_1,s)}_{\in H} = \underbrace{(g_2g_1\bullet s)}_{\in S}\underbrace{h_S(g_2,g_1\bullet s)h_S(g_1,s)}_{\in H}.$$

However, this already implies the cocycle equation

$$h_S(g_2g_1,s) = h_S(g_2,g_1\bullet s)h_S(g_1,s)$$

for $g_1,g_2\in G$, $s\in S$ and the function $h_S\colon G\times S\to H$.

Given a unitary representation τ of H we also obtain the cocycle $C_{S,\tau}$ as in the lemma. As τ is continuous and h is measurable, $C_{S,\tau}$ is also measurable. □

With this lemma to hand, we are finally able to give the definition.

Definition 5.46 (Unitarily normalized induced representations). Assume that H is a closed subgroup of the group G, and let $S\subseteq G$ be a cross-section for H. Let ν be a quasi-invariant σ-finite measure on $S\cong G/H$. Let τ be a unitary representation, and let $C = C_{S,\tau} = \tau\circ h_S\colon G\times S\to H$ be the exact cocycle from Lemma 5.45. The *unitarily normalized unitary representation* $\pi = \operatorname{Ind}_H^G\tau$ is defined as the unitary representation $\pi^{\nu,C}$ of Proposition 5.33 on $\mathcal{H}_\pi = L^2_\nu(S,\mathcal{H}_\tau)$ when applied to the action of G on S and the cocycle C.

We note that the definition depends on the cross-section S and the quasi-invariant measure ν. However, we now show that different choices will give rise to unitarily equivalent representations of G. This justifies the notation $\operatorname{Ind}_H^G\tau$ for the induced representation.

Corollary 5.47. *Let H be a closed subgroup of the group G and let τ be a unitary representation of H. Let $S \subseteq G$ be a cross-section for H and let ν, ν' be two quasi-invariant σ-finite measures on S. Then the unitarily normalized induced representations defined by S and ν or S and ν' are unitarily isomorphic. Moreover, $\pi = \operatorname{Ind}_H^G \tau$ can also be defined without a choice of a cross-section similarly to the discussion in Sections* 5.5.1 *and* 5.5.2.

PROOF. We first suppose that ν, ν' are two quasi-invariant σ-finite measures on S. By Lemma 5.42 ν and ν' define the same measure class on $S \cong G/H$. By Lemma 5.31 we have $J_\nu(g, s) = F(g{\bullet}s) J_{\nu'}(g, s) F(s)^{-1}$ for $g \in G$, $s \in S$, and $F = \frac{d\nu'}{d\nu}$. We define $U(f)(s) = F(s)^{\frac{1}{2}} f(s)$ for $f \in L^2_{\nu'}(S, \mathcal{H}_\tau)$ and $s \in S$ so that

$$\|U(f)\|^2_{L^2_\nu(S,\mathcal{H}_\tau)} = \int_S \|f(s)\|^2_{\mathcal{H}_\tau} \underbrace{F(s)\,d\nu}_{d\nu'(s)} = \|f\|^2_{L^2_{\nu'}(S,\mathcal{H}_\tau)}.$$

This implies that $U\colon L^2_{\nu'}(S, \mathcal{H}_\tau) \to L^2_\nu(S, \mathcal{H}_\tau)$ is a unitary isomorphism. For the induced unitary representations $\pi^{\nu,C}$ and $\pi^{\nu',C}$ as in Definition 5.46 this gives

$$\begin{aligned}
\big(U\pi_g^{\nu',C} U^{-1} f\big)(s) &= F(s)^{\frac{1}{2}} \big(\pi_g^{\nu',C} U^{-1} f\big)(s) \\
&= F(s)^{\frac{1}{2}} J_{\nu'}(g, g^{-1}{\bullet}s)^{\frac{1}{2}} \tau(h_S(g, g^{-1}{\bullet}s))(U^{-1}f)(g^{-1}{\bullet}s) \\
&= \big(F(s) J_{\nu'}(g, g^{-1}{\bullet}s) F(g^{-1}{\bullet}s)^{-1}\big)^{\frac{1}{2}} \tau(h_S(g, g^{-1}{\bullet}s)) f(g^{-1}{\bullet}s) \\
&= J_\nu(g, g^{-1}{\bullet}s)^{\frac{1}{2}} \tau(h_S(g, g^{-1}{\bullet}s)) f(g^{-1}{\bullet}s) \\
&= (\pi_g^{\nu,C} f)(s)
\end{aligned}$$

for $g \in G$, $f \in L^2_\nu(S, \mathcal{H}_\tau)$, and $s \in S$. This proves the first part of the corollary.

For the second part of the corollary we let ν be a quasi-invariant σ-finite measure on G/H. We define the vector space V_G to consist of all measurable functions $F\colon G \to \mathcal{H}_\tau$ with

(a) $F(gh) = \tau_h^{-1} F(g)$ for all $g \in G$ and $h \in H$, and
(b) $\|F\|^2_{V_G} = \displaystyle\int_{G/H} \|F(g)\|^2_{\mathcal{H}_\tau} \, d\nu(gH) < \infty$ (where the integral is well-defined for $gH \in G/H$ by (a)).

This defines a Hilbert space naturally isomorphic to $\mathcal{H}_\pi$ for the induced representation $\pi = \operatorname{Ind}_H^G \tau$ defined by a cross-section $S \subseteq G$ and the measure ν considered on $S \cong G/H$. Indeed, a function $F \in V_G$ may simply be restricted to S to obtain $f = F|_S \in \mathcal{H}_\pi$ with $\|f\|_{L^2(S,\mathcal{H}_\tau)} = \|F\|_{V_G}$ by the definition in (b). Similarly, given some $f \in L^2(S, \mathcal{H}_\tau)$ we may define $F(sh) = \tau_h^{-1} f(s)$ for $(s, h) \in S \times H$ which satisfies

$$F(sh_0h) = \tau_{h_0h}^{-1} f(s) = \tau_h^{-1} \tau_{h_0}^{-1} f(s) = \tau_h^{-1} F(sh_0)$$

for all $sh_0 \in SH = G$ and $h \in H$ as required in (a), and $\|F\|_{V_G} = \|f\|_{L^2(S,\mathcal{H}_\tau)}$.

The representation on V_G is a 'normalized regular representation' defined for $g_0 \in G$ by sending $F \in V_G$ to the function

$$G \ni g \mapsto J_\nu(g_0, g_0^{-1}gH)^{\frac{1}{2}} F(g_0^{-1}g).$$

We evaluate this at $s \in S$ and obtain $J_\nu(g_0, g_0^{-1}sH)^{\frac{1}{2}} F(g_0^{-1}s)$, where we may use the properties of F to calculate

$$\begin{aligned} F(g_0^{-1}s) &= F(g_0^{-1}\bullet s h_S(g_0^{-1}, s)) \\ &= \tau(h_S(g_0^{-1}, s)^{-1}) F(g_0^{-1}\bullet s) \\ &= \tau(h_S(g_0, g_0^{-1}\bullet s)) F(g_0^{-1}\bullet s). \end{aligned}$$

In the last line we used the cocycle property

$$h_S(g_0, g_0^{-1}\bullet s) h_S(g_0^{-1}, s) = h_S(e, s) = e$$

to express the inverse of $h_S(g_0^{-1}, s)$. Hence the image of F under g_0 when restricted to S has the form

$$S \ni s \longmapsto J_\nu(g_0, g_0^{-1}sH)^{\frac{1}{2}} \tau(h_S(g_0, g_0^{-1}\bullet s)) f(g_0^{-1}\bullet s)$$

for $f = F|_S$. This matches our definition of $\pi = \mathrm{Ind}_H^G \tau$ using S and ν. Hence our representation on V_G is isomorphic to $\pi = \mathrm{Ind}_H^G \tau$ (and is unitary). □

Exercise 5.48. Let $\tau = \mathbb{1}_H$ be the trivial representation of the trivial subgroup $H = \{e\}$ of the group G. Show that the unitarily normalized induced representation $\mathrm{Ind}_{\{e\}}^G \mathbb{1}_{\{e\}}$ is isomorphic to the left regular representation of G on $L^2(G)$.

Exercise 5.49. Verify that the unitary representations discussed in Sections 5.2.1–5.2.4 are special cases of the unitarily normalized induced representations.

Exercise 5.50. Find a group G, a subgroup $H < G$, and a unitary representation τ of H such that $\mathcal{H}_\pi$ for $\pi = \mathrm{Ind}_H^G \tau$ does not contain a closed H-invariant subspace on which the restriction is isomorphic to τ.

5.5.4 The Cross-Section

To complete our general definition of the induced representation, we return to the construction of the cross-section. We first show how to find a 'global' cross-section assuming a 'local' cross-section is given. This serves as a warm-up but also applies trivially if H is a discrete subgroup, if H is a closed subgroup of a Lie group, and will be useful in the general case.

Exercise 5.51. Let G be a σ-compact Lie group and $H < G$ a closed subgroup. Use the exponential map and a linear complement to the Lie algebra of H inside the Lie algebra of G to find a 'local' cross-section satisfying the assumptions of the following lemma.

Lemma 5.52 (Upgrading from a local cross-section). *Let H be a closed subgroup of the group G and let $p\colon G \to G/H$ be the canonical quotient map sending $g \in G$ to $gH \in G/H$. Suppose there exists a compact neighbourhood K of the identity in G and a measurable subset $S_{\mathrm{local}} \subseteq K$ so that $|gH \cap S_{\mathrm{local}}| = 1$ for all $g \in K$. Suppose also that the selection map*

$$G/H \supseteq p(K) \longrightarrow S_{\mathrm{local}}$$

sending $kH \in p(K)$ to the unique $s \in S_{\mathrm{local}}$ with $kH = sH$ is measurable. Then there exists a cross-section $S \subseteq G$ for H.

PROOF. As G is σ-compact and K has non-empty interior we can find a countable cover $G = \bigcup_{n=1}^{\infty} g_n K$. Note that

$$p^{-1}\bigl(p(g_n K)\bigr) = g_n KH = \{g_n kh \mid k \in K \text{ and } h \in H\}$$

are closed subsets for $n \in \mathbb{N}$. Using this we define the measurable sets

$$\begin{aligned} S_1 &= g_1 S_{\mathrm{local}} \\ S_2 &= g_2 S_{\mathrm{local}} \smallsetminus g_1 KH \\ &\vdots \\ S_n &= g_n S_{\mathrm{local}} \smallsetminus \bigl(g_1 KH \cup \cdots \cup g_{n-1} KH\bigr) \end{aligned}$$

for $n \geqslant 2$ and the measurable set

$$S = \bigcup_{n=1}^{\infty} S_n.$$

We will show that S is a cross-section for H. Given $g \in G$ there exists a minimal $n \in \mathbb{N}$ with $gH \in p(g_n K)$. Note that $g_n^{-1} g \in KH$ and that by applying our assumption on S_{local} we find some $s \in S_{\mathrm{local}}$ with $g_n^{-1} gH = sH$. It follows that $g_n s \in S_n \subseteq S$ as n was chosen minimally with $gH \in p(g_n K)$. Moreover, $gH = g_n sH$.

Suppose now that $gH = s_1 H = s_2 H$ for some $s_1, s_2 \in S$. Without loss of generality we may assume $s_1 \in S_{n_1}$ and $s_2 \in S_{n_2}$ with $n_1 \leqslant n_2$. If $n_1 < n_2$ then $gH = s_1 H \in g_{n_1} KH$ is contained in the set that is removed from S_{n_2} in its definition, which contradicts the assumption that $s_2 \in S_{n_2}$. Hence we have $n_1 = n_2 = n$, $g_n^{-1} s_1 H = g_n^{-1} s_2 H$, and $s_1, s_2 \in S_{\mathrm{local}}$ implies $s_1 = s_2$ by the assumed properties of S_{local}.

Now let $B \subseteq S$ be a measurable subset. Then $B_n = g_n^{-1}(B \cap S_n)$ is a measurable subset of S_{local} for all $n \in \mathbb{N}$. By the assumed properties of S_{local} this implies that the image $p(B_n)$ in G/H is measurable for all $n \in \mathbb{N}$. Multiplying again with g_n and taking the union, we see that

$$p(B) = \bigcup_{n=1}^{\infty} p(g_n B_n) = \bigcup_{n=1}^{\infty} g_n p(B_n)$$

is measurable. Note that continuity of the map $S \ni s \mapsto sH \in G/H$ implies the converse statement that the pre-image of a measurable set in G/H is measurable in S. □

Depending on the interests of the reader, Exercise 5.51 and Lemma 5.52 might be sufficient, and so the reader may skip the following proof. However, for completeness we also construct a cross-section without any additional assumptions.

PROOF OF LEMMA 5.44. It is sufficient to find a local cross-section satisfying the assumptions of Lemma 5.52. To do this let $K = K_1 \supseteq K_2 \supseteq \cdots$ be a nested sequence of compact sets that form a basis of the neighbourhoods of the identity in G. Starting with $S_1 = K_1$ we will iteratively construct a sequence of measurable sets $S_1 \supseteq S_2 \supseteq \cdots$ satisfying $p(S_n) = p(K_1)$ for all $n \in \mathbb{N}$. In fact the sets S_n will be defined as the union of the sets in a finite list

$$B_{n,1}, \ldots, B_{n,M_n} \subseteq K_1. \tag{5.13}$$

The construction will consist of two sub-steps applied to the sets in (5.13). In the first sub-step we will apply compactness to split $B_{n,m}$ into a union of smaller sets contained in translates of K_{n+1}. In the second sub-step we use the order in the list to trim the sets and remove unnecessary duplications in the images of the sets (much as we did in the proof of Lemma 5.52). Doing this all carefully will result in the set $S = \bigcap_{n=1}^{\infty} S_n$ being a local cross-section as in Lemma 5.52.

For $n = 1$ we set $M_1 = 1$ and $B_{1,1} = K_1$, which trivially satisfies all the claims in the inductive hypothesis below.

Suppose now that $n \geqslant 1$ and that we have already constructed a list of sets as in (5.13) for some $M_n \in \mathbb{N}$ satisfying the following properties:

(i) (Smallness) $B_{n,m}$ is a subset of a left translate of K_n for $m = 1, \ldots, M_n$.
(ii) (Disjoint images) The images $p(B_{n,1}), \ldots, p(B_{n,M_n})$ in G/H are pairwise disjoint.
(iii) (No point left behind) The union $S_n = B_{n,1} \cup \cdots \cup B_{n,M_n}$ satisfies

$$p(S_n) = p(K_1).$$

(iv) (Closure) For $m = 1, \ldots, M_n$ we have

$$p\big(\overline{B_{n,m}} \smallsetminus B_{n,m}\big) \subseteq p\big(B_{n,1} \cup \cdots \cup B_{n,m-1}\big).$$

(v) (Compact unions) For $m = 1, \ldots, M_n$ the image

$$p\big(B_{n,1} \cup \cdots \cup B_{n,m}\big) \subseteq G/H$$

is compact.
(vi) (Measurability) The sets $B_{n,m} \subseteq G$ and their images $p(B_{n,m}) \subseteq G/H$ are measurable for $m = 1, \ldots, M_n$. In particular, $S_n \subseteq G$ is measurable.

As mentioned before, we first apply to the inductively given list in (5.13) a splitting sub-step and then a trimming sub-step. As K_1 is compact and K_{n+1} is a neighbourhood of $e \in G$ there exist a finite cover of the form

$$K_1 \subseteq \bigcup_{j=1}^{J} g_j K_{n+1}$$

for some $J = J_n$ and $g_1, \ldots, g_J \in K_1$. Now fix some $k \in \{1, \ldots, M_n\}$. Using the cover we *split* $B_{n,k}$ into the sets $C_j = B_{n,k} \cap g_j K_{n+1}$ for $j = 1, \ldots, J$ satisfying

$$B_{n,k} = \bigcup_{j=1}^{J} C_j,$$

where we have suppressed the dependence of C_j on n and k to simplify the notation. Next we *trim* the resulting sets by defining

$$D_j = C_j \smallsetminus p^{-1}\big(p(D_1 \cup \cdots \cup D_{j-1})\big)$$

for $j = 1, \ldots, J$.

Recall that we apply these two steps to each of the sets in the list (5.13). We define the new list

$$B_{n+1,1}, \ldots, B_{n+1,M_{n+1}}$$

with $M_{n+1} = JM_n$ by concatenating the lists of sets obtained from splitting and trimming each of the sets $B_{n,k}$ in their natural order. Thus, for example, the first block of J sets $B_{n+1,1}, \ldots, B_{n+1,J}$ are precisely the sets $D_1, \ldots, D_J$ obtained from splitting and trimming the first set $B_{n,1}$ in the inductively given old list (5.13), and similarly for each of the following blocks of J sets.†

Notice that by construction each of the sets $B_{n+1,m}$ (is equal to a set of the form $D_j \subseteq C_j$ from one of the blocks and hence) is a subset of a left translate of K_{n+1}. This proves the smallness property claimed in (i).

To verify (ii) for the new list, pick two indices m_1 and m_2 with

$$m_1 < m_2 \leqslant M_{n+1}.$$

If $m_1 < m_2$ belong to different blocks in the construction, then we must have $B_{n+1,m_1} \subseteq B_{n,k_1}$ and $B_{n+1,m_2} \subseteq B_{n,k_2}$ for some $k_1 < k_2 \leqslant M_n$. In this case (ii) for the old list in (5.13) implies that $p(B_{n,k_1}) \cap p(B_{n,k_2}) = \emptyset$ and so $p(B_{n+1,m_1}) \cap p(B_{n+1,m_2}) = \emptyset$ as well. If $m_1 < m_2$ belong to the same block then the corresponding sets B_{n+1,m_1} and B_{n+1,m_2} were constructed as D_{j_1} and D_{j_2} for $1 \leqslant j_1 < j_2 \leqslant J$ in the splitting and trimming construction starting with $B = B_{n,k}$ for some $k \leqslant M_n$. However, in this process the trimming definition of D_{j_2} ensures that $p(B_{n+1,m_1}) = p(D_{j_1})$ and $p(B_{n+1,m_2}) = p(D_{j_2})$ are disjoint. This proves the disjointness property in (ii) for the new list.

† To keep the indexing simple, we ignore the fact that many of the sets constructed will be empty.

Now fix some $k \in \{1, \dots, M_n\}$ and $g \in B_{n,k}$. As $B_{n,k} = \bigcup_{j=1}^{J} C_j$ for the sets in our splitting step there exists a minimal j for which $p(g) \in p(C_j)$. Pick some $g' \in C_j$ with $p(g') = p(g)$. However, minimality of j implies that

$$p(g) = p(g') \notin p(D_1 \cup \cdots \cup D_{j-1})$$

and so $g' \in D_j$. Together this shows that

$$p(B_{n,k}) = \bigcup_{j=1}^{J} p(D_j). \tag{5.14}$$

We combine this for $k = 1, \dots, M_n$ with the surjectivity claim in (iii) for the old list in (5.13). Together we obtain for $S_{n+1} = \bigcup_{m=1}^{M_{n+1}} B_{n+1,m}$ that

$$p(S_{n+1}) = \bigcup_{m=1}^{M_{n+1}} p(B_{n+1,m}) = \bigcup_{k=1}^{M_n} p(B_{n,k}) = p(K_1)$$

as claimed in (iii).

To prove the closure property in (iv), fix $m \in \{1, \dots, M_{n+1}\}$. Suppose also that m corresponds to j in the kth block so that $B_{n+1,m} = D_j$ if splitting and trimming is applied to $B_{n,k}$ as above. Let $g \in \overline{B_{n+1,m}} \smallsetminus B_{n+1,m} = \overline{D_j} \smallsetminus D_j$. Notice that

$$\overline{B_{n+1,m}} \subseteq \overline{B_{n,k}}.$$

If $g \notin B_{n,k}$, then (iv) for the old list in (5.13) gives

$$p(g) \in p(B_{n,1} \cup \cdots \cup B_{n,k-1}).$$

Together with (5.14) applied to the sets $B_{n,1}, \dots, B_{n,k-1}$ this implies that $p(g)$ belongs to the union of the images of the new sets in the previous blocks.

So we now assume that $g \in B_{n,k}$. By the splitting construction the subsets $C_1, \dots, C_j \subseteq B_{n,k}$ are relatively closed. As $g \in \overline{D_j}$, $D_j \subseteq C_j$, and $g \in B_{n,k}$, this implies that $g \in C_j$. As $g \notin D_j$ the definition in the trimming sub-step shows that $p(g) \in p(D_1 \cup \cdots \cup D_{j-1})$.

In any case we have shown that

$$g \in \overline{B_{n+1,m}} \smallsetminus B_{n+1,m} \implies p(g) \in p(B_{n+1,1} \cup \cdots \cup B_{n+1,m-1}).$$

This proves the closure property claimed in (iv) for the new list.

We now use (iv) inductively to prove the compact union property in (v). For $m = 1$ the statement (iv) means that $B_{n+1,1} \subseteq K_1$ is closed and so $p(B_{n+1,1})$ is compact. Suppose now that (v) already holds for $n+1$ and $m < M_{n+1}$. Then by continuity of p we have

$$p\big(\overline{B_{n+1,1} \cup \cdots \cup B_{n+1,m}}\big) = p\big(B_{n+1,1} \cup \cdots \cup B_{n+1,m}\big) \tag{5.15}$$

as the latter is compact. It follows that

$$\begin{aligned}
p\big(\overline{B_{n+1,1}\cup\cdots\cup B_{n+1,m+1}}\big) &= p\big(\overline{B_{n+1,1}\cup\cdots\cup B_{n+1,m}}\cup\overline{B_{n+1,m+1}}\big)\\
&= p\big(B_{n+1,1}\cup\cdots\cup B_{n+1,m}\cup\overline{B_{n+1,m+1}}\big)\\
&= p\big(B_{n+1,1}\cup\cdots\cup B_{n+1,m+1}\cup\overline{B_{n+1,m+1}}\smallsetminus B_{n+1,m+1}\big)\\
&= p\big(B_{n+1,1}\cup\cdots\cup B_{n+1,m+1}\big),
\end{aligned}$$

where we used the fact that the closure of a finite union is the finite union of the closures, the equation in (5.15), simply decomposed $\overline{B_{n+1,m+1}}$ into two parts, and finally used (iv) for $B_{n+1,m+1}$. It follows that the compact union property in (v) holds for the new list for all $m\in\{1,\dots,M_{n+1}\}$.

Using (v) and the disjointness property in (ii) it now also follows that the images $p(B_{n+1,m})\subseteq G/H$ are measurable for $m=1,\dots,M_{n+1}$. Fix again some $k\in\{1,\dots,M_n\}$. By the measurability property in (vi) for the old list in (5.13) the set $B_{n,k}$ is measurable, which also implies that the sets $C_1,\dots,C_J$ obtained from splitting $B_{n,k}$ are measurable. We now also know that the images $p(D_1),\dots,p(D_J)$ are measurable (as they are part of the new list). However, this implies that their pre-images, and hence also the sets $D_1,\dots,D_J$ obtained in the trimming sub-step, are measurable. As this holds for all blocks we obtain the measurability claimed in (vi) for the new list. This concludes the induction.

Now define the measurable set

$$S=\bigcap_{n=1}^{\infty}S_n\subseteq S_1=K_1.$$

We claim that S is a local cross-section.

Suppose $g_1,g_2\in S$ satisfy $p(g_1)=p(g_2)$. As $g_1,g_2\in S_n$ for $n\in\mathbb{N}$ it follows by the disjointness in (ii) that $g_1,g_2\in B_{n,m}$ for the same $m\in\{1,\dots,M_n\}$. By the smallness in (i) this shows that g_1,g_2 belong to the same left translate of K_n. As this holds for all n and the sets $K_1\supseteq K_2\supseteq\cdots$ form a basis of the neighbourhoods of $e\in G$, we must have $g_1=g_2$.

Now suppose $g\in K_1$. Applying the surjectivity claim in (iii) for every $n\in\mathbb{N}$ gives a sequence (g_n) with $g_n\in S_n$ and $p(g_n)=p(g)$ for all $n\in\mathbb{N}$. Let g_∞ be a limit of a convergent subsequence of (g_n) (and so, in particular, $p(g_\infty)=p(g)$ also). We claim that g_∞ must lie in S. To see the claim fix some $n_0\in\mathbb{N}$ and let $m_0\in\{1,\dots,M_{n_0}\}$ be chosen with $g_{n_0}\in B_{n_0,m_0}$. Then for $n\geqslant n_0$ we have $g_n\in S_n\subseteq S_{n_0}$ and

$$p(g_n)=p(g)=p(g_{n_0})\in p(B_{n_0,m_0}). \tag{5.16}$$

By the disjointness in (ii) this shows that $g_n\in B_{n_0,m_0}$ also. It follows that g_∞ lies in $\overline{B_{n_0,m_0}}$ which, together with (5.16), the closure property in (iv), and the disjointness in (ii) for n_0 implies that $g_\infty\in B_{n_0,m_0}$. In other words we have shown that $g_\infty\in S_{n_0}$. As $n_0\in\mathbb{N}$ was arbitrary, we deduce the claim.

To summarize: $S \subseteq K_1$ is measurable and satisfies $|S \cap gH| = 1$ for all g in K_1. It remains to show that the selection map sending $p(k) \in p(K_1)$ to the unique $s \in S$ with $p(k) = p(s)$ is measurable. For this it is sufficient to show that the image $p(O)$ of a relatively open set $O \subseteq S$ is measurable in G/H.

For $n \in \mathbb{N}$ and $m \in \{1, \dots, M_n\}$ we claim that

$$p(B_{n,m} \cap S) = p(B_{n,m}),$$

which implies measurability of this image by (vi). To prove the claim, let g lie in $B_{n,m} \subseteq K_1$. By the argument above there exists some $s \in S$ with $p(s) = p(g)$. Since $S \subseteq S_n$, the disjointness property in (ii) for n implies that $s \in B_{n,m}$ also, which already gives $p(g) \in p(B_{n,m} \cap S)$.

Suppose $O \subseteq S$ is relatively open and let $g \in O$. Using the smallness in (i) and the fact that the compact sets $K_1 \supseteq K_2 \supseteq \cdots$ form a basis of the neighbourhoods of $e \in G$, we can find some $n \in \mathbb{N}$ and $m \in \{1, \dots, M_n\}$ so that $g \in B_{n,m} \cap S \subseteq O$. It follows that O is a countable union of sets of the form $B_{n,m} \cap S$ for various n, m, which shows that $p(O)$ is measurable. □

5.6 Two Tools for Classification Theorems

We now present the main technical results of the chapter which can be helpful in describing the unitary dual of groups with a non-trivial closed normal abelian subgroup. This will, in particular, apply to the affine group in one dimension in Section 5.2.3, the Heisenberg group in Section 5.2.4, and many other groups.

Proposition 5.53 (Construction of irreducible representations). *Assume that $A \lhd G$ is a closed normal abelian subgroup of the group G, let $t_0 \in \widehat{A}$, and let $H = \mathrm{Stab}_G(t_0)$ with respect to the natural action of G on $\widehat{A}$. We suppose that τ is an irreducible unitary representation of H such that $\tau|_A$ is multiplication by χ_{t_0}. Then $\pi = \mathrm{Ind}_H^G \tau$ is an irreducible unitary representation of G. Moreover, $\pi|_A$ is isomorphic to a multiplication representation for a quasi-invariant measure on $G \centerdot t_0$ with multiplicity $\dim \mathcal{H}_\tau$. Finally, for $\tau, \tau' \in \widehat{H}$ we have that $\mathrm{Ind}_H^G \tau$ is isomorphic to $\mathrm{Ind}_H^G \tau'$ if and only if τ is isomorphic to τ'.*

We note that in the case where $t_0 \in \widehat{A}$ (for example, if $t_0 = 0$) is fixed by all of G the proposition is tautologically true: In such a situation $H = G$ and $\mathrm{Ind}_G^G \tau$ is isomorphic to τ.

The complicated—and, in a sense, negative—examples in Sections 5.2.5–5.2.7 show that a 'completeness' result, giving a converse to Proposition 5.53, must require further assumptions.

Definition 5.54. Let the group G act on the space X. We say that the action has *countable separation of orbits* if there exists a countable collection $\mathcal{C} \subseteq \mathcal{B}_X$ so that

- any $B \in \mathcal{C}$ is G-invariant, meaning that $g \cdot B = B$ for all $g \in G$, and
- for any x_1, x_2 with $G \cdot x_1 \neq G \cdot x_2$ there exists some $B \in \mathcal{C}$ with $x_1 \in B$ but $x_2 \notin B$.

With this we can formulate a partial converse to Proposition 5.53.

Theorem 5.55 (Classification tool for irreducible representations). *Assume that $A \lhd G$ is a closed normal abelian subgroup and suppose that the G-action on $\widehat{A}$ has countable separation of orbits. Let π be an irreducible unitary representation of G. Then there exists some $t_0 \in \widehat{A}$ so that π is isomorphic to $\mathrm{Ind}_H^G \tau$ for some irreducible unitary representation τ of $H = \mathrm{Stab}_G(t_0)$ with $\tau|_A = \chi_{t_0} I$.*

We first show how these two results are useful in concrete situations.

5.6.1 The Euclidean Isometry Groups

For a semi-direct product $G = K \ltimes A$ of a compact group K and an abelian group A the technical assumption of countable separation is always satisfied. Indeed, as K is compact every G-orbit (or, equivalently, K-orbit) is in fact compact and we may equip the space $G \backslash \widehat{A}$ of orbits with the Hausdorff distance

$$\mathsf{d}_{G\backslash\widehat{A}}(G \cdot t_1, G \cdot t_2) = \min_{g_1, g_2 \in K} \mathsf{d}_{\widehat{A}}(g_1 \cdot t_1, g_2 \cdot t_2).$$

The map $\widehat{A} \to G\backslash\widehat{A}$ is continuous and $G\backslash\widehat{A}$ is a locally compact σ-compact metric space. This implies that its Borel σ-algebra is countably generated and hence the G-action is countably separated. If the unitary dual of K and its subgroups (arising as stabilizers of points in $\widehat{A}$) is known, then Proposition 5.53 and Theorem 5.55 lead to a complete description of $\widehat{G}$.

This applies in particular to the isometry groups $G = \mathrm{SO}_d(\mathbb{R}) \ltimes \mathbb{R}^d$ for any $d \in \mathbb{N}$. We briefly discuss a few special cases and refer to Fulton and Harris [32] for a discussion of $\widehat{\mathrm{SO}_d(\mathbb{R})}$ for $d \geqslant 5$.

Example 5.56. The case $d = 2$ was discussed in Section 5.2.2 but now also follows trivially from our general setup.

Example 5.57. Let $d = 3$. The old representations of $G/A \cong \mathrm{SO}_3(\mathbb{R})$ are described in Section 4.2.2. Note that the orbits of G on $\widehat{A} \cong \mathbb{R}^3$ are the spheres $r\mathbb{S}^2$ with $r \in [0, \infty)$. For $r > 0$ we may take the initial point $t_0 = re_1$ and define its stabilizer subgroup

$$H = \mathrm{Stab}_G(t_0) \cong \mathbb{T} \ltimes \mathbb{R}^3$$

for $\mathbb{T} \cong \mathrm{SO}_2(\mathbb{R})$. For an integer $n \in \mathbb{Z}$ we define the character $\tau_{r,n} \in \widehat{H}$ obtained as the product of the nth character on $\mathbb{T}$ and the character defined by $t_0 = re_1$ on $\mathbb{R}^3$. Using Proposition 5.53 we now obtain the new representation $\pi^{(r,n)} \in \widehat{G}$

for $(r, n) \in (0, \infty) \times \mathbb{Z}$. Moreover Theorem 5.55 shows that we obtain in this way all irreducible unitary representations of G.

Exercise 5.58. Use the low-dimension accidental isomorphism

$$\mathrm{SO}_4(\mathbb{R}) \cong \mathrm{SU}_2(\mathbb{R}) \times \mathrm{SU}_2(\mathbb{R}) / \{(I, I), (-I, -I)\}$$

and the description of $\widehat{\mathrm{SU}_2(\mathbb{R})}$ to find a complete description of $\widehat{\mathrm{SO}_4(\mathbb{R}) \ltimes \mathbb{R}^4}$. Show in particular that for any $n \in \mathbb{N}$ there exist infinitely many $\pi \in \widehat{\mathrm{SO}_4(\mathbb{R}) \ltimes \mathbb{R}^4}$ so that the spectral multiplicity of $\pi|_{\mathbb{R}^4}$ is n.

5.6.2 The Affine Group in One Dimension

We now use the completeness result for the affine group $G = \mathbb{R}_{>0} \ltimes \mathbb{R}$ in one dimension from Section 5.2.3.

Proof of Proposition 5.22. It remains to show that any irreducible unitary representation of G is one of the representations found in Section 5.2.3. If $\pi|_A$ is trivial, then π must be a character on $G/A \cong \mathbb{R}_{>0}$. So suppose that $\pi|_A$ is non-trivial and apply Theorem 5.55. For this we note that the action of G on $\widehat{A} \cong \mathbb{R}$ has precisely three orbits, namely $(-\infty, 0)$, $\{0\}$, and $(0, \infty)$. Hence the G-action trivially has countable separation of orbits. It follows that π is isomorphic to $\mathrm{Ind}_H^G \tau$ for some irreducible unitary representation τ of $H = \mathrm{Stab}_G(t_0)$ and $t_0 \in \widehat{A}$. As $\pi|_A$ is non-trivial, we see from Proposition 5.53 that $t_0 \neq 0$ which implies in our case that $H = A$ and $\tau = \chi_{t_0}$. Depending on the sign of t_0 we obtain that π is isomorphic to π^+ or π^- (see also Lemma 5.42, Corollary 5.47, and Exercise 5.49). $\square$

Exercise 5.59 (Other fields). Let $\mathbb{K}$ be a local field (that is, $\mathbb{R}$, $\mathbb{C}$, a finite extension of $\mathbb{Q}_p$ for some prime p, or a finite extension of $\mathbb{F}_p((t))$ for some prime p). Extend the results of Section 5.2.3 and this section to describe the relationship between the unitary dual of $\mathbb{K}^\times$ and of the affine group

$$G = \left\{ \begin{pmatrix} s & a \\ 0 & 1 \end{pmatrix} \;\middle|\; s \in \mathbb{K}^\times, a \in \mathbb{K} \right\}.$$

Exercise 5.60 (The group Sol). Define the group $\mathrm{Sol} = \mathbb{R}_{>0} \ltimes \mathbb{R}^2$ as the matrix group

$$\mathrm{Sol} = \left\{ \begin{pmatrix} s & 0 & a_1 \\ 0 & s^{-1} & a_2 \\ 0 & 0 & 1 \end{pmatrix} \;\middle|\; s > 0, a = \begin{pmatrix} a_1 \\ a_2 \end{pmatrix} \in \mathbb{R}^2 \right\}.$$

Describe $\widehat{\mathrm{Sol}}$, and prove that your description is complete.

5.6.3 The Heisenberg Group

Next we complete our discussion from Section 5.2.4 concerning the $(2d+1)$-dimensional Heisenberg group.

PROOF OF THEOREM 5.23. Let

$$\begin{aligned} A &= \{(0, y, z) \mid y \in \mathbb{R}^d, z \in \mathbb{R}\}, \\ S &= \{(x, 0, 0) \mid x \in \mathbb{R}^d\} \end{aligned}$$

so that $G = S \ltimes A = \{(x, y, z) \mid x, y \in \mathbb{R}^d, z \in \mathbb{R}\}$ as in Section 5.2.4 In that section we already discussed the old and constructed the new representations and know that they are irreducible (relying on Lemmas 5.8 and 5.10; see also Exercise 5.49). Moreover, it is clear that different $\xi \in \mathbb{R}^\times$ give rise to inequivalent irreducible representations as ξ is uniquely determined as being the character obtained by restriction to the centre $C(G)$.

Moreover, the S-orbits on $\widehat{A} \cong \mathbb{R}^{d+1}$ have the form

$$S \bullet \begin{pmatrix} t \\ \xi \end{pmatrix} = \left\{ \begin{pmatrix} t - x\xi \\ \xi \end{pmatrix} \mid x \in \mathbb{R}^d \right\}$$

for $\begin{pmatrix} t \\ \xi \end{pmatrix} \in \mathbb{R}^{d+1}$. For $\xi \neq 0$ this gives $\mathbb{R}^d \times \{\xi\}$ and for $\xi = 0$ we instead obtain the singleton $\{(t, 0)\}$. To see that this action has countable separation of orbits consider the sets $\mathbb{R}^d \times (a, b)$ for rationals $a < b$ and $B_{\frac{1}{n}}(t) \times \{0\}$ for $n \in \mathbb{N}$ and $t \in \mathbb{Q}^d$. Hence we may apply Theorem 5.55 to $\pi \in \widehat{G}$ and deduce that $\pi \cong \mathrm{Ind}_H^G(\tau)$, where $\tau|_A$ is equal to multiplication by the character defined by $\begin{pmatrix} t \\ \xi \end{pmatrix} \in \mathbb{R}^{d+1}$. If $\xi = 0$ or, equivalently, if π restricted to the centre is trivial, then π is an old representation corresponding to a character on $G/C(G) \cong \mathbb{R}^{2d}$. So suppose $\xi \neq 0$ in which case the S-orbit is $\mathbb{R}^d \times \{\xi\}$ and $H = A$. Using the Lebesgue measure on $\mathbb{R}^d \times \{\xi\}$ (see Lemma 5.42) it follows that $\pi \cong \mathrm{Ind}_H^G(\tau)$ is isomorphic to the new representation π^ξ constructed in Section 5.2.4.

It remains to show the last part of the theorem giving a more concrete formula for π^ξ. We fix $\xi \neq 0$ and identify $t \in \mathbb{R}^d$ with $\begin{pmatrix} t \\ \xi \end{pmatrix} \in \mathbb{R}^{d+1}$. For an element $(x, y, z) \in G$ note that

$$(x, y, z) = (0, y, z) \underbrace{(x, 0, 0)}_{=s}$$

and $s^{-1} \bullet \begin{pmatrix} t \\ \xi \end{pmatrix} = \widehat{\theta_x} \begin{pmatrix} t \\ \xi \end{pmatrix} = \begin{pmatrix} t + x\xi \\ \xi \end{pmatrix}$. Putting this into the definition of π^ξ in Lemma 5.8 we obtain for $f \in L^2(\mathbb{R}^d)$ and $t \in \mathbb{R}^d$ that

$$\left(\pi^\xi_{(x,y,z)} f\right)(t) = \left(\pi^\xi_{(0,y,z)} \pi^\xi_{(x,0,0)} f\right)(t) = \mathrm{e}^{2\pi \mathrm{i}(\langle y, t\rangle + z\xi)} f(t + \xi x)$$

as claimed. □

5.6.4 Product Groups

We outline in this section a recipe for describing the dual of products of groups in which one of them has a normal abelian subgroup. Suppose therefore that G has a normal abelian subgroup A so that the G-action on $\widehat{A}$ has countable separation of orbits. Suppose also that L is a locally compact σ-compact metric group. Then $\widetilde{G} = G \times L$ still has (an isomorphic copy of) A as a normal abelian subgroup. Moreover the $\widetilde{G}$-action on $\widehat{A}$ also has countable separation of orbits (as $\widetilde{G}$-orbits are just G-orbits). This allows us to apply Theorem 5.55 to show that any irreducible unitary representation π of $\widetilde{G}$ has the form $\operatorname{Ind}_{\widetilde{H}}^{\widetilde{G}} \tau$. Here $\widetilde{H} = H \times L$ with $H = \operatorname{Stab}_G(t_0)$ for some $t_0 \in \widehat{A}$ and τ is an irreducible unitary representation of $\widetilde{H}$ so that $\tau|_A$ is multiplication by χ_{t_0}. If now $H < G$ is a proper subgroup for which we can already describe $\widehat{H \times L}$ this leads to a description of π.

Exercise 5.61. Use the scheme outlined above to prove that $\widehat{G \times L} \cong \widehat{G} \times \widehat{L}$ in the sense of Proposition 3.15 where G is the isometry group of the plane, the affine group in one dimension, or the Heisenberg group.

5.6.5 Constructing Irreducible Unitary Representations

Having seen the power of our two main results from this chapter, we now turn to their proofs.

Lemma 5.62 (A Fubini characterisation of null sets). *Let $H < G$ be a closed subgroup and ν a quasi-invariant measure on G/H. Then a measurable subset $B \subseteq G$ is a null set for m_G if and only if $m_H(H \cap g_0^{-1}B) = 0$ for ν-almost every $g_0H \in G/H$.*

Proof. By Lemma 5.42 it does not matter which quasi-invariant measure ν on G/H we consider here. So we again choose some integrable $\rho > 0$ and define ν as the push-forward of $\rho \, \mathrm{d}m_G$. Note that this implies

$$\int_{G/H} f \, \mathrm{d}\nu = \int_G f(g_0H)\rho(g_0) \, \mathrm{d}m_G \tag{5.17}$$

for any measurable $f \geqslant 0$ on G/H, first for simple functions and then more generally.

We now let $B \subseteq G$ be measurable and define

$$f\colon G/H \ni g_0H \longmapsto m_H(H \cap g_0^{-1}B) = \int_H \mathbb{1}_B(g_0h) \, \mathrm{d}m_H(h).$$

The left-invariance of m_H implies that f is well-defined. Moreover, considering G measurably as $S \times H$ for a cross-section $S \cong G/H$ as in Lemma 5.44 the measurability of f follows from Fubini's theorem. We now apply (5.17) for this function and obtain (using Fubini's theorem again) that

$$\begin{aligned}\int_{G/H} m_H(H\cap g_0^{-1}B)\,\mathrm{d}\nu(g_0H) &= \int_G\int_H \mathbb{1}_B(g_0h)\,\mathrm{d}m_H(h)\rho(g_0)\,\mathrm{d}m_G(g_0)\\ &= \int_H\int_G \mathbb{1}_{Bh^{-1}}(g_0)\rho(g_0)\,\mathrm{d}m_G(g_0)\,\mathrm{d}m_H(h).\end{aligned}$$

Recall from Section 1.2.3 that right translation preserves the measure class of m_G. It follows that B is a null set for m_G if and only if Bh^{-1} is a null set for m_G for all $h\in H$. Hence B is a null set if and only if the integral above vanishes, which gives the lemma. □

Proof of Proposition 5.53. Let $A\lhd H<G$, $t_0\in\widehat{A}$, and $\tau\in\widehat{H}$ be as in the proposition. By Definition 5.46 (relying, *inter alia*, on Proposition 5.33) we know that $\pi=\operatorname{Ind}_H^G\tau$ is a unitary representation of G. Recall that $\mathcal{H}_\pi=L^2_\nu(S,\mathcal{H}_\tau)$ for a quasi-invariant measure ν on a cross-section S for H. Moreover,

$$S\cong G/H\cong G{\centerdot}t_0$$

by the properties of a cross-section and the definition of $H=\operatorname{Stab}_G(t_0)$. We note that also the second isomorphism is bi-measurable: As G/H is equipped with the quotient topology continuity of the G-action (from Lemma 5.4) implies that the map $G/H\ni gH\mapsto g{\centerdot}t_0\in G{\centerdot}t_0\subseteq\widehat{A}$ is continuous. This also implies that the restriction to a compact subset of G/H is a homeomorphism, which together with σ-compactness implies measurability of the inverse map.

This allows us to transport measurable functions from one space to the other, and to think of ν also as a quasi-invariant measure on $G{\centerdot}t_0$. In particular, we have $\mathcal{H}_\pi\cong L^2_\nu(G{\centerdot}t_0,\mathcal{H}_\tau)$.

We now describe $\pi|_A$ by unfolding our definition of $\pi_a=\pi_a^{\nu,C}$ for $a\in A$. For $s\in S$ we have $as=s(s^{-1}as)$ with $s\in S$ and $s^{-1}as\in A<H$, which gives

$$h_S(a,s)=s^{-1}as$$

for the exact cocycle with values in H from Lemma 5.45. Moreover, this also shows that the action of $a\in A$ on $s\in S$ is trivial and so the derivative cocycle satisfies $J(a,s)=1$ for almost every $s\in S$. Combining these with the definition in Proposition 5.33, we have

$$\begin{aligned}(\pi_af)(s) &= 1\cdot\tau\bigl(h_S(a,\underbrace{a^{-1}{\centerdot}s}_{=s})\bigr)f(\underbrace{a^{-1}{\centerdot}s}_{=s})\\ &= \tau(s^{-1}as)f(s)\\ &= \chi_{t_0}(s^{-1}as)f(s)\\ &= \chi_{s{\centerdot}t_0}f(s)\end{aligned}$$

for $f\in L^2_\nu(S,\mathcal{H}_\tau)$ and $s\in S$. Using the isomorphisms $S\ni s\mapsto s{\centerdot}t_0\in G{\centerdot}t_0$ and

$$\mathcal{H}_\pi = L^2_\nu(S, \mathcal{H}_\tau) \cong L^2_\nu(G\boldsymbol{\cdot}t_0, \mathcal{H}_\tau)$$

we can also write this in the form

$$(\pi_a f)(t) = \chi_t(a) f(t)$$

for $a \in A$, $f \in L^2_\nu(G\boldsymbol{\cdot}t_0, \mathcal{H}_\tau)$, and $t \in G\boldsymbol{\cdot}t_0 \subseteq \widehat{A}$. In other words, $\pi|_A$ simply takes the form of the multiplication representation appearing in the spectral theorems of Chapter 2.

It remains to prove the irreducibility of π and the isomorphism claim. We will prove both by almost the same argument. To this end we let τ, τ' lie in $\widehat{H}$, define $\pi = \operatorname{Ind}_H^G \tau$ and $\pi' = \operatorname{Ind}_H^G \tau'$ (using the same measure ν and cross-section $S \subseteq G$ for H). We suppose that the operator

$$B \colon L^2_\nu(S, \mathcal{H}_\tau) \longrightarrow L^2_\nu(S, \mathcal{H}_{\tau'})$$

is intertwining for π and π'. As B is also intertwining for $\pi|_A$ we wish to apply Proposition 2.72. If $\mathcal{H}_\tau = \mathcal{H}_{\tau'}$ this is literally allowed. If not we consider instead the map $(f, f') \mapsto (0, Bf)$ on $L^2_\nu(G\boldsymbol{\cdot}t_0, \mathcal{H}_\tau) \times L^2_\nu(G\boldsymbol{\cdot}t_0, \mathcal{H}_{\tau'})$ together with the isomorphism

$$L^2_\nu(G\boldsymbol{\cdot}t_0, \mathcal{H}_\tau) \times L^2_\nu(G\boldsymbol{\cdot}t_0, \mathcal{H}_{\tau'}) \cong L^2_\nu(G\boldsymbol{\cdot}t_0, \mathcal{H}_\tau \times \mathcal{H}_{\tau'}),$$

which allows us to apply Proposition 2.72. This shows that B admits a pointwise description: For ν-almost every $t \in G\boldsymbol{\cdot}t_0$ there exists a bounded operator $B_t \colon \mathcal{H}_\tau \to \mathcal{H}_{\tau'}$ (depending measurably on t) so that

$$(Bf)(t) = B_t(f(t)) \tag{5.18}$$

for $f \in L^2_\nu(G\boldsymbol{\cdot}t_0, \mathcal{H}_\tau)$ and almost every $t \in G\boldsymbol{\cdot}t_0$. As B is not just intertwining for $\pi|_A$ and $\pi'|_A$ but in fact for all of π and π', the operators B_t will satisfy a compatibility condition for different $t \in G\boldsymbol{\cdot}t_0$. This condition, together with transitivity of the action on the orbit and irreducibility of τ, will lead to the desired conclusions.

For this we again work with the original definition $\mathcal{H}_\pi = L^2_\nu(S, \mathcal{H}_\tau)$ and think of $B_{(\cdot)}$ as a measurable function $B_{(\cdot)} \colon S \ni s \mapsto B_s \in \mathrm{B}(\mathcal{H}_\tau)$. Note that

$$B\pi_g = \pi'_g B$$

for all $g \in G$. Expanding the definition of $B\pi_g$ using (5.18) we have

$$\begin{aligned}\big(B\pi_g f\big)(s) &= B_s\big((\pi_g f)(s)\big)\\ &= J(g, g^{-1}\boldsymbol{\cdot}s)^{\frac{1}{2}} B_s \tau(h_S(g, g^{-1}\boldsymbol{\cdot}s)) f(g^{-1}\boldsymbol{\cdot}s)\end{aligned}$$

for $f \in L^2_\nu(S, \mathcal{H}_\tau)$ and ν-almost every $s \in S$. Expanding $\pi'_g B$ similarly, we have

$$\begin{aligned}(\pi'_g Bf)(s) &= J(g, g^{-1}\bullet s)^{\frac{1}{2}}\tau'(h_S(g, g^{-1}\bullet s))(Bf)(g^{-1}\bullet s)\\ &= J(g, g^{-1}\bullet s)^{\frac{1}{2}}\tau'(h_S(g, g^{-1}\bullet s))B_{g^{-1}\bullet s}f(g^{-1}\bullet s).\end{aligned}$$

Setting $s = g\bullet s_0$ it follows that

$$B_{g\bullet s_0}\tau(h_S(g, s_0)) = \tau'(h_S(g, s_0))B_{s_0} \tag{5.19}$$

for all $g \in G$ and ν-almost every $s_0 \in S$ (with the null set in S potentially depending on $g \in G$). We note that the required uniqueness of the pointwise operators is a consequence of Proposition 2.72(3).

By Fubini's theorem we may fix some $s_0 \in S$ (from the complement of a ν-null set) so that (5.19) will hold for s_0 and m_G-almost every $g \in G$. Recall that m_G is quasi-invariant under right multiplication and (from Lemma 5.62) that $N \subseteq G$ is a null set if and only if $m_H(s^{-1}N \cap H) = 0$ for ν-almost every $s \in S$. Combining these two statements we set $g = shs_0^{-1}$ and deduce that (5.19) holds for g for ν-almost every s and (depending on s) for m_H-almost every $h \in H$. We fix one such $s \in S$. Note that $gs_0 = sh$, which gives $g\bullet s_0 = s$ and

$$h_S(g, s_0) = h.$$

Together with (5.19) we obtain

$$B_s\tau_h = \tau'_h B_{s_0} \tag{5.20}$$

for m_H-almost every $h \in H$, say for $h \in H\smallsetminus N$ with $m_H(N) = 0$. Equivalently, we have $\tau'_{h^{-1}}B_s = B_{s_0}\tau_{h^{-1}}$ for $h \in H\smallsetminus N$. If we pick elements $h_1, h_2 \in H\smallsetminus N$ we obtain

$$\tau'_h B_{s_0} = \tau'_{h_2^{-1}h_1}B_{s_0} = \tau'_{h_2^{-1}}B_s\tau_{h_1} = B_{s_0}\tau_{h_2^{-1}h_1} = B_{s_0}\tau_h$$

for $h = h_2^{-1}h_1$. However, any $h \in H$ can be written in the form $h = h_2^{-1}h_1$ with $h_1, h_2 \in H\smallsetminus N$ (by picking $h_1 \in H\smallsetminus(N \cup Nh)$). It follows that the operator $B_{s_0} \in \mathrm{B}(\mathcal{H}_\tau)$ is intertwining for τ and τ'. Using this together with (5.20) gives $B_s = B_{s_0}$ for ν-almost every $s \in S$.

Let us now assume that $\tau = \tau'$ is an irreducible representation of H. Then $B_{s_0} = \lambda I_{\mathcal{H}_\tau}$ for some $\lambda \in \mathbb{C}$ by Schur's lemma (Theorem 1.29). Therefore $B_s = B_{s_0} = \lambda I_{\mathcal{H}_\tau}$ for ν-almost every $s \in S$ and we see that $B = \lambda I_{\mathcal{H}_\pi}$. As this holds for any intertwining operator $B \in \mathrm{B}(\mathcal{H}_\pi)$ we deduce that π is irreducible.

Assume now that $\tau, \tau' \in \widehat{H}$ are irreducible unitary representations of H and $B\colon \mathcal{H}_\pi \to \mathcal{H}_{\pi'}$ is a non-trivial intertwining operator for π and π'. Then the map $B_{s_0}\colon \mathcal{H}_\tau \to \mathcal{H}_{\tau'}$ is a non-trivial intertwining operator for τ and τ' and so τ and τ' are also isomorphic. The converse follows easily as well. □

5.6.6 Identifying Representations as Induced

For the proof of Theorem 5.55 we start with an unknown irreducible unitary representations π and wish to show that it is obtained from an induced representation. The first step for this relies on our assumptions concerning the action of G on $\widehat{A}$.

Lemma 5.63 (Ergodicity and countable separation). *Let the group G act on the space X and suppose that the G-action has countable separation of orbits. Then any ergodic measure μ is concentrated on a single orbit, meaning that $\mu(X \smallsetminus G{\bullet}x_0) = 0$ for some $x_0 \in X$.*

PROOF. Let μ be a non-zero ergodic measure on X and let $B \in \mathcal{B}_X$ be a G-invariant set. By ergodicity we have $\mu(B) = 0$ or $\mu(X \smallsetminus B) = 0$. We apply this for all sets in $\mathcal{C}$ as in the definition of countable separation of orbits in Definition 5.54. By taking the union of these sets we obtain a single G-invariant null set $N \subseteq X$ so that for any $B \in \mathcal{C}$ we have $B \subseteq N$ or $X \smallsetminus B \subseteq N$. Equivalently, we have $X \smallsetminus N \subseteq X \smallsetminus B$ or $X \smallsetminus N \subseteq B$ for every $B \in \mathcal{C}$. That is, all points in $X \smallsetminus N$ belong to the same sets in $\mathcal{C}$. As N is a null set we can pick some $x_0 \in X \smallsetminus N$. The definition of countable separation now implies that we must have $X \smallsetminus N = G{\bullet}x_0$ and $\mu(X \smallsetminus G{\bullet}x_0) = \mu(N) = 0$ follows. □

In the following we will work under the assumptions of Theorem 5.55: $A \lhd G$ is a closed normal abelian subgroup so that the G-action on $\widehat{A}$ has countable separation of orbits, and $\pi \in \widehat{G}$ is an irreducible unitary representation.

Let $\nu = \mu_{\max}$ be the maximal spectral measure on $\widehat{A}$ for $\pi|_A$. By Lemma 5.5 the measure ν is quasi-invariant under the action of G. By Proposition 5.7 irreducibility of π implies ergodicity of ν. By our assumption on the G-action on $\widehat{A}$ and Lemma 5.63, it follows that ν is concentrated on a single orbit $G{\bullet}t_0$ for some $t_0 \in \widehat{A}$. By Lemma 5.42 this identifies the measure class of ν. Moreover, Proposition 5.7 also shows that $\pi|_A$ can be described as the multiplication representation on $L^2_\nu(G{\bullet}t_0, \mathcal{H}) \cong \mathcal{H}_\pi$, where $\mathcal{H} = \mathcal{V}_n$ for some $n \in \mathbb{N} \cup \{\infty\}$. By using this isomorphism we will also think of π as a unitary representation on $L^2_\nu(G{\bullet}t_0, \mathcal{H})$.

Our next step is to find a cocycle describing π.

Lemma 5.64 (Finding a cocycle). *There exists a cocycle C for the G-action on $G{\bullet}t_0$ with values in $\mathscr{U}(\mathcal{H})$ so that, on denoting the representation on $L^2_\nu(G{\bullet}t_0, \mathcal{H})$ again by π, we have*

$$\pi_g(f)(t) = J(g, g^{-1}{\bullet}t)^{\frac{1}{2}} C(g, g^{-1}{\bullet}t) f(g^{-1}{\bullet}t)$$

for $g \in G$, $f \in L^2_\nu(G{\bullet}t_0, \mathcal{H})$, and $t \in G{\bullet}t_0$.

PROOF. We start by noting that on $L^2_\nu(G{\bullet}t_0, \mathcal{H})$ we also have an action-induced unitary representation. In fact G/A acts on $\widehat{A}$ and ν is quasi-invariant by Lemma 5.5. Hence we may apply Proposition 5.33 for the trivial cocycle for G

acting on $\widehat{A}$ taking on the value $I_{\mathcal{H}}$ only, to obtain the unitary representation ρ with

$$(\rho_g f)(t) = J(g, g^{-1}{\cdot}t)^{\frac{1}{2}} f(g^{-1}{\cdot}t)$$

for $g \in G$, $f \in L^2_\nu(G{\cdot}t_0, \mathcal{H})$, and $t \in G{\cdot}t_0$.

We note that $\pi_g \pi_a \pi_{g^{-1}} = \pi_{gag^{-1}}$ and $\rho_a = I$ for $g \in G$ and $a \in A$. We claim that we also have $\rho_g \pi_a \rho_{g^{-1}} = \pi_{gag^{-1}}$. Indeed

$$\begin{aligned}(\rho_g \pi_a \rho_{g^{-1}} f)(t) &= J(g, g^{-1}{\cdot}t)^{\frac{1}{2}} (\pi_a \rho_{g^{-1}} f)(g^{-1}{\cdot}t) \\ &= J(g, g^{-1}{\cdot}t)^{\frac{1}{2}} \langle a, g^{-1}{\cdot}t\rangle (\rho_{g^{-1}} f)(g^{-1}{\cdot}t) \\ &= J(g, g^{-1}{\cdot}t)^{\frac{1}{2}} J(g^{-1}, t)^{\frac{1}{2}} \langle a, \widehat{\theta}_g(t)\rangle f(t) \\ &= \langle gag^{-1}, t\rangle f(t) = (\pi_{gag^{-1}} f)(t)\end{aligned}$$

for $f \in L^2(G{\cdot}t_0, \mathcal{H})$ and $t \in G{\cdot}t_0$. Together this implies that $\rho_{g^{-1}}\pi_g$ commutes with π_a for $a \in A$ as $\rho_{g^{-1}}\pi_g \pi_a \pi_{g^{-1}} \rho_f = \rho_{g^{-1}} \pi_{gag^{-1}} \rho_g = \pi_a$. Therefore, Proposition 2.72 concerning the centralizer of $\pi|_A$ applies and shows that $\rho_{g^{-1}}\pi_g$ has a pointwise definition. That is, for every $g \in G$ there exists a measurable function $C(g, \cdot)\colon \widehat{A} \to \mathrm{B}(\mathcal{H})$ so that

$$(\rho_{g^{-1}}\pi_g f)(t) = C(g, t) f(t)$$

for $f \in L^2_\nu(G{\cdot}t_0, \mathcal{H})$ and $t \in G{\cdot}t_0$. Combined with the definition of ρ_g this gives

$$\begin{aligned}(\pi_g f)(t) &= (\rho_g \rho_{g^{-1}} \pi_g)(t) \\ &= J(g, g^{-1}{\cdot}t)^{\frac{1}{2}} (\rho_{g^{-1}} \pi_g f)(g^{-1}t) \\ &= J(g, g^{-1}{\cdot}t)^{\frac{1}{2}} C(g, g^{-1}{\cdot}t) f(g^{-1}{\cdot}t)\end{aligned}$$

for $f \in L^2_\nu(G{\cdot}t_0, \mathcal{H})$ and $t \in G{\cdot}t_0$. We also note that by Proposition 2.72(3) we have $\operatorname{ess\,sup} \|C(g, \cdot)\|_{\mathrm{op}} = \|\rho_{g-1}\pi_g\|_{\mathrm{op}} = 1$. Applying Proposition 2.72(3) also to $(\rho_{g^{-1}}\pi_g)^{-1} = \pi_{g^{-1}}\rho_g$, we see that $C(g, \cdot)$ almost surely takes values in $\mathscr{U}(\mathcal{H})$.

Now let $g_1, g_2 \in G$ and calculate

$$\begin{aligned}(\pi_{g_1}\pi_{g_2})(t) &= J(g_1, g_1^{-1}{\cdot}t)^{\frac{1}{2}} C(g_1, g_1^{-1}{\cdot}t)(\pi_{g_2} f)(g_1^{-1}{\cdot}t) \\ &= \underbrace{J(g_1, g_1^{-1}{\cdot}t)^{\frac{1}{2}} J(g_2, g_2^{-1} g_1^{-1}{\cdot}t)^{\frac{1}{2}}}_{=J(g_1 g_2, g_2^{-1} g_1^{-1}{\cdot}t)^{\frac{1}{2}}} C(g_1, g_1^{-1}{\cdot}t) C(g_2, g_2^{-1} g_1^{-1}{\cdot}t) f(g_2^{-1} g_1^{-1}{\cdot}t)\end{aligned}$$

for $f \in L^2_\nu(G{\cdot}t_0, \mathcal{H})$ and $t \in G{\cdot}t_0$. Comparing this with the formula for $\pi_{g_1 g_2} f$ we obtain for $g_1, g_2 \in G$ the cocycle equation

$$C(g_1 g_2, g_2^{-1} g_1^{-1}{\cdot}t) = C(g_1, g_1^{-1}{\cdot}t) C(g_2, g_2^{-1} g_1^{-1}{\cdot}t)$$

for ν-almost every $t \in G{\cdot}t_0$.

It remains to see that $C(\cdot, \cdot)$ is really measurable as a function of two variables. For this let $(\mathcal{P}_n)$ be a sequence of finite partitions of $\widehat{A}$ increasing in the sense that $\mathcal{P}_n \subseteq \sigma(\mathcal{P}_{n+1})$ for all $n \geqslant 1$ and with

$$\mathcal{B}_{\widehat{A}} = \bigvee_{n=1}^{\infty} \sigma(\mathcal{P}_n).$$

Recall that ν was defined as a maximal spectral measure and so is finite. We assume without loss of generality that ν is a probability measure. Also recall from Lemma 5.36 that measurability of a $\mathscr{U}(\mathcal{H})$-valued map can be defined using inner products. Fixing $v_1, v_2 \in \mathcal{H}$ and $P \in \mathcal{P}_n$ with $\nu(P) > 0$ for some $n \in \mathbb{N}$, we define $f_j(t) = v_j \mathbb{1}_P(t)$ for $t \in G\bullet t_0$ and $j = 1, 2$. Then we have $f_j \in L^2_\nu(G\bullet t_0, \mathcal{H})$ and

$$\int_P \langle C(g,t)v_1, v_2\rangle \,\mathrm{d}\nu = \langle \rho_{g^{-1}}\pi_g f_1, f_2\rangle = \langle \pi_g f_1, \rho_g f_2\rangle$$

depends continuously on $g \in G$ by continuity of the unitary representations. Varying $P \in \mathcal{P}_n$ it follows that

$$E_\nu \left(\langle C(g,t)v_1, v_2\rangle \mid \sigma(\mathcal{P}_n)\right) \tag{5.21}$$

depends measurably on $(g,t) \in G \times G\bullet t_0$. Applying the increasing martingale convergence theorem we see that for every $g \in G$ the limit of (5.21) exists ν-almost surely and is equal to $\langle C(g,t)v_1, v_2\rangle$ ν-almost surely. After possibly modifying the definition of $C(g,t)$ for $g \in G$ and t in a ν-null set (depending on g) we obtain that $C(\cdot,\cdot)$ is measurable. □

The discussion above, including Lemma 5.64, has already achieved a good portion of the proof of Theorem 5.55. Starting with $\pi \in \widehat{G}$ we have found a quasi-invariant measure ν on a single orbit of the form $G\bullet t_0 \subseteq \widehat{A}$ and a cocycle describing π completely. What remains is to describe the cocycle in terms of a unitary representation of $H = \mathrm{Stab}_G(t_0)$ on $\mathcal{H}$. The next lemma paves the way by finding a 'better version' of the cocycle.

Lemma 5.65 (A better cocycle). *There exists a measurable set $W \subseteq G\bullet t_0$ of full ν-measure and a cocycle C_W for the G-action on $G\bullet t_0$ with the following properties.*

- *$C_W(g,t)$ is defined whenever $g \in G$ and $t, g\bullet t \in W$.*
- *The cocycle equation $C_W(g_1g_2,t) = C_W(g_1, g_2\bullet t)C_W(g_2,t)$ holds for any g_1 and g_2 in G whenever $t, g_2\bullet t, g_1g_2\bullet t \in W$.*
- *For every $g \in G$ we have $C_W(g,\cdot) = C(g,\cdot)$ ν-almost surely and so*

$$\pi_g(f)(t) = J(g, g^{-1}\bullet t)^{\frac{1}{2}} C_W(g, g^{-1}\bullet t) f(g^{-1}\bullet t)$$

for $g \in G$, $f \in L^2_\nu(G\bullet t_0)$, and $t \in G\bullet t_0$.

Proof. We continue using the notation introduced in Lemma 5.64. In particular, let C be the cocycle for the action of G on the orbit $G\bullet t_0 \subseteq \widehat{A}$ so that

- $C(g,t)$ is defined for every $g \in G$ and (depending on $g \in G$) for ν-almost every $t \in G\bullet t_0$; and

- the cocycle equation

$$C(g_1 g_2, t) = C(g_1, g_2{\cdot}t)C(g_2, t) \tag{5.22}$$

holds for $g_1, g_2 \in G$ and (depending on g_1, g_2) for ν-almost every $t \in G{\cdot}t_0$.

DEFINING W. Using Fubini's theorem on $(G \times G) \times (G{\cdot}t_0)$ we can also turn this around and obtain a measurable subset $W \subseteq G{\cdot}t_0$ of full ν-measure such that for every $t \in W$ we have

- $C(g, t)$ is defined for m_G-almost every $g \in G$, and
- the cocycle equation (5.22) holds for $m_G \times m_G$-a.e. (g_1, g_2) in $G \times G$.

We will say that t is *well-connected* if $t \in W$.

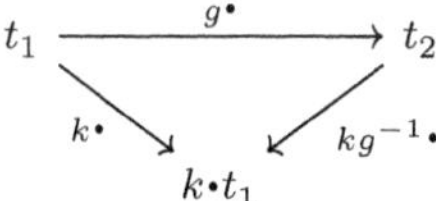

Fig. 5.2: The value $C(g, t_1)$ may not be defined (or may not satisfy the cocycle equation). To obtain a (better) definition, we introduce an arbitrary intermediate point $k{\cdot}t_1$ and use a 'majority vote' for choosing the correct value for $C_W(g, t_1)$.

DEFINING C_W. We now use the set W and the cocycle C to define C_W. For this let $g \in G$ and $t_1 \in W$ so that $t_2 = g{\cdot}t_1 \in W$ also. We claim that

$$G \ni k \longmapsto C(kg^{-1}, t_2)^{-1}C(k, t_1) \tag{5.23}$$

is almost surely independent of $k \in G$ (see Figure 5.2).

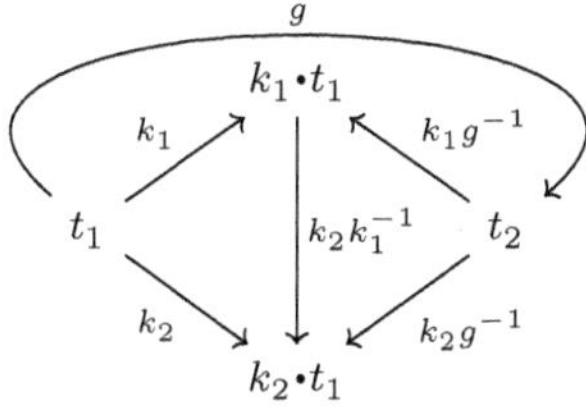

Fig. 5.3: If the cocycle equation holds for the left-hand and right-hand triangle then using the top two straight arrows or using the bottom two arrows to define $C_W(g, t_1)$ leads to the same value.

Figure 5.3 is a guide for the proof of the claim concerning (5.23). We will choose $k_1 \in G$ so that the cocycle equation holds for both the left-hand and the right-hand triangle for almost every $k_2 \in G$, which will give the claim in (5.23) quickly. We first note that by the properties of well-connected points $C(\cdot, t_1)$ and $C(\cdot, t_2)$ are defined m_G-almostwhere. This implies that (5.23) is defined for almost every $k \in G$. As $t_1 \in W$ is well-connected we may apply Fubini's theorem on $G \times G$ to find $k_1 \in G$ (from a subset of full measure) so that the cocycle equation (5.22) holds for $t = t_1$, $g_2 = k_1$, and almost every $g_1 \in G$. We set $g_1 = k_2 k_1^{-1}$ and obtain for almost every $k_2 \in G$ the cocycle equation

$$C(k_2 k_1^{-1} k_1, t_1) = C(k_2 k_1^{-1}, k_1 \cdot t_1) C(k_1, t_1)$$

corresponding to the left-hand triangle in Figure 5.3.

We proceed similarly with the right-hand triangle. As $t_2 \in W$ is well-connected we may suppose in addition that $k_1 \in G$ is from a subset of full measure so that the cocycle equation also holds for $t = t_2$, $g_2 = k_1 g^{-1}$, and $g_1 = k_2 k_1^{-1}$ for almost every k_2, which gives

$$C(k_2 k_1^{-1} k_1 g^{-1}, t_2) = C(k_2 k_1^{-1}, k_1 g^{-1} \cdot t_2) C(k_1 g^{-1}, t_2).$$

Simplifying these two equations and using the equality $g^{-1} \cdot t_2 = t_1$ we obtain

$$\begin{aligned} C(k_2, t_1) &= C(k_2 k_1^{-1}, k_1 \cdot t_1) C(k_1, t_1); \\ C(k_2 g^{-1}, t_2) &= C(k_2 k_1^{-1}, k_1 \cdot t_1) C(k_1 g^{-1}, t_2). \end{aligned}$$

However, on multiplying the inverse of the second line by the first line we obtain

$$C(k_2 g^{-1}, t_2)^{-1} C(k_2, t_1) = C(k_1 g^{-1}, t_2)^{-1} C(k_1, t_1)$$

for almost every $k_2 \in G$ as claimed in (5.23).

The claim in (5.23) now allows us to define $C_W(g, t)$ for $t \in W$ and $g \in G$ with $g \cdot t \in W$ as the almost sure value of the function

$$G \ni k \longmapsto C(k g^{-1}, g \cdot t)^{-1} C(k, t).$$

Fubini's theorem also implies that $C_W(g, t)$ is measurable.

Cocycle Equation. Let $t \in W$ and $g_1, g_2 \in G$ be such that $g_2 \cdot t \in W$ and $g_1 g_2 \cdot t \in W$. Then $C_W(g_2, t)$ is the almost sure value of

$$k \longmapsto C(k g_2^{-1}, g_2 \cdot t)^{-1} C(k, t) \tag{5.24}$$

and $C_W(g_1, g_2 \cdot t)$ is the almost sure value of

$$k \longmapsto C(k g_1^{-1}, g_1 g_2 \cdot t)^{-1} C(k, g_2 \cdot t)$$

or, equivalently, of

$$k \longmapsto C(k g_2^{-1} g_1^{-1}, g_1 g_2 \cdot t)^{-1} C(k g_2^{-1}, g_2 \cdot t). \tag{5.25}$$

Taking the product of (5.25) and (5.24) we obtain the cocycle equation

$$C_W(g_1g_2,t) = C_W(g_1,g_2\cdot t)C_W(g_2,t) \tag{5.26}$$

for all $t \in W$ and $g_1, g_2 \in G$ with $g_2\cdot t \in W$ and $g_1g_2\cdot t \in W$.

COMPARING TO OLD COCYCLE. Now fix some $g \in G$ and recall that (5.22) originally held for all $(g_1, g_2) \in G$ and ν-almost every $t \in G\cdot t_0$. We fix $g_2 = g$ and apply Fubini's theorem only on $(g_1, t) \in G \times G\cdot t_0$. Hence for ν-almost every t we have $t, g\cdot t \in W$ and that (5.22) holds for $g_2 = g$ and m_G-almost every $g_1 = kg^{-1}$. However, this gives that

$$C(g,t) = C(kg^{-1}, g\cdot t)^{-1}C(k,t)$$

for ν_G-almost every $t \in W$ and m_G-almost every $k \in G$. Hence

$$C(g,t) = C_W(g,t) \tag{5.27}$$

for every $g \in G$ and ν-almost every $t \in W$. □

To summarize, Lemma 5.65 allows us to replace the initial cocycle C with the cocycle C_W. The gain from this argument is that the domain of definition of C_W is more clearly spelled out, and that the cocycle equation (5.26) in fact holds whenever the original point t, the intermediate point $g_2\cdot t$ and the final endpoint $g_1g_2\cdot t$ are well-connected (that is, belong to W). This is crucial for our next step.

PROOF OF THEOREM 5.55. As $t_0 \in \widehat{A}$ was thus far only used to describe the orbit $G\cdot t_0$ we may suppose without loss of generality that $t_0 \in W$ is well-connected. We define H to be $\mathrm{Stab}_G(t_0)$ and

$$\tau\colon H \ni h \longmapsto C_W(h,t_0) \in \mathscr{U}(\mathcal{H}). \tag{5.28}$$

Lemma 5.65 now actually shows that τ is a homomorphism. By Lemma 5.38 measurability of τ implies continuity, so τ is in fact a unitary representation of H on $\mathcal{H}_\tau = \mathcal{H}$.

Now let $S \subseteq G$ be a cross-section as in Lemma 5.44. Without loss of generality we may assume that $e \in S$. Note that $S \cong G/H \cong G\cdot t_0$ and hence we also have $\mathcal{H}_\pi \cong L^2_\nu(S,\mathcal{H})$ and may identify W with a subset of S containing e. Writing Lemma 5.65 in this notation gives

$$\pi_g(f)(s) = J_\nu(g,g^{-1}\cdot s)^{\frac{1}{2}}C_W(g,g^{-1}\cdot s)f(g^{-1}\cdot s) \tag{5.29}$$

for $g \in G$, $f \in L^2_\nu(S,\mathcal{H})$, and $s \in S$ satisfying $s, g^{-1}\cdot s \in W$.

We wish to relate the cocycle C_W to the exact cocycle $C_{S,\tau}$ defined in Lemma 5.45 by the cross-section S and the unitary representation τ. Note that by definition of the G-action on S and the properties of the exact cocycle we have

$$g(g^{-1}\cdot s) = sh_S(g,g^{-1}\cdot s).$$

Assuming for $g \in G$ that $g^{-1}\bullet s, s \in W$, Lemma 5.65 and $e \in W$ (corresponding to t_0 in (5.28)) give the relation

$$\begin{aligned} C_W(g(g^{-1}\bullet s), e) &= C_W(g, g^{-1}\bullet s) C_W(g^{-1}\bullet s, e) \\ &= C_W(s, e) \underbrace{C_W(h_S(g, g^{-1}\bullet s), e)}_{=\tau(h_S(g, g^{-1}\bullet s))}. \end{aligned}$$

Therefore we obtain the 'coboundary equation'

$$C_W(s, e)^{-1} C_W(g, g^{-1}\bullet s) C_W(g^{-1}\bullet s, e) = C_{S,\tau}(g, g^{-1}\bullet s) \tag{5.30}$$

for $g \in G$ and $s, g^{-1}\bullet s \in W$, which will lead to the desired conclusion.

Let $B\colon L^2_\nu(S, \mathcal{H}) \to L^2_\nu(S, \mathcal{H})$ be the unitary operator defined by

$$(Bf)(s) = C_W(s, e) f(s) \tag{5.31}$$

for $f \in L^2_\nu(S, \mathcal{H})$ and $s \in W \subseteq S$. We calculate

$$\begin{aligned} (B^{-1}\pi_g Bf)(s) &= C_W(s, e)^{-1} (\pi_g Bf)(s) && \text{(by (5.31))} \\ &= C_W(s, e)^{-1} J(g, g^{-1}\bullet s)^{\frac{1}{2}} C_W(g, g^{-1}\bullet s)(Bf)(g^{-1}\bullet s) && \text{(by (5.29))} \\ &= J(g, g^{-1}\bullet s)^{\frac{1}{2}} C_W(s, e)^{-1} C_W(g, g^{-1}\bullet s) C_W(g^{-1}\bullet s, e) f(g^{-1}\bullet s) \\ & && \text{(by (5.31))} \\ &= J(g, g^{-1}\bullet s)^{\frac{1}{2}} C_{S,\tau}(g, g^{-1}\bullet s) f(g^{-1}\bullet s) && \text{(by (5.30))} \end{aligned}$$

for $g \in G$, $f \in L^2(S, \mathcal{H})$, and $s \in W$ with $g^{-1}\bullet s \in W$. In other words, we see that $B^{-1}\pi B$ is the induced unitary representation $\mathrm{Ind}_H^G \tau$. □

5.7 Summary and Outlook

Our discussions regarding the interplay between $\pi|_H$ and the G-action on $\widehat{H}$ for a normal abelian subgroup $H \lhd G$ in Section 5.1 will be important in, for example, Chapter 7.

The complete description of $\widehat{G}$ for some concrete metabelian groups will provide interesting examples for the discussion of the Fell topology in the next chapter.

Our examples considered in this chapter all fall into one of two categories. For the isometry groups, the affine group, and the Heisenberg group we obtained a classification of the unitary dual but for the lamplighter group and some others this turned out to be an unreasonable goal. The first type are known as groups of type I or tame groups and the second as groups of type II or wild groups.[(15)]

We note that Proposition 5.53 and Theorem 5.55 in fact allow the treatment of any connected nilpotent Lie group.

Chapter 6
Weak Containment and the Fell Topology

Recall from Definition 1.22 that a unitary representation π is contained in a unitary representation ρ, written as $\pi < \rho$, if there exists a ρ-invariant closed subspace $\mathcal{V} \subseteq \mathcal{H}_\rho$ such that the restriction $\rho|_\mathcal{V}$ of ρ to this subspace is isomorphic to π. In this chapter we will introduce the weaker, and often more useful, notion of weak containment. We also recall that in Corollary 1.78 we have shown that irreducible representations can be used to approximate any other cyclic representation. In fact, the approximation could be done uniformly on compact sets in G, which opens the door to many other discussions. Our discussions of weak containment and the Fell topology will both use the compact-open topology on $\mathcal{P}(G)$. We will also discuss these notions in detail for abelian groups as in Chapter 2, and for the examples of metabelian groups as in Chapter 5.

6.1 Weak Containment

We start with the following important notion.

Definition 6.1 (Weak containment). Let π and ρ be unitary representations of the group G. We say that π is *weakly contained in* ρ and write $\pi \prec \rho$ if every diagonal π-matrix coefficient can be approximated, uniformly within compact subsets of G, by finite sums of diagonal ρ-matrix coefficients. That is, if for every $v \in \mathcal{H}_\pi$, every compact $Q \subseteq G$, and every $\varepsilon > 0$ there exists some $n \geqslant 1$ and $w_1, \ldots, w_n \in \mathcal{H}_\rho$ such that

$$\Big\| \varphi_v^\pi - \sum_{j=1}^n \varphi_{w_j}^\rho \Big\|_{Q,\infty} = \max_{g\in Q} \Big| \langle \pi_g v, v\rangle - \sum_{j=1}^n \langle \rho_g w_j, w_j\rangle \Big| \leqslant \varepsilon. \tag{6.1}$$

Notice that by allowing sums of ρ-matrix coefficients we are able to ignore any multiplicity questions for the representations π or ρ (see also Exercise 6.2).

Exercise 6.2 (Elementary observations). Let π and ρ be unitary representations of G.

M. Einsiedler and T. Ward, *Unitary Representations and Unitary Duals*,
Graduate Texts in Mathematics 308, https://doi.org/10.1007/978-3-032-03899-9_6

(a) Show that $\pi < \rho$ implies $\pi \prec \rho$ (where in fact we could set $Q = G$ and $\varepsilon = 0$ in the definition).
(b) Now let ρ be any unitary representation, and define $\pi = \rho \oplus \rho$ to be the direct sum of ρ with itself. Show that $\pi \prec \rho$ (where again we could set $Q = G$ and $\varepsilon = 0$ in the definition). However, if $\dim \mathcal{H}_\rho < \infty$ or ρ is irreducible, then we cannot have $\pi < \rho$.
(c) Let π and ρ be unitary representations of G. Show that $\pi \prec \rho$, $\pi^\infty \prec \rho$, and $\pi \prec \rho^\infty$ are mutually equivalent.

For an abelian group G we again write $\widehat{G}$ for its Pontryagin dual (using additive notation as in Chapter 2), $t \in \widehat{G}$ for its elements, and χ_t for the associated character and unitary representation on $\mathbb{C}$.

Exercise 6.3 (Abelian regular representation). Let λ be the regular representation of an abelian group G and let $t \in \widehat{G}$. Show that the character χ_t is weakly contained in λ. Show moreover that χ_t is not contained in λ if G is non-compact.

We hope that the abelian case above will help to motivate the notion of weak containment, and now return to the general case.

Essential Exercise 6.4 (Transitivity). Let π, ρ, γ be unitary representations of G with $\pi \prec \rho$ and $\rho \prec \gamma$. Show that $\pi \prec \gamma$.

Exercise 6.5 (Twisting weak containment). Let π and ρ be unitary representations of G and let χ be a unitary character of G. Show that $\pi \prec \rho$ implies that $\chi \otimes \pi \prec \chi \otimes \rho$ (see Lemma 1.28).

Exercise 6.6 (Restricting weak containment). Let $H < G$ be a closed subgroup and let π and ρ be be unitary representations of G with $\pi \prec \rho$. Show that $\pi|_H \prec \rho|_H$.

We start with the following simple reduction.

Lemma 6.7 (Normalization of vector and sum). *Let π and ρ be unitary representations of G. In the definition of weak containment $\pi \prec \rho$ it suffices to consider unit vectors $v \in \mathcal{H}_\pi$. Moreover, if $\pi \prec \rho$ and $v \in \mathcal{H}_\pi$ has $\|v\| = 1$, then for every compact subset $Q \subseteq G$ and $\varepsilon > 0$, there exist vectors $w_j \in \mathcal{H}_\rho$ for $j = 1, \ldots, n$ with $\sum_{j=1}^n \|w_j\|^2 = 1$ satisfying* (6.1).

Proof. The first claim in the lemma follows simply from normalizing a given vector $v \in \mathcal{H}_\pi \smallsetminus \{0\}$. So assume now that $\pi \prec \rho$ and $v \in \mathcal{H}_\pi$ has $\|v\| = 1$. We suppose without loss of generality that $e \in Q$ and $\varepsilon \in (0,1)$. In this case (6.1) implies that $|1 - s| = \left|\varphi_v^\pi(e) - \sum_{j=1}^n \varphi_{w_j}^\rho(e)\right| < \varepsilon$ for $s = \sum_{j=1}^n \|w_j\|^2$, and so $s > 0$. This gives, with $\widetilde{w}_j = s^{-\frac{1}{2}} w_j$ for $j = 1, \ldots, n$, that

$$\left\|\varphi_v^\pi - \sum_{j=1}^n \varphi_{\widetilde{w}_j}^\rho\right\|_{Q,\infty} \leqslant \varepsilon + \left\|\sum_{j=1}^n \varphi_{w_j}^\rho - \sum_{j=1}^n \varphi_{\widetilde{w}_j}^\rho\right\|_{Q,\infty}$$
$$\leqslant \varepsilon + |1 - s^{-1}| \underbrace{\left\|\sum_{j=1}^n \varphi_{w_j}^\rho\right\|_{Q,\infty}}_{=s} \leqslant \varepsilon + |s - 1| < 2\varepsilon.$$

Since $\sum_{j=1}^{n} \|\widetilde{w}_j\|^2 = 1$, this implies the lemma. □

We now show that for irreducible representations of G we do not have to consider sums of matrix coefficients in the definition of weak containment. For this, recall from Proposition 1.79 that the compact-open and the weak* topologies on $\mathcal{P}^1(G)$ agree. For this reason, we will not specify the topology on $\mathcal{P}^1(G)$ whenever we use either of the two.

Proposition 6.8 (Irreducible representations). *Let π be an irreducible unitary representation of G. Suppose that, for a unit vector $v \in \mathcal{H}_\pi$, there exists a sequence (ϕ_n) in $\mathcal{P}^1(G)$ of the form*

$$\phi_n = \sum_{j=1}^{J(n)} \phi_{n,j}$$

with $\phi_{n,j} \in \mathcal{P}^{\leqslant 1}(G) \smallsetminus \{0\}$ for all $j \in \{1, \dots, J(n)\}$, so that (ϕ_n) converges to φ_v^π as $n \to \infty$. Then there exists a sequence $(j(n))$ with $j(n)$ in $\{1, \dots, J(n)\}$ for all $n \geqslant 1$ with the property that the sequence $(\widetilde{\phi}_{n,j(n)})$ in $\mathcal{P}^1(G)$ converges to φ_v^π as $n \to \infty$, where

$$\widetilde{\phi}_{n,j(n)} = \phi_{n,j(n)}(e)^{-1} \phi_{n,j(n)} \in \mathcal{P}^1(G)$$

for all $n \geqslant 1$.

In particular, if $\pi \prec \rho$ for a unitary representation ρ of G, then there exists for every unit vector $v \in \mathcal{H}_\pi$, every compact subset $Q \subseteq G$, and every $\varepsilon > 0$, a unit vector $w \in \mathcal{H}_\rho$ such that $\|\varphi_v^\pi - \varphi_w^\rho\|_{Q,\infty} < \varepsilon$.

PROOF. Suppose first that $\pi \prec \rho$ for a unitary representation ρ of G, and that $v \in \mathcal{H}_\pi$ is a unit vector. Then, by Lemma 6.7, for every compact $Q \subseteq G$ and $\varepsilon > 0$ we can find vectors $w_1, \dots, w_J \in \mathcal{H}_\rho$ with corresponding matrix coefficients $\phi_j = \varphi_{w_j}^\rho$ for $j = 1, \dots, J$ so that

$$\Big\| \varphi_v^\pi - \sum_{j=1}^{J} \phi_j \Big\|_{Q,\infty} < \varepsilon,$$

satisfying

$$\sum_{j=1}^{J} \phi_j(e) = \sum_{j=1}^{J} \|w_j\|^2 = 1$$

and hence

$$\phi = \sum_{j=1}^{J} \phi_j \in \mathcal{P}^1(G)$$

approximates φ_v^π, as in the first part of the proposition. Using $\varepsilon = \frac{1}{n}$ and σ-compactness of G, it follows that it suffices to prove the first part of the proposition.

Let $v \in \mathcal{H}_\pi$ have $\|v\| = 1$ and recall from Proposition 1.73 that $\varphi_v^\pi \in \mathcal{P}^1(G)$ is extremal. Recall from Section 1.7 that

$$\mathcal{P}^{\leqslant 1}(G) = [0,1]\mathcal{P}^1(G)$$

is weak* compact and that φ_v^π is also extremal when viewed as an element of $\mathcal{P}^{\leqslant 1}(G)$. By the assumption, there exists a sequence (ϕ_n) with

$$\phi_n = \sum_{j=1}^{J(n)} \phi_{n,j} \in \mathcal{P}^1(G)$$

with $\phi_{n,j} \in \mathcal{P}^{\leqslant 1}(G) \smallsetminus \{0\}$ for all n, so that $\phi_n \to \varphi_v^\pi$ as $n \to \infty$ (in the weak* or compact-open topology). We define $s_{n,j} = \phi_{n,j}(e)$ and

$$\widetilde{\phi}_{n,j} = s_{n,j}^{-1} \phi_{n,j} \in \mathcal{P}^1(G)$$

for $j = 1, \ldots, J(n)$. Using this notation, we have $\sum_{j=1}^{J(n)} s_{n,j} = \phi(e) = 1$ and that the convex combination $\phi_n = \sum_{j=1}^{J(n)} s_{n,j} \widetilde{\phi}_{n,j}$ converges to φ_v^π as $n \to \infty$.

This is the reason why the proposition should be true: We approximate an extremal point by convex combinations, which should imply that most of the vectors used in the convex combination approximate the extremal vector. The proposition follows from the following more general lemma applied to the case $V = L^\infty(G) = L^1(G)'$ with the weak* topology and $K = \mathcal{P}^{\leqslant 1}(G)$. For this, recall that $K \subseteq \overline{B_1^V(0)}$ is metrizable in the weak* topology, since $L^1(G)$ is separable. □

Lemma 6.9. *Let V be a locally convex vector space and let $K \subseteq V$ be a metrizable compact convex subset. Suppose that $x_0 \in K$ is an extremal point, that $y_{n,1}, \ldots, y_{n,J(n)} \in K$ and $s_{n,1}, \ldots, s_{n,J(n)} > 0$ satisfy $\sum_{j=1}^{J(n)} s_{n,j} = 1$, and that the convex combination $\sum_{j=1}^{J(n)} s_{n,j} y_{n,j}$ converges to x_0 as $n \to \infty$. Then there exists a sequence (j_n) so that y_{n,j_n} converges to x_0 as $n \to \infty$.*

PROOF. We define the finitely supported probability measure

$$\mu_n = \sum_{j=1}^{J(n)} s_{n,j} \delta_{y_{n,j}} \tag{6.2}$$

on K.

We now show that μ_n has barycentre

$$y_n = \sum_{j=1}^{J(n)} s_{n,j} y_{n,j}.$$

To see this, let $L \in V'$ be a continuous linear functional and calculate

$$\int_K L\,\mathrm{d}\mu_n = \sum_{j=1}^{J(n)} s_{n,j} \underbrace{\int_K L\,\mathrm{d}\delta_{y_{n,j}}}_{=L(y_{n,j})}$$
$$= L\left(\sum_{j=1}^{J(n)} s_{n,j} y_{n,j}\right) = L(y_n).$$

Now recall that the barycentre y_n converges to x_0 as $n \to \infty$. Therefore we obtain a sequence of probability measures (μ_n) whose barycentres y_n converge to x_0 as $n \to \infty$. As K is a compact metric space by assumption, the same holds for the space of Borel probability measures on K (see [25, Prop. 8.27]). Hence we can choose a subsequence (μ_{n_ℓ}) with $\lim_{\ell\to\infty} \mu_{n_\ell} = \nu$ for some probability measure ν on K. Let $L \in V'$. By continuity of L on K, we have

$$\int_K L\,\mathrm{d}\nu = \lim_{\ell\to\infty} \int_K L\,\mathrm{d}\mu_{n_\ell} = \lim_{\ell\to\infty} L(y_{n_\ell}) = L(x_0).$$

It follows that the barycentre of ν is given by x_0.

We claim that, by extremality of x_0, we have $\nu = \delta_{x_0}$. Assuming the claim, it follows that we even have

$$\lim_{n\to\infty} \mu_n = \delta_{x_0}. \tag{6.3}$$

Indeed, in a compact metric space, if every convergent subsequence has the same limit, then this limit is the limit of the whole sequence. Moreover, we conclude now from the definition of μ_n in (6.2), from (6.3), and Urysohn's lemma, that there exists a sequence $(j(n))$ with $1 \leqslant j(n) \leqslant J(n)$ for all $n \in \mathbb{N}$ such that $x_{n,j(n)}$ converges to x_0 as $n \to \infty$.

To prove the claim, we suppose that $\nu \neq \delta_{x_0}$ (for the purposes of a contradiction). Then there exists some $z \in \operatorname{supp}\nu \smallsetminus \{x_0\}$ and some $L_0 \in V'$ with $L_0(z) \neq L_0(x_0)$. Multiplying L_0 by -1 or, in the case that V is a complex vector space, by $\pm\mathrm{i}$ if necessary, we assume that

$$\Re L_0(z) < s < \Re L_0(x_0)$$

for some $s \in \mathbb{R}$. We define the neighbourhood

$$O = \{y \in K \mid \Re L_0(y) < s\}.$$

Note that $\nu(O) > 0$ as $z \in \operatorname{supp}\nu \cap O$. Also $\nu(O) < 1$ as otherwise $x_0 \in O$. Hence we can use O to split ν into the convex combination

$$\nu = \nu(O)\big(\tfrac{1}{\nu(O)}\nu|_O\big) + \nu(O^c)\big(\tfrac{1}{\nu(O^c)}\nu|_{O^c}\big).$$

However, this implies that x_0 is, as the barycentre of ν, the non-trivial convex combination of the barycentre of $\frac{1}{\nu(O)}\nu|_O$ that belongs to O and another barycentre in $K \smallsetminus O$ (which exist by [25, Lem. 8.88]). This contradicts extremality of x_0, which proves the claim and finishes the proof of the lemma. □

By Proposition 6.8, the statement $\pi \prec \rho$ for an irreducible unitary representation π is equivalent to the statement that for any $v \in \mathcal{H}_\pi$ there is a $w \in \mathcal{H}_\rho$ with the property that v and w behave very similarly on a large compact subset (described by φ_v^π). Moreover, in the case that $\pi = \chi$ is a character, we may call the vector $w \in \mathcal{H}_\rho$ with this behaviour an *approximate eigenvector* for the character χ. If π is not irreducible, then the same can be said, but only using a vector in the Hilbert space $\mathcal{H}_\rho^\infty$.

Exercise 6.10 (Approximate eigenvectors). Let π be a unitary representation of G, and let χ be a unitary character of G. Show that $\chi \prec \pi$ if and only if for any compact set $Q \subseteq G$ and $\varepsilon > 0$ there exists a vector w such that $\|\pi_g w - \chi(g)w\| < \varepsilon$ for all $g \in Q$.

The following example shows that weak containment has surprising properties for non-abelian groups.

Example 6.11. Let $G = \mathbb{R}_{>0} \ltimes \mathbb{R}$ be the affine group, let $A = \{h_a \mid a \in \mathbb{R}\}$ be the normal abelian subgroup, and let $S = \{g_s \mid s \in \mathbb{R}_{>0}\}$ be the complementary subgroup as in Section 5.2.3. Recall that G has the old representations defined by characters on $S = G/A$ and the two new representations π^+ and π^-, where π^+ is defined by the measure μ_+ on $\widehat{A} \cong \mathbb{R}$ with support $[0, \infty)$. We claim that $\mathbb{1} \prec \pi^+$. To see this define, for $\varepsilon \in (0, 1)$,

$$v_\varepsilon = \frac{1}{\sqrt{|\log \varepsilon|}} \mathbb{1}_{[\varepsilon^2, \varepsilon]} \in \mathcal{H}_+ = L^2_{\mu_+}(\mathbb{R})$$

with

$$\|v_\varepsilon\|^2_{\mathcal{H}_+} = \frac{1}{|\log \varepsilon|} \int_{\varepsilon^2}^{\varepsilon} \frac{\mathrm{d}t}{t} = 1.$$

For $a \in \mathbb{R}$ we have

$$\big|\langle \pi^+(h_a)v_\varepsilon, v_\varepsilon\rangle - 1\big| \leqslant \frac{1}{|\log \varepsilon|} \int_{\varepsilon^2}^{\varepsilon} \underbrace{\big|\mathrm{e}^{2\pi i a t} - 1\big|}_{\leqslant 2\pi|a|\varepsilon} \frac{\mathrm{d}t}{t} \leqslant 2\pi|a|\varepsilon$$

by the mean value theorem, which shows that $\varphi_{v_\varepsilon}^{\pi^+}(h_a) = \langle \pi^+(h_a)v_\varepsilon, v_\varepsilon\rangle$ converges to 1 uniformly on compact subsets of $\mathbb{R}$. For $s \in (\varepsilon, \frac{1}{\varepsilon})$, we also have

$$\pi^+(g_s)v_\varepsilon = \mathbb{1}_{[\varepsilon^2, \varepsilon]} \circ \widehat{\theta_{g_s}} = \mathbb{1}_{[\varepsilon^2/s, \varepsilon/s]}$$

and so

$$\big|\langle \pi^+(g_s)v_\varepsilon, v_\varepsilon\rangle - 1\big| = \left|\frac{1}{|\log \varepsilon|} \int_{\max(\varepsilon^2, \varepsilon^2/s)}^{\min(\varepsilon, \varepsilon/s)} \frac{\mathrm{d}t}{t} - 1\right| = \frac{|\log s|}{|\log \varepsilon|} \longrightarrow 0$$

uniformly on compact subsets of $\mathbb{R}_{>0}$. Therefore both $\|\pi^+(h_a)v_\varepsilon - v_\varepsilon\| \to 0$ and $\|\pi^+(g_s)v_\varepsilon - v_\varepsilon\| \to 0$ uniformly on compact subsets of $\mathbb{R}$, resp. $\mathbb{R}_{>0}$. This implies that $\mathbb{1} \prec \pi^+$ (see Exercise 6.12) even though $\mathbb{1}, \pi^+ \in \widehat{G}$.

Exercise 6.12. (a) Complete the argument above to show that $\mathbb{1} \prec \pi^+$.
(b) Extend the argument above by showing that $\chi \prec \pi^+$ and $\chi \prec \pi^-$ for any character χ of $G/A \cong \mathbb{R}_{>0}$.

6.1.1 Weak Containment for Compact Groups

For a compact group the notion of weak containment $\rho_1 \prec \rho_2$ for unitary representations ρ_1 and ρ_2 from Section 6.1 has the following special properties.

Corollary 6.13 (Characterization of weak containment). *Let ρ_1 and ρ_2 be unitary representations of the compact group G. Then the following are equivalent:*

(1) $\rho_1 \prec \rho_2$;
(2) $\pi < \rho_1$ *implies that* $\pi < \rho_2$ *for any* $\pi \in \widehat{G}$; *and*
(3) $\rho_1 < \rho_2^\infty$.

Moreover, for $\pi \in \widehat{G}$ and a unitary representation ρ of G, we have $\pi \prec \rho$ if and only if $\pi < \rho$.

PROOF. We begin by proving the final claim in the corollary. Suppose that π is in $\widehat{G}$ and ρ is a unitary representation of G. Clearly $\pi < \rho$ implies $\pi \prec \rho$, so suppose for the converse that π is not contained in ρ. By the decomposition theorem (Theorem 3.19) this gives

$$\mathcal{H}_\rho = \bigoplus_{\ell \in I} \mathcal{V}_\ell$$

for a finite or countable index set I and closed subspaces $\mathcal{V}_\ell$ for $\ell \in I$ with $\rho|_{\mathcal{V}_\ell} \cong \pi_\ell \in \widehat{G} \smallsetminus \{\pi\}$. Let $v \in \mathcal{H}_\pi$ be a unit vector and let φ_v^π be its principal matrix coefficient. For any $w \in \mathcal{H}_{\pi_0}$ with $\pi_0 \in \widehat{G} \smallsetminus \{\pi\}$ we obtain from Schur orthogonality (Theorem 3.30) that $\varphi_w^{\pi_0} \perp \varphi_v^\pi$, where it does not matter whether w is normalized to unit length or not.

If now $w = \sum_{\ell \in I} w_\ell \in \mathcal{H}_\rho$ with $w_\ell \in \mathcal{V}_\ell$, then

$$\varphi_w^\rho(g) = \langle \rho_g w, w \rangle = \sum_{\ell \in I} \langle \rho_g w_\ell, w_\ell \rangle = \sum_{\ell \in I} \varphi_{w_\ell}^\rho(g)$$

for all $g \in G$. Since $\|\varphi_{w_\ell}^\rho\| = \|w_\ell\|^2$ this series converges uniformly. Together with the above, this implies that $\varphi_w^\rho \perp \varphi_v^\pi$. Clearly this extends to finite sums as in the definition of weak containment (Definition 6.1), and shows that π is not weakly contained in ρ.

Suppose now that ρ_1 and ρ_2 are unitary representations of G and $\rho_1 \prec \rho_2$ as in (1). If $\pi < \rho_1$, then by transitivity of weak containment (see Exercise 6.4 and its hint on page 535) we have $\pi \prec \rho_2$, which by the above also implies that $\pi < \rho_2$ as required.

If now ρ_1 and ρ_2 satisfy (2), then

$$\operatorname{mult}(\pi, \rho_1) \leqslant \infty \cdot \operatorname{mult}(\pi, \rho_2) = \operatorname{mult}(\pi, \rho_2^\infty),$$

and (3) follows from the characterization of containment in Corollary 3.25.

Now suppose that $\rho_1 < \rho_2^\infty$ as in (3). Any $v \in \mathcal{H}_{\rho_1}$ then corresponds under the assumed unitary intertwining isomorphism to some $w = (w_n)_n$ lying in $\bigoplus_{n=1}^\infty \mathcal{H}_{\rho_2}$. This implies that $\varphi_v^{\rho_1} = \varphi_w^{\rho_2^\infty} = \sum_{n=1}^\infty \varphi_{w_n}^{\rho_2}$, and the series converges uniformly. As $v \in \mathcal{H}_\rho$ was arbitrary, this implies that $\rho_1 \prec \rho_2$ by definition of weak containment. □

6.2 Amenability, Property (T), and Spectral Gap

Using the notion of weak containment, we can formulate two important properties that the group G might have. For this, the following weakening of the notion of invariant vectors from Section 1.1.3 will be useful.

Definition 6.14 (Almost invariant vectors). A unitary representation π of G has *almost invariant vectors* if for every compact set $Q \subseteq G$ and $\varepsilon > 0$ there exists a unit vector $u \in \mathcal{H}_\pi$ with $\|\pi_g u - u\| < \varepsilon$ for all $g \in Q$. We also say, for a compact subset $Q \subseteq G$ and $\varepsilon > 0$, that a unit vector $u \in \mathcal{H}_\pi$ is (Q, ε)-*invariant* if $\|\pi_g u - u\| < \varepsilon$ for all $g \in Q$.

The following corollary to Proposition 6.8 links this notion to weak containment.

Corollary 6.15 (Almost invariant vectors). *Let π be a unitary representation of G. Then $\mathbb{1}_G \prec \pi$ if and only if π has almost invariant vectors.*

PROOF. Suppose that π has almost invariant vectors. Let $Q \subseteq G$ be compact and $\varepsilon > 0$. Then there exists a unit vector $v \in \mathcal{H}_\pi$ with $\|\pi_g v - v\| < \varepsilon$ for all $g \in Q$. However, this implies that

$$|1 - \varphi_v^\pi(g)| = |\|v\|^2 - \langle \pi_g v, v \rangle| = |\langle v - \pi_g v, v \rangle| < \varepsilon$$

for all $g \in Q$. As the compact subset Q and $\varepsilon > 0$ were arbitrary, we see that $\mathbb{1}_G \prec \pi$.

For the converse, suppose that $\mathbb{1}_G \prec \pi$. By Proposition 6.8 there exists, for every compact subset $Q \subseteq G$ and $\varepsilon > 0$, a unit vector $v \in \mathcal{H}_\pi$ with

$$\|1 - \varphi_v^\pi\|_{Q,\infty} < \varepsilon^2.$$

For $g \in Q$ we now have

$$\|\pi_g v - v\|^2 = \langle \pi_g v, \pi_g v\rangle - 2\Re\langle \pi_g v, v\rangle + \|v\|^2 = 2 - 2\Re\varphi_v^\pi(g) < 2\varepsilon^2.$$

Taking the square root, we obtain $\|\pi_g v - v\| < \sqrt{2}\varepsilon$ for $g \in Q$. It follows that π has almost invariant vectors. □

Definition 6.16 (Amenability). We say that G is *amenable* if the trivial representation $\mathbb{1}_G$ is weakly contained in the regular representation λ_G, or equivalently if the regular representation λ_G has almost invariant vectors.

In particular, we see from Exercise 6.3 that any abelian group is amenable.

Exercise 6.17. Show that the solvable groups considered in Chapter 5 are amenable.

Definition 6.18 (Kazhdan's property (T)). We say that G has *property* (T) if, for any unitary representation ρ of G, we have that $\mathbb{1}_G \prec \rho$ holds only if $\mathbb{1}_G < \rho$, or equivalently if any representation that has almost invariant vectors in fact has invariant vectors.

For non-compact locally compact groups, amenability and property (T) are in some sense opposite properties. The first is a 'soft' property permitting many different behaviours for the actions of the group, the second a 'rigidity' property that sharply constrains the possible behaviours of actions. Some groups have neither property, including finitely-generated free groups and, as we will show in Chapter 8, $\mathrm{SL}_2(\mathbb{R})$. Exercise 6.19 identifies the very limited class of groups with both properties.

Exercise 6.19. Show that G is compact if and only if G is amenable and has property (T).

We will briefly encounter property (T) again. The notion will be formulated in terms of a topology on $\widehat{G}$ in Section 6.4, where we will also explain the notation (T), and discussed further in Sections 7.3 and 7.4. We refer to [25, Ch. 10] for a discussion of both amenability and property (T), and to the monograph of Bekka, de la Harpe, and Valette [5] for an extensive treatment of property (T).

Exercise 6.20. Let $\mathbb{F}_2 = \langle a, b\rangle$ be the free group generated by a and b. Show that $\mathbb{F}_2$ is not amenable, and that $\mathbb{F}_2$ does not have property (T).

6.2.1 (Uniform) Spectral Gap

Definition 6.21 (Spectral gap). Let π be a unitary representation of G, and write

$$\mathcal{H}_\pi^G = \{v \in \mathcal{H}_\pi \mid \pi_g v = v \text{ for all } g \in G\}$$

for the subspace of invariant vectors. We say that π has *spectral gap* if π restricted to $(\mathcal{H}_\pi^G)^\perp$ does not have almost invariant vectors. Equivalently, π has spectral gap if there exists a compact subset $Q \subseteq G$ and some $\varepsilon > 0$ with the property that there are no (Q, ε)-invariant unit vectors in $(\mathcal{H}_\pi^G)^\perp$.

Comparing Definitions 6.14, 6.18, and 6.21, we see that a unitary representation of a group with property (T) always has a spectral gap. The next lemma shows that more is true.

Definition 6.22 (Uniform spectral gap). A collection S of unitary representations of G has *uniform spectral gap* if there exist $\varepsilon > 0$ and a compact subset $Q \subseteq G$ such that for every $\pi \in S$ and any $v \in (\mathcal{H}_\pi^G)^\perp$ there is some $g \in Q$ with $\|\pi_g v - v\| \geqslant \varepsilon \|v\|$.

Lemma 6.23 (Uniform spectral gap). *Suppose G has property* (T)*. Then all its unitary representations have uniform spectral gap.*

PROOF. By our standing assumption, G can be written as $G = \bigcup_{n=1}^\infty Q_n$ for compact sets Q_n with $Q_n \subseteq Q_{n+1}^o$ for all $n \geqslant 1$. Suppose now that the lemma does not hold. Then for every $n \geqslant 1$ there is a unitary representation π_n of G on $\mathcal{H}_{\pi_n}$ so that π_n fails Definition 6.22 for Q_n and $\varepsilon = \frac{1}{n}$. Restricting to $(\mathcal{H}_{\pi_n}^G)^\perp$, we may assume that π_n has no non-trivial invariant vectors but that there exists a vector v_n with $\|v_n\| = 1$ and with

$$\sup_{g \in Q_n} \|\pi_n(g)v_n - v_n\| < \tfrac{1}{n}.$$

Now define $\mathcal{H}_\pi = \bigoplus_{n \geqslant 1} \mathcal{H}_{\pi_n}$ with the unitary representation $\pi = \bigoplus_n \pi_n$ of G on $\mathcal{H}_\pi$. It follows that π has no non-zero G-invariant vectors, but does have almost invariant vectors. This contradicts the assumption that G has property (T). □

Exercise 6.24. Let G be a discrete group with property (T). Show that G is finitely generated.

The phrase 'spectral gap' and, in particular, the word 'gap' is made more clear in the next result.

Proposition 6.25 (Spectral gap in terms of convolution operators). *A unitary representation π of G has spectral gap if and only if there exists some $\delta > 0$ and some non-negative function $f \in L^1(G)$ with $\int f \,\mathrm{d}m_G = 1$ and with*

$$\left\| \pi_*(f)|_{(\mathcal{H}_\pi^G)^\perp} \right\|_{\mathrm{op}} \leqslant 1 - \delta < 1. \tag{6.4}$$

That is, the norm of the bounded operator $\pi_*(f)$, when restricted to the orthogonal complement of the space of invariant vectors, is strictly less than one.

Moreover, f can be chosen in $C_c(G)$ with $f = f^* \geqslant 0$, or alternatively equal to $\frac{1}{m(P)} \mathbb{1}_P$ for some compact $P \subseteq G$ with $P^o \supseteq Q \cup \{e\}$, where $Q \subseteq G$ is as in the definition of spectral gap. Finally, if a collection S of unitary representations of G has uniform spectral gap, then f and δ as in (6.4) *can be chosen uniformly across S.*

Notice that the operator $\pi_*(f)$ in (6.4) satisfies $\pi_*(f)v = v$ for all $v \in \mathcal{H}_\pi^G$. Therefore, (6.4) does indeed give a gap in the spectrum of $\pi_*(f)$.

Proof of Proposition 6.25. One direction of the equivalence is straightforward. Indeed, suppose that π is a unitary representation of G and $f \in L^1(G)$ satisfies $f \geqslant 0$, $\int f \, dm_G = 1$, and

$$\left\|\pi_*(f)|_{(\mathcal{H}_\pi^G)^\perp}\right\|_{\mathrm{op}} \leqslant \lambda = 1 - \delta < 1. \tag{6.5}$$

Then there exists a compact set $Q \subseteq G$ with

$$\int_{G \smallsetminus Q} f \, dm_G < \tfrac{\delta}{3}.$$

This implies that, for any $v \in (\mathcal{H}_\pi^G)^\perp$, there exists some $g \in Q$ with

$$\|\pi_g v - v\| \geqslant \tfrac{\delta}{3}\|v\|.$$

Indeed, if this were not true, then we would have $\|\pi_g v - v\| < \frac{\delta}{3}\|v\|$ for all g in Q, and hence, by Exercise 1.59 (see the hint on page 527) applied to $\nu = f|_Q \, dm_G$, we have

$$\|\pi_*(f)v - v\| \leqslant \underbrace{\|\pi_*(f|_Q)v - \int_Q f \, dm_G v\|}_{<\frac{\delta}{3}\int_Q f \, dm_G\|v\|} + \underbrace{\|\pi_*(f|_{G\smallsetminus Q})v\|}_{<\frac{\delta}{3}\|v\|} + \underbrace{\int_{G\smallsetminus Q} f \, dm_G}_{<\frac{\delta}{3}} \|v\| < \delta\|v\|,$$

which contradicts (6.5).

Suppose now, for the converse, that the unitary representation π of G has spectral gap in the sense of Definition 6.21, and let (Q, ε) be chosen with the property that there are no (Q, ε)-invariant unit vectors in $(\mathcal{H}_\pi^G)^\perp$. We will show, starting with $Q \subseteq G$, $\varepsilon > 0$, and a compact set

$$P \supseteq P^o \supseteq Q \cup \{e\} \tag{6.6}$$

as in the proposition, that $f = \frac{1}{m(P)}\mathbb{1}_P$ satisfies (6.4).

At the heart of the argument, we will use strict convexity of the norm in $\mathcal{H}_\pi$. For this, a uniform continuity property for applying π_g will be useful. To achieve this, we define $w = \pi_*(f^*)v$ for $v \in \left(\mathcal{H}_\pi^G\right)^\perp \smallsetminus \{0\}$. Thus

$$\begin{aligned}
\|\pi_g w - \pi_h w\| &= \|\pi_g \pi_*(f^*)v - \pi_h \pi_*(f^*)v\| \\
&= \|\pi_*(\lambda_g f^* - \lambda_h f^*)v\| \\
&\leqslant \|\lambda_g f^* - \lambda_h f^*\|_1 \|v\| = \|f^* - \lambda_{g^{-1}h} f^*\|_1 \|v\|.
\end{aligned}$$

This shows that there exists a neighbourhood U of $e \in G$ (independent of π, $\mathcal{H}_\pi$, and v) such that

$$\|\pi_g w - \pi_h w\| < \tfrac{\varepsilon}{4}\|v\| \tag{6.7}$$

whenever $g^{-1}h \in U$. Using compactness of Q, the assumption that

$$Q \cup \{e\} \subseteq P^o,$$

and decreasing U if necessary, we can also assume that

$$gU \subseteq P$$

for all $g \in Q \cup \{e\}$.

If

$$\|w\| = \|\pi_*(f^*)v\| < \tfrac{3}{4}\|v\|, \tag{6.8}$$

then we have basically achieved our goal. Hence we will assume for now that

$$\|w\| = \|\pi_*(f^*)v\| \geqslant \tfrac{3}{4}\|v\|.$$

Since $w \in \left(\mathcal{H}_\pi^G\right)^\perp$ is non-zero, there exists some $g_1 \in Q$ satisfying

$$\|\pi_{g_1} w - w\| \geqslant \varepsilon \|w\|.$$

Then, by the above assumption on w, we also have

$$\|\pi_{g_1} w - w\| \geqslant \tfrac{3\varepsilon}{4}\|v\|. \tag{6.9}$$

Moreover, the uniform continuity estimate in (6.7) gives

$$\|\pi_{h_0} w - w\| < \tfrac{\varepsilon}{4}\|v\| \tag{6.10}$$

for all $h_0 \in B_0 = U \subseteq P$, and

$$\|\pi_{h_1} w - \pi_{g_1} w\| < \tfrac{\varepsilon}{4}\|v\| \tag{6.11}$$

for all $h_1 \in B_1 = g_1 U \subseteq P$. In particular, by combining (6.9)–(6.11) with the triangle inequality, we obtain

$$\|\pi_{h_0} w - \pi_{h_1} w\| > \tfrac{\varepsilon}{4}\|v\|$$

for $h_0 \in B_0$ and $h_1 \in B_1$, and so $B_0 \cap B_1 = \emptyset$. With these preparations, and setting $s = \frac{m(B_0)}{m(P)}$, we now obtain that

$$\pi_*\left(\frac{1}{m(P)}\mathbb{1}_P\right)w = \underbrace{\frac{m(B_0)}{m(P)}}_{=s}\underbrace{\pi_*\left(\frac{1}{m(B_0)}\mathbb{1}_{B_0}\right)w}_{=w_0} + \underbrace{\frac{m(B_1)}{m(P)}}_{=s}\underbrace{\pi_*\left(\frac{1}{m(B_1)}\mathbb{1}_{B_1}\right)w}_{=w_1} + \underbrace{\frac{m(\widetilde{P})}{m(P)}}_{=1-2s}\underbrace{\pi_*\left(\frac{1}{m(\widetilde{P})}\mathbb{1}_{\widetilde{P}}\right)w}_{=\widetilde{w}}$$

where $\widetilde{P} = P\smallsetminus(B_0 \cup B_1)$. Using (6.10) and (6.11) gives $\|w_0 - w\| \leqslant \frac{\varepsilon}{4}\|v\|$ and $\|w_1 - \pi_{g_1}w\| \leqslant \frac{\varepsilon}{4}\|v\|$ by Exercise 1.59. Together with (6.9), we have

$$\|w_0 - w_1\| \geqslant \tfrac{\varepsilon}{4}\|v\|. \tag{6.12}$$

We now show the desired estimate by a detailed calculation. First note that

$$\begin{aligned}\left\|\pi_*(\tfrac{1}{m(P)}\mathbb{1}_P)w\right\|^2 &= \langle sw_0+sw_1+(1-2s)\widetilde{w}, sw_0+sw_1+(1-2s)\widetilde{w}\rangle\\ &= s^2\|w_0\|^2+s^2\|w_1\|^2+(1-2s)^2\|\widetilde{w}\|^2+2s^2\Re\langle w_0,w_1\rangle\\ &\quad + 2s(1-2s)\Re\langle w_0,\widetilde{w}\rangle+2s(1-2s)\Re\langle w_1,\widetilde{w}\rangle\\ &\leqslant \left(2s^2+(1-2s)^2+2s^2+2s(1-2s)+2s(1-2s)\right)\|w\|^2\\ &= \|w\|^2 \leqslant \|v\|^2\end{aligned}$$

by the Cauchy–Schwarz inequality and the bounds $\|w_0\|, \|w_1\|, \|\widetilde{w}\| \leqslant \|w\|$. Hence it is sufficient to improve the inequality for one term only, and we will then obtain (6.4).

For this, notice that (6.12) implies

$$\begin{aligned}\left(\tfrac{\varepsilon}{4}\right)^2\|v\|^2 \leqslant \|w_0 - w_1\|^2 &= \|w_0\|^2 + \|w_1\|^2 - 2\Re\langle w_0, w_1\rangle\\ &\leqslant 2\|w\|^2 - 2\Re\langle w_0, w_1\rangle\\ &\leqslant 2\|v\|^2 - 2\Re\langle w_0, w_1\rangle,\end{aligned}$$

which gives

$$\Re\langle w_0, w_1\rangle \leqslant \left(1 - \tfrac{\varepsilon^2}{32}\right)\|v\|^2.$$

Combining this with the above gives, for an arbitrary $v \in (\mathcal{H}_\pi^G)^\perp$, either (6.8) or for $w = \pi_*(f^*)v$ with $f = \frac{1}{m(P)}\mathbb{1}_P$, that

$$\left\|\pi_*\left(\tfrac{1}{m(P)}\mathbb{1}_P\right)w\right\|^2 \leqslant \left(1 - s^2\tfrac{\varepsilon^2}{32}\right)\|v\|^2.$$

In either case, we have shown that $\pi_*(f)\pi_*(f^*)$ is a contraction on $(\mathcal{H}_\pi^G)^\perp$. Since $\pi_*(f^*) = \pi_*(f)^*$ by Section 1.5.3, $(\mathcal{H}_\pi^G)^\perp$ is invariant, and

$$\|AA^*\| = \|A\|^2$$

for any operator A on a Hilbert space, we have

$$\|\pi_*(f)\pi_*(f^*)|_{(\mathcal{H}_\pi^G)^\perp}\| = \|\pi_*(f)|_{(\mathcal{H}_\pi^G)^\perp}\|^2.$$

To summarize, we have shown that for P as in (6.6) the function

$$f = \frac{1}{m(P)}\mathbb{1}_P$$

satisfies (6.4) and that δ can be chosen to depend only on Q, ε, and P. Now note that we have $f * f^* \in C_c(G)$ by Exercise 1.49, $(f * f^*)^* = f * f^*$, and that $f * f^*$ also satisfies (6.4). Hence the second half of the equivalence, and all of the additional claims in the second part of the proposition, follow from this. □

6.2.2 (Uniform) Spectral Gap and Ergodic Theory

In ergodic theory, the existence of a spectral gap has many interesting consequences. The first of these is an unexpected and striking quantitative 'random walk ergodic theorem'.

Suppose that G acts on a locally compact space X and μ is an invariant and ergodic probability measure on X such that the induced unitary representation has spectral gap. We note that, using the notation of Definition 6.21, the G-action is ergodic if and only if $L^2_\mu(X)^G = \mathbb{C}\mathbb{1}$. Now let $f \in C_c(G)$ be as in Proposition 6.25. Then

$$v - \int v \,\mathrm{d}\mu \in \left(L^2_\mu(X)^G\right)^\perp$$

and hence

$$\left\|\pi_*(f)^n v - \int v \,\mathrm{d}\mu\right\|_{L^2_\mu} \leqslant (1-\delta)^n \|v\|_{L^2_\mu} \tag{6.13}$$

for $v \in L^2_\mu(X)$. That is, the average

$$\begin{aligned}\pi_*(f)^n v(x) &= \pi_*(f^{*n})v(x)\\ &= \int_G \cdots \int_G f(g_n)\cdots f(g_1)v(g_n\cdots g_1{\boldsymbol\cdot}x)\,\mathrm{d}m(g_1)\cdots \mathrm{d}m(g_n)\end{aligned}$$

converges exponentially fast in the square mean norm to the integral $\int v \,\mathrm{d}\mu$. This is an odd kind of ergodic theorem, corresponding to a 'random walk'

$$g_1, g_2g_1, \ldots, g_n\cdots g_1$$

on G, and, via

$$g_1\boldsymbol{\cdot}x,\ (g_2g_1)\boldsymbol{\cdot}x,\ldots,\ (g_n\cdots g_1)\boldsymbol{\cdot}x,$$

also on X. In fact, we have to use a particular density f^{*n} on the group to form the averages, but it is striking to obtain an exponential rate for all L^2-functions.† We will refine this phenomenon in Section 8.7 for particular representations of $\mathrm{SL}_2(\mathbb{R})$, and for more general groups we refer to the monograph of Gorodnik and Nevo [39].

Another consequence of the existence of a uniform spectral gap concerns the topology of the space of ergodic measures for the action. For actions of $\mathbb{Z}$, one can find many examples of continuous actions for which the set of ergodic invariant measures is dense in the much larger convex set of invariant probability measures (examples include the full shift on $\mathbb{F}_2^{\mathbb{Z}}$ discussed in Section 5.2.5, hyperbolic toral automorphisms, or the time-one map in an Anosov flow). As was discovered by Glasner and Weiss [35] (see also work of Burger and Sarnak [9] for a related argument in a different context), uniform spectral gap implies that the structure of the set of ergodic measures is more well-behaved.

Corollary 6.26 (Ergodic measures with uniform spectral gap). *Let G act continuously on a locally compact σ-compact metric space X. Suppose that (μ_n) is a sequence of invariant ergodic probability measures on X such that the induced unitary representations have uniform spectral gap. If $\mu_n \to \mu$ as $n \to \infty$ in the weak* topology for some probability measure μ, then μ is also G-invariant and ergodic, with the same spectral gap. In particular, if G has property* (T)*, then the set of ergodic measures is a closed subset of the convex set of invariant measures.*

PROOF. By the hypotheses and Proposition 6.25 there exists $f = f^* \in C_c(X)$ with $f \geqslant 0$, $\int f \,\mathrm{d}m_G = 1$ such that

$$\left\| \pi_*(f)v - \int v \,\mathrm{d}\mu_n \right\|_{L^2_{\mu_n}} \leqslant (1-\delta)\|v\|_{L^2_{\mu_n}} \tag{6.14}$$

for all $v \in L^2_{\mu_n}(X)$, and so in particular for all $v \in C_c(X)$. Now notice that

$$\left(\pi_*(f)v\right)(x) = \int_{\operatorname{supp} f} f(g)v(g^{-1}\boldsymbol{\cdot}x)\,\mathrm{d}m(g)$$

for $x \in X$ defines another continuous function of compact support. Squaring both sides of (6.14) and expanding the definition shows it is an inequality between quantities derived from integrals of continuous functions (for example, $|v|^2$ on the right-hand side) against μ_n. Thus for every fixed $v \in C_c(X)$ the inequality (6.14) depends continuously on μ_n, and so also holds for the limit measure μ.

† This is in complete contrast to the case of a single transformation, where Rokhlin towers (see [24, Sec. 2.9]) can be used to show that no uniform rate occurs in the convergence of ergodic averages; see Krengel [58].

Moreover, (6.14) now extends by density of $C_c(X) \subseteq L^2_\mu(X)$ and continuity (with respect to $\|\cdot\|_{L^2_\mu}$) to all functions $v \in L^2_\mu(X)$.

If now $v \in L^2_\mu(X)$ is G-invariant then, by applying (6.14) to the function

$$v - \int v \, \mathrm{d}\mu,$$

we get

$$\left\| v - \int v \, \mathrm{d}\mu \right\|_{L^2_\mu} \leqslant (1-\delta) \left\| v - \int v \, \mathrm{d}\mu \right\|_{L^2_\mu}$$

which shows that

$$v = \int v \, \mathrm{d}\mu,$$

and so the limit measure μ is ergodic.

The final claim follows from the above, together with Lemma 6.23. □

Exercise 6.27. Square and expand (6.14) to verify that both sides depend continuously on μ_n for a fixed continuous function $v \in C_c(X)$.

Exercise 6.28. Let G act continuously on X, preserving a probability measure μ. Suppose the G-action is ergodic, and the induced representation has spectral gap. Deduce a 'random walk' pointwise ergodic theorem for $v \in L^2_\mu(X)$ with exponential rate of convergence at almost every $x \in X$.

6.2.3 Spectral Gap on Compact Quotients

Proposition 6.29. *Let $\Gamma < G$ be a uniform lattice in a connected unimodular group G, so that $X = \Gamma\backslash G$ is compact. Then the action-associated representation π^X defined by*

$$\pi^X_g(f)(x) = f(xg)$$

for $g \in G$, $f \in L^2(X)$, and $x \in X$ has spectral gap. Moreover, for any function $\psi \in C_c(G)$ the convolution operator $\pi^X_*(\psi)$ is compact with image in $C(X)$.

Proof. By Section C.1 of Appendix C, we may assume that G is equipped with a left-invariant metric d_G. As explained in Section C.3, since $\Gamma <$ is a uniform lattice we can use d_G to induce a metric d_X on X with the property that there is a uniform $r > 0$ for which

$$B^G_r(e) \ni g \longmapsto xg \in B^X_r(x) \tag{6.15}$$

is an isometry for all $x \in X$. We may also assume that $\overline{B^G_r(e)}$ is compact and that (6.15) is measure-preserving with respect to the measures m_G on $B^G_r(e)$ and m_X on $B^X_r(x)$.

Let ψ be a function in $C_c(G)$. By compactness of $\operatorname{supp}\psi$, there exist finitely many elements $g_1, \dots, g_n \in G$ with

$$\overline{B_r^G(e)}\operatorname{supp}\psi \subseteq \bigcup_{j=1}^n g_j B_r^G(e).$$

Then for any $f \in L^2(X)$ and $x \in X$ we have

$$\begin{aligned}\int_{\operatorname{supp}\psi} |\psi(g)f(xg)|\,\mathrm{d}m_G(g) &\leqslant \sum_{j=1}^n \int_{g_j B_r^G(e)} |\psi(g)f(xg)|\,\mathrm{d}m_G(g)\\ &\leqslant \sum_{j=1}^n \|\psi\|_\infty \left\|B_r^G(e) \ni h \mapsto f(xg_jh)\right\|_1\\ &\leqslant n\|\psi\|_\infty\|f\|_1 \leqslant nm_X(X)^{\frac12}\|\psi\|_\infty\|f\|_2, \end{aligned}\qquad (6.16)$$

where we have used the fact that (6.15) is measure-preserving. This shows that the map $G \ni\mapsto \psi(g)f(xg)$ is integrable for all $x \in X$. Using Fubini's theorem, it follows that the convolution operator

$$\pi_*(\psi)f(x) = \int_G \psi(g)f(xg)\,\mathrm{d}m_G(g)$$

can be defined by a Lebesgue integral (instead of just a weak integral) and that $\pi_*(\psi)f$ is a bounded function on X.

To see continuity of $\pi_*(\psi)f$, let $x_1, x_2 \in X$ have $\mathsf{d}_X(x_1,x_2) < r$. Since (6.15) is an isometry, there exists $h \in B_r^G(e)$ with $x_2 = x_1h$ and

$$\mathsf{d}_X(x_1,x_2) = \mathsf{d}_X(x_1,x_1h) = \mathsf{d}_G(e,h).$$

Therefore

$$\begin{aligned}\left|\pi_*(\psi)f(x_1)-\pi_*(\psi)f(x_2)\right| &= \left|\int\psi(g)f(x_1g)\,\mathrm{d}m_G(g)-\int\psi(g)f(x_1hg)\,\mathrm{d}m_G(g)\right|\\ &= \left|\int\psi(g)f(x_1g)\,\mathrm{d}m_G(g)-\int\psi(h^{-1}g)f(x_1g)\,\mathrm{d}m_G(g)\right|\\ &\leqslant \int_{B_r^G(e)\operatorname{supp}\psi} |\psi(g)-\psi(h^{-1}g)||f(x_1g)|\,\mathrm{d}m_G(g).\end{aligned}$$

Consider the continuous function $\overline{B_r^G(e)} \times G \ni (h,g) \mapsto \psi(h^{-1}g)$. As this function is also compactly supported, it is uniformly continuous. Hence there exists for every $\varepsilon > 0$ some $\delta \in (0,r)$ so that

$$h \in B_\delta^G(e) \implies \left|\psi(g) - \psi(h^{-1}g)\right| < \varepsilon$$

for all $g \in G$. Using this estimate together with the argument leading to (6.16), we obtain

$$\left|\pi_*(\psi)f(x_1) - \pi_*(\psi)f(x_2)\right| < \varepsilon C\|f\|_2$$

for all $x_1, x_2 \in X$ with $\mathsf{d}_X(x_1, x_2) < \delta$ and a constant C depending only on X and ψ. In particular, we deduce that $\pi_*(\psi)f$ is a continuous function on X.

Moreover, our argument above shows that $\pi_*(\psi)B_1^{L^2(X)}(0)$ is a bounded and equicontinuous subset of $C(X)$. By the Arzela–Ascoli theorem, this implies that the closure of $\pi_*(\psi)B_1^{L^2(X)}(0)$ is compact in $C(X)$, which by definition means that $\pi_*(\psi)\colon L^2(X) \to C(X)$ is a compact operator. Composing this map with the continuous embedding $C(X) \to L^2(X)$, we may view $\pi_*(\psi)$ also as a compact operator from $L^2(X)$ to $L^2(X)$.

To prove the spectral gap as claimed in the first part of the proposition, we wish to employ Proposition 6.25. So suppose that the function $\psi \in C_c(G)$ satisfies $\psi \geqslant 0$, $\psi(e) > 0$, $\|\psi\|_1 = 1$, and $\psi^* = \psi$. By the discussion above we know that $\pi_*(\psi)$ is compact. As it is also a self-adjoint operator of norm at most 1, we deduce that $\pi_*(\psi)$ is diagonalizable with eigenvalues in $[-1, 1]$ and converging to 0 (if there are infinitely many eigenvalues). Note that $\operatorname{supp}\psi$ contains a neighbourhood of $e \in G$, which implies that G is generated by $\operatorname{supp}\psi$ since G is connected. Note that $\pi_*(\psi)f = f$ if and only if $f \in C(X)$ and we have

$$\int \psi(g) f(xg) \,\mathrm{d}m_G(g) = f(x). \tag{6.17}$$

We claim that this holds if and only if f is invariant under G. It is straightforward to see that invariance of f under G implies (6.17). So suppose for the converse that (6.17) holds. If necessary we may replace f by $\Re f$ and $\Im f$ and assume that f is real-valued. As X is compact there exists some $x_{\max} \in X$ so that

$$f(x_{\max}) = \max_{x \in X} f(x).$$

Applying (6.17) for $x_{\max}$ shows that we must have $f(x_{\max} g) = f(x_{\max})$ for all $g \in \operatorname{supp}\psi$, which can be iterated to give the same for all $g \in G = \langle \operatorname{supp}\psi \rangle$. As the G-action is transitive on $X = \Gamma\backslash G$ we have that the eigenspace for $\pi_*(\psi)$ and eigenvalue 1 is precisely the space $\mathbb{C}\mathbb{1}_X$ of constant functions on X.

Moreover suppose $f \in L^2(X)$ satisfies $\pi_*(\psi)f = -f$. Then

$$\pi_*(\psi * \psi)f = \pi_*(\psi)^2 f = f,$$

and so by the above applied to $\psi * \psi$, we deduce that f is constant. However, together with the assumption $\pi_*(\psi)f = -f$ we see that $f = 0$.

It follows that all eigenvalues of $\pi_*(\psi)$ restricted to $(\mathbb{C}\mathbb{1}_X)^\perp = \left(L^2(X)^G\right)^\perp$ have absolute value less than one. By the spectral theorem for compact self-adjoint operators, this is equivalent to

$$\left\| \pi_*(\psi)|_{\left(L^2(X)^G\right)^\perp} \right\|_{\mathrm{op}} < 1.$$

It follows that the action-associated representation has spectral gap by Proposition 6.25. □

6.3 Characterizing Weak Containment

To avoid cumbersome notation we also introduce a shorthand for positive-definite functions appearing in Definition 6.1 and the results of Section 6.1.

Definition 6.30 (Positive-definite functions associated to π). For a unitary representation π of G we define the set

$$\mathcal{P}_\pi^1 = \left\{ \phi^\pi = \sum_{j=1}^n \varphi_{v_j}^\pi \;\middle|\; n \in \mathbb{N}, v_1, \ldots, v_n \in \mathcal{H}_\pi \text{ and } \sum_{j=1}^n \|v_j\|^2 = 1 \right\},$$

and refer to its elements $\phi^\pi \in \mathcal{P}_\pi^1$ as *positive-definite functions associated* to π. Moreover, we define $\overline{\mathcal{P}_\pi^1}$ to be the closure of $\mathcal{P}_\pi^1$ in the compact-open topology, and will refer to its elements ϕ^π as positive-definite functions *weakly associated* to π.

By comparing this with the definition of weak containment (see Definition 6.1 and Lemma 6.7), we see that a representation π of G is weakly contained in another representation ρ of G if and only if for any unit vector $v \in \mathcal{H}_\pi$ we have $\varphi_v^\pi \in \overline{\mathcal{P}_\rho^1}$.

We defined weak containment as a weakening of actual containment (Definition 1.22), and we have seen in the abelian setting examples where weak containment cannot be replaced by containment (see Exercise 6.3). We will see that weak containment will become more useful and important after we have obtained equivalent formulations. This motivates the following result, which goes back to the work of Fell [29] and Eymard [28].

Theorem 6.31 (Weak containment). *Let π and ρ be unitary representations of G. Then the following are equivalent:*

($\pi \prec_{\mathrm{diag}} \rho$) *For every unit vector $v \in \mathcal{H}_\pi$ we have $\varphi_v^\pi \in \overline{\mathcal{P}_\rho^1}$.*
($\pi \prec_{\mathrm{op}} \rho$) *We have $\|\pi_*(f)\|_{\mathrm{op}} \leqslant \|\rho_*(f)\|_{\mathrm{op}}$ for all $f \in L^1(G)$ (or $C_c(G)$).*
($\pi \prec_{\mathrm{mc}} \rho$) *For every $v, w \in \mathcal{H}_\pi$, compact $Q \subseteq G$ and $\varepsilon > 0$ there exist $a_1, \ldots, a_n, b_1, \ldots, b_n \in \mathcal{H}_\rho$ such that we again have the uniform approximation of the matrix coefficients*

$$\left| \langle \pi_g v, w \rangle - \sum_{j=1}^n \langle \rho_g a_j, b_j \rangle \right| < \varepsilon$$

for all $g \in Q$ with the additional constraint[†] *that*

$$\sum_{j=1}^n \|a_j\| \|b_j\| \leqslant \|v\| \|w\|.$$

[†] The reader should recall from the proof of Lemma 6.7 that if $e \in Q$ then this additional constraint is—up to ε—automatic in Definition 6.1. This is unclear in the first estimate of ($\pi \prec_{\mathrm{mc}} \rho$).

If any (and hence all) of these conditions are satisfied then we again say that π is weakly contained *in ρ, and write $\pi \prec \rho$.*

We will prove the theorem in three steps.

PROOF THAT $(\pi \prec_{\text{diag}} \rho)$ IMPLIES $(\pi \prec_{\text{op}} \rho)$. Let $f \in C_c(G)$ and define

$$f_0 = f^* * f \in C_c(G).$$

As this implies (by the work in Section 1.5.3) that both $\pi_*(f_0) = \pi_*(f)^*\pi_*(f)$ and $\rho_*(f_0) = \rho_*(f)^*\rho_*(f)$ are self-adjoint positive operators, we have

$$\|\pi_*(f_0)\|_{\text{op}} = \sup_{\substack{v \in \mathcal{H}_\pi \\ \|v\|=1}} \langle \pi_*(f_0)v, v\rangle$$

and

$$\|\rho_*(f_0)\|_{\text{op}} = \sup_{\substack{w \in \mathcal{H}_\rho \\ \|w\|=1}} \langle \rho_*(f_0)w, w\rangle.$$

Fix some unit vector $v \in \mathcal{H}_\pi$, let $Q = \operatorname{supp} f_0$ and $\varepsilon > 0$. By assumption, we can find $\phi^\rho = \sum_{j=1}^n \varphi^\rho_{w_j} \in \mathcal{P}^1_\rho$ with $w_1, \dots, w_n \in \mathcal{H}_\rho$ and $\sum_{j=1}^n \|w_j\|^2 = 1$ such that

$$\|\varphi^\pi_v - \phi^\rho\|_{Q,\infty} < \varepsilon.$$

We multiply $\varphi^\pi_v(g) - \phi^\rho(g)$ by $f_0(g)$ and integrate over $Q = \operatorname{supp} f_0$ to obtain

$$\left| \langle \pi_*(f_0)v, v\rangle - \sum_{j=1}^n \langle \rho_*(f_0)w_j, w_j\rangle \right| < \varepsilon \|f_0\|_1.$$

Therefore

$$\begin{aligned}\langle \pi_*(f_0)v, v\rangle &\leqslant \sum_{j=1}^n \langle \rho_*(f_0)w_j, w_j\rangle + \varepsilon\|f_0\|_1 \\ &\leqslant \sum_{j=1}^n \|\rho_*(f_0)\|_{\text{op}}\|w_j\|^2 + \varepsilon\|f_0\|_1 = \|\rho_*(f_0)\|_{\text{op}} + \varepsilon\|f_0\|_1.\end{aligned}$$

As $\varepsilon > 0$ and the unit vector $v \in \mathcal{H}_\pi$ were arbitrary, we deduce that

$$\|\pi_*(f_0)\|_{\text{op}} \leqslant \|\rho_*(f_0)\|_{\text{op}}.$$

Now recall that

$$\pi_*(f_0) = \pi_*(f^* * f) = \pi_*(f)^*\pi_*(f),$$

and similarly

$$\rho_*(f_0) = \rho_*(f)^*\rho_*(f),$$

which together with the identity $\|A^*A\|_{\text{op}} = \|A\|^2_{\text{op}}$ for any operator A on a Hilbert space gives $\|\pi_*(f)\|_{\text{op}} \leqslant \|\rho_*(f)\|_{\text{op}}$. By approximating $f \in L^1(G)$ by elements in $C_c(G)$ this inequality extends to $f \in L^1(G)$, and hence $(\pi \prec_{\mathrm{op}} \rho)$ follows. □

PROOF THAT $(\pi \prec_{\mathrm{op}} \rho)$ IMPLIES $(\pi \prec_{\mathrm{mc}} \rho)$. For this step we are going to use the weak* topology on the closed unit ball $B_1 = \overline{B_1^{L^\infty(G)}(0)}$. Define $\mathcal{M}_\rho^{\leqslant 1} \subseteq B_1$ as the set consisting of all finite sums

$$\sum_{j=1}^{n} \varphi_{a_j,b_j}^{\rho} = \sum_{j=1}^{n} \langle \rho_g a_j, b_j \rangle$$

of ρ-matrix coefficients φ_{a_i,b_j}^{ρ} with the side constraint that

$$\sum_{j=1}^{n} \|a_j\| \|b_j\| \leqslant 1.$$

Let $\overline{\mathcal{M}_\rho^{\leqslant 1}}$ be the weak* closure of $\mathcal{M}_\rho^{\leqslant 1}$. Notice that the nature of the definition of $\mathcal{M}_\rho^{\leqslant 1}$ implies that $\mathcal{M}_\rho^{\leqslant 1}$ and $\overline{\mathcal{M}_\rho^{\leqslant 1}}$ are also convex subsets.

We claim that for a given $v, w \in \mathcal{H}_\pi$ with $\|v\| = \|w\| = 1$ the π-matrix coefficient

$$\varphi_{v,w}^{\pi} = \langle \pi_g v, w \rangle$$

belongs to $\overline{\mathcal{M}_\rho^{\leqslant 1}}$. Suppose this is not the case. Then, by a corollary of the Hahn–Banach lemma (see [25, Th. 8.73]), and since $\overline{\mathcal{M}_\rho^{\leqslant 1}}$ is closed and convex, this implies that there is an $\mathbb{R}$-valued $\mathbb{R}$-linear functional L and some $s \in \mathbb{R}$ such that

$$L(\phi) \leqslant s < L(\varphi_{v,w}^{\pi})$$

for all $\phi \in \overline{\mathcal{M}_\rho^{\leqslant 1}}$. It is easy[†] to verify that

$$L_{\mathbb{C}}(\phi) = L(\phi) - \mathrm{i}L(\mathrm{i}\phi)$$

for $\phi \in L^\infty(G)$ defines a $\mathbb{C}$-valued $\mathbb{C}$-linear functional on $L^\infty(G)$ with the property that $L = \Re L_{\mathbb{C}}$. However, since $\overline{\mathcal{M}_\rho^{\leqslant 1}}$ is closed in the weak* topology we may ensure that L and $L_{\mathbb{C}}$ are continuous in the weak* topology, which implies that $L_{\mathbb{C}}$ is given by an evaluation map (see [25, Lem. 8.13]). Together, it follows that there exists some function $f \in L^1(G)$ such that

$$L(\phi) = \Re \int_G f \phi \,\mathrm{d}m$$

for all $\phi \in L^\infty(G)$. Therefore,

[†] Indeed, any $\mathbb{R}$-valued $\mathbb{R}$-linear functional $L_{\mathbb{R}}$ on a complex vector space V has $L_{\mathbb{R}} = \Re L_{\mathbb{C}}$ for a $\mathbb{C}$-valued $\mathbb{C}$-linear functional $L_{\mathbb{C}}$ defined by $L_{\mathbb{C}}(v) = L_{\mathbb{R}}(v) - \mathrm{i}L_{\mathbb{R}}(\mathrm{i}v)$ for $v \in V$.

$$\Re \int_G f \varphi_{a,b}^{\rho} \, dm \leqslant s < \Re \int_G f \varphi_{v,w}^{\pi} \, dm$$

for all $a, b \in \mathcal{H}_\rho$ with $\|a\|, \|b\| \leqslant 1$. Using the definition of $\rho_*(f)$ and $\pi_*(f)$ this gives

$$\Re \langle \rho_*(f)a, b \rangle \leqslant s < | \langle \pi_*(f)v, w \rangle | \leqslant \|\pi_*(f)\|.$$

As $a, b \in \mathcal{H}_\rho$ are arbitrary with $\|a\|, \|b\| \leqslant 1$ this gives

$$\|\rho_*(f)\| \leqslant s < \|\pi_*(f)\|,$$

which contradicts the assumption ($\pi \prec_{\text{op}} \rho$). It follows that $\varphi_{v,w}^{\pi}$ can be approximated in the weak* topology by elements of $\mathcal{M}_\rho^{\leqslant 1}$. To see that the matrix coefficient $\varphi_{v,w}^{\pi}$ can be approximated in the compact-open topology by finite sums, as claimed in ($\pi \prec_{\text{mc}} \rho$), we will need Lemma 6.32 below. □

Lemma 6.32 (Approximation in the compact-open topology). *Suppose π and ρ are unitary representations of G and that elements $v, w \in \mathcal{H}_\pi$ have the property that the matrix coefficient $\varphi_{v,w}^{\pi} \in C_b(G)$ can be approximated arbitrarily well in the weak* topology by finite sums*

$$\phi = \sum_{j=1}^{n} \varphi_{a_j, b_j}^{\rho}$$

of matrix coefficients defined by vectors $a_1, \dots, a_n, b_1, \dots, b_n \in \mathcal{H}_\rho$ that satisfy the constraint

$$\sum_{j=1}^{n} \|a_j\| \|b_j\| \leqslant \|v\| \|w\|.$$

Then $\varphi_{v,w}^{\pi}$ can also be approximated arbitrarily well by such sums in the compact-open topology.

As we will see, the proof is an adaptation of the method of proof of Proposition 1.79

PROOF OF LEMMA 6.32. Let π be a unitary representation, let $v, w \in \mathcal{H}_\pi$, and let us use the shorthand $\varphi_{v,w} = \varphi_{v,w}^{\pi}$. We fix some $\varepsilon > 0$ and some compact $K \subseteq G$. By continuity of the representation there exists a compact neighbourhood U of $e \in G$ such that

$$\left| \varphi_{v,w}(gh) - \varphi_{v,w}(g) \right| = \left| \left\langle \pi_g(\pi_h v - v), w \right\rangle \right| < \varepsilon \tag{6.18}$$

for all $g \in G$ and $h \in U$. We define $f_0 = \frac{1}{m(U)} \mathbb{1}_U$ and choose some open neighbourhood $V = V^{-1} \subseteq U$ of $e \in G$ such that

$$\left\| \lambda_h f_0 - f_0 \right\|_1 < \varepsilon$$

for all $h \in V$. Moreover, we use translates of V to cover K, so that

$$\bigcup_{\ell=1}^{n} g_\ell V \supseteq K, \tag{6.19}$$

and use the elements $g_1, \dots, g_m$ to define $f_\ell = \frac{1}{m(U)} \mathbb{1}_{g_\ell U}$ for $\ell = 1, \dots, m$ and the weak* neighbourhood

$$\mathcal{N}_{f_0,\dots,f_m;\varepsilon} = \bigcap_{\ell=0,\dots,m} \left\{ \phi \in L^\infty(G) \mid \left| \int (\phi - \varphi_{v,w}) f_\ell \,\mathrm{d}m \right| < \varepsilon \right\}$$

of $\varphi_{v,w}$. Suppose now that

$$\phi = \sum_{j=1}^{n} \varphi^{\rho}_{a_j,b_j} \in \mathcal{N}_{f_0,\dots,f_m;\varepsilon} \tag{6.20}$$

for some unitary representation ρ and vectors $a_1, \dots, a_n, b_1, \dots, b_n \in \mathcal{H}_\rho$ with $\sum_{j=1}^{n} \|a_j\| \|b_j\| \leqslant \|v\| \|w\|$. The pairing of ϕ and f_ℓ can be expressed differently as

$$\int_G \phi f_\ell \,\mathrm{d}m = \frac{1}{m(U)} \sum_{j=1}^{n} \int_U \langle \rho_{g_\ell h} a_j, b_j \rangle \,\mathrm{d}m(h) = \sum_{j=1}^{n} \Big\langle \rho_{g_\ell} \underbrace{\rho_*(f_0) a_j}_{=\widetilde{a_j}}, b_j \Big\rangle.$$

Using the continuity properties in (6.18) twice we also have, for any $g \in g_\ell U$,

$$\int_G \varphi_{v,w} f_\ell \,\mathrm{d}m = \frac{1}{m(U)} \int_{g_\ell U} \varphi_{v,w} \,\mathrm{d}m = \varphi_{v,w}(g) + \mathrm{O}(\varepsilon).$$

Combining the last two formulas with the assumption (6.20), we obtain

$$\sum_{j=1}^{n} \langle \rho_{g_\ell} \widetilde{a_j}, b_j \rangle = \varphi_{v,w}(g) + \mathrm{O}(\varepsilon) \tag{6.21}$$

for $\ell = 0, \dots, m$.

Finally, for $\widetilde{a_j} = \rho_*(f_0) a_j$ we will need the following uniform continuity property, namely that $h \in V$ implies

$$\|\rho_h \widetilde{a_j} - \widetilde{a_j}\| = \|\rho_h \rho_*(f_0) a_j - \rho_*(f_0) a_j\| \leqslant \|\lambda_h f_0 - f_0\|_1 \|a_j\| \leqslant \varepsilon \|a_j\| \tag{6.22}$$

for $j = 1, \dots, n$. Therefore, combining the above for $g = g_\ell h \in g_\ell V$ we obtain

$$\begin{aligned}\sum_{j=1}^{n}\langle\rho_g\widetilde{a_j},b_j\rangle &= \sum_{j=1}^{n}\langle\rho_{g_\ell h}\widetilde{a_j},b_j\rangle\\ &= \sum_{j=1}^{n}\left(\langle\rho_{g_\ell}\widetilde{a_j},b_j\rangle + \mathrm{O}(\varepsilon\|a_j\|\|b_j\|)\right) && \text{(by (6.22))}\\ &= \varphi_{v,w}(g) + \mathrm{O}_{v,w}(\varepsilon) && \text{(by (6.21))}\end{aligned}$$

and

$$\sum_{j=1}^{n}\|\widetilde{a_j}\|\|b_j\| \leqslant \sum_{j=1}^{n}\|a_j\|\|b_j\| \leqslant \|v\|\|w\|.$$

We now can replace the sum of matrix coefficients in (6.20) by the sum

$$\widetilde{\phi} = \sum_{j=1}^{n}\varphi^{\rho}_{\widetilde{a}_j,b_j}$$

and, using the cover in (6.19), we obtain the desired statement. □

Proof that $(\pi\prec_{\mathrm{mc}}\rho)$ implies $(\pi\prec_{\mathrm{diag}}\rho)$. Let $v\in\mathcal{H}_\pi$ have $\|v\| = 1$, and let $Q\subseteq G$ be a compact subset containing e. Fix some $\varepsilon\in(0,1]$. Applying $(\pi\prec_{\mathrm{mc}}\rho)$ we find $a_j,b_j\in\mathcal{H}_\rho$ with

$$\left|\langle\pi_g v,v\rangle - \sum_{j=1}^{n}\langle\rho_g a_j,b_j\rangle\right| < \varepsilon^4$$

for all $g\in Q$ and

$$\sum_{j=1}^{n}\|a_j\|\|b_j\| \leqslant \|v\|^2 = 1.$$

Since $e\in Q$ we also have

$$\left|\sum_{j=1}^{n}\langle a_j,b_j\rangle - \|v\|^2\right| < \varepsilon^4. \tag{6.23}$$

Together these inequalities will show that a_j and b_j are close to each other for 'most' j, which will allow us to switch to diagonal matrix coefficients.

Indeed, let

$$\mathrm{Bad} = \{j \mid \Re\langle a_j,b_j\rangle \leqslant (1-\varepsilon^2)\|a_j\|\|b_j\|\}.$$

Then using (6.23) we obtain

$$\begin{aligned}
1-\varepsilon^4 &< \Re\sum_{j=1}^{n}\langle a_j,b_j\rangle \\
&\leqslant \sum_{j\in\mathrm{Bad}}(1-\varepsilon^2)\|a_j\|\|b_j\| + \sum_{j\notin\mathrm{Bad}}\|a_j\|\|b_j\| \\
&\leqslant 1-\varepsilon^2\sum_{j\in\mathrm{Bad}}\|a_j\|\|b_j\|,
\end{aligned}$$

which gives

$$\sum_{j\in\mathrm{Bad}}\|a_j\|\|b_j\| < \varepsilon^2.$$

Dropping those bad indices altogether, we may worsen our estimate by ε^2 to

$$\left|\langle\pi_g v,v\rangle - \sum_{j=1}^{n}\langle\rho_g a_j,b_j\rangle\right| < 2\varepsilon^2,$$

but may assume without loss of generality that

$$\Re\langle a_j,b_j\rangle > (1-\varepsilon^2)\|a_j\|\|b_j\| \tag{6.24}$$

for all j. Multiplying a_j by a scalar and b_j by its inverse, we may also assume that $\|a_j\| = \|b_j\|$. In that case (6.24) gives

$$\|a_j-b_j\|^2 = 2\|a_j\|^2 - 2\Re\langle a_j,b_j\rangle < 2\varepsilon^2\|a_j\|^2,$$

and so

$$|\langle\rho_g a_j,a_j\rangle - \langle\rho_g a_j,b_j\rangle| = |\langle\rho_g a_j,a_j-b_j\rangle| \leqslant \sqrt{2}\varepsilon\|a_j\|^2.$$

Thus we may replace b_j by a_j and worsen the approximation only by another 2ε. However, this means that we have obtained a uniform approximation of $\langle\pi_g v,v\rangle$ on Q by a sum of diagonal matrix coefficients for ρ. Applying Lemma 6.7 gives $(\pi\prec_{\mathrm{diag}}\rho)$. □

Exercise 6.33. Show that the following condition is also equivalent to weak containment:

$(\pi\prec_{\mathrm{measure}}\rho)$ We have $\|\pi_*(\nu)\| \leqslant \|\rho_*(\nu)\|$ for all $\nu\in\mathcal{M}(G)$.

6.3.1 Weak Containment for Abelian Groups

Using the characterization $(\pi\prec_{\mathrm{op}}\rho)$ of weak containment in Theorem 6.31 it is quite convenient to characterize weak containment for abelian groups using the support of unitary representations as in Definition 2.67.

Corollary 6.34 (Weak containment for abelian groups). *Let G be an abelian group, and let π and ρ be unitary representations of G. Then $\pi\prec\rho$ if* *and only if* $\operatorname{supp}\pi \subseteq \operatorname{supp}\rho$. *In particular, for any* $t \in \widehat{G}$, *we have* $\chi_t \prec \rho$ *if and only if* $t \in \operatorname{supp}\rho$.

PROOF. By the spectral theorem (Corollaries 2.13 and 2.65) we have

$$\|\pi_*(f)\|_{\mathrm{op}} = \|\check{f}\|_{\operatorname{supp}\pi,\infty}$$

for $f \in L^1(G)$, and similarly for ρ. If now $\operatorname{supp}\pi \subseteq \operatorname{supp}\rho$, then we obtain

$$\|\pi_*(f)\|_{\mathrm{op}} = \|\check{f}\|_{\operatorname{supp}\pi,\infty} \leqslant \|\check{f}\|_{\operatorname{supp}\rho,\infty} = \|\rho_*(f)\|_{\mathrm{op}}$$

for all $f \in L^1(G)$, which implies $\pi \prec \rho$ by Theorem 6.31.

Suppose now that $\operatorname{supp}\pi$ is not contained in $\operatorname{supp}\rho$, and choose an element $t_0 \in \operatorname{supp}\pi \smallsetminus \operatorname{supp}\rho$. By Urysohn's lemma, there exists some $F \in C_c(\widehat{G})$ with $F(t_0) = 1$ and $F|_{\operatorname{supp}\rho} = 0$. By Corollary 2.5, there exists some function $f \in L^1(G)$ with $\|\check{f} - F\|_\infty < \frac{1}{2}$. Therefore

$$\|\check{f}\|_{\operatorname{supp}\rho,\infty} = \|\check{f} - F\|_{\operatorname{supp}\rho,\infty} < \tfrac{1}{2}$$

and

$$\|\check{f}\|_{\operatorname{supp}\pi,\infty} \geqslant |\check{f}(t_0)| > \tfrac{1}{2},$$

which gives $\|\pi_*(f)\|_{\mathrm{op}} > \|\rho_*(f)\|_{\mathrm{op}}$ and, by Theorem 6.31, that π is not weakly contained in ρ. □

6.3.2 Cyclic Representations

Using the characterization of weak containment in Theorem 6.31, we can characterize weak containment in terms of the generator, and explain the meaning of $\overline{\mathcal{P}^1_\rho}$ more clearly. For this the following lemma will be useful.

Lemma 6.35 (Convolution formula for operator norm). *Let* π *be a unitary representation of* G, *and let* $f \in C_c(G)$. *Then*

$$\begin{aligned}\|\pi_*(f)\|_{\mathrm{op}} &= \sup_{v \in \mathcal{H}_\pi} \lim_{n\to\infty} \left(\langle \pi_*(f^* * f)^{*n} v, v\rangle\right)^{\frac{1}{2n}} \\ &= \sup_{v \in \mathcal{H}_\pi} \lim_{n\to\infty} \left(\int_G (f^* * f)^{*n} \varphi_v^\pi \,\mathrm{d}m\right)^{\frac{1}{2n}},\end{aligned}$$

where the supremum could also be taken over a dense subset of $\mathcal{H}_\pi$.

PROOF. Let $f \in C_c(G)$ and define $f_0 = f^* * f \in C_c(G)$ so that

$$T = \pi_*(f_0) = \pi_*(f)^* \pi_*(f)$$

is a positive self-adjoint operator. Fix some $v \in \mathcal{H}_\pi$ and let μ_v^T denote the spectral measure on $[0,\infty) \subseteq \mathbb{R}$ for the self-adjoint operator T (see [25, Sec. 12.4]) so that, in particular,

$$\langle \pi_*(f_0)v, v\rangle = \int_0^\infty x \,\mathrm{d}\mu_v^T(x).$$

We apply Hölder's inequality for the conjugate pair of exponents $(n, \frac{n}{n-1})$ to the last expression and obtain

$$\int_0^\infty x \cdot 1 \,\mathrm{d}\mu_v^T(x) \leqslant \underbrace{\left(\int_0^\infty x^n \,\mathrm{d}\mu_v^T(x)\right)^{\frac{1}{n}}}_{\longrightarrow S_v} \underbrace{\left(\int_{\mathbb{R}} \mathrm{d}\mu_v^T\right)^{1-\frac{1}{n}}}_{\longrightarrow \|v\|^2}$$

as $n \to \infty$, where[†] the limit S_v can be expressed using the continuous functional calculus for T (see [25, Sec. 12.4]) as

$$S_v = \sup \operatorname{supp} \mu_v^T = \lim_{n\to\infty} \left(\int_0^\infty x^n \,\mathrm{d}\mu_v^T(x)\right)^{\frac{1}{n}} = \lim_{n\to\infty} \langle \pi_*(f_0)^n v, v\rangle^{\frac{1}{n}}.$$

Combining these, we obtain that

$$\langle \pi_*(f_0)v, v\rangle \leqslant S_v \|v\|^2.$$

Taking the supremum over all $v \in \mathcal{H}_\pi$ (or v from a dense subset of $\mathcal{H}_\pi$) with $\|v\| \leqslant 1$ gives

$$\|\pi_*(f_0)\|_{\mathrm{op}} = \sup_{\|v\|\leqslant 1} \langle \pi_*(f_0)v, v\rangle \leqslant \sup_{v\in\mathcal{H}_\pi} S_v,$$

as $\pi_*(f_0)$ is positive and self-adjoint.

Also note that

$$S_v = \lim_{n\to\infty} \langle \pi_*(f_0)^n v, v\rangle^{\frac{1}{n}} \leqslant \lim_{n\to\infty} \left(\|\pi_*(f_0)^n\|_{\mathrm{op}} \|v\|^2\right)^{\frac{1}{n}} = \|\pi_*(f_0)\|_{\mathrm{op}}$$

and so

$$\sup_{v\in\mathcal{H}_\pi} S_v \leqslant \|\pi_*(f_0)\|_{\mathrm{op}},$$

which implies that

$$\|\pi_*(f_0)\|_{\mathrm{op}} = \sup_{v\in\mathcal{H}_\pi} \lim_{n\to\infty} \langle \pi_*(f_0)^n v, v\rangle^{\frac{1}{n}} = \sup_{v\in\mathcal{H}_\pi} \lim_{n\to\infty} \left(\int f_0^{*n} \varphi_v^\pi \,\mathrm{d}m\right)^{\frac{1}{n}}$$

[†] We note that the limits exist for the following simple reason: if $F \in L^\infty(X,\mu)$ is a non-negative bounded function on a finite measure space (X,μ), then $\left(\int_X F^n \,\mathrm{d}\mu\right)^{\frac{1}{n}}$ converges to $\|F\|_\infty$ as $n \to \infty$.

Finally, since $f_0 = f^* * f$ for $f \in C_c(G)$, the lemma follows from the identity $\|\pi_*(f^* * f)\|_{\mathrm{op}} = \|\pi_*(f)\|_{\mathrm{op}}^2$. □

Corollary 6.36 (Generators suffice). *Let ρ be a unitary representation of G. Then, for any cyclic representation π with generator $v_0 \in \mathcal{H}_\pi$ of norm $\|v_0\| = 1$, we have $\pi \prec \rho$ if and only if $\varphi_{v_0}^\pi \in \overline{\mathcal{P}_\rho^1}$. Moreover, we have*

$$\begin{aligned}[0,1]\overline{\mathcal{P}_\rho^1} &= \textit{the closure of } [0,1]\mathcal{P}_\rho^1 \textit{ in the compact-open topology}\\ &= \textit{the closure of } [0,1]\mathcal{P}_\rho^1 \textit{ in the weak* topology}\\ &= \{0\} \cup \big\{\varphi_{v_0}^\pi \;\big|\; \pi \prec \rho \textit{ is cyclic with generator } v_0 \textit{ of norm } \|v_0\| \leqslant 1\big\}.\end{aligned}$$

Proof. Recall that $\overline{\mathcal{P}_\rho^1}$ is defined in Definition 6.30 as the closure of $\mathcal{P}_\rho^1$ in the compact-open topology. We start by proving the equality of the four sets. Continuity of scalar multiplication implies for $\phi^\rho \in \overline{\mathcal{P}_\rho^1}$ and $s \in [0,1]$ that $s\phi^\rho \in s\overline{\mathcal{P}_\rho^1}$ belongs to the closure of $[0,1]\mathcal{P}_\rho^1$ in the compact-open topology. By Proposition 1.79, the compact-open topology on $\overline{B_1^{L^\infty(G)}(0)}$ is stronger than the weak* topology, so the compact-open closure of $[0,1]\mathcal{P}_\rho^1$ is contained in its weak* closure. This already gives two inclusions.

Suppose now that ϕ^ρ belongs to the weak* closure of $[0,1]\mathcal{P}_\rho^1$. By the GNS construction in Theorem 1.76 and Corollary 1.78, there exists a cyclic unitary representation π with generator v_0, so that $\phi^\rho = \varphi_{v_0}^\pi$. Since $\|\phi^\rho\|_\infty \leqslant 1$, we also have $\|v_0\| \leqslant 1$. We need to show that $\pi \prec \rho$ by using the assumption that the matrix coefficient $\varphi_{v_0}^\pi$ of the generator $v_0 \in \mathcal{H}_\pi$ belongs to the weak* closure of $[0,1]\mathcal{P}_\rho^1$. Equivalently, we assume that the matrix coefficient

$$\varphi_{v_0}^\pi = \lim_{k\to\infty} \varphi_{w_k}^{\rho^\infty}$$

is equal to the weak* limit of matrix coefficients of vectors $w_k \in \mathcal{H}_\rho^\infty$ for the representation ρ^∞ with $\|w_k\| \leqslant 1$ for all $k \geqslant 1$.

For this we will apply the convolution formula in Lemma 6.35 and the notion $(\prec_{\mathrm{op}})$ from Theorem 6.31. First note that if $f_0 \in C_c(G)$, then

$$\begin{aligned}\left|\int_G f_0 \varphi_{v_0}^\pi \,\mathrm{d}m\right| &= \lim_{k\to\infty} \left|\int_G f_0 \varphi_{w_k}^{\rho^\infty} \,\mathrm{d}m\right|\\ &= \lim_{k\to\infty} \left|\langle \rho_*^\infty(f_0) w_k, w_k\rangle\right| \leqslant \|\rho_*^\infty(f_0)\|_{\mathrm{op}} = \|\rho_*(f_0)\|_{\mathrm{op}}.\end{aligned}$$

To obtain other vectors from our generator v_0, we let $v = \pi_*(\psi)v_0$ for some function $\psi \in C_c(G)$. Now apply the above for

$$f_n = \psi^* * (f^* * f)^{*n} * \psi \in C_c(G)$$

for $f \in C_c(G)$ and $n \in \mathbb{N}$. This gives

$$\left|\int_G (f^* * f)^{*n} \varphi_v^{*n} \,\mathrm{d}m\right| = \left|\left\langle \pi_*(f^* * f)^{*n} \pi_*(\psi) v_0, \pi_*(\psi) v_0 \right\rangle\right|$$
$$= \left|\left\langle \pi_*(f_n) v_0, v_0 \right\rangle\right|$$
$$= \left|\int f_n \varphi_{v_0}^\pi \,\mathrm{d}m\right| \leqslant \|\rho_*(f_n)\| \leqslant \|\rho_*(\psi)\|^2 \|\rho_*(f)\|^{2n}.$$

Taking the $2n$th root and the limit as $n \to \infty$, we arrive at

$$\lim_{n\to\infty} \left(\int_G (f^* * f)^{*n} \varphi_v^\pi \,\mathrm{d}m\right)^{\frac{1}{2n}} \leqslant \|\rho_*(f)\|_{\mathrm{op}}.$$

As v_0 is a generator of π by assumption, the subspace $\pi_*\left(C_c(G)\right) v_0$ is dense by Corollary 1.53, and hence the convolution formula in Lemma 6.35 implies $\|\pi_*(f)\|_{\mathrm{op}} \leqslant \|\rho_*(f)\|_{\mathrm{op}}$. The characterization ($\prec_{\mathrm{op}}$) of weak containment in Theorem 6.31 now gives $\pi \prec \rho$, as $f \in C_c(G)$ was arbitrary. Hence the set in the second line in the statement of the corollary is contained in the set in the third line.

Suppose now that $\pi \prec \rho$ is cyclic with generator v_0 of norm $\|v_0\| \leqslant 1$. Then $\widetilde{v}_0 = \|v_0\|^{-1} v_0$ satisfies $\varphi_{\widetilde{v}_0}^\pi \in \overline{\mathcal{P}_\rho^1}$ by definition of weak containment, and hence $\varphi_{v_0}^\pi$ belongs to $[0,1]\overline{\mathcal{P}_\rho^1}$.

If π is cyclic with generator v_0 satisfying $\|v_0\| = 1$ and $\varphi_{v_0}^\pi \in \overline{\mathcal{P}_\rho^1}$, then the argument above shows that $\pi \prec \rho$. As the converse of the first claim in the corollary is clear by definition, this gives the corollary. □

6.3.3 Approximation by Irreducible Representations

The following strengthens Corollary 1.78 by restricting the collection of irreducible representations needed for approximating the matrix coefficient of a given unitary representation.

Proposition 6.37 (Weakly contained irreducible representations). *Let ρ be a unitary representation of the group G and let $v \in \mathcal{H}_\rho$ have $\|v\| = 1$. Then φ_v^ρ can be approximated in the compact-open topology by sums $\sum_{j=1}^n \varphi_{v_j}^{\pi_j}$ with $v_j \in \mathcal{H}_{\pi_j}$ for $j = 1, \ldots, n$ and $\sum_{j=1}^n \|v_j\|^2 = 1$ for irreducible unitary representations $\pi_1, \ldots, \pi_n$ weakly contained in ρ.*

Proof. We claim that the extremal points of the convex set $\overline{\mathcal{P}_\rho^1}$ are also extremal in $\mathcal{P}^1(G)$, and so correspond to irreducible representations by Proposition 1.73. For this, let ϕ be an extremal point of $\overline{\mathcal{P}_\rho^1} \subseteq \mathcal{P}^1(G)$, and suppose that π is a cyclic unitary representation of G with generator v such that $\phi = \varphi_v^\pi$ (see Theorem 1.76) and $\|v_0\|^2 = \|\phi\| = 1$. By the characterization of weak containment for cyclic representations in Corollary 6.36 we also have $\pi \prec \rho$. If now ϕ is not an extremal point of $\mathcal{P}^1(G)$, then Proposition 1.73 shows that π is not irreducible and we can find non-trivial subspaces $\mathcal{V}_1$, $\mathcal{V}_2$ such that $\mathcal{H}_\pi = \mathcal{V}_1 \oplus \mathcal{V}_2$, non-zero vectors $v_1 \in \mathcal{V}_1$, $v_2 \in \mathcal{V}_2$ with $v = v_1 + v_2$ and hence $\phi = \varphi_v^\pi = \varphi_{v_1}^\pi + \varphi_{v_2}^\pi$. Moreover, we have $\|v_1\|^{-2}\varphi_{v_1}^\pi \neq \|v_2\|^{-2}\varphi_{v_2}^\pi$ since otherwise $v = v_1 + v_2 \in \mathcal{V}_1 \oplus \mathcal{V}_2$ would not generate $\mathcal{H}_\pi = \mathcal{V}_1 \oplus \mathcal{V}_2$, but instead the graph of an isomorphism between the cyclic representation $\pi|_{\mathcal{V}_1}$ and $\pi|_{\mathcal{V}_2}$ (see Proposition 1.65). It follows that $\pi|_{\mathcal{V}_j} < \pi \prec \rho$ for $j = 1, 2$ and hence $\|v_1\|^{-2}\varphi_{v_1}^\pi, \|v_2\|^{-2}\varphi_{v_2}^\pi \in \overline{\mathcal{P}_\rho^1}$ which, together with the identity $\phi = \varphi_v^\pi = \varphi_{v_1}^\pi + \varphi_{v_2}^\pi$ contradicts extremality of $\phi \in \overline{\mathcal{P}_\rho^1}$. This proves the claim that the extremal points of $\overline{\mathcal{P}_\rho^1}$ are also extremal in $\mathcal{P}^1(G)$, and so correspond to irreducible unitary representations of G (see Proposition 1.73).

By Corollary 6.36, the weak* closure of the convex set $[0,1]\mathcal{P}_\rho^1$ is given by $\mathcal{C} = [0,1]\overline{\mathcal{P}_\rho^1}$. Hence $\mathcal{C} \subseteq \mathcal{P}^{\leqslant 1}(G)$ is weak* compact by the Banach–Alaoglu theorem. Moreover, the extreme points of $\mathcal{C}$ are 0 and the extreme points of $\overline{\mathcal{P}^1(G)}$ belonging to $\mathcal{C}$. Applying now the proof of Corollary 1.78 with this in mind and replacing $\mathcal{P}^{\leqslant 1}(G)$ there with $\mathcal{C}$ gives the proposition. □

Exercise 6.38 (Characterizing weak containment using $\widehat{G}$). Let ρ and γ be unitary representations of G. Show that $\rho \prec \gamma$ if and only if for any $\pi \in \widehat{G}$ with $\pi \prec \rho$ we also have $\pi \prec \gamma$.

6.3.4 Weak Containment for some Metabelian Groups*

In the context of metabelian groups as in Chapter 5, the following corollary to Proposition 6.37 is of interest as it helps to analyze which irreducible representations are weakly contained in a given unitary representation.

Corollary 6.39 (Weak containment and abelian subgroups). *Let A be a closed abelian subgroup of the group G. Then*

$$\overline{\bigcup_{\substack{\pi \in \widehat{G} \\ \pi \prec \rho}} \operatorname{supp}(\pi|_A)} = \operatorname{supp}(\rho|_A) \tag{6.25}$$

for any unitary representation ρ of G.

Proof. If $\pi \in \widehat{G}$ is weakly contained in ρ, then we may restrict the approximation claim in Definition 6.1 to A, and obtain $\pi|_A \prec \rho|_A$. This shows that $\operatorname{supp} \pi|_A \subseteq \operatorname{supp} \rho|_A$ by Corollary 6.34 and the first inclusion in (6.25) follows.

For the converse, we apply Propositions 6.8 and 6.37 (surprisingly often). Suppose now that $t \in \operatorname{supp} \rho|_A$, so that $\chi_t \prec \rho|_A$ by Corollary 6.34. By Proposition 6.8 applied to the irreducible representation $\chi_t \prec \rho|_A$, there exists a sequence of vectors (w_n) in $\mathcal{H}_\rho$ so that $\varphi_{w_n}^{\rho|_A}$ approximates χ_t in the compact-open topology. Applying Proposition 6.37 (for G) to each $\varphi_{w_n}^\rho$, there exist $\pi_j \in \widehat{G}$

weakly contained in ρ and $v_j \in \mathcal{H}_{\pi_j}$ for $j = 1, \ldots, J(n)$ so that $\sum_{j=1}^{J(n)} \|v_j\|^2 = 1$ and $\sum_{j=1}^{J(n)} \varphi_{v_j}^{\pi_j}$ approximates $\varphi_{w_n}^{\rho}$ arbitrarily well. Restricting to A, we may also assume that $\sum_{j=1}^{J(n)} \varphi_{v_j}^{\pi_j|_A}$ approximates χ_t. Applying the first part of Proposition 6.8 (again for A), it follows that χ_t can also be approximated by matrix coefficients $\varphi_{v_n}^{\pi_n|_A}$ with $\pi_n \in \widehat{G}$ weakly contained in ρ and $v_n \in \mathcal{H}_{\pi_n}$ a unit vector.

Once more we apply Proposition 6.37 (for A) and see that $\varphi_{v_n}^{\pi_n|_A}$ can be approximated arbitrarily well by a sum

$$\sum_{j=1}^{J(n)} c_j \chi_{t_{n,j}}$$

with $\chi_{t_{n,j}} \prec \pi_n|_A$ and $c_j \geqslant 0$ for $j = 1, \ldots, J(n)$ and $\sum_{j=1}^{J(n)} c_j = 1$. Combining this with the above, we see that the same is true for χ_t. Finally, we apply Proposition 6.8 again to see that χ_t can be approximated in the compact-open topology by characters χ_{t_n} with $\chi_{t_n} \prec \pi_n|_H$. However, by Corollary 6.34 this shows that $t_n \in \operatorname{supp} \pi_n|_H$ and, by Corollary 2.5, that $t = \lim_{n\to\infty} t_n$ also belongs to the closure appearing on the left-hand side of (6.25). □

Let us study again the question of whether an irreducible representation can be weakly contained in other irreducible representations.

- For abelian groups, Corollary 6.34 shows that $\chi_{t_0} \prec \chi_{t_1}$ if and only if $t_0 = t_1$.
- Let $G = \mathrm{SO}_2(\mathbb{R}) \ltimes \mathbb{R}^2$ as in Section 5.2.2. Restricting the irreducible representations χ_m, χ_n, π^r, and π^s for $m, n \in \mathbb{Z}$ and $r, s \in (0, \infty)$ to the normal abelian subgroup $A \cong \mathbb{R}^2$, we see from Corollary 6.39 that $\pi^r \prec \pi^s$ implies $r = s$, that $\chi_m \not\prec \pi^r$, and that $\pi^r \not\prec \chi_m$. For χ_m and χ_n we apply Corollary 6.34 to $\mathrm{SO}_2(\mathbb{R}) \cong G/A$ and see that $\chi_m \prec \chi_n$ implies $m = n$.
- For the affine group $G = \mathbb{R}_{>0} \ltimes \mathbb{R}$ from Section 5.2.3, we saw in Example 6.11 and Exercise 6.12 that $\chi \prec \pi^{\pm}$ for any character χ on $S = \mathbb{R}_{>0}$. Corollary 6.39 for the normal abelian subgroup $A \cong \mathbb{R}$ shows, however, that $\pi^- \not\prec \pi^+$ and $\pi^+ \not\prec \pi^-$. Moreover, Corollary 6.34 (applied to $S \cong G/A$) shows that no character is weakly contained in another.

Exercise 6.40. (a) Analyze the question of weak containment for all pairs of irreducible unitary representations of the Heisenberg group in Section 5.2.4.
(b) Repeat this for the solvable group Sol in Exercise 5.60.

Exercise 6.41 (Weak containment for the isometry group of the plane). Let ρ be a unitary representation of the isometry group $\mathrm{SO}_2(\mathbb{R}) \ltimes \mathbb{R}^2$ of the plane, and let X be the set $\operatorname{supp} \rho|_{\mathbb{R}^2}$. Prove the following characterizations of weak containment for irreducible unitary representations.
(a) Show that $\pi^r \prec \rho$ if and only if $r\mathbb{S}^1 \subseteq X$.
(b) Show that if 0 is an accumulation point of X, then $\chi_n \prec \rho$ for all $n \in \mathbb{Z}$.
(c) Show that if 0 is not an accumulation point of X, then for every $n \in \mathbb{Z}$ we have $\chi_n \prec \rho$ if and only if $\chi_n < \rho$.

6.4 Fell Topology*

The Fell topology on the unitary dual $\widehat{G}$ of G is useful to give $\widehat{G}$ additional structure, which can be helpful to better understand and visualize $\widehat{G}$. We already saw this in Chapter 2, since the topology on the Pontryagin dual for a locally compact abelian group will turn out to be a special case of the Fell topology (see Sections 2.4 and 6.5.1). However, unlike our discussion above of weak containment, we will only need the Fell topology for non-abelian groups when we want to discuss its properties in a construction or an example.

A word of warning before we get started: The Fell topology can have, much like the Zariski topology in algebraic geometry, unusual properties—and in particular, for non-abelian non-compact groups it is often not Hausdorff. Despite this, we will speak of convergence and limits even if the limit of a sequence may not be uniquely determined by the sequence.

At times it is also useful to have the Fell topology defined on a larger class of unitary representations. For this we recall our standing assumptions that we only wish to consider separable Hilbert spaces, so that it is sufficient to consider unitary representations on $\mathcal{V}_n = \mathbb{C}^n$ for $n \in \mathbb{N}$ or $\mathcal{V}_\infty = \ell^2(\mathbb{N})$.

Definition 6.42 (The set of unitary representations). For the group G, we define the set

$$\mathscr{U}(G) = \bigcup_{n \in \mathbb{N} \cup \{\infty\}} \big\{\pi \mid \pi \text{ is a unitary representation on } \mathcal{V}_n\big\}.$$

For $\pi \in \mathscr{U}(G)$, we will again write $\mathcal{H}_\pi = \mathcal{V}_n$ if π is a unitary representation on $\mathcal{V}_n$ for $n \in \mathbb{N} \cup \{\infty\}$.

As we will see, the Fell topology will have a very strong relationship with the notion of weak containment in Definition 6.1 and Theorem 6.31. Due to the equivalent notions of weak containment in Theorem 6.31, we also have the choice of defining the Fell topology using compact-open approximation of diagonal matrix coefficients as in $(\pi \prec_{\mathrm{diag}} \rho)$, or operator norms for convolution operators as in $(\pi \prec_{\mathrm{op}} \rho)$ from Theorem 6.31. The former may seem more natural, but is notationally slightly heavier and at times requires longer arguments, so we will use operator norms more prominently.

Definition 6.43 (Fell open sets). For any $f \in L^1(G)$ and $\alpha \geqslant 0$, we define the *principal Fell open subset*

$$\mathscr{FO}(f, \alpha) = \big\{\pi \in \mathscr{U}(G) \mid \|\pi_*(f)\|_{\mathrm{op}} > \alpha\big\}.$$

We declare the principal Fell open sets to be a sub-basis of the *Fell topology* on $\mathscr{U}(G)$. That is, a Fell open set is, by definition, a union of finite intersections of principal Fell open sets as above.

Lemma 6.44 (Fell topology). *The Fell topology is second countable. In fact, if $D \subseteq L^1(G)$ is dense, then the sets $\mathscr{FO}(f, r)$ for $f \in D$ and $r \in \mathbb{Q}$ with $r > 0$ form a sub-basis for the Fell topology.*

Proof. Let $D \subseteq L^1(G)$ be dense in $L^1(G)$. Also fix $F \in L^1(G)$ and $\alpha \geqslant 0$. We claim that this implies

$$\mathscr{FO}(F,\alpha) = \bigcup_{\substack{\|F-f\|_1 < r-\alpha \\ r\in\mathbb{Q}\cap(0,\infty);\, f\in D}} \mathscr{FO}(f,r). \tag{6.26}$$

Suppose first that $\pi \in \mathscr{FO}(f,r)$ for some $f \in D$ and rational $r > 0$ with

$$\|F - f\|_1 < r - \alpha.$$

By definition $\pi \in \mathscr{FO}(f,r)$ means that $\|\pi_*(f)\|_{\mathrm{op}} > r$. Hence we have

$$\|\pi_*(F)\|_{\mathrm{op}} \geqslant \|\pi_*(f)\|_{\mathrm{op}} - \|\pi_*(F-f)\|_{\mathrm{op}} > r - \|F-f\|_1 > \alpha$$

by the triangle inequality or, equivalently, $\pi \in \mathscr{FO}(F,\alpha)$.

So suppose now that $\pi \in \mathscr{FO}(F,\alpha)$, so $\|\pi_*(F)\|_{\mathrm{op}} > \alpha$. Then there exists some $r \in \mathbb{Q}$ with $\|\pi_*(F)\|_{\mathrm{op}} > r > \alpha \geqslant 0$ and some $f \in D$ with

$$\|F - f\|_1 < \|\pi_*(F)\|_{\mathrm{op}} - r.$$

This implies that $\|\pi_*(F-f)\|_{\mathrm{op}} \leqslant \|F-f\|_1 < \|\pi_*(F)\|_{\mathrm{op}} - r$, and so

$$\|\pi_*(f)\|_{\mathrm{op}} \geqslant \|\pi_*(F)\|_{\mathrm{op}} - \|\pi_*(F-f)\|_{\mathrm{op}} > r,$$

or equivalently that $\pi \in \mathscr{FO}(f,r)$ also belongs to the right-hand side of (6.26).

Taking all possible finite intersections of principal Fell open sets of the form $\mathscr{FO}(f,r)$ with $f \in D$ and rational $r > 0$ we obtain, because of (6.26) and the definition of the Fell topology, a basis of the Fell topology. Since $L^1(G)$ is separable, we also obtain a countable basis of the Fell topology. □

The following is a corollary to the characterization of weak containment in Theorem 6.31 and the above definition of the Fell topology.

Corollary 6.45 (Fell topology and weak containment). *For unitary representations $\pi, \rho \in \mathscr{U}(G)$ we have $\pi \prec \rho$ if and only if $\pi \in \overline{\{\rho\}}$, where the closure is taken with respect to the Fell topology. Moreover, for a sequence (ρ_n) in $\mathscr{U}(G)$ and $\pi \in \mathscr{U}(G)$ the following conditions are equivalent:*

(i) *ρ_n converges to π as $n \to \infty$ in the Fell topology.*

(ii) *For any strictly monotonely increasing sequence (n_k) in $\mathbb{N}$ we have*

$$\pi \prec \bigoplus_{k=1}^{\infty} \rho_{n_k}.$$

(iii) *For any $f \in L^1(G)$ (or $C_c(G)$), we have*

$$\liminf_{n\to\infty} \|\rho_{n,*}(f)\|_{\mathrm{op}} \geqslant \|\pi_*(f)\|_{\mathrm{op}}.$$

PROOF. We start by proving the equivalences, and assume first that (ρ_n) converges to π as in (i). For any $f \in L^1(G)$ and $\varepsilon > 0$ the set

$$\mathscr{FO}(f, \|\pi_*(f)\|_{\text{op}} - \varepsilon)$$

is an open neighbourhood of π, and so there exists some $N \in \mathbb{N}$ with

$$\rho_n \in \mathscr{FO}(f, \|\pi_*(f)\|_{\text{op}} - \varepsilon)$$

for all $n \geqslant N$. This implies that

$$\liminf_{n\to\infty} \|\rho_{n,*}(f)\|_{\text{op}} \geqslant \|\pi_*(f)\|_{\text{op}} - \varepsilon.$$

Since $\varepsilon > 0$ and $f \in L^1(G)$ were arbitrary, condition (iii) follows.

Suppose now that (iii) holds, (n_k) is a strictly increasing sequence in $\mathbb{N}$, and define

$$\rho = \bigoplus_{k=1}^{\infty} \rho_{n_k}. \tag{6.27}$$

Then

$$\|\rho_*(f)\|_{\text{op}} = \sup_{k\in\mathbb{N}} \|\rho_{n_k,*}(f)\|_{\text{op}} \geqslant \liminf_{n\to\infty} \|\rho_{n,*}(f)\|_{\text{op}} \geqslant \|\pi_*(f)\|_{\text{op}}$$

for any $f \in L^1(G)$ (or $C_c(G)$). By the condition ($\pi \prec_{\text{op}} \rho$) of Theorem 6.31, this implies that $\pi \prec \rho$, as claimed in (ii).

So assume now that (ii) holds. If (ρ_n) does not converge to π, then there exists a neighbourhood of π such that for infinitely many $n \in \mathbb{N}$ the representation ρ_n does not belong to this neighbourhood. Without loss of generality we may assume that this neighbourhood of π is a principal Fell open set $\mathscr{FO}(f, \alpha)$ for some $f \in L^1(G)$ and $\alpha \geqslant 0$. Hence there exists a strictly increasing sequence (n_k) in $\mathbb{N}$ with $\|\rho_{n_k,*}(f)\|_{\text{op}} \leqslant \alpha < \|\pi_*(f)\|_{\text{op}}$. Defining ρ as in (6.27) gives

$$\|\rho_*(f)\| \leqslant \alpha < \|\pi_*(f)\|$$

and contradicts the assumed weak containment in (ii) by Theorem 6.31.

For the first claim in the corollary, we apply the above to the constant sequence defined by $\rho_n = \rho$ for all $n \in \mathbb{N}$ (see also Exercise 6.2(c)). □

We note that the Fell topology (by definition) does not distinguish between equivalent representations. More precisely, if $\pi, \rho \in \mathscr{U}(G)$ are equivalent representations, then $\pi \in \mathscr{O}$ if and only if $\rho \in \mathscr{O}$ for any Fell open set $\mathscr{O}$ in $\mathscr{U}(G)$. In particular, when restricting to irreducible representations the Fell topology can also be considered as a topology on $\widehat{G}$. We will also consider the Fell topology on other subsets of $\mathscr{U}(G)$.

Exercise 6.46. Show that a sequence (ρ_n) in $\mathscr{U}(G)$ converges to $\pi \in \mathscr{U}(G)$ if and only if ρ_n converges to $\pi|_{\langle v\rangle_\pi}$ as $n \to \infty$ for any unit vector $v \in \mathcal{H}_\pi$ (or any vector in a generating set $\{v_k \mid k \in \mathbb{N}\}$ satisfying $\mathcal{H}_\pi = \bigoplus_{k=1}^{\infty} \langle v_k \rangle_\pi$).

Exercise 6.47. Show that the operation $\oplus\colon \mathscr{U}(G) \times \mathscr{U}(G) \longrightarrow \mathscr{U}(G)$ is continuous with respect to the Fell topology. (Formally, we use for every $m, n \in \mathbb{N}$ some fixed isomorphisms $\mathcal{V}_m \oplus \mathcal{V}_n \cong \mathcal{V}_{m+n}$, $\mathcal{V}_n \oplus \mathcal{V}_\infty \cong \mathcal{V}_\infty \cong \mathcal{V}_\infty \oplus \mathcal{V}_n$, and $\mathcal{V}_\infty \oplus \mathcal{V}_\infty \cong \mathcal{V}_\infty$ to interpret $\oplus$ as an actual map from $\mathscr{U}(G) \times \mathscr{U}(G)$ to $\mathscr{U}(G)$.)

Exercise 6.48. Suppose G is compact. Show that the Fell topology on $\widehat{G}$ is discrete.

6.4.1 Explaining the 'Property (T)' Notation

The next proposition relates property (T) as in Definition 6.18 to the Fell topology.

Proposition 6.49 (Isolation of $\mathbb{1}_G$). *The group G has property (T) if and only if the trivial representation of G is an isolated point of $\widehat{G}$ with respect to the Fell topology.*

We note that in the sense of this proposition, '(T)' is a visual signifier for a neighbourhood of the trivial representation that only contains the trivial representation.

PROOF OF PROPOSITION 6.49. Suppose first that G has property (T). By Lemma 6.23, this implies that all unitary representations of G have uniform spectral gap. By Proposition 6.25 we find some $f \in L^1(G)$ and $\delta > 0$ so that the norm estimate (6.4) holds for any unitary representation π of G. In particular, $\|\pi_*(f)\| \leqslant 1 - \delta$ for all $\pi \in \widehat{G} \smallsetminus \{\mathbb{1}_G\}$, and hence

$$\mathscr{FO}(f, 1-\delta) \cap \widehat{G} = \{\mathbb{1}_G\}.$$

In other words, $\mathbb{1}_G$ is an isolated point of $\widehat{G}$.

We now suppose that the trivial representation is isolated within $\widehat{G}$ with respect to the Fell topology. By Lemma 6.44 applied to $C_c(G) \subseteq L^1(G)$, we see that the sets $\mathscr{FO}(f, \alpha)$ for $f \in C_c(G)$ and $\alpha > 0$ form a sub-basis of the Fell topology. Hence, by our assumption, there exist $f_1, \ldots, f_n \in C_c(G)$ and $\alpha_1, \ldots, \alpha_n > 0$ so that

$$\bigcap_{j=1}^{n} \mathscr{FO}(f_j, \alpha_j) \cap \widehat{G} = \left\{\pi \in \widehat{G} \;\middle|\; \|\pi_*(f_j)\| > \alpha_j \text{ for } j = 1, \ldots, n\right\} = \{\mathbb{1}_G\}. \tag{6.28}$$

We claim that (6.28) implies that $\widehat{G} \smallsetminus \{\mathbb{1}_G\}$ has uniform spectral gap as in Definition 6.22, which we will use to prove property (T) for G.

For this, we first note that $\mathbb{1}_{G,*}(f_j)$ is equal to multiplication by $\int_G f_j \,\mathrm{d}m$ on $\mathbb{C}$, so that (6.28) implies $\left\|\mathbb{1}_{G,*}(f_j)\right\| = \left|\int_G f_j \,\mathrm{d}m\right| > \alpha_j$ for $j = 1, \ldots, n$. We define

$$\varepsilon = \min_{j=1,\ldots,n} \left(\frac{\left|\int_G f_j \,\mathrm{d}m\right| - \alpha_j}{\|f_j\|_1} \right) > 0$$

and

$$Q = \bigcup_{j=1}^{n} \operatorname{supp} f_j.$$

Let $\pi \in \widehat{G} \smallsetminus \{\mathbb{1}_G\}$ and let $v \in \mathcal{H}_\pi$ be a unit vector. We assume now for the purposes of a contradiction that v is (Q, ε)-almost invariant. By (6.28) there exists some $j \in \{1, \ldots, n\}$ so that $\pi \notin \mathscr{FO}(f_j, \alpha_j)$ or, equivalently,

$$\|\pi_*(f_j)\|_{\text{op}} \leqslant \alpha_j. \tag{6.29}$$

Combining the almost invariance with Exercise 1.59 (see also the hint on page 527) we obtain

$$\left\| \pi_*(f_j)v - \left(\int_G f_j \, \mathrm{d}m \right) v \right\| < \varepsilon \|f_j\|_1.$$

Using the triangle inequality, (6.29), and the definition of ε this gives

$$\left| \int_G f_j \, \mathrm{d}m \right| < \|\pi_*(f_j)\|_{\text{op}} + \varepsilon \|f_j\|_1 \leqslant \alpha_j + \left(\left| \int_G f_j \, \mathrm{d}m \right| - \alpha_j \right).$$

This contradiction shows that $\mathcal{H}_\pi$ does not contain any (Q, ε)-almost invariant vectors, and we deduce that $\widehat{G} \smallsetminus \{\mathbb{1}_G\}$ has uniform spectral gap.

By Proposition 6.25 there exists some function $f_0 \in C_c(G)$ and $\delta > 0$ with $f_0 \geqslant 0$, $\int_G f_0 \, \mathrm{d}m = \int_G f_0^* * f_0 \, \mathrm{d}m = 1$, and

$$\|\pi_*(f_0)\|_{\text{op}} \leqslant 1 - \delta$$

for all $\pi \in \widehat{G} \smallsetminus \{\mathbb{1}_G\}$. For any $\pi \in \widehat{G} \smallsetminus \{\mathbb{1}_G\}$ and any unit vector $w \in \mathcal{H}_\pi$, this shows

$$\int_G f_0^* * f_0 \varphi_w^\pi \, \mathrm{d}m = \langle \pi_*(f_0^* * f_0) w, w \rangle = \|\pi_*(f_0) w\|^2 \leqslant (1 - \delta)^2.$$

Equivalently, we have

$$\int_G f_0^* * f_0 \phi \, \mathrm{d}m \leqslant (1 - \delta)^2 \tag{6.30}$$

for all $\phi \in \mathcal{E}^1(G) \smallsetminus \{\mathbb{1}\}$.

Suppose that ρ is a unitary representation of G with almost invariant vectors. In particular, there exists a unit vector $v \in \mathcal{H}_\rho$ with

$$\int_G f_0^* * f_0(g) \varphi_v^\rho(g) \, \mathrm{d}m(g) = \langle \rho_*(f_0^* * f_0) v, v \rangle = \|\rho_*(f_0) v\|^2 > (1 - \delta)^2. \tag{6.31}$$

We claim that

$$v_1 = \lim_{n \to \infty} \rho_*(f_0)^n v$$

exists in $\mathcal{H}_\rho$, and is a non-zero invariant vector.

To see the claim, we apply Corollary 1.83 to find a probability measure μ on the set $\mathcal{E}^1(G)$ of extremal points of $\mathcal{P}^1(G)$ satisfying

$$\varphi_v^\rho(g) = \int_{\mathcal{E}^1(G)} \phi(g) \, \mathrm{d}\mu(\phi).$$

Combining this with (6.30)–(6.31), we see that $s = \mu(\{\mathbb{1}_G\}) > 0$, which we will use to prove the claim. If $s = 1$, then $\varphi_v^\rho = \mathbb{1}_G$ and v is already invariant. So assume that $s < 1$ and set

$$\phi_0 = \frac{1}{1-s} \int_{\mathcal{E}^1(G) \smallsetminus \{\mathbb{1}_G\}} \phi \, \mathrm{d}\mu$$

so that $\varphi_v^\rho = (1-s)\phi_0 + s$. Applying Theorem 1.76, we find a cyclic unitary representation π^0 on a Hilbert space $\mathcal{H}_0 = \langle v_0 \rangle$ so that $\phi_0 = \varphi_{v_0}^{\pi^0}$. We define $\mathcal{H}_\tau = \mathcal{H}_0 \oplus \mathbb{C}$, and $w = \big((1-s)^{\frac{1}{2}} v_0, s^{\frac{1}{2}}\big)$. For the unitary representation $\tau = \pi^0 \oplus \mathbb{1}$ on $\mathcal{H}_\tau$ we have

$$\varphi_w^\tau = (1-s)\varphi_{v_0}^{\pi^0} + s = (1-s)\phi_0 + s = \varphi_v^\rho.$$

Hence $\langle v \rangle_\pi \subseteq \mathcal{H}_\rho$ is isomorphic to $\langle w \rangle_\tau \subseteq \mathcal{H}_\tau$, and so it suffices for the claim to show that

$$\lim_{n \to \infty} \tau_*(f_0)^n w = (0, s^{\frac{1}{2}}).$$

Indeed, note that by the definition of $\tau = \pi^0 \oplus \mathbb{1}_G$ the vector $(0, s^{\frac{1}{2}})$ is invariant, and the convergence in the second component holds trivially. For the first component, we let $n \geqslant 1$ and apply Fubini's theorem to calculate

$$\begin{aligned}
(1-s) \left\| \pi^0_*(f_0)^n v_0 \right\|^2 &= (1-s) \int_G \left(f_0^{*n}\right)^* * f_0^{*n} \underbrace{\varphi_{v_0}^{\pi^0}}_{=\phi_0} \, \mathrm{d}m \\
&= \int_{\mathcal{E}^1(G) \smallsetminus \{\mathbb{1}_G\}} \int_G \left(f_0^{*n}\right)^* * f_0^{*n} \phi \, \mathrm{d}m \, \mathrm{d}\mu(\phi) \\
&= \int_{\mathcal{E}^1(G) \smallsetminus \{\mathbb{1}_G\}} \left\| \pi^\phi_*(f_0)^n v_\phi \right\|^2 \mathrm{d}\mu(\phi) \longrightarrow 0
\end{aligned}$$

as $n \to \infty$. This shows the claim, and hence shows that $\mathcal{H}_\rho$ contains invariant vectors. As ρ was an arbitrary unitary representation with almost invariant vectors, it follows that G has property (T). □

6.4.2 The Support of a Unitary Representation

Definition 6.50 (Support of a unitary representation). For a unitary representation ρ of G, we define the *support of* ρ as

$$\operatorname{supp}\rho = \left\{\pi \in \widehat{G} \;\middle|\; \pi \prec \rho\right\}.$$

We note that by Corollary 6.45 we have $\operatorname{supp}\rho = \widehat{G} \cap \overline{\{\rho\}}$, where the closure is taken in the Fell topology, which shows in particular that the support is a Fell closed subset of $\widehat{G}$. We also note that Corollary 6.34 shows that the above definition of the support of a unitary representation agrees with the definition of the support in the case of an abelian group.

Exercise 6.51 (Weak containment and support). Let ρ_1, ρ_2 be unitary representations of G. Show that $\rho_1 \prec \rho_2$ if and only if $\operatorname{supp}\rho_1 \subseteq \operatorname{supp}\rho_2$.

With Definition 6.50 we can give a generalization of Corollary 6.39.

Exercise 6.52 (Support of restriction). Let $H < G$ be a closed subgroup, and let ρ be a unitary representation of G. Then

$$\overline{\bigcup_{\substack{\pi \in \widehat{G} \\ \pi \prec \rho}} \operatorname{supp}(\pi|_H)} = \operatorname{supp}(\rho|_H).$$

Exercise 6.53. Show that π has spectral gap if and only if $\operatorname{supp}\pi|_{(\mathcal{H}_\pi^G)^\perp}$ does not contain the trivial representation $\mathbb{1}_G$.

6.4.3 The Fell Topology Using Matrix Coefficients

We outline an alternative definition of the Fell topology using the compact-open topology on $\mathcal{P}^1(G)$, but will leave some steps as an exercise.

Definition 6.54 (Neighbourhoods in the Fell topology). For a compact set $Q \subseteq G$, $\varepsilon > 0$, and a continuous positive-definite function $\phi \in \mathcal{P}^1(G)$, we define the neighbourhood

$$CO(\phi, Q, \varepsilon) = \left\{F \in \mathcal{P}^1(G) \;\middle|\; \|F - \phi\|_{Q,\infty} < \varepsilon\right\}$$

in the compact-open topology. The *principal Fell open set* associated to Q, ε, and ϕ is defined by

$$\mathscr{FO}_{\mathrm{diag}}(\phi, Q, \varepsilon) = \left\{\pi \in \mathscr{U}(G) \;\middle|\; \text{there exists } \phi_\pi \in \mathcal{P}^1_\pi \text{ with } \phi_\pi \in CO(\phi, Q, \varepsilon)\right\}.$$

Proposition 6.55 (Fell topology using matrix coefficients). *By varying $\phi \in \mathcal{P}^1(G)$, the compact subset $Q \subseteq G$, and $\varepsilon > 0$ we obtain another sub-basis of the Fell topology on $\mathscr{U}(G)$.*

In the proof of Proposition 6.55 we will make use of the phrase 'diag-Fell topology' to indicate the topology defined by the sub-basis in Definition 6.54. We will also need the following elementary observations.

Lemma 6.56 (Inclusions). *Let $\phi_1, \phi_2 \in \mathcal{P}^1(G)$, Q_1, Q_2 compact subsets of G, and $\varepsilon_1, \varepsilon_2 > 0$. Suppose that*

$$CO(\phi_1, Q_1, \varepsilon_1) \subseteq CO(\phi_2, Q_2, \varepsilon_2).$$

Then

$$\mathscr{FO}_{\mathrm{diag}}(\phi_1, Q_1, \varepsilon_1) \subseteq \mathscr{FO}_{\mathrm{diag}}(\phi_2, Q_2, \varepsilon_2).$$

PROOF. The proof consists of checking the definitions. Indeed, suppose that π is in $\mathscr{FO}_{\mathrm{diag}}(\phi_1, Q_1, \varepsilon_1)$. Then there exists a positive-definite function ϕ_π associated to π with $\phi_\pi \in CO(\phi_1, Q_1, \varepsilon_1)$. By our assumption, this implies that

$$\phi_\pi \in CO(\phi_2, Q_2, \varepsilon_2)$$

and so $\pi \in \mathscr{FO}_{\mathrm{diag}}(\phi_2, Q_2, \varepsilon_2)$. □

Lemma 6.57 (Sub-bases of neighbourhoods). *Let ρ be a unitary representation of G. By definition, every neighbourhood of ρ in the diag-Fell topology contains a finite intersection of sets of the form $\mathscr{FO}_{\mathrm{diag}}(\phi, Q, \varepsilon)$, where ϕ is in $\mathcal{P}^1(G)$, $Q \subseteq G$ is compact, and $\varepsilon > 0$. The same is true if we restrict ϕ to either of the following classes of functions:*

- *$\phi = \varphi_v^\rho$ for unit vectors $v \in \mathcal{H}_\rho$;*
- *$\phi = \varphi_w^\pi$ for unit vectors $w \in \mathcal{H}_\pi$ and irreducible unitary representation π weakly contained in ρ.*

PROOF. Let $\phi \in \mathcal{P}^1(G)$, let $Q \subseteq G$ be a compact subset, and $\varepsilon > 0$. We suppose that $\rho \in \mathscr{FO}_{\mathrm{diag}}(\phi, Q, \varepsilon)$. By definition of the principal Fell open set $\mathscr{FO}_{\mathrm{diag}}(\phi, Q, \varepsilon)$, this means that there exist some positive-definite function $\phi_\rho \in \mathcal{P}_\rho^1$ associated to ρ with

$$\|\phi_\rho - \phi\|_{Q,\infty} < \varepsilon.$$

By definition of $\phi_\rho \in \mathcal{P}_\rho^1$ there exist vectors $v_1, \dots, v_n \in \mathcal{H}_\rho$ with the property that $\sum_{j=1}^n \|v_j\|^2 = 1$ and $\phi_\rho = \sum_{j=1}^n \varphi_{v_j}^\rho$. We suppose without loss of generality that $s_j = \|v_j\| > 0$ for $j = 1, \dots, n$. We set $\delta = \varepsilon - \|\phi_\rho - \phi\|_{Q,\infty} > 0$, define $\widetilde{v}_j$ to be $s_j^{-1} v_j \in \mathcal{H}_\rho$ for $j = 1, \dots, n$, and claim that

$$\mathscr{FO}_{\mathrm{diag}}(\varphi_{\widetilde{v}_1}^\rho, Q, \delta) \cap \cdots \cap \mathscr{FO}_{\mathrm{diag}}(\varphi_{\widetilde{v}_n}^\rho, Q, \delta) \subseteq \mathscr{FO}_{\mathrm{diag}}(\phi, Q, \varepsilon),$$

which will prove the first case in the lemma. To see the claim, suppose that π belongs to the left-hand side. Then there exists, for every $j \in \{1, \dots, n\}$, a positive-definite function $\phi_j \in \mathcal{P}_\pi^1$ associated to π with $\|\phi_j - \varphi_{\widetilde{v}_j}^\rho\| < \delta$. We define $\phi_\pi = \sum_{j=1}^n s_j^2 \phi_j \in \mathcal{P}_\pi^1$, and note that $\varphi_{v_j}^\rho = s_j^2 \varphi_{\widetilde{v}_j}^\rho$ for $j = 1, \dots, n$. Together, we obtain

$$\|\phi_\pi - \phi_\rho\|_{Q,\infty} \leqslant \sum_{j=1}^n s_j^2 \|\phi_j - \varphi_{\widetilde{v}_j}^\rho\|_{Q,\infty} < \delta.$$

By the triangle inequality, this gives $\|\phi_\pi - \phi\|_{Q,\infty} < \varepsilon$, and the claim follows from Lemma 6.56.

To see the second case of the lemma, it now suffices to consider a neighbourhood of the form $\mathscr{F}\mathscr{O}_{\mathrm{diag}}(\varphi_v^\rho, Q, \varepsilon)$ for a unit vector $v \in \mathcal{H}_\rho$, a compact $Q \subseteq G$, and some $\varepsilon > 0$. We set $\delta = \frac{\varepsilon}{2}$ and apply Proposition 6.37 to find irreducible unitary representations $\pi_1, \dots, \pi_n \prec \rho$ and vectors $v_j \in \mathcal{H}_{\pi_j}$ for $j = 1, \dots, n$ so that $\sum_{j=1}^n \|v_j\|^2 = 1$ and $\|\varphi_v^\rho - \sum_{j=1}^n \varphi_{v_j}^{\pi_j}\|_{Q,\infty} < \frac{\varepsilon}{2}$. We again suppose that $s_j = \|v_j\| > 0$ and set $\widetilde{v}_j = s_j^{-1} v_j \in \mathcal{H}_{\pi_j}$ for $j = 1, \dots, n$, and use the same argument as above to show that

$$\mathscr{F}\mathscr{O}_{\mathrm{diag}}(\varphi_{\widetilde{v}_1}^{\pi_1}, Q, \tfrac{\varepsilon}{2}) \cap \dots \cap \mathscr{F}\mathscr{O}_{\mathrm{diag}}(\varphi_{\widetilde{v}_n}^{\pi_n}, Q, \tfrac{\varepsilon}{2}) \subseteq \mathscr{F}\mathscr{O}_{\mathrm{diag}}(\varphi_v^\rho, Q, \varepsilon).$$

This gives the second case in the lemma. □

Essential Exercise 6.58. Use Lemma 6.56 to show that the diag-Fell topology is first-countable.

Proof of Proposition 6.55. By Exercise 6.58 the diag-Fell topology is first-countable. Recall that, by Lemma 6.44, this also holds for the Fell topology. Hence for both topologies the closed subsets can be characterized by limits of converging sequences (see Exercise 6.59). We claim that the characterization of convergence in terms of weak containment in Corollary 6.45(ii) also holds for the diag-Fell topology. Hence the Fell topology and the diag-Fell topology have the same convergent sequences with the same sets of possible limits. This implies that the Fell topology and the diag-Fell topology have the same closed sets, and so agree.

For the proof of the claim, we first assume that (ρ_n) is a sequence in $\mathscr{U}(G)$ converging to $\pi \in \mathscr{U}(G)$ in the diag-Fell topology. This implies quite directly (using the original definition of weak containment in Definition 6.1) that

$$\pi \prec \bigoplus_{k=1}^{\infty} \rho_{n_k} \tag{6.32}$$

for any increasing sequence $(n_k)_k$ in $\mathbb{N}$.

Finally, we suppose for the converse implication in the claim that (ρ_n) is a sequence in $\mathscr{U}(G)$ that does not converge in the diag-Fell topology. Hence there exists a neighbourhood U of π in the diag-Fell topology for which $\rho_{n_k} \notin U$ for all $k \in \mathbb{N}$ for some increasing sequence (n_k) in $\mathbb{N}$. By Lemma 6.57, we may assume that

$$U = \mathscr{F}\!\mathscr{O}_{\mathrm{diag}}(\varphi^{\pi_1}_{w_1}, Q_1, \varepsilon_1) \cap \cdots \cap \mathscr{F}\!\mathscr{O}_{\mathrm{diag}}(\varphi^{\pi_\ell}_{w_\ell}, Q_\ell, \varepsilon_\ell)$$

for some irreducible unitary representations $\pi_1, \dots, \pi_\ell \prec \pi$, some unit vectors $w_j \in \mathcal{H}_{\pi_j}$ for $j = 1, \dots, \ell$, compact subsets $Q_1, \dots, Q_\ell \subseteq G$, and numbers $\varepsilon_1, \dots, \varepsilon_\ell > 0$. Choosing a subsequence of $(n_k)_k$ (which we will still denote by $(n_k)_k$ for convenience), we may even suppose that

$$U = \mathscr{F}\!\mathscr{O}_{\mathrm{diag}}(\varphi^{\widetilde{\pi}}_{w}, Q, \varepsilon)$$

for some irreducible unitary representation $\widetilde{\pi} \prec \pi$, a unit vector $w \in \mathcal{H}_{\widetilde{\pi}}$, a compact set $Q \subseteq G$, and some $\varepsilon > 0$. We now show that π is not weakly contained in $\bigoplus_{k=1}^{\infty} \rho_{n_k}$, which will prove the claim above and hence also the proposition.

Suppose, for the purpose of a contradiction, that (6.32) holds. By the transitivity in Exercise 6.4 we also obtain $\widetilde{\pi} \prec \bigoplus_{k=1}^{\infty} \rho_{n_k}$. Therefore $\varphi^{\widetilde{\pi}}_{w}$ can be obtained as a limit of positive-definite functions associated to $\bigoplus_{k=1}^{\infty} \rho_{n_k}$. Since each matrix coefficient for $\bigoplus_{k=1}^{\infty} \rho_{n_k}$ is itself a converging sum of matrix coefficients of ρ_{n_k} for $k \in \mathbb{N}$, we deduce that $\varphi^{\widetilde{\pi}}_{w}$ can be approximated by finite sums of matrix coefficients of ρ_{n_k} for $k \in \mathbb{N}$. Since $\widetilde{\pi}$ is irreducible, we may apply Proposition 6.8 and obtain that $\varphi^{\widetilde{\pi}}_{w}$ is a limit (in the compact-open topology) of matrix coefficients of ρ_{n_k} for $k \in \mathbb{N}$. However, this is a contradiction of our construction of the subsequence $(n_k)_k$ since no positive-definite function associated to ρ_{n_k} for $k \in \mathbb{N}$ can approximate $\varphi^{\widetilde{\pi}}_{w}$ on Q with an error smaller than ε. □

Exercise 6.59. Let $\mathcal{T}$ be a topology on a set X satisfying the first countability axiom (that is, that every point in X has a countable basis of neighbourhoods). Show that $A \subseteq X$ is closed with respect to $\mathcal{T}$ if and only if it contains every limit point of every sequence of elements of A.

Exercise 6.60. Let $H < G$ be a closed subgroup, and let (ρ_n) be a sequence in $\mathscr{U}(G)$ converging to $\pi \in \mathscr{U}(G)$. Show that $(\rho_n|_H)$ converges to $\pi|_H$.

6.4.4 The Fell Topology as a Quotient Topology

As discussed in Section 1.6 (specifically, in Propositions 1.65 and 1.73), cyclic representations with a chosen generator are, up to unitary isomorphism, in one-to-one correspondence with non-zero positive-definite functions. Moreover, pairs (π, v) where $\pi \in \widehat{G}$ is an irreducible representation and $v \in \mathcal{H}_\pi$ is a unit vector are in a one-to-one correspondence with extremal elements $\varphi^{\pi}_{v} \in \mathcal{E}^1(G)$, where we write $\mathcal{E}^1(G)$ for the extreme elements of $\mathcal{P}^1(G)$. This, and Corollary 6.36, gives another interpretation of the Fell topology on $\widehat{G}$. For $\phi \in \mathcal{E}^1(G)$ we write π^ϕ for the irreducible unitary representation from the GNS construction in Theorem 1.76, and write $[\pi^\phi] \in \widehat{G}$ for its equivalence class under unitary isomorphism.

Proposition 6.61 (Quotient topology of compact-open topology). *Using the compact-open topology on $\mathcal{E}^1(G)$, the map*

$$p\colon \mathcal{E}^1(G) \ni \phi \longmapsto [\pi^\phi] \in \widehat{G}$$

is both continuous and open with respect to the Fell topology on $\widehat{G}$. In particular, the Fell topology on $\widehat{G}$ is equal to the quotient topology of the compact-open topology on $\mathcal{E}^1(G)$ with respect to the map p.

Proof. Suppose first that (ϕ_n) is a sequence in $\mathcal{E}^1(G)$ that converges in the compact-open topology to $\phi \in \mathcal{E}^1(G)$. Let $[\pi_n] = p(\phi_n) \in \widehat{G}$ and $[\pi] = p(\phi) \in \widehat{G}$ with generators $v_n \in \mathcal{H}_{\pi_n}$ and $v \in \mathcal{H}_\pi$ respectively, so that $\varphi_{v_n}^{\pi_n} = \phi_n$ for $n \in \mathbb{N}$ and $\varphi_v^\pi = \phi$. To prove that (π_n) converges to π in the Fell topology, we use the equivalent condition in Corollary 6.45(ii). So let (n_k) be an increasing sequence in $\mathbb{N}$. Since $\phi_{n_k} \to \phi$ as $k \to \infty$ it follows from Corollary 6.36 that

$$\pi \prec \rho = \bigoplus_{k=1}^{\infty} \pi_{n_k}.$$

As the subsequence (π_{n_k}) was arbitrary, it follows that (π_n) converges to π in the Fell topology. This gives continuity of the map p with respect to the Fell topology on $\widehat{G}$.

For the proof that p is an open map, we suppose now that $\pi = p(\phi) \in \widehat{G}$ with $\phi \in \mathcal{E}^1(G)$, and that the sequence (π_n) in $\widehat{G}$ converges to π in the Fell topology. We claim that this implies the existence of unit vectors $v_n \in \mathcal{H}_{\pi_n}$ so that $\phi = \lim_{n\to\infty} \varphi_{v_n}^{\pi_n}$ in the compact-open topology. Assume the claim for now, and let $O \subseteq \mathcal{E}^1(G)$ be open. Applying the claim to $\pi = p(\phi)$ for some $\phi \in O$ and any sequence (π_n) converging to π in the Fell topology, we find some N so that $\varphi_{v_n}^{\pi_n} \in O$, and hence $\pi_n \in p(O)$, for $n \geqslant N$. As the Fell topology is first-countable by Lemma 6.44, and the converging sequence with limit in $p(O)$ was arbitrary, this implies that $p(O)$ is open. Finally, note that continuity, openness, and surjectivity of p together show that a subset $U \subseteq \widehat{G}$ is open in the Fell topology if and only if $O = p^{-1}(U) \subseteq \mathcal{E}^1(G)$ is open in the compact-open topology, which makes the Fell topology the quotient topology on $\widehat{G}$.

So let $\pi = p(\phi)$ with $\phi \in \mathcal{E}^1(G)$ and $(\pi_n)_n$ converging to π, as in the claim. Moreover, let $Q \subseteq G$ be compact and let $\varepsilon > 0$. Suppose, for the purpose of a contradiction, that for infinitely many $n \in \mathbb{N}$ there is no unit vector $v_n \in \mathcal{H}_{\pi_n}$ with $\varphi_{v_n}^{\pi_n} \in CO(\phi, Q, \varepsilon)$. We choose an increasing sequence (n_k) of such integers. By our assumption that π_n converges to π, and by Corollary 6.45(ii), we have $\pi \prec \rho = \bigoplus_{k=1}^\infty \pi_{n_k}$. Since the matrix coefficients of ρ are uniformly converging sums of matrix coefficients of π_{n_k} for $k \in \mathbb{N}$, any positive-definite function associated to ρ can be approximated by finite sums of matrix coefficients of π_{n_k} for $k \in \mathbb{N}$. Applying the definition of $\pi \prec \rho$ to the generator $v \in \mathcal{H}_\pi$ with matrix coefficient ϕ and Proposition 6.8, it follows that ϕ can be approximated by matrix coefficients of unit vectors $v_{n_k} \in \mathcal{H}_{\pi_{n_k}}$ for some $k \in \mathbb{N}$. However, this contradicts our construction of the sequence (n_k) in $\mathbb{N}$. This contradiction shows that, for a fixed compact subset $Q \subseteq G$ and a fixed $\varepsilon > 0$, for all but finitely many $n \in \mathbb{N}$ there exists a unit vector v_n with $\varphi_{v_n}^{\pi_n} \in CO(\phi, Q, \varepsilon)$. Using the fact that G is σ-compact and taking $\varepsilon = \frac{1}{\ell}$ for $\ell \in \mathbb{N}$, this implies the claim and hence the proposition. □

6.4.5 Continuity of the Contragredient Map

Proposition 6.62 (Continuity of contragredient). *The contragredient map defined by*

$$\mathscr{U}(G) \ni \pi \longmapsto \overline{\pi} \in \mathscr{U}(G)$$

is continuous with respect to the Fell topology on $\mathscr{U}(G)$.

PROOF. We note that for any $\phi \in \mathcal{P}^1(G)$, compact $K \subseteq G$, $\varepsilon > 0$, unitary representation $\pi \in \mathscr{U}(G)$, and unit vector $v \in \mathcal{H}_\pi$ we have

$$\varphi_{\overline{\pi}}^{v'} = \overline{\varphi_\pi^v} \in CO(\phi, K, \varepsilon)$$

if and only if $\varphi_\pi^v \in CO(\overline{\phi}, K, \varepsilon)$, which implies that the image of $\mathscr{FO}_{\text{diag}}(\phi, K, \varepsilon)$ is precisely $\mathscr{FO}_{\text{diag}}(\overline{\phi}, K, \varepsilon)$. □

Exercise 6.63. Prove Proposition 6.62 using the principal Fell open sets $\mathscr{FO}(f, \alpha)$ for all values $\alpha \in \mathbb{R}$ and $f \in L^1(G)$.

6.4.6 Compatibility and Continuity of Tensor Products

Proposition 6.64 (Compatibility and continuity). *Let π_1 and π_2 be unitary representations of G and let ρ_1 and ρ_2 be unitary representations of H. If $\pi_1 \prec \pi_2$ and $\rho_1 \prec \rho_2$ then the outer tensor products also satisfy*

$$\pi_1 \otimes \rho_1 \prec \pi_2 \otimes \rho_2.$$

Moreover, the tensor product construction

$$\mathscr{U}(G) \times \mathscr{U}(H) \ni (\pi, \rho) \longmapsto \pi \otimes \rho \in \mathscr{U}(G \times H)$$

is continuous. If $G = H$ then both claims also hold for the inner tensor products.

PROOF. Suppose first that π_1 and ρ_1 are cyclic and have generators $v \in \mathcal{H}_{\pi_1}$ and $w \in \mathcal{H}_{\rho_1}$ respectively. Then $\pi_1 \prec \pi_2$ and $\rho_1 \prec \rho_2$ imply that, for any compact subsets $K \subseteq G$ and $L \subseteq H$ and any $\varepsilon > 0$, there exist $\phi^{\pi_2} \in \mathcal{P}_{\pi_2}^1$ and $\phi^{\rho_2} \in \mathcal{P}_{\rho_2}^1$ with $\|\varphi_v^{\pi_1} - \phi^{\pi_2}\|_{K,\infty} < \varepsilon$ and $\|\varphi_w^{\rho_1} - \phi^{\rho_2}\|_{L,\infty} < \varepsilon$. We recall that $\varphi_{v\otimes w}^{\pi_1\otimes\rho_1} = \varphi_v^{\pi_1} \otimes \varphi_w^{\rho_1}$ (that is, $\varphi_{v\otimes w}^{\pi_1\otimes\rho_1}(g,h) = \varphi_v^{\pi_1}(g)\varphi_w^{\rho_1}(h)$ for all (g,h) in $G \times H$), which extends by linearity to $\phi = \phi^{\pi_2} \otimes \phi^{\rho_2} \in \mathcal{P}^1_{\pi_2\otimes\rho_2}$. Multiplying the approximation claims above with $\|\varphi_w^{\rho_1}\|_{L,\infty}$ and $\|\phi^{\pi_2}\|_{K,\infty}$ respectively gives

$$\begin{aligned}\left\|\varphi_{v\otimes w}^{\pi_1\otimes\rho_1} - \phi\right\|_{K\times L,\infty} &\leqslant \left\|\varphi_v^{\pi_1} - \phi^{\pi_2}\right\|_{K,\infty}\|\varphi_w^{\rho_1}\|_{L,\infty} + \left\|\phi^{\pi_2}\right\|_{K,\infty}\left\|\varphi_w^{\rho_1} - \phi^{\rho_2}\right\|_{L,\infty}\\ &< 2\varepsilon.\end{aligned}$$

We also note that $v\otimes w \in \mathcal{H}_{\pi_1} \otimes \mathcal{H}_{\rho_1}$ is a generator for $\pi_1 \otimes \rho_1$ by Exercise 3.12. As the compact subsets and $\varepsilon > 0$ were arbitrary, we obtain from Corollary 6.36 that $\pi_1 \otimes \rho_1 \prec \pi_2 \otimes \rho_2$.

If now π_1 (resp. ρ_1) is not cyclic, we write $\pi_1 = \bigoplus_{m=1}^{\infty} \pi_{1,m}$ (and, respectively, $\rho_1 = \bigoplus_{n=1}^{\infty} \rho_{1,n}$) as a direct sum of cyclic representations. This gives

$$\pi_1 \otimes \rho_1 = \bigoplus_{m,n=1}^{\infty} \pi_{1,m} \otimes \rho_{1,n} \prec \left(\pi_2 \otimes \rho_2\right)^{\infty} \prec \pi_2 \otimes \rho_2,$$

and hence the first part of the proposition.

The proof of the second part is similar. Suppose that (π_n) is a sequence in $\mathscr{U}(G)$ converging to $\pi \in \mathscr{U}(G)$, and (ρ_n) is a sequence in $\mathscr{U}(H)$ converging to $\rho \in \mathscr{U}(H)$. We claim that $\pi \otimes \rho \prec \bigoplus_{n=1}^{\infty} \pi_n \otimes \rho_n$.

For the proof of the claim, we first assume that $v \in \mathcal{H}_\pi$ and $w \in \mathcal{H}_\rho$ are generators so that $v \otimes w$ is a generator for $\pi \otimes \rho$ by Exercise 3.12. By Proposition 6.55 we can find, for compact subsets $K \subseteq G$ and $L \subseteq H$, $\varepsilon > 0$ and every large enough $n \in \mathbb{N}$, positive-definite functions $\phi^{\pi_n} \in \mathcal{P}^1_{\pi_n}$ and $\phi^{\rho_n} \in \mathcal{P}^1_{\rho_n}$ with $\|\varphi_v^\pi - \phi^{\pi_n}\|_{K,\infty} < \varepsilon$ and $\|\varphi_w^\rho - \phi^{\rho_n}\|_{L,\infty} < \varepsilon$. As in the first part of the proof, this implies that

$$\|\varphi_{v\otimes w}^{\pi\otimes\rho} - \phi^{\pi_n} \otimes \phi^{\rho_n}\|_{K\times L,\infty} < 2\varepsilon.$$

Together with Corollary 6.36, we see that $\pi \otimes \rho \prec \bigoplus_{n=1}^{\infty} \pi_n \otimes \rho_n$ as claimed.

As the claim holds for any sequences (π_n) and (ρ_n) converging to π and ρ respectively, we can also apply it to subsequences. By Corollary 6.45, this shows that $\pi_n \otimes \rho_n \to \pi \otimes \rho$ as $n \to \infty$. If π (resp. ρ) is not cyclic, we can apply the above to all its cyclic subspaces, and conclude the argument as in the first part of the proof.

By restricting the uniform approximation in the weak containment statement or the convergence statement to compact subsets of

$$\Delta_G = \{(g,g) \mid g \in G\} \subseteq G \times G,$$

we obtain the two statements also for the inner tensor product representations. □

Exercise 6.65. Let G have property (T) and let $\pi \in \widehat{G}$ be a finite-dimensional irreducible representation. Show that $\pi \prec \rho$ implies $\pi < \rho$ for any unitary representation ρ of G.

6.5 Examples of the Fell Topology*

We will describe in this section the Fell topology on the unitary duals of the three metabelian groups discussed in detail in Chapter 5. Before doing so, we completely describe the abelian case.

6.5.1 Abelian Groups

Corollary 6.66 (Fell topology for abelian groups). *Let G be abelian. Then the Fell topology on $\widehat{G}$ agrees with the locally compact, σ-compact, metric topology of Corollary 2.5. Moreover, a sequence (ρ_n) in $\mathscr{U}(G)$ converges to $\pi \in \mathscr{U}(G)$ if and only if for any $t \in \operatorname{supp} \pi$ there exists $t_n \in \operatorname{supp} \rho_n$ for all $n \in \mathbb{N}$ such that $t_n \to t$ as $n \to \infty$.*

PROOF. Recall that the topology on the Pontryagin dual in Corollary 2.5 can be defined as the compact-open topology for the characters viewed as elements of $C_b(G)$. Also note that all diagonal matrix coefficients for the unitary representation χ_t defined by $t \in \widehat{G}$ are positive multiples of χ_t and hence the only positive-definite function $\phi \in \mathcal{P}^1_{\chi_t}$ associated to χ_t is $\phi = \chi_t$ itself. Therefore

$$\mathscr{F}\!\mathscr{O}_{\mathrm{diag}}(\phi, Q, \varepsilon) \cap \widehat{G} = \left\{t \in \widehat{G} \mid \|\chi_t - \phi\|_{Q,\infty} < \varepsilon\right\}$$

for all $\phi \in \mathcal{P}^1(G)$, compact $Q \subseteq G$, and $\varepsilon > 0$. By Proposition 6.55, this shows that the Fell topology on the Pontryagin dual $\widehat{G}$ agrees with the compact-open topology.

For the last claim assume, initially for any $t \in \operatorname{supp} \pi$, that there exists some choice of $t_n \in \operatorname{supp} \rho_n$ for all $n \in \mathbb{N}$ with $t_n \to t$ as $n \to \infty$. Let (n_k) be a strictly increasing sequence in $\mathbb{N}$. We then obtain from the assumption and the definition of the support of a unitary representation for an abelian group that

$$\operatorname{supp} \pi \subseteq \overline{\bigcup_{k=1}^{\infty} \operatorname{supp} \rho_{n_k}} = \operatorname{supp} \bigoplus_{k=1}^{\infty} \rho_{n_k}. \tag{6.33}$$

By Corollary 6.34, this implies that $\pi \prec \bigoplus_{k=1}^{\infty} \rho_{n_k}$. As the strictly increasing sequence (n_k) was arbitrary, Corollary 6.45 gives that $\rho_n \to \pi$ as $n \to \infty$.

Assume now that (ρ_n) is a sequence in $\mathscr{U}(G)$ converging to π. By Corollary 6.45 this shows that $\pi \prec \bigoplus_{n=1}^{\infty} \rho_{n_k}$ for any strictly increasing sequence (n_k) in $\mathbb{N}$. Hence by Corollary 6.34 and the definition of the support for unitary representations of abelian groups, we also obtain (6.33). Now let $t \in \operatorname{supp} \pi$. It follows that for any neighbourhood O of t we have $O \cap \operatorname{supp} \rho_n \neq \emptyset$ for all but finitely many $n \in \mathbb{N}$, as otherwise we could construct a subsequence (n_k) that contradicts

$$t \in \overline{\bigcup_{k=1}^{\infty} \operatorname{supp} \rho_{n_k}}.$$

Now let $\widehat{G} = O_1 \supseteq O_2 \supseteq \cdots$ be a countable basis of the neighbourhoods of t, and choose recursively $1 = N_1 \leqslant N_2 \leqslant \cdots$ in $\mathbb{N}$ such that for every $\ell \in \mathbb{N}$ and $n \geqslant N_\ell$ the intersection $\operatorname{supp} \rho_n \cap O_\ell$ is non-empty. Finally choose for every $\ell \in \mathbb{N}$ and $n \in \mathbb{N}$ with $N_\ell \leqslant n < N_{\ell+1}$ the element $t_n \in \operatorname{supp} \rho_n$ such that t_n lies in $\operatorname{supp} \rho_n \cap O_\ell$. It follows that $t_n \to t$ as $n \to \infty$, which concludes the proof of the last statement in the corollary. □

6.5.2 The Isometry Group of the Plane

Let $G = \mathrm{SO}_2(\mathbb{R}) \ltimes \mathbb{R}^2$ be the group of isometries on $\mathbb{R}^2$. By Section 5.2.2, and using the same terminology, we have $\widehat{G} \cong \mathbb{Z} \sqcup (0, \infty)$, where the first set corresponds to all possible old representations χ_n for $n \in \mathbb{Z}$, and the second set to new representations π^r for all possible radii $r > 0$. We will show that on each of these sets separately, the Fell topology is the usual topology. More precisely, the Fell topology restricted to the old set $\mathbb{Z} \subseteq \widehat{G}$ is the discrete topology, and the Fell topology restricted to the Fell open new set $(0, \infty)$ is the standard metric topology. However, every old representation χ_n for $n \in \mathbb{Z}$ has a neighbourhood basis consisting of sets of the form $\{\chi_n\} \sqcup \{\pi^r \mid r \in (0, \delta)\}$ for $\delta > 0$. This is difficult to visualize, but Figure 6.1 is an attempt to do this. In particular, the Fell topology on $\widehat{G}$ is not Hausdorff but is T_1, as for every two points there is a neighbourhood of the first that does not contain the second.

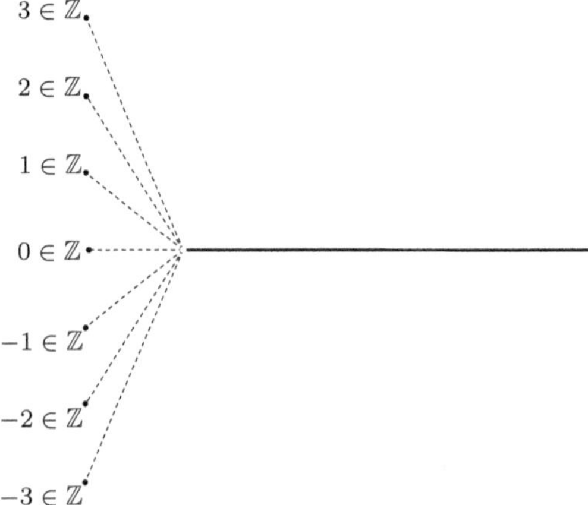

Fig. 6.1: The dotted lines from the elements $n \in \mathbb{Z}$ to the end point of the open ray indicates that these points are all limit points of π^r for $r \to 0$, while they themselves have no limit points.

Let us first discuss the Fell topology on the old set $\mathbb{Z} \subseteq \widehat{G}$. Here the old unitary representation is defined on $\mathcal{V}_1 = \mathbb{C}$ for every $n \in \mathbb{Z}$ by the character χ_n on $K \cong G/A$ and so there is only one choice of associated matrix coefficients to define a neighbourhood of χ_n (namely χ_n itself). If we choose the compact subset $Q = K$ (in Definition 6.54), it follows from the discreteness of $\widehat{K} \cong \mathbb{Z}$ that we can define a Fell neighbourhood of χ_n that does not contain any χ_m for $m \in \mathbb{Z} \smallsetminus \{n\}$.

To understand the Fell topology on all of $\widehat{G}$, we define the map

$$F\colon \begin{cases} \chi_n \longmapsto 0 & \text{for } n \in \mathbb{Z}, \\ \pi^r \longmapsto r & \text{for } r \in (0,\infty) \end{cases} \tag{6.34}$$

and note that the last claim in Corollary 6.66 (applied to the restriction to the subgroup $A \cong \mathbb{R}^2$) implies continuity of $F : \widehat{G} \to \mathbb{R}$ with respect to the Fell topology of $\widehat{G}$ (see Exercise 6.67). In particular, we can combine this with the above and deduce that for any $n \in \mathbb{Z}$ the set $\{\chi_n\} \sqcup \{\pi^r \mid r \in (0,\delta)\}$ is a neighbourhood of χ_n.

Exercise 6.67. Use Corollary 6.66 to prove continuity of the function F defined by (6.34).

The converse statement that for every Fell neighbourhood U of χ_n there exists some $\delta > 0$ with $\{\chi_n\} \sqcup (0,\delta) \subseteq U$ is easy to establish, as we may use the function $e_{-n} \in \mathcal{H}_r$ for sufficiently small $r > 0$ to approximate the matrix coefficient of $1 \in \mathbb{C}$ with respect to the representation χ_n of G. Informally, π^r degenerates for $r \searrow 0$ to the regular representation of K on $L^2(K)$ with A acting trivially, which should help in explaining the discussion above.

The description of a Fell neighbourhood of π^r for $r \in (0,\infty)$ is similar. By continuity of F, we have that

$$O = \{\pi^s \mid s \in (r-\delta, r+\delta)\}.$$

is a Fell open subset of $\widehat{G}$. On the other hand, it is easy to approximate the matrix coefficient of any $v_r \in \mathcal{H}_r$ by some $v_s \in \mathcal{H}_s$ if only s is sufficiently close to r. This shows that for any Fell neighbourhood U of π^r in $\widehat{G}$, there exists some $\eta > 0$ with

$$\{\pi^s \mid s \in (r-\eta, r+\eta)\} \subseteq U.$$

This concludes the description of the Fell topology on $\widehat{G}$ for $G = \mathrm{SO}_2(\mathbb{R}) \ltimes \mathbb{R}^2$.

Exercise 6.68. Describe the Fell topology on $\widehat{G_d}$, where G_d is as in Exercise 5.14.

6.5.3 The Affine Group in One Dimension

We now let $G = \mathbb{R}_{>0} \ltimes \mathbb{R}$ be the affine group from Section 5.2.3. By Proposition 5.22, the unitary dual of G consists of the old characters on the multiplicative diagonal subgroup $S = \mathbb{R}_{>0}$ and two new irreducible representations π^+, π^-

corresponding to the S-orbit $S{\cdot}1 = (0,\infty)$ and the S-orbit $S{\cdot}(-1) = (-\infty,0)$ within the dual $\widehat{A} \cong \mathbb{R}$ of the additive subgroup $A \cong \mathbb{R}$.

By Corollary 6.66 (applied to the restrictions to A), we see that old representations can only converge to old representations, that π^+ (and, similarly, π^-) can only converge to itself and to old representations. In fact we have already shown in Section 6.3.4 that every old representation is weakly contained in π^+ (and similarly in π^-).

Finally, on the set $\widehat{S} \cong \mathbb{R}$ of old characters we have the standard metric topology. We summarize this in Figure 6.2.

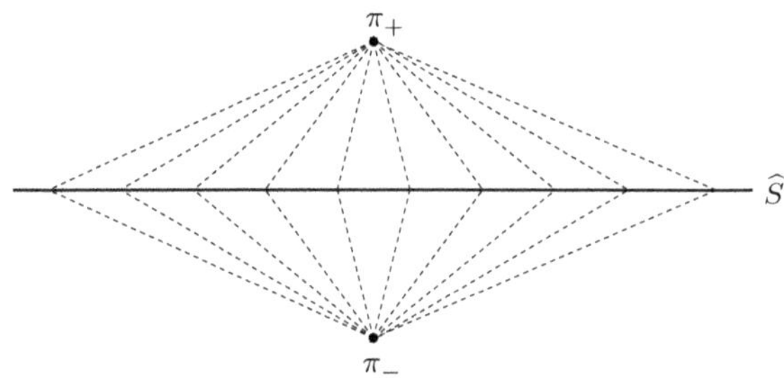

Fig. 6.2: The unitary dual of the affine group consists of the real line $\widehat{S} \cong \mathbb{R}$ and two 'nearly generic' points each of whose closures contain the real line but not the other of the two points.

6.5.4 The Heisenberg Group

We now fix $d \geqslant 1$ and let G be the $(2d+1)$-dimensional Heisenberg group as in Section 5.2.4. By the Stone–von Neumann theorem (Theorem 5.23), the unitary dual consists of the set $\widehat{\mathbb{R}^{2d}}$ of old representations arising from characters on $G/C(G) \cong \mathbb{R}^{2d}$ and one irreducible representation π^ξ for every non-trivial character on the centre defined by some $\xi \in \mathbb{R}^\times$.

By the last statement in Corollary 6.66 (applied to the restriction to the centre $C(G)$), the function $F\colon \widehat{G} \to \mathbb{R}$ defined by

$$F\colon \begin{cases} \chi \longmapsto 0 & \text{for } \chi \in \widehat{\mathbb{R}^{2d}}, \\ \pi^\xi \longmapsto \xi & \text{for } \xi \in \mathbb{R}^\times \end{cases}$$

is continuous. In particular, the set $\widehat{G/C(G)}$ of old characters is closed as a subset of $\widehat{G}$. Also, the Fell topology restricted to $\widehat{G/C(G)}$ becomes the standard metric topology on $\mathbb{R}^{2d}$.

Moreover, by the concrete description of π^ξ in (5.5) for $\xi \in \mathbb{R}^\times$ on the Hilbert space $L^2(\mathbb{R}^d)$ (independent of ξ) it is easy to see that if the sequence (ξ_n) in $\mathbb{R}^\times$ converges to $\xi \in \mathbb{R}^\times$, then the sequence (π^{ξ_n}) converges to π^ξ in the Fell topology. In the case of $\xi_n \to 0$ as $n \to \infty$ we also see that the definition of π^{ξ_n} degenerates to the multiplication representation of $\mathbb{R}^d$ on $L^2(\mathbb{R}^d)$. In particular, it is easy to find a sequence (v_n) in $L^2(\mathbb{R}^d)$ so that the matrix coefficient $\varphi_{v_n}^{\pi^{\xi_n}}$ converges to 1 in the compact-open topology as $n \to \infty$. This implies that π^{ξ_n} converges to the trivial character $\mathbb{1} \in \widehat{G/C(G)}$ as $n \to \infty$. For a non-trivial representation $\chi \in \widehat{G/C(G)}$ we can twist π^{ξ_n} as in Lemma 1.28 to obtain from Proposition 6.55 that $\pi^{\xi_n} \otimes \chi$ converges to $\chi \in \widehat{G/C(G)}$ as $n \to \infty$. By Theorem 5.23 we have that $\pi^{\xi_n} \otimes \chi$ is isomorphic to π^{ξ_n} (since $\chi(C(G)) = 1$), and so π^{ξ_n} converges to χ as $n \to \infty$ also. We again try to depict this (for the case $d = 1$) in Figure 6.3.

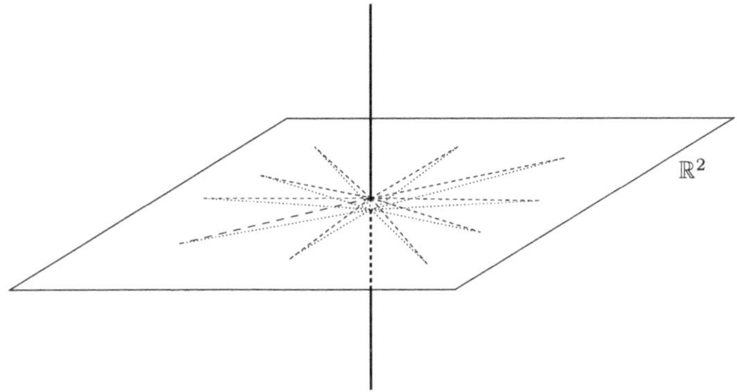

Fig. 6.3: The plane $\mathbb{R}^2$ consisting of old characters is closed, and has the standard topology. The vertical line parameterizes the new representations π^ξ with non-trivial central characters, and the dotted lines indicate that π^ξ converges as $\xi \to 0$ to all points in the plane simultaneously.

Exercise 6.69. Describe the Fell topology on the unitary dual of the group $G = \mathrm{Sol}$ from Exercise 5.60.

6.6 Summary and Outlook

As our examples may convey, the notion of weak containment is an important generalization of the notion of containment in Section 1.2.5. We note that weak containment will become crucial for the discussion of temperedness in Chapter 8.

Chapter 7
Smooth Vectors and Decay for $\mathrm{SL}_3(\mathbb{R})$

In this chapter we introduce smooth vectors for a unitary representation. This notion will be important for the study of unitary representations of Lie groups in the following chapters. Already in this chapter we will use smooth vectors in our study of *effective decay* of matrix coefficients, and for an effective Howe–Moore theorem for the group $\mathrm{SL}_3(\mathbb{R})$ (see also Theorem 1.90).

It may seem strange to discuss the group $\mathrm{SL}_3(\mathbb{R})$ before we treat $\mathrm{SL}_2(\mathbb{R})$ in Chapters 8 and 9. The reason for this is that one of our goals for this volume was to explain decay of matrix coefficients, which is a much more general phenomenon for $\mathrm{SL}_3(\mathbb{R})$. In fact with the preparations from the previous chapters the proof for $\mathrm{SL}_3(\mathbb{R})$ is surprisingly straightforward and takes about seven pages. However, to understand this for $\mathrm{SL}_2(\mathbb{R})$ requires the classification of its unitary dual and involves understanding the so-called complementary series representations. Both of these topics are treated in Chapter 9 and require much more work. We complete our discussion in this direction in Section 9.8, where we relate the decay properties to the first non-trivial eigenvalue of the Laplace operator on compact hyperbolic surfaces.

7.1 Smooth Vectors

7.1.1 Differential Operators and Smooth Vectors

Throughout this section we assume that G is a Lie group with Lie algebra $\mathfrak{g}$.

Definition 7.1 (Partial derivative). Let π be a unitary representation of the Lie group G. A vector $v \in \mathcal{H}_\pi$ has a *partial derivative* $\pi_\partial(\mathbf{a})v \in \mathcal{H}_\pi$ *in the direction* $\mathbf{a} \in \mathfrak{g}$ if

$$\pi_\partial(\mathbf{a})v = \frac{\mathrm{d}}{\mathrm{d}t}\bigg|_{t=0} \big(\pi_{\exp(t\mathbf{a})}\big)v = \lim_{t\to 0} \frac{1}{t}\big(\pi_{\exp(t\mathbf{a})}v - v\big)$$

M. Einsiedler and T. Ward, *Unitary Representations and Unitary Duals*,
Graduate Texts in Mathematics 308, https://doi.org/10.1007/978-3-032-03899-9_7

exists in $\mathcal{H}_\pi$. We say that v is C^1-*smooth* if $\pi_\partial(\mathbf{a})v$ exists for all $\mathbf{a} \in \mathfrak{g}$, is C^r-*smooth* for some $r \geqslant 1$ if $\pi_\partial(\mathbf{a}_1) \cdots \pi_\partial(\mathbf{a}_r)v$ exists for all $\mathbf{a}_1, \ldots, \mathbf{a}_r \in \mathfrak{g}$, and is *smooth* if v is C^r-smooth for all $r \geqslant 1$.

These notions will become more familiar after we see an example and establish some standard properties of derivatives and integrals in this context.

Essential Exercise 7.2. Let $r \geqslant 1$ and let π be a unitary representation of the Lie group G. Show that the set $\mathcal{V} \subseteq \mathcal{H}_\pi$ of C^r-smooth vectors is a subspace, and that the map $\mathcal{V} \ni v \mapsto \pi_\partial(\mathbf{a}_1) \cdots \pi_\partial(\mathbf{a}_r)v \in \mathcal{H}_\pi$ is linear for any $\mathbf{a}_1, \ldots, \mathbf{a}_r \in \mathfrak{g}$.

Example 7.3 (Smooth vectors for unitary representations of $\mathrm{SO}_2(\mathbb{R})$*).* Let

$$G = \mathrm{SO}_2(\mathbb{R}) = \left\{ k_\theta = \begin{pmatrix} \cos\theta & -\sin\theta \\ \sin\theta & \cos\theta \end{pmatrix} \;\middle|\; \theta \in [0, 2\pi) \right\}$$

and let

$$\mathbf{w} = \begin{pmatrix} 0 & -1 \\ 1 & 0 \end{pmatrix} \in \mathfrak{g} = \mathfrak{so}_2(\mathbb{R}).$$

Furthermore, let π be a unitary representation of $\mathrm{SO}_2(\mathbb{R})$. Suppose first that v in $\mathcal{H}_\pi$ is an eigenvector of weight $n \in \mathbb{Z}$ (that is, $\pi_{k_\theta} v = \mathrm{e}^{\mathrm{i}n\theta} v$ for all k_θ in $\mathrm{SO}_2(\mathbb{R})$). Then v is smooth, since

$$\pi_\partial(\mathbf{w})v = \lim_{t \to 0} \frac{1}{t} \big(\pi_{\exp(t\mathbf{w})} v - v \big) = \lim_{t \to 0} \frac{1}{t} \big(\mathrm{e}^{\mathrm{i}nt} - 1 \big) v = \mathrm{i}nv.$$

More generally, $v \in \mathcal{H}_\pi$ with eigenvector decomposition $v = \sum_{n \in \mathbb{Z}} v_n$ has a partial derivative $\pi_\partial(\mathbf{w})v$ if and only if $\sum_{n \in \mathbb{Z}} n^2 \|v_n\|^2 < \infty$, and in this case

$$\pi_\partial(\mathbf{w})v = \sum_{n \in \mathbb{Z}} \mathrm{i}nv_n.$$

Indeed, suppose first that $\sum_{n \in \mathbb{Z}} n^2 \|v_n\|^2 < \infty$, which gives that

$$\pi_\partial(\mathbf{w})v = \lim_{t \to 0} \sum_{n \in \mathbb{Z}} \underbrace{\frac{1}{t} \big(\mathrm{e}^{\mathrm{i}nt} - 1 \big)}_{|\cdot| \leqslant n} v_n = \sum_{n \in \mathbb{Z}} \mathrm{i}nv_n$$

by (a trivial form of) dominated convergence. Suppose that $\pi_\partial(\mathbf{w})v = \widetilde{v}$ exists, and assume that $\widetilde{v} = \sum_{n \in \mathbb{Z}} \widetilde{v}_n$ is the eigenvalue decomposition. For any $n \in \mathbb{Z}$ and $u \in \mathcal{H}_\pi$ with eigenvector decomposition $u = \sum_{m \in \mathbb{Z}} u_m$, we then have

$$\begin{aligned}\langle \widetilde{v}_n, u\rangle = \langle \widetilde{v}, u_n\rangle &= \lim_{t\to 0}\frac{1}{t}\langle \pi_{\exp(t\mathbf{w})}v - v, u_n\rangle \\ &= \lim_{t\to 0}\frac{1}{t}\big(\langle \pi_{\exp(t\mathbf{w})}v, u_n\rangle - \langle v, u_n\rangle\big) \\ &= \lim_{t\to 0}\big(\langle v, \pi_{\exp(-t\mathbf{w})}u_n\rangle - \langle v, u_n\rangle\big) \\ &= \lim_{t\to 0}\frac{1}{t}(\mathrm{e}^{int}-1)\langle v_n, u_n\rangle = \langle inv_n, u\rangle.\end{aligned}$$

As this holds for all $u \in \mathcal{H}_\pi$, we see that $\widetilde{v}_n = inv_n$ and hence

$$\|\widetilde{v}\|^2 = \sum_{n\in\mathbb{Z}} n^2\|v_n\|^2 < \infty$$

as claimed.

7.1.2 Fundamental Properties

The following will be a convenient tool for establishing basic properties of partial derivatives.

Lemma 7.4 (Fundamental theorem). *Let π be a unitary representation of the Lie group G. If for $v \in \mathcal{H}_\pi$ the derivative $\pi_\partial(\mathbf{a})v$ exists for some $\mathbf{a} \in \mathfrak{g}$, then*

$$\pi_{\exp(t\mathbf{a})}v - v = \int_0^t \pi_{\exp(s\mathbf{a})}\pi_\partial(\mathbf{a})v \,\mathrm{d}s \tag{7.1}$$

for all $t \in \mathbb{R}$ (with the usual sign conventions for Riemann integrals). Conversely, if $v, \widetilde{v} \in \mathcal{H}_\pi$ satisfy

$$\pi_{\exp(t\mathbf{a})}v - v = \int_0^t \pi_{\exp(s\mathbf{a})}\widetilde{v} \,\mathrm{d}s \tag{7.2}$$

for $t \in \mathbb{R}$, then the partial derivative $\pi_\partial(\mathbf{a})v = \widetilde{v}$ exists.

Proof. First notice that $\pi_{\exp(s\mathbf{a})}\pi_\partial(\mathbf{a})v$ depends continuously on s, which implies that the $\mathcal{H}_\pi$-valued weak integral on the right-hand side of (7.1) exists. Now fix some vector $w \in \mathcal{H}_\pi$ and notice that the derivative of the map $s \mapsto \big\langle \pi_{\exp(s\mathbf{a})}v, w\big\rangle$ is given by

$$\begin{aligned}\lim_{t\to 0}&\frac{\big\langle \pi_{\exp((s+t)\mathbf{a})}v, w\big\rangle - \big\langle \pi_{\exp(s\mathbf{a})}v, w\big\rangle}{t} \\ &= \lim_{t\to 0}\frac{1}{t}\big\langle \pi_{\exp(t\mathbf{a})}v - v, \pi_{\exp(-s\mathbf{a})}w\big\rangle \\ &= \big\langle \pi_\partial(\mathbf{a})v, \pi_{\exp(-s\mathbf{a})}w\big\rangle \\ &= \big\langle \pi_{\exp(s\mathbf{a})}\pi_\partial(\mathbf{a})v, w\big\rangle,\end{aligned}$$

and so is also continuous in s. Hence, by the fundamental theorem of calculus (for $\mathbb{C}$-valued functions),

$$\begin{aligned}\left\langle \pi_{\exp(t\mathbf{a})}v, w\right\rangle - \langle v, w\rangle &= \int_0^t \left\langle \pi_{\exp(s\mathbf{a})}\pi_\partial(\mathbf{a})v, w\right\rangle \mathrm{d}s \\ &= \left\langle \int_0^t \pi_{\exp(s\mathbf{a})}\pi_\partial(\mathbf{a})v \,\mathrm{d}s, w\right\rangle.\end{aligned}$$

As this holds for all $w \in \mathcal{H}_\pi$, we see that (7.1) holds for $\mathbf{a}$ (assuming only that $\pi_\partial(\mathbf{a})v$ exists).

Suppose now for the converse that $v, \widetilde{v} \in \mathcal{H}_\pi$ satisfy (7.2). Then

$$\lim_{t\to 0}\frac{1}{t}\left(\pi_{\exp t\mathbf{a}}v - v\right) = \lim_{t\to 0}\frac{1}{t}\int_0^t \pi_{\exp(s\mathbf{a})}\widetilde{v}\,\mathrm{d}s = \widetilde{v}$$

follows from continuity of the representation, as in the proof of Proposition 1.52. □

Exercise 7.5. Suppose that π is a unitary representation of a Lie group G with Lie algebra $\mathfrak{g}$. Suppose $v \in \mathcal{H}_\pi$ has $v_{\mathbf{a}} \in \mathcal{H}_\pi$ as a weak derivative in the direction $\mathbf{a} \in \mathfrak{g}$ in the sense that

$$\frac{\mathrm{d}}{\mathrm{d}t}\Big|_{t=0}\left\langle \pi_{\exp(t\mathbf{a})}v, w\right\rangle = \langle v_{\mathbf{a}}, w\rangle$$

for all $w \in \mathcal{H}_\pi$ (or just for a dense set of vectors). Show in this case that $v_{\mathbf{a}} = \pi_\partial(\mathbf{a})v$ is in fact the derivative of v in the sense of Definition 7.1.

We note that the following linearity claim comes as no surprise. However, our assumptions regarding the existence of partial derivatives is significantly different to the standard lemma from multi-dimensional analysis.

Lemma 7.6 (Linearity). *Let $\mathbf{b}_1, \ldots, \mathbf{b}_{\dim\mathfrak{g}}$ be a basis of the Lie algebra $\mathfrak{g}$ of the Lie group G. Let π be a unitary representation of the Lie group G, and suppose that $v \in \mathcal{H}_\pi$ has the property that $\pi_\partial(\mathbf{b}_j)v$ exists for all j from 1 to $\dim\mathfrak{g}$. Then $\pi_\partial(\mathbf{a})v$ exists for all $\mathbf{a} \in \mathfrak{g}$, and depends linearly on $\mathbf{a}$.*

PROOF. By assumption, $\pi_\partial(\mathbf{b}_j)v$ exists for $j = 1, \ldots, d = \dim\mathfrak{g}$ so that (7.1) holds, in particular, already for $\mathbf{a} = \mathbf{b}_j$. We will combine (7.1) with the coordinate system of the second kind defined by

$$\Psi\colon \mathbb{R}^d \ni (t_1, t_2, \ldots, t_d) \longmapsto \exp(t_1\mathbf{b}_1)\exp(t_2\mathbf{b}_2)\cdots\exp(t_d\mathbf{b}_d) \in G.$$

Since the derivative of Ψ at 0 is the map $(s_1, \ldots, s_d) \mapsto s_1\mathbf{b}_1 + \cdots + s_d\mathbf{b}_d \in \mathfrak{g}$ and so is invertible, Ψ indeed defines a local diffeomorphism. For some $\mathbf{a} \in \mathfrak{g}$ we define smooth functions $t_j = t_j(t)$ for t close to 0 and $j = 1, \ldots, d$ by

$$(t_1(t), t_2(t), \ldots, t_d(t)) = \Psi^{-1}\left(\exp(t\mathbf{a})\right),$$

or equivalently by

$$\exp(t\mathbf{a}) = \exp(t_1(t)\mathbf{b}_1) \cdots \exp(t_d(t)\mathbf{b}_d). \tag{7.3}$$

Recall that the derivative of $t \mapsto \exp(t\mathbf{a})$ at $t = 0$ is $\mathbf{a} \in \mathfrak{g}$. Hence, taking the derivative of (7.3), we obtain from the chain rule in multi-dimensional analysis that

$$\mathbf{a} = s_1\mathbf{b}_1 + \cdots + s_d\mathbf{b}_d$$

where

$$\lim_{t\to 0} \frac{t_j(t)}{t} = s_j \tag{7.4}$$

for $j = 1, \ldots, d$.

We now express $\pi_{\exp(t\mathbf{a})}v - v$ as the telescoping sum

$$\sum_{j=1}^{d} \left(\pi_{\exp(t_1\mathbf{b}_1)\cdots\exp(t_j\mathbf{b}_j)}v - \pi_{\exp(t_1\mathbf{b}_1)\cdots\exp(t_{j-1}\mathbf{b}_{j-1})}v\right)$$
$$= \sum_{j=1}^{d} \pi_{\exp(t_1\mathbf{b}_1)\cdots\exp(t_{j-1}\mathbf{b}_{j-1})}\left(\pi_{\exp(t_j\mathbf{b}_j)}v - v\right)$$

and apply (7.1) for the directions $\mathbf{b}_1, \ldots, \mathbf{b}_d$. This shows that $\pi_{\exp(t\mathbf{a})}v - v$ equals

$$\sum_{j=1}^{d} \pi_{\exp(t_1\mathbf{b}_1)\cdots\exp(t_{j-1}\mathbf{b}_{j-1})} \int_0^{t_j} \pi_{\exp(s\mathbf{b}_j)}\pi_\partial(\mathbf{b}_j)v \,\mathrm{d}s$$
$$= \sum_{j=1}^{d} \int_0^{t_j} \pi_{\exp(t_1\mathbf{b}_1)\cdots\exp(t_{j-1}\mathbf{b}_{j-1})\exp(s\mathbf{b}_j)}\pi_\partial(\mathbf{b}_j)v \,\mathrm{d}s.$$

We now divide by t and use (7.4) together with continuity of the representation to obtain

$$\lim_{t\to 0} \frac{1}{t}\left(\pi_{\exp(t\mathbf{a})}v - v\right) = \sum_{j=1}^{d} s_j\pi_\partial(\mathbf{b}_j)v.$$

This proves that

$$\pi_\partial(\mathbf{a})v = s_1\pi_\partial(\mathbf{b}_1)v + \cdots + s_d\pi_\partial(\mathbf{b}_d)v$$

exists and depends linearly on $\mathbf{a} \in \mathfrak{g}$. □

We will bring this into connection with the *adjoint* representation $\mathrm{Ad}_g : \mathfrak{g} \to \mathfrak{g}$ for $g \in G$, satisfying $\exp(\mathrm{Ad}_g(\mathbf{a})) = g\exp(\mathbf{a})g^{-1}$ for $g \in G$ and $\mathbf{a} \in \mathfrak{g}$.

Proposition 7.7 (Chain rule). *Let π be a unitary representation of the Lie group with Lie algebra $\mathfrak{g}$. Let $v \in \mathcal{H}_\pi$ be C^1-smooth and $g \in G$. Then $\pi_g v$ is C^1-smooth and*

$$\pi_g\pi_\partial(\mathbf{a})v = \pi_\partial(\mathrm{Ad}_g\,\mathbf{a})\pi_g v \tag{7.5}$$

for all $\mathbf{a} \in \mathfrak{g}$. In particular, the vector space of C^r-smooth vectors is invariant under π_g for every $g \in G$ and $r \geqslant 1$.

PROOF. By Lemma 7.4 we have

$$\pi_{\exp(t\mathbf{a})}v - v = \int_0^t \pi_{\exp(s\mathbf{a})}\pi_\partial(\mathbf{a})v \, \mathrm{d}s$$

for all $t \in \mathbb{R}$. We apply π_g on both sides. On the left-hand side this gives

$$\pi_g\bigl(\pi_{\exp(t\mathbf{a})}\bigr) \underbrace{\pi_{g^{-1}}\pi_g v}_{=v} - \pi_g v = \pi_{\exp(t \operatorname{Ad}_g \mathbf{a})}\pi_g v - \pi_g v.$$

Applying the same trick on the right-hand side gives

$$\pi_{\exp(t \operatorname{Ad}_g \mathbf{a})}\pi_g v - \pi_g v = \int_0^t \pi_{\exp(s \operatorname{Ad}_g \mathbf{a})}\pi_g \pi_\partial(\mathbf{a})v \, \mathrm{d}s$$

for all $t \in \mathbb{R}$. However, the second part of Lemma 7.4 applies and gives

$$\pi_\partial(\operatorname{Ad}_g \mathbf{a})\pi_g v = \pi_g \pi_\partial(\mathbf{a})v.$$

As this holds for all $\mathbf{a} \in \mathfrak{g}$ and $\operatorname{Ad}_g \colon \mathfrak{g} \to \mathfrak{g}$ is bijective, we also obtain that $\pi_g v$ is C^1-smooth as claimed. □

Proposition 7.8 (Existence of smooth vectors). *Let π be a unitary representation of the Lie group G with Lie algebra $\mathfrak{g}$, $v \in \mathcal{H}_\pi$, and $\psi \in C_c^\infty(G)$. Then $\pi_*(\psi)v$ is smooth, and $\pi_\partial(\mathbf{a})\pi_*(\psi)v = \pi_*\bigl(\lambda_\partial(\mathbf{a})\psi\bigr)v$ for any $\mathbf{a} \in \mathfrak{g}$, where*[†]

$$\lambda_\partial(\mathbf{a})\psi(g) = \frac{\partial}{\partial t}\Big|_{t=0} \psi\left(\exp(-t\mathbf{a})g\right)$$

is the partial derivative with respect to the left regular representation. Moreover, for a smooth approximate identity (ψ_n) in $C_c^\infty(G)$ (see Proposition 1.46) we have

$$v = \lim_{n \to \infty} \pi_*(\psi_n)v$$

for any $v \in \mathcal{H}_\pi$ and so, in particular, the smooth vectors in $\mathcal{H}_\pi$ are dense.

PROOF. Using the definition of the convolution operator we see that

$$\begin{aligned} \pi_{\exp(t\mathbf{a})}\pi_*(\psi)v &= \int_G \pi_{\exp(t\mathbf{a})}\psi(h)\pi_h v \, \mathrm{d}m(h) \\ &= \int_G \psi\left(\exp(-t\mathbf{a})g\right)\pi_g v \, \mathrm{d}m(g) \end{aligned}$$

[†] This formula may look a bit unusual, as it corresponds to a right-invariant vector field on G (rather than a left-invariant vector field).

by using the substitution $g = \exp(t\mathbf{a})h$. This gives

$$\frac{1}{t}\big(\pi_{\exp(t\mathbf{a})}\pi_*(\psi)v - \pi_*(\psi)v\big) = \frac{1}{t}\int_G \big(\psi\left(\exp(-t\mathbf{a})g\right) - \psi(g)\big)\pi_g v \, \mathrm{d}m(g)$$
$$= \int_G \frac{\psi(\exp(-t\mathbf{a})g) - \psi(g)}{t}\pi_g v \, \mathrm{d}m(g)$$

for all $t \in \mathbb{R} \smallsetminus \{0\}$. As $\psi \in C_c^\infty(G)$, we know that

$$\frac{\psi(\exp(-t\mathbf{a})g) - \psi(g)}{t} \longrightarrow \lambda_\partial(\mathbf{a})\psi(g)$$

as $t \to 0$, that this convergence is uniform in g, and that this convergence takes place inside a compact subset of G in the sense that the left-hand side vanishes for all $t \in [-1,1] \smallsetminus \{0\}$ outside the compact subset $\exp\left([-1,1]\mathbf{a}\right)\operatorname{supp}(\psi)$. In particular, the convergence also takes place in $L^1(G)$, and it follows that

$$\pi_\partial(\mathbf{a})\pi_*(\psi)v = \pi_*(\lambda_\partial(\mathbf{a})\psi)v$$

exists. Applying this inductively to expressions of the form

$$\pi_\partial(\mathbf{a}_n)\cdots\pi_\partial(\mathbf{a}_1)\pi_*(\psi)v$$

for $n \geqslant 1$ shows that $\pi_*(\psi)v$ is smooth.

Using an approximate identity in $C_c^\infty(G) \subseteq L^1(G)$ the proposition follows from Proposition 1.52. □

Definition 7.9 (Sobolev norm). Let π be a unitary representation of G, let $\mathbf{b}_1, \ldots, \mathbf{b}_{\dim \mathfrak{g}}$ be a basis of $\mathfrak{g} = \operatorname{Lie} G$, and let $r \geqslant 0$ be an integer. The *degree r Sobolev norm* of a C^r-smooth vector $v \in \mathcal{H}_\pi$ (with respect to the fixed basis) is defined by

$$\mathcal{S}(v)^2 = \mathcal{S}_r(v)^2 = \|v\|^2 + \sum_{s=1}^{r} \sum_{j_1,\ldots,j_s=1}^{\dim \mathfrak{g}} \|\pi_\partial(\mathbf{b}_{j_1})\cdots\pi_\partial(\mathbf{b}_{j_s})v\|^2.$$

Essential Exercise 7.10 (Lipshitz bound). Let π be a unitary representation of the Lie group G and v a C^1-smooth vector. Show that

$$\|\pi_{\exp \mathbf{a}} v - v\| \leqslant \|\mathbf{a}\| \, \mathcal{S}(v)$$

for all $\mathbf{a} \in \mathfrak{g}$, where $\mathcal{S}$ is a degree-one Sobolev norm defined by an orthonormal basis of $\mathfrak{g}$.

Exercise 7.11. Extend Proposition 7.8, and show that $\mathcal{S}_r(\pi_*(\psi)v) \ll_\psi \|v\|$ (and express the implicit constant in terms of ψ).

Essential Exercise 7.12. Let π be a unitary representation of a Lie group G with Lie algebra $\mathfrak{g}$. Let $r \geqslant 1$ and $v \in \mathcal{H}_\pi$ be a C^r-smooth vector. Let $\mathcal{S}$ denote the degree r Sobolev norm. Show that $\mathcal{S}(\pi_g v) \ll_{r,g} \mathcal{S}(v)$, and that the implicit constant can be chosen to be uniformly bounded on compact subsets of G.

Exercise 7.13. Let π be a unitary representation of G, let $\mathbf{b}_1, \ldots, \mathbf{b}_{\dim \mathfrak{g}}$ be a basis of $\mathfrak{g}$, the Lie algebra of G, and let $\widetilde{\mathbf{b}}_1, \ldots, \widetilde{\mathbf{b}}_{\dim \mathfrak{g}}$ be another basis of $\mathfrak{g}$. Let $r \geqslant 1$ and let $\mathcal{S}$ (respectively $\widetilde{\mathcal{S}}$) be the degree r Sobolev norm defined by $\mathbf{b}_1, \ldots, \mathbf{b}_{\dim \mathfrak{g}}$ (resp. by $\widetilde{\mathbf{b}}_1, \ldots, \widetilde{\mathbf{b}}_{\dim \mathfrak{g}}$). Show that we have $\mathcal{S}(v) \ll \widetilde{\mathcal{S}}(v) \ll \mathcal{S}(v)$ for any C^r-smooth $v \in \mathcal{H}_\pi$. Show also that if $r = 1$ and both bases are orthonormal with respect to an inner product on $\mathfrak{g}$, then $\mathcal{S}(v) = \widetilde{\mathcal{S}}(v)$ for any C^1-smooth $v \in \mathcal{H}_\pi$.

Exercise 7.14. Let π and ρ be unitary representations of G, let $B\colon \mathcal{H}_\pi \to \mathcal{H}_\rho$ be an intertwining operator and let $r \geqslant 1$. Show that B maps C^r-smooth vectors to C^r-smooth vectors and that $B\pi_\partial(a) \subseteq \rho_\partial(a)B$ for all $a \in \mathfrak{g}$.

7.1.3 Smooth Vectors for Unitary Flows

Example 7.15 (Derivatives and smooth vectors for $\mathbb{R}^d$). We let $G = \mathbb{R}^d$ for some $d \in \mathbb{N}$, and will use the standard basis $\mathbf{e}_1, \ldots, \mathbf{e}_d$ of its Lie algebra $\mathfrak{g}$ (which is also $\mathbb{R}^d$). Let π be a unitary representation. Applying the spectral theorem (Corollary 2.13), we assume that $\mathcal{H}_\pi = L^2_\mu(X)$ for a finite (or σ-finite) measure μ on $X = \mathbb{R}^d \times \mathbb{N}$ and that π is defined by the multiplication representation[†]

$$\big(\pi_x v\big)(t, n) = \mathrm{e}^{2\pi\mathrm{i}(x\cdot t)} v(t, n)$$

for all $x \in \mathbb{R}^d$, $v \in L^2_\mu(X)$, and $(t, n) \in \mathbb{R}^d \times \mathbb{N}$. In this case we obtain, for a C^1-smooth vector v, that

$$\pi_\partial(\mathbf{e}_j)v = \lim_{s\to 0} \underbrace{\frac{\mathrm{e}^{2\pi\mathrm{i}st_j} - 1}{s}}_{|\cdot|\leqslant 2\pi|t_j|} v(t, n) = 2\pi\mathrm{i}t_j v(t, n) \tag{7.6}$$

for $j = 1, \ldots, d$, so that

$$\mathcal{S}_1(v)^2 = \|v\|^2 + \sum_{j=1}^{d} \|M_{2\pi\mathrm{i}t_j} v\| \tag{7.7}$$

where $M_{2\pi\mathrm{i}t_j}$ is the multiplication operator on $L^2_\mu(X)$ defined by

$$\big(M_{2\pi\mathrm{i}t_j} v\big)(t, n) = 2\pi\mathrm{i}t_j v(t, n)$$

for all $v \in L^2_\mu(X)$, $t = (t_1, \ldots, t_d) \in \mathbb{R}^d$, and $n \in \mathbb{N}$. Conversely, if $v \in L^2_\mu(X)$ and $j \in \{1, \ldots, d\}$ have the property that $M_{2\pi\mathrm{i}t_j}(v)$ belongs to $L^2_\mu(X)$, then, by

[†] In Section 2.2 we used a simplified notation and wrote M_g for the multiplication operator defined by the function $\widehat{G} \ni t \mapsto \langle g, t\rangle \in \mathbb{S}^1$. In the case of $x \in \mathbb{R}^d$ and $t \in \widehat{\mathbb{R}^d} \cong \mathbb{R}^d$ this function corresponds to $\mathbb{R}^d \ni t \mapsto \mathrm{e}^{2\pi\mathrm{i}(x\cdot t)}$ by Exercise 2.6 or Proposition 2.42.

applying dominated convergence in (7.6), we see that $\pi_\partial(\mathbf{e}_j)(v)$ exists. If this holds for all $j \in \{1,\dots,d\}$, then v is C^1-smooth and the degree-one Sobolev norm is given by (7.7).

Now let $r \in \mathbb{N}$. Applying the above recursively to the partial derivatives, we see that $v \in \mathcal{H}_\pi$ is C^r-smooth if and only if $pv \in L^2_\mu(X)$ where p is any polynomial in $\mathbb{C}[t_1,\dots,t_d]$ of degree at most r.

Finally, we wish to apply this to the regular representation λ of $\mathbb{R}^d$ on $L^2(\mathbb{R}^d)$, which will reveal the connection to Sobolev spaces (see, for example, [25, Ch. 5]). By the Plancherel formula (Theorem 2.17), the regular representation is isomorphic to the multiplication representation as above for the Lebesgue measure $\mu = m_{\mathbb{R}^d}$ on $X = \mathbb{R}^d$. Applying the above, we see that $v \in L^2(\mathbb{R}^d)$ is smooth for the regular representation if and only if $\breve{v} \in L^2(\mathbb{R}^d)$ satisfies $p\breve{v} \in L^2(\mathbb{R}^d)$ for any polynomial $p \in \mathbb{C}[t_1,\dots,t_d]$. We define the polynomial

$$p(t) = \prod_{j=1}^{d}(t_j^2+1)$$

and note that $\frac{1}{p} \in L^2(\mathbb{R}^d)$. It follows that $\breve{v} = \frac{1}{p}(p\breve{v}) \in L^1(\mathbb{R}^d)$ and hence, by Theorem 2.17 and Corollary 2.5 applied to the Fourier transform instead of the Fourier back transform, that $v = \widehat{(\breve{v})} \in C_0(\mathbb{R}^d)$.

We claim that a smooth vector $v \in L^2(\mathbb{R}^d)$ actually belongs to $C^\infty(\mathbb{R}^d)$. For this, fix some index j in $\{1,\dots,d\}$, and note that the above also applies to $\lambda_\partial(\mathbf{e}_j)v \in C_0(\mathbb{R}^d)$. Here

$$\lambda_\partial(\mathbf{e}_j)v = \lim_{s\to 0}\frac{1}{s}(\lambda_{s\mathbf{e}_j}v - v)$$

is defined as a limit in L^2, which, for the isometric Fourier transform on L^2, becomes

$$M_\partial(\mathbf{e}_j)\breve{v}(t) = \lim_{s\to 0}\frac{1}{s}(\mathrm{e}^{2\pi\mathrm{i}st_j}-1)\breve{v}(t) = 2\pi\mathrm{i}t_j\breve{v}(t) \tag{7.8}$$

in $L^2(\mathbb{R}^d)$ and for almost any $t \in \mathbb{R}^d$ as above. We now multiply this once more by $p(t)$ and apply dominated convergence (by relying on the fact that the map $t \mapsto t_j p(t)\breve{v}(t)$ lies in $L^2(\mathbb{R}^d)$) to see that

$$\lim_{s\to 0} p(t)\underbrace{\frac{1}{s}(\mathrm{e}^{2\pi\mathrm{i}st_j}-1)}_{|\cdot|\leqslant 2\pi|t_j|}\breve{v}(t) = p(t)2\pi\mathrm{i}t_j\breve{v}(t)$$

converges in $L^2(\mathbb{R}^d)$. Multiplying by $\frac{1}{p} \in L^2(\mathbb{R}^d)$ gives convergence of (7.8) in $L^1(\mathbb{R}^d)$ by the Cauchy–Schwarz inequality. However, this gives

$$\left\|\frac{1}{s}(\lambda_{s\mathbf{e}_j}v - v) - \lambda_\partial(\mathbf{e}_j)v\right\|_\infty \leqslant \left\|\frac{1}{s}(\mathrm{e}^{2\pi\mathrm{i}st_j}-1)\breve{v}(t) - \widecheck{\lambda_\partial(\mathbf{e}_j)v}\right\|_1 \longrightarrow 0$$

as $s \to 0$ by the continuity bound in Corollary 2.5. As this holds for all j in $\{1, \dots, d\}$ and can be applied recursively to the partial derivatives of v, it follows that $v \in C^\infty(\mathbb{R}^d)$.

Once again the reasoning above can be reversed to see that $v \in L^2(\mathbb{R}^d)$ is smooth with respect to the regular representation if and only if $v \in C^\infty(\mathbb{R}^d)$ and its partial derivatives $\partial^\alpha v$ belong to $L^2(\mathbb{R}^d)$ for all $\alpha \in \mathbb{N}_0^d$ (see Exercise 7.16).

Exercise 7.16. Complete the proof of the last claim in Example 7.15.

Exercise 7.17. (a) Let $G = \mathrm{SO}_2(\mathbb{R}) \ltimes \mathbb{R}^2$ be the isometry group of the plane as in Section 5.2.2. Let $\pi \in \widehat{G}$ be an irreducible representation. Find and prove a description of the space of smooth vectors in $\mathcal{H}_\pi$. Also show that any $v \in \mathcal{H}_\pi$ is smooth for the restriction of π to $H = \mathbb{R}^2$.
(b) Let G be the affine group as in Section 5.2.3, and let $\pi^+ \in \widehat{G}$ be the irreducible representation corresponding to the set $(0, \infty) \subseteq \mathbb{R} \cong \widehat{\mathbb{R}}$. Show that any $f \in C_c^\infty\big((0, \infty)\big)$ is a smooth vector. Can you again characterize smoothness with an appropriate moment condition?
(c) Let G be the Heisenberg group as in Section 5.2.4, and let $\pi^\xi \in \widehat{G}$ be the irreducible representation corresponding to the central character χ_ξ determined by $\xi \in \mathbb{R}^\times$. Show that any $f \in C_c^\infty(\mathbb{R})$ is a smooth vector. Can you again characterize smoothness?

7.2 The Total Derivative*

We wish to study the functorial properties of partial derivatives, which will lead to some interesting results, for example, for $G = \mathrm{SU}_2(\mathbb{R})$.

7.2.1 Definition and Basic Properties

Definition 7.18 (Total derivative). Let π be a unitary representation of G. The *total derivative* of π is defined on every C^1-smooth vector $v \in \mathcal{H}_\pi$ as the linear map $T_\pi(v)$ in $\mathrm{Hom}(\mathfrak{g}, \mathcal{H}_\pi)$ given by

$$T_\pi(v)\colon \mathbf{a} \longmapsto \pi_\partial(\mathbf{a})v$$

for $\mathbf{a} \in \mathfrak{g}$. After fixing a basis $\mathbf{b}_1, \dots, \mathbf{b}_{\dim \mathfrak{g}}$ of $\mathfrak{g}$ we can identify $T_\pi(v)$ with the tuple

$$\big(\pi_\partial(\mathbf{b}_1)v, \dots, \pi_\partial(\mathbf{b}_{\dim \mathfrak{g}})v\big) \in \mathcal{H}_\pi^{\dim \mathfrak{g}} \cong \mathrm{Hom}(\mathfrak{g}, \mathcal{H}_\pi).$$

Lemma 7.19 (Closed operator). *Let π be a unitary representation of G. Then the total derivative T_π with domain*

$$D_{T_\pi} = \{v \in \mathcal{H}_\pi \mid v \textit{ is } C^1\textit{-smooth}\}$$

is a densely defined closed operator.

PROOF. Suppose that (v_n) in D_{T_π} is a sequence with

$$(v_n, T_\pi(v_n)) \longrightarrow (v, L) \in \mathcal{H}_\pi \times \mathrm{Hom}(\mathfrak{g}, \mathcal{H}_\pi)$$

as $n \to \infty$, and let $\mathbf{a} \in \mathfrak{g}$. By Lemma 7.4 this implies that

$$\pi_{\exp(t\mathbf{a})} v_n - v_n = \int_0^t \pi_{\exp(s\mathbf{a})} \underbrace{\pi_\partial(\mathbf{a}) v_n}_{=T_\pi(v_n)\mathbf{a}} \, \mathrm{d}s$$

for any $t \in \mathbb{R}$. Since $T_\pi(v_n) \to L$ in $\mathrm{Hom}(\mathfrak{g}, \mathcal{H}_\pi)$ as $n \to \infty$ we have

$$T_\pi(v_n)(\mathbf{a}) \longrightarrow L(\mathbf{a})$$

as $n \to \infty$. Moreover, since the integral defines a continuous operator on $\mathcal{H}_\pi$ we also obtain from this that

$$\pi_{\exp(t\mathbf{a})} v - v = \int_0^t \pi_{\exp(s\mathbf{a})} L(\mathbf{a}) \, \mathrm{d}s.$$

By the second part of Lemma 7.4, this gives $\pi_\partial(\mathbf{a}) v = L(\mathbf{a})$ for any $\mathbf{a} \in \mathfrak{g}$. However, this implies that $v \in D_{T_\pi}$ and $T_\pi(v) = L$, and hence the lemma. □

Lemma 7.20 (Chain rule). *Let π be a unitary representation of G. Then*

$$T_\pi(\pi_g v) = D\pi_g(T_\pi(v))$$

for every C^1-smooth vector $v \in \mathcal{H}_\pi$, where $D\pi$ is the continuous representation on $\mathrm{Hom}(\mathfrak{g}, \mathcal{H}_\pi)$ defined by $D\pi_g(L) = \pi_g \circ L \circ \mathrm{Ad}_{g^{-1}}$ for any L in $\mathrm{Hom}(\mathfrak{g}, \mathcal{H}_\pi)$.

We note that continuity of the representation is defined as in Definition 1.1(3) but that we did not claim unitarity of the representation $D\pi$ (see Section 7.2.2).

PROOF OF LEMMA 7.20. Let $g \in G$, $\mathbf{a} \in \mathfrak{g}$, and let $v \in \mathcal{H}_\pi$ be a C^1-smooth vector. Using the fact that π_g is bounded we have

$$\begin{aligned}
\pi_\partial(\mathbf{a}) \pi_g v &= \frac{\partial}{\partial t}\Big|_{t=0} \pi_{\exp(t\mathbf{a})g} v \\
&= \lim_{t \to 0} \frac{1}{t} \pi_g \left(\pi_{\exp(t \, \mathrm{Ad}_g^{-1} \mathbf{a})} v - v \right) = \pi_g \pi_\partial(\mathrm{Ad}_g^{-1} \mathbf{a}) v.
\end{aligned}$$

As this holds for any $\mathbf{a} \in \mathfrak{g}$ we see that $T_\pi \circ \pi_g = D\pi_g \circ T_\pi$ (where defined). We note that the formulation in (7.5) is obtained by replacing $\mathbf{a}$ with $\mathrm{Ad}_g \mathbf{a}$.

To see that $D\pi$ defines a representation on $\mathrm{Hom}(\mathfrak{g}, \mathcal{H}_\pi)$ let $g, h \in G$ and let $L \in \mathrm{Hom}(\mathfrak{g}, \mathcal{H}_\pi)$, and calculate

$$\begin{aligned}
D\pi_g\left(D\pi_h(L)\right) &= \pi_g \circ D\pi_h(L) \circ \mathrm{Ad}_{g^{-1}} \\
&= \pi_g \circ \pi_h \circ L \circ \mathrm{Ad}_{h^{-1}} \circ \mathrm{Ad}_{g^{-1}} = D\pi_{gh}(L).
\end{aligned}$$

Moreover, notice that $D\pi_e(L) = L$.

As noted after Definition 7.18, we make the identification of $\mathrm{Hom}(\mathfrak{g}, \mathcal{H}_\pi)$ with $\mathcal{H}_\pi^{\dim \mathfrak{g}}$ using a fixed basis of $\mathfrak{g} = \mathrm{Lie}\, G$. This identification gives the vector space $\mathrm{Hom}(\mathfrak{g}, \mathcal{H}_\pi)$ the structure of a Hilbert space. With this, we also have

$$\begin{aligned}\left\|D\pi_g(L)\right\|^2 &= \sum_{j=1}^{\dim \mathfrak{g}} \|\pi_g\big(L(\mathrm{Ad}_g^{-1} \mathbf{b}_j)\big)\|^2 = \sum_{j=1}^{\dim \mathfrak{g}} \|L(\mathrm{Ad}_g^{-1} \mathbf{b}_j)\|^2 \\ &\leqslant \sum_{j=1}^{\dim \mathfrak{g}} \left\| \sum_{k=1}^{\dim \mathfrak{g}} \left[\mathrm{Ad}_g^{-1}\right]_{kj} (L(\mathbf{b}_k)) \right\|^2 \ll_g \|L\|^2,\end{aligned}$$

where $\left[\mathrm{Ad}_g^{-1}\right]_{kj}$ denotes the matrix entry of the matrix representing the linear map

$$\mathrm{Ad}_g^{-1} \colon \mathfrak{g} \longrightarrow \mathfrak{g}$$

in the basis $\mathbf{b}_1, \ldots, \mathbf{b}_{\dim \mathfrak{g}}$. In other words, $D\pi_g$ is a bounded operator on the space $\mathrm{Hom}(\mathfrak{g}, \mathcal{H}_\pi)$. To see the continuity of the representation $D\pi$, fix some $j \in \{1, \ldots, \dim \mathfrak{g}\}$, let $L \in \mathrm{Hom}(\mathfrak{g}, \mathcal{H}_\pi)$, and suppose (g_n) is a sequence in G with $g_n \to g$ as $n \to \infty$. Then

$$\begin{aligned}D\pi_{g_n}(L)(\mathbf{b}_j) = \pi_{g_n}\left(L\left(\mathrm{Ad}_{g_n}^{-1} \mathbf{b}_j\right)\right) &= \sum_{k=1}^{\dim \mathfrak{g}} \left[\mathrm{Ad}_{g_n}^{-1}\right]_{kj} \pi_{g_n}(L(\mathbf{b}_k)) \\ \longrightarrow \sum_{k=1}^{\dim \mathfrak{g}} \left[\mathrm{Ad}_g^{-1}\right]_{kj} \pi_g(L(\mathbf{b}_k)) &= \pi_g\left(L\, \mathrm{Ad}_g^{-1} \mathbf{b}_j\right) = D\pi_g(L)(\mathbf{b}_j)\end{aligned}$$

as $n \to \infty$. This gives $D\pi_{g_n}(L) \to D\pi_g(L)$ as $n \to \infty$, as required. □

We finish this subsection with an interesting exercise, which requires the following definition.

Definition 7.21 (Adjoint operator). Let T be a densely defined closed operator from $\mathcal{H}_1$ to $\mathcal{H}_2$. The adjoint operator T^* is defined on the domain

$$D_{T^*} = \left\{ w \in \mathcal{H}_2 \;\middle|\; D_T \ni v \longmapsto \langle Tv, w\rangle_{\mathcal{H}_2} \text{ is bounded} \right\}$$

and satisfies

$$\langle Tv, w\rangle_{\mathcal{H}_2} = \langle v, T^* w\rangle_{\mathcal{H}_1}$$

for all $v \in D_T$ and $w \in D_{T^*}$.

We refer to [25, Lemma 13.3] for the properties of the adjoint operator.

Exercise 7.22. Show that $\pi_\partial(\mathbf{a})^*$ agrees with $-\pi_\partial(\mathbf{a})$ for any unitary representation π of G and element $\mathbf{a} \in \mathfrak{g}$.

7.2.2 Unitarity of the Derivative Representation

In this section we prove the following proposition which gives unitarity of the total derivative in some interesting cases.

Proposition 7.23 (Unitarity of D and Equivariance of $T_\pi^* T_\pi$). *Let G be a Lie group with Lie algebra $\mathfrak{g} = \operatorname{Lie} G$. Suppose that $\mathfrak{g}$ is equipped with an inner product with the property that Ad_g is orthogonal for any $g \in G$, and let π be a unitary representation of G. Use an orthonormal basis of $\mathfrak{g}$ to define the isomorphism $\operatorname{Hom}(\mathfrak{g}, \mathcal{H}_\pi) \cong \mathcal{H}_\pi^{\dim \mathfrak{g}}$, and hence a Hilbert space structure on $\operatorname{Hom}(\mathfrak{g}, \mathcal{H}_\pi)$. Then the derivative representation $D\pi$ on $\operatorname{Hom}(\mathfrak{g}, \mathcal{H}_\pi)$ is unitary, and hence $\Omega = T_\pi^* T_\pi$ is a densely defined closed intertwining operator from $\mathcal{H}_\pi$ to $\mathcal{H}_\pi$. If $\mathbf{b}_1, \ldots, \mathbf{b}_{\dim G}$ is an orthonormal basis of $\mathfrak{g}$ and $v \in \mathcal{H}_\pi$ is C^2-smooth, then*

$$T_\pi^* T_\pi v = -\sum_{j=1}^{\dim G} \pi_\partial(\mathbf{b}_j)^2 v = \Omega v. \tag{7.9}$$

If π is irreducible, then there exists some $\alpha_\pi \geqslant 0$ with $\Omega v = \alpha_\pi v$ for all v in $\mathcal{H}_\pi$.

Notice that the Hilbert space structure of $\operatorname{Hom}(\mathfrak{g}, \mathcal{H}_\pi)$ and the representation D becomes clearer after noting that $\operatorname{Hom}(\mathfrak{g}, \mathcal{H}_\pi) \cong \mathcal{H}_\pi \otimes_{\mathbb{R}} \mathfrak{g}$ and that the latter carries a unitary representation since $\mathcal{H}_\pi$ carries a unitary representation and the real Hilbert space $\mathfrak{g}$ carries a natural representation of G that is assumed to be 'orthogonal'.

Allowing ourselves to consider formal products of Lie algebra elements (giving elements of the so-called universal enveloping algebra; see Section 9.1), we may write the differential operator Ω on $\mathcal{H}_\pi$ also as $-\pi_\partial\left(\sum_{j=1}^{\dim G} \mathbf{b}_j \circ \mathbf{b}_j\right)$.

Proof of Proposition 7.23. Let $\mathfrak{g} = \operatorname{Lie} G$ and π be as in the proposition and suppose that $\mathbf{b}_1, \ldots, \mathbf{b}_{\dim G}$ is an orthonormal basis with respect to the assumed inner product on $\mathfrak{g}$. For $g \in G$ and $L \in \operatorname{Hom}(\mathfrak{g}, \mathcal{H}_\pi)$ we have, by unitarity of π_g, that $\left\|D\pi_g L\right\|^2_{\operatorname{Hom}(\mathfrak{g}, \mathcal{H}_\pi)}$ is equal to

$$\begin{aligned}
\sum_{j=1}^{\dim \mathfrak{g}} \left\|\pi_g L(\operatorname{Ad}_g^{-1} \mathbf{b}_j)\right\|^2_{\mathcal{H}_\pi} &= \sum_{j=1}^{\dim \mathfrak{g}} \left\|L(\operatorname{Ad}_g^{-1} \mathbf{b}_j)\right\|^2_{\mathcal{H}_\pi} \\
&= \sum_{j=1}^{\dim \mathfrak{g}} \left\| \sum_{k=1}^{\dim \mathfrak{g}} \left[\operatorname{Ad}_g^{-1}\right]_{kj} L(\mathbf{b}_k) \right\|^2_{\mathcal{H}_\pi} \\
&= \sum_{j=1}^{\dim \mathfrak{g}} \sum_{k,\ell=1}^{\dim \mathfrak{g}} \left[\operatorname{Ad}_g^{-1}\right]_{kj} \left[\operatorname{Ad}_g^{-1}\right]_{\ell j} \langle L(\mathbf{b}_k), L(\mathbf{b}_\ell)\rangle_{\mathcal{H}_\pi},
\end{aligned}$$

where $[\operatorname{Ad}_g^{-1}]_{kj}$ again denotes the entries of the matrix representation of Ad_g^{-1} in the basis $\mathbf{b}_1, \ldots, \mathbf{b}_{\dim \mathfrak{g}}$. Reordering the summation in the last expression above and using the fact that $\left[\mathrm{Ad}_g^{-1}\right]$ is an orthogonal matrix we obtain

$$\left\|D\pi_g L\right\|^2_{\mathrm{Hom}(\mathfrak{g},\mathcal{H}_\pi)} = \sum_{k,\ell=1}^{\dim\mathfrak{g}} \underbrace{\sum_{j=1}^{\dim\mathfrak{g}} \left[\mathrm{Ad}_g^{-1}\right]_{kj}\left[\mathrm{Ad}_g^{-1}\right]_{\ell j}}_{\delta_{k,\ell}} \langle L(\mathbf{b}_k), L(\mathbf{b}_\ell)\rangle_{\mathcal{H}_\pi}$$

$$= \sum_{k=1}^{\dim\mathfrak{g}} \|L(\mathbf{b}_k)\|^2_{\mathcal{H}_\pi} = \|L\|_{\mathrm{Hom}(\mathfrak{g},\mathcal{H}_\pi)}$$

as required.

This implies that $T_\pi^* T_\pi$ is intertwining where it is defined. To see this, we first recall that for $g \in G$ we have

$$T_\pi \pi_g = D\pi_g \circ T_\pi$$

by Lemma 7.20. Suppose now that v is in the domain of $T_\pi^* T_\pi$ and $g \in G$. Then, for all w in the domain of T_π, we have

$$\begin{aligned}\langle \pi_g T_\pi^* T_\pi v, w\rangle_{\mathcal{H}_\pi} = \langle T_\pi^* T_\pi v, \pi_g^{-1} w\rangle_{\mathcal{H}_\pi} &= \langle T_\pi v, T_\pi \pi_{g^{-1}} w\rangle_{\mathrm{Hom}(\mathfrak{g},\mathcal{H}_\pi)}\\ &= \langle T_\pi v, D_{\pi_{g^{-1}}} T_\pi w\rangle_{\mathrm{Hom}(\mathfrak{g},\mathcal{H}_\pi)}\\ &= \langle D_{\pi_g} T_\pi v, T_\pi w\rangle_{\mathrm{Hom}(\mathfrak{g},\mathcal{H}_\pi)}\\ &= \langle T_\pi \pi_g v, T_\pi w\rangle_{\mathrm{Hom}(\mathfrak{g},\mathcal{H}_\pi)}.\end{aligned}$$

However, this, by definition of the adjoint, implies that $T_\pi \pi_g v$ belongs to the domain of T_π^* and that

$$T_\pi^* T_\pi \pi_g v = \pi_g T_\pi^* T_\pi v.$$

This shows that the domain of $T_\pi^* T_\pi$ is invariant under π_g, and that

$$\pi_g T_\pi^* T_\pi \supseteq T_\pi^* T_\pi \pi_g$$

for all $g \in G$. Applying this for g^{-1} together with the invariance of the domain of $T_\pi^* T_\pi$, we actually obtain

$$\pi_g T_\pi^* T_\pi = T_\pi^* T_\pi \pi_g$$

for $g \in G$, as required. For the proof that $T_\pi^* T_\pi$ is densely defined and closed, we refer to [25, Th. 13.10] (see also Section 1.3.4 for similar arguments).

Now let $w_1, \ldots, w_{\dim\mathfrak{g}} \in \mathcal{H}_\pi$ be C^1-smooth vectors. We claim that the linear map L defined by

$$L\left(\sum_{j=1}^{\dim\mathfrak{g}} s_j \mathbf{b}_j\right) = \sum_{j=1}^{\dim\mathfrak{g}} s_j w_j$$

belongs to the domain $D_{T_\pi^*}$ of T_π^*, and

$$T_\pi^*(L) = -\sum_{j=1}^{\dim \mathfrak{g}} \pi_\partial(\mathbf{b}_j) w_j.$$

For this, let $v \in \mathcal{H}_\pi$ be C^1-smooth (that is, in the domain of T_π) and calculate

$$\begin{aligned}
\left\langle -\sum_{j=1}^{\dim \mathfrak{g}} \pi_\partial(\mathbf{b}_j) w_j, v \right\rangle_{\mathcal{H}_\pi} &= -\sum_{j=1}^{\dim \mathfrak{g}} \langle \pi_\partial(\mathbf{b}_j) w_j, v \rangle_{\mathcal{H}_\pi} \\
&= -\sum_{j=1}^{\dim \mathfrak{g}} \lim_{t\to 0} \frac{1}{t} \left\langle \pi_{\exp(t\mathbf{b}_j)} w_j - w_j, v \right\rangle_{\mathcal{H}_\pi} \\
&= -\sum_{j=1}^{\dim \mathfrak{g}} \lim_{t\to 0} \frac{1}{t} \left\langle w_j, \pi_{\exp(-t\mathbf{b}_j)} v - v \right\rangle_{\mathcal{H}_\pi} \\
&= \sum_{j=1}^{\dim \mathfrak{g}} \langle w_j, \pi_\partial(\mathbf{b}_j) v \rangle_{\mathcal{H}_\pi} = \langle L, T_\pi v \rangle_{\mathrm{Hom}(\mathfrak{g}, \mathcal{H}_\pi)}.
\end{aligned}$$

As $v \in D_{T_\pi}$ was arbitrary this gives the claim, which implies (7.9). The final claim in the proposition follows from Schur's lemma (Corollary 1.38). □

7.2.3 The Casimir Operator for $\mathrm{SU}_2(\mathbb{R})$

We wish to study an example of Proposition 7.23 and explicitly calculate the constants α_π for all irreducible representations of $G = \mathrm{SU}_2(\mathbb{R})$ (which we already classified in Section 4.2). For this we will also use the basis $\mathbf{b}_1, \mathbf{b}_2, \mathbf{b}_3 \in \mathrm{SU}_2(\mathbb{R})$ in (4.5).

Moreover, in the so-called universal enveloping algebra $\mathfrak{E}$ of $\mathfrak{su}_2(\mathbb{R})$, we will also consider formal products $\mathbf{a} \circ \mathbf{b} \in \mathfrak{E}$, squares $\mathbf{a}^{\circ 2} = \mathbf{a} \circ \mathbf{a} \in \mathfrak{E}$, the formal identity $\mathbb{1}_{\mathfrak{E}}$ of $\mathfrak{E}$, and the rules

$$\begin{cases} \pi_\partial(\mathbf{a} \circ \mathbf{b}) = \pi_\partial(\mathbf{a}) \pi_\partial(\mathbf{b}), \\ \quad \pi_\partial(\mathbb{1}_{\mathfrak{E}}) = I \end{cases}$$

for all Lie algebra elements $\mathbf{a}, \mathbf{b} \in \mathfrak{su}_2(\mathbb{R})$ (see also Section 9.1).

Corollary 7.24 (Casimir operator on $\mathrm{Sym}^n(\mathbb{C}^2)$). *For every $n \in \mathbb{N}_0$ the so-called*[(16)] Casimir element $\Omega = \mathbb{1}_{\mathfrak{E}} - (\mathbf{b}_1 \circ \mathbf{b}_1 + \mathbf{b}_2 \circ \mathbf{b}_2 + \mathbf{b}_3 \circ \mathbf{b}_3)$ *acts on* $\mathrm{Sym}^n(\mathbb{C}^2)$ *by differentiation, and equals the scalar multiplication*

$$\pi_\partial(\Omega) = I - \pi_\partial(\mathbf{b}_1^{\circ 2} + \mathbf{b}_2^{\circ 2} + \mathbf{b}_3^{\circ 2}) = (n+1)^2 I$$

by the square of the dimension of $\mathrm{Sym}^n(\mathbb{C}^2)$.

We note that we added $\mathbb{1}_{\mathfrak{C}}$ in the definition of Ω to make the conclusion of the corollary easier to remember.

PROOF OF COROLLARY 7.24. We first note that $\mathbf{b}_1, \mathbf{b}_2, \mathbf{b}_3 \in \mathfrak{su}_2(\mathbb{R})$ as defined in (4.5) form an orthonormal basis for the inner product defined by the quadratic form det. Moreover, as discussed in Section 4.2.2, $\mathrm{SU}_2(\mathbb{R})$ acts via the adjoint representation by orthogonal matrices on $\mathfrak{su}_2(\mathbb{R})$ with respect to this inner product. Thus $G = \mathrm{SU}_2(\mathbb{R})$ and $\mathfrak{su}_2(\mathbb{R})$ equipped with this inner product satisfy the assumptions in Proposition 7.23.

Let $n \in \mathbb{N}_0$. As the representation π on $\mathrm{Sym}^n(\mathbb{C}^2)$ is an irreducible representation of $\mathrm{SU}_2(\mathbb{R})$ by Theorem 4.6, we obtain from Proposition 7.23 that

$$-\pi_\partial(\mathbf{b}_1^{\circ 2} + \mathbf{b}_2^{\circ 2} + \mathbf{b}_3^{\circ 2}) = \alpha_n I$$

for some $\alpha_n \geqslant 0$.

To calculate α_n, we use the basis vectors $e_1^{\odot n} \in \mathrm{Sym}^n(\mathbb{C}^2)$. For $t \in \mathbb{R}$ we have

$$\exp(t\mathbf{b}_1) = \exp\begin{pmatrix} it & \\ & -it \end{pmatrix} = \begin{pmatrix} \mathrm{e}^{it} & \\ & \mathrm{e}^{-it} \end{pmatrix}$$

and

$$\pi_{\exp t\mathbf{b}_1} e_1^{\odot n} = \mathrm{e}^{int} e_1^{\odot n},$$

which implies that

$$\pi_\partial(\mathbf{b}_1) e_1^{\odot n} = ine_1^{\odot n}$$

and

$$\pi_\partial(\mathbf{b}_1^{\circ 2}) e_1^{\odot n} = -n^2 e_1^{\odot n}. \tag{7.10}$$

For $\mathbf{b}_2, \mathbf{b}_3$ we similarly have

$$\exp(t\mathbf{b}_2) = \exp\begin{pmatrix} & it \\ it & \end{pmatrix} = \begin{pmatrix} \cos t & \mathrm{i}\sin t \\ \mathrm{i}\sin t & \cos t \end{pmatrix},$$

$$\exp(t\mathbf{b}_3) = \exp\begin{pmatrix} & -t \\ t & \end{pmatrix} = \begin{pmatrix} \cos t & -\sin t \\ \sin t & \cos t \end{pmatrix},$$

$$\pi_{\exp t\mathbf{b}_2} e_1^{\odot n} = (\cos t\, e_1 + \mathrm{i}\sin t\, e_2)^{\odot n},$$

and

$$\pi_{\exp t\mathbf{b}_3} e_1^{\odot n} = (\cos t\, e_1 + \sin t\, e_2)^{\odot n}.$$

Expanding the latter expressions using the binomial theorem, we can take the derivative with respect to t at $t = 0$ and notice that only one term is relevant to obtain

$$\pi_\partial(\mathbf{b}_2) e_1^{\odot n} = ine_1^{\odot(n-1)} \odot e_2$$

and

$$\pi_\partial(\mathbf{b}_3) e_1^{\odot n} = ne_1^{\odot(n-1)} \odot e_2.$$

We repeat this step and obtain

$$\begin{aligned}\pi_{\exp t\mathbf{b}_2}\pi_\partial(\mathbf{b}_2)e_1^{\odot n} &= \mathrm{i}n(\cos t\, e_1 + \mathrm{i}\sin t\, e_2)^{\odot(n-1)} \odot (\mathrm{i}\sin t\, e_1 + \cos t\, e_2)\\ &= \mathrm{i}n\big(\cos^{n-1} t\, e_1^{\odot(n-1)} + \mathrm{i}(n-1)\cos^{n-2} t \sin t\, e_1^{\odot(n-2)} \odot e_2 + \cdots\big)\\ &\qquad\qquad \odot (\mathrm{i}\sin t\, e_1 + \cos t\, e_2)\\ &= \mathrm{i}n\big(\mathrm{i}\sin t\cos^{n-1} t\, e_1^{\odot n} + \mathrm{i}(n-1)\sin t\cos^{n-1} t\, e_1^{\odot(n-2)} \odot e_2^{\odot 2} + \cdots\big)\end{aligned}$$

and

$$\begin{aligned}\pi_{\exp t\mathbf{b}_3}\pi_\partial(\mathbf{b}_3)e_1^{\odot n} &= n\big(\cos t\, e_1 + \sin t\, e_2\big)^{\odot(n-1)} \odot \big(-\sin t\, e_1 + \cos t\, e_2\big)\\ &= n\big(\cos^{n-1} t\, e_1^{\odot(n-1)} + (n-1)\cos^{n-2} t \sin t\, e_1^{\odot(n-2)} \odot e_2 + \cdots\big)\\ &\qquad\qquad \odot \big(-\sin t\, e_1 + \cos t\, e_2\big)\\ &= n\big(-\sin t\cos^{n-1} t\, e_1^{\odot n} + (n-1)\sin t\cos^{n-1} t\, e_1^{\odot(n-2)} \odot e_2^{\odot 2} + \cdots\big),\end{aligned}$$

which implies that

$$\pi_\partial(\mathbf{b}_2^{\circ 2})e_1^{\odot n} = -ne_1^{\odot n} - n(n-1)e_1^{\odot(n-2)} \odot e_2^{\odot 2}$$

and

$$\pi_\partial(\mathbf{b}_3^{\circ 2})e_1^{\odot n} = -ne_1^{\odot n} + n(n-1)e_1^{\odot(n-2)} \odot e_2^{\odot 2}.$$

Together with (7.10) this gives

$$-\pi_\partial(\mathbf{b}_1^{\circ 2} + \mathbf{b}_2^{\circ 2} + \mathbf{b}_3^{\circ 2})e_1^{\odot n} = (n^2 + 2n)e_1^{\odot n}.$$

The corollary follows by adding $e_1^{\odot n}$. □

7.3 Effective Decay, Definitions, and First Results

In the following we will be interested in concrete examples of closed linear groups. By a *closed linear group* we mean a closed subgroup G of $\mathrm{SL}_d(\mathbb{R})$ for some $d \geqslant 1$. The assumption that $G < \mathrm{SL}_d(\mathbb{R})$ instead of the seemingly more general $G < \mathrm{GL}_d(\mathbb{R})$ is harmless, as we can consider $\mathrm{GL}_d(\mathbb{R})$ itself as a closed subgroup of $\mathrm{SL}_{d+1}(\mathbb{R})$. One reason for the assumption is that it gives the Hilbert–Schmidt norm $\|\cdot\|_{\mathrm{HS}}$ on $\mathrm{Mat}_{d,d}(\mathbb{R})$ for elements of

$$G < \mathrm{SL}_d(\mathbb{R}) \subseteq \mathrm{Mat}_{d,d}(\mathbb{R})$$

more meaning. We also note that for the purposes of establishing effective decay, the notion of degree r Sobolev norms for unitary representations from Definition 7.9 will be important.

Definition 7.25 (Effective decay of matrix coefficients). Let G be a closed linear group with Lie algebra $\mathfrak{g}$, let π be a unitary representation of G, let $r \geqslant 0$, denote the degree r Sobolev norm on C^r-smooth vectors in $\mathcal{H}_\pi$ by $\mathcal{S}(\cdot)$, and write

$$\mathcal{H}_\pi^G = \{v \in \mathcal{H}_\pi \mid \pi_g v = v \text{ for all } g \in G\}$$

for the subspace of fixed vectors. We say that π has *effective decay of matrix coefficients* if there exists some $\kappa > 0$ such that

$$\left|\langle \pi_g v, w\rangle\right| \ll \|g\|_{\mathrm{HS}}^{-\kappa} \mathcal{S}(v)\, \mathcal{S}(w)$$

for all C^r-smooth $v, w \in (\mathcal{H}_\pi^G)^\perp$ and for all $g \in G$, where the implicit constant is allowed to depend on π, κ, and $r \geqslant 0$. We will call κ a *decay exponent* of the unitary representation by.

For semi-simple groups effective decay of matrix coefficients as defined above gives a formulation of effectiveness of the Howe–Moore theorem (see Section 1.8 and Theorem 1.90). Our aim is to show that many natural actions have this property, and we will give an example of a decay exponent in Section 7.4.

We emphasize that the above notions as defined depend on the fact that we consider closed subgroups $G \leqslant \mathrm{SL}_d(\mathbb{R})$. The reader troubled by this may fix a Riemannian metric on an abstract Lie group G and use instead of the norm $\|g\|_{\mathrm{HS}}$ for $g \in G$ the expression $\mathrm{e}^{-\mathsf{d}(g,e)}$ for $g \in G$, and use this to define a notion of *exponential decay of matrix coefficients.* We note, however, that this notion will now depend on the choice of the Riemannian metric (instead of the particular embedding). We have chosen the terminology above as it is much easier to generalize Definition 7.25 in its formulation for closed linear subgroups of $\mathrm{SL}_d(\mathbb{Q}_p)$ or $\mathrm{SL}_d\big(\mathbb{F}_p((t))\big)$ for a prime p. We also note that for the simply connected Lie group $\widetilde{\mathrm{SL}_2(\mathbb{R})}$ with infinite centre (any kind of) decay of matrix coefficients cannot hold for irreducible representations due to Corollary 1.32. Finally, in this volume we will only be interested in $\mathrm{SL}_d(\mathbb{R})$ for $d \in \{2, 3\}$, and here there should be no doubt that the Hilbert–Schmidt norm $\|\cdot\|_{\mathrm{HS}}$ is a meaningful measuring tool for the size of the group elements.

7.3.1 Relationship to Spectral Gap

We will show here that effective decay of matrix coefficients implies spectral gap (see Section 6.2.1). We will also see in Section 7.4 that $\mathrm{SL}_3(\mathbb{R})$ satisfies the following stronger property.

Definition 7.26 (Uniform decay exponent). We say that a closed linear group G has a *uniform decay exponent* $\kappa > 0$ if κ is a decay exponent for any unitary representation of G, and both the Sobolev degree r and the implicit constant in Definition 7.25 can be chosen absolute.

Proposition 7.27 (Effective decay implies spectral gap). *A unitary representation of a closed linear group with effective decay of matrix coefficients has spectral gap. Moreover, any closed linear group G for which there exists a uniform decay exponent $\kappa > 0$ has property* (T).

PROOF. Let G be a closed linear group, let $\kappa > 0$ be a decay exponent for a unitary representation π of G, and let $\mathcal{S}$ be the Sobolev norm as in Definition 7.25. Also let $\psi \in C_c^\infty(G)$ satisfy $\|\psi\|_1 = 1$ and $\psi \geqslant 0$. We define

$$A = \pi_*(\psi)^* \pi_g \pi_*(\psi)$$

for some $g \in G$ to be determined.

For $v, w \in \mathcal{H}_\pi$ we apply Exercise 7.11 to see that $\mathcal{S}\bigl(\pi_*(\psi)v\bigr) \ll_\psi \|v\|$ and $\mathcal{S}\bigl(\pi_*(\psi)w\bigr) \ll_\psi \|w\|$. By assumption, we also have

$$\bigl|\langle Av, w\rangle\bigr| = \bigl|\langle \pi_g \pi_*(\psi)v, \pi_*(\psi)w\rangle\bigr| \ll_\psi \|g\|_{\mathrm{HS}}^{-\kappa} \|v\| \|w\|$$

for all $v, w \in \bigl(\mathcal{H}_\pi^G\bigr)^\perp$. We now choose $g \in G$ sufficiently large to ensure that†

$$\bigl|\langle Av, w\rangle\bigr| \leqslant \tfrac{1}{2}\|v\|\|w\|$$

for all $v, w \in \bigl(\mathcal{H}_\pi^G\bigr)^\perp$, or, equivalently, so that

$$\|A\|_{(\mathcal{H}_\pi^G)^\perp} \leqslant \tfrac{1}{2}.$$

Now recall that by Section 1.5.3—and (1.19) in particular—we have

$$A = \pi_*(\psi^*)\pi_g\pi_*(\psi) = \pi_*\bigl(\psi^* * \lambda_g \psi\bigr).$$

Thus Proposition 6.25 implies that π has spectral gap.

If G has a uniform decay exponent, then any unitary representation has spectral gap by the above. However, this implies that G has property (T). □

The following exercise shows that property (T) and possessing a positive uniform decay exponents are *not* equivalent in general.

Exercise 7.28. For the purpose of this exercise, use the fact that $\mathrm{SL}_3(\mathbb{Z})$ has property (T), which follows from the fact that $\mathrm{SL}_3(\mathbb{R})$ has property (T) (see Theorem 7.30), and the fact that $\mathrm{SL}_3(\mathbb{Z})$ is a lattice in $\mathrm{SL}_3(\mathbb{R})$ (see [25, Sec. 10.3]). Show that the natural action of $\mathrm{SL}_3(\mathbb{Z})$ on $\mathbb{T}^3 \cong \mathbb{R}^3/\mathbb{Z}^3$ is ergodic, and that it does not have (effective) decay of matrix coefficients.

7.3.2 Eigenvectors of $\mathrm{SO}_2(\mathbb{R})$ for Representations of $\mathrm{SL}_2(\mathbb{R})$

In this section we study unitary representations of $\mathrm{SL}_2(\mathbb{R})$, obtaining some technical results for later use. For this we define the subgroups

$$A = \left\{ a_t = \begin{pmatrix} \mathrm{e}^t & \\ & \mathrm{e}^{-t} \end{pmatrix} \;\middle|\; t \in \mathbb{R} \right\} < \mathrm{SL}_2(\mathbb{R})$$

and

† Note that if G is compact, then G has property (T) by Exercise 6.19 in any case.

$$K = \mathrm{SO}_2(\mathbb{R}) = \left\{ k_\theta = \begin{pmatrix} \cos\theta & -\sin\theta \\ \sin\theta & \cos\theta \end{pmatrix} \;\middle|\; \theta \in [0, 2\pi) \right\} < \mathrm{SL}_2(\mathbb{R}),$$

and will frequently use the following terminology.

For a unitary representation π of $\mathrm{SL}_2(\mathbb{R})$, we say that a vector $v \in \mathcal{H}_\pi$ is a *K-eigenvector* if there exists some $n \in \mathbb{Z}$ with

$$\pi_{k_\theta} v = \mathrm{e}^{\mathrm{i}n\theta} v$$

for all $k_\theta \in K$. We will also refer to n as the *weight* of the K-eigenvector. Moreover, in the cases where $\mathcal{H}_\pi$ is clearly a space of functions, we will also call any K-eigenvector a K-eigenfunction.

Using 'Fourier series' we now show that for establishing effective decay of matrix coefficients for $\mathrm{SL}_2(\mathbb{R})$, it suffices to study K-eigenvectors. We will use this observation repeatedly.

Proposition 7.29 (Upgrade to smooth vectors for $\mathrm{SL}_2(\mathbb{R})$). *Let π be a unitary representation of $\mathrm{SL}_2(\mathbb{R})$, $c > 0$, and $\kappa > 0$ so that*

$$\left|\langle \pi_{a_t} v, w\rangle\right| \leqslant c\mathrm{e}^{-\kappa t} \|v\| \|w\|$$

for all $t \in \mathbb{R}$ and all K-eigenvectors $v, w \in \mathcal{H}_\pi$. Suppose also that $B \in \mathrm{B}(\mathcal{H}_\pi)$ is a bounded operator that commutes with π_k for all $k \in \mathrm{SO}_2(\mathbb{R})$. Then we have

$$\left|\langle \pi_g B v, w\rangle\right| \ll c\|B\|_{\mathrm{op}} \|g\|_{\mathrm{HS}}^{-\kappa}\, \mathcal{S}(v)\, \mathcal{S}(w)$$

for all $g \in \mathrm{SL}_2(\mathbb{R})$ and all C^1-smooth vectors $v, w \in \mathcal{H}_\pi$, where the implicit constant is absolute.

Proof. If $v \in \mathcal{H}_\pi$ is C^1-smooth, then the decomposition of $v = \sum_{m\in\mathbb{Z}} v_m$ into K-eigenvectors not only converges in $\mathcal{H}_\pi$ (which it always does) but in fact converges absolutely. To see this, let $\mathbf{w} \in \mathfrak{sl}_2(\mathbb{R})$ denote the element in the Lie algebra of $\mathrm{SL}_2(\mathbb{R})$ corresponding to $\mathrm{SO}_2(\mathbb{R})$ such that

$$\pi_\partial(\mathbf{w}) v_m = \lim_{\theta\to 0} \frac{\pi_{k_\theta} v_m - v_m}{\theta} = \lim_{\theta\to 0} \frac{\mathrm{e}^{\mathrm{i}m\theta} - 1}{\theta} v_m = \mathrm{i}m v_m$$

and

$$\pi_\partial(\mathbf{w}) v = \sum_{m\in\mathbb{Z}} \mathrm{i}m v_m$$

by Example 7.3. Let $\mathbf{b}_1, \mathbf{b}_2, \mathbf{b}_3$ be a basis of $\mathfrak{sl}_2(\mathbb{R})$, so that $\pi_\partial(\mathbf{w})v$ can be expressed as a sum of $\pi_\partial(\mathbf{b}_1)v$, $\pi_\partial(\mathbf{b}_2)v$, and $\pi_\partial(\mathbf{b}_3)v$ (see Lemma 7.4). Therefore,

$$\begin{aligned}\sum_{m\in\mathbb{Z}} m^2 \|v_m\|^2 = \|\pi_\partial(\mathbf{w})v\|^2 &\ll \left(\|\pi_\partial(\mathbf{b}_1)v\| + \|\pi_\partial(\mathbf{b}_2)v\| + \|\pi_\partial(\mathbf{b}_3)v\|\right)^2 \\ &\ll \mathcal{S}(v)^2 < \infty\end{aligned}$$

by the triangle inequality, and

$$\begin{aligned}\sum_{m\in\mathbb{Z}}\|v_m\| &= \|v_0\| + \sum_{m\in\mathbb{Z}\smallsetminus\{0\}}\tfrac{1}{m}m\|v_m\|\\ &\leqslant \|v_0\| + \Biggl(\sum_{m\in\mathbb{Z}\smallsetminus\{0\}}\tfrac{1}{m^2}\Biggr)^{\frac12}\Biggl(\sum_{m\in\mathbb{Z}\smallsetminus\{0\}}m^2\|v_m\|^2\Biggr)^{\frac12} \ll \mathcal{S}(v)\end{aligned}$$

by the Cauchy–Schwarz inequality in $\ell^2(\mathbb{Z}\smallsetminus\{0\})$ and the fact that the sequence $\bigl(\frac{1}{m^2}\bigr)_{m\in\mathbb{N}}$ is summable.

Also note that B and π_{k_θ} for $k_\theta\in K$ map any K-eigenvector to a K-eigenvector with the same weight.

With this, we can now finish the proof using the Cartan decomposition

$$g = k_\theta a_t k_{\theta'}$$

of $g\in\mathrm{SL}_2(\mathbb{R})$ with $t\geqslant 0$ and $k_\theta,k_{\theta'}\in K$ satisfying $\|g\|_{\mathrm{HS}} = \|a_t\|_{\mathrm{HS}}\asymp \mathrm{e}^t$. Indeed, we obtain from our assumption applied separately to each summand below that

$$\begin{aligned}\bigl|\langle\pi_g Bv,w\rangle\bigr| &= \Bigl|\sum_{m,n}\langle\pi_{a_t}\pi_{k_{\theta'}}Bv_m,\pi_{k_\theta}^*w_n\rangle\Bigr|\\ &\ll c\,\underbrace{\mathrm{e}^{-\kappa t}}_{\ll\|g\|_{\mathrm{HS}}^{-\kappa}}\sum_{m,n}\underbrace{\|Bv_m\|}_{\leqslant\|B\|_{\mathrm{op}}\|v_m\|}\;\|w_n\|\\ &\ll c\|B\|_{\mathrm{op}}\|g\|_{\mathrm{HS}}^{-\kappa}\,\mathcal{S}(v)\,\mathcal{S}(w)\end{aligned}$$

as claimed. □

7.4 A Uniform Decay Exponent for $\mathrm{SL}_3(\mathbb{R})$*

As indicated earlier, our goal in this chapter is uniform effective decay for $\mathrm{SL}_3(\mathbb{R})$. We note that the decay exponent $\frac38$ below is not optimal, and refer to work of Oh [76] for more general and sharp results.

Theorem 7.30 (Effective decay for $\mathrm{SL}_3(\mathbb{R})$). *The group $\mathrm{SL}_3(\mathbb{R})$ has a uniform decay exponent: If π is a unitary representation of $\mathrm{SL}_3(\mathbb{R})$ and the vectors $v,w\in\bigl(\mathcal{H}_\pi^{\mathrm{SL}_3(\mathbb{R})}\bigr)^\perp$ are C^1-smooth, then*

$$\bigl|\langle\pi_g v,w\rangle\bigr| \ll \|g\|_{\mathrm{HS}}^{-\frac38}\,\mathcal{S}(v)\,\mathcal{S}(w)$$

for all $g\in\mathrm{SL}_3(\mathbb{R})$, where the implicit constant is absolute and $\mathcal{S}$ is a degree-one Sobolev norm. In particular, $\mathrm{SL}_3(\mathbb{R})$ has property (T).

7.4.1 Eigenvectors of $\mathrm{SO}_2(\mathbb{R})$

We will use the terminology of Section 7.3.2 for the restriction of a unitary representation π of $\mathrm{SL}_3(\mathbb{R})$ to the subgroup

$$\mathrm{ASL}_2(\mathbb{R}) = \left\{ \begin{pmatrix} g & x \\ 0 & 1 \end{pmatrix} \;\middle|\; g \in \mathrm{SL}_2(\mathbb{R}), x \in \mathbb{R}^2 \right\} < \mathrm{SL}_3(\mathbb{R}),$$

containing

$$\mathrm{SL}_2(\mathbb{R}) = \left\{ \begin{pmatrix} g & 0 \\ & 1 \end{pmatrix} \;\middle|\; g \in \mathrm{SL}_2(\mathbb{R}) \right\}$$

and so also $K = \mathrm{SO}_2(\mathbb{R})$. We will also make use of the normal abelian closed subgroup

$$H \lhd \mathrm{ASL}_2(\mathbb{R})$$

defined by

$$H = \left\{ h_x = \begin{pmatrix} I & x \\ & 1 \end{pmatrix} \;\middle|\; x \in \mathbb{R}^2 \right\} \cong \mathbb{R}^2,$$

as well as the elements

$$a_t = \begin{pmatrix} \mathrm{e}^t & 0 & 0 \\ 0 & \mathrm{e}^{-t} & 0 \\ & & 1 \end{pmatrix} \in \mathrm{SL}_2(\mathbb{R}) \tag{7.11}$$

for $t \in \mathbb{R}$. Since $H \lhd \mathrm{ASL}_2(\mathbb{R})$, we may use the results in Section 5.1 for $\pi|_{\mathrm{ASL}_2(\mathbb{R})}$ for the proof of the following first step towards Theorem 7.30. For this, we first note that for $g \in \mathrm{SL}_2(\mathbb{R})$ we will denote the inner automorphism of $\mathrm{ASL}_2(\mathbb{R})$ defined by

$$\begin{pmatrix} g & 0 \\ & 1 \end{pmatrix} \in \mathrm{ASL}_2(\mathbb{R})$$

by θ_g, so

$$\theta_g(h_x) = \begin{pmatrix} g & 0 \\ & 1 \end{pmatrix} \begin{pmatrix} I & x \\ & 1 \end{pmatrix} \begin{pmatrix} g^{-1} & 0 \\ & 1 \end{pmatrix} = \begin{pmatrix} I & gx \\ & 1 \end{pmatrix} = h_{gx}$$

for all $x \in \mathbb{R}^2$. In particular, the dual automorphism $\widehat{\theta}_g$ on $\widehat{H} \cong \mathbb{R}^2$ is given by the linear map defined by g^{t}, and the action of $g \in \mathrm{SL}_2(\mathbb{R})$ on this dual group $\mathbb{R}^2$ from Section 5.1 is defined by $(g^{\mathrm{t}})^{-1}$.

Lemma 7.31 (Eigenvectors). *Let π be as in Theorem 7.30, and suppose that $v, w \in \left(\mathcal{H}_\pi^{\mathrm{SL}_3(\mathbb{R})}\right)^\perp$ are K-eigenvectors with*

$$\mathrm{SO}_2(\mathbb{R}) = K < \mathrm{ASL}_2(\mathbb{R}) < \mathrm{SL}_3(\mathbb{R}).$$

Then the diagonal matrices in (7.11) satisfy

$$\left|\langle \pi_{a_t} v, w \rangle\right| \ll \mathrm{e}^{-\frac{|t|}{2}} \|v\| \|w\|$$

for all $t \in \mathbb{R}$, where the implicit constant is absolute.

PROOF. Let $v \in \mathcal{H}$ be a K-eigenvector of weight $n \in \mathbb{Z}$, so that

$$\pi_{k_\theta} v = \mathrm{e}^{\mathrm{i}n\theta} v \tag{7.12}$$

for all $k_\theta \in \mathrm{SO}_2(\mathbb{R}) < \mathrm{ASL}_2(\mathbb{R})$. Now notice that (7.12) and sesqui-linearity of the inner product implies that $\mu_{\pi_{k_\theta} v} = \mu_v$ for the spectral measures of $\pi|_H$ (also see Proposition 2.52(2)). By Proposition 5.1, this implies that μ_v is invariant under the rotation k_θ for all $\theta \in \mathbb{R}$.

Since $v \in (\mathcal{H}_\pi^{\mathrm{SL}_3(\mathbb{R})})^\perp$, we have $\mu_v(\{0\}) = 0$ by Exercise 1.88 (see also the hint on page 529) and the same holds for μ_w. We assume that $t > 0$ (switching the roles of v and w then gives the other case). We define the subsets

$$B_{\text{horizontal}} = \{(x_1, x_2) \mid \left|\tfrac{x_2}{x_1}\right| \leqslant \mathrm{e}^{-t}\}$$

and

$$B_{\text{vertical}} = \{(x_1, x_2) \mid \left|\tfrac{x_2}{x_1}\right| \geqslant \mathrm{e}^{t}\},$$

which are the two sectors in $\mathbb{R}^2$ illustrated in Figure 7.1.

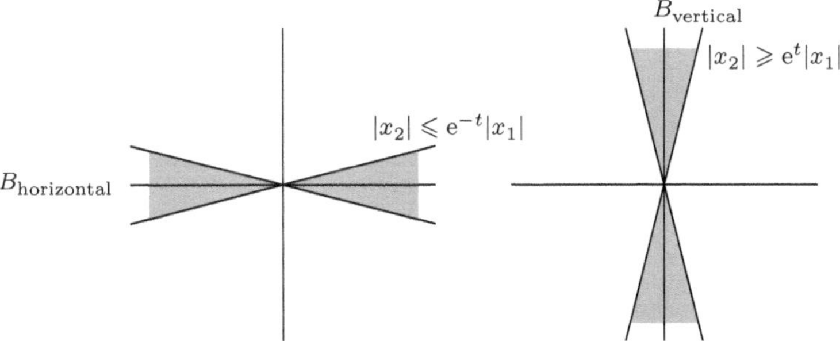

Fig. 7.1: The sets used for decomposing v and w.

Next we use these and the functional calculus for $\pi|_H$ to define the associated projection operator Π_B for

$$B \in \{B_{\text{horizontal}}, \mathbb{R}^2 \smallsetminus B_{\text{horizontal}}, B_{\text{vertical}}, \mathbb{R}^2 \smallsetminus B_{\text{vertical}}\}$$

and split v and w into two parts according to these two sectors in $\mathbb{R}^2$,

$$v = v_{\text{main}} + v_{\text{horizontal}}$$

with $v_{\text{main}} = \Pi_{\mathbb{R}^2 \smallsetminus B_{\text{horizontal}}}(v)$, $v_{\text{horizontal}} = \Pi_{B_{\text{horizontal}}}(v)$ and

$$w = w_{\text{main}} + w_{\text{vertical}},$$

with $w_{\mathrm{main}} = \Pi_{\mathbb{R}^2 \smallsetminus B_{\mathrm{vertical}}}(w)$ and $w_{\mathrm{vertical}} = \Pi_{B_{\mathrm{vertical}}}(w)$.

We now use the fact that μ_v is invariant under rotation, $\mu_v(\{0\}) = 0$, and $B_{\mathrm{horizontal}}$ consists of two sectors with internal angle $\ll \mathrm{e}^{-t}$. Using Exercise 2.60 (see the hint on page 530) we obtain from these the bound

$$\|v_{\mathrm{horizontal}}\|^2 = \mu_v(B_{\mathrm{horizontal}}) \ll \mathrm{e}^{-t}\|v\|^2, \tag{7.13}$$

and similarly

$$\|w_{\mathrm{vertical}}\|^2 = \mu_w(B_{\mathrm{vertical}}) \ll \mathrm{e}^{-t}\|w\|^2. \tag{7.14}$$

Also, by Exercise 2.60 we have

$$\mu_{v_{\mathrm{main}}} = \mu_v|_{\mathbb{R}^2 \smallsetminus B_{\mathrm{horizontal}}},$$

which together with Proposition 5.1 gives

$$\mu_{\pi_{a_t} v_{\mathrm{main}}} = (a_t^{-1})_* \mu_v|_{\mathbb{R}^2 \smallsetminus B_{\mathrm{horizontal}}}$$

since the transpose of a_t is a_t itself. A simple calculation now reveals that the set $a_t^{-1}(\mathbb{R}^2 \smallsetminus B_{\mathrm{horizontal}})$ agrees with B_{vertical} with the exception of the boundaries, which are null sets. It follows that

$$\mu_{\pi_{a_t} v_{\mathrm{main}}} \perp \mu_{w_{\mathrm{main}}}$$

which implies that $\pi_{a_t} v_{\mathrm{main}} \perp w_{\mathrm{main}}$ by Proposition 2.52(7). Together with (7.13) and (7.14), this gives

$$\begin{aligned} \left|\langle \pi_{a_t} v, w\rangle\right| &\leqslant \underbrace{\left|\langle \pi_{a_t} v_{\mathrm{main}}, w_{\mathrm{main}}\rangle\right|}_{=0} + \left|\langle \pi_{a_t} v_{\mathrm{main}}, w_{\mathrm{vertical}}\rangle\right| + \left|\langle \pi_{a_t} v_{\mathrm{horizontal}}, w\rangle\right| \\ &\ll \mathrm{e}^{-\frac{t}{2}}\|v\|\|w\| \end{aligned}$$

by the Cauchy–Schwarz inequality. □

7.4.2 Bootstrapping to the General Case

Lemma 7.31 will allow us to use the following lemma with the value $\kappa = \frac{1}{2}$.

Lemma 7.32 (Smooth vectors for SL_3). *Let π be a unitary representation of $\mathrm{SL}_3(\mathbb{R})$ and assume that*

$$\left|\langle \pi_{a_t} v, w\rangle\right| \leqslant c\mathrm{e}^{-\kappa t}\|v\|\|w\| \tag{7.15}$$

holds for some constants $c, \kappa > 0$ and all $\mathrm{SO}_2(\mathbb{R})$-eigenfunctions v, w lying in $\left(\mathcal{H}_\pi^{\mathrm{SL}_3(\mathbb{R})}\right)^\perp$. Then

$$|\langle \pi_a v, w\rangle| \ll c\mathrm{e}^{-\frac{1}{2}\kappa|t_2-t_1|}\,\mathcal{S}(v)\,\mathcal{S}(w)$$

for all C^1-smooth vectors $v, w \in \left(\mathcal{H}_\pi^{\mathrm{SL}_3(\mathbb{R})}\right)^\perp$, where

$$a = \begin{pmatrix} \mathrm{e}^{t_1} & & \\ & \mathrm{e}^{t_2} & \\ & & \mathrm{e}^{t_3} \end{pmatrix} \tag{7.16}$$

for $t_1, t_2, t_3 \in \mathbb{R}$ with $t_1 + t_2 + t_3 = 0$.

Proof. We will use Proposition 7.29 and its notation for π restricted to

$$\left(\mathcal{H}_\pi^{\mathrm{SL}_3(\mathbb{R})}\right)^\perp.$$

For this, we notice that

$$a = \begin{pmatrix} \mathrm{e}^{t_1} & & \\ & \mathrm{e}^{t_2} & \\ & & \mathrm{e}^{t_3} \end{pmatrix} = \underbrace{\begin{pmatrix} \mathrm{e}^{t_1+\frac{1}{2}t_3} & & \\ & \mathrm{e}^{t_2+\frac{1}{2}t_3} & \\ & & 1 \end{pmatrix}}_{=a_{t_1+\frac{1}{2}t_3}} \underbrace{\begin{pmatrix} \mathrm{e}^{-\frac{1}{2}t_3} & & \\ & \mathrm{e}^{-\frac{1}{2}t_3} & \\ & & \mathrm{e}^{t_3} \end{pmatrix}}_{=b}$$

where $a_{t_1+\frac{1}{2}t_3} \in \mathrm{ASL}_2(\mathbb{R})$ is as defined in (7.11) and $b \in \mathrm{SL}_3(\mathbb{R})$ commutes with K. We set $B = \pi_b$ and apply Proposition 7.29. We note that its conclusion holds even if $\mathcal{S}(\cdot)$ denotes the Sobolev norm with respect to the unitary action π of $\mathrm{SL}_3(\mathbb{R})$. Indeed, we may include a basis of $\mathfrak{sl}_2(\mathbb{R})$ in the basis of $\mathfrak{sl}_3(\mathbb{R})$ in the definition of the Sobolev norm, and then use Exercise 7.13 to see that this assumption does not matter since we allow ourselves to change the implicit constant. Therefore

$$|\langle \pi_a v, w\rangle| = \left|\left\langle \pi_{a_{t_1+\frac{1}{2}t_3}} Bv, w\right\rangle\right| \ll c\mathrm{e}^{-\kappa|t_1+\frac{1}{2}t_3|}\,\mathcal{S}(v)\,\mathcal{S}(w)$$

for all C^1-smooth vectors $v, w \in \left(\mathcal{H}_\pi^{\mathrm{SL}_3(\mathbb{R})}\right)^\perp$. Since $t_1 + \frac{1}{2}t_3 = \frac{1}{2}(t_1 - t_2)$, this gives the lemma. □

Proof of Theorem 7.30. By Lemma 7.31, we have (7.15) for $\kappa = \frac{1}{2}$ and obtain the conclusion of Lemma 7.32. By conjugation with permutation matrices, we have that Lemma 7.32 also shows

$$|\langle \pi_a v, w\rangle| \ll \mathrm{e}^{-\frac{1}{4}|t_3-t_1|}\,\mathcal{S}(v)\,\mathcal{S}(w) \tag{7.17}$$

and

$$|\langle \pi_a v, w\rangle| \ll \mathrm{e}^{-\frac{1}{4}|t_3-t_2|}\,\mathcal{S}(v)\,\mathcal{S}(w)$$

for all diagonal matrices a, in the notation of (7.16). Hence we may choose the best of these three estimates. Suppose without loss of generality that

$$t_3 \geqslant t_2 \geqslant t_1,$$

so that the best estimate is given by (7.17). Now notice that $\|a\|_{\mathrm{HS}}$ is essentially equal to e^{t_3} in the sense that

$$\mathrm{e}^{t_3} \leqslant \|a\|_{\mathrm{HS}} \ll \mathrm{e}^{t_3},$$

and $t_1 + t_2 + t_3 = 0$ gives

$$\mathrm{e}^{t_1 - t_3} \leqslant \mathrm{e}^{\frac{1}{2}(t_1+t_2)-t_3} = \mathrm{e}^{-\frac{3}{2}t_3} \ll \|a\|_{\mathrm{HS}}^{-\frac{3}{2}}.$$

Combining this with (7.17), it follows that Theorem 7.30 holds for diagonal matrices.

To generalize the estimate to an arbitrary $g \in \mathrm{SL}_3(\mathbb{R})$ we use the Cartan decomposition $g = kak'$ for $k, k' \in \mathrm{SO}_3(\mathbb{R})$ and a diagonal a (see the footnote on page 65). Notice that $\|g\|_{\mathrm{HS}} = \|a\|_{\mathrm{HS}}$, and that $\mathcal{S}(\pi_k v) \ll \mathcal{S}(v)$ uniformly for $k \in \mathrm{SO}_3(\mathbb{R})$ (see Exercise 7.12). Therefore

$$\begin{aligned}\left|\langle \pi_g v, w\rangle\right| &= \left|\langle \pi_a(\pi_{k'} v), \pi_k^{-1} w\rangle\right| \\ &\ll \|a\|_{\mathrm{HS}}^{-\frac{3}{8}}\, \mathcal{S}(\pi_{k'} v)\, \mathcal{S}(\pi_k^{-1} w) \\ &\ll \|g\|_{\mathrm{HS}}^{-\frac{3}{8}}\, \mathcal{S}(v)\, \mathcal{S}(w),\end{aligned}$$

as required. Going through the implicit constants appearing in the proof, starting with Lemma 7.31 and Proposition 7.29, one verifies that the implicit constant above is also absolute. The last claim now follows from Proposition 7.27. □

Exercise 7.33. Extend the argument used in the proof of Theorem 7.30 to find a uniform decay exponent for $G = \mathrm{SL}_d(\mathbb{R})$ with $d \geqslant 3$ (for example by again using subgroups of the form $\mathrm{ASL}_2(\mathbb{R}) = \mathrm{SL}_2(\mathbb{R}) \ltimes \mathbb{R}^2$ leading to the exponent $\frac{1}{4}\frac{d}{d-1}$).

Exercise 7.34. Recall that $\mathrm{SL}_3(\mathbb{R})$ has a double cover $\widetilde{\mathrm{SL}_3(\mathbb{R})}$ (since the fundamental group of $\mathrm{SO}_3(\mathbb{R})$ is $\mathbb{Z}/2\mathbb{Z}$). Generalize Definition 7.25 and Theorem 7.30 to $\widetilde{\mathrm{SL}_3(\mathbb{R})}$ by using the isogeny $\imath\colon \widetilde{\mathrm{SL}_3(\mathbb{R})} \to \mathrm{SL}_3(\mathbb{R})$ to define the size of $g \in \widetilde{\mathrm{SL}_3(\mathbb{R})}$ by $\|g\|_{\mathrm{HS}} = \|\imath(g)\|_{\mathrm{HS}}$.

7.5 An Effective Pointwise Ergodic Theorem*

In this section we outline through guided exercises how to prove an 'effective pointwise ergodic theorem'. For ergodic systems the pointwise ergodic theorem states that for almost every point in the space the time average of a function sampled along the orbit of the point converges to the space average of the function (see [24] for example).† Here we describe how effective decay of matrix coefficients for smooth functions can lead to an effective ergodic theorem.

† As mentioned in the footnote to page 271, for a single transformation there is no general rate of convergence in the ergodic theorem.

An effective form of the ergodic theorem giving a concrete error estimate will require us to remove a little more of the space than just a null set. However, how much of the space needs to be removed should again have a concrete estimate.

As mixing implies ergodicity and ergodicity implies the pointwise ergodic theorem, it stands to reason that an effective version of mixing should imply an effective pointwise ergodic theorem. We explain now that this is indeed the case. The argument used is quite flexible and can be adapted and generalized in various ways. By Theorem 7.30 it should be clear that the following formulation applies, for example, directly to finite volume quotients of the form $X = \Gamma\backslash \mathrm{SL}_3(\mathbb{R})$, smooth compactly supported functions on X, and the action of both diagonalizable or unipotent one-parameter subgroups of $\mathrm{SL}_3(\mathbb{R})$. Our discussions in Chapter 9 will lead to further instances (see Corollary 9.84).

Assumptions

We let X be a locally compact, σ-compact metric space and suppose that $\mathbb{R}$ acts on X, with the action written as $x \mapsto g_t {\boldsymbol\cdot} x$ for $t \in \mathbb{R}$ and $x \in X$. Let μ be a probability measure on X preserved by the action. We suppose that the Koopman representation π of $\mathbb{R}$ on $L^2_\mu(X)$ and a function $f \in L^\infty(X)$ satisfies effective decay of matrix coefficients in the form

$$\left| \langle \pi_t f, f \rangle - \left| \int_X f \,\mathrm{d}\mu \right|^2 \right| \leqslant t^{-\kappa} S(f)^2 \tag{7.18}$$

for a constant $\kappa \in (0,1)$ and a constant $S(f) \geqslant \|f\|_\infty \geqslant \|f\|_2$ depending only on f.

Conclusion

There exists a constant $\delta > 0$ depending only on κ with the following property: For every $T \geqslant 1$ there exists an exceptional set $X_T \subseteq X$ satisfying

$$\mu(X_T) \ll T^{-\delta}$$

and for all $x \in X \smallsetminus X_T$ and $t \geqslant T$ we have

$$\left| \frac{1}{t} \int_0^t f(g_s {\boldsymbol\cdot} x) \,\mathrm{d}s - \int_X f \,\mathrm{d}\mu \right| \ll t^{-\delta} S(f).$$

The proof is given in the form of three exercises as follows.

Exercise 7.35 (Discrepancy estimate). Define the *discrepancy* of f at $x \in X$ for $t \geqslant 1$ by

$$D_t f(x) = \left| \frac{1}{t} \int_0^t f(g_s {\boldsymbol\cdot} x) \,\mathrm{d}s - \int_X f \,\mathrm{d}\mu \right|.$$

(a) Show that $\|D_t f\|_2^2 \ll_\kappa t^{-\kappa} S(f)^2$.
(b) Conclude that

$$\mu\Big(\{x \in X \mid D_t f(x) \geqslant t^{-\delta} S(f)\}\Big) \ll t^{2\delta-\kappa}$$

for any $\delta > 0$.

We now fix $\delta = \frac{\kappa}{4}$, $\alpha = \frac{3}{\kappa}$ and define

$$X_T = \bigcup_{n \geqslant \lfloor T^{\frac{1}{\alpha}} \rfloor} \{x \in X \mid D_{n^\alpha} f(x) > n^{-\delta\alpha} S(f)\}$$

for $T \geqslant 1$.

Exercise 7.36 (Effective null set). Show that $\mu(X_T) \ll_\kappa T^{-\frac{\kappa}{6}}$ for $T \geqslant 1$.

Exercise 7.37 (Conclusion). Show that

$$\left| \frac{1}{t} \int_0^t f(g_s \boldsymbol{\cdot} x)\,\mathrm{d}s - \int_X f\,\mathrm{d}\mu \right| \ll t^{-\delta} S(f)$$

for $T \geqslant 1$, $x \in X \smallsetminus X_T$ and $t \geqslant T$.

We note that in the case of a diagonal subgroup $A < \mathrm{SL}_3(\mathbb{R})$ one could replace the polynomial decay in (7.18) by an exponential decay, but that this does not lead to a stronger form of the effective pointwise ergodic theorem because of the contribution of the 'diagonal' in the estimate for $\|D_t f\|^2$.

It is also possible to render effective the notion of generic point. Here the pointwise ergodic theorem holds for all smooth functions with a concrete error estimate and independently of the function for all points except for a set of small measure. We refer to the work of Einsiedler, Margulis, and Venkatesh [23, Sec. 9] for the details.

7.6 Summary and Outlook

The main purpose of this chapter was to introduce smoothness to our discussion of unitary representations. As we have already seen in the discussion of effective decay of matrix coefficients, smoothness may be a required assumption in certain applications of the theory. We will study this notion further in Chapter 9, where it will also become an important tool for describing $\widehat{\mathrm{SL}_2(\mathbb{R})}$ completely.

Chapter 8
Discrete Series Representations and Temperedness for $\mathrm{SL}_2(\mathbb{R})$

In Chapter 2 we have seen that for abelian non-compact groups no irreducible representation can be contained in the regular representation.† For compact groups, we have seen in Chapter 3 that every irreducible representation is contained in the regular representation.

However, as we will see in Section 8.4, for non-abelian non-compact groups it is possible for certain irreducible unitary representations to be contained in the regular representation. Before we get to these examples, we classify such representations from an abstract point of view, and relate weak containment in the regular representation to almost square integrability of diagonal matrix coefficients.

However, our main purpose in this chapter will be to start our discussion of the group $\mathrm{SL}_2(\mathbb{R})$, and discuss its discrete series and tempered representations.

8.1 Abstract Discrete Series Representations

Definition 8.1. An irreducible unitary representation π of the group G is called a *discrete series representation* if π is contained in the regular representation of G.

Theorem 8.2 (Characterization of discrete series). *Suppose G is unimodular, and let π be an irreducible unitary representation of G. Then the following are equivalent.*[17]

(1) *π is a discrete series representation (that is, $\pi < \lambda$).*
(2) *There exist vectors $u, v \in \mathcal{H}_\pi \smallsetminus \{0\}$ such that the matrix coefficient $\varphi^\pi_{u,v}$ is square integrable (that is, $\varphi^\pi_{u,v} \in L^2(G)$).*
(3) *The matrix coefficient $\varphi^\pi_{u,v}$ is square integrable for any pair $u, v \in \mathcal{H}_\pi$.*

† For this, see Exercise 1.19 and its hint on page 525, Exercise 2.14, Lemma 2.19, and the characterization of containment in Proposition 2.57.

M. Einsiedler and T. Ward, *Unitary Representations and Unitary Duals*,
Graduate Texts in Mathematics 308, https://doi.org/10.1007/978-3-032-03899-9_8

PROOF. We first show that (1) implies (2). Suppose that $\pi < \lambda$ and so $\mathcal{H}_\pi = \mathcal{V}$ for an invariant supspace $\mathcal{V} \subseteq L^2(G)$ as in (1). Let $v \in \mathcal{V}\smallsetminus\{0\}$ and let u be the projection $P(f)$ where $P\colon L^2(G) \to \mathcal{V} \subseteq L^2(G)$ is the orthogonal projection operator and f is a function in $C_c(G)$. For $f \in C_c(G)$ and $g \in G$, we also define $\widetilde{f}(g) = f(g^{-1})$. Since $\mathcal{V}$ is invariant, P is intertwining and so

$$\begin{aligned}\varphi^\pi_{u,v}(g) = \langle \lambda_g Pf, v\rangle = \langle \lambda_g f, v\rangle &= \int f(g^{-1}h)\overline{v}(h)\,\mathrm{d}m(h)\\ &= \int \overline{v}(h)\widetilde{f}(h^{-1}g)\,\mathrm{d}m(h) = \overline{v} * \widetilde{f}(g)\end{aligned}$$

for all $g \in G$. Now recall that $\overline{v} * \widetilde{f} \in L^2(G)$ (see [25, Lem. 3.75] or generalize Lemma 2.18), and note that density of $C_c(G) \subseteq L^2(G)$ implies that there exists some $f \in C_c(G)$ such that $u = Pf \in \mathcal{V}\smallsetminus\{0\}$. This shows (2).

Notice that (3) clearly implies (2) as $\mathcal{H}_\pi \neq \{0\}$. Hence the following step will finish the proof of the theorem.

We will now show that (2) implies (1) and (3). So suppose that u_0, v_0 in $\mathcal{H}_\pi\smallsetminus\{0\}$ have $\varphi^\pi_{u_0,v_0} \in L^2(G)$. We fix u_0 and consider the subspace

$$D_{u_0} = \{v \in \mathcal{H}_\pi \mid \varphi^\pi_{u_0,v} \in L^2(G)\}$$

as the domain of the (possibly unbounded) operator T from $\mathcal{H}_\pi$ into $L^2(G)$ defined by

$$T(v) = \overline{\varphi^\pi_{u_0,v}} \in L^2(G)$$

for $v \in D_{u_0}$. As the main step of the remaining proof, we claim that T is a densely defined closed intertwining operator.

Equivariance follows from the properties of the matrix coefficients. Indeed, for $v \in D_{u_0}$ and $g_0 \in G$ we have

$$\begin{aligned}\lambda_{g_0}\overline{\varphi^\pi_{u_0,v}}(g) = \overline{\varphi^\pi_{u_0,v}(g_0^{-1}g)} &= \overline{\langle \pi_{g_0^{-1}g}u_0, v\rangle}\\ &= \overline{\langle \pi_g u_0, \pi_{g_0}v\rangle} = \overline{\varphi^\pi_{u_0,\pi_{g_0}v}}(g)\end{aligned}$$

for every $g \in G$, equivalently $\lambda_{g_0}T(v) = T(\pi_{g_0}(v)) \in L^2(G)$ and $\pi_{g_0}v \in D_{u_0}$.

Since $v_0 \in D_{u_0}\smallsetminus\{0\}$ and D_{u_0} is invariant under π by the above, we see that $\overline{D_{u_0}} = \mathcal{H}_\pi$ by irreducibility of π. In other words, T is densely defined. To complete the proof of the claim, it remains to show that T is a closed operator. Suppose therefore that (v_n) is a sequence in D_{u_0} with

$$(v_n, Tv_n) \longrightarrow (v, f) \in \mathcal{H}_\pi \times L^2(G)$$

as $n \to \infty$. Choosing a subsequence if necessary we may upgrade the L^2 convergence and suppose without loss of generality that $Tv_n \to f$ almost everywhere. Furthermore, notice that $v_n \to v \in \mathcal{H}_\pi$ implies

$$(Tv_n)(g) = \overline{\varphi^{\pi}_{u_0,v_n}(g)} = \overline{\langle \pi_g u_0, v_n \rangle} \longrightarrow \overline{\langle \pi_g u_0, v \rangle}$$

as $n \to \infty$ for all $g \in G$. Together we see that

$$f(g) = \lim_{n\to\infty} (Tv_n)(g) = \overline{\langle \pi_g u_0, v \rangle}$$

for almost every $g \in G$. However, since $f \in L^2(G)$ we see that $v \in D_{u_0}$ belongs to the domain of T and $f = Tv$. In particular, $(v, f) \in \operatorname{Graph}(T)$ as required.

The claim above, the assumed irreducibility of π, and Schur's lemma in the form of Corollary 1.38 together imply that $D_{u_0} = \mathcal{H}_\pi$ and either $T = 0$ or T is a scalar multiple of a unitary isomorphism. Using $u_0 \in D_{u_0} = \mathcal{H}_\pi$ we see that $Tu_0 = \overline{\varphi^{\pi}_{u_0,u_0}} \neq 0$ and obtain $\pi < \lambda$ as in (1).

To prove (3), let $u_1, v_1 \in \mathcal{H}_\pi$ be any vectors. By the argument above we already have $\varphi_{u_0,v_1} \in L^2(G)$ and repeat the argument as follows. Define

$$\widetilde{D}_{v_1} = \{u \in \mathcal{H}_\pi \mid \varphi^{\pi}_{u,v_1} \in L^2(G)\}$$

and $\widetilde{T}(u) = \varphi^{\pi}_{u,v_1}$ for $u \in \widetilde{D}_{v_1}$. Now recall that

$$\rho_{g_0}(\widetilde{T}(u))(g) = \langle \pi_{gg_0} u, v_1 \rangle = \langle \pi_g \pi_{g_0} u, v_1 \rangle = \widetilde{T}(\pi_{g_0} u)(g)$$

for $g, g_0 \in G$ since G is assumed to be unimodular. We now apply the argument above to deduce that $\widetilde{T}$ is a densely defined closed intertwining operator from $\mathcal{H}_\pi$ to $L^2(G)$ with the latter equipped with the right regular representation. As above, this implies by Schur's lemma (Corollary 1.38) that $D_{v_1} = \mathcal{H}_\pi$, and in particular $\varphi^{\pi}_{u_1,v_1} \in L^2(G)$. As $u_1, v_1 \in \mathcal{H}_\pi$ were arbitrary, (3) follows. □

To see an example of a discrete series representation, the reader may continue with Section 8.4 (which only builds on this section and some preparations concerning hyperbolic geometry in Section 8.3). As one of our motivations for this volume is to better understand decay of matrix coefficients, Theorem 8.2 is of interest to us. Indeed, the requirement to belong to $L^2(G)$ is a requirement on the average decay of the matrix coefficients. This observation will become very important for us after replacing the strong condition of containment by weak containment as in the next section.

8.2 Almost Square Integrable Matrix Coefficients

We now present a theorem of Cowling, Haagerup and Howe [19] that will allow us to characterize in Section 8.5.5 the so called *tempered representations* for the group $\mathrm{SL}_2(\mathbb{R})$.

Definition 8.3 (Tempered representations). Let π be a unitary representation of G. We say that π is *tempered* if it is weakly contained in the regular representation λ of G.

Our interest in tempered representations comes from our wish to understand decay of matrix coefficients. We will see the connection explicitly in Section 8.5.5, where the following theorem will be a crucial ingredient.

The reader may also wonder whether non-tempered representations actually exist, since so far we have mostly encountered tempered representations (without pointing this out). However, we will see that (for example) the trivial representation is not always tempered (which relates to the definition of amenability in Section 6.2 and Exercise 6.20).

Definition 8.4 (Almost square integrability). Let π be a unitary representation of G. We say that π is *almost square integrable* if there exists a dense subset $\mathcal{V} \subseteq \mathcal{H}_\pi$ such that $\varphi_v^\pi \in L^{2+\varepsilon}(G)$ for all $v \in \mathcal{V}$ and $\varepsilon > 0$.

Theorem 8.5 (Almost square integrable matrix coefficients [19]). *Let G be a locally compact group and π a unitary representation. Suppose that $v \in \mathcal{H}_\pi$ has the property that the diagonal matrix coefficient*

$$\varphi_v^\pi(g) = \langle \pi_g v, v \rangle$$

for $g \in G$ belongs to $L^{2+\varepsilon}(G)$ for all $\varepsilon > 0$. Then the cyclic representation generated by v is weakly contained in the regular representation of G. Moreover, if π is almost square integrable, then π is tempered.

Along the way, and as a warm-up for the proof of the theorem, we will obtain the following result which is of independent interest.

Proposition 8.6 (Sub-exponential growth). *Suppose that G has sub-exponential growth, meaning that for every compact subset $K \subseteq G$ we have*

$$\lim_{n\to\infty} m\left(K^n\right)^{\frac{1}{n}} = 1. \tag{8.1}$$

Then every unitary representation is tempered.

For the proof of the theorem and the proposition we are going to use Lemma 6.35 and its twin for the regular representation below and the equivalent definition ($\pi \prec_{\rm op} \rho$) of weak containment from Theorem 6.31.

Lemma 8.7 (2-norm formula for operator norm). *Let $f \in C_c(G)$. Then for the left regular representation λ of G we have*

$$\|\lambda_*(f)\|_{\rm op} = \lim_{n\to\infty} \|(f^* * f)^{*n}\|_2^{\frac{1}{2n}}.$$

PROOF. By Lemma 6.35 we have

$$\|\lambda_*(f)\| = \sup_{f_0\in L^2(G)} \lim_{n\to\infty} \left(\int_G (f^* * f)^{*n}\varphi_{f_0}\,\mathrm{d}m\right)^{\frac{1}{2n}}$$

where $\varphi_{f_0}(g) = \langle\lambda_g f_0, f_0\rangle$ for $f_0 \in L^2(G)$. Also, by Lemma 6.35, we may also restrict to functions $f_0 \in C_c(G)$ in which case $\varphi_{f_0} \in C_c(G)$. Noticing that for $f_0 \in C_c(G)$

$$\int_G (f^* * f)^{*n}\varphi_{f_0}\,\mathrm{d}m \leqslant \|(f^* * f)^{*n}\|_2\|\varphi_{f_0}\|_2,$$

we get

$$\|\lambda_*(f)\|_{\mathrm{op}} \leqslant \liminf_{n\to\infty} \|(f^* * f)^{*n}\|_2^{\frac{1}{2n}}. \tag{8.2}$$

On the other hand, setting $f_0 = f^* * f \in C_c(G)$ gives

$$\begin{aligned}(f^* * f)^{*n}(g) &= (f^* * f)^{*(n-1)} * f_0(g)\\ &= \int_G (f^* * f)^{*(n-1)}(h)f_0(h^{-1}g)\,\mathrm{d}m(h)\\ &= \left(\lambda_*\left((f^* * f)^{*(n-1)}\right)f_0\right)(g)\end{aligned}$$

for all $g \in G$, which leads to

$$\begin{aligned}\|(f^* * f)^{*n}\|_2 &\leqslant \|\lambda_*(f^* * f)^{*(n-1)}\|_{\mathrm{op}}\|f_0\|_2\\ &\leqslant \|\lambda_*(f)\|_{\mathrm{op}}^{2(n-1)}\|f_0\|_2\end{aligned}$$

and in turn to

$$\limsup_{n\to\infty} \|(f^* * f)^{*n}\|_2^{\frac{1}{2n}} \leqslant \|\lambda_*(f)\|_{\mathrm{op}}.$$

Together with (8.2), this gives the lemma. □

Proof of Proposition 8.6. Let G have sub-exponential growth, let π be a unitary representation of G, and f a function in $C_c(G)$. We wish to show that $\|\pi_*(f)\|_{\mathrm{op}} \leqslant \|\lambda_*(f)\|_{\mathrm{op}}$, and will use Lemmas 6.35 and 8.7 to prove this. So set $f_0 = f^* * f$ and $K = \operatorname{supp} f_0$. Then $\operatorname{supp} f_0^{*n} \subseteq K^n$, and we may use Cauchy–Schwarz for any $v \in \mathcal{H}_\pi$ to obtain

$$\left|\int_G f_0^{*n}\varphi_v^\pi\,\mathrm{d}m\right|^2 \leqslant \int_{K^n} |f_0^{*n}|^2\,\mathrm{d}m \int_{K^n} |\varphi_v^\pi|^2\,\mathrm{d}m.$$

Taking the $4n$th root and using $\|\varphi_v^\pi\|_\infty \leqslant \|v\|^2$, this gives

$$\lim_{n\to\infty}\left|\int_G f_0^{*n}\varphi_v^\pi\,\mathrm{d}m\right|^{\frac{1}{2n}} \leqslant \lim_{n\to\infty} \|f_0^{*n}\|_2^{\frac{1}{2n}} \lim_{n\to\infty} \left(\|v\|^4 m(K^n)\right)^{\frac{1}{4n}} = \|\lambda_*(f)\|_{\mathrm{op}}$$

by (8.1) and Lemma 8.7. Taking the supremum over $v \in \mathcal{H}_\pi$, Lemma 6.35 shows that $\|\pi_*(f)\|_{\mathrm{op}} \leqslant \|\lambda_*(f)\|_{\mathrm{op}}$. As $f \in C_c(G)$ was arbitrary, Theorem 6.31 shows that $\pi \prec \lambda$. □

In general, groups may not satisfy sub-exponential growth (see Exercise 8.9 below), but the following is a general estimate.

Lemma 8.8 (At most exponential growth). *Let $K \subseteq G$ be a compact subset with non-empty interior. Then there exist constants $c > 0$, $M > 1$ (depending on K) with $m_G(K^n) \leqslant cM^n$ for all $n \geqslant 1$.*

PROOF. By compactness, there exists a finite collection $x_1, \dots, x_M \in G$ with the property that

$$K^2 \subseteq \bigcup_{j=1}^{M} x_j K.$$

By induction, this implies that

$$K^n \subseteq \bigcup_{j_1,\dots,j_{n-1}=1}^{M} x_{j_1} \cdots x_{j_{n-1}} K,$$

which gives the lemma with $c = \frac{m_G(K)}{M}$. □

We are now ready to prove the main result of this section.

PROOF OF THEOREM 8.5. Suppose that π is a unitary representation of G and $v \in \mathcal{H}_\pi$ has the property that the matrix coefficient $\varphi_v^\pi(g) = \langle \pi_g v, v \rangle$ belongs to $L^{2+\varepsilon}(G)$ for all $\varepsilon > 0$. Let us now bound the expressions appearing in Lemma 6.35. To do this, fix $\varepsilon > 0$, let $f \in C_c(G)$, write $f_0 = f^* * f$ and set $K = \operatorname{supp} f_0$. This implies that $\operatorname{supp} f_0^{*n} \subseteq K^n$, and so

$$\begin{aligned}
\left| \int_{K^n} f_0^{*n}(g) \varphi_v^\pi \,\mathrm{d}m \right|^2 &\leqslant \int_{K^n} |f_0^{*n}|^2 \,\mathrm{d}m \int_{K^n} |\varphi_v^\pi|^2 \,\mathrm{d}m \\
&\leqslant \|f_0^{*n}\|_2^2 \left(\int_{K^n} |\varphi_v^\pi|^{2+\varepsilon} \,\mathrm{d}m_G \right)^{\frac{2}{2+\varepsilon}} \left(\int_{K^n} \mathrm{d}m_G \right)^{\frac{\varepsilon}{2+\varepsilon}} \\
&\leqslant \|f_0^{*n}\|_2^2 \, \|\varphi_v^\pi\|_{2+\varepsilon}^2 \, (cM^n)^{\frac{\varepsilon}{2+\varepsilon}},
\end{aligned}$$

where we have used the Cauchy–Schwarz and Hölder inequalities with the conjugate pair of exponents $\left(\frac{2+\varepsilon}{2}, \frac{2+\varepsilon}{\varepsilon}\right)$ and Lemma 8.8. Taking the $4n$th root and letting n go to infinity, it follows that

$$\begin{aligned}
\lim_{n\to\infty} \left(\int_G (f^* * f)^{*n} \varphi_v^\pi \,\mathrm{d}m \right)^{\frac{1}{2n}} &\leqslant \lim_{n\to\infty} \|(f^* * f)^{*n}\|_2^{\frac{1}{2n}} M^{\frac{\varepsilon}{4(2+\varepsilon)}} \\
&= \|\lambda_*(f)\|_{\mathrm{op}} M^{\frac{\varepsilon}{4(2+\varepsilon)}}
\end{aligned}$$

by Lemma 8.7. Since this holds for all $\varepsilon > 0$, we also obtain

$$\lim_{n\to\infty}\left(\int_G (f^* * f)^{*n}\varphi_v^\pi \,\mathrm{d}m\right)^{\frac{1}{2n}} \leqslant \|\lambda_*(f)\|_{\mathrm{op}}.$$

If there is a dense set of vectors $v \in \mathcal{H}_\pi$ for which the matrix coefficients are almost square integrable, then Lemma 6.35 gives $\|\pi(f)\|_{\mathrm{op}} \leqslant \|\lambda_*(f)\|_{\mathrm{op}}$ for all $f \in C_c(G)$ and Theorem 6.31 gives $\pi \prec \lambda$, proving the last part of the theorem.

Now suppose that $v \in \mathcal{H}_\pi$ has almost square integrable matrix coefficients. To prove the first part of the theorem assume without loss of generality that $\mathcal{H}_\pi$ is a cyclic representation with generator v. Hence vectors of the form

$$w = \sum_{j=1}^{n} \alpha_j \pi_{g_j} v$$

for $\alpha_1, \dots, \alpha_n \in \mathbb{C}$ and $g_1, \dots, g_n \in G$ are dense in $\mathcal{H}_\pi$. We claim that the matrix coefficients of such vectors also belong to $L^{2+\varepsilon}(G)$ for all $\varepsilon > 0$. To see this let w be as above and calculate

$$\begin{aligned}\varphi_w^\pi(g) = \langle \pi_g w, w\rangle &= \sum_{j,k=1}^{n} \alpha_j \overline{\alpha_k} \left\langle \pi_g \pi_{g_j} v, \pi_{g_k} v\right\rangle \\ &= \sum_{j,k=1}^{n} \alpha_j \overline{\alpha_k} \left\langle \pi_{g_k^{-1} g g_j} v, v\right\rangle \\ &= \sum_{j,k=1}^{n} \alpha_j \overline{\alpha_k} \varphi_v^\pi(g_k^{-1} g g_j).\end{aligned}$$

Clearly the left regular representation preserves the p-norm for $p \in [1, \infty]$ and so $\|\varphi_v^\pi(g_k^{-1} \cdot g_j)\|_p = \|\varphi_v^\pi(\cdot g_j)\|_p$. Furthermore,

$$\|\varphi_v^\pi(\cdot g_j)\|_p^p = \int_G |\varphi_v^\pi(g g_j)|^p \,\mathrm{d}m(g) = \int |\varphi_v^\pi(g)|^p \,\mathrm{d}m(g) \Delta(g_j)^{-1} < \infty$$

by Lemma 1.15. It follows that $\varphi_w^\pi \in L^{2+\varepsilon}(G)$ for all $\varepsilon > 0$, and the theorem follows from the case considered above. □

It is clear that, unlike the statement for discrete series representations in Theorem 8.2, the condition of almost square integrability is not a characterization for temperedness (for example, for $G = \mathbb{R}$). However, as we will see in Section 8.5.5, it is a characterization for $G = \mathrm{SL}_2(\mathbb{R})$, for example.

Exercise 8.9. Show that the affine group from Section 5.2.3 does not have sub-exponential growth, but that all its irreducible unitary representations (as in Proposition 5.22) are tempered.

8.3 The Group $\mathrm{SL}_2(\mathbb{R})$

To give interesting examples of discrete series representations and of tempered representations, we need to study non-compact semi-simple Lie groups. The principal example—and hence the first group we should try to understand—is $\mathrm{SL}_2(\mathbb{R})$.

8.3.1 The Action of $\mathrm{SL}_2(\mathbb{R})$ on $\mathbb{H}$

As we will see in the discussions below, and in the following chapter, the group $\mathrm{SL}_2(\mathbb{R})$ and its unitary representations are intimately linked to hyperbolic geometry. We briefly review the link between $\mathrm{SL}_2(\mathbb{R})$ and hyperbolic geometry, and refer to [24, Ch. 9] or the survey by Cannon *et al.* [12] for example, for more details.

The upper half-plane model $\mathbb{H}$ of the hyperbolic plane is defined by

$$\mathbb{H} = \{z \in \mathbb{C} \mid \Im(z) > 0\}$$

together with the hyperbolic metric and volume. The hyperbolic Riemannian metric is defined by

$$\frac{\mathrm{d}x^2 + \mathrm{d}y^2}{y^2}$$

and the hyperbolic area measure is defined by

$$\mathrm{dvol} = \frac{\mathrm{d}x\,\mathrm{d}y}{y^2}$$

at any point $z = x + \mathrm{i}y \in \mathbb{H}$ with $x = \Re(z) \in \mathbb{R}$ and $y = \Im(z) > 0$. The group $\mathrm{SL}_2(\mathbb{R}) = \{g \in \mathrm{Mat}_{2,2} \mid \det g = 1\}$ acts on $\mathbb{H}$ via Möbius transformations as

$$g\cdot z = \frac{az+b}{cz+d} \tag{8.3}$$

where

$$g = \begin{pmatrix} a & b \\ c & d \end{pmatrix} \in \mathrm{SL}_2(\mathbb{R}).$$

A calculation reveals that

$$\Im(g\cdot z) = \frac{\Im(z)}{|cz+d|^2}, \tag{8.4}$$

$$\frac{\mathrm{d}}{\mathrm{d}z}(g\cdot z) = \frac{a(cz+d)-(az+b)c}{(cz+d)^2} = \frac{1}{(cz+d)^2}, \tag{8.5}$$

and that g preserves both the hyperbolic Riemannian metric and the hyperbolic area measure. We also note that the action of $\mathrm{SL}_2(\mathbb{R})$ is transitive and that

$$\mathrm{Stab}_{\mathrm{SL}_2(\mathbb{R})}(\mathrm{i}) = \mathrm{SO}_2(\mathbb{R}),$$

so that $\mathbb{H} \cong \mathrm{SL}_2(\mathbb{R})/\mathrm{SO}_2(\mathbb{R})$.

Essential Exercise 8.10. (a) Show (8.4) and conclude that $g\boldsymbol{\cdot}z \in \mathbb{H}$ for any $g \in \mathrm{SL}_2(\mathbb{R})$ and $z \in \mathbb{H}$.
(b) Show that the Möbius transformations in (8.3) define a continuous action of $\mathrm{SL}_2(\mathbb{R})$ on $\mathbb{H}$.
(c) Show that the Möbius transformation (8.3) preserves the hyperbolic Riemannian metric and the hyperbolic area measure for any $g \in \mathrm{SL}_2(\mathbb{R})$.
(d) Show that $\mathrm{SO}_2(\mathbb{R}) = \mathrm{Stab}_{\mathrm{SL}_2(\mathbb{R})}(\mathrm{i})$ is the stabilizer of the point $\mathrm{i} \in \mathbb{H}$.
(e) Show that the action of the subgroup

$$B = \left\{ \begin{pmatrix} a & b \\ 0 & a^{-1} \end{pmatrix} \;\middle|\; a > 0, b \in \mathbb{R} \right\}$$

on $\mathbb{H}$ is simply transitive and deduce that the action of $\mathrm{SL}_2(\mathbb{R})$ is transitive.

By (8.5) and the chain rule, the action of $\mathrm{SL}_2(\mathbb{R})$ on $\mathbb{H}$ extends via taking the derivative to an action of $\mathrm{SL}_2(\mathbb{R})$ on the tangent bundle $\mathrm{T}\mathbb{H}$. Since the action of $\mathrm{SL}_2(\mathbb{R})$ on $\mathrm{T}\mathbb{H} = \mathbb{H} \times \mathbb{C}$ preserves the Riemannian metric, this action restricts to an action of $\mathrm{SL}_2(\mathbb{R})$ on the unit tangent bundle

$$\mathrm{T}^1\mathbb{H} = \left\{ (z, v) \mid z \in \mathbb{H}, \|v\|_z = \tfrac{\|v\|_2}{\Im(z)} = 1 \right\}.$$

Moreover, this action is transitive, $-I \in \mathrm{SL}_2(\mathbb{R})$ acts trivially,[†] and the action descends to a simply transitive action of $\mathrm{PSL}_2(\mathbb{R}) = \mathrm{SL}_2(\mathbb{R})/\{\pm I\}$ on $\mathrm{T}^1\mathbb{H}$.

As we have already seen in Exercise 8.10(e), the action is transitive on $\mathbb{H}$ and so it suffices to consider all unit tangent vectors at a given point in $\mathbb{H}$. Hence we fix attention on the point $\mathrm{i} \in \mathbb{H}$. Here the action of $K = \mathrm{SO}_2(\mathbb{R})$ and Exercise 8.10(d) help. In fact, for

$$k_\theta = \begin{pmatrix} \cos\theta & -\sin\theta \\ \sin\theta & \cos\theta \end{pmatrix} \in K$$

and $z \in \mathbb{H}$ we have

$$k_\theta \boldsymbol{\cdot} z = \frac{\cos\theta z - \sin\theta}{\sin\theta z + \cos\theta}.$$

Using (8.5) at $z = \mathrm{i}$, this gives

$$\left(\frac{\mathrm{d}}{\mathrm{d}z}(k_\theta \boldsymbol{\cdot} z) \right)\bigg|_{z=\mathrm{i}} = \frac{1}{(\cos\theta + \mathrm{i}\sin\theta)^2} = \mathrm{e}^{-2\mathrm{i}\theta},$$

which in words means that the action of $\mathrm{SO}_2(\mathbb{R})$ fixes i but rotates the unit tangent vector at i at double speed clockwise. This implies the simple transitivity of the action of $\mathrm{PSL}_2(\mathbb{R})$ on $\mathrm{T}^1\mathbb{H}$.

[†] Note that $(-I)\boldsymbol{\cdot}z = \frac{-z}{-1} = z$ for all $z \in \mathbb{H}$, so $-I$ acts trivially on $\mathbb{H}$ and on $\mathrm{T}^1\mathbb{H}$.

8.3.2 The Action of $\mathrm{SU}_{1,1}(\mathbb{R})$ on $\mathbb{D}$

The Poincaré disk model $\mathbb{D}$ of the hyperbolic plane is defined by

$$\mathbb{D} = \{w \in \mathbb{C} \mid |w| < 1\}$$

together with the hyperbolic metric and volume. In the disk model, the hyperbolic Riemannian metric is defined by

$$4\frac{\mathrm{d}x^2 + \mathrm{d}y^2}{(1 - |w|^2)^2} \tag{8.6}$$

for $w = x + \mathrm{i}y$, and the associated hyperbolic area measure is defined by

$$\mathrm{dvol} = 4\frac{\mathrm{d}x\,\mathrm{d}y}{(1 - |w|^2)^2}.$$

For this model it is more convenient to use the special unitary group

$$\mathrm{SU}_{1,1}(\mathbb{R}) = \left\{ \begin{pmatrix} \alpha & \overline{\beta} \\ \beta & \overline{\alpha} \end{pmatrix} \;\middle|\; \alpha, \beta \in \mathbb{C} \text{ with } |\alpha|^2 - |\beta|^2 = 1 \right\},$$

which is defined as the stabilizer inside $\mathrm{SL}_2(\mathbb{C})$ of the Hermitian form

$$\mathbb{C}^2 \ni (z_1, z_2) \longmapsto |z_1|^2 - |z_2|^2$$

of signature $(1, 1)$. In fact

$$h = \begin{pmatrix} \alpha & \overline{\beta} \\ \beta & \overline{\alpha} \end{pmatrix} \in \mathrm{SU}_{1,1}(\mathbb{R})$$

with $\alpha, \beta \in \mathbb{C}$ and $|\alpha|^2 - |\beta|^2 = 1$ acts on $w \in \mathbb{D}$ by the Möbius transformation

$$h \bullet w = \frac{\alpha w + \overline{\beta}}{\beta w + \overline{\alpha}}. \tag{8.7}$$

A calculation reveals that

$$1 - |h \bullet w|^2 = \frac{1 - |w|^2}{|\beta w + \overline{\alpha}|^2}, \tag{8.8}$$

and that h preserves both the hyperbolic Riemannian metric and the hyperbolic area measure.

Essential Exercise 8.11. (a) Show (8.8) and deduce that $h \bullet w \in \mathbb{D}$ for any element $h \in \mathrm{SU}_{1,1}(\mathbb{R})$ and $w \in \mathbb{D}$.
(b) Show that the Möbius transformations in (8.3) define a continuous action of $\mathrm{SU}_{1,1}(\mathbb{R})$ on $\mathbb{D}$.

(c) Show that the Möbius transformation (8.7) preserves the hyperbolic Riemannian metric and the hyperbolic area measure on $\mathbb{D}$ for any $h \in \mathrm{SU}_{1,1}(\mathbb{R})$.
(d) Show that

$$K' = \left\{ k'_\theta = \begin{pmatrix} \mathrm{e}^{-\mathrm{i}\theta} & 0 \\ 0 & \mathrm{e}^{\mathrm{i}\theta} \end{pmatrix} \;\middle|\; \theta \in [0, 2\pi) \right\} = \mathrm{Stab}_{\mathrm{SU}_{1,1}(\mathbb{R})}(0)$$

is the stabilizer of the point $0 \in \mathbb{D}$. Show also that the action of $k'_\theta \in K'$ rotates unit tangent vectors at 0 at double speed clockwise.

8.3.3 Equivalence of the Models

These two models of the hyperbolic plane and their associated isometry groups are in fact equivalent. We let $\overline{\mathbb{C}} = \mathbb{C} \cup \{\infty\}$ be the Riemann sphere and note that any $g \in \mathrm{GL}_2(\mathbb{C})$ acts as a Möbius transformation on $\overline{\mathbb{C}}$. The map

$$\Phi \colon \overline{\mathbb{C}} \ni w \longmapsto \begin{pmatrix} 1 & \mathrm{i} \\ \mathrm{i} & 1 \end{pmatrix} \bullet w = \frac{w + \mathrm{i}}{\mathrm{i}w + 1} \in \overline{\mathbb{C}}$$

satisfies, for $w = x + \mathrm{i}y \in \mathbb{C}$, the equation

$$\Im \Phi(x + \mathrm{i}y) = \Im \frac{(x + \mathrm{i}y + \mathrm{i})(-\mathrm{i}x - y + 1)}{(\mathrm{i}x - y + 1)(-\mathrm{i}x - y + 1)} = \frac{1 - x^2 - y^2}{x^2 + (y-1)^2} = \frac{1 - \|w\|^2}{\|\mathrm{i}w + 1\|^2},$$

and so $\Phi(\mathbb{D}) \subseteq \mathbb{H}$. Conversely,

$$\Psi \colon \overline{\mathbb{C}} \ni z \longmapsto \begin{pmatrix} 1 & -\mathrm{i} \\ -\mathrm{i} & 1 \end{pmatrix} \bullet z = \frac{z - \mathrm{i}}{-\mathrm{i}z + 1} \in \overline{\mathbb{C}}$$

satisfies for $z = x + \mathrm{i}y \in \mathbb{H}$ that

$$|\Psi(x + \mathrm{i}y)|^2 = \left| \frac{x + \mathrm{i}y - \mathrm{i}}{-\mathrm{i}x + y + 1} \right|^2 = \frac{x^2 + (y-1)^2}{x^2 + (y+1)^2}$$

and so $\Psi(\mathbb{H}) \subseteq \mathbb{D}$.

We also note that for

$$g_1 = \begin{pmatrix} a_1 & b_1 \\ c_1 & d_1 \end{pmatrix}, g_2 = \begin{pmatrix} a_2 & b_2 \\ c_2 & d_2 \end{pmatrix} \in \mathrm{GL}_2(\mathbb{C})$$

and $z \in \overline{\mathbb{C}}$ we have

$$\begin{aligned} g_1 \cdot (g_2 \cdot z) = g_1 \cdot \frac{a_2 z + b_2}{c_2 z + d_2} &= \frac{a_1 \frac{a_2 z + b_2}{c_2 z + d_2} + b_1}{c_1 \frac{a_2 z + b_2}{c_2 z + d_2} + d_1} \\ &= \frac{a_1(a_2 z + b_2) + b_1(c_2 z + d_2)}{c_1(a_2 z + b_2) + d_1(c_2 z + d_2)} \\ &= \frac{(a_1 a_2 + b_1 c_2) z + (a_1 b_2 + b_1 d_2)}{(c_1 a_2 + d_1 c_2) z + (c_1 b_2 + d_1 d_2)} = (g_1 g_2) \cdot z, \end{aligned}$$

and for

$$g = \begin{pmatrix} a & \\ & a \end{pmatrix} \in \mathrm{GL}_2(\mathbb{C})$$

we have $g \cdot z = z$. Together with

$$\begin{pmatrix} 1 & \mathrm{i} \\ \mathrm{i} & 1 \end{pmatrix} \begin{pmatrix} 1 & -\mathrm{i} \\ -\mathrm{i} & 1 \end{pmatrix} = \begin{pmatrix} 2 & 0 \\ 0 & 2 \end{pmatrix}$$

this implies that $\Phi \circ \Psi = \Psi \circ \Phi = I$, and hence that $\mathbb{D}$ and $\mathbb{H}$ are biholomorphic under these maps.

Let us also verify that the Riemannian metric on $\mathbb{D}$ and on $\mathbb{H}$ are mapped to each other with respect to Φ and Ψ. For this we let $w = x + \mathrm{i}y \in \mathbb{D}$ and $v \in \mathrm{T}_w\mathbb{D}$ be a tangent vector and calculate

$$\Phi'(w) = \frac{2}{(\mathrm{i}w + 1)^2}.$$

In other words, the complex derivative $\Phi'(w)$ corresponds as a derivative from $\mathbb{R}^2$ to $\mathbb{R}^2$ to a rotation and a scalar multiplication by

$$\frac{2}{|\mathrm{i}w + 1|^2} = \frac{2}{x^2 + (y-1)^2}.$$

Therefore, the squared length of the image vector $\Phi'(w)v$ as an element of $\mathrm{T}_{\Phi(w)}\mathbb{H}$ is given by

$$\frac{|\Phi'(w)v|^2}{(\Im \Phi(w))^2} = \frac{\frac{4|v|^2}{(x^2+(y-1)^2)^2}}{\left(\frac{1-x^2-y^2}{x^2+(y-1)^2}\right)^2} = \frac{4|v|^2}{(1-|w|^2)^2}.$$

Therefore the two notions of hyperbolic Riemannian metric, and so[†] also the two notions of hyperbolic volume, are equivalent.

Finally, we calculate for

$$g = \begin{pmatrix} a & b \\ c & d \end{pmatrix} \in \mathrm{SL}_2(\mathbb{R})$$

the product

[†] This follows either abstractly as the volume form can be obtained from the Riemannian metric or via a similarly short direct calculation.

$$\frac{1}{2}\begin{pmatrix}1 & -\mathrm{i}\\ -\mathrm{i} & 1\end{pmatrix}\begin{pmatrix}a & b\\ c & d\end{pmatrix}\begin{pmatrix}1 & \mathrm{i}\\ \mathrm{i} & 1\end{pmatrix}=\frac{1}{2}\begin{pmatrix}a+d+\mathrm{i}(b-c) & b+c-\mathrm{i}(d-a)\\ b+c+\mathrm{i}(d-a) & a+d-\mathrm{i}(b-c)\end{pmatrix}=\begin{pmatrix}\alpha & \overline{\beta}\\ \beta & \overline{\alpha}\end{pmatrix}$$

where $\alpha=\frac{a+d+\mathrm{i}(b-c)}{2}$ and $\beta=\frac{b+c+\mathrm{i}(d-a)}{2}$ satisfy

$$|\alpha|^2-|\beta|^2=\tfrac{1}{4}\left[(a+d)^2+(b-c)^2-(b+c)^2-(d-a)^2\right]=ad-bc=1.$$

Conversely, for

$$h=\begin{pmatrix}\alpha & \overline{\beta}\\ \beta & \overline{\alpha}\end{pmatrix}\in\mathrm{SU}_{1,1}(\mathbb{R})$$

we can use the formulas

$$\begin{aligned}a&=\Re\alpha-\Im\beta,\\ b&=\Im\alpha+\Re\beta,\\ c&=-\Im\alpha+\Re\beta,\\ d&=\Re\alpha+\Im\beta\end{aligned}$$

to define

$$g=\begin{pmatrix}a & b\\ c & d\end{pmatrix}\in\mathrm{SL}_2(\mathbb{R}),$$

giving back h under the conjugation above. It follows that the groups $\mathrm{SL}_2(\mathbb{R})$ and $\mathrm{SU}_{1,1}(\mathbb{R})$ are conjugated, and that the action of $\mathrm{SL}_2(\mathbb{R})$ on $\mathbb{H}$ corresponds to the action of $\mathrm{SU}_{1,1}(\mathbb{R})$ on $\mathbb{D}$. Moreover $\Psi(\mathrm{i})=0$, $\Phi(0)=\mathrm{i}$, and the compact subgroup $\mathrm{SO}_2(\mathbb{R})\subseteq\mathrm{SL}_2(\mathbb{R})$ is conjugated to the diagonal compact subgroup

$$K'=\left\{k'_\theta=\begin{pmatrix}\mathrm{e}^{-\mathrm{i}\theta} & \\ & \mathrm{e}^{\mathrm{i}\theta}\end{pmatrix}\;\middle|\;\theta\in[0,2\pi)\right\}<\mathrm{SU}_{1,1}(\mathbb{R}).$$

In fact for $g=k_\theta$ we have

$$\tfrac{1}{2}\begin{pmatrix}1 & -\mathrm{i}\\ -\mathrm{i} & 1\end{pmatrix}\begin{pmatrix}\cos\theta & -\sin\theta\\ \sin\theta & \cos\theta\end{pmatrix}\begin{pmatrix}1 & \mathrm{i}\\ \mathrm{i} & 1\end{pmatrix}=\begin{pmatrix}\mathrm{e}^{-\mathrm{i}\theta} & \\ & \mathrm{e}^{\mathrm{i}\theta}\end{pmatrix}$$

and obtain that $k_\theta\in K$ is mapped to $k'_\theta\in K'$ with matching parametrization.

Exercise 8.12. The subgroup $K=\mathrm{SO}_2(\mathbb{R})<\mathrm{SL}_2(\mathbb{R})$ (and, similarly, the diagonal subgroup K' of $\mathrm{SU}_{1,1}(\mathbb{R})$) is often called a *maximal* compact subgroup. Justify this by showing that if $H<\mathrm{SL}_2(\mathbb{R})$ is a subgroup properly containing $\mathrm{SO}_2(\mathbb{R})$, then $H=\mathrm{SL}_2(\mathbb{R})$.

8.3.4 Two Geometric Decompositions of $\mathrm{SL}_2(\mathbb{R})$

The action of $\mathrm{SL}_2(\mathbb{R})$ on the hyperbolic plane is fundamental to hyperbolic geometry, and is also a valuable tool for understanding the group $\mathrm{SL}_2(\mathbb{R})$ itself, and (as we will see) for understanding its unitary representations. This geometric viewpoint allows us to decompose $\mathrm{SL}_2(\mathbb{R})$ in two different ways.

The first decomposition that will be useful is the *Iwasawa decomposition* of $\mathrm{SL}_2(\mathbb{R})$, using, in addition to the maximal compact subgroup $K = \mathrm{SO}_2(\mathbb{R})$, the diagonal subgroup

$$A = \left\{ a_t = \begin{pmatrix} \mathrm{e}^t & \\ & \mathrm{e}^{-t} \end{pmatrix} \;\middle|\; t \in \mathbb{R} \right\}$$

and the upper unipotent[†] subgroup

$$U = \left\{ u_x = \begin{pmatrix} 1 & x \\ & 1 \end{pmatrix} \;\middle|\; x \in \mathbb{R} \right\}.$$

With this notation, the Iwasawa decomposition states that every $g \in \mathrm{SL}_2(\mathbb{R})$ can be written in a unique way as a product

$$g = u_x a_t k_\theta,$$

where $u_x \in U$, $a_t \in A$, and $k_\theta \in K$ for $x, t \in \mathbb{R}$ and $\theta \in [0, 2\pi)$. The geometric meaning of this decomposition is revealed in Exercise 8.10(d)–(e), which also provides a proof. Indeed, $g{\bullet}\mathrm{i} \in \mathbb{H}$ and by transitivity of the action of the subgroup $B = UA$ on $\mathbb{H}$ there exists $x \in \mathbb{R}$ and $t > 0$ with

$$g{\bullet}\mathrm{i} = \underbrace{\begin{pmatrix} 1 & x \\ & 1 \end{pmatrix} \begin{pmatrix} \mathrm{e}^t & \\ & \mathrm{e}^{-t} \end{pmatrix}}_{=b} {\bullet}\mathrm{i} = x + \mathrm{e}^{2t}\mathrm{i}.$$

Applying b^{-1} and using the fact that the stabilizer of i is $\mathrm{SO}_2(\mathbb{R})$, we obtain $b^{-1}g{\bullet}\mathrm{i} = \mathrm{i}$ and hence $b^{-1}g = k_\varphi$ for some φ, giving $g = u_x a_t k_\varphi$ as required. Moreover, it follows that $x = \Re(g{\bullet}\mathrm{i})$, $t = \frac{1}{2}\log \Im(g{\bullet}\mathrm{i})$, and that the angle of the vector $g{\bullet}(\mathrm{i}, \mathrm{i})$ is determined by 2θ measuring clockwise from the vector pointing upwards. Taking inverses, we see that the Iwasawa decomposition also holds in the order KAU. We also refer to Exercise 4.4 for a different proof of the Iwasawa decomposition.

The second decomposition of $\mathrm{SL}_2(\mathbb{R})$ that we will use is the *Cartan decomposition*, which states that every $g \in \mathrm{SL}_2(\mathbb{R})$ can be written as a product $g = k_\theta a_t k_\psi$ with $k_\theta, k_\psi \in K$ and $a_t \in A$. Moreover, we may assume that $t \geqslant 0$, and with this assumption the element $a_t \in A$ is uniquely determined by g. If $t > 0$ then the only ambiguity in the choice of k_φ and k_ψ is that we can multiply both simultaneously by the element $-I$ of the centre.

The existence of the Cartan decomposition follows from linear algebra (see the footnote on page 65). The claimed uniqueness of a_t follows, since left and right multiplication by $k_\theta, k_\psi \in K = \mathrm{SO}_2(\mathbb{R})$ preserves the Hilbert–Schmidt norm $\|\cdot\|_{\mathrm{HS}}$, which gives

[†] A subgroup of $\mathrm{SL}_d(\mathbb{R})$ for $d \geqslant 2$ is called unipotent if all its elements have 1 as their only eigenvalue.

$$\|g\|_{\mathrm{HS}} = \|k_\theta a_t k_\psi\|_{\mathrm{HS}} = \|a_t\|_{\mathrm{HS}} = \sqrt{\mathrm{e}^{2t} + \mathrm{e}^{-2t}} = \sqrt{2\cosh(2t)},$$

and so uniquely determines $t \geqslant 0$.

To practise using the disk model a little more, we explain the geometric meaning of the Cartan decomposition of $\mathrm{SU}_{1,1}(\mathbb{R}) = K'AK'$ using $\mathbb{D}$. Here we use

$$K' = \left\{ k'_\theta = \begin{pmatrix} \mathrm{e}^{-\mathrm{i}\theta} & \\ & \mathrm{e}^{\mathrm{i}\theta} \end{pmatrix} \;\middle|\; \theta \in [0, 2\pi) \right\}$$

as in Exercise 8.11(d) and

$$A' = \left\{ a'_t = \begin{pmatrix} \cosh t & \sinh t \\ \sinh t & \cosh t \end{pmatrix} \;\middle|\; t \in \mathbb{R} \right\}.$$

Note that this subgroup $A' < \mathrm{SU}_{1,1}(\mathbb{R}) \cap \mathrm{SL}_2(\mathbb{R})$, which is conjugated to our previous choice $A < \mathrm{SL}_2(\mathbb{R})$ by an element of K, could also have been used in the Cartan decomposition for $\mathrm{SL}_2(\mathbb{R})$, and that our isomorphism between $\mathrm{SL}_2(\mathbb{R})$ and $\mathrm{SU}_{1,1}(\mathbb{R})$ fixes A'. So suppose that $g \in \mathrm{SU}_{1,1}(\mathbb{R})$ and that

$$w = g \bullet 0 = \rho \mathrm{e}^{-2\mathrm{i}\theta} \in \mathbb{D}$$

for some $\rho \in [0, 1)$ and $\theta \in [0, \pi)$. Applying the inverse $k'_{-\theta}$ of k'_θ to w gives

$$k'_{-\theta} g \bullet 0 = k'_{-\theta} \bullet w = \rho \geqslant 0.$$

We define $t = \operatorname{arctanh} \rho \in \mathbb{R}_{\geqslant 0}$ and obtain

$$k'_{-\theta} g \bullet 0 = \tanh t = \begin{pmatrix} \cosh t & \sinh t \\ \sinh t & \cosh t \end{pmatrix} \bullet 0.$$

Applying a'_{-t}, we obtain $a_t k'_\theta g \bullet 0 = 0$. Using Exercise 8.11(d), we find $k'_\psi \in K'$ with $\psi \in [0, 2\pi)$ and

$$a'_{-t} k'_{-\theta} g = k'_\psi$$

so that

$$g = k'_\theta \begin{pmatrix} \cosh t & \sinh t \\ \sinh t & \cosh t \end{pmatrix} k'_\psi$$

as claimed. In this way, $k'_\theta \in K'$ and $t \geqslant 0$ correspond to the position of the point $g \bullet 0 \in \mathbb{D}$. In fact $t \geqslant 0$ is uniquely determined and if $t > 0$ then the coset $k'_\theta\{\pm I\} \in K'/\{\pm I\}$ is uniquely determined by g. Moreover, $k'_{\theta+\psi}\{\pm I\}$ corresponds to the direction of the unit tangent vector when g is applied to the tangent vector 1 at 0.

8.3.5 Volume Calculations

As we will see, it will be important to be able to explicitly calculate integrals over $\mathrm{SL}_2(\mathbb{R})$. For instance, to prove or disprove that certain irreducible unitary representations are a discrete series representation in the sense of Definition 8.1. Moreover, some crucial results of this chapter will depend on playing the description of the Haar measure on G expressed in terms of the Iwasawa and Cartan decompositions against each other. Hence we wish to describe the Haar measure m of $\mathrm{SL}_2(\mathbb{R})$ using both decompositions.

For the Iwasawa decomposition, this is achieved using the following general fact applied to the subgroups $L = B$ and $R = K$ of $G = \mathrm{SL}_2(\mathbb{R})$.

Lemma 8.13 (Product decomposition of Haar measure). *Let G be a locally compact, σ-compact, metric, unimodular group and suppose that L, R are closed subgroups of G such that LR is a neighbourhood of the identity and the map $\Phi\colon L \times R \ni (\ell, r) \mapsto \ell r \in LR \subseteq G$ is a homeomorphism. Then the Haar measure m_G restricted to LR is proportional to the push-forward of the product measure $m_L \times m_R^{(\mathrm{r})}$ of the left Haar measure m_L on L and the right Haar measure $m_R^{(\mathrm{r})}$ on R.*

We refer to [24, Lem. 11.31] for the details (or [25, Lem. 10.57] for the case of interest here), but point out the main idea of the proof: Indeed, a consequence of the fact that m_G is left and right invariant is that $\Psi_*^{-1}(m_G|_{LR})$ is a left Haar measure on $L \times R$, where $\Psi\colon L \times R \ni (\ell, r) \mapsto \ell r^{-1} \in G$.

Exercise 8.14. Show that $\mathrm{SL}_d(\mathbb{R})$ is unimodular for all $d \geqslant 2$.

To describe the Haar measure m of $\mathrm{SL}_2(\mathbb{R})$ using Lemma 8.13, we need to calculate the left Haar measure of B and the right Haar measure of K. For the former we use Exercise 8.10(e) to identify B with $\mathbb{H}$ and move the hyperbolic area measure from $\mathbb{H}$ to B. We use the coordinates $z = x + \mathrm{i}y \in \mathbb{H}$ and

$$b = u_x a_t = \begin{pmatrix} 1 & x \\ & 1 \end{pmatrix} \begin{pmatrix} \mathrm{e}^t & \\ & \mathrm{e}^{-t} \end{pmatrix} \in B$$

which gives, with $z = b{\centerdot}\mathrm{i}$ that $x = \Re z$ and $y = \Im z = \mathrm{e}^{2t}$. With $\mathrm{dvol} = \frac{\mathrm{d}x\,\mathrm{d}y}{y^2}$, and $\mathrm{d}y = 2\mathrm{e}^{2t}\,\mathrm{d}t$ this leads to the description

$$\mathrm{d}m_B = 2\,\mathrm{d}x \mathrm{e}^{-2t}\,\mathrm{d}t. \tag{8.9}$$

Using now the coordinates $g = u_x a_t k_\psi \in \mathrm{SL}_2(\mathbb{R})$ and Lemma 8.13, we obtain for the Haar measure m on $\mathrm{SL}_2(\mathbb{R})$ that

$$\mathrm{d}m = 2\,\mathrm{d}x \mathrm{e}^{-2t}\,\mathrm{d}t\,\mathrm{d}m_K(k_\psi)$$

by using $\mathrm{d}m_K(k_\psi) = \frac{1}{2\pi}\,\mathrm{d}\psi$ as the Haar measure on K. Note that with this normalization the Haar measure on $\mathrm{SL}_2(\mathbb{R})$ induces the hyperbolic volume measure on $\mathrm{SL}_2(\mathbb{R})/K \cong \mathbb{H}$.

We now wish to calculate the Haar measure in terms of the Cartan decomposition, using once more the conjugated group $\mathrm{SU}_{1,1}(\mathbb{R})$. For this, we make the choices

$$g = k'_\theta a'_t k'_\psi \tag{8.10}$$

with $\theta \in [0, \pi)$, $t > 0$, $\psi \in [0, 2\pi)$ to uniquely parametrize $\mathrm{SU}_{1,1}(\mathbb{R}) \smallsetminus K'$. Using again the hyperbolic plane, but this time in the disk model, we have that

$$w = x + \mathrm{i}y = g\cdot 0 = k'_\theta a'_t \cdot 0 = k'_\theta \cdot \tanh t = \mathrm{e}^{-2\theta \mathrm{i}} \tanh t.$$

Using Euclidean polar coordinates with radius $r = |w| = \tanh t$ and angle $\gamma = -2\theta$, the hyperbolic area on $\mathbb{D}$ is therefore given by

$$\begin{aligned} \mathrm{dvol} = 4\frac{\mathrm{d}x\,\mathrm{d}y}{(1-r^2)^2} = 4r\frac{\mathrm{d}r\,\mathrm{d}\gamma}{(1-r^2)^2} &= 4\tanh t \frac{(\tanh t)'\,\mathrm{d}t\,2\,\mathrm{d}\theta}{(1-\tanh^2 t)^2} \\ &= 8 \sinh t \cosh t\,\mathrm{d}t\,\mathrm{d}\theta \\ &= 4 \sinh 2t\,\mathrm{d}t\,\mathrm{d}\theta, \end{aligned}$$

since

$$\begin{aligned} (\tanh t)' &= \frac{\cosh^2 t - \sinh^2 t}{\cosh^2 t} = \frac{1}{\cosh^2 t}, \\ (1 - \tanh^2 t)^2 &= \left(\frac{\cosh^2 t - \sinh^2 t}{\cosh^2 t}\right)^2 = \frac{1}{\cosh^4 t}, \end{aligned}$$

and

$$\sinh 2t = 2 \sinh t \cosh t.$$

We again use the normalized Haar measure

$$\mathrm{d}m_{K'}(k'_\psi) = \frac{1}{2\pi}\,\mathrm{d}\psi$$

on K' and use the calculation above for the hyperbolic area measure dvol on the space $\mathrm{SU}_{1,1}(\mathbb{R})/K' \cong \mathbb{D}$. Hence the Haar measure on $\mathrm{SU}_{1,1}(\mathbb{R})$, and equivalently on $\mathrm{SL}_2(\mathbb{R})$, is given (in the Cartan decomposition (8.10)) by

$$\mathrm{d}m = \frac{2}{\pi} \sinh 2t\,\mathrm{d}\theta\,\mathrm{d}t\,\mathrm{d}\psi, \tag{8.11}$$

where as before $g = k_\theta a_t k_\psi$ with $\theta \in [0, \pi)$, $t > 0$, and $\psi \in [0, 2\pi)$.

To practice using the above formulas, we calculate some volumes in the next lemma.

Lemma 8.15 (Volume of hyperbolic and norm balls). *We have*

$$B_R^{\mathbb{D}}(0) = \left\{ w \in \mathbb{D} \mid |w| < \tanh\left(\tfrac{R}{2}\right) \right\} \tag{8.12}$$

and

$$\mathrm{vol}\big(B_R^{\mathbb{H}}(z)\big) = \mathrm{vol}\big(B_R^{\mathbb{D}}(w)\big) = 2\pi\,(\cosh(R) - 1)$$

for all $R > 0$, $z \in \mathbb{H}$, *and* $w \in \mathbb{D}$. *Moreover,*

$$m_{\mathrm{SL}_2(\mathbb{R})}\big(B_r^{\|\cdot\|_{\mathrm{HS}}}\big) = \pi\big(r^2 - 2\big)$$

for all $r \geqslant \sqrt{2}$, *where*

$$B_r^{\|\cdot\|_{\mathrm{HS}}} = \{g \in \mathrm{SL}_2(\mathbb{R}) \mid \|g\|_{\mathrm{HS}} \leqslant r\}.$$

PROOF. By the discussion in Section 8.3.3, the volume calculation may be carried out using the disk model $\mathbb{D} = \{w \in \mathbb{C} \mid |w| < 1\}$ of the hyperbolic plane, which carries the Riemannian metric

$$4\frac{\mathrm{d}x^2 + \mathrm{d}y^2}{(1 - r^2)^2}$$

at the point $w = x + \mathrm{i}y$, where $r^2 = x^2 + y^2$. Moreover, the transitive action of $\mathrm{SU}_{1,1}(\mathbb{R})$ on $\mathbb{D}$ preserves the hyperbolic Riemannian metric, and hence also the hyperbolic area. Therefore it suffices to consider the point $w = 0$ and calculate the volume of the ball of (hyperbolic) radius R around 0, which by invariance of $B_R^{\mathbb{D}}(0)$ under the diagonal subgroup K' is a disk of some radius ρ around 0 in the Euclidean metric as well. The Euclidean radius ρ may be calculated using the relation

$$R = 2\int_0^\rho \frac{\mathrm{d}r}{1 - r^2} = \log\left(\frac{1 + r}{1 - r}\right)\Big|_0^\rho = \log\left(\frac{1 + \rho}{1 - \rho}\right)$$

or, equivalently,

$$\rho = \frac{\mathrm{e}^R - 1}{\mathrm{e}^R + 1} = \frac{\mathrm{e}^{\frac{R}{2}} - \mathrm{e}^{-\frac{R}{2}}}{\mathrm{e}^{\frac{R}{2}} + \mathrm{e}^{-\frac{R}{2}}} = \tanh\big(\tfrac{R}{2}\big).$$

This gives (8.12). Therefore, the volume of the hyperbolic ball of radius R around $w = 0$ is given by

$$\begin{aligned}
\mathrm{vol}\left(B_R^{\mathbb{D}}(0)\right) &= 4\int_0^\rho \int_0^{2\pi} \frac{r\,\mathrm{d}r\,\mathrm{d}\gamma}{(1 - r^2)^2} = 4\pi \int_0^{\rho^2} \frac{\mathrm{d}u}{(1 - u)^2} \\
&= 4\pi \frac{1}{1 - u}\Big|_0^{\rho^2} = 4\pi\left(\frac{1}{1 - \rho^2} - 1\right) \\
&= 2\pi\left(\frac{2(\mathrm{e}^R + 1)^2}{(\mathrm{e}^R + 1)^2 - (\mathrm{e}^R - 1)^2} - 2\right) = 2\pi\,(\cosh(R) - 1)\,.
\end{aligned}$$

We note that $g = k_\theta a_t k_\psi \in B_r^{\|\cdot\|_{\mathrm{HS}}}$ if and only if $\mathrm{e}^{2t} + \mathrm{e}^{-2t} \leqslant r^2$ (that is, $\cosh 2t \leqslant \frac{r^2}{2}$). Hence we can use (8.11) and calculate

$$\begin{aligned}m\big(B_r^{\|\cdot\|_{\mathrm{HS}}}\big) &= \int_0^{2\pi}\int_0^{\frac{1}{2}\operatorname{arcosh}\frac{r^2}{2}}\int_0^{\pi}\frac{2}{\pi}\sinh 2t\,\mathrm{d}\theta\,\mathrm{d}t\,\mathrm{d}\psi\\ &= 2\pi\int_0^{\operatorname{arcosh}\frac{r^2}{2}}\sinh u\,\mathrm{d}u\\ &= 2\pi\big(\tfrac{r^2}{2}-1\big) = \pi(r^2-2).\end{aligned}$$

Alternatively, we could have put $R = \operatorname{arcosh}\frac{r^2}{2}$ into the calculation of the hyperbolic R-ball to obtain $m\big(B_r^{\|\cdot\|_{\mathrm{HS}}}\big) = \operatorname{vol}\big(B_R^{\mathbb{D}}(0)\big) = \pi\big(r^2-2\big)$. □

8.3.6 The Hyperbolic Metric on $\mathbb{H}$

For some of our later discussions (specifically, those in Section 9.4.5), we will need to describe the hyperbolic metric more concretely than we did in Section 8.3.1.

Lemma 8.16 (Hyperbolic metric). *The hyperbolic distance between the points $z_1 = x_1 + \mathrm{i}y_2$ and $z_2 = x_2 + \mathrm{i}y_2$ in $\mathbb{H}$ is given by*

$$\mathsf{d}_{\mathrm{hyp}}(z_1,z_2) = \operatorname{arcosh}\left(1+\frac{|z_1-z_2|^2}{2y_1y_2}\right).$$

Proof. For the proof, we will again use the isometric action of $\mathrm{SL}_2(\mathbb{R})$ on $\mathbb{H}$. In fact, we claim that the expression

$$\frac{|z_1-z_2|^2}{y_1y_2} \tag{8.13}$$

is also invariant under the simultaneous action of $g \in \mathrm{SL}_2(\mathbb{R})$ on z_1 and z_2. Assuming the claim (which will follow from a calculation) for the moment, we now prove the lemma. By transitivity of the isometric action of $\mathrm{SL}_2(\mathbb{R})$ we may assume that $z_1 = \mathrm{i}$. Without moving z_1 we may also apply an element of $\mathrm{SO}_2(\mathbb{R})$, allowing us to assume that $z_2 = \mathrm{i}y_2$. Indeed, the action of $\mathrm{SO}_2(\mathbb{R})$ fixes i and acts transitively on unit tangent vectors in $\mathrm{T}_{\mathrm{i}}\mathbb{H}$, while the geodesic through z_1 and z_2 is uniquely determined by z_1 and the unit tangent vector in $\mathrm{T}_{z_1}\mathbb{H}$ in the direction of the geodesic. Since the action is isometric, this does not change $\mathsf{d}_{\mathrm{hyp}}(z_1,z_2)$, nor the expression in (8.13) by the claim. We now calculate the latter for $z_1 = \mathrm{i}$ and $z_2 = \mathrm{i}y_2$, and obtain

$$\frac{|z_1-z_2|^2}{y_1y_2} = \frac{(y_2-1)^2}{y_2} = \frac{y_2^2-2y_2+1}{y_2}.$$

For the term $\operatorname{arcosh}\big(1+\frac{|z_1-z_2|^2}{2y_1y_2}\big)$, this gives

$$\operatorname{arcosh}\left(\frac{y_2+y_2^{-1}}{2}\right)=|\log y_2|.$$

On the other hand, the geodesic path from $z_1 = \mathrm{i}$ to $z_2 = \mathrm{i}y_2$ (for $y_2 \neq 1$) is parametrized by $[0,1] \ni t \mapsto \mathrm{i}y_2^t$, which gives

$$\mathsf{d}_{\mathrm{hyp}}(\mathrm{i},\mathrm{i}y_2)=\int_0^1 \frac{|\log y_2| y_2^t}{y_2^t}\,dt=|\log y_2|.$$

Thus the claim that the expression in (8.13) is invariant under the action of $\mathrm{SL}_2(\mathbb{R})$ implies the lemma.

For the claim, we first notice that the action of $u_x = \begin{pmatrix} 1 & x \\ & 1 \end{pmatrix} \in U$ on $z \in \mathbb{H}$ satisfies $u_x \cdot z = z + x$, does not change $y = \Im(z)$, nor the difference $z_1 - z_2$ if applied to both z_1 and z_2. This already gives part of the claim. Also notice that the expression (8.13) is homogeneous in z_1, z_2 of degree 0. Therefore it is also invariant under the action of $a_t = \begin{pmatrix} \mathrm{e}^t & \\ & \mathrm{e}^{-t} \end{pmatrix} \in A$ satisfying $a_t \cdot z = \mathrm{e}^{2t} z$ for $z \in \mathbb{H}$.

Finally, we wish to show that the expression in (8.13) is also invariant under K which, together with the above, implies the claim. To see the invariance under K equally well as invariance under U and A, we express the expression in (8.13) for $w_1, w_2 \in \mathbb{D}$. With $z_j = \Phi(w_j) = \frac{w_j+\mathrm{i}}{\mathrm{i}w_j+1} \in \mathbb{H}$ for $j = 1, 2$ as in Section 8.3.3, we obtain

$$\begin{aligned}
\frac{|z_1-z_2|^2}{y_1y_2} &= \frac{\left|\frac{w_1+\mathrm{i}}{\mathrm{i}w_1+1}-\frac{w_2+\mathrm{i}}{\mathrm{i}w_2+1}\right|^2}{\frac{1-|w_1|^2}{|\mathrm{i}w_1+1|^2}\frac{1-|w_2|^2}{|\mathrm{i}w_2+1|^2}} \\
&= \frac{|\mathrm{i}w_1+1|^2|\mathrm{i}w_2+1|^2\left|\frac{(w_1+\mathrm{i})(\mathrm{i}w_2+1)-(w_2+\mathrm{i})(\mathrm{i}w_1+1)}{(\mathrm{i}w_1+1)(\mathrm{i}w_2+1)}\right|^2}{\left(1-|w_1|^2\right)\left(1-|w_2|^2\right)} \\
&= \frac{\left|\left(w_1\mathrm{i}w_2+w_1-w_2+\mathrm{i}\right)-\left(w_2\mathrm{i}w_1+w_2-w_1+\mathrm{i}\right)\right|^2}{\left(1-|w_1|^2\right)\left(1-|w_2|^2\right)} \\
&= \frac{|2(w_1-w_2)|^2}{\left(1-|w_1|^2\right)\left(1-|w_2|^2\right)}.
\end{aligned}$$

Since the action of $k'_\theta \in K'$ on $w \in \mathbb{D}$ is by multiplication by $\mathrm{e}^{-2\mathrm{i}\theta}$, we now obtain the desired invariance under K as well. □

We will also use the terminology introduced in Section 7.3.2 for the compact subgroup $K' < \mathrm{SU}_{1,1}(\mathbb{R})$ and unitary representations of $\mathrm{SU}_{1,1}(\mathbb{R})$. We note that in the following arguments we will always study either $\mathrm{SL}_2(\mathbb{R})$ or $\mathrm{SU}_{1,1}(\mathbb{R})$, but not both at the same time. Hence we will simplify the notation for

$$K = \left\{ k_\theta = \begin{pmatrix} \mathrm{e}^{-\mathrm{i}\theta} & \\ & \mathrm{e}^{\mathrm{i}\theta} \end{pmatrix} \,\middle|\, \theta \in [0, 2\pi) \right\} < \mathrm{SU}_{1,1}(\mathbb{R}).$$

8.4 (Mock) Discrete Series Representations for $\mathbf{SL_2}(\mathbb{R})$

We will describe in this section the first and second types of irreducible unitary representations for the group $G = \mathrm{SL}_2(\mathbb{R})$. As we will see, these are intimately connected to complex analysis and will also give examples of discrete series representations (in the sense of Definition 8.1) and tempered almost square integrable representations (in the sense of Definition 8.3 and Definition 8.4).

8.4.1 The Discrete Series Representation

The discussion in Section 8.3.3 shows that we can work either with the upper half-plane model or the disk model of the hyperbolic plane. As the description of a particular important orthogonal basis will be more convenient in the disk model, we will work here with $\mathbb{D}$ and the group $\mathrm{SU}_{1,1}(\mathbb{R})$ instead of $\mathbb{H}$ and the isomorphic group $\mathrm{SL}_2(\mathbb{R})$.

We fix an integer $n \geqslant 2$, and define the Hilbert space

$$V_n = L^2\left(\mathbb{D}, (1 - |z|^2)^n \,\mathrm{dvol}\right) \tag{8.14}$$

where $\mathrm{dvol} = \frac{4\,\mathrm{d}x\,\mathrm{d}y}{(1-|z|^2)^2}$ denotes the hyperbolic volume on $\mathbb{D}$. We note that, unlike the infinite measure vol, the finite measure

$$(1 - |z|^2)^n \,\mathrm{dvol} = 4(1 - |z|^2)^{n-2} \,\mathrm{d}x\,\mathrm{d}y$$

is not invariant under $\mathrm{SU}_{1,1}(\mathbb{R})$, but is instead only quasi-invariant. We define the representation π^n of $\mathrm{SU}_{1,1}(\mathbb{R})$ on V_n by the formula

$$\left(\pi^n_g f\right)(z) = (-\beta z + \alpha)^{-n} f(g^{-1}\bullet z) \tag{8.15}$$

where

$$g = \begin{pmatrix} \alpha & \overline{\beta} \\ \beta & \overline{\alpha} \end{pmatrix} \in \mathrm{SU}_{1,1}(\mathbb{R}),$$

and $f \in V_n$, $z \in \mathbb{D}$, and the factor $(-\beta z + \alpha)$ is the denominator of

$$g^{-1}\bullet z = \frac{\overline{\alpha} z - \overline{\beta}}{-\beta z + \alpha}.$$

As indicated in Section 8.3.2, this factor $-\beta z + \alpha$ also appears in the formula for the complex derivative of $z \mapsto g^{-1}\bullet z$ given by

$$(g^{-1}\bullet z)' = \frac{\overline{\alpha}(-\beta z+\alpha)-(\overline{\alpha}z-\overline{\beta})(-\beta)}{(-\beta z+\alpha)^2} = \frac{1}{(-\beta z+\alpha)^2}$$

and hence also in the Jacobian determinant $|-\beta z+\alpha|^{-4}$ of the map $z \mapsto g^{-1}\bullet z$ considered as a real differentiable map from an open subset of $\mathbb{R}^2$ to itself.

Using the measure-preserving substitution $w = g^{-1}\bullet z$ for the measure vol, we wish to calculate

$$\|\pi_g^n f\|_{V_n}^2 = \int_{\mathbb{D}} |-\beta z+\alpha|^{-2n} |f(g^{-1}\bullet z)|^2 \left(1-|z|^2\right)^n \,\mathrm{dvol}(z).$$

Moreover (8.8) for $h = g^{-1}$ and the initial point z has the form

$$1-|w|^2 = 1-|g^{-1}\bullet z| = \frac{1-|z|^2}{|-\beta z+\alpha|^2},$$

and so we obtain

$$\begin{aligned}
\|\pi_g^n f\|_{V_n}^2 &= \int_{\mathbb{D}} |f(w)|^2 (1-|w|^2)^n \underbrace{\left(\left(\frac{1-|z|^2}{1-|w|^2}\right)|-\beta z+\alpha|^{-2}\right)^n}_{=1} \mathrm{dvol}(w) \\
&= \int_{\mathbb{D}} |f(w)|^2 \left(1-|w|^2\right)^n \,\mathrm{dvol}(w) = \|f\|_{V_n}^2.
\end{aligned}$$

We also need to verify that

$$\pi_g^n \pi_h^n = \pi_{gh}^n \tag{8.16}$$

for all $g, h \in \mathrm{SU}_{1,1}(\mathbb{R})$. By definition, we have

$$\pi_h^n f(w) = (-\gamma w+\eta)^{-n} f(h^{-1}\bullet w)$$

for

$$h = \begin{pmatrix} \eta & \overline{\gamma} \\ \gamma & \overline{\eta} \end{pmatrix} \in \mathrm{SU}_{1,1}(\mathbb{R})$$

and $f \in V_n$ and $w \in \mathbb{D}$.

For $g = \begin{pmatrix} \alpha & \overline{\beta} \\ \beta & \overline{\alpha} \end{pmatrix}$ we now have for $z \in \mathbb{D}$ and $w = g^{-1}\bullet z \in \mathbb{D}$ that

$$\begin{aligned}
\left(\pi_g^n \pi_h^n(f)\right)(z) &= (-\beta z+\alpha)^{-n} \pi_h^n f(g^{-1}\bullet z) \\
&= (-\beta z+\alpha)^{-n} (-\gamma w+\eta)^{-n} f(h^{-1}\bullet(g^{-1}\bullet z)) \\
&= (-\beta z+\alpha)^{-n} \left(-\gamma \frac{\overline{\alpha}z-\overline{\beta}}{-\beta z+\alpha}+\eta\right)^{-n} f\left((gh)^{-1}\bullet z\right) \\
&= \left(-\gamma(\overline{\alpha}z-\overline{\beta})+\eta(-\beta z+\alpha)\right)^{-n} f\left((gh)^{-1}\bullet z\right) \\
&= \left(-(\gamma\overline{\alpha}+\eta\beta)z+(\gamma\overline{\beta}+\eta\alpha)\right)^{-n} f\left((gh)^{-1}\bullet z\right).
\end{aligned}$$

Since

$$\begin{pmatrix} \alpha & \overline{\beta} \\ \beta & \overline{\alpha} \end{pmatrix} \begin{pmatrix} \eta & \overline{\gamma} \\ \gamma & \overline{\eta} \end{pmatrix} = \begin{pmatrix} \alpha\eta + \overline{\beta}\gamma & \alpha\overline{\gamma} + \overline{\beta}\overline{\eta} \\ \beta\eta + \overline{\alpha}\gamma & \beta\overline{\gamma} + \overline{\alpha}\overline{\eta} \end{pmatrix},$$

it follows that (8.16) holds for all $g, h \in \mathrm{SU}_{1,1}(\mathbb{R})$. Therefore π^n is a unitary representation of $\mathrm{SU}_{1,1}(\mathbb{R})$ (see also Exercise 8.17).

Exercise 8.17. Either mimic the argument from Proposition 1.6 to prove the continuity requirement for π^n, or show that the above is a special case of a twisted normalized representation in Proposition 1.6.

Definition 8.18 (Bergman space & discrete series representations). Let n be a natural number with $n \geqslant 2$, and let

$$\mathrm{Hol}(\mathbb{D}) = \{f\colon \mathbb{D} \longrightarrow \mathbb{C} \mid f \text{ is holomorphic}\}.$$

The subspace $A_n(\mathbb{D}) = V_n \cap \mathrm{Hol}(\mathbb{D}) < V_n$ is called a *weighted Bergman space*. The restriction of π^n to $A_n(\mathbb{D})$ is called the *holomorphic discrete series representation* $\delta^{n,+}$.

Lemma 8.19 (Holomorphic discrete series representations). *Let $n \geqslant 2$ be a natural number. Then the weighted Bergman space $A_n(\mathbb{D})$ is closed in V_n and is invariant under π^n.*

PROOF. We first show that $A_n(\mathbb{D}) \subseteq V_n$ is closed. For this, we claim that if a sequence in $A_n(\mathbb{D})$ converges with respect to $\|\cdot\|_{V_n}$ to some element of V_n, then the convergence also holds uniformly on compact subsets of D. The claim implies that the limit is holomorphic.

Define the compact subset $K = \overline{B_r(0)}$ for $r \in (0,1)$. Then

$$f(z_0) = \frac{1}{2\pi \mathrm{i}} \oint_{|z|=R} \frac{f(z)}{z - z_0}\, \mathrm{d}z$$

for $f \in A_n(\mathbb{D})$, $z_0 \in K$, and $R \in (r,1)$ by the Cauchy residue formula, where $\oint$ denotes the positively oriented complex path integral along the circle defined by $|z| = R$. Equivalently, we have

$$f(z_0) = \frac{1}{2\pi \mathrm{i}} \int_0^{2\pi} \frac{f(R\mathrm{e}^{\mathrm{i}\theta})}{R\mathrm{e}^{\mathrm{i}\theta} - z_0} R\mathrm{i}\mathrm{e}^{\mathrm{i}\theta}\, \mathrm{d}\theta.$$

We now fix two additional real parameters s, t with $0 < r < s < t < 1$, and vary R within $[s,t]$ to obtain

$$\begin{aligned} f(z_0) &= \frac{1}{2\pi(t-s)} \int_s^t \int_0^{2\pi} \frac{f(R\mathrm{e}^{\mathrm{i}\theta})}{R\mathrm{e}^{\mathrm{i}\theta} - z_0} \mathrm{e}^{\mathrm{i}\theta} R\, \mathrm{d}\theta\, \mathrm{d}R \\ &= \frac{1}{2\pi(t-s)} \int_{s\leqslant|z|\leqslant t} \frac{f(z)}{z - z_0} \frac{z}{|z|(1-|z|^2)^{n-2}} \underbrace{\left(1-|z|^2\right)^{n-2} \mathrm{d}x\, \mathrm{d}y}_{=\frac{1}{4}(1-|z|^2)^n\, \mathrm{dvol}(z)} \end{aligned}$$

by using polar coordinates for $z = x + \mathrm{i}y = R\mathrm{e}^{\mathrm{i}\theta}$. This shows that the map

$$A_n(\mathbb{D}) \ni f \longmapsto f(z_0) \in \mathbb{C}$$

is continuous with respect to $\|\cdot\|_{V_n}$ for all $z_0 \in \overline{B_r(0)}$, and in fact also gives

$$\left\| f|_{\overline{B_r(0)}} \right\|_\infty \ll_{s,t} \|f\|_{V_n}.$$

However, this proves the claim and hence shows that $A_n(\mathbb{D})$ is a closed subspace of V_n.

For the invariance, let $f \in A_n(\mathbb{D})$, $g \in \mathrm{SU}_{1,1}(\mathbb{R})$ and recall from above that we already know that $\pi_g^n f \in V_n$. However, the definition of $\pi_g^n f$ also shows that $\pi_g^n f$ is holomorphic on $\mathbb{D}$, which gives $\pi_g^n f \in A_n(\mathbb{D})$ as required. □

Definition 8.20 (Anti-holomorphic discrete series representations). Assume that $n \geqslant 2$. The *anti-holomorphic discrete series representation* $\delta^{n,-}$ is defined as the restriction of the unitary representation $\overline{\pi}^n$ defined by

$$\overline{\pi}_g^n(f)(z) = (-\overline{\beta}\overline{z} + \overline{\alpha})^{-n} f(g^{-1} \bullet z)$$

for

$$g = \begin{pmatrix} \alpha & \overline{\beta} \\ \beta & \overline{\alpha} \end{pmatrix} \in \mathrm{SU}_{1,1}(\mathbb{R})$$

and $f \in V_n$, $z \in \mathbb{D}$ to the conjugated subspace

$$\overline{A_n(\mathbb{D})} = V_n \cap \overline{\mathrm{Hol}(\mathbb{D})}$$

of anti-holomorphic functions in V_n.

Exercise 8.21. Show that $\delta^{n,+}$ is the contragredient representation to $\delta^{n,-}$ for any integer $n \geqslant 2$.

What follows holds similarly for the anti-holomorphic discrete series representations, but we will only consider the holomorphic discrete series representations.

Lemma 8.22 (K-weights in $\delta^{n,+}$). *Let $n \geqslant 2$ be a natural number. The discrete series representation $\delta^{n,+}$ contains K-eigenfunctions for the compact diagonal subgroup*

$$K = \left\{ k_\theta = \begin{pmatrix} \mathrm{e}^{-\mathrm{i}\theta} & \\ & \mathrm{e}^{\mathrm{i}\theta} \end{pmatrix} \;\middle|\; \theta \in [0, 2\pi) \right\} < \mathrm{SU}_{1,1}(\mathbb{R})$$

for the weights $n, n+2, n+4, \ldots$ each with multiplicity one, and only these. In fact the functions $e_\ell \colon \mathbb{D} \ni z \mapsto z^\ell$ for $\ell \in \mathbb{N}_0$ are mutually orthogonal, span $A_n(\mathbb{D})$, and

$$\delta_{k_\theta}^{n,+} e_\ell = \mathrm{e}^{\mathrm{i}(n+2\ell)\theta} e_\ell \tag{8.17}$$

for all $k_\theta \in K$ and $\ell \in \mathbb{N}_0$.

PROOF. For any $\ell \geqslant 0$ and $k_\theta \in K$, we note that

$$\begin{aligned}\delta^{n,+}_{k_\theta} e_\ell(z) &= (-0z + \mathrm{e}^{-\mathrm{i}\theta})^{-n} e_\ell\left(\frac{\mathrm{e}^{\mathrm{i}\theta} z}{\mathrm{e}^{-\mathrm{i}\theta}}\right) \\ &= \mathrm{e}^{\mathrm{i}n\theta}(\mathrm{e}^{2\mathrm{i}\theta} z)^\ell \\ &= \mathrm{e}^{\mathrm{i}(n+2\ell)\theta} z^\ell = \mathrm{e}^{\mathrm{i}(n+2\ell)\theta} e_\ell(z)\end{aligned}$$

for all $z \in \mathbb{D}$, which gives (8.17). Since the weights are mutually distinct, we see that the non-zero vectors $e_0, e_1, e_2, \ldots \in A_n(\mathbb{D})$ are mutually orthogonal.

It remains to show that the closed linear hull of $\{e_\ell \mid \ell \in \mathbb{N}_0\}$ is $A_n(\mathbb{D})$, as the other claims in the lemma then follow from this. So suppose that f lies in $A_n(\mathbb{D})$. Then f is holomorphic on $\mathbb{D}$, and complex analysis therefore implies that

$$f(z) = \sum_{\ell=0}^{\infty} a_\ell z^\ell$$

for all $z \in \mathbb{D}$, and that this convergence is uniform on compact subsets of $\mathbb{D}$. This gives, with $\mathrm{dvol} = 4\frac{r\,\mathrm{d}r\,\mathrm{d}\theta}{(1-r^2)^2}$, that

$$\begin{aligned}\|f\|^2_{A_n(\mathbb{D})} &= 4\int_0^1 (1-r^2)^{n-2} \int_0^{2\pi} |f(r\mathrm{e}^{\mathrm{i}\theta})|^2 \,\mathrm{d}\theta\, r\,\mathrm{d}r \\ &= 4\int_0^1 (1-r^2)^{n-2} r \underbrace{\int_0^{2\pi} \left|\sum_{\ell=0}^{\infty} a_\ell r^\ell \mathrm{e}^{\mathrm{i}\ell\theta}\right|^2 \mathrm{d}\theta}_{=\sum_{\ell=0}^{\infty} |a_\ell r^\ell|^2 \cdot 2\pi} \,\mathrm{d}r \\ &= 4\sum_{\ell=0}^{\infty} \int_0^1 (1-r^2)^{n-2} r \int_0^{2\pi} |a_\ell r^\ell \mathrm{e}^{\mathrm{i}\ell\theta}|^2 \,\mathrm{d}\theta\,\mathrm{d}r = \sum_{\ell=0}^{\infty} \|a_\ell e_\ell\|^2_{A_n(\mathbb{D})}\end{aligned}$$

by using the fact that the functions $[0, 2\pi] \ni \theta \mapsto \mathrm{e}^{\mathrm{i}\ell\theta}$ for $\ell \in \mathbb{N}_0$ are mutually orthogonal. Therefore, the series $\sum_{\ell=0}^{\infty} a_\ell e_\ell$ also converges in $A_n(\mathbb{D})$, and the integral formulas in the proof of Lemma 8.19 show that the limit of the series in $A_n(\mathbb{D})$ equals f. □

Due to the conclusion of the previous lemma, we will also say for $n \geqslant 2$ that n is the *lowest K-weight* for $\delta^{n,+}$. Similarly, $-n$ is the *highest K-weight* for $\delta^{n,-}$. If we want to speak about the representations $\delta^{n,\pm}$ simultaneously, we also say that $\pm n$ is the *terminal K-weight.*

Theorem 8.23 (The discrete series representation $\delta^{n,+}$). *Let $n \geqslant 2$. Then the holomorphic discrete series representations $\delta^{n,+}$ is an irreducible unitary representation of $\mathrm{SU}_{1,1}(\mathbb{R})$ that is also a discrete series representation (in the sense of Definition 8.1). Moreover, $\delta^{n,+}$ has K-eigenfunctions with weights in $n + 2\mathbb{N}_0$, each with multiplicity one, and only these. This holds similarly for $\delta^{n,-}$ and weights in $-(n + 2\mathbb{N}_0)$.*

It remains to prove irreducibility and square integrability of at least one non-trivial principal matrix coefficient. For the former, we will use the Lie algebra element

$$\mathbf{d} = \begin{pmatrix} & 1 \\ 1 & \end{pmatrix} \in \mathfrak{su}_{1,1}(\mathbb{R})$$

satisfying

$$a_t = \exp(t\mathbf{d}) = \begin{pmatrix} \cosh t & \sinh t \\ \sinh t & \cosh t \end{pmatrix} \tag{8.18}$$

for $t \in \mathbb{R}$.

PROOF OF IRREDUCIBILITY OF $\delta^{n,+}$. Suppose that $\mathcal{V} \subseteq A_n(\mathbb{D})$ is a non-trivial $\delta^{n,+}$-invariant closed subspace. As $K \cong \mathbb{S}^1$ is compact and abelian there exists some non-trivial eigenvector $v \in \mathcal{V}$ for $\delta^{n,+}|_K$ and some weight. By Lemma 8.22, this shows that $e_\ell \in \mathcal{V}$ for some $\ell \in \mathbb{N}_0$. We will now show that $\delta^{n,+}_\partial(\mathbf{d})e_\ell$ exists and conclude that $\delta^{n,+}_\partial(\mathbf{d})e_\ell \in \mathcal{V}$. By definition,

$$\delta^{n,+}_\partial(\mathbf{d})e_\ell = \lim_{t\to 0} \frac{1}{t}\left(\delta^{n,+}(\exp(t\mathbf{d}))e_\ell - e_\ell\right). \tag{8.19}$$

Now

$$\begin{aligned}\delta^{n,+}(\exp(t\mathbf{d}))e_\ell(z) &= (-z\sinh t + \cosh t)^{-n}\left(\frac{z\cosh t - \sinh t}{-z\sinh t + \cosh t}\right)^\ell \\ &= (z\cosh t - \sinh t)^\ell\,(-z\sinh t + \cosh t)^{-n-\ell}\end{aligned}$$

and

$$\begin{aligned}\lim_{t\to 0}\frac{1}{t}&\left((z\cosh t - \sinh t)^\ell(-z\sinh t + \cosh t)^{-n-\ell} - z^\ell\right) \\ &= \frac{\partial}{\partial t}\bigg|_{t=0} (z\cosh t - \sinh t)^\ell\,(-z\sinh t + \cosh t)^{-n-\ell} \\ &= \ell z^{\ell-1}\cdot(-1)\cdot(1) + z^\ell\,(-n-\ell)\cdot 1\cdot(-z) \\ &= -\ell z^{\ell-1} + (n+\ell)z^{\ell+1}.\end{aligned}$$

The latter convergence holds for any $z \in \mathbb{C}$, uniformly on $\overline{\mathbb{D}}$. Moreover, the limit in (8.19) also exists with respect to $\|\cdot\|_{V_n}$, since

$$\int_{\mathbb{D}} (1-|z|^2)^n \,\mathrm{dvol} = 4\int_{\mathbb{D}} (1-|z|^2)^{n-2}\,\mathrm{d}x\,\mathrm{d}y < \infty.$$

In fact, we obtain

$$\delta^{n,+}_\partial(\mathbf{d})e_\ell = -\ell e_{\ell-1} + (n+\ell)e_{\ell+1}.$$

As $e_\ell \in \mathcal{V}$ and $\mathcal{V}$ is assumed to be invariant, we obtain $\delta^{n,+}_\partial(\mathbf{d})e_\ell \in \mathcal{V}$. As the vectors $e_0, e_1, e_2, \ldots$ are K-eigenfunctions with different eigenvalues by Lemma 8.22, we also obtain $e_{\ell-1} \in \mathcal{V}$ if $\ell > 0$ and $e_{\ell+1} \in \mathcal{V}$ (since $n + \ell$

is non-zero). Iterating this argument, we obtain $e_\ell \in \mathcal{V}$ for all $\ell \in \mathbb{N}_0$, and so $\mathcal{V} = A_n(\mathbb{D})$ by Lemma 8.22. It follows that $\delta^{n,+}$ is irreducible. □

To prove that $\delta^{n,+}$ is a discrete series representation in the sense of Definition 8.1, it suffices by Theorem 8.2 to find one non-zero vector in $A_n(\mathbb{D})$ for which the associated matrix coefficient belongs to $L^2(\mathrm{SU}_{1,1}(\mathbb{R}))$. For this reason we calculate the matrix coefficient of $e_0 \in A_n(\mathbb{D})$. For this, we first consider a_t as in (8.18) for $t \in \mathbb{R}$, and calculate

$$\begin{aligned}\varphi_{e_0}(a_t) &= \left\langle \delta^{n,+}_{a_t} e_0, e_0 \right\rangle_{A_n(\mathbb{D})} \\ &= \int_{\mathbb{D}} (-z \sinh t + \cosh t)^{-n} \left(1 - |z|^2\right)^n \,\mathrm{dvol}(z) \\ &= 4(\cosh t)^{-n} \int_0^1 (1-r^2)^{n-2} r \int_0^{2\pi} \left(1 - \mathrm{e}^{\mathrm{i}\theta} r \tanh t\right)^{-n} \,\mathrm{d}\theta \,\mathrm{d}r.\end{aligned}$$

Since $|r \tanh t| < 1$ for all $r \in [0,1)$ and $t \in \mathbb{R}$, we can simply expand the innermost term in a power series. After integration with respect to $\theta \in [0, 2\pi]$, we therefore obtain the constant 2π for the inner integral, and hence

$$\varphi_{e_0}(a_t) = 4\pi (\cosh t)^{-n} \int_0^1 \left(1 - r^2\right)^{n-2} 2r \,\mathrm{d}r = \frac{4\pi}{n-1} (\cosh t)^{-n}.$$

Proof of square integrability. We recall from (8.10) that the Cartan decomposition of $\mathrm{SU}_{1,1}(\mathbb{R})$ is given by KAK, meaning that any $g \in \mathrm{SU}_{1,1}(\mathbb{R})$ can be written as $g = k_\theta a_t k_\psi$, where a_t is as above and

$$k_\theta = \begin{pmatrix} \mathrm{e}^{-\mathrm{i}\theta} & \\ & \mathrm{e}^{\mathrm{i}\theta} \end{pmatrix} \in K$$

for $\theta \in [0, \pi)$ and $\psi \in [0, 2\pi)$. With this, the matrix coefficient φ_{e_0} satisfies

$$|\varphi_{e_0}(g)| = \left| \left\langle \delta^{n,+}_{a_t k_\psi} e_0, \delta^{n,+}_{k_\theta^{-1}} e_0 \right\rangle \right| = \left| \mathrm{e}^{\mathrm{i}n\psi} \mathrm{e}^{\mathrm{i}n\theta} \underbrace{\left\langle \delta^{n,+}_{a_t} e_0, e_0 \right\rangle}_{=\varphi_{e_0}(a_t)} \right| = \frac{4\pi}{n-1} (\cosh t)^{-n}$$

by the calculation above.

To see that φ_{e_0} belongs to $L^2(\mathrm{SU}_{1,1}(\mathbb{R}))$, we need the description of the Haar measure m in the coordinates of the Cartan decomposition. By (8.11), this has the form $\mathrm{d}m = \frac{2}{\pi} \sinh 2t \,\mathrm{d}\theta \,\mathrm{d}t \,\mathrm{d}\psi$, and so we obtain

$$\begin{aligned}\int_{\mathrm{SU}_{1,1}(\mathbb{R})} |\varphi_{e_0}(g)|^2 \,\mathrm{d}m &= \frac{2}{\pi} \int_0^{2\pi} \int_0^\infty \int_0^\pi |\varphi_{e_0}(a_t)|^2 \sinh(2t) \,\mathrm{d}\theta \,\mathrm{d}t \,\mathrm{d}\psi \\ &= 4\pi \left(\frac{4\pi}{n-1}\right)^2 \int_0^\infty (\cosh t)^{-2n} \sinh(2t) \,\mathrm{d}t.\end{aligned}$$

Since $(\cosh t)^{-1} \ll \mathrm{e}^{-t}$ and $\sinh 2t \ll \mathrm{e}^{2t}$, and since $n \geqslant 2$ we deduce that $\varphi_{e_0} \in L^2(\mathrm{SU}_{1,1}(\mathbb{R}))$. By Theorem 8.2, this implies that $\delta^{n,+}$ is a discrete series representation of $\mathrm{SU}_{1,1}(\mathbb{R})$. □

8.4.2 Mock Discrete Series Representations

The mock discrete series representation corresponds in a way to setting $n = 1$ in the discussions above. However, if we simply set $n = 1$ in the definition of V in (8.14), then the underlying measure would be an infinite measure on $\mathbb{D}$. This would mean that the most basic holomorphic functions e_ℓ defined by $e_\ell(z) = z^\ell$ would not any longer be square-integrable for $\ell \in \mathbb{N}_0$. To guess how we could give a meaningful definition of a Hilbert space for $n = 1$, we allow for a moment n to lie in $(1, 2)$ and consider $L^2(\mathbb{D}, (1 - |z|^2)^{n-2}\,\mathrm{d}x\,\mathrm{d}y)$ for $n \searrow 1$. A calculation reveals that $(1 - |z|^2)^{n-2}\,\mathrm{d}x\,\mathrm{d}y$ defines in this case a finite measure with more and more total mass as n decreases to 1. If we normalize this measure to be a probability measure for each n, then those probability measures have a weak* limit in the space of probability measures on $\overline{\mathbb{D}} = \mathbb{D} \cup \mathbb{S}^1$ as $n \searrow 1$. In fact, this limit is given by the normalized arc length measure on $\partial\mathbb{D} = \mathbb{S}^1$. This motivates the following definition and the alternative name 'limit discrete series representation' for the 'mock discrete series representations' discussed here. The definition could be chosen closer to the discussion above, but the following will be easier to work with.

Definition 8.24 (Hardy space). The *Hardy space* $H(\mathbb{D})$ for $\mathbb{D}$ is defined to consist of all holomorphic functions $f\colon \mathbb{D} \to \mathbb{C}$ for which the Hardy norm $\|f\|_{H(\mathbb{D})}$ is finite, where

$$\|f\|_{H(\mathbb{D})}^2 = \sup_{0 \leqslant r < 1} \int_0^1 |f(r\mathrm{e}^{2\pi \mathrm{i}\theta})|^2\,\mathrm{d}\theta. \tag{8.20}$$

Lemma 8.25 (Norm on Hardy space). *For elements $f \in H(\mathbb{D})$ the functions $\mathbb{S}^1 \ni z \mapsto f(rz)$ have a limit as $r \nearrow 1$ in $L^2(\mathbb{S}^1)$. Extending f to almost every $z \in \mathbb{S}^1$ by this limit, we also have*

$$\|f\|_{H(\mathbb{D})} = \big\|f|_{\partial\mathbb{D}}\big\|_{L^2(\mathbb{S}^1)},$$

where we use the normalized arc length measure to define $L^2(\mathbb{S}^1)$. In this way $H(\mathbb{D})$ has the functions $e_\ell\colon \mathbb{D} \ni z \mapsto z^\ell$ for $\ell \in \mathbb{N}_0$ as an orthonormal basis and can be identified with a subspace of $L^2(\mathbb{S}^1)$. In fact for a holomorphic function defined by

$$f(z) = \sum_{k=0}^{\infty} a_k z^k \tag{8.21}$$

we have

$$\|f\|_{H(\mathbb{D})}^2 = \sum_{k=0}^{\infty} |a_k|^2. \tag{8.22}$$

PROOF. We first prove the last claim concerning $f \in \mathrm{Hol}(\mathbb{D})$ in (8.21)–(8.22). For this notice that

$$\int_0^1 |f(r\mathrm{e}^{2\pi\mathrm{i}\theta})|^2 \,\mathrm{d}\theta = \int_0^1 \left| \sum_{\ell=0}^{\infty} a_\ell r^\ell \mathrm{e}^{2\pi\mathrm{i}\ell\theta} \right|^2 \mathrm{d}\theta = \sum_{k=0}^{\infty} |a_k|^2 r^{2k}$$

by orthogonality of $\mathrm{e}^{2\pi\mathrm{i}\ell\theta}$ for $\ell \in \mathbb{N}_0$. It follows that the integral appearing in (8.20) increases monotonically as $r \nearrow 1$ to $\sum_{k=0}^{\infty} |a_k|^2$, proving (8.22).

For $f \in H(\mathbb{D})$ the above shows for the coefficients of f that

$$\sum_{\ell=0}^{\infty} |a_\ell|^2 < \infty,$$

which implies that

$$\mathbb{S}^1 \ni z \longmapsto \sum_{\ell=0}^{\infty} a_\ell r^\ell z^\ell$$

converges for $r \nearrow 1$ in $L^2(\mathbb{S}^1)$ to the function

$$\mathbb{S}^1 \ni z \longmapsto \sum_{\ell=0}^{\infty} a_\ell z^\ell.$$

The remaining claims now follow from this. □

Essential Exercise 8.26. Show that for any function $f \in H(\mathbb{D})$ the values $f(z)$ for $z \in \mathbb{D}$ are determined by the extension of f to $\mathbb{S}^1 \subseteq \overline{\mathbb{D}}$ through the Cauchy integral formula

$$f(z) = \frac{1}{2\pi\mathrm{i}} \oint_{\mathbb{S}^1} \frac{f(w)}{w - z} \,\mathrm{d}w.$$

Definition 8.27 (Mock discrete series representations). We define the mock discrete series representation $\delta^{1,+}$ on $H(\mathbb{D})$ by

$$\delta_g^{1,+} f(z) = (-\beta z + \alpha)^{-1} f\left(\frac{\overline{\alpha} z - \beta}{-\overline{\beta} z + \alpha}\right) \tag{8.23}$$

for $g = \begin{pmatrix} \alpha & \overline{\beta} \\ \beta & \overline{\alpha} \end{pmatrix} \in \mathrm{SU}_{1,1}(\mathbb{R})$, $f \in H(\mathbb{D})$, and $z \in \mathbb{D}$.

Lemma 8.28 (Holomorphic mock discrete series representation). *The mock discrete series representation $\delta^{1,+}$ is a unitary representation on the Hardy space $H(\mathbb{D})$.*

PROOF. From (8.8) we know that the Möbius transformation $\mathbb{C} \ni z \mapsto g \cdot z$ for $g \in \mathrm{SU}_{1,1}(\mathbb{R})$ maps $\mathbb{D}$ to $\mathbb{D}$ and $\mathbb{S}^1$ to $\mathbb{S}^1$. We define $\pi^{1,+}$ on $L^2(\mathbb{S}^1)$ by

$$\pi_g^{1,+}f(z) = (-\beta z + \alpha)^{-1} f\left(\frac{\overline{\alpha}z - \beta}{-\beta z + \alpha}\right)$$

for $g = \begin{pmatrix} \alpha & \overline{\beta} \\ \beta & \overline{\alpha} \end{pmatrix} \in \mathrm{SU}_{1,1}(\mathbb{R})$, $f \in L^2(\mathbb{S}^1)$, and $z \in \mathbb{S}^1$. Note that this matches (8.15) and (8.23) except that we define $\pi^{1,+}$ for functions on the space $\mathbb{S}^1 = \partial\mathbb{D}$.

Once more the complex derivative of $z \mapsto g^{-1}\bullet z$ is given by $(-\beta z + \alpha)^{-2}$. This in turn tells us the derivative of the normalized arc length $\psi \in [0, 1]$ of the point $g^{-1}\bullet z = \mathrm{e}^{2\pi \mathrm{i}\psi}$ with respect to the normalized arc length $\theta \in [0, 1]$ of $z = \mathrm{e}^{2\pi \mathrm{i}\theta}$. As $\mathrm{SU}_{1,1}(\mathbb{R})$ is connected, this is given by $\frac{\mathrm{d}\psi}{\mathrm{d}\theta} = |-\beta z + \alpha|^{-2}$. Therefore

$$\begin{aligned} \left\|\pi_g^{1,+}f\right\|_{L^2(\mathbb{S}^1)}^2 &= \int_0^1 \underbrace{|-\beta \mathrm{e}^{2\pi \mathrm{i}\theta} + \alpha|^{-2}}_{\frac{\mathrm{d}\psi}{\mathrm{d}\theta}} |f(g^{-1}\bullet \mathrm{e}^{2\pi \mathrm{i}\theta})|^2 \,\mathrm{d}\theta \\ &= \int_0^1 |f(\mathrm{e}^{2\pi \mathrm{i}\psi})|^2 \,\mathrm{d}\psi = \|f\|_{L^2(\mathbb{S}^1)}^2. \end{aligned}$$

The argument proving (8.16) works unchanged here and shows the homomorphism property $\pi_{gh}^{1,+} = \pi_g^{1,+}\pi_h^{1,+}$ for $g, h \in \mathrm{SU}_{1,1}(\mathbb{R})$. Therefore $\pi^{1,+}$ is a unitary representation of $\mathrm{SU}_{1,1}(\mathbb{R})$ on $L^2(\mathbb{S}^1)$ (see Exercise 8.29).

By Lemma 8.25 the Hardy space $H(\mathbb{D})$ can be identified with a closed subspace of $L^2(\mathbb{S}^1)$ containing the functions e_ℓ for $\ell \in \mathbb{N}_0$ as an orthonormal basis. For one of the basis vectors $e_\ell \in H(\mathbb{D})$ with $\ell \in \mathbb{N}_0$ and $g \in \mathrm{SU}_{1,1}(\mathbb{R})$, we also have that $\delta_g^{1,+}e_\ell$ is given by

$$\begin{aligned} \delta_g^{1,+}e_\ell(z) &= (-\beta z + \alpha)^{-1}\left(\frac{\overline{\alpha}z - \overline{\beta}}{-\beta z + \alpha}\right)^\ell \\ &= \alpha^{-\ell-1}\left(1 - \frac{\beta}{\alpha}z\right)^{-(\ell+1)} \left(\overline{\alpha}z - \overline{\beta}\right)^\ell. \end{aligned}$$

Recalling that $|\alpha|^2 - |\beta|^2 = 1$ we see that $|\beta| < |\alpha|$, and that this defines a holomorphic function on an open set O_g containing $\overline{\mathbb{D}}$. In particular, we see that $\delta_g^{1,+}e_\ell \in H(\mathbb{D})$ can be identified with $\pi_g^{1,+}e_\ell$. Since this holds for all elements $g \in \mathrm{SU}_{1,1}(\mathbb{R})$ and $\ell \in \mathbb{N}_0$, it follows that $H(\mathbb{D}) \subseteq L^2(\mathbb{S}^1)$ is invariant under $\pi^{1,+}$.

Moreover, by Exercise 8.26 we know that L^2-convergence of a sequence in $H(\mathbb{D}) \subseteq L^2(\mathbb{S}^1)$ implies compact-open convergence in $\mathbb{D}$. Together with linearity of $\delta^{1,+}$ and $\pi^{1,+}$ and the case of the basis vectors, it follows that $\delta^{1,+}$ can be identified with the restriction of $\pi^{1,+}$ to $H(\mathbb{D})$. □

Exercise 8.29. Prove the continuity of $\pi^{1,+}$ either directly or by applying Proposition 1.6.

Definition 8.30 (Anti-holomorphic mock discrete series representation). The *anti-holomorphic mock discrete series representation* $\delta^{1,-}$ is defined to be the contragredient representation to the holomorphic mock discrete series representation.

Theorem 8.31 (Mock discrete series representation). *The holomorphic and anti-holomorphic mock discrete series representations are irreducible tempered representations but not discrete series representations. The holomorphic mock discrete series representation $\delta^{1,+}$ has K-eigenfunctions with weights in $1 + 2\mathbb{N}_0$, and only those, each with multiplicity one. Similarly, the anti-holomorphic mock discrete series representation $\delta^{1,-}$ has K-eigenfunctions with weights $-1 - 2\mathbb{N}_0$, and only these, each with multiplicity one.*

We will again say that 1 is the *lowest K-weight* of $\delta^{1,+}$, -1 is the *highest K-weight* of $\delta^{1,-}$, and that ± 1 are the *terminal K-weights* of $\delta^{1,\pm}$.

The last part of Theorem 8.31 follows immediately from studying the orthonormal basis of $H(\mathbb{D})$ comprising e_ℓ for $\ell \in \mathbb{N}_0$. Indeed,

$$\delta^{1,+}_{k_\theta} e_\ell(z) = (-0z + \mathrm{e}^{-\mathrm{i}\theta})^{-1} \left(\frac{\mathrm{e}^{\mathrm{i}\theta} z + 0}{0 + \mathrm{e}^{-\mathrm{i}\theta}} \right)^\ell = \mathrm{e}^{\mathrm{i}(1+2\ell)\theta} e_\ell(z)$$

for $\theta \in [0, 2\pi)$ shows that e_ℓ is a K-eigenfunction with weight $1 + 2\ell$. For the contragredient representation, this also implies that $e'_\ell \in H(\mathbb{D})'$ is a K-eigenfunction with weight $-1 - 2\ell$.

Proof of irreducibility. We again use the differential operator associated to

$$\mathbf{d} = \begin{pmatrix} 0 & 1 \\ 1 & 0 \end{pmatrix}$$

and the basis vectors e_ℓ for $\ell \in \mathbb{N}_0$. Indeed, let us suppose $\mathcal{V} \subseteq H(\mathbb{D})$ is a non-trivial closed invariant subspace. As $K < \mathrm{SU}_{1,1}(\mathbb{R})$ is a compact abelian subgroup, $\mathcal{V}$ must contain a K-eigenfunction for some weight $k \in \mathbb{Z}$. As the basis vectors are K-eigenfunctions of mutually different weights, it follows that $e_\ell \in \mathcal{V}$ for some $\ell \in \mathbb{N}_0$. We recall that

$$\delta^{1,+}_\partial(\mathbf{d})(e_\ell) = \lim_{t \to 0} \frac{1}{t} \left(\delta^{1,+} \left(\exp(t\mathbf{d}) \right)(e_\ell) - e_\ell \right).$$

Repeating the calculation after (8.19) but with $n = 1$ we obtain that the pointwise derivative is equal to $-\ell e_{\ell-1} + (\ell+1) e_{\ell+1}$. In fact the underlying convergence is uniform on (for example) the ball $B_2^{\mathbb{C}}$, and it follows that the convergence also takes place in $H(\mathbb{D})$. Since $\mathcal{V}$ is closed, it follows that

$$\delta^{1,+}_\partial(\mathbf{d})(e_\ell) = -\ell e_{\ell-1} + (\ell + 1) e_{\ell+1} \in \mathcal{V}.$$

Since $e_{\ell-1}$ and $e_{\ell+1}$ have different weights, and $\mathcal{V}$ is T-invariant, we see that $e_{\ell+1} \in \mathcal{V}$ and if $\ell > 1$ we also have $e_{\ell-1} \in \mathcal{V}$. Therefore $\mathcal{V}$ contains all the basis vectors, so $\mathcal{V} = H(\mathbb{D})$, and the mock discrete series representation is irreducible. This implies the same statement for its contragredient representation. □

Proof of temperedness. As the matrix coefficients of the contragredient representation are the conjugates of the matrix coefficients of the original representation, it suffices to study the mock discrete series representation $\delta^{1,+}$. We again calculate the matrix coefficient φ_{e_0} first along the one-parameter subgroup corresponding to $\mathbf{d}$. This gives

$$\varphi_{e_0}(a_t) = \langle \delta^{1,+}_{a_t} e_0, e_0 \rangle = \int_0^1 \bigl(-\sinh t e^{2\pi i\theta} + \cosh t\bigr)^{-1} \cdot 1 \cdot 1 \, d\theta = (\cosh t)^{-1}$$

for all $t \in \mathbb{R}$, where we once more expanded the expression

$$\bigl(-\sinh t e^{2\pi i\theta} + \cosh t\bigr)^{-1}$$

into a geometric series to calculate the integrals. Using the formula (8.11) for the Haar measure in the KAK coordinates, and the fact that e_0 is a K-eigenfunction for K with weight 1, we obtain

$$\begin{aligned}\int_G |\varphi_{e_0}(g)|^{2+\varepsilon} \, dm(g) &= \frac{2}{\pi} \int_0^{2\pi} \int_0^\infty \int_0^\pi \underbrace{\bigl|\varphi_{e_0}(k_\theta a_t k_\psi)\bigr|}_{=(\cosh t)^{-1}}{}^{2+\varepsilon} \sinh 2t \, d\theta \, dt \, d\psi \\ &= 4\pi \int_0^\infty (\cosh t)^{-(2+\varepsilon)} \sinh 2t \, dt \begin{cases} = \infty & \text{if } \varepsilon = 0; \\ < \infty & \text{if } \varepsilon > 0. \end{cases}\end{aligned}$$

Using the case $\varepsilon = 0$, it follows from Theorem 8.2 that the mock discrete series representation is not a discrete series representation. Using $\varepsilon > 0$, we can use Theorem 8.5 to see that the mock discrete series representation is tempered (that is, weakly contained in the regular representation). □

8.5 Effective Decay for the Regular Representation of $\mathrm{SL}_2(\mathbb{R})$

The goal in this section is to prove the following effective decay property for the regular representation of $\mathrm{SL}_2(\mathbb{R})$. This material comes from work of Harish-Chandra [41]. We again use the terminology introduced in Section 7.3.2.

Theorem 8.32 (Effective decay for the regular representation). *Let f_1 and f_2 in $L^2(\mathrm{SL}_2(\mathbb{R}))$ be K-eigenfunctions for the regular representation. Then*

$$\bigl|\langle \lambda_g f_1, f_2 \rangle\bigr| \leqslant \|f_1\| \|f_2\| \Xi(g),$$

where $\Xi\colon \mathrm{SL}_2(\mathbb{R}) \to \mathbb{R}$ is the Harish-Chandra spherical function, which satisfies

$$\Xi(g) \ll \|g\|_{\mathrm{HS}}^{-1} \log \|g\|_{\mathrm{HS}} \ll_\varepsilon \|g\|_{\mathrm{HS}}^{-1+\varepsilon}$$

for all $g \in \mathrm{SL}_2(\mathbb{R})$ *and all* $\varepsilon > 0$.

8.5.1 Subgroups and the Modular Character

Let us recall once more the notation for elements and subgroups of $\mathrm{SL}_2(\mathbb{R})$ from Section 8.3. We write

$$K = \mathrm{SO}_2(\mathbb{R}) = \left\{ k_\theta = \begin{pmatrix} \cos\theta & -\sin\theta \\ \sin\theta & \cos\theta \end{pmatrix} \;\middle|\; \theta \in [0, 2\pi) \right\}$$

for the usual maximal compact subgroup, and

$$A = \left\{ a_t = \begin{pmatrix} \mathrm{e}^t & \\ & \mathrm{e}^{-t} \end{pmatrix} \;\middle|\; t \in \mathbb{R} \right\}$$

for the connected diagonal subgroup. Note that the centre $M = \{\pm I\}$ of $\mathrm{SL}_2(\mathbb{R})$ is also the maximal compact subgroup of the centralizer AM of A, and that AM consists of all diagonal matrices (with positive and negative eigenvalues). We also define

$$U = \left\{ u_x = \begin{pmatrix} 1 & x \\ & 1 \end{pmatrix} \;\middle|\; x \in \mathbb{R} \right\}$$

to be the upper unipotent matrix. Finally, we define the connected upper Borel subgroup

$$B = \left\{ \begin{pmatrix} \mathrm{e}^t & x \\ & \mathrm{e}^{-t} \end{pmatrix} \;\middle|\; t \in \mathbb{R}, x \in \mathbb{R} \right\} = AU < \mathrm{SL}_2(\mathbb{R}).$$

We will often denote elements of the full upper triangular group BM by $u_x a_t m$ or $a_t m u_x$, with the implicit assumptions $t \in \mathbb{R}$, $x \in \mathbb{R}$, and $m \in M$.

We now give a concrete description of the modular character Δ_B. Notice that every element $b \in B = AU$ can be written uniquely as a product $b = u_x a_t$ of $u_x \in U$ and $a_t \in A$. We recall from (8.9) that in this coordinate system (that is, order of writing A and U) the Haar measure is given by

$$\mathrm{d}m_B = 2\mathrm{e}^{-2t}\,\mathrm{d}x\,\mathrm{d}t,$$

which we obtained by identifying $b \in B$ with $b\boldsymbol{\cdot}\mathrm{i} \in \mathbb{H}$. More precisely, we have

$$\int_B f\,\mathrm{d}m_B = 2\int_{\mathbb{R}^2} f(u_x a_t)\mathrm{e}^{-2t}\,\mathrm{d}x\,\mathrm{d}t$$

for any integrable function f on B.

To calculate Δ_B, we first let $b_0 = u_{x_0} \in B$ and let $f \geqslant 0$ be any integrable function on B. Then we get for the right translated function that

$$\begin{aligned}\int_B f(bb_0^{-1})\,\mathrm{d}m_B(b) &= 2\int_{\mathbb{R}^2} f\big(u_x a_t u_{-x_0}\big)\mathrm{e}^{-2t}\,\mathrm{d}x\,\mathrm{d}t\\ &= 2\int_{\mathbb{R}^2} f\big(u_{x'} a_t\big)\mathrm{e}^{-2t}\,\mathrm{d}x'\,\mathrm{d}t\end{aligned}$$

by using the calculation

$$a_t u_{-x_0} = \begin{pmatrix}\mathrm{e}^t & \\ & \mathrm{e}^{-t}\end{pmatrix}\begin{pmatrix}1 & -x_0\\ & 1\end{pmatrix} = \begin{pmatrix}\mathrm{e}^t & -\mathrm{e}^t x_0\\ & \mathrm{e}^{-t}\end{pmatrix} = u_{-\mathrm{e}^{2t}x_0} a_t$$

and the substitution $x' = x - \mathrm{e}^{2t}x_0$ for the Lebesgue measure on $\mathbb{R}$. Therefore $\Delta_B(u_x) = 1$.

Suppose now that $b_0 = a_{t_0} \in B$ and that $f \geqslant 0$ is again an integrable function on B. Then we may use the substitution $t' = t - t_0$ to get

$$\begin{aligned}\int_B f(bb_0^{-1})\,\mathrm{d}m_B(b) &= 2\int_{\mathbb{R}^2} f(u_x a_t a_{-t_0})\mathrm{e}^{-2t}\,\mathrm{d}x\,\mathrm{d}t\\ &= 2\int_{\mathbb{R}^2} f(u_x a_{t'})\mathrm{e}^{-2t'-2t_0}\,\mathrm{d}x\,\mathrm{d}t'\\ &= \mathrm{e}^{-2t_0}\int_B f(b)\,\mathrm{d}m_B(b).\end{aligned}$$

Therefore the modular character Δ_B of B (which is characterized by this formula; see Lemma 1.15) is given by

$$\Delta_B\big(a_{t_0}u_{x_0}\big) = \Delta_B\big(a_{t_0}\big) = \mathrm{e}^{-2t_0} \tag{8.24}$$

for all $b_0 = a_{t_0}u_{x_0} \in B$.

For the right Haar measure $m_B^{(\mathrm{r})}$ we also have

$$\int_B f(b_0 b)\,\mathrm{d}m_B^{(\mathrm{r})}(b) = \Delta_B(b_0)\int_B f\,\mathrm{d}m_B^{(\mathrm{r})} = \mathrm{e}^{-2t_0}\int_B f\,\mathrm{d}m_B^{(\mathrm{r})}. \tag{8.25}$$

Indeed, applying Lemma 1.16 twice and the above, we obtain

$$\begin{aligned}\int_B f(b_0 b)\,\mathrm{d}m_B^{(\mathrm{r})}(b) &= \int_B f(b_0 b^{-1})\,\mathrm{d}m_B(b) = \int_B f\big((bb_0^{-1})^{-1}\big)\,\mathrm{d}m_B(b)\\ &= \Delta_B(b_0)\int_B f(b^{-1})\,\mathrm{d}m_B(b)\\ &= \mathrm{e}^{-2t_0}\int_B f\,\mathrm{d}m_B^{(\mathrm{r})}.\end{aligned}$$

We recall that we normalize the Haar measure m_K on K so that $m_K(K) = 1$.

8.5.2 A Particular Induced Representation

In order to define the Harish-Chandra spherical function we will need a simple case of induction for representations. Given any abstract representation π_B on a (complex) vector space V_B of a subgroup B of a group G, we can define the *induced representation* to be the left regular representation of G on the vector space

$$\Big\{f\colon G \longrightarrow V_B \;\Big|\; f(gb) = \pi_B(b)^{-1}f(g) \text{ for all } g \in G, b \in B\Big\}.$$

We refer to Section 5.5.1 for an introduction to this concept.

We will be particularly interested in the case where π_B is the trivial representation, in which case the representation space above is simply the space

$$\Big\{f\colon G/B \longrightarrow \mathbb{C}\Big\}.$$

In the context of unitary represenation it is of course natural to require in addition some square-integrability conditions as in Section 5.5.3. Thus, for example, if B is the upper Borel subgroup in $\mathrm{SL}_2(\mathbb{R})$ recall that $\mathrm{SL}_2(\mathbb{R})/B \cong K$, and we may consider the space $L^2(K)$. However, this definition together with the left regular representation has a fundamental flaw: There is no $\mathrm{SL}_2(\mathbb{R})$-invariant measure on $\mathrm{SL}_2(\mathbb{R})/B \cong K$. To rectify this and obtain a unitary representation, we make the following definition.†

Definition 8.33 (Unitary induction for $B < \mathrm{SL}_2(\mathbb{R})$). Let B be the connected upper Borel subgroup of $G = \mathrm{SL}_2(\mathbb{R})$ as above. Let π_B be a unitary representation of $B < G = \mathrm{SL}_2(\mathbb{R})$ on some Hilbert space $\mathcal{H}_B$. The *unitary (or normalized) induced representation*

$$(\mathcal{H}_G, \pi_G) = \mathrm{Ind}_B^G(\mathcal{H}_B, \pi_B)$$

is defined to be the left regular representation on the space of those functions $f\colon G \to \mathcal{H}_B$ with the following properties:

(1) f is measurable,
(2) $f(gb) = \Delta_B(b)^{\frac{1}{2}}\pi_B(b)^{-1}f(g)$ for all $g \in G$ and $b \in B$, and
(3) $\|f|_K\|_{L^2(K)} < \infty$,

where $\Delta_B(a_t u_x) = \Delta_B(a_t)$ is the modular character of B, and

$$\|f|_K\|^2_{L^2(K)} = \int_K \|f(k)\|^2_{\mathcal{H}_B}\,\mathrm{d}m_K(k)$$

† Our definitions here are special cases of the unitarily normalized representation in Proposition 1.6 or the unitary induction in Section 5.5. However, in order to make some of the arguments to come more convenient, we use different formulas to obtain unitarity (see Exercise 8.35).

defines the norm on the Hilbert space $\mathcal{H}_G$. As usual, we identify any function of semi-norm zero with the zero function.

Instead of concentrating on the technical details of the representation constructed above in general, we will for now only discuss the following crucial special case.

We let $\mathbb{1}_B$ denote the trivial representation of B on $\mathbb{C}$, and define

$$(\mathcal{H}_0, \pi^0) = \operatorname{Ind}_B^G(\mathbb{C}, \mathbb{1}_B)$$

by setting $\mathcal{H}_0$ to be the space of functions $f\colon G \to \mathbb{C}$ with the properties

(1) f is measurable;
(2) $f(gb_0) = \Delta_B(b_0)^{\frac{1}{2}} f(g)$ for all $g \in G, b_0 \in B$; and
(3) $\|f\|_{\mathcal{H}_0} = \|f|_K\|_{L^2(K)} < \infty$.

We also write π^0 for the left regular representation on $\mathcal{H}_0$, that is

$$\pi^0(g)(f)(x) = f(g^{-1}x)$$

for all $g, x \in G$, and write $\langle \cdot, \cdot \rangle_{\mathcal{H}_0}$ for the inner product on $\mathcal{H}_0$ compatible with the norm $\|\cdot\|_{\mathcal{H}_0}$, as in (3).

We will see in Corollary 8.37 that $f \in \mathcal{H}_0$ and $g \in G$ implies $\pi^0(g)(f) \in \mathcal{H}_0$, and that π^0 is indeed a unitary representation. Using only the definition, we now show that

$$\mathcal{H}_0 \ni f \longmapsto f|_K \in L^2(K) \tag{8.26}$$

is an isomorphism of Hilbert spaces. To see that the map is onto, suppose that f_K is a square-integrable function on K and define $f\colon G \to \mathbb{C}$ by

$$f(kb) = f_K(k)\Delta_B(b)^{\frac{1}{2}}$$

for $kb \in G = KB$. We only have to verify (2), so let $g = kb \in G$ and $b_0 \in B$. Then

$$f(gb_0) = f(kbb_0) = f_K(k)\Delta_B(bb_0)^{\frac{1}{2}} = \Delta_B(b_0)^{\frac{1}{2}} f(g)$$

as required. Similarly, any $f \in \mathcal{H}_0$ is uniquely determined by $f|_K$, because of $G = KB$ and property (2). Finally, the map in (8.26) is a well-defined isometry by the definition of $\|\cdot\|_{\mathcal{H}_0}$ in (3).

We also note that two functions $f_1, f_2 \in \mathcal{H}_0$ satisfy $\|f_1 - f_2\|_{\mathcal{H}_0} = 0$ if and only if f_1 and f_2 agree m_G-almost everywhere on G. For this, recall that we may define the Haar measure m_G on $G = KB$ as the product measure of m_K and $m_B^{(\mathrm{r})}$ (see Lemma 8.13). Suppose now that $f_1, f_2 \in \mathcal{H}_0$. Then

$$\|f_1 - f_2\|_{\mathcal{H}_0} = 0$$

if and only if $f_1(k) = f_2(k)$ for m_K-almost every $k \in K$, which by (2) is equivalent to $f_1(kb) = f_2(kb)$ for m_K-almost every $k \in K$ and all $b \in B$, and hence also to $f_1(g) = f_2(g)$ for m_G-almost every $g \in G$.

We note that our implicit requirement in the discussion above was that f as in (1)–(3) be defined on all of G. As is common with other function spaces, it will be convenient to loosen this at times by permitting f to not be defined (or to be equal to ∞) on a right B-invariant null set $S = SB \subseteq G$. Setting

$$\widetilde{f}(g) = \begin{cases} f(g) & \text{for } g \in G \setminus S, \\ 0 & \text{for } g \in S \end{cases}$$

then defines an equivalent function $\widetilde{f} \in \mathcal{H}_0$ (since $S = SB$ is a null set in G if and only if $S \cap K$ is a null set in K).

Exercise 8.34. (a) Show directly that $\pi^0(g)\mathcal{H}_0 = \mathcal{H}_0$, and that $\pi^0(g)$ is unitary.
(b) Show that $(\mathcal{H}_0, \pi^0)$ is a unitary representation of G.

Exercise 8.35. Show that the above notion of unitary induction $\mathrm{Ind}_B^G \pi$ of a unitary representation π of B is isomorphic to the representation defined in Section 5.5.3.

8.5.3 The Hertz Domination Principle

The link between the regular representation and the representation π^0 introduced above is revealed in the next fundamental result.

Proposition 8.36 (Hertz domination principle). *Let $G = \mathrm{SL}_2(\mathbb{R})$. To any $f \in L^2(G)$ we can associate an element $\breve{f} \in \mathcal{H}_0$ defined by*

$$\breve{f}(g) = \left(\int_B |f(gb)|^2 \, \mathrm{d}m_B^{(\mathrm{r})}(b) \right)^{\frac{1}{2}}$$

for all $g \in G$. Then

(1) $\breve{f} \in \mathcal{H}_0$,
(2) $\breve{cf} = |c|\,\breve{f}$ for $c \in \mathbb{C}$,
(3) $\|\breve{f}\|_{\mathcal{H}_0} = \|f\|_2$,
(4) $\breve{\lambda_g f} = \pi_g^0 \breve{f}$ for all $g \in G$, and finally
(5) $|\langle \lambda_g f_1, f_2 \rangle| \leqslant \langle \pi_g^0 \breve{f_1}, \breve{f_2} \rangle$ for all $g \in G$.

PROOF. Let $f \in L^2(G)$, define $\breve{f}\colon G \to [0,\infty]$ as in the proposition, and let $g \in G$ and $b_0 \in B$. Using the definition of $\breve{f}$ and $\Delta_B\colon B \to (0,\infty)$ we get from (8.25) that

$$\begin{aligned} \breve{f}(gb_0) &= \left(\int_B |f(gb_0 b)|^2 \, \mathrm{d}m_B^{(\mathrm{r})}(b) \right)^{\frac{1}{2}} \\ &= \left(\Delta_B(b_0) \int_B |f(gb)|^2 \, \mathrm{d}m_B^{(\mathrm{r})}(b) \right)^{\frac{1}{2}} = \Delta_B(b_0)^{\frac{1}{2}} \breve{f}(g) \end{aligned}$$

whenever at least one side is defined. Notice that $\overbrace{cf} = |c|\overbrace{f}$ for $c \in \mathbb{C}$ follows from the definition. Moreover,

$$\begin{aligned}\|\overbrace{f}\|_{\mathcal{H}_0}^2 &= \|\overbrace{f}|_K\|_{L^2(K)}^2 \\ &= \int_K \int_B |f(kb)|^2 \mathrm{d}m_B^{(\mathrm{r})}(b)\,\mathrm{d}m_K(k) = \int_G |f(g)|^2 \mathrm{d}m(g) = \|f\|_2^2 < \infty,\end{aligned}$$

since the Haar measure $\mathrm{d}m_G$ decomposes as $\mathrm{d}m_K\,\mathrm{d}m_B^{(\mathrm{r})}$ in the coordinates of the Iwasawa decomposition. In particular, we have $\overbrace{f}(k) < \infty$ for m_K-almost every $k \in K$, and $N = \{g \in G \mid \overbrace{f}(g) = \infty\}$ is a right B-invariant null set in G. Formally, we may modify $\overbrace{f}$ and set it equal to 0 on N, but instead we will simply ignore this null set. In that sense, $\overbrace{f} \in \mathcal{H}_0$ and we obtain the statements (1), (2), and (3).

Now fix some $g_0 \in G$. Just using the definitions, we see that

$$\begin{aligned}\overbrace{\lambda_{g_0} f}(g) &= \left(\int_B |(\lambda_{g_0} f)(gb)|^2\,\mathrm{d}m_B^{(\mathrm{r})}(b)\right)^{\frac{1}{2}} \\ &= \left(\int_B |f(g_0^{-1}gb)|^2\,\mathrm{d}m_B^{(\mathrm{r})}(b)\right)^{\frac{1}{2}} \\ &= \overbrace{f}(g_0^{-1}g) = (\pi_{g_0}^0 \overbrace{f})(g)\end{aligned}$$

for all $g \in G$, which gives claim (4) in the proposition.

As we now show, the last statement of the proposition is a simple corollary of the decomposition of the Haar measure on G already used above and the Cauchy–Schwarz inequality applied in the right way. Indeed, let f_1, f_2 be functions in $L^2(G)$. Then

$$\begin{aligned}|\langle \lambda_g(f_1), f_2\rangle| &= \left|\int_K \int_B f_1(g^{-1}kb) f_2(kb)\,\mathrm{d}m_B^{(\mathrm{r})}(b)\,\mathrm{d}m_K(k)\right| \\ &\leqslant \int_K \int_B |f_1(g^{-1}kb) f_2(kb)|\,\mathrm{d}m_B^{(\mathrm{r})}(b)\,\mathrm{d}m_K(k) \\ &\leqslant \int_K \left(\int_B |f_1(g^{-1}kb)|^2\,\mathrm{d}m_B^{(\mathrm{r})}(b)\right)^{\frac{1}{2}} \\ &\qquad\qquad \left(\int_B |f_2(kb)|^2\,\mathrm{d}m_B^{(\mathrm{r})}(b)\right)^{\frac{1}{2}}\,\mathrm{d}m_K(k) \\ &= \int_K \left(\pi_g^0 \overbrace{f_1}\right)(k)\,\overbrace{f_2}(k)\,\mathrm{d}m_K(k) \\ &= \langle \pi_g^0 \overbrace{f_1}, \overbrace{f_2}\rangle_{\mathcal{H}_0}\end{aligned}$$

as claimed. □

With the link between λ and π^0 provided by Proposition 8.36, it is now straightforward to check that π^0 is a unitary representation.

Corollary 8.37 (Unitarity of π^0). *π^0 defines a unitary representation of the group $\mathrm{SL}_2(\mathbb{R})$ on $\mathcal{H}_0$.*

Proof. Let $f_0 \in \mathcal{H}_0$. For $g_0, g \in G$ and $b \in B$ we then have

$$\pi^0_{g_0}(f_0)(gb) = f_0(g_0^{-1}gb) = \Delta(b)^{\frac{1}{2}} f_0(g_0^{-1}g) = \Delta(b)^{\frac{1}{2}} \pi^0_{g_0}(f_0)(g),$$

as is required for elements of $\mathcal{H}_0$. To see that $\pi^0_{g_0}(f_0) \in \mathcal{H}_0$, we need to show that it is square-integrable on K. In fact we will show that

$$\|\pi^0_{g_0}(f_0)|_K\|_{L^2(K)} = \|\pi^0_{g_0} f_0\|_{\mathcal{H}_0} = \|f_0\|_{\mathcal{H}_0}.$$

For this, we fix some measurable $S \subseteq B$ with $m_B^{(\mathrm{r})}(S) = 1$. We now define a measurable function $f\colon G \to \mathbb{C}$ by

$$f(kb) = |f_0(k)| \mathbb{1}_S(b).$$

Since $f_0|_K \in L^2(K)$ by definition of $\mathcal{H}_0$, we see that $f \in L^2(G)$ (by the product structure of the Haar measure). Applying the definition in Proposition 8.36 we also get

$$\breve{f}(g) = |f_0(g)| \tag{8.27}$$

first for all $g = k \in K$, but since both $\breve{f}$ and f_0 belong to $\mathcal{H}_0$ the defining property of $\mathcal{H}_0$ extends this equality to all $g \in G$. In particular, we obtain $\|f_0\|_{\mathcal{H}_0} = \|f\|_2$. Now fix some $g \in G$ and use (8.27), Proposition 8.36(4), (3), unitarity of the regular representation λ on $L^2(G)$, and (8.27) again to obtain

$$\begin{aligned}\|\pi^0_g f_0\|_{\mathcal{H}_0} = \|\pi^0_g \breve{f}\|_{\mathcal{H}_0} = \|\breve{\lambda_g f}\|_{\mathcal{H}_0} &= \|\lambda_g f\|_2 \\ &= \|f\|_2 = \|\breve{f}\|_{\mathcal{H}_0} = \|f_0\|_{\mathcal{H}_0}.\end{aligned}$$

This shows that $\pi^0_g f_0 \in \mathcal{H}_0$, and that π^0_g is unitary. Continuity of π^0 follows from Lemma 1.13 considering $C(K) \subseteq L^2(K)$, from Proposition 1.6, or from Section 5.5. We leave the choice and the details to the reader. □

We note that π^0 is not irreducible, since $M = \{\pm I\}$ does not act as a scalar. In fact, splitting $\mathcal{H}_0$ into even and odd components for M defines two unitary representations that will be studied more carefully in the next chapter. As we will see in Chapter 9, the even subspace is irreducible and the odd subspace is a sum of two irreducible subspaces.

Exercise 8.38 (Hertz domination for $L^2(\mathbb{R}^2)$). For $f \in L^2(\mathbb{R}^2)$ define

$$\breve{f}(g) = \left(\int |f(rge_1)|^2 r \,\mathrm{d}r\right)^{\frac{1}{2}}$$

for $g \in \mathrm{SL}_2(\mathbb{R})$. Consider the action-associated representation $\pi^{\mathbb{R}^2}$ of $\mathrm{SL}_2(\mathbb{R})$ on $L^2(\mathbb{R}^2)$, and show properties (1)–(5) of Proposition 8.36 (replacing the regular representation λ by $\pi^{\mathbb{R}^2}$).

8.5.4 The Harish-Chandra Spherical Function

Definition 8.39 (Harish-Chandra spherical function for $\mathrm{SL}_2(\mathbb{R})$). After extending the constant function $\mathbb{1}_K$ in $L^2(K)$ to an element f_0 in $\mathcal{H}_0$ with the property that $f_0|_K \equiv \mathbb{1}_K$, we define the Harish-Chandra spherical function for $\mathrm{SL}_2(\mathbb{R})$ by

$$\Xi(g) = \langle \pi_g^0 f_0, f_0 \rangle_{\mathcal{H}_0}$$

for $g \in \mathrm{SL}_2(\mathbb{R})$.

Proposition 8.40 (Estimate for HC-spherical function for $\mathrm{SL}_2(\mathbb{R})$). *The Harish-Chandra spherical function is continuous and bi-K-invariant, meaning that*

$$\Xi(kgk') = \Xi(g)$$

for all $g \in G = \mathrm{SL}_2(\mathbb{R})$ and $k, k' \in K$. Moreover, $\Xi \notin L^2(G)$ but

$$\Xi(g) \asymp \|g\|_{\mathrm{HS}}^{-1} \log \|g\|_{\mathrm{HS}} \ll_\varepsilon \|g\|_{\mathrm{HS}}^{-1+\varepsilon} \tag{8.28}$$

for all $g \in G$, and $\Xi \in L^{2+\varepsilon}(G)$ for all $\varepsilon > 0$.

Proof. First note that

$$f_0(ka_t u_x) = \Delta_B(a_t)^{\frac{1}{2}} = \mathrm{e}^{-t}$$

defines a continuous function f_0 on $ka_t u_x \in G$. For $g \in G$ and $k, k' \in K$ we have

$$\Xi(kgk') = \langle \pi_g^0 \pi_{k'}^0 f_0, \pi_{k^{-1}}^0 f_0 \rangle_{\mathcal{H}_0} = \Xi(g)$$

since $\pi_{k^{-1}}^0 f_0 = \pi_{k'}^0 f_0 = f_0$.

It follows that it is enough to consider

$$g = a_t = \begin{pmatrix} \mathrm{e}^t & \\ & \mathrm{e}^{-t} \end{pmatrix}$$

with $t \geqslant 0$ in the proof of the estimate (8.28). We now calculate

$$\begin{aligned} \Xi(a_t) &= \langle \pi_{a_t}^0 f_0, f_0 \rangle_{\mathcal{H}_0} \\ &= \int_K f_0(a_t^{-1}k) \underbrace{f_0(k)}_{=1} \,\mathrm{d}m_K(k) = \frac{1}{2\pi} \int_0^{2\pi} f_0(a_t^{-1}k_\theta) \,\mathrm{d}\theta \end{aligned}$$

and arrive at the problem of writing $a_t^{-1}k_\theta$ in the desired form $k_\psi a_{t_0} u_{x_0}$ for some $k_\psi \in K$, $a_{t_0} \in A$, and $u_{x_0} \in U$. More precisely, we only need to calculate $t_0 = t_0(t, \theta)$, since we have

$$f_0(a_t^{-1}k_\theta) = f_0(k_\psi a_{t_0} u_{x_0}) = \Delta_B(a_{t_0})^{\frac{1}{2}} = \mathrm{e}^{-t_0}$$

by definition of $f_0 \in \mathcal{H}_0$. Notice that e^{t_0} is the length of the first column of $a_t^{-1}k_\theta$, since

$$a_t^{-1}k_\theta\begin{pmatrix}1\\0\end{pmatrix} = k_\psi a_{t_0}u_{x_0}\begin{pmatrix}1\\0\end{pmatrix} = \mathrm{e}^{t_0}k_\psi\begin{pmatrix}1\\0\end{pmatrix},$$

which gives

$$\mathrm{e}^{t_0} = \left\|\begin{pmatrix}\mathrm{e}^{-t} & \\ & \mathrm{e}^{t}\end{pmatrix}\begin{pmatrix}\cos\theta\\ \sin\theta\end{pmatrix}\right\| = \sqrt{\mathrm{e}^{-2t}\cos^2\theta + \mathrm{e}^{2t}\sin^2\theta}.$$

Therefore

$$\Xi(a_t) = \frac{4}{2\pi}\int_0^{\frac{\pi}{2}}\left(\mathrm{e}^{-2t}\cos^2\theta + \mathrm{e}^{2t}\sin^2\theta\right)^{-\frac{1}{2}}\,\mathrm{d}\theta.$$

Next notice that

$$\mathrm{e}^{-2t}\cos^2\theta + \mathrm{e}^{2t}\sin^2\theta \asymp \max\left\{\mathrm{e}^{-2t}\cos^2\theta, \mathrm{e}^{2t}\sin^2\theta\right\}.$$

Moreover, the maximum is $\mathrm{e}^{2t}\sin^2\theta$ unless θ is close to 0, specifically unless $\tan^2\theta < \mathrm{e}^{-4t}$. This gives

$$\begin{aligned}\Xi(a_t) &\asymp \int_0^{\arctan \mathrm{e}^{-2t}}\underbrace{\left(\mathrm{e}^{-2t}\cos^2\theta\right)^{-\frac{1}{2}}}_{\asymp \mathrm{e}^{t}}\,\mathrm{d}\theta + \int_{\arctan \mathrm{e}^{-2t}}^{\frac{\pi}{2}}\underbrace{\left(\mathrm{e}^{2t}\sin^2\theta\right)^{-\frac{1}{2}}}_{\asymp \mathrm{e}^{-t}\frac{1}{\theta}}\,\mathrm{d}\theta\\ &\asymp \mathrm{e}^{t}\arctan \mathrm{e}^{-2t} + \mathrm{e}^{-t}|\log\arctan \mathrm{e}^{-2t}|\\ &\asymp \mathrm{e}^{-t} + \mathrm{e}^{-t}t \ll_\varepsilon \mathrm{e}^{-t(1-\varepsilon)}\end{aligned}$$

for all $\varepsilon > 0$. Also recall that $\|a_t\|_{\mathrm{HS}} \asymp \mathrm{e}^{t}$ for $t \geqslant 0$, which proves the estimate for Ξ in the proposition.

For the last claim, we need the description of the Haar measure on $\mathrm{SL}_2(\mathbb{R})$ in the coordinates of the Cartan decomposition from (8.11). Using the coordinates given by $g = k_\theta a_t k_\psi$ with $\theta \in [0, 2\pi)$, $t \in [0, \infty)$, and $\psi \in [0, \pi)$, we have

$$\mathrm{d}m_G \propto \sinh 2t\,\mathrm{d}\theta\,\mathrm{d}t\,\mathrm{d}\psi$$

by (8.11). Therefore,

$$\begin{aligned}\int_G (\Xi(g))^{2+\varepsilon}\,\mathrm{d}m_G(g) &\asymp \int_0^\infty (\Xi(a_t))^{2+\varepsilon}\sinh 2t\,\mathrm{d}t\\ &\asymp \int_0^\infty \left(\mathrm{e}^{-t} + \mathrm{e}^{-t}t\right)^{2+\varepsilon}\underbrace{\sinh 2t}_{\ll \mathrm{e}^{2t}}\,\mathrm{d}t\\ &\ll_\varepsilon 1 + \int_1^\infty \mathrm{e}^{-\varepsilon t}t^{2+\varepsilon}\,\mathrm{d}t < \infty\end{aligned}$$

for any $\varepsilon > 0$. However, for $\varepsilon = 0$ the integral diverges. □

Proof of Theorem 8.32. Let $f_1, f_2 \in L^2(G)$ be K-eigenfunctions for the regular representation, and let $k_\theta \in K$. Then

$$\lambda_{k_\theta}(f_j) = \mathrm{e}^{2\pi \mathrm{i} n_j \theta} f_j,$$

and Proposition 8.36(4) imply that

$$\pi^0_{k_\theta}(\widetilde{f_j}) = \widetilde{\lambda_{k_\theta} f_j} = \widetilde{\mathrm{e}^{2\pi \mathrm{i} n_j \theta} f_j} = \widetilde{f_j}$$

or, in other words, that $\widetilde{f_j}$ is invariant under K. We know that $\pi^0|_K$ is equal to the regular representation on $\mathcal{H}_0 \cong L^2(K)$. Therefore, there exists up to scalar multiples only one K-invariant function in $\mathcal{H}_0 \cong L^2(K)$. Since $\|f_0\|_{\mathcal{H}_0} = 1$ (by our choice that $m_K(K) = 1$), it follows that

$$\widetilde{f_j} = \|f_j\| f_0$$

for $j = 1, 2$. Therefore, Proposition 8.36(5) gives

$$|\langle \lambda_g f_1, f_2 \rangle| \leqslant \langle \pi^0_g \|f_1\| f_0, \|f_2\| f_0 \rangle_{\mathcal{H}_0} = \Xi(g) \|f_1\| \|f_2\|.$$

The estimates for $\Xi(g)$ are given by Proposition 8.40. □

Exercise 8.41 (Decay of action-associated representation on $L^2(\mathbb{R}^2)$). Use Exercise 8.38 and the results above to prove the same decay estimates as in Theorem 8.32 for the action-associated representation of $\mathrm{SL}_2(\mathbb{R})$ on $L^2(\mathbb{R}^2)$.

8.5.5 Tempered Decay for $\mathrm{SL}_2(\mathbb{R})$

We now present the companion result to Theorem 8.5, also due to Cowling, Haagerup and Howe [19], for this case. For a unitary representation π of a group K, we say that $v \in \mathcal{H}_\pi$ is K-finite if $\dim\langle \pi(K)v \rangle < \infty$. Below, we will again set

$$K = \mathrm{SO}_2(\mathbb{R}) < \mathrm{SL}_2(\mathbb{R}).$$

Theorem 8.42 (Tempered representations of $\mathrm{SL}_2(\mathbb{R})$). *For any unitary representation π of $\mathrm{SL}_2(\mathbb{R})$ the following are equivalent:*

(1) *π is tempered (that is, $\pi \prec \lambda$);*
(2) *for any two K-finite vectors (or for any K-eigenvalues) $v, w \in \mathcal{H}_\pi$ we have*

$$\left|\langle \pi_g v, w \rangle\right| \leqslant \left(\dim\langle \pi(K)v \rangle\right)^{\frac{1}{2}} \left(\dim\langle \pi(K)w \rangle\right)^{\frac{1}{2}} \|v\| \|w\| \Xi(g);$$

(3) *for any two C^1-smooth vectors $v, w \in \mathcal{H}_\pi$ we have*

$$\left|\langle \pi_g v, w \rangle\right| \ll_\varepsilon \mathcal{S}(v)\, \mathcal{S}(w) \|g\|_{\mathrm{HS}}^{-1+\varepsilon},$$

where $\mathcal{S}(\cdot)$ denotes a degree-one Sobolev norm, and the implied constant depends on ε but not on the representation π; and

(4) *π is almost square integrable (that is, on a dense set of vectors the matrix coefficients belong to $L^{2+\varepsilon}$ for all $\varepsilon > 0$).*

PROOF (OF ALL BUT ONE IMPLICATION). The fact that (2) restricted to K-eigenvectors implies (3) follows from Proposition 7.29 and the estimate concerning the Harish-Chandra spherical function in Proposition 8.40.

Assume now that (3) holds. Then (8.11) shows (as in the last step of the proof of Proposition 8.40) that the matrix coefficients of smooth vectors of $\mathcal{H}_\pi$ belong to $L^{2+\varepsilon}(G)$ for all $\varepsilon > 0$. This implies (4), since the smooth vectors are dense by Proposition 7.8.

Next note that Theorem 8.5 shows that (4) implies (1). Hence it remains to show that (1) implies (2). □

For the remaining implication, we are going to use the following lemma.

Lemma 8.43. *Let π and ρ be unitary representations of $G = \mathrm{SL}_2(\mathbb{R})$. Suppose that $\pi \prec \rho$ and that $v, w \in \mathcal{H}_\pi$ are K-eigenvectors. Then, for any bi-K-invariant compact subset $\Omega = K\Omega K \subseteq G$ and $\varepsilon > 0$, there exist K-eigenvectors $v_1, \dots, v_J, w_1, \dots, w_J$ in $\mathcal{H}_\rho$ such that*

$$\sum_{j=1}^{J} \|v_j\|\|w_j\| \leqslant \|v\|\|w\|$$

and

$$\left\| \varphi^\pi_{v,w} - \sum_{j=1}^{J} \varphi^\rho_{v_j,w_j} \right\|_{\Omega,\infty} < \varepsilon.$$

PROOF. Let $v, w \in \mathcal{H}_\pi$. By our assumption and condition ($\prec_{\mathrm{mc}}$) in Theorem 6.31, there exist some $J \geqslant 1$ and vectors $\widetilde{v}_j, \widetilde{w}_j \in \mathcal{H}_\rho$ for $j = 1, \dots, J$ such that

$$\sum_{j=1}^{J} \|\widetilde{v}_j\|\|\widetilde{w}_j\| \leqslant \|v\|\|w\| \tag{8.29}$$

and

$$\left\| \varphi^\pi_{v,w} - \sum_{j=1}^{J} \varphi^\rho_{\widetilde{v}_j,\widetilde{w}_j} \right\|_{\Omega,\infty} < \varepsilon. \tag{8.30}$$

Assume now that $v, w \in \mathcal{H}_\pi$ have K-weight $m, n \in \mathbb{Z}$, so that

$$\pi_{k_\theta} v = \mathrm{e}^{\mathrm{i}m\theta} v = \chi_m(k_\theta) v$$

and

$$\pi_{k_\theta} w = \mathrm{e}^{\mathrm{i}n\theta} w = \chi_n(k_\theta) w$$

for all $k_\theta \in K$. We wish to use this to improve (8.30) to an approximation that only uses K-eigenvectors. For this, we define the projected vectors

$$v_j = (\rho|_K)_*(\overline{\chi_m})\widetilde{v}_j$$

and

$$w_j = (\rho|_K)_*(\overline{\chi_n})\widetilde{w}_j$$

for $j = 1, \ldots, J$, so that

$$\begin{aligned}\rho_{k_\psi} v_j &= \int_K \overline{\chi_m}(k_\theta)\rho_{k_\psi k_\theta}\widetilde{v}_j \,\mathrm{d}m_K(k_\theta)\\ &= \chi_m(k_\psi)\int_K \overline{\chi_m}(k_\psi k_\theta)\rho_{k_\psi k_\theta}\widetilde{v}_j \,\mathrm{d}m_K(k_\theta) = \chi_m(k_\psi)v_j,\end{aligned}$$

and similarly

$$\rho_{k_\psi} w_j = \chi_n(k_\psi)w_j$$

for all $k_\psi \in K$. Moreover, we have

$$\begin{aligned}\|v_j\| &\leqslant \|\widetilde{v}_j\|,\\ \|w_j\| &\leqslant \|\widetilde{w}_j\|.\end{aligned} \tag{8.31}$$

Using the definition of v_j and w_j and the estimate (8.30), we obtain that

$$\begin{aligned}\sum_{j=1}^J \langle \rho_g v_j, w_j\rangle &= \int_K\int_K \overline{\chi_m}(k_\theta)\chi_n(k_\psi)\sum_{j=1}^J \underbrace{\langle \rho_{gk_\theta}\widetilde{v}_j, \rho_{k_\psi}\widetilde{w}_j\rangle}_{=\langle \rho_{k_\psi^{-1}gk_\theta}\widetilde{v}_j,\widetilde{w}_j\rangle} \mathrm{d}m_K(k_\theta)\mathrm{d}m_K(k_\psi)\\ &= \int_K\int_K \overline{\chi_m}(k_\theta)\chi_n(k_\psi)\langle \pi_{gk_\theta}v, \pi_{k_\psi}w\rangle \,\mathrm{d}m_K(k_\theta)\mathrm{d}m_K(k_\psi) + \mathrm{O}(\varepsilon)\\ &= \langle \pi_g v, w\rangle + \mathrm{O}(\varepsilon)\end{aligned}$$

for $g \in \Omega = K\Omega K$, where we used the fact that v has K-weight m and w has K-weight n. Together with (8.29) and (8.31), this gives the lemma. □

PROOF THAT (1) $\Longrightarrow$ (2) IN THEOREM 8.42. We suppose that $\pi \prec \lambda$ and let $v, w \in \mathcal{H}_\pi$ be K-eigenvectors. Let $\Omega = K\Omega K \subseteq G$ be a bi-K-invariant subset, and $\varepsilon > 0$. Applying Theorem 6.31 and Lemma 8.43, we find K-eigenfunctions

$$f_{v,1}, \ldots, f_{v,J}, f_{w,1}, \ldots, f_{w,J}$$

in $L^2(G)$ satisfying

$$\sum_{j=1}^J \|f_{v,j}\|\|f_{w,j}\| \leqslant \|v\|\|w\|$$

and

$$\left\| \varphi_{v,w}^{\pi} - \sum_{j=1}^{J} \varphi_{f_{v,j}, f_{w,j}}^{\lambda} \right\|_{\Omega,\infty} < \varepsilon.$$

Next we apply the decay properties of K-eigenfunctions for the regular representation in Theorem 8.32 and obtain

$$\begin{aligned}
\left|\left\langle \varphi_{v,w}^{\pi}(g)\right\rangle\right| &\leqslant \sum_{j=1}^{J} \left|\left\langle \varphi_{f_{v,j}, f_{w,j}}^{\lambda}(g)\right\rangle\right| + \varepsilon \\
&\leqslant \sum_{j=1}^{J} \|f_{v,j}\| \|f_{w,j}\| \Xi(g) + \varepsilon \leqslant \|v\| \|w\| \Xi(g) + \varepsilon
\end{aligned}$$

for all $g \in \Omega$. Varying the bi-K-invariant compact subset $\Omega \subseteq G$ and $\varepsilon > 0$, we deduce that

$$\left|\left\langle \pi_g v, w \right\rangle\right| \leqslant \|v\| \|w\| \Xi(g) \tag{8.32}$$

for all K-eigenvectors $v, w \in \mathcal{H}_\pi$ and $g \in G$, which is property (2) in the special case of K-eigenvectors.

Suppose now that $v, w \in \mathcal{H}_\pi$ are K-finite. Applying the projections

$$(\pi|_K)_*(\overline{\chi}_m),$$

we can write v as a sum of K-eigenvectors $v_m \in \overline{\langle \pi(K) v \rangle}$ for $m \in \mathbb{Z}$. Since v is K-finite, it follows that

$$v = \sum_{m \in M} v_m,$$

and similarly

$$w = \sum_{n \in N} w_n$$

with $M, N \subseteq \mathbb{Z}$ satisfying $|M| = \dim(\langle \pi(K) v \rangle)$ and $|N| = \dim(\langle \pi(K) w \rangle)$. Applying (8.32) for each $m \in M$ and $n \in N$, we obtain

$$\left|\left\langle \pi_g v, w \right\rangle\right| \leqslant \sum_{\substack{m \in M \\ n \in N}} \left|\left\langle \pi_g v_m, w_n \right\rangle\right| \leqslant \Bigl(\sum_{m \in M} \|v_m\|\Bigr) \Bigl(\sum_{n \in N} \|w_n\|\Bigr) \Xi(g).$$

On the other hand,

$$\sum_{m \in M} \|v_m\| \leqslant |M|^{\frac{1}{2}} \left(\sum_{m \in M} \|v_m\|^2 \right)^{\frac{1}{2}} = \left(\dim \langle \pi(K) v \rangle\right)^{\frac{1}{2}} \|v\|$$

by Cauchy–Schwarz and orthogonality of the decomposition of v into its different K-eigenvectors. Putting this together gives

$$\left|\left\langle \pi_g v, w \right\rangle\right| \leqslant \left(\dim \langle \pi(K) v \rangle\right)^{\frac{1}{2}} \left(\dim \langle \pi(K) w \rangle\right)^{\frac{1}{2}} \|v\| \|w\| \Xi(g)$$

for all $g \in G$ and K-finite vectors $v, w \in \mathcal{H}_\pi$, as required. □

8.6 (Non-Uniform) Integrability and Decay Exponents*

In the Howe–Moore theorem (Theorem 1.90), we have seen that matrix coefficients for unitary representations of $\mathrm{SL}_2(\mathbb{R})$ without fixed vectors decay at infinity. In Section 7.4 we went much further for $\mathrm{SL}_3(\mathbb{R})$, and showed the existence of a uniform decay exponent. This raises the question of whether $\mathrm{SL}_2(\mathbb{R})$ also has a uniform decay exponent. We explain in this section using geometry and dynamics why there is no uniform decay exponent as in Definition 7.26 for $\mathrm{SL}_2(\mathbb{R})$. As this will become much clearer in our detailed study of $\widehat{\mathrm{SL}_2(\mathbb{R})}$ in Chapter 9 we will leave the details as exercises. However, we begin by defining and studying the notion of integrability exponents.

8.6.1 Integrability Exponents

The following definition is closely related to the notion of almost square integrability (Definition 8.4) and effective decay of matrix coefficients (Definition 7.25).

Definition 8.44 (Integrability exponents). Let G be a locally compact σ-compact metric group and π a unitary representation of G. For $p > 0$ we say that π is *p-integrable* or that p is an *integrability exponent* of π if there exists a dense set of vectors $\mathcal{V} \subseteq \left(\mathcal{H}_\pi^G\right)^\perp$ such that the matrix coefficients

$$G \ni g \longmapsto \varphi_{v,w}(g) = \left\langle \pi_g v, w \right\rangle$$

lie in $L^p(G)$ for $v, w \in \mathcal{V}$.

Essential Exercise 8.45. Let G be a locally compact σ-compact metric group and π a cyclic representation with generator $v_0 \in \mathcal{H}_\pi$. If $\varphi_{v_0}^\pi \in L^p(G)$ for some $p > 0$, then p is an integrability exponent of π.

As a first preliminary relationship between p_π and the parameter κ_π in Definition 7.25, we prove the following.

Lemma 8.46 (Decay implies integrability). *For a unitary representation* π *of* $\mathrm{SL}_2(\mathbb{R})$ *with decay exponent* $\kappa > 0$ *any* $p > \frac{2}{\kappa}$ *is an integrability exponent.*

PROOF. For simplicity of notation we assume that π has no fixed vectors. By assumption

$$|\varphi_{v,w}(g)| = |\langle \pi_g v, w \rangle| \ll \|g\|_{\mathrm{HS}}^{-\kappa}\, \mathcal{S}(v)\, \mathcal{S}(w)$$

for all C^r-smooth $v, w \in \mathcal{H}_\pi$ and some $r > 0$. Using (8.11) we now calculate

$$\begin{aligned}\|\varphi_{v,w}\|_p^p &= \int_{\mathrm{SL}_2(\mathbb{R})} |\varphi_{v,w}(g)|^p \,\mathrm{d}m(g)\\ &\ll_{v,w} \int_{KAK} \|k_\varphi a_t k_\psi\|_{\mathrm{HS}}^{-\kappa p} \sinh 2t \,\mathrm{d}\varphi\,\mathrm{d}t\,\mathrm{d}\psi\\ &\ll_{v,w} \int_0^\infty \mathrm{e}^{-\kappa p t}\mathrm{e}^{2t}\,\mathrm{d}t < \infty.\end{aligned}$$

□

The results of Chapter 9 will lead to a much more satisfactory converse statement for any unitary representation of $\mathrm{SL}_2(\mathbb{R})$ (see Section 9.7). For now we make the following weaker claims.

Essential Exercise 8.47 (Integrability implies decay). Let π be a unitary representation of G with integrability exponent $p \in (0,\infty)$, and let $d \in \mathbb{N}$ satisfy $\frac{p}{d} \leqslant 2$.
(a) If $\mathcal{H}_\pi$ has no invariant vectors, show that $\pi^{\otimes d}$ is tempered.
(b) If $G = \mathrm{SL}_2(\mathbb{R})$, show that π has exponential decay of matrix coefficients with decay exponent $\kappa = \frac{1}{d} - \varepsilon$ for all $\varepsilon > 0$. Moreover, show that it suffices to consider degree-one Sobolev norms, and the implicit constant in Definition 7.25 only depends on d and ε.

8.6.2 Dynamics on Quotients of $\mathbf{PSL_2(\mathbb{R})}$

Our construction below of unitary representations with non-uniform decay exponents is quite geometric. As explained in Section 8.3, the group

$$\mathrm{PSL}_2(\mathbb{R}) = \mathrm{SL}_2(\mathbb{R})/\{\pm I\}$$

acts transitively via Möbius transformations on the hyperbolic plane $\mathbb{H}$. Moreover, after equipping $\mathbb{H}$ with the hyperbolic Riemannian metric, this extends to a simple transitive action of $\mathrm{PSL}_2(\mathbb{R})$ on the unit tangent bundle $\mathrm{T}^1\mathbb{H}$ of the hyperbolic surface. Choosing a reference vector in $\mathrm{T}^1\mathbb{H}$ (usually the upward pointing vector at the point $\mathrm{i} \in \mathbb{H}$) this gives an identification between $\mathrm{PSL}_2(\mathbb{R})$ and $\mathrm{T}^1\mathbb{H}$.

For discrete subgroups $\Gamma < \mathrm{PSL}_2(\mathbb{R})$ (the so-called *Fuchsian groups*[(18)]), we may use the discussions above to identify the quotient $X = \Gamma\backslash\mathrm{PSL}_2(\mathbb{R})$ with the unit tangent bundle of the hyperbolic surface[†] $M = \Gamma\backslash\mathbb{H}$. Using a measurable fundamental domain $F_{\Gamma,\mathbb{H}}$ for the action of Γ on $\mathbb{H}$, one can define a measurable fundamental domain F_Γ in $\mathrm{PSL}_2(\mathbb{R})$ for the action of Γ (so that, in particular, $|\Gamma g \cap F_\Gamma| = 1$ for all $g \in \mathrm{PSL}_2(\mathbb{R})$). The discrete subgroup Γ is called a *lattice* if $m(F_\Gamma) < \infty$ (equivalently, if $\mathrm{vol}(F_{\Gamma,\mathbb{H}}) < \infty$). In this case one

† If M is a manifold, then this is the unit tangent bundle in the usual sense. If Γ has torsion elements, then M is an orbifold with non-smooth points, and this should be viewed simply as a definition of the unit tangent bundle of M.

can use the Haar measure m restricted to F_Γ to induce a measure m_X on X, which will be finite and invariant under the action of $g \in G$ on $x = \Gamma h \in X$ defined by right multiplication,

$$g \cdot x = g \cdot (\Gamma h) = \Gamma h g^{-1}.$$

We will call m_X the Haar measure on X (see Exercise 8.48). Moreover, this gives rise to the action-associated representation defined by

$$(\pi_g^X f)(x) = f(xg)$$

for $g \in \mathrm{PSL}_2(\mathbb{R})$ (or $g \in \mathrm{SL}_2(\mathbb{R})$), $f \in L^2_{m_X}(X)$, and $x \in X$. We refer any reader not familiar with these constructions to [24, Ch. 9].

Exercise 8.48 (Haar measure on X). Let $\Gamma < \mathrm{PSL}_2(\mathbb{R})$ be discrete, let

$$\pi\colon G \ni g \longmapsto \Gamma g \in X = \Gamma \backslash \mathrm{PSL}_2(\mathbb{R})$$

denote the canonical quotient map, and let F_Γ be a measurable fundamental domain for Γ. Show that $m_X = \pi_*(m|_{F_\Gamma})$ is a G-invariant measure independent of the choice of the fundamental domain F_Γ. Prove that m_X is ergodic for the action of $\mathrm{PSL}_2(\mathbb{R})$.

8.6.3 Non-uniform Decay for $\mathrm{SL}_2(\mathbb{R})$

Proposition 8.49 (A free lattice in $\mathrm{PSL}_2(\mathbb{R})$). *We let*

$$\alpha = \begin{pmatrix} 1 & 2 \\ 0 & 1 \end{pmatrix}, \ \beta = \begin{pmatrix} 1 & 0 \\ 2 & 1 \end{pmatrix}$$

considered in $\mathrm{PSL}_2(\mathbb{R})$. The Sanov[(19)] *subgroup $\Gamma = \langle \alpha, \beta \rangle$ generated by α and β is a free group on two generators, and is a lattice in $\mathrm{PSL}_2(\mathbb{R})$.*

As explained in Section 8.6.2, the lattice statement can be checked in the hyperbolic plane by exhibiting a fundamental domain of finite hyperbolic area for the action on $\mathbb{H}$. We will ignore the null sets arising from the boundaries of hyperbolic polygons in our discussions and split the proof of the proposition into a series of exercises.

Essential Exercise 8.50. Verify that

$$\left.\begin{aligned} \alpha \cdot (\mathbb{H} \smallsetminus D_\alpha^-) &\subseteq D_\alpha^+ \\ \alpha^{-1} \cdot (\mathbb{H} \smallsetminus D_\alpha^+) &\subseteq D_\alpha^- \end{aligned}\right\} \tag{8.33}$$

and

$$\left.\begin{aligned} \beta \cdot (\mathbb{H} \smallsetminus D_\beta^-) &\subseteq D_\beta^+, \\ \beta^{-1} \cdot (\mathbb{H} \smallsetminus D_\beta^+) &\subseteq D_\beta^- \end{aligned}\right\} \tag{8.34}$$

in the notation from Figure 8.1 (and, as promised, ignoring null sets).

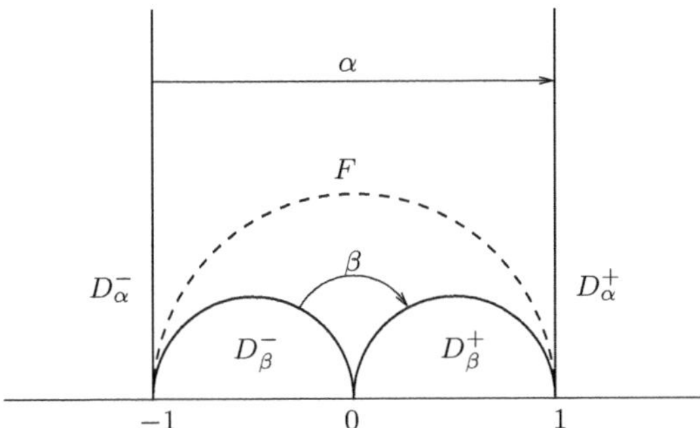

Fig. 8.1: A fundamental domain F for Γ in $\mathrm{SL}_2(\mathbb{R})$, with the unit circle shown in a dashed line. The sets D_α^+, D_α^-, D_β^+, and D_β^- together with F form a partition of $\mathbb{H}$, which can be used to show that α and β generate a free group.

Essential Exercise 8.51. The subset $F \subseteq \mathbb{H}$ and the pair of relations (8.33) and (8.34) allow the following simple 'ping-pong' argument to be made. Let

$$\gamma = \alpha^{m_1}\beta^{n_1}\alpha^{m_2}\beta^{n_2}\cdots\alpha^{m_k}\beta^{n_k}$$

be a reduced word in the generators α and β, where $n_1, n_2, \dots, n_{k-1}, m_2, \dots, m_k$ are in $\mathbb{N}$ but m_1, n_k are in $\mathbb{N}_0$. Show that

- $\gamma \cdot F \subseteq D_\alpha^+$ if and only if $m_1 > 0$;
- $\gamma \cdot F \subseteq D_\alpha^-$ if and only if $m_1 < 0$;
- $\gamma \cdot F \subseteq D_\beta^+$ if and only if $m_1 = 0$ and $n_1 > 0$; and
- $\gamma \cdot F \subseteq D_\beta^-$ if and only if $m_1 = 0$ and $n_1 < 0$.

Conclude that α and β generate a free group.

Essential Exercise 8.52. Show that F is a fundamental domain.

Essential Exercise 8.53. Show that Γ is a lattice.

Since Γ is a free group, we may define a homomorphism

$$\phi_n \colon \Gamma \longrightarrow \Gamma/[\Gamma, \Gamma] \cong \mathbb{Z}^2 \longrightarrow \mathbb{Z}/n\mathbb{Z}$$

for every $n \in \mathbb{N}$, by sending $\alpha \in \Gamma$ to the generator $1 \in \mathbb{Z}/n\mathbb{Z}$ and $\beta \in \Gamma$ to zero. Clearly the kernel $\Lambda_n = \ker \phi_n < \Gamma$ has index n in Γ, and so is again a lattice in $\mathrm{PSL}_2(\mathbb{R})$. We note that $X_n = \Lambda_n \backslash \mathrm{PSL}_2(\mathbb{R})$ is called a $\mathbb{Z}/n\mathbb{Z}$-cover of $\Gamma \backslash \mathrm{PSL}_2(\mathbb{R})$. In fact, the fundamental domain $F_n = F_{\Lambda_n}$ of Λ_n consists of n copies of the fundamental domain F_Γ of Γ that can be chosen to be adjacent, as in Figure 8.2. Note that $\alpha^n \in \Lambda_n$ will identify the left boundary of F_n with the right boundary of F_n, so that, roughly speaking and from far away, the geometries of $\Lambda_n \backslash \mathbb{H}$ or of $X_n = \Lambda_n \backslash \mathrm{PSL}_2(\mathbb{R})$ look like that of a circle. This can be used to define, for each $n \in \mathbb{N}$, the unitary action-associated representation π^{X_n} on $L^2(X_n)$ and as a result to prove the following result.

Proposition 8.54 (Non-uniformity). *The decay exponents for the unitary representation π^{X_n} of $\mathrm{SL}_2(\mathbb{R})$ on $L^2(\Lambda_n\backslash\mathrm{PSL}_2(\mathbb{R}))$ go to zero as $n\to\infty$. Equivalently, the integrability exponents of π^{X_n} go to infinity as $n\to\infty$.*

The idea of the proof relies on the 'circle-like shape' of $\Lambda_n\backslash\mathrm{PSL}_2(\mathbb{R})$ mentioned above. This allows us to find two subsets K_1 and K_2 that are in measure almost $\frac{1}{4}$ of the whole space but have large distance from each other as in Figure 8.2. However, if the decay exponent of π^{X_n} is not close to zero, then it should be possible to move from K_1 to K_2 using a uniformly bounded element of $\mathrm{SL}_2(\mathbb{R})$.

We will split the formal argument into exercises. For this we suppose for the purposes of a contradiction that there exists an integrability exponent p_n of π^{X_n} satisfying the bound $p_n \leqslant 2d$. By Exercise 8.47 it follows that π^{X_n} has decay exponent

$$\kappa = \frac{1}{d} - \frac{1}{2d} = \frac{1}{2d},$$

and that it suffices to consider degree-one Sobolev norms and a uniform (that is, independent of n) implicit constant.

We define the sets K_1 and K_2 of $\Lambda_n\backslash\mathrm{PSL}_2(\mathbb{R})$ as in Figure 8.2. Formally, let $K_\Gamma\subseteq F_\Gamma$ be a fixed compact subset with non-empty interior, $\ell=\lceil\frac{n}{4}\rceil$, and define

$$\left.\begin{aligned} K_1 &= \Lambda_n \bigcup_{j=0}^{\ell-1}\alpha^j K_\Gamma,\\ K_2 &= \Lambda_n \bigcup_{j=0}^{\ell-1}\alpha^{2\ell+j} K_\Gamma. \end{aligned}\right\}\tag{8.35}$$

We note that

$$v_n = m_{X_n}(X_n) = \mathrm{vol}(F_n) = n\mathrm{vol}(F_\Gamma).$$

Let $c_0=\frac{\mathrm{vol}(K_\Gamma)}{\mathrm{vol}(F_\Gamma)}>0$ be the ratio of the volume of the compact region K_Γ within one copy of F_Γ and $\mathrm{vol}(F_\Gamma)$, which is independent of n. The construction of K_1 and K_2 (using $\ell=\lceil\frac{n}{4}\rceil$) then implies that

$$m_{X_n}(K_1) = m_{X_n}(K_2) = \ell c_0\mathrm{vol}(F_\Gamma) \geqslant \frac{c_0}{4}v_n.$$

Essential Exercise 8.55. Prove that there exists a constant C independent of n with the property that for every $n\in\mathbb{N}$ with $p_n<2d$ there exists some element $g\in\mathrm{SL}_2(\mathbb{R})$ with $\|g\|_{\mathrm{HS}}\leqslant C$ and $K_1g^{-1}\cap K_2\neq\emptyset$.

Essential Exercise 8.56. Use the definition of K_j for $j=1,2$ and g as in Exercise 8.55 to find $\lambda\in\Lambda_n$ and $j_1,j_2\in\{0,1,\ldots,\ell-1\}$ with

$$\gamma = \alpha^{-2\ell-j_2}\lambda\alpha^{j_1}\in\Gamma\cap(K_\Gamma g K_\Gamma^{-1}).$$

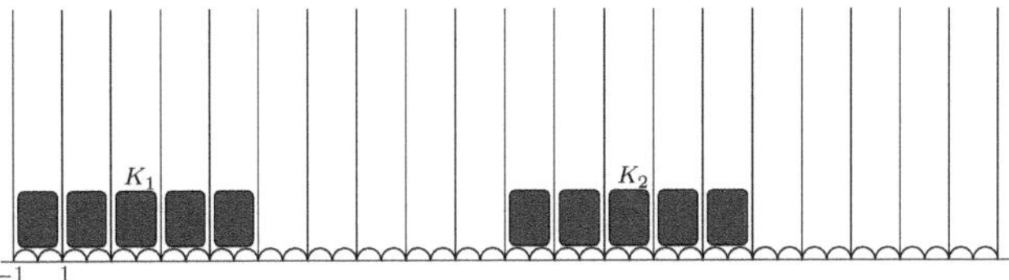

Fig. 8.2: We let $\ell = \lceil \frac{n}{4} \rceil$, define the compact subset K_1 using the first ℓ copies of F in F_n, and the compact subset K_2 using the third ℓ copies of F in F_n. This forces K_1 and K_2 to have a large distance between them once n is large.

Essential Exercise 8.57. Conclude that there are only finitely many n for which the integrability exponent p_n is less than $2d$.

We note that the action-associated representation on a quotient

$$X = \Gamma \backslash \mathrm{PSL}_2(\mathbb{R})$$

by a lattice $\Gamma < \mathrm{PSL}_2(\mathbb{R})$ always has a positive decay exponent, equivalently a finite integrability exponent, or spectral gap. Moreover for the so-called 'congruence lattices' one even has a uniform decay exponent. Our discussions show that 'arithmetic lattices' do not have uniform decay exponents.

8.7 A Special Case of the Kunze–Stein Phenomenon*

We again let $G = \mathrm{SL}_2(\mathbb{R})$, and will explain a special case of a surprising result that was first discovered by Kunze and Stein [60] and extended to semi-simple groups by Cowling [18] (after other results in the same spirit; we refer to Cowling's work for references).

Recall that any function $\psi \in L^1(G)$ acts as a convolution operator $\pi_*(\psi)$ on $\mathcal{H}_\pi$ for any unitary representation π, so that we have the trivial operator bound

$$\|\pi_*(\psi)\|_{\mathrm{op}} \leqslant \|\psi\|_1.$$

In the context of ergodic theory, one often sets $\psi = \frac{1}{m(S)} \mathbb{1}_S$ for some measurable set $S \subseteq G$ of large finite measure. Unfortunately, the trivial bound is, in this context, quite useless. However, the operator bound in the following theorem and its corollary can be powerful in ergodic theory and its applications to number theory (see the monograph of Gorodnik and Nevo [39] for more on this).

Theorem 8.58 (Spherical Kunze–Stein for $\mathrm{SL}_2(\mathbb{R})$). *Let $G = \mathrm{SL}_2(\mathbb{R})$ and $K = \mathrm{SO}_2(\mathbb{R}) < G$. Let $\varepsilon > 0$ and let $\psi \in L^1(G) \cap L^{2-\varepsilon}(G)$ be bi-K-invariant in the sense that $\psi(kgk') = \psi(g)$ for all $k, k' \in K$ and $g \in G$. Then we have*

$$\|\pi_*(\psi)\|_{\mathrm{op}} \ll_\varepsilon \|\psi\|_{2-\varepsilon}$$

for any tempered representation π of G.

We note that the restriction to bi-K-invariant functions is not needed, but it does dramatically simplify the proof of the theorem. The general case is a corollary of the Plancherel formula for $\mathrm{SL}_2(\mathbb{R})$, which we will not discuss in this volume.

PROOF OF THEOREM 8.58. Let π be a unitary representation of G, and let ψ be a function in $L^1(G)$ satisfying $\psi(kgk') = \psi(g)$ for all $k, k' \in K$ and $g \in G$. Let $P_K \colon \mathcal{H}_\pi \to \mathcal{H}_\pi^K$ denote the orthogonal projection onto the subspace of K-invariant vectors in $\mathcal{H}_\pi$. Note that $P_K = (\pi|_K)_*(\mathbb{1})$.

We claim that our assumption on ψ implies that $\pi_*(\psi) = P_K \pi_*(\psi) P_K$. Indeed, using abbreviated and slightly informal notation, we have

$$\begin{aligned}
\pi_*(\psi) &= \int_G \underbrace{\psi(g)}_{=\psi(k_1^{-1} g k_2^{-1})} \pi_g \,\mathrm{d}m(g) \\
&= \int_G \psi(h) \pi_{k_1} \pi_h \pi_{k_2} \,\mathrm{d}m(h) \\
&= \pi_{k_1} \circ \pi_*(\psi) \circ \pi_{k_2} \\
&= (\pi|_K)_* (\mathbb{1}) \circ \pi_*(\psi) \circ (\pi|_K)_* (\mathbb{1}) = P_K \pi_*(\psi) P_K
\end{aligned}$$

by using the substitution $h = k_1^{-1} g k_2^{-1}$, and then integrating with respect to the normalized Haar measure over $k_1, k_2 \in K$.

We now specialize to the regular representation. For $f_1, f_2 \in L^2(G)$ we then obtain

$$\begin{aligned}
\big|\langle \lambda_*(\psi) f_1, f_2 \rangle\big| &= \big|\langle \lambda_*(\psi) P_K f_1, P_K f_2 \rangle\big| \\
&= \left| \int_G \psi(g) \langle \lambda_g P_K f_1, P_K f_2 \rangle \,\mathrm{d}m(g) \right| \\
&\leqslant \int_G |\psi(g)| \underbrace{\big|\langle \lambda_g P_K f_1, P_K f_2 \rangle\big|}_{\leqslant \langle \pi_g^0 \overline{P_K f_1}, \overline{P_K f_2} \rangle} \,\mathrm{d}m(g) \\
&\leqslant \int_G |\psi(g)| \left\langle \pi_g^0 \overline{P_K f_1}, \overline{P_K f_2} \right\rangle \mathrm{d}m(g)
\end{aligned}$$

by the Hertz domination principle in Proposition 8.36. Using the other properties in Proposition 8.36, more can be said. In fact, since $P_K f_j$ is K-invariant and $\overline{\lambda_g f} = \pi^0(g) \overline{f}$ for all $g \in G$ and $f \in L^2(G)$, we see that $\overline{P_K f_j}$ is K-invariant also. However, in $\mathcal{H}_0 \cong L^2(K)$, there is only one K-invariant function up to scalars. It follows that

$$\overline{P_K f_j} = \| \overline{P_K f_j} \|_{\mathcal{H}_0} f_0$$

with

$$\| \overline{P_K f_j} \|_{\mathcal{H}_0} = \| \overline{P_K f_j} \|_{L^2(K)} \leqslant \|f_j\|_2$$

for $j = 1, 2$.

Recalling the definition of the Harish-Chandra spherical function

$$\Xi(g) = \langle \pi_g^0 f_0, f_0 \rangle_{\mathcal{H}_0},$$

and putting this into the above estimate, we obtain

$$\begin{aligned} |\langle \lambda_*(\psi) f_1, f_2 \rangle| &\leqslant \int_G |\psi(g)| \Xi(g) \, dm(g) \|f_1\|_2 \|f_2\|_2 \\ &\leqslant \|\psi\|_{2-\varepsilon} \|\Xi\|_q \|f_1\|_2 \|f_2\|_2 \end{aligned}$$

by assuming $\psi \in L^{2-\varepsilon}(G)$ for some $\varepsilon > 0$ and letting $q \in (2, \infty)$ be the Hölder conjugate of $p = 2 - \varepsilon$. This shows that

$$\|\lambda_*(\psi)\|_{\mathrm{op}} \leqslant \underbrace{\|\Xi\|_q}_{<\infty} \|\psi\|_{2-\varepsilon}$$

by the integrability properties of Ξ in Proposition 8.40.

Finally, let π be a tempered representation. Then we also have

$$\|\pi_*(\psi)\| \leqslant \|\lambda_*(\psi)\|_{\mathrm{op}} \leqslant \|\Xi\|_q \|\psi\|_{2-\varepsilon}$$

by the characterization ($\prec_{\mathrm{op}}$) of weak containment in Theorem 6.31. □

8.7.1 Kunze–Stein Phenomenon as an Ergodic Theorem

We now suppose that $G = \mathrm{SL}_2(\mathbb{R})$ acts continuously on X, preserving a locally finite measure μ. We write π for the induced action-associated representation on $L^2_\mu(X)$ (see Proposition 1.3). Recall that in this case we have

$$\pi_*(\psi) f(x) = \int_G \psi(g) f(g^{-1} \cdot x) \, dm(g)$$

for almost every $x \in X$. Moreover, if $\psi = \frac{1}{m(S)} \mathbb{1}_S$ for a measurable set $S \subseteq G$ with finite measure, then

$$\pi_*(\psi) f(x) = \frac{1}{m(S)} \int_S f(g^{-1} \cdot x) \, dm(g)$$

is called an *ergodic average*.

Corollary 8.59 (Kunze–Stein in ergodic theory). *Let π be an action-associated unitary representation of $G = \mathrm{SL}_2(\mathbb{R})$ as above. Assume in addition that π has integrability exponent $p_\pi < \infty$. Let $\ell \in \mathbb{N}$ satisfy $\frac{p_\pi}{2\ell} < 2$. Then we have*

$$\left\| \pi_*(\psi)|_{(\mathcal{H}_\pi^G)^\perp} \right\|_{\mathrm{op}} \ll_\varepsilon \|\psi\|_{2-\varepsilon}^{\frac{1}{2\ell}}.$$

for any $\varepsilon > 0$ *and* $\psi \in L^1(G) \cap L^{2-\varepsilon}(G)$ *with*

- $\psi \geqslant 0$;
- $\int \psi \, \mathrm{d}m = 1$;
- $\psi(kgk') = \psi(g)$ *for all* $k, k' \in K$ *and* $g \in G$.

Moreover, for a measurable set $S = KSK \subseteq G$ *with* $m(S) < \infty$, *we have*

$$\left\| \frac{1}{m(S)} \int_S f(g^{-1} \mathbin{\bullet} x) \, \mathrm{d}m(g) - I_f \right\|_2 \ll_\varepsilon m(S)^{-\frac{1}{4\ell}+\varepsilon} \|f\|_2$$

for all $\varepsilon > 0$. *Here* $I_f \in L^2_\mu(X)$ *is given by*

$$I_f = \begin{cases} \left(\int f \, \mathrm{d}\mu\right) \mathbb{1}_X & \text{if } \mu(X) = 1, \\ 0 & \text{if } \mu(X) = \infty. \end{cases}$$

PROOF. We let $d = 2\ell$. Proposition 8.47 shows that $\pi|^{\otimes d}_{(\mathcal{H}_\pi^G)^\perp}$ is tempered. Let $f_1, f_2 \in (\mathcal{H}_\pi^G)^\perp$ be real-valued. Then

$$\begin{aligned} |\langle \pi_*(\psi) f_1, f_2 \rangle|^d &= \left| \int_G \langle \pi_g f_1, f_2 \rangle \, \psi(g) \, \mathrm{d}m(g) \right|^d \\ &\leqslant \left| \int_G \underbrace{\langle \pi_g f_1, f_2 \rangle^d}_{= \langle \pi_g^{\otimes d} f_1^{\otimes d}, f_2^{\otimes d} \rangle} \psi(g) \, \mathrm{d}m(g) \right| \end{aligned}$$

by Jensen's inequality for the real-valued function $g \mapsto \langle \pi_g f_1, f_2 \rangle$, the convexity of the function $\mathbb{R} \ni t \mapsto t^d$ (for $d = 2\ell$ even), and the probability measure $\psi \, \mathrm{d}m$ on G. Moreover, Theorem 8.58 gives

$$\left| \langle \pi_*^{\otimes d}(\psi) f_1^{\otimes d}, f_2^{\otimes d} \rangle \right| \ll_p \|\psi\|_{2-\varepsilon} \|f_1^{\otimes d}\| \|f_2^{\otimes d}\|.$$

Taking the dth root, and combining the two estimates, we obtain

$$|\langle \pi_*(\psi) f_1, f_2 \rangle| \ll_p \|\psi\|_{2-\varepsilon}^{\frac{1}{d}} \|f_1\|_2 \|f_2\|_2.$$

Using linearity, this extends to complex-valued functions, and implies the first part of the corollary.

If $\mu(X) = \infty$ and the action is ergodic, then $\mathcal{H}_\pi^G = \{0\}$. We apply the above to $\psi = \frac{1}{m(S)} \mathbb{1}_S$. Since

$$\|\psi\|_{2-\varepsilon}^{\frac{1}{2\ell}} = \left(m(S)^{-(2-\varepsilon)} m(S) \right)^{\frac{1}{(2-\varepsilon)2\ell}} = m(S)^{\frac{-1+\varepsilon}{(2-\varepsilon)2\ell}}$$

and $\varepsilon > 0$ is arbitrary, this implies the infinite volume case. If $\mu(X) = 1$, then $\mathcal{H}_\pi^G = \mathbb{C}\mathbb{1}_X$ and we can apply the above to

$$f - \Big(\int f \,\mathrm{d}\mu\Big)\mathbb{1}_X \in \left(\mathcal{H}_\pi^G\right)^\perp.$$

□

We refer to the monograph of Gorodnik and Nevo [39] for a more thorough discussion of the applications of the Kunze–Stein phenomenon in ergodic theory, and its connection to number theory.

Exercise 8.60 (Counting lattice points). Suppose $\Gamma < \mathrm{SL}_2(\mathbb{R})$ is a lattice, and let X be the homogeneous space $\Gamma\backslash\mathrm{SL}_2(\mathbb{R})$. Suppose also that the unitary representation π^X on $L^2(X)$ has finite integrability exponent. Use Corollary 8.59 to prove an asymptotic counting result for $\{\gamma \in \Gamma \mid \|g\|_{\mathrm{HS}} \leqslant R\}$ with a concrete error term depending on κ.

8.8 The Weil Representation*

In this section we introduce a curious unitary representation known as the Segal–Shale–Weil or metaplectic representation. Starting with the irreducible unitary representation $\pi = \pi^1$ of the three-dimensional Heisenberg group H in Theorem 5.23 on $\mathcal{H}_\pi = L^2(\mathbb{R})$ we will create a new representation π^Φ of H on $L^2(\mathbb{R})$ by composing π with an automorphism Φ of H. However, the new representation π^Φ will actually be isomorphic to the old representation π and the unitary operator $U = U_\Phi$ of $L^2(\mathbb{R})$ with $U\pi_h U^{-1} = \pi_h^\Phi = \pi_{\Phi(h)}$ for all $h \in H$ is then associated to the automorphism Φ of H.

As we will see, $\mathrm{SL}_2(\mathbb{R})$ acts by automorphisms on H. Thus one could hope to obtain in this way a unitary representation ρ of $\mathrm{SL}_2(\mathbb{R})$ on $L^2(\mathbb{R})$, and this is almost—but not quite—what will happen. Apart from the curious interplay between H, $\mathrm{SL}_2(\mathbb{R})$, and $L^2(\mathbb{R})$ that will follow, it will also be interesting to see that the Fourier transform will (essentially) be part of this Weil representation.

8.8.1 Automorphisms of H

Let $H = \{(x, y, z) \mid x, y, z \in \mathbb{R}\}$ be the three-dimensional Heisenberg group using the notation from Section 5.2.4. The Lie algebra of H is given by

$$\mathfrak{h} = \left\{ (x, y, z) = \begin{pmatrix} 0 & x & z \\ 0 & 0 & y \\ 0 & 0 & 0 \end{pmatrix} \mid x, y, z \in \mathbb{R} \right\}$$

and is spanned by the matrices

$$\mathbf{e}_x = \begin{pmatrix} 0 & 1 & 0 \\ 0 & 0 & 0 \\ 0 & 0 & 0 \end{pmatrix}, \mathbf{e}_y = \begin{pmatrix} 0 & 0 & 0 \\ 0 & 0 & 1 \\ 0 & 0 & 0 \end{pmatrix}, \mathbf{e}_z = \begin{pmatrix} 0 & 0 & 1 \\ 0 & 0 & 0 \\ 0 & 0 & 0 \end{pmatrix}.$$

We note the commutation relations

$$[\mathbf{e}_x, \mathbf{e}_y] = \mathbf{e}_z$$

and

$$[\mathbf{e}_x, \mathbf{e}_z] = [\mathbf{e}_y, \mathbf{e}_z] = 0.$$

Moreover,

$$\log(x, y, z) = x\mathbf{e}_x + y\mathbf{e}_y + \bigl(z - \tfrac{1}{2}xy\bigr)\mathbf{e}_z$$

for $(x, y, z) \in H$ and

$$\exp(x\mathbf{e}_x + y\mathbf{e}_y + z\mathbf{e}_z) = (x, y, z + \tfrac{1}{2}xy)$$

for $x, y, z \in \mathbb{R}$.

A very special case of the Baker–Campbell–Hausdorff formula[20] arises here as

$$\log(h_1 h_2) = \log h_1 + \log h_2 + \tfrac{1}{2}[\log h_1, \log h_2] \tag{8.36}$$

for $h_1, h_2 \in H$. This formula shows that the group multiplication in H is uniquely determined by the Lie bracket in $\mathfrak{h}$ and by using the universal map $\log\colon H \to \mathfrak{h}$ and its inverse $\exp\colon \mathfrak{h} \to H$. For completeness we quickly verify (8.36) by a direct calculation. Let $h_j = (x_j, y_j, z_j) \in H$ for $j = 1, 2$. Then

$$\begin{aligned} \log(h_1 h_2) &= \log\bigl(x_1 + x_2, y_1 + y_2, z_1 + z_2 + x_1 y_2\bigr) \\ &= (x_1 + x_2)\mathbf{e}_x + (y_1 + y_2)\mathbf{e}_y + \bigl(z_1 + z_2 + x_1 y_2 - \tfrac{1}{2}(x_1 + x_2)(y_1 + y_2)\bigr)\mathbf{e}_z \\ &= (x_1 + x_2)\mathbf{e}_x + (y_1 + y_2)\mathbf{e}_y + \bigl(z_1 + z_2 + \tfrac{1}{2}(x_1 y_2 - x_1 y_1 - x_2 y_1 - x_2 y_2)\bigr)\mathbf{e}_z \end{aligned}$$

and

$$\begin{aligned} \log h_1 + \log h_2 + \tfrac{1}{2}[\log h_1, \log h_2] = x_1\mathbf{e}_x + y_1\mathbf{e}_y + \bigl(z_1 - \tfrac{1}{2}x_1 y_1\bigr)\mathbf{e}_z& \\ + x_2\mathbf{e}_x + y_2\mathbf{e}_y + \bigl(z_2 - \tfrac{1}{2}x_2 y_2\bigr)\mathbf{e}_z& \\ + \tfrac{1}{2}(x_1 y_2 - x_2 y_1)\mathbf{e}_z& \end{aligned}$$

together give (8.36).

We now let $g = \begin{pmatrix} a & b \\ c & d \end{pmatrix} \in \mathrm{SL}_2(\mathbb{R})$ and define $\phi_g\colon \mathfrak{h} \to \mathfrak{h}$ by the formulas

$$\begin{aligned} \phi(\mathbf{e}_x) &= a\mathbf{e}_x + c\mathbf{e}_y, \\ \phi(\mathbf{e}_y) &= b\mathbf{e}_x + d\mathbf{e}_y, \\ \phi(\mathbf{e}_z) &= \mathbf{e}_z, \end{aligned}$$

and linear extension. We then have

$$\begin{aligned}[\phi_g(\mathbf{e}_x), \phi_g(\mathbf{e}_y)] &= [a\mathbf{e}_x + c\mathbf{e}_y, b\mathbf{e}_x + d\mathbf{e}_y] \\ &= (ad - bc)[\mathbf{e}_x, \mathbf{e}_y] = \mathbf{e}_z = \phi(\mathbf{e}_z) = \phi([\mathbf{e}_x, \mathbf{e}_z]), \\ [\phi_g(\mathbf{e}_x), \phi_g(\mathbf{e}_z)] &= 0 = \phi_g([\mathbf{e}_x, \mathbf{e}_z]),\end{aligned}$$

and similarly for $\mathbf{e}_y$ and $\mathbf{e}_z$. This shows that $\phi_g\colon \mathfrak{h} \to \mathfrak{h}$ is a Lie algebra automorphism. Using (8.36) we obtain that

$$\Phi_g = \exp\circ\phi_g \circ \log\colon H \longrightarrow H$$

is a group automorphism for any $g \in \mathrm{SL}_2(\mathbb{R})$.

Exercise 8.61. Verify that $\Phi_{g_1} \circ \Phi_{g_2} = \Phi_{g_1 g_2}$ for $g_1, g_2 \in \mathrm{SL}_2(\mathbb{R})$.

8.8.2 A Projective Representation of $\mathrm{SL}_2(\mathbb{R})$

Let π be the irreducible unitary representation of H on $L^2(\mathbb{R})$ associated to the central character $\chi((0,0,z)) = \mathrm{e}^{2\pi \mathrm{i} z}$ for $z \in \mathbb{R}$. It will be more convenient to present the unitary representation in a slightly different way. Indeed, we define

$$\big(\pi_{(x,y,z)}f\big)(t) = \mathrm{e}^{2\pi\mathrm{i}(xt+z-xy)} f(t-y) \tag{8.37}$$

for $(x,y,z) \in H$, $f \in L^2(\mathbb{R})$, and $t \in \mathbb{R}$.

Exercise 8.62. Verify that π as in (8.37) defines a unitary representation isomorphic to π^1 as in Theorem 5.23.

For $g \in \mathrm{SL}_2(\mathbb{R})$ we let $\Phi_g\colon H \to H$ be the group automorphism of H defined in Section 8.8.1. Using π and Φ_g we now define a new unitary representation π^g of H on $L^2(\mathbb{R})$ by the formula $\pi^g_h = \pi_{\Phi_g(h)}$ for $h \in H$. For $z \in \mathbb{R}$ we then have that

$$\pi^g_{(0,0,z)} = \pi_{\Phi_g((0,0,z))} = \pi_{(0,0,z)} = \chi(z) I$$

is simply multiplication by $\chi(z)$. In other words π^g is an irreducible unitary representation of H with central character χ. By Theorem 5.23 (and Exercise 8.62) this shows that π^g must be isomorphic to π. That is, there exists a unitary operator $\rho_g\colon L^2(\mathbb{R}) \to L^2(\mathbb{R})$ with the property that

$$\pi^g_h = \pi_{\Phi_g(h)} = \rho_g \pi_h \rho_g^{-1} \tag{8.38}$$

for all $h \in H$.

By Schur's lemma (Theorem 1.29) ρ_g is almost, but not quite, uniquely determined by g: All other operators satisfying (8.38) belong to $\mathbb{S}^1\rho_g$. In this sense our discussions so far define a so-called *projective unitary representation* of $\mathrm{SL}_2(\mathbb{R})$ on $L^2(\mathbb{R})$. For any $g \in \mathrm{SL}_2(\mathbb{R})$ we have a collection $\mathbb{S}^1\rho_g$ of scalar multiples of unitary operators on $L^2(\mathbb{R})$ such that

$$\mathbb{S}^1\rho_{g_1 g_2} = \mathbb{S}^1\rho_{g_1}\rho_{g_2} \tag{8.39}$$

for all $g_1, g_2 \in \mathrm{SL}_2(\mathbb{R})$.

Exercise 8.63. Use Exercise 8.61 to verify (8.39).

8.8.3 Four Special Cases

We now wish to find the operator ρ_g for some choices of $g \in \mathrm{SL}_2(\mathbb{R})$.

THE WEYL ELEMENT. We start by considering the *Weyl element* $W = \begin{pmatrix} 0 & -1 \\ 1 & 0 \end{pmatrix}$ corresponding to a rotation by $\frac{\pi}{2}$. The corresponding Lie algebra automorphism ϕ_W is given by

$$\phi_W \colon \mathfrak{h} \ni x\mathbf{e}_x + y\mathbf{e}_y + z\mathbf{e}_z \longmapsto -y\mathbf{e}_x + x\mathbf{e}_y + z\mathbf{e}_z,$$

which gives the group automorphism $\Phi_W \colon H \to H$ defined for $(x, y, z) \in H$ by

$$\begin{aligned} \Phi_W((x,y,z)) &= \exp\bigl(\phi_W(x\mathbf{e}_x + y\mathbf{e}_y + (z - \tfrac{1}{2}xy)\mathbf{e}_z)\bigr) \\ &= \exp\bigl(-y\mathbf{e}_x + x\mathbf{e}_y + (z - \tfrac{1}{2}xy)\mathbf{e}_z\bigr) \\ &= (-y, x, z - xy). \end{aligned}$$

Composing with π now defines π^W. By (8.37) we have

$$\begin{aligned} \bigl(\pi^W_{(x,y,z)} f\bigr)(t) &= \bigl(\pi_{(-y,x,z-xy)} f\bigr)(t) \\ &= \mathrm{e}^{2\pi\mathrm{i}(-yt+(z-xy)+xy)} f(t-x) \\ &= \mathrm{e}^{2\pi\mathrm{i}(-yt+z)} f(t-x) \end{aligned}$$

for $(x, y, z) \in H$, $f \in L^2(\mathbb{R})$, and $t \in \mathbb{R}$. We claim that the associated unitary operator ρ_W is given by the Fourier transform (or rather by any of its multiples)

$$\rho_W f = \alpha \widehat{f} \tag{8.40}$$

for $f \in L^2(\mathbb{R})$ and some fixed $\alpha \in \mathbb{S}^1$. To see this we have to calculate for $U \colon L^2(\mathbb{R}) \ni f \mapsto \widehat{f}$ the composition $U\pi_{(x,y,z)}U^{-1}$ for $(x, y, z) \in H$. Note that for $f \in L^2(\mathbb{R})$ we have $U^{-1} f = \check{f}$, the Fourier back transform of f. Therefore by (8.37) we have

$$\bigl(\pi_{(x,y,z)} U^{-1} f\bigr)(s) = \mathrm{e}^{2\pi\mathrm{i}(xs+z-xy)} \check{f}(s-y)$$

for $s \in \mathbb{R}$ and so

$$\begin{aligned}
\big(U\pi_{(x,y,z)}U^{-1}f\big)(t) &= \int_{\mathbb{R}} \mathrm{e}^{2\pi\mathrm{i}(xs+z-xy)}\check{f}(\underbrace{s-y}_{=s'})\mathrm{e}^{-2\pi\mathrm{i}st}\,\mathrm{d}s\\
&= \int_{\mathbb{R}} \mathrm{e}^{2\pi\mathrm{i}(x(s'+y)+z-xy-(s'+y)t)}\check{f}(s')\,\mathrm{d}s'\\
&= \mathrm{e}^{2\pi\mathrm{i}(-yt+z)}\int_{\mathbb{R}}\check{f}(s')\mathrm{e}^{-2\pi\mathrm{i}(t-x)s'}\,\mathrm{d}s'\\
&= \mathrm{e}^{2\pi\mathrm{i}(-yt+z)}f(t-x)\\
&= \big(\pi^{W}_{(x,y,z)}f\big)(t)
\end{aligned}$$

for $t\in\mathbb{R}$, which proves (8.40). At this point $\alpha\in\mathbb{S}^1$ is arbitrary, and we will find the 'best' choice for it later.

THE DIAGONAL SUBGROUPS. Next we consider elements $g_a=\begin{pmatrix}a&0\\0&a^{-1}\end{pmatrix}\in A$ for some $a>0$ of the connected diagonal subgroup A. The corresponding automorphism is given by

$$\Phi_{g_a}: H\ni(x,y,z)\mapsto(ax,\tfrac{1}{a}y,z).$$

Exercise 8.64. Verify the claimed formula for Φ_{g_a}.

Using (8.37) it follows that

$$\big(\pi^{g_a}_{(x,y,z)}f\big)(t)=\mathrm{e}^{2\pi\mathrm{i}(axt+z-xy)}f(t-\tfrac{1}{a}y) \tag{8.41}$$

for $(x,y,z)\in H$, $f\in L^2(\mathbb{R})$, and $t\in\mathbb{R}$. We claim that the associated unitary operator ρ_{g_a} can be chosen to be

$$\big(\rho_{g_a}f\big)(t)=a^{\frac{1}{2}}f(at)$$

for $f\in L^2(\mathbb{R})$ and $t\in\mathbb{R}$. To see this we calculate

$$\begin{aligned}
\big(\rho_{g_a}\pi_{(x,y,z)}\rho_{g_a^{-1}}f\big)(t) &= a^{\frac{1}{2}}\big(\pi_{(x,y,z)}\rho_{g_a^{-1}}f\big)(at)\\
&= a^{\frac{1}{2}}\mathrm{e}^{2\pi\mathrm{i}(xat+z-xy)}\big(\rho_{g_a^{-1}}f\big)(at-y)\\
&= \mathrm{e}^{2\pi\mathrm{i}(axt+z-xy)}f(t-\tfrac{1}{a}y),
\end{aligned}$$

which matches (8.41) and so justifies the claimed choice.

Our goal is to obtain a unitary representation ρ. For this we note that the map $A\ni g_a\mapsto\rho_{g_a}$ already defines a unitary representation of $A\cong\mathbb{R}$. As a multiple by some element of $\mathbb{S}^1$ of the unitary operator ρ_{g_a} would also satisfy the desired property, we could equally well consider the twisted unitary representation $A\ni g_a\mapsto a^{\beta\mathrm{i}}\rho_{g_a}$. However, we claim that the 'best' choice of β is in fact 0 by the following exercise.

Exercise 8.65. (a) Verify that $A\ni g_a\mapsto\rho_{g_a}$ is a unitary representation.
(b) Show that W normalizes A and maps an element of A to its inverse.
(c) Show that

$$\langle W, A\rangle \ni W^j g_a \longmapsto \rho_W^j a^{\beta \mathrm{i}} \rho_{g_a}$$

defines a unitary representation of the group $\langle W, A\rangle$ generated by W and A if and only if $\beta = 0$ and $\alpha^4 = 1$.

THE UPPER UNIPOTENT SUBGROUP. We now repeat this analysis for elements $u_b = \begin{pmatrix} 1 & b \\ 0 & 1 \end{pmatrix}$ of the upper unipotent subgroup $U \leqslant \mathrm{SL}_2(\mathbb{R})$. The corresponding automorphism of H is given by

$$\begin{aligned} \Phi_{u_b} \colon H \ni (x, y, z) &\stackrel{\log}{\longmapsto} (x, y, z - \tfrac{1}{2}xy) \\ &\stackrel{\phi_{u_b}}{\longmapsto} (x + by, y, z - \tfrac{1}{2}xy) \\ &\stackrel{\exp}{\longmapsto} \big(x + by, y, z - \tfrac{1}{2}xy + \tfrac{1}{2}(x + by)y\big) = \big(x + by, y, z + \tfrac{1}{2}by^2\big). \end{aligned}$$

By (8.37) the unitary representation π^{u_b} is therefore given by

$$\big(\pi^{u_b}_{(x,y,z)} f\big)(t) = \big(\pi_{(x+by,y,z+\frac{1}{2}by^2)} f\big)(t) = \mathrm{e}^{2\pi \mathrm{i}((x+by)t+z+\frac{1}{2}by^2-(x+by)y)} f(t-y) \tag{8.42}$$

for $(x, y, z) \in H$, $f \in L^2(\mathbb{R})$, and $t \in \mathbb{R}$.

Notice that for $y = 0$ this agrees with the original unitary representation $\pi_{(x,0,z)}$. Roughly speaking, this shows that the spectral significance of the parameter $t \in \mathbb{R}$ is the same for π and for π^{u_b}. Therefore we might want to consider the ansatz

$$\big(\rho_{u_b} f\big)(t) = \mathrm{e}^{2\pi \mathrm{i}\gamma(b,t)} f(t) \tag{8.43}$$

for $f \in L^2(\mathbb{R})$, $t \in \mathbb{R}$, and a yet to be determined real-valued function γ of b and t.

This leads to

$$\begin{aligned} \big(\rho_{u_b}\pi_{(x,y,z)}\rho_{u_b}^{-1} f\big)(t) &= \mathrm{e}^{2\pi \mathrm{i}\left(\gamma(b,t)+xt+z-xy\right)} \big(\rho_{u_b}^{-1} f\big)(t-y) \\ &= \mathrm{e}^{2\pi \mathrm{i}\left(\gamma(b,t)-\gamma(b,t-y)+xt+z-xy\right)} f(t-y) \end{aligned} \tag{8.44}$$

for $(x, y, z) \in H$, $f \in L^2(\mathbb{R})$, and $b, t \in \mathbb{R}$. Comparing (8.42) with (8.44) we would like to have

$$\gamma(b,t) - \gamma(b,t-y) + xt + z - xy = (x + by)t + z + \tfrac{1}{2}by^2 - (x + by)y$$

and thus

$$\begin{aligned} \gamma(b,t) - \gamma(b,t-y) &= byt - \tfrac{1}{2}by^2 \\ &= \tfrac{1}{2}bt^2 - \tfrac{1}{2}b(t-y)^2 \end{aligned}$$

by completing the square. Therefore

$$\gamma(b,t) = \gamma_0 b + \tfrac{1}{2}bt^2 \tag{8.45}$$

for a constant γ_0 could be used in (8.43) to define ρ_{u_b} satisfying

$$\rho_{u_b}\pi_{(x,y,z)}\rho_{u_b}^{-1} = \pi_{(x,y,z)}^{u_b} \tag{8.46}$$

for all $(x,y,z) \in H$. By insisting on linear dependence of $\gamma(b,t)$ on b we also obtain that $\mathbb{R} \ni b \mapsto \rho_{u_b}$ is a unitary representation. The 'best' choice of the constant γ_0 turns out to be 0 as explained in the following exercise. Therefore we define

$$\big(\rho_{u_b} f\big)(t) = \mathrm{e}^{\pi i b t^2} f(t),$$

which satisfies (8.46).

Exercise 8.66. Let $g_a = \begin{pmatrix} a & 0 \\ 0 & a^{-1} \end{pmatrix} \in A$ with $a > 1$. Define ρ_{u_b} for $b \in \mathbb{R}$ using (8.43) and (8.45). Calculate both $g_a u_b g_a^{-1}$ and $\rho_{g_a}\rho_{u_b}\rho_{g_a}^{-1}$ and conclude that the latter coincides with $\rho_{g_a u_b g_a^{-1}}$ for all $b \in \mathbb{R}$ if and only if $\gamma_0 = 0$.

The lower unipotent subgroup. Finally we calculate

$$W u_{-c} W^{-1} = \begin{pmatrix} 0 & -1 \\ 1 & 0 \end{pmatrix}\begin{pmatrix} 1 & -c \\ 0 & 1 \end{pmatrix}\begin{pmatrix} 0 & 1 \\ -1 & 0 \end{pmatrix} = \begin{pmatrix} 1 & 0 \\ c & 1 \end{pmatrix}.$$

Hence we define for a lower unipotent matrix $u_c' = \begin{pmatrix} 1 & 0 \\ c & 1 \end{pmatrix}$ the unitary operator $\rho_{u_c'}$ that sends f to the Fourier transform of

$$\mathbb{R} \ni s \longmapsto \mathrm{e}^{-\pi i c s^2} \check{f}(s)$$

since the unknown scalar α in our definition of ρ_W cancels.

8.8.4 A Brief Summary

By the discussion in Section 8.8.2 we can associate to any $g \in \mathrm{SL}_2(\mathbb{R})$ a unitary representation ρ_g on $L^2(\mathbb{R})$, which, up to a scalar in $\mathbb{S}^1$, is unique. We calculated these operators in Section 8.8.3 and made—whenever possible—some reasonable choices for the scalar. We summarize this discussion in Table 8.1.

Hoping to obtain a unitary representation of $\mathrm{SL}_2(\mathbb{R})$ we wish to calculate the 'best' value of α. As we already obtained good definitions for upper and lower unipotent matrices the factorization

$$W = \begin{pmatrix} 1 & -1 \\ 0 & 1 \end{pmatrix}\begin{pmatrix} 1 & 0 \\ 1 & 1 \end{pmatrix}\begin{pmatrix} 1 & -1 \\ 0 & 1 \end{pmatrix} \tag{8.47}$$

might be useful. This suggests that we should try to define ρ_W as the composition of $\rho_{u_{-1}}$, $\rho_{u_1'}$, and $\rho_{u_{-1}}$ again. This involves daunting expressions for a general function $f \in L^2(\mathbb{R})$.

Table 8.1: The operators constructed so far.

$g \in \mathrm{SL}_2(\mathbb{R})$	associated unitary operator on $L^2(\mathbb{R})$
$W = \begin{pmatrix} 0 & -1 \\ 1 & 0 \end{pmatrix}$	$f \longmapsto \alpha \widehat{f}$
$g_a = \begin{pmatrix} a & 0 \\ 0 & a^{-1} \end{pmatrix}$	$f \longmapsto a^{\frac{1}{2}} f(a\cdot)$
$u_b = \begin{pmatrix} 1 & b \\ 0 & 1 \end{pmatrix}$	$f \longmapsto \mathrm{e}^{\pi \mathrm{i} b(\cdot)^2} f(\cdot)$
$u'_c = \begin{pmatrix} 1 & 0 \\ c & 1 \end{pmatrix}$	$f \longmapsto \widehat{\mathrm{e}^{-\pi \mathrm{i} c(\cdot)^2} \check{f}(\cdot)}$

However, we already know that the outcome will be a multiple of $\widehat{f}$ and we only wish to calculate the constant involved. For this reason it suffices to consider one specific non-zero L^2 function for which the calculation is more manageable.

8.8.5 A Concrete Class of Functions

We define the modified Gaussian function

$$f_z \colon \mathbb{R} \ni t \longmapsto \mathrm{e}^{\pi \mathrm{i} z t^2}$$

for a fixed $z \in \mathbb{C}$ with $\Im(z) > 0$ and wish to calculate the images of f_z under the various group elements considered earlier. For $a > 0$ we have

$$\big(\rho_{g_a} f_z\big)(t) = a^{\frac{1}{2}} f_z(at) = a^{\frac{1}{2}} \mathrm{e}^{\pi \mathrm{i} z a^2 t^2} = a^{\frac{1}{2}} f_{a^2 z}(t) \tag{8.48}$$

for $t \in \mathbb{R}$. Similarly we have for $b \in \mathbb{R}$ that

$$\big(\rho_{u_b} f_z\big)(t) = \mathrm{e}^{\pi \mathrm{i} b t^2} f_z(t) = \mathrm{e}^{\pi \mathrm{i}(z+b)t^2} = f_{z+b}(t) \tag{8.49}$$

for $t \in \mathbb{R}$. To calculate $\rho_{u'_c} f_z$ we need to know the Fourier transform of f_z.

Lemma 8.67 (Fourier transform). *For $z \in \mathbb{C}$ with $\Im(z) > 0$ we have*

$$\widehat{f_z} = \check{f_z} = I_z f_{-\frac{1}{z}}$$

where $I_z = \int_{\mathbb{R}} f_z(t)\,\mathrm{d}t$ satisfies $I_z^2 = \frac{\mathrm{i}}{z}$.

PROOF. We start by calculating

$$I_z^2 = \iint\limits_{\mathbb{R}\times\mathbb{R}} \mathrm{e}^{\pi \mathrm{i} z(t_1^2+t_2^2)} \underbrace{\mathrm{d}t_1\,\mathrm{d}t_2}_{r\,\mathrm{d}r\,\mathrm{d}\theta} = \int_0^{2\pi} \mathrm{d}\theta \int_0^{\mathbb{R}} \mathrm{e}^{\pi \mathrm{i} z r^2} r\,\mathrm{d}r = \int_0^{\infty} \mathrm{e}^{\mathrm{i} z u}\,\mathrm{d}u = -\frac{1}{\mathrm{i}z} = \frac{\mathrm{i}}{z}$$

by using the substitution $u = \pi r^2$ with $\mathrm{d}u = 2\pi r\,\mathrm{d}r$ and since $\Im(z) > 0$. Moreover, by completing the square we have

$$\widehat{f_z}(t) = \int_{\mathbb{R}} \mathrm{e}^{\pi \mathrm{i} z s^2} \mathrm{e}^{-2\pi \mathrm{i} s t}\,\mathrm{d}s = \int_{\mathbb{R}} \mathrm{e}^{\pi \mathrm{i} z (s^2 - 2s\frac{t}{z} + \frac{t^2}{z^2} - \frac{t^2}{z^2})} = \underbrace{\mathrm{e}^{-\pi \mathrm{i} \frac{1}{z} t^2}}_{f_{-\frac{1}{z}}(t)} \int_{\mathbb{R}} \mathrm{e}^{\pi \mathrm{i} z (s - \frac{t}{z})^2}\,\mathrm{d}s.$$

By exploiting the rapid decay of the entire function $u \mapsto \mathrm{e}^{\pi \mathrm{i} z u^2}$ for $|\Re(u)| \to \infty$ with $|\Im(u)|$ bounded and shifting the path integral we see that the remaining integral is equal to I_z. Hence $\widehat{f_z} = \check{f_z} = I_z f_{-\frac{1}{z}}$ as claimed. □

Exercise 8.68. Verify the argument involving shifting the path integral more carefully.

Recall our definition $\rho_{u'_c} f = \widehat{\rho_{u_{-c}} \check{f}}$. Therefore by Lemma 8.67 we have

$$f_z \overset{\smile}{\longmapsto} \check{f_z} = I_z f_{-\frac{1}{z}} \overset{\rho_{u_{-c}}}{\longmapsto} I_z f_{-\frac{1}{z}-c} = I_z f_{-\frac{cz+1}{z}} \overset{\frown}{\longmapsto} I_z I_{-\frac{cz+1}{z}} f_{\frac{z}{cz+1}}$$

for $z \in \mathbb{C}$ with $\Im(z) > 0$. Moreover

$$\left(I_z I_{-\frac{cz+1}{z}}\right)^2 = \frac{\mathrm{i}}{z} \frac{-z\mathrm{i}}{(cz+1)} = \frac{1}{cz+1}.$$

Combining the two we obtain

$$\rho_{u'_c} f_z = \left(\sqrt{cz+1}\right)^{-1} f_{\frac{z}{cz+1}}, \tag{8.50}$$

where the sign of the square root is determined by requiring continuous dependence on $c \in \mathbb{R}$. More concretely, we may use a holomorphic square root on $\mathbb{C} \smallsetminus (-\infty, 0]$ with values in $\{z \in \mathbb{H} \mid \Re(z) > 0\}$.

8.8.6 Calculation of the Constant

We define ρ_g for g equal to W, a diagonal, or an upper or lower unipotent as in Table 8.1. Together with the formula for W in (8.47) we can now calculate the 'best' choice for α (or, more precisely, for α^2). Indeed for $z \in \mathbb{C}$ with $\Im(z) > 0$ we have by Lemma 8.67, (8.49), and (8.50) for $c = 1$ and $z - 1$ instead of z that

$$\begin{aligned} \alpha I_z f_{-\frac{1}{z}} = \alpha \widehat{f_z} = \rho_W f_z &\underset{(*)}{=} \rho_{u_{-1}} \rho_{u'_1} \rho_{u_{-1}} f_z \\ &= \rho_{u_{-1}} \rho_{u'_1} f_{z-1} = \pm \rho_{u_{-1}} z^{-\frac{1}{2}} f_{\frac{z-1}{z}} = \pm z^{-\frac{1}{2}} f_{-\frac{1}{z}}, \end{aligned}$$

where the step $(*)$ is wishful thinking motivated by our desire to obtain a unitary representation of $\mathrm{SL}_2(\mathbb{R})$. With this we conclude that $\alpha I_z = \pm z^{-\frac{1}{2}}$. Taking the square and using Lemma 8.67 again we obtain $\alpha^2 \frac{\mathrm{i}}{z} = \frac{1}{z}$ and so $\alpha^2 = -\mathrm{i}$. In other words, we should have $\rho_W f = \alpha \widehat{f}$ for all $f \in L^2(\mathbb{R})$ and $\alpha \in \mathbb{S}^1$ with $\alpha^2 = -\mathrm{i}$.

Note that $\widehat{\widehat{f}}$ is the function $t \mapsto f(-t)$ and so the Fourier transform operator has order 4. For ρ_W (defined using a square root of $-\mathrm{i}$ as α) this implies that

$$\rho_W^4 = \alpha^4 I = -I.$$

As $W^4 = I$ we see that ρ cannot be a unitary representation of $\mathrm{SL}_2(\mathbb{R})$. However, since $\rho_W^8 = I$ we can repair this problem by considering the double cover of $\mathrm{SL}_2(\mathbb{R})$.

8.8.7 The Metaplectic Group

Recall that

$$\mathrm{SL}_2(\mathbb{R}) = KAU \cong \mathbb{S}^1 \times \mathbb{R}^2$$

as a topological space, where $K = \mathrm{SO}_2(\mathbb{R}) \cong \mathbb{S}^1$. Therefore the fundamental group of $\mathrm{SL}_2(\mathbb{R})$ is $\mathbb{Z}$ and the universal cover $\widetilde{\mathrm{SL}_2(\mathbb{R})}$ of $\mathrm{SL}_2(\mathbb{R})$ is an infinite cover.

We define[(21)] the *metaplectic group* $\mathrm{Mp}_2(\mathbb{R})$ as the double cover of $\mathrm{SL}_2(\mathbb{R})$: The simple closed loop in $\mathrm{SL}_2(\mathbb{R})$ defined by

$$[0, 2\pi] \ni \theta \longmapsto k_\theta = \begin{pmatrix} \cos\theta & -\sin\theta \\ \sin\theta & \cos\theta \end{pmatrix} \tag{8.51}$$

corresponding to $\mathrm{SO}_2(\mathbb{R}) \cong \mathbb{S}^1$ lifts to a path in $\mathrm{Mp}_2(\mathbb{R})$ that connects the identity in $\mathrm{Mp}_2(\mathbb{R})$ to the second element in the pre-image of I_2. However, lifting the closed path in $\mathrm{SL}_2(\mathbb{R})$ defined by $[0, 4\pi] \ni \theta \mapsto k_\theta$ gives a closed path in $\mathrm{Mp}_2(\mathbb{R})$.

It will be convenient to have a more concrete definition. In fact we may define $\mathrm{Mp}_2(\mathbb{R})$ as the set of tuples $\widetilde{g} = (g, \varepsilon(\cdot))$, where $g = \begin{pmatrix} a & b \\ c & d \end{pmatrix} \in \mathrm{SL}_2(\mathbb{R})$ and $\varepsilon\colon \mathbb{H} \to \mathbb{C}$ is a holomorphic square root of $\mathbb{H} \ni z \mapsto cz + d$. As $\mathbb{H}$ is simply connected and $cz + d \neq 0$ for $g = \begin{pmatrix} a & b \\ c & d \end{pmatrix} \in \mathrm{SL}_2(\mathbb{R})$ and $z \in \mathbb{H}$, it follows from analytic continuation that for every $g \in \mathrm{SL}_2(\mathbb{R})$ there are precisely two choices of $\varepsilon\colon \mathbb{H} \to \mathbb{C}$ with $\widetilde{g} = (g, \varepsilon) \in \mathrm{Mp}_2(\mathbb{R})$. In other words the canonical projection

$$\mathrm{Mp}_2(\mathbb{R}) \ni \widetilde{g} = (g, \varepsilon) \longmapsto g \in \mathrm{SL}_2(\mathbb{R})$$

is two-to-one.

Essential Exercise 8.69. (a) Verify that the binary operation

$$(g_1,\varepsilon_1)\cdot(g_2,\varepsilon_2)=\bigl(g_1g_2,z\mapsto\varepsilon_1(g_2\bullet z)\varepsilon_2(z)\bigr)$$

for $(g_1,\varepsilon_1),(g_2,\varepsilon_2)\in \mathrm{Mp}_2(\mathbb{R})$ makes $\mathrm{Mp}_2(\mathbb{R})$ into a group, where we again use the Möbius transformation from (8.3).
(b) Use the compact-open topology on the second coordinate to define a topology on $\mathrm{Mp}_2(\mathbb{R})$. Show that this makes $\mathrm{Mp}_2(\mathbb{R})$ into a topological group locally isomorphic to $\mathrm{SL}_2(\mathbb{R})$.
(c) Define a path

$$[0,4\pi]\ni\theta\longmapsto(k_\theta,\varepsilon_\theta)$$

where ε_θ is the holomorphic square root defined by analytic continuation starting with $\varepsilon_\theta(\mathrm{i})=\mathrm{e}^{\mathrm{i}\frac{\theta}{2}}$. Show that this defines a simple closed loop in $\mathrm{Mp}_2(\mathbb{R})$ that projects to the loop in (8.51) traversed twice. Conclude that $\mathrm{Mp}_2(\mathbb{R})$ is connected.

As $\mathrm{Mp}_2(\mathbb{R})$ is locally isomorphic to $\mathrm{SL}_2(\mathbb{R})$ we can uniquely lift the previously studied one-parameter subgroups of $\mathrm{SL}_2(\mathbb{R})$ to one-parameter subgroups of $\mathrm{Mp}_2(\mathbb{R})$. In the case of A and U these are simply given by

$$\widetilde{A}=\{(g_a,1)\mid g_a\in A\}$$

and

$$\widetilde{U}=\{(u_b,1)\mid u_b\in U\}.$$

For the lift of U' note that $c\mathbb{H}+1\subseteq\mathbb{C}\smallsetminus(-\infty,0]$ for all $c\in\mathbb{R}$. Hence we may let $\sqrt{\cdot}$ denote the holomorphic square root defined on $\mathbb{C}\smallsetminus(-\infty,0]$ and taking values in $\{z\in\mathbb{H}\mid\Re(z)>0\}$. With this the lift of U' is then given by

$$\widetilde{U'}=\bigl\{\bigl(u'_c,\mathbb{H}\ni z\longmapsto\sqrt{cz+1}\bigr)\mid u'_c\in U'\bigr\}.\tag{8.52}$$

8.8.8 The Weil Representation

As $\mathrm{Mp}_2(\mathbb{R})$ is locally isomorphic to $\mathrm{SL}_2(\mathbb{R})$ and connected by Exercise 8.69 it follows that $\mathrm{Mp}_2(\mathbb{R})$ is generated by the one-parameter subgroups $\widetilde{A}$, $\widetilde{U}$, and $\widetilde{U'}$. We use the discussions in Section 8.8.3 as our guide and declare the summary in Table 8.1 to give the definition of $\rho_{\widetilde{g}}$ for $\widetilde{g}\in\widetilde{A}\cup\widetilde{U}\cup\widetilde{U'}$. Moreover, if

$$\widetilde{g}=\widetilde{g_1}\cdots\widetilde{g_k}\in\mathrm{Mp}_2(\mathbb{R})$$

with $\widetilde{g_1},\ldots,\widetilde{g_k}\in\widetilde{A}\cup\widetilde{U}\cup\widetilde{U'}$ we would like to define $\rho_{\widetilde{g}}$ by

$$\rho_{\widetilde{g}}=\rho_{\widetilde{g_1}}\cdots\rho_{\widetilde{g_k}}.\tag{8.53}$$

Using all our earlier discussions and, in particular, the link to Möbius transformations on $\mathbb{H}$ and the term $(cz+d)$ appearing so often in hyperbolic geometry and analytic number theory we are now ready to prove the following result.

Proposition 8.70 (Weil representation). *Table* 8.1 *for* $\widetilde{g} \in \widetilde{A} \cup \widetilde{U} \cup \widetilde{U'}$ *and* (8.53) *for any* $\widetilde{g} \in \mathrm{Mp}_2(\mathbb{R})$ *define a unitary representation of* $\mathrm{Mp}_2(\mathbb{R})$ *on* $\mathcal{H}_\rho = L^2(\mathbb{R})$.

Proof. The main issue is to show that (8.53) gives a well-defined unitary operator. So suppose that

$$\widetilde{g} = \widetilde{g_1} \cdots \widetilde{g_k} = \widetilde{h_1} \cdots \widetilde{h_\ell}$$

for some $k, \ell \geqslant 1$ and $\widetilde{g_1}, \dots, \widetilde{g_k}, \widetilde{h_1}, \dots, \widetilde{h_\ell} \in \widetilde{A} \cup \widetilde{U} \cup \widetilde{U'}$. We claim that

$$\rho_{\widetilde{g_1}} \cdots \rho_{\widetilde{g_k}} = \rho_{\widetilde{h_1}} \cdots \rho_{\widetilde{h_\ell}}. \tag{8.54}$$

By Sections 8.8.2 and 8.8.3 our operators are part of a projective unitary representation satisfying (8.39), which gives

$$\mathbb{S}^1 \rho_{\widetilde{g_1}} \cdots \rho_{\widetilde{g_k}} = \mathbb{S}^1 \rho_{\widetilde{h_1}} \cdots \rho_{\widetilde{h_\ell}}.$$

Hence to prove the desired equality (8.54) we only need to verify equality on a simple function (or on a class of functions).

Define

$$\mathscr{C} = \{\gamma f_z \mid \gamma \in \mathbb{C}, z \in \mathbb{H}\}$$

using the modified Gaussian functions from Section 8.8.5 and define an action of $\mathrm{Mp}_2(\mathbb{R})$ on $\mathscr{C}$ by

$$(g, \varepsilon)\boldsymbol{\cdot}(\gamma f_z) = \varepsilon(z)^{-1} \gamma f_{g\boldsymbol{\cdot} z}$$

for $(g, \varepsilon) \in \mathrm{Mp}_2(\mathbb{R})$ and $\gamma f_z \in \mathscr{C}$. This is indeed an action since

$$\begin{aligned}
(g_1, \varepsilon_1)\boldsymbol{\cdot}\big((g_2, \varepsilon_2)\boldsymbol{\cdot}\gamma f_z\big) &= (g_1, \varepsilon_1)\boldsymbol{\cdot}\big(\varepsilon_2(z)^{-1} \gamma f_{g_2\boldsymbol{\cdot} z}\big) \\
&= \varepsilon_1(g_2\boldsymbol{\cdot} z)^{-1} \varepsilon_2(z)^{-1} \gamma f_{g_1 g_2\boldsymbol{\cdot} z} \\
&= \big((g_1, \varepsilon_1) \cdot (g_2, \varepsilon_2)\big)\boldsymbol{\cdot}\gamma f_z
\end{aligned}$$

for $\gamma f_z \in \mathscr{C}$ and $(g_1, \varepsilon_1), (g_2, \varepsilon_2) \in \mathrm{Mp}_2(\mathbb{R})$.

Our discussions in Section 8.8.5 and (8.48)–(8.50) in particular can be summarized as

$$\rho_{\widetilde{g}}(\gamma f_z) = \widetilde{g}\boldsymbol{\cdot}(\gamma f_z)$$

for $\widetilde{g} \in \widetilde{A} \cup \widetilde{U} \cup \widetilde{U'}$ and $\gamma f_z \in \mathscr{C}$. In the case of $\widetilde{U'}$ this holds as we use the same square root in (8.50) and (8.52). For the compositions in (8.54) this implies

$$\rho_{\widetilde{g_1}} \cdots \rho_{\widetilde{g_k}}(\gamma f_z) = (\widetilde{g_1} \cdots \widetilde{g_k})\boldsymbol{\cdot}(\gamma f_z) = (\widetilde{h_1} \cdots \widetilde{h_\ell})\boldsymbol{\cdot}(\gamma f_z) = \rho_{\widetilde{h_1}} \cdots \rho_{\widetilde{h_\ell}}(\gamma f_z)$$

for $\gamma f_z \in \mathscr{C}$, which proves (8.54).

Our definitions of $\rho_{\widetilde{g}}$ for $\widetilde{g} \in \widetilde{A} \cup \widetilde{U} \cup \widetilde{U'}$ show that the restrictions to these three one-parameter subgroups are continuous in the strong operator topology. Note that every $\widetilde{g}$ close to the identity in $\mathrm{Mp}_2(\mathbb{R})$ has a unique representation in the form $\widetilde{g} = \widetilde{g}_a \widetilde{u}_b \widetilde{u}'_c$ with $g_a \in \widetilde{A}$, $\widetilde{u}_b \in \widetilde{U}$, and $\widetilde{u}'_c \in \widetilde{U'}$ close to the identity. Continuity of $\widetilde{g} \mapsto \rho_{\widetilde{g}}$ follows from these facts. □

Exercise 8.71. (a) Fill in the details of the proof of continuity above.
(b) Show that the Weil representation is not irreducible.

8.9 Summary and Outlook

In this chapter we started to uncover some of the mysteries of $\mathrm{SL}_2(\mathbb{R})$ and its (irreducible) unitary representations. By building on this, we will be able to completely describe $\widehat{\mathrm{SL}_2(\mathbb{R})}$ in Chapter 9.

The Weil representation provides another connection between representation theory and harmonic analysis, and plays many roles ranging from the study of modular forms of half-integral weight in number theory to quantum mechanics. It generalizes to the two-to-one covers (also known as the metaplectic group $\mathrm{Mp}_{2\mathrm{n}}(\mathbb{R})$) of the symplectic group $\mathrm{Sp}_{2n}(\mathbb{R})$, using the $(2n+1)$-dimensional Heisenberg group.

Chapter 9
Unitary Representations of $\mathrm{SL}_2(\mathbb{R})$

In this chapter we will again focus our attention on $\mathrm{SL}_2(\mathbb{R})$ and its unitary representations. For this, we recall some of our previous discussions:

- We already classified all finite-dimensional representations of $\mathrm{SL}_2(\mathbb{R})$ in Section 4.1. This result, and even more so the method of proof, will be of interest here.
- In Section 8.3 we recalled the geometric significance of $\mathrm{SL}_2(\mathbb{R}) \cong \mathrm{SU}_{1,1}(\mathbb{R})$ by connecting it to the hyperbolic plane $\mathbb{H} \cong \mathbb{D}$.
- In Section 8.4 we already found our first two types of non-trivial irreducible unitary representations of $\mathrm{SL}_2(\mathbb{R})$, namely the discrete series and the mock discrete series representations.
- In Section 8.5 we studied the regular representation of $\mathrm{SL}_2(\mathbb{R})$.
- This allowed us to characterize temperedness for $\mathrm{SL}_2(\mathbb{R})$ in terms of integrability and decay of matrix coefficients in Section 8.5.

We will extend these results here, to obtain a complete description of $\widehat{\mathrm{SL}_2(\mathbb{R})}$. Moreover, we will decompose natural unitary representations into irreducible representations.[(22)] In particular, we will study the action-associated representation of $\mathrm{SL}_2(\mathbb{R})$ on $L^2(\mathbb{H})$, and see that the 'Fourier transform on $\mathbb{H}$' is intimately related to the principal series representations of $\mathrm{SL}_2(\mathbb{R})$. Finally, the complementary series representation will allow a better understanding of (the lack of) spectral gap, decay rates, and integrability exponents for $\mathrm{SL}_2(\mathbb{R})$.

9.1 The Universal Enveloping Algebra and Smooth Vectors

We recall that the Casimir operator for $\mathrm{SU}_2(\mathbb{R})$ in Proposition 7.23 and Corollary 7.24 commutes with $\mathrm{SU}_2(\mathbb{R})$. Because of the connection between $\mathrm{SU}_2(\mathbb{R})$ and $\mathrm{SL}_2(\mathbb{R})$ developed in Section 4.1.2, it stands to reason that $\mathrm{SL}_2(\mathbb{R})$ should also possess a Casimir operator. However, the Casimir operator of $\mathrm{SL}_2(\mathbb{R})$

M. Einsiedler and T. Ward, *Unitary Representations and Unitary Duals*,
Graduate Texts in Mathematics 308, https://doi.org/10.1007/978-3-032-03899-9_9

will have 'mixed signature', and will not arise from an application of Proposition 7.23.

After developing the necessary abstract machinery in this section, it will also be relatively straightforward to define the raising and lowering operators for any unitary representation of $\mathrm{SL}_2(\mathbb{R})$, which will lead to the description of $\widehat{\mathrm{SL}_2(\mathbb{R})}$ in Section 9.2.

9.1.1 The Universal Enveloping Algebra

We briefly introduce the algebra $\mathfrak{E}$ containing the Lie algebra $\mathfrak{g}$ of a Lie group G as well as higher-order terms like the Casimir element. We refer to Knapp [54, Ch. 3] for a more careful introduction to this concept.

Definition 9.1 (The algebra $\mathfrak{E}$). The *universal enveloping algebra* $\mathfrak{E}$ of a Lie algebra $\mathfrak{g}$ is the linear hull of all formal multi-linear associative non-commuting products $\mathbf{b}_1 \circ \mathbf{b}_2 \circ \cdots \circ \mathbf{b}_n$ for $n \in \mathbb{N}_0$ and $\mathbf{b}_1, \ldots, \mathbf{b}_n \in \mathfrak{g}$ modulo the ideal generated by the expressions $\mathbf{a} \circ \mathbf{b} - \mathbf{b} \circ \mathbf{a} - [\mathbf{a}, \mathbf{b}]$ for $\mathbf{a}, \mathbf{b} \in \mathfrak{g}$. By also allowing the empty product $\mathbb{1}_{\mathfrak{E}}$ (corresponding to $n = 0$) the algebra $\mathfrak{E}$ is also unital. We will write $\mathbf{b}^{\circ n} = \mathbf{b} \circ \mathbf{b} \circ \cdots \circ \mathbf{b}$ for powers of $\mathbf{b} \in \mathfrak{E}$ in the algebra $\mathfrak{E}$ for all $n \in \mathbb{N}$, and define $\mathbf{b}^{\circ 0}$ to be $\mathbb{1}_{\mathfrak{E}}$ for all $\mathbf{b} \in \mathfrak{E}$.

At first sight this definition might look very much like abstract nonsense. However, as we will see, we should think of $\mathfrak{E}$ as the algebra of all partial differential operators that can be obtained by composition from the first order differential operators that correspond to elements of $\mathfrak{g}$.

Because of its definition, $\mathfrak{E}$ is not a graded algebra (that is, there is no good definition of homogeneous degree in $\mathfrak{E}$), since the generators of the ideal in its definition have terms of different degree. However, it can be written as an increasing union

$$\mathfrak{E} = \bigcup_{d=0}^{\infty} \mathfrak{E}_{\leqslant d},$$

where $\mathfrak{E}_{\leqslant 0} = \mathbb{C}\mathbb{1}_{\mathfrak{E}}$, $\mathfrak{E}_{\leqslant 1} = \mathfrak{E}_{\leqslant 0} + \mathfrak{g}$, and $\mathfrak{E}_{\leqslant d}$ is the subspace of $\mathfrak{E}$ generated by all products $\mathbf{b}_1 \circ \mathbf{b}_2 \circ \cdots \circ \mathbf{b}_n$ with $n \leqslant d$ and $\mathbf{b}_1, \ldots, \mathbf{b}_n \in \mathfrak{g}$. We say that $\mathbf{e} \in \mathfrak{E}$ has degree $d \in \mathbb{N}$ if $\mathbf{e} \in \mathfrak{E}_{\leqslant d} \smallsetminus \mathfrak{E}_{\leqslant d-1}$, and has degree 0 if $\mathbf{e} \in \mathfrak{E}_{\leqslant 0}$.

We note that the algebra $\mathfrak{E}$ is *enveloping* in the sense that it contains the Lie algebra $\mathfrak{g}$, and that it is *universal* in the sense that any Lie algebra representation of $\mathfrak{g}$ (sending Lie brackets to Lie brackets as in (4.8)) can be extended to an algebra representation of $\mathfrak{E}$ (sending products to products). This functorial property holds essentially by definition of $\mathfrak{E}$.

However, let us start by extending the adjoint representation to $\mathfrak{E}$. We define $\mathrm{Ad}_g(\mathbb{1}_{\mathfrak{E}}) = \mathbb{1}_{\mathfrak{E}}$,

$$\mathrm{Ad}_g(\mathbf{b}_1 \circ \cdots \circ \mathbf{b}_n) = (\mathrm{Ad}_g \mathbf{b}_1) \circ \cdots \circ (\mathrm{Ad}_g \mathbf{b}_n)$$

for all $g \in G$, $n \in \mathbb{N}$, and $\mathbf{b}_1, \ldots, \mathbf{b}_n \in \mathfrak{g}$ and extend Ad_g linearly to all of $\mathfrak{E}$. For this we need to point out that

$$\begin{aligned}\mathrm{Ad}_g(\mathbf{a} \circ \mathbf{b} - \mathbf{b} \circ \mathbf{a} - [\mathbf{a}, \mathbf{b}]) &= (\mathrm{Ad}_g \mathbf{a}) \circ (\mathrm{Ad}_g \mathbf{b}) - (\mathrm{Ad}_g \mathbf{b}) \circ (\mathrm{Ad}_g \mathbf{a}) \\ &\quad - \mathrm{Ad}_g([\mathbf{a}, \mathbf{b}]) \\ &= (\mathrm{Ad}_g \mathbf{a}) \circ (\mathrm{Ad}_g \mathbf{b}) - (\mathrm{Ad}_g \mathbf{b}) \circ (\mathrm{Ad}_g \mathbf{a}) \\ &\quad - [\mathrm{Ad}_g(\mathbf{a}), \mathrm{Ad}_g(\mathbf{b})]\end{aligned}$$

for $g \in G$ and $\mathbf{a}, \mathbf{b} \in \mathfrak{g}$. Hence the adjoint action of G on the algebra of formal products sends the ideal appearing in Definition 9.1 to itself, and we obtain a well-defined representation of G on $\mathfrak{E}$ and, by restriction, also on $\mathfrak{E}_{\leqslant d}$ for all $d \in \mathbb{N}_0$.

By the discussion in Section 4.1.3, we may also take the derivative of the adjoint representation of G on $\mathfrak{E}$ (or, said more carefully, on $\mathfrak{E}_{\leqslant d}$ for all $d \in \mathbb{N}_0$) to obtain a representation of $\mathfrak{g}$ on $\mathfrak{E}$. We will again call this representation the adjoint representation, and denote it by

$$\begin{aligned}\mathrm{ad}\colon \mathfrak{g} &\longrightarrow \mathrm{End}(\mathfrak{E}) \\ \mathbf{c} &\longmapsto \mathrm{ad}_{\mathbf{c}}.\end{aligned}$$

In fact, for $n \in \mathbb{N}$, $\mathbf{b}_1, \ldots, \mathbf{b}_n, \mathbf{c} \in \mathfrak{g}$ the adjoint representation satisfies

$$\begin{aligned}\mathrm{ad}_{\mathbf{c}}(\mathbf{b}_1 \circ \cdots \circ \mathbf{b}_n) &= \frac{\mathrm{d}}{\mathrm{d}t}\bigg|_{t=0} \big(\mathrm{Ad}_{\exp(t\mathbf{c})} \mathbf{b}_1\big) \circ \cdots \circ \big(\mathrm{Ad}_{\exp(t\mathbf{c})} \mathbf{b}_n\big) \\ &= (\mathrm{ad}_{\mathbf{c}} \mathbf{b}_1) \circ \mathbf{b}_2 \circ \cdots \circ \mathbf{b}_n + \cdots + \mathbf{b}_1 \circ \mathbf{b}_2 \circ \cdots \circ (\mathrm{ad}_{\mathbf{c}} \mathbf{b}_n). \qquad (9.1)\end{aligned}$$

This generalized product rule follows, for example, by restricting to $\mathfrak{E}_{\leqslant d}$ for some $d \geqslant n$, applying the approximation formula

$$\mathrm{Ad}_{\exp(t\mathbf{c})}(\mathbf{b}_j) = \mathbf{b}_j + t\,\mathrm{ad}_{\mathbf{c}}(\mathbf{b}_j) + \mathrm{O}(t^2)$$

for $t \to 0$ and $j \in \{1, \ldots, n\}$, using multi-linearity to expand the product above, and letting t go to zero.

We note that substituting the relation

$$\mathrm{ad}_{\mathbf{c}}(\mathbf{b}_j) = [\mathbf{c}, \mathbf{b}_j] = \mathbf{c} \circ \mathbf{b}_j - \mathbf{b}_j \circ \mathbf{c}$$

for $j = 1, \ldots, n$ into (9.1) gives the telescoping sum

$$\begin{aligned}\mathrm{ad}_{\mathbf{c}}(\mathbf{b}_1 \circ \cdots \circ \mathbf{b}_n) &= (\mathbf{c} \circ \mathbf{b}_1 \circ \mathbf{b}_2 \circ \cdots \circ \mathbf{b}_n - \mathbf{b}_1 \circ \mathbf{c} \circ \mathbf{b}_2 \circ \cdots \circ \mathbf{b}_n) \\ &\quad + (\mathbf{b}_1 \circ \mathbf{c} \circ \mathbf{b}_2 \circ \cdots \circ \mathbf{b}_n - \mathbf{b}_1 \circ \mathbf{b}_2 \circ \mathbf{c} \circ \cdots \circ \mathbf{b}_n) \\ &\quad + \cdots \\ &\quad + (\mathbf{b}_1 \circ \mathbf{b}_2 \circ \cdots \circ \mathbf{c} \circ \mathbf{b}_n - \mathbf{b}_1 \circ \mathbf{b}_2 \circ \cdots \circ \mathbf{b}_n \circ \mathbf{c}) \\ &= \mathbf{c} \circ \mathbf{b}_1 \circ \cdots \circ \mathbf{b}_n - \mathbf{b}_1 \circ \cdots \circ \mathbf{b}_n \circ \mathbf{c},\end{aligned}$$

which by linearity shows that

$$\mathrm{ad}_{\mathbf{c}}(\mathbf{e}) = \mathbf{c} \circ \mathbf{e} - \mathbf{e} \circ \mathbf{c} \tag{9.2}$$

for $\mathbf{c} \in \mathfrak{g}$ and $\mathbf{e} \in \mathfrak{E}$.

To get a better feeling for $\mathfrak{E}$, we describe how to obtain a basis for it. For this, suppose that $\mathbf{b}_1, \dots, \mathbf{b}_n \in \mathfrak{g}$ form a basis of the Lie algebra $\mathfrak{g}$. Then, by the Poincaré–Birkhoff–Witt Theorem (see Knapp [54, Th. 3.8]) the products

$$\mathbf{b}_1^{\circ k_1} \circ \cdots \circ \mathbf{b}_n^{\circ k_n} \tag{9.3}$$

for $(k_1, \dots, k_n) \in \mathbb{N}_0^n$ form a basis of $\mathfrak{E}$. We will not need or prove this in detail, but wish to indicate briefly why this should be true.

Firstly, because of the assumed multi-linearity of the products appearing in Definition 9.1, it is clear that $\mathfrak{E}$ is the linear hull of $\mathbb{1}_{\mathfrak{E}}$, $\mathbf{b}_1,\dots,\mathbf{b}_n$, and all non-commuting products of $\mathbf{b}_1,\dots,\mathbf{b}_n$ in any order and multiplicity. However, using the ideal appearing in Definition 9.1 we may swap two basis elements in such a product, possibly at the cost of adding a term of lower degree. This allows us to order the basis elements in a product so that they are of the form in (9.3). Using this and an induction on the degree, we arrive at the statement that $\mathfrak{E}$ is the linear hull of products as in (9.3) for $(k_1, \dots, k_n) \in \mathbb{N}_0^n$. In a way the ideal appearing in Definition 9.1 precisely allows us to do this re-ordering of arbitrary products of basis elements to transform them to the shape of (9.3), but does not allow anything else. This suggests the second half of the Poincaré–Birkhoff–Witt Theorem, namely the statement that the products in (9.3) are all linearly independent within $\mathfrak{E}$.

9.1.2 The Casimir Element for $\mathfrak{sl}_2(\mathbb{R})$

As it is our goal in this chapter to describe the unitary dual $\widehat{\mathrm{SL}_2(\mathbb{R})}$, and the notion of universal enveloping algebra is meant to be a tool for that goal, it is only natural that we want to study the universal enveloping algebra $\mathfrak{E}$ of the Lie algebra $\mathfrak{sl}_2(\mathbb{R})$. This will, in particular, reveal that $\mathfrak{E}$ has an interesting central element.

Hence we let $\mathfrak{g} = \mathfrak{sl}_2(\mathbb{R})$. We will use the basis elements $\mathbf{a}, \mathbf{e}, \mathbf{f}$ as in (4.2). These form an $\mathfrak{sl}_2$-triple as they satisfy the relations $[\mathbf{a}, \mathbf{e}] = 2\mathbf{e}$, $[\mathbf{a}, \mathbf{f}] = -2\mathbf{f}$, and $[\mathbf{e}, \mathbf{f}] = \mathbf{a}$ by (4.3). In addition to these, we will also use the Lie algebra elements

$$\mathbf{d} = \mathbf{e} + \mathbf{f} = \begin{pmatrix} 0 & 1 \\ 0 & 0 \end{pmatrix} + \begin{pmatrix} 0 & 0 \\ 1 & 0 \end{pmatrix} = \begin{pmatrix} 0 & 1 \\ 1 & 0 \end{pmatrix} \in \mathfrak{sl}_2(\mathbb{R})$$

and

$$\mathbf{k} = -\mathbf{e} + \mathbf{f} = \begin{pmatrix} 0 & -1 \\ 0 & 0 \end{pmatrix} + \begin{pmatrix} 0 & 0 \\ 1 & 0 \end{pmatrix} = \begin{pmatrix} 0 & -1 \\ 1 & 0 \end{pmatrix} \in \mathfrak{sl}_2(\mathbb{R}).$$

We will define the central element Ω of the universal enveloping algebra $\mathfrak{E}$ of $\mathfrak{sl}_2(\mathbb{R})$ directly below. However, we first wish to explain an abstract argument (relying on the Poincaré–Birkhoff–Witt Theorem and the theory of finite-dimensional representations of $\mathfrak{sl}_2(\mathbb{R})$ in Section 4.1) showing that such a central element of degree two has to exist.

For this, note first that by the Poincaré–Birkhoff–Witt Theorem $\mathfrak{E}_{\leqslant 2}$ has the 10 elements

$$\mathbb{1}_{\mathfrak{E}}, \mathbf{a}, \mathbf{e}, \mathbf{f}, \mathbf{a}^{\circ 2}, \mathbf{a} \circ \mathbf{e}, \mathbf{a} \circ \mathbf{f}, \mathbf{e}^{\circ 2}, \mathbf{e} \circ \mathbf{f}, \mathbf{f}^{\circ 2}$$

as a basis. To understand $\mathfrak{E}_{\leqslant 2}$ as a finite-dimensional representation for the Lie algebra $\mathfrak{sl}_2(\mathbb{R})$, we have to find its highest weight vectors. Clearly $\mathbb{1}_{\mathfrak{E}}$ is a highest weight vector corresponding to weight 0 and spanning a trivial sub-representation. Next we note that $\mathbf{e}$ is a highest weight vector (since we know that $\mathrm{ad}_{\mathbf{a}}(\mathbf{e}) = 2\mathbf{e}$ and $\mathrm{ad}_{\mathbf{e}}(\mathbf{e}) = 0$). Also note that $\mathbf{e}$ generates the sub-representation $\mathfrak{g} \subseteq \mathfrak{E}_{\leqslant 2}$. Moreover, $\mathbf{e}^{\circ 2}$ is another highest weight vector, since

$$\mathrm{ad}_{\mathbf{a}}(\mathbf{e} \circ \mathbf{e}) = \mathrm{ad}_{\mathbf{a}}(\mathbf{e}) \circ \mathbf{e} + \mathbf{e} \circ \mathrm{ad}_{\mathbf{a}}(\mathbf{e}) = 4\mathbf{e} \circ \mathbf{e}$$

and

$$\mathrm{ad}_{\mathbf{e}}(\mathbf{e} \circ \mathbf{e}) = \mathrm{ad}_{\mathbf{e}}(\mathbf{e}) \circ \mathbf{e} + \mathbf{e} \circ \mathrm{ad}_{\mathbf{e}}(\mathbf{e}) = 0.$$

Since $\mathbf{e}^{\circ 2}$ has weight 4, it generates a 5-dimensional irreducible subspace $\mathcal{V}_5$ inside $\mathfrak{E}_{\leqslant 2}$. Together we see that the sub-representations $\mathbb{C}\mathbb{1}_{\mathfrak{E}}$, $\mathfrak{g}$, and $\mathcal{V}_5$ found so far have dimension 1, 3, and 5 respectively. As $\mathfrak{E}_{\leqslant 2}$ has dimension 10 we see that there exists a one-dimensional invariant complement $\mathbb{C}\Omega$ to

$$\mathbb{C}\mathbb{1}_{\mathfrak{E}} \oplus \mathfrak{g} \oplus \mathcal{V}_5 = \mathfrak{E}_{\leqslant 1} \oplus \mathcal{V}_5,$$

on which $\mathfrak{sl}_2(\mathbb{R})$ acts trivially. This shows the existence of an element Ω of degree 2 with $\mathrm{ad}_{\mathbf{b}}(\Omega) = 0$ for all $\mathbf{b} \in \mathfrak{sl}_2(\mathbb{R})$. By (9.2) we may also write this as

$$\mathbf{b} \circ \Omega = \Omega \circ \mathbf{b}$$

for all $\mathbf{b} \in \mathfrak{sl}_2(\mathbb{R})$. Since $\mathfrak{E}$ is generated by $\mathbb{1}_{\mathfrak{E}}$ and $\mathfrak{sl}_2(\mathbb{R})$ as an algebra, we find that Ω belongs to the centre of $\mathfrak{E}$.

In order to be logically independent of the Poincaré–Birkhoff–Witt Theorem and, more importantly, to be able to do concrete calculations with Ω, we now give a direct definition of Ω.

Lemma 9.2 (The Casimir element for $\mathfrak{sl}_2(\mathbb{R})$). *Let* $\mathfrak{g} = \mathfrak{sl}_2(\mathbb{R})$ *and let* $\mathbf{a}$, $\mathbf{e}$, $\mathbf{f}$ *be the* $\mathfrak{sl}_2$*-triple from* (4.2)*. Define* $\mathbf{d} = \mathbf{e} + \mathbf{f}$ *and* $\mathbf{k} = -\mathbf{e} + \mathbf{f}$*. Then*

$$\Omega = \mathbb{1}_{\mathfrak{E}} + \mathbf{a}^{\circ 2} + \mathbf{d}^{\circ 2} - \mathbf{k}^{\circ 2} = \mathbb{1}_{\mathfrak{E}} + \mathbf{a}^{\circ 2} + 2\mathbf{e} \circ \mathbf{f} + 2\mathbf{f} \circ \mathbf{e}$$

is a degree two element in the centre of $\mathfrak{E}$, which we will refer to as the Casimir element.

PROOF. We use the definitions of $\mathbf{d}$ and $\mathbf{k}$ to see that

$$\begin{aligned}
&\mathbb{1}_{\mathfrak{E}} + \mathbf{a}^{\circ 2} + (\mathbf{e}+\mathbf{f})^{\circ 2} - (-\mathbf{e}+\mathbf{f})^{\circ 2} \\
&= \mathbb{1}_{\mathfrak{E}} + \mathbf{a}^{\circ 2} + \mathbf{e}^{\circ 2} + \mathbf{e}\circ\mathbf{f} + \mathbf{f}\circ\mathbf{e} + \mathbf{f}^{\circ 2} - \left(\mathbf{e}^{\circ 2} - \mathbf{e}\circ\mathbf{f} - \mathbf{f}\circ\mathbf{e} + \mathbf{f}^{\circ 2}\right) \\
&= \mathbb{1}_{\mathfrak{E}} + \mathbf{a}^{\circ 2} + 2\mathbf{e}\circ\mathbf{f} + 2\mathbf{f}\circ\mathbf{e}.
\end{aligned}$$

Hence the two expressions in the lemma define the same element Ω. Next we note that $\mathrm{ad}_{\mathbf{b}}(\mathbb{1}_{\mathfrak{E}}) = 0$ for all $\mathbf{b}\in\mathfrak{g}$. Hence we obtain, by the product rule in (9.1), that

$$\begin{aligned}
\mathrm{ad}_{\mathbf{a}}(\Omega) &= \mathrm{ad}_{\mathbf{a}}\left(\mathbf{a}^{\circ 2} + 2\mathbf{e}\circ\mathbf{f} + 2\mathbf{f}\circ\mathbf{e}\right) \\
&= 2(\mathrm{ad}_{\mathbf{a}}\,\mathbf{e})\circ\mathbf{f} + 2\mathbf{e}\circ(\mathrm{ad}_{\mathbf{a}}\,\mathbf{f}) + 2(\mathrm{ad}_{\mathbf{a}}\,\mathbf{f})\circ\mathbf{e} + 2\mathbf{f}\circ\mathrm{ad}_{\mathbf{a}}(\mathbf{e}) \\
&= 2(2\mathbf{e})\circ\mathbf{f} + 2\mathbf{e}\circ(-2\mathbf{f}) + 2(-2\mathbf{f})\circ\mathbf{e} + 2\mathbf{f}\circ(2\mathbf{e}) = 0
\end{aligned}$$

and

$$\begin{aligned}
\mathrm{ad}_{\mathbf{e}}(\Omega) &= \mathrm{ad}_{\mathbf{e}}\left(\mathbf{a}^{\circ 2} + 2\mathbf{e}\circ\mathbf{f} + 2\mathbf{f}\circ\mathbf{e}\right) \\
&= \mathrm{ad}_{\mathbf{e}}(\mathbf{a})\circ\mathbf{a} + \mathbf{a}\circ\mathrm{ad}_{\mathbf{e}}(\mathbf{a}) + 2\mathbf{e}\circ(\mathrm{ad}_{\mathbf{e}}\,\mathbf{f}) + 2(\mathrm{ad}_{\mathbf{e}}\,\mathbf{f})\circ\mathbf{e} \\
&= -2\mathbf{e}\circ\mathbf{a} - 2\mathbf{a}\circ\mathbf{e} + 2\mathbf{e}\circ\mathbf{a} + 2\mathbf{a}\circ\mathbf{e} = 0.
\end{aligned}$$

The calculation $\mathrm{ad}_{\mathbf{f}}(\Omega) = 0$ is similar, but also follows from the properties of finite-dimensional representations of $\mathfrak{sl}_2(\mathbb{R})$ in Section 4.1.4 applied to ad on $\mathfrak{E}_{\leqslant 2}$: Since $\mathrm{ad}_{\mathbf{a}}(\Omega) = \mathrm{ad}_{\mathbf{e}}(\Omega) = 0$ it follows that $\Omega\in\mathfrak{E}_{\leqslant 2}$ is a highest weight vector with weight zero, and so generates the trivial representation of $\mathfrak{sl}_2(\mathbb{R})$.

Applying (9.2) and using the fact that $\mathfrak{E}$ is generated by $\mathbb{1}_{\mathfrak{E}}$ and $\mathfrak{sl}_2(\mathbb{R})$ as an algebra, this implies (as discussed just before the lemma) that Ω belongs to the centre of $\mathfrak{E}$.

It remains to prove that Ω has degree two. Assume the opposite, meaning that $\Omega\in\mathfrak{E}_{\leqslant 1}$. Since $\mathfrak{E}_{\leqslant 1} = \mathbb{1}_{\mathfrak{E}}\oplus\mathfrak{g}$ and $\mathfrak{g}$ has no centre, this would imply that $\Omega = \alpha\mathbb{1}_{\mathfrak{E}}$ for some $\alpha\in\mathbb{C}$. To derive a contradiction from this, we apply the universal property of $\mathfrak{E}$: Any Lie algebra representation ρ of $\mathfrak{sl}_2(\mathbb{R})$ on a real vector space $\mathcal{V}$ can be extended to an algebra representation of $\mathfrak{E}$. Indeed, we may simply define ρ by $\rho(\mathbb{1}_{\mathfrak{E}}) = I$, composition as in

$$\rho(\mathbf{b}_1\circ\cdots\circ\mathbf{b}_n) = \rho(\mathbf{b}_1)\cdots\rho(\mathbf{b}_n)$$

for $\mathbf{b}_1,\dots,\mathbf{b}_n\in\mathfrak{g}$ and $n\in\mathbb{N}$, and linear extension. For this, note that the ideal appearing in Definition 9.1 is sent to 0 due to the definition of a Lie algebra homomorphism (see (4.8)).

We first apply this universal property to the trivial representation ρ on $\mathbb{C}$. Here $\rho(\mathbf{b}) = 0$ for all $\mathbf{b}\in\mathfrak{g}$, which implies that $\rho(\Omega) = \rho(\mathbb{1}_{\mathfrak{E}}) = 1$.

Next we apply the universal property to the standard representation ρ on the space $\mathcal{V} = \mathbb{C}^2$. In this case

$$\rho(\Omega) = I + \mathbf{a}^2 + \mathbf{d}^2 - \mathbf{k}^2 = 4I,$$

where $\mathbf{a}^2 = I$, $\mathbf{d}^2 = I$, $\mathbf{k}^2 = -I$ are just the matrix squares. It follows that the central element $\Omega \in \mathfrak{E}_{\leqslant 2}$ is not a multiple of $\mathbb{1}_{\mathfrak{E}}$, does not belong to $\mathfrak{E}_{\leqslant 1}$, and hence has degree two as claimed. □

The reader who hopes to find more central elements in $\mathfrak{E}_{\leqslant d}$ for larger values of $d \in \mathbb{N}$, is invited to solve the following exercise.

Essential Exercise 9.3. Let $\mathfrak{g} = \mathfrak{sl}_2(\mathbb{R})$, and assume the Poincaré–Birkhoff–Witt Theorem for its universal enveloping algebra.
(a) For every $d \in \mathbb{N}$, calculate the weight of $\mathbf{e}^{\circ d} \in \mathfrak{E}_{\leqslant d}$. Show that all other eigenvectors of $\mathfrak{E}_{\leqslant d}$ have smaller weight. Conclude that $\mathfrak{E}_{\leqslant d}$ contains an irreducible invariant subspace $\mathcal{V}_{2d+1}$ of dimension $(2d+1)$ that is not contained in $\mathfrak{E}_{\leqslant d-1}$.
(b) Calculate the dimension of $\mathfrak{E}_{\leqslant 3}$ and $\mathfrak{E}_{\leqslant 4}$, analyze the representation appearing, and show that the centre of $\mathfrak{E}$ intersected with $\mathfrak{E}_{\leqslant 4}$ is the linear hull of $\mathbb{1}, \Omega, \Omega^{\circ 2}$.
(c) Repeat (b) for all $\mathfrak{E}_{\leqslant d}$ with $d \in \mathbb{N}$ to see that the centre of $\mathfrak{E}$ is the linear hull of $\mathbb{1}_{\mathfrak{E}}$ and $\Omega^{\circ n}$ for $n \in \mathbb{N}$.

9.1.3 Higher-order Differential Operators

We return to the general case and now show, as promised, that $\mathfrak{E}$ can be thought of as the algebra of differential operators arising by composition from $\mathfrak{g}$.

Proposition 9.4 (Differential operators coming from $\mathfrak{E}$). *Let G be a Lie group with Lie algebra $\mathfrak{g}$ and let π be a unitary representation of G. Then the representation of $\mathfrak{g}$ via π_∂ on smooth vectors extends to a representation of the universal enveloping algebra $\mathfrak{E}$ of $\mathfrak{g}$ on smooth vectors in such a way that*

$$\pi_\partial(\mathbf{b}_1 \circ \cdots \circ \mathbf{b}_n) = \pi_\partial(\mathbf{b}_1) \cdots \pi_\partial(\mathbf{b}_n) \tag{9.4}$$

for all $n \in \mathbb{N}$ and $\mathbf{b}_1, \ldots, \mathbf{b}_n \in \mathfrak{g}$. Furthermore,

$$\pi_g \pi_\partial(\mathbf{e}) \pi_{g^{-1}} = \pi_\partial\big(\mathrm{Ad}_g(\mathbf{e})\big) \tag{9.5}$$

for all $\mathbf{e} \in \mathfrak{E}$ and $g \in G$.

Proof. Let $v \in \mathcal{H}_\pi$ be smooth. Recall that by Lemma 7.4 the vector $\pi_\partial(\mathbf{b})v$ depends linearly on $\mathbf{b} \in \mathfrak{g}$. This extends to multi-linear dependence of

$$\pi_\partial(\mathbf{b}_1) \cdots \pi_\partial(\mathbf{b}_n) v$$

on $\mathbf{b}_1, \ldots, \mathbf{b}_n \in \mathfrak{g}$. Therefore π_∂ extends from $\mathfrak{g}$ to the algebra of formal multi-linear non-commuting products of elements of $\mathfrak{g}$ as appearing in Definition 9.1.

However, to see that π_∂ extends to $\mathfrak{E}$ we have to show that π_∂ sends the ideal appearing in Definition 9.1 to zero, or equivalently that

$$\pi_\partial\big([\mathbf{a},\mathbf{b}]\big)v = \big(\pi_\partial(\mathbf{a})\pi_\partial(\mathbf{b}) - \pi_\partial(\mathbf{b})\pi_\partial(\mathbf{a})\big)v \tag{9.6}$$

for all $\mathbf{a},\mathbf{b} \in \mathfrak{g}$ and all smooth $v \in \mathcal{H}_\pi$. To see this, we fix a vector $w \in \mathcal{H}_\pi$ and look at the matrix coefficient $\varphi = \varphi^\pi_{w,v}$. Let us use left-translation by elements of G to define a vector field $\lambda_\partial(\mathbf{m})$ on G for every $\mathbf{m} \in \mathfrak{g}$, as in Proposition 7.8. Smoothness of v and Lemma 7.20 show that

$$\begin{aligned}\lambda_\partial(\mathbf{m})\varphi(g) = \frac{\mathrm{d}}{\mathrm{d}t}\Big|_{t=0}\varphi^\pi_{w,v}(\exp(-t\mathbf{m})g) &= \frac{\mathrm{d}}{\mathrm{d}t}\Big|_{t=0}\big\langle \pi_{\exp(-t\mathbf{m})}\pi_g w, v\big\rangle \\ &= \frac{\mathrm{d}}{\mathrm{d}t}\Big|_{t=0}\big\langle \pi_g w, \pi_{\exp(t\mathbf{m})}v\big\rangle \\ &= \big\langle \pi_g w, \pi_\partial(\mathbf{m})v\big\rangle = \varphi^\pi_{w,\pi_\partial(\mathbf{m})v}(g)\end{aligned}$$

exists for all $\mathbf{m} \in \mathfrak{g}$ and $g \in G$, which, when iterated, shows that $\varphi \in C^\infty(G)$. However, for smooth functions on G, the formula

$$\lambda_\partial([\mathbf{a},\mathbf{b}])\varphi = \lambda_\partial(\mathbf{a})\lambda_\partial(\mathbf{b})\varphi - \lambda_\partial(\mathbf{b})\lambda_\partial(\mathbf{a})\varphi \tag{9.7}$$

is the definition of the Lie bracket for general Lie groups (see Exercise 9.5). At $g = e$, together with the above, this becomes

$$\begin{aligned}\big\langle w, \pi_\partial\big([\mathbf{a},\mathbf{b}]\big)v\big\rangle &= \lambda_\partial([\mathbf{a},\mathbf{b}])\varphi^\pi_{w,v}(e) \\ &= \big(\lambda_\partial(\mathbf{a})\lambda_\partial(\mathbf{b})\varphi^\pi_{w,v} - \lambda_\partial(\mathbf{b})\lambda_\partial(\mathbf{a})\varphi^\pi_{w,v}\big)(e) \\ &= \big(\lambda_\partial(\mathbf{a})\varphi^\pi_{w,\pi_\partial(\mathbf{b})v} - \lambda_\partial(\mathbf{b})\varphi^\pi_{w,\pi_\partial(\mathbf{a})v}\big)(e) \\ &= \langle w, \pi_\partial(\mathbf{a})\pi_\partial(\mathbf{b})v\rangle - \langle w, \pi_\partial(\mathbf{b})\pi_\partial(\mathbf{a})v\rangle\,.\end{aligned}$$

As this holds for all $w \in \mathcal{H}_\pi$, we obtain (9.6).

For $g \in G$ and $\mathbf{e} \in \mathfrak{g}$, the identity in (9.5) is simply the chain rule in Proposition 7.7. Moreover, if $\mathbf{e} = \mathbf{a}_1 \circ \mathbf{a}_2 \circ \cdots \circ \mathbf{a}_n \in \mathfrak{E}$ then we also have

$$\begin{aligned}\pi_g\pi_\partial(\mathbf{e})\pi_{g^{-1}} &= \pi_g\pi_\partial(\mathbf{a}_1)\pi_\partial(\mathbf{a}_2)\cdots\pi_\partial(\mathbf{a}_n)\pi_{g^{-1}} \\ &= \pi_g\pi_\partial(\mathbf{a}_1)\pi_{g^{-1}}\pi_g\pi_\partial(\mathbf{a}_2)\pi_{g^{-1}}\cdots\pi_g\pi_\partial(\mathbf{a}_n)\pi_{g^{-1}} \\ &= \pi_\partial\big(\mathrm{Ad}_g(\mathbf{a}_1)\big)\pi_\partial\big(\mathrm{Ad}_g(\mathbf{a}_2)\big)\cdots\pi_\partial\big(\mathrm{Ad}_g(\mathbf{a}_n)\big) \\ &= \pi_\partial\big(\mathrm{Ad}_g(\mathbf{a}_1)\circ\cdots\circ\mathrm{Ad}_g(\mathbf{a}_n)\big) = \pi_\partial\big(\mathrm{Ad}_g(\mathbf{e})\big)\end{aligned}$$

which, together with linearity, proves (9.5) for $\mathbf{e} \in \mathfrak{E}$. □

Essential Exercise 9.5 (Lie brackets for linear groups). Prove (9.7) for closed linear groups, where the Lie bracket is defined by (4.1) using matrix products.

Corollary 9.6 (Adjoint). *Let G be a Lie group with Lie algebra $\mathfrak{g}$ and universal enveloping algebra $\mathfrak{E}$. Then there exists a linear anti-homomorphism*

$$^*\colon \mathfrak{E} \longrightarrow \mathfrak{E}$$

satisfying $\mathbf{a}^* = -\mathbf{a}$ *and* $(\mathbf{e} \circ \mathbf{f})^* = \mathbf{f}^* \circ \mathbf{e}^*$ *for all* $\mathbf{a} \in \mathfrak{g}$, *and* $\mathbf{e}, \mathbf{f} \in \mathfrak{E}$. *Moreover, we have*

$$\langle \pi_\partial(\mathbf{e})u, v\rangle = \langle u, \pi_\partial(\mathbf{e}^*)v\rangle$$

for all $\mathbf{e} \in \mathfrak{E}$ *whenever* π *is a unitary representation of* G *and* $u, v \in \mathcal{H}_\pi$ *are smooth.*

We will refer to $*$ on $\mathfrak{E}$ as the *formal adjoint.*

PROOF OF COROLLARY 9.6. We define $^*\colon \mathfrak{g} \ni \mathbf{a} \mapsto -\mathbf{a} \in \mathfrak{g}$ and extend * to a linear map from $\mathfrak{E}$ to $\mathfrak{E}$ by requiring

$$\big(\mathbf{e}_1 \circ \mathbf{e}_2\big)^* = \mathbf{e}_2^* \circ \mathbf{e}_1^* \tag{9.8}$$

for all $\mathbf{e}_1, \mathbf{e}_2 \in \mathfrak{E}$.

To see that * is well-defined, we verify that the relations appearing in Definition 9.5 are preserved. For $\mathbf{a}, \mathbf{b} \in \mathfrak{g}$ we have $\mathbf{a}^* = -\mathbf{a}$, $\mathbf{b}^* = -\mathbf{b}$, and that $[\mathbf{a}, \mathbf{b}] \in \mathfrak{g}$ satisfies $[\mathbf{a}, \mathbf{b}]^* = -[\mathbf{a}, \mathbf{b}]$. Together with (9.8), we obtain that

$$\begin{aligned}\big(\mathbf{a} \circ \mathbf{b} - \mathbf{b} \circ \mathbf{a} - [\mathbf{a}, \mathbf{b}]\big)^* &= (-\mathbf{b}) \circ (-\mathbf{a}) - (-\mathbf{a}) \circ (-\mathbf{b}) + [\mathbf{a}, \mathbf{b}]\\ &= \mathbf{b} \circ \mathbf{a} - \mathbf{a} \circ \mathbf{b} - [\mathbf{b}, \mathbf{a}]\end{aligned}$$

indeed is once more in the ideal appearing in the definition of $\mathfrak{E}$.

Now let $u, v \in \mathcal{H}_\pi$ be smooth and $\mathbf{a} \in \mathfrak{g}$. Then

$$\begin{aligned}\langle \pi_\partial(\mathbf{a})u, v\rangle &= \lim_{t\to 0} \left\langle \tfrac{1}{t}\big(\pi_{\exp(t\mathbf{a})}u - u\big), v\right\rangle\\ &= \lim_{t\to 0} \tfrac{1}{t}\big(\langle u, \pi_{\exp(-t\mathbf{a})}v\rangle - \langle u, v\rangle\big)\\ &= \lim_{t\to 0} \left\langle u, \tfrac{1}{t}\big(\pi_{\exp(-t\mathbf{a})}v - v\big)\right\rangle\\ &= \langle u, -\pi_\partial(\mathbf{a})v\rangle = \langle u, \pi_\partial(\mathbf{a}^*)v\rangle\,.\end{aligned}$$

Moreover, if $\mathbf{b} \in \mathfrak{g}$ then

$$\langle \pi_\partial(\mathbf{a} \circ \mathbf{b})u, v\rangle = \langle \pi_\partial(\mathbf{b})u, \pi_\partial(\mathbf{a}^*)v\rangle = \langle u, \pi_\partial(\mathbf{b}^* \circ \mathbf{a}^*)v\rangle = \big\langle u, \pi_\partial\big((\mathbf{a} \circ \mathbf{b})^*\big)v\big\rangle\,.$$

The corollary follows by iterating this, and combining it with sesqui-linearity of the inner product. □

9.1.4 Central Elements of the Universal Enveloping Algebra

We have seen in Section 9.1.2 that the universal enveloping algebra $\mathfrak{E}$ of $\mathfrak{sl}_2(\mathbb{R})$ has a central element, to which we will apply the following general results.

Essential Exercise 9.7. Let G be a connected closed linear group with Lie algebra $\mathfrak{g}$ and universal enveloping algebra $\mathfrak{E}$. Suppose that $\Omega \in \mathfrak{E}$ is central in the sense that $\mathbf{a} \circ \Omega = \Omega \circ \mathbf{a}$ for all $\mathbf{a} \in \mathfrak{g}$ (or, equivalently, $\mathbf{a} \in \mathfrak{E}$). Show that $\mathrm{Ad}_g(\Omega) = \Omega$ for all $g \in G$.

Proposition 9.8 (Operators coming from the centre of $\mathfrak{E}$). *Let G be a closed linear group with Lie algebra $\mathfrak{g}$, and let Ω be a central element of the universal enveloping algebra $\mathfrak{E}$ of $\mathfrak{g}$ satisfying $\mathrm{Ad}_g(\Omega) = \Omega$ for all $g \in G$. Then, for any unitary representation π of G, the closure of $\pi_\partial(\Omega)$ (defined on smooth vectors) is a well-defined closed intertwining operator. If π is irreducible, then the closure is multiplication by a scalar $\alpha_{\Omega,\pi} \in \mathbb{C}$ (respectively $\alpha_{\Omega,\pi} \in \mathbb{R}$ if $\Omega^* = \Omega$ also).*

PROOF. Let $\Omega \in \mathfrak{E}$ be central and let π be a unitary representation as in the proposition. We define T_π as the closure of $\pi_\partial(\Omega)$ acting on smooth vectors. More formally, we have that $v \in D_{T_\pi}$ belongs to the domain of T_π, and $T_\pi v = w$ is the image, if there exists a sequence (v_n) in $\mathcal{H}_\pi$ of smooth vectors with

$$\big(v_n, \pi_\partial(\Omega)v_n\big) \in \mathrm{Graph}\big(\pi_\partial(\Omega)\big)$$

converging to (v, w) as $n \to \infty$. To see that this defines a well-defined operator, we need to show that

$$(v, w) = \lim_{n\to\infty} \big(v_n, \pi_\partial(\Omega)v_n\big) \in \overline{\mathrm{Graph}\big(\pi_\partial(\Omega)\big)}$$

with $v = 0$ implies that $w = 0$. To see this, suppose $u \in \mathcal{H}_\pi$ is smooth, then

$$\langle w, u\rangle = \lim_{n\to\infty} \langle \pi_\partial(\Omega)v_n, u\rangle = \lim_{n\to\infty} \langle v_n, \pi_\partial(\Omega^*)u\rangle\,,$$

where we applied Corollary 9.6. Since $\lim_{n\to\infty} v_n = 0$ by assumption, we obtain $\langle w, u\rangle = 0$ for all smooth $u \in \mathcal{H}_\pi$. Since smooth vectors are dense by Proposition 7.8, it follows that $w = 0$. Hence T_π is a well-defined closed operator.

Suppose now that $v \in D_{T_\pi}$ so that, by definition,

$$(v, T_\pi v) = \lim_{n\to\infty} (v_n, \pi_\partial(\Omega)v_n) \in \mathrm{Graph}(T_\pi)$$

for a sequence (v_n) of smooth vectors. Fix some $g \in G$. Then

$$\pi_g \pi_\partial(\Omega)v_n = \pi_\partial\big(\underbrace{\mathrm{Ad}_g(\Omega)}_{=\Omega}\big)\pi_g v = \pi_\partial(\Omega)\pi_g v_n$$

by Corollary 9.6 and our assumptions on Ω. Now $v_n \to v$ and $\pi_\partial(\Omega)v_n \to T_\pi v$ as $n \to \infty$, so $\pi_g v_n \to \pi_g v$ and

$$\pi_\partial(\Omega)\pi_g v_n = \pi_g \pi_\partial(\Omega)v_n \longrightarrow \pi_g T_\pi v$$

as $n \to \infty$. Therefore

$$\left(\pi_g v, \pi_g T_\pi v\right) = \lim_{n\to\infty} \left(\pi_g v_n, \pi_\partial(\Omega)\pi_g v_n\right),$$

which shows that $T_\pi \pi_g v = \pi_g T_\pi v$ for all $g \in G$ and $v \in D_{T_\pi}$. In other words, T_π is a well-defined closed intertwining operator.

We note that if π is irreducible then Schur's lemma in the form of Corollary 1.38 implies that T_π is simply multiplication by some $\alpha_{\pi,\Omega} \in \mathbb{C}$. If, in addition, $\Omega^* = \Omega$ and $v \in \mathcal{H}_\pi$ is a smooth unit vector, then

$$\alpha_{\pi,\Omega} = \alpha_{\pi,\Omega} \langle v, v\rangle = \langle \pi_\partial(\Omega)v, v\rangle = \langle v, \pi_\partial(\Omega^*)v\rangle = \overline{\alpha_{\pi,\Omega}} \langle v, v\rangle = \overline{\alpha_{\pi,\Omega}}$$

shows that $\alpha_{\pi,\Omega} \in \mathbb{R}$. □

9.1.5 Complexification of the Universal Enveloping Algebra

For some of our discussions, the following extension of Definition 9.1 will be important.

Definition 9.9 (The complexification $\mathfrak{E}_\mathbb{C}$). Let $\mathfrak{g}$ be a Lie algebra with universal enveloping algebra $\mathfrak{E}$. We define the *complexification of* $\mathfrak{E}$ by

$$\mathfrak{E}_\mathbb{C} = \mathfrak{E} \otimes_\mathbb{R} \mathbb{C}.$$

For $\mathbf{a}_1 + \mathrm{i}\mathbf{b}_1, \mathbf{a}_2 + \mathrm{i}\mathbf{b}_2 \in \mathfrak{E}_\mathbb{C}$ with $\mathbf{a}_1, \mathbf{b}_1, \mathbf{a}_2, \mathbf{b}_2 \in \mathfrak{E}$ and $g \in G$, we also define multiplication by

$$(\mathbf{a}_1 + \mathrm{i}\mathbf{b}_1) \circ (\mathbf{a}_2 + \mathrm{i}\mathbf{b}_2) = \mathbf{a}_1 \circ \mathbf{a}_2 - \mathbf{b}_1 \circ \mathbf{b}_2 + \mathrm{i}(\mathbf{a}_1 \circ \mathbf{b}_2 + \mathbf{b}_1 \circ \mathbf{a}_2),$$

the adjoint operator by

$$\mathrm{Ad}_g(\mathbf{a}_1 + \mathrm{i}\mathbf{b}_1) = \mathrm{Ad}_g(\mathbf{a}_1) + \mathrm{i}\,\mathrm{Ad}_g(\mathbf{b}_2),$$

conjugation by

$$\overline{\mathbf{a}_1 + \mathrm{i}\mathbf{b}_1} = \mathbf{a}_1 - \mathrm{i}\mathbf{b}_1,$$

and the formal adjoint by $(\mathbf{a}_1 + \mathrm{i}\mathbf{b}_1)^* = \mathbf{a}_1^* - \mathrm{i}\mathbf{b}_1^*$.

We verify that multiplication is actually bilinear over $\mathbb{C}$ on $\mathfrak{E}_\mathbb{C}$. To see this, note first that multiplication is clearly linear over $\mathbb{R}$. Moreover,

$$\begin{aligned}
\big(\mathrm{i}(\mathbf{a}_1 + \mathrm{i}\mathbf{b}_1)\big) \circ (\mathbf{a}_2 + \mathrm{i}\mathbf{b}_2) &= (-\mathbf{b}_1 + \mathrm{i}\mathbf{a}_1) \circ (\mathbf{a}_2 + \mathrm{i}\mathbf{b}_2) \\
&= -\mathbf{b}_1 \circ \mathbf{a}_2 - \mathbf{a}_1 \circ \mathbf{b}_2 + \mathrm{i}(-\mathbf{b}_1 \circ \mathbf{b}_2 + \mathbf{a}_1 \circ \mathbf{a}_2) \\
&= \mathrm{i}\big(\mathbf{a}_1 \circ \mathbf{a}_2 - \mathbf{b}_1 \circ \mathbf{b}_2 + \mathrm{i}(\mathbf{a}_1 \circ \mathbf{b}_2 + \mathbf{b}_1 \circ \mathbf{a}_2)\big) \\
&= \mathrm{i}\big((\mathbf{a}_1 + \mathrm{i}\mathbf{b}_1) \circ (\mathbf{a}_2 + \mathrm{i}\mathbf{b}_2)\big).
\end{aligned}$$

Together with linearity over $\mathbb{R}$ in the first argument, linearity over $\mathbb{C}$ in the first argument follows. The proof for the second argument is identical. Finally, the proof of conjugate-linearity of conjugation and the formal adjoint on $\mathfrak{E}_\mathbb{C}$ are similar.

Proposition 9.10 (Differentials operators coming from $\mathfrak{E}_\mathbb{C}$). *Let G be a Lie group with Lie algebra $\mathfrak{g}$, and let π be a unitary representation of G. Then the representation of $\mathfrak{g}$ via π_∂ on smooth vectors extends to a complex representation of the complexification $\mathfrak{E}_\mathbb{C}$ of the universal enveloping algebra $\mathfrak{E}$ of $\mathfrak{g}$ on smooth vectors in $\mathcal{H}_\pi$ satisfying the properties of Proposition* 9.4 *and Corollary* 9.6.

PROOF. For $\mathbf{a}, \mathbf{b} \in \mathfrak{E}$, we define $\pi_\partial(\mathbf{a} + \mathrm{i}\mathbf{b}) = \pi_\partial(\mathbf{a}) + \mathrm{i}\pi_\partial(\mathbf{b})$. This gives an extension of the above representation to $\mathfrak{E}_\mathbb{C}$, satisfying linearity over $\mathbb{C}$. Moreover, if $\mathbf{a}_1, \mathbf{a}_2, \mathbf{b}_1, \mathbf{b}_2 \in \mathfrak{E}$, then, by definition of $\circ$ on $\mathfrak{E}_\mathbb{C}$,

$$\begin{aligned}\pi_\partial\big((\mathbf{a}_1 + \mathrm{i}\mathbf{b}_1)\circ(\mathbf{a}_2 + \mathrm{i}\mathbf{b}_2)\big) &= \pi_\partial\big(\mathbf{a}_1\circ\mathbf{a}_2 - \mathbf{b}_1\circ\mathbf{b}_2\big) + \mathrm{i}\pi_\partial\big(\mathbf{a}_1\circ\mathbf{b}_2 + \mathbf{b}_1\circ\mathbf{a}_2\big)\\ &= \pi_\partial(\mathbf{a}_1)\pi_\partial(\mathbf{a}_2) - \pi_\partial(\mathbf{b}_1)\pi_\partial(\mathbf{b}_2)\\ &\qquad + \mathrm{i}\big(\pi_\partial(\mathbf{a}_1)\pi_\partial(\mathbf{b}_2) - \pi_\partial(\mathbf{b}_1)\pi_\partial(\mathbf{a}_2)\big)\\ &= \pi_\partial(\mathbf{a}_1 + \mathrm{i}\mathbf{b}_1)\pi_\partial(\mathbf{a}_2 + \mathrm{i}\mathbf{b}_2),\end{aligned}$$

so (9.4) also holds for $\mathfrak{E}_\mathbb{C}$.

The extension of (9.5) follows simply by linearity of both sides over $\mathbb{C}$. Finally, the complex version of Corollary 9.6 follows from its real counterpart and sesquilinearity of the inner product on $\mathcal{H}_\pi$. □

To conclude, we wish to emphasize that the operators $\pi_\partial(\mathbf{e})$ for

$$\mathbf{e} \in \mathfrak{g}_\mathbb{C} = \mathfrak{g} \otimes_\mathbb{R} \mathbb{C} \subseteq \mathfrak{E}_\mathbb{C}$$

(or $\mathbf{e} \in \mathfrak{E}_\mathbb{C}$) do not in general directly correspond to first (or higher order) partial derivatives, but instead to complex linear combinations of these.

We also note that the complexification $\mathfrak{E}_\mathbb{C}$ of the universal enveloping algebra of $\mathfrak{sl}_2(\mathbb{R})$ reveals that our definition of Ω as in Lemma 9.2 formally matches the definition of the Casimir operator in Corollary 7.24. Indeed, there we used the basis vectors

$$\mathbf{b}_1 = \begin{pmatrix} \mathrm{i} & 0 \\ 0 & -\mathrm{i} \end{pmatrix} = \mathrm{i}\mathbf{a}, \mathbf{b}_2 = \begin{pmatrix} 0 & \mathrm{i} \\ \mathrm{i} & 0 \end{pmatrix} = \mathrm{i}\mathbf{d}, \mathbf{b}_3 = \begin{pmatrix} 0 & -1 \\ 1 & 0 \end{pmatrix} = \mathbf{k}$$

of $\mathfrak{su}_2(\mathbb{R}) \subseteq \mathfrak{sl}_2(\mathbb{C}) = \mathfrak{sl}_2(\mathbb{R}) \otimes_\mathbb{R} \mathbb{C}$, and hence

$$\Omega = \mathbb{1}_\mathfrak{E} + \mathbf{a}^{\circ 2} + \mathbf{d}^{\circ 2} - \mathbf{k}^{\circ 2} = \mathbb{1}_\mathfrak{E} - \mathbf{b}_1^{\circ 2} - \mathbf{b}_2^{\circ 2} - \mathbf{b}_3^{\circ 2}$$

within the complex universal enveloping algebra $\mathfrak{E}_\mathbb{C}$.

9.2 Raising, Lowering, and the Dual of $\mathrm{SL}_2(\mathbb{R})$

We now specialize and extend the material from the previous section in the case of $\mathrm{SL}_2(\mathbb{R})$. Our goal for the section is a description of $\widehat{\mathrm{SL}_2(\mathbb{R})}$, although the concrete construction of the principal and complementary series representations will be postponed to later sections.

9.2.1 The Casimir Operator

It follows from the discussion in Section 9.1.2 that Proposition 9.8 applies to unitary representations of $\mathrm{SL}_2(\mathbb{R})$ as in Corollary 9.11 below (see also Exercise 9.7). Indeed, the Casimir element $\Omega = \mathbb{1}_{\mathfrak{E}} + \mathbf{a}^{\circ 2} + \mathbf{d}^{\circ 2} - \mathbf{k}^{\circ 2}$ of $\mathfrak{sl}_2(\mathbb{R})$ satisfies $\Omega^* = \Omega$ by its definition in Lemma 9.2.

Corollary 9.11 (Casimir eigenvalue). *Let π be a unitary representation of $\mathrm{SL}_2(\mathbb{R})$ and let $\Omega \in \mathfrak{E}$ be the Casimir element in the centre of the enveloping algebra $\mathfrak{E}$ of $\mathfrak{sl}_2(\mathbb{R})$. Then the closure T_π of $\pi_\partial(\Omega)$ is a closed intertwining operator. If π is in addition irreducible, then $T_\pi = \alpha_\pi I$ for some $\alpha_\pi \in \mathbb{R}$.*

We will refer to $\pi_\partial(\Omega)$ (and T_π) as the *Casimir operator* and to α_π as the *Casimir eigenvalue* of the irreducible representation π. Note that for the trivial representation $\mathbb{1}$ of $\mathrm{SL}_2(\mathbb{R})$ the Casimir eigenvalue is $\alpha_\pi = 1$.

9.2.2 The Raising and Lowering Operators

We define[†] the elements $\mathbf{r}^+ = \frac{1}{2}(\mathbf{a} - \mathrm{i}\mathbf{d})$ and $\mathbf{r}^- = \frac{1}{2}(\mathbf{a} + \mathrm{i}\mathbf{d})$ of $\mathfrak{sl}_2(\mathbb{C})$, and note that $\overline{\mathbf{r}^+} = \mathbf{r}^-$. We calculate

$$\mathrm{ad}_{\mathbf{k}}(\mathbf{a}) = [-\mathbf{e} + \mathbf{f}, \mathbf{a}] = [\mathbf{a}, \mathbf{e}] - [\mathbf{a}, \mathbf{f}] = 2\mathbf{e} + 2\mathbf{f} = 2\mathbf{d}$$

and

$$\mathrm{ad}_{\mathbf{k}}(\mathbf{d}) = [-\mathbf{e} + \mathbf{f}, \mathbf{e} + \mathbf{f}] = -[\mathbf{e}, \mathbf{f}] + [\mathbf{f}, \mathbf{e}] = -2[\mathbf{e}, \mathbf{f}] = -2\mathbf{a},$$

which gives

$$\mathrm{ad}_{\mathbf{k}}(\mathbf{r}^+) = \mathbf{d} - \mathrm{i}(-\mathbf{a}) = \mathrm{i}(\mathbf{a} - \mathrm{i}\mathbf{d}) = 2\mathrm{i}\mathbf{r}^+$$

and, by conjugation,

$$\mathrm{ad}_{\mathbf{k}}(\mathbf{r}^-) = -2\mathrm{i}\mathbf{r}^-$$

also. With $\mathrm{Ad}_{\exp(\theta\mathbf{k})} = \exp(\mathrm{ad}_{\theta\mathbf{k}})$ for $\theta \in \mathbb{R}$, this also implies

[†] It is an unfortunate coincidence that $\mathbf{d} \in \mathfrak{sl}_2(\mathbb{R})$ multiplied by $\mathrm{i} \in \mathbb{C}$, giving $\mathrm{i}\mathbf{d}$, is notationally close to a familiar 'identity' notation, id.

$$\begin{cases} \mathrm{Ad}_{k_\theta}(\mathbf{r}^+) = \mathrm{e}^{2\mathrm{i}\theta}\mathbf{r}^+ \\ \mathrm{Ad}_{k_\theta}(\mathbf{r}^-) = \mathrm{e}^{-2\mathrm{i}\theta}\mathbf{r}^- \end{cases} \tag{9.9}$$

for all $k_\theta \in K$.

For a unitary representation π of $\mathrm{SL}_2(\mathbb{R})$, we will call $\pi_\partial(\mathbf{r}^+)$ the *raising operator* and $\pi_\partial(\mathbf{r}^-)$ the *lowering operator.*

Proposition 9.12 (Raising and lowering operators). *Let π be a unitary representation of $\mathrm{SL}_2(\mathbb{R})$. For any smooth K-eigenvector $v \in \mathcal{H}_\pi$ of weight n, the vector $\pi_\partial(\mathbf{r}^\pm)v$ has weight $n \pm 2$ (but either or both might be zero).*

Proof. Let $k_\theta \in K$ and let $v \in \mathcal{H}_\pi$ be a smooth vector of K-weight n. Then the chain rule in Proposition 9.4 and (9.9) imply that

$$\begin{aligned} \pi_{k_\theta}\pi_\partial(\mathbf{r}^+)v &= \pi_\partial\bigl(\mathrm{Ad}_{k_\theta}(\mathbf{r}^+)\bigr)\pi_{k_\theta}v \\ &= \pi_\partial\bigl(\mathrm{e}^{2\mathrm{i}\theta}\mathbf{r}^+\bigr)\mathrm{e}^{\mathrm{i}n\theta}v \\ &= \mathrm{e}^{\mathrm{i}(n+2)\theta}\pi_\partial(\mathbf{r}^+)v, \end{aligned}$$

and the calculation for $\pi_\partial(\mathbf{r}^-)v$ is identical. □

We also wish to relate the elements $\mathbf{r}^\pm$ to the Casimir operator in Lemma 9.2. By definition of $\mathbf{r}^+$ and $\mathbf{r}^-$, we have

$$\begin{aligned} \mathbf{a} &= \mathbf{r}^+ + \mathbf{r}^-, \\ \mathbf{d} &= \mathrm{i}(\mathbf{r}^+ - \mathbf{r}^-), \end{aligned}$$

and

$$[\mathbf{r}^+, \mathbf{r}^-] = \tfrac{1}{4}[\mathbf{a} - \mathrm{i}\mathbf{d}, \mathbf{a} + \mathrm{i}\mathbf{d}] = \tfrac{1}{4}\bigl(\mathrm{i}[\mathbf{a}, \mathbf{d}] - \mathrm{i}[\mathbf{d}, \mathbf{a}]\bigr) = \tfrac{\mathrm{i}}{2}[\mathbf{a}, \mathbf{d}] = -\mathrm{i}\mathbf{k} \tag{9.10}$$

since

$$[\mathbf{a}, \mathbf{d}] = [\mathbf{a}, \mathbf{e} + \mathbf{f}] = 2\mathbf{e} - 2\mathbf{f} = -2\mathbf{k}.$$

Using the definition of Ω in Lemma 9.2, we now obtain

$$\begin{aligned} \Omega &= \mathbb{1}_{\mathfrak{E}} + (\mathbf{r}^+ + \mathbf{r}^-) \circ (\mathbf{r}^+ + \mathbf{r}^-) - (\mathbf{r}^+ - \mathbf{r}^-) \circ (\mathbf{r}^+ - \mathbf{r}^-) - \mathbf{k}^{\circ 2} \\ &= \mathbb{1}_{\mathfrak{E}} + (\mathbf{r}^+ \circ \mathbf{r}^+ + \mathbf{r}^+ \circ \mathbf{r}^- + \mathbf{r}^- \circ \mathbf{r}^+ + \mathbf{r}^- \circ \mathbf{r}^-) \\ &\qquad - (\mathbf{r}^+ \circ \mathbf{r}^+ - \mathbf{r}^+ \circ \mathbf{r}^- - \mathbf{r}^- \circ \mathbf{r}^+ + \mathbf{r}^- \circ \mathbf{r}^-) - \mathbf{k}^{\circ 2} \\ &= \mathbb{1}_{\mathfrak{E}} + 2\mathbf{r}^+ \circ \mathbf{r}^- + 2\mathbf{r}^- \circ \mathbf{r}^+ - \mathbf{k}^{\circ 2}. \end{aligned}$$

We can also write this in the form

$$\Omega = \mathbb{1}_{\mathfrak{E}} + 4\mathbf{r}^+ \circ \mathbf{r}^- - 2[\mathbf{r}^+, \mathbf{r}^-] - \mathbf{k}^{\circ 2}.$$

Using (9.10), this gives

$$\Omega = 4\mathbf{r}^+ \circ \mathbf{r}^- + \underbrace{\mathbb{1}_{\mathfrak{E}} + 2\mathrm{i}\mathbf{k} - \mathbf{k}^{\circ 2}}_{(\mathbb{1}_{\mathfrak{E}} + \mathrm{i}\mathbf{k})^{\circ 2}}.$$

Therefore

$$\Omega = 4\mathbf{r}^+ \circ \mathbf{r}^- + (\mathbb{1}_{\mathfrak{E}} + \mathrm{i}\mathbf{k})^{\circ 2} = 4\mathbf{r}^- \circ \mathbf{r}^+ + (\mathbb{1}_{\mathfrak{E}} - \mathrm{i}\mathbf{k})^{\circ 2}, \tag{9.11}$$

where the second formula follows from the first by conjugation.

Corollary 9.13 (Norm of raised and lowered vectors). *Let π be a unitary representation of $\mathrm{SL}_2(\mathbb{R})$. Then for any smooth eigenvector $v \in \mathcal{H}_\pi$ for K and Ω with K-weight n and Casimir eigenvalue α, we have*

$$\begin{cases} \|\pi_\partial(\mathbf{r}^+)v\|^2 &= \frac{1}{4}\big((n+1)^2 - \alpha_\pi\big)\|v\|^2 \\ \|\pi_\partial(\mathbf{r}^-)v\|^2 &= \frac{1}{4}\big((n-1)^2 - \alpha_\pi\big)\|v\|^2. \end{cases}$$

To make these two formulas more memorable, we note that in both cases $n \pm 1$ is precisely the weight that lies in between the weight n of v and the weight $n \pm 2$ of the vector $\pi_\partial(\mathbf{r}^\pm)v$.

Proof of Corollary 9.13. We note that $(\mathbf{r}^+)^* = -\overline{\mathbf{r}^+} = -\mathbf{r}^-$ by Corollary 9.6 and Definition 9.9. Using Proposition 9.10 and the identity (9.11) in the form

$$\mathbf{r}^- \circ \mathbf{r}^+ = \tfrac{1}{4}\left(\Omega - (\mathbb{1}_{\mathfrak{E}} - \mathrm{i}\mathbf{k})^{\circ 2}\right)$$

gives

$$\begin{aligned} \|\pi_\partial(\mathbf{r}^+)v\|^2 &= \left\langle \pi_\partial(\mathbf{r}^+)v, \pi_\partial(\mathbf{r}^+)v \right\rangle \\ &= -\left\langle \pi_\partial(\mathbf{r}^- \circ \mathbf{r}^+)v, v \right\rangle \\ &= -\tfrac{1}{4}\left\langle \pi_\partial(\Omega - (\mathbb{1}_{\mathfrak{E}} - \mathrm{i}\mathbf{k})^{\circ 2})v, v \right\rangle. \end{aligned}$$

Since v has Casimir eigenvalue α and since

$$\pi_\partial(\mathbb{1}_{\mathfrak{E}} - \mathrm{i}\mathbf{k})v = v - \mathrm{i}\pi_\partial(\mathbf{k})v = v + nv = (n+1)v,$$

we obtain

$$\|\pi_\partial(\mathbf{r}^+)v\|^2 = -\tfrac{1}{4}\left(\alpha_\pi - (n+1)^2\right)\|v\|^2.$$

Similarly,

$$\begin{aligned} \|\pi_\partial(\mathbf{r}^-)v\|^2 &= \left\langle \pi_\partial(\mathbf{r}^-)v, \pi_\partial(\mathbf{r}^-)v \right\rangle \\ &= -\left\langle \pi_\partial(\mathbf{r}^+ \circ \mathbf{r}^-)v, v \right\rangle \\ &= -\tfrac{1}{4}\left\langle \pi_\partial(\Omega - (\mathbb{1}_{\mathfrak{E}} + \mathrm{i}\mathbf{k})^{\circ 2})v, v \right\rangle \\ &= -\tfrac{1}{4}\left(\alpha_\pi - (n-1)^2\right)\|v\|^2, \end{aligned}$$

completing the proof. □

We note that the raising and lowering operators are useful for proving irreducibility. In fact we already used these operators implicitly in the proofs of irreducibility of the discrete and mock discrete series representations of the group $\mathrm{SL}_2(\mathbb{R}) \cong \mathrm{SU}_{1,1}(\mathbb{R})$ in Section 8.4. More importantly, we will use the more abstract framework above involving the Casimir operator to classify all irreducible unitary representations of $\mathrm{SL}_2(\mathbb{R})$. In fact Corollary 9.13 already implies important restrictions for the Casimir eigenvalue α_π for an irreducible unitary representation π: If v as in the corollary is non-zero with K-weight n, then we must have

$$\alpha_\pi \leqslant (n \pm 1)^2. \tag{9.12}$$

Applying our above discussion to the irreducible unitary representation of Section 8.4, we obtain their Casimir eigenvalue.

Corollary 9.14 (Casimir for discrete and mock discrete series). *For every integer $n \geqslant 2$, we have*

$$\alpha_{\delta^{n,+}} = \alpha_{\delta^{n,-}} = (n-1)^2$$

for the discrete series representations $\delta^{n,\pm}$. Similarly, we have

$$\alpha_{\delta^{1,+}} = \alpha_{\delta^{1,-}} = 0$$

for the two mock discrete series representations $\delta^{1,\pm}$.

PROOF. Let $n \geqslant 2$ and write e_0 for the constant function in $A_n(\mathbb{D})$ which has K-weight n by Lemma 8.22. The proof of irreducibility in Theorem 8.23 shows that e_0 is smooth (see also Lemma 9.16). By Proposition 9.12 we know that $\delta_\partial^{n,+}(\mathbf{r}^-)e_0$ has weight $n-2$, and by Corollary 9.13 we have

$$\|\delta_\partial^{n,+}(\mathbf{r}^-)e_0\| = \tfrac{1}{4}\left((n-1)^2 - \alpha_{\delta^{n,+}}\right)\|e_0\|.$$

However, by Theorem 8.23, there is no vector of weight $n-2$. Hence $\delta_\partial^{n,+}(\mathbf{r}^-)e_0$ must be trivial, and so

$$\alpha_{\delta^{n,+}} = (n-1)^2,$$

as claimed. The argument for the mock discrete series representation $\delta^{1,+}$ uses Theorem 8.31, but is otherwise identical.

This implies the same formulas for the contragredient representations $\delta^{n,-}$ for $n \geqslant 2$ and for $\delta^{1,-}$. □

Exercise 9.15. Explain the last step in the proof of Corollary 9.14 more carefully.

9.2.3 Smooth K-finite Vectors

For a unitary representation π of $\mathrm{SL}_2(\mathbb{R})$ and a vector $v \in \mathcal{H}_\pi$, we say that v is K-finite if $\dim\langle\pi(K)v\rangle < \infty$ (or, equivalently, if v is a finite sum of K-eigenfunctions).

Because of the discussions above, the following lemma is useful for the study of general unitary representations of $\mathrm{SL}_2(\mathbb{R})$.

Lemma 9.16 (Smooth K-finite vectors). *Let π be a unitary representation of $\mathrm{SL}_2(\mathbb{R})$. Then the subspace of smooth K-finite vectors is dense in $\mathcal{H}_\pi$. Moreover, if for some $n \in \mathbb{Z}$ the space of K-eigenvectors in $\mathcal{H}_\pi$ of weight n is finite-dimensional, then every K-eigenvector of weight n is smooth and in the image of a convolution operator $\pi_*(\psi)$ for some $\psi \in C_c^\infty(\mathrm{SL}_2(\mathbb{R}))$.*

For the proof we will use a special kind of approximate identity for $\mathrm{SL}_2(\mathbb{R})$ as in the next lemma.

Lemma 9.17 (K-invariant approximate identity). *There exists an approximate identity on $\mathrm{SL}_2(\mathbb{R})$ consisting of functions $\psi \in C_c^\infty(\mathrm{SL}_2(\mathbb{R}))$ satisfying*

$$\psi(kgk^{-1}) = \psi(g) \tag{9.13}$$

for $k \in K$ and $g \in \mathrm{SL}_2(\mathbb{R})$. For $\psi \in C_c^\infty(\mathrm{SL}_2(\mathbb{R}))$ satisfying (9.13) and a unitary representation π the operator $\pi_*(\psi)$ satisfies

$$\pi_k\pi_*(\psi) = \pi_*(\psi)\pi_k$$

for $k \in K$. Moreover, $\pi_*(\psi)$ maps a K-eigenvector of weight $n \in \mathbb{Z}$ to a smooth K-eigenvector of weight n.

PROOF. For $\psi \in C_c^\infty(\mathrm{SL}_2(\mathbb{R}))$ we define

$$\widetilde{\psi}(g) = \int_K \psi(\ell g\ell^{-1})\,\mathrm{d}m_K(\ell).$$

Then, by a simple substitution,

$$\widetilde{\psi}(kgk^{-1}) = \int_K \psi(\ell kgk^{-1}\ell^{-1})\,\mathrm{d}m_K(\ell) = \widetilde{\psi}(g)$$

for all $k \in K$ and $g \in \mathrm{SL}_2(\mathbb{R})$. Moreover, as the map $g \mapsto \psi(\ell g\ell^{-1})$ is smooth for every $\ell \in K$, differentiation under the integral sign implies that $\widetilde{\psi}$ is in $C_c^\infty(\mathrm{SL}_2(\mathbb{R}))$. Finally, if (ψ_n) is an approximate identity in $L^1(\mathrm{SL}_2(\mathbb{R}))$ as in Proposition 1.46, then $(\widetilde{\psi_n})$ is also an approximate identity.

Suppose now that π is a unitary representation of $\mathrm{SL}_2(\mathbb{R})$, $v \in \mathcal{H}_\pi$, and that $\psi \in C_c^\infty(\mathrm{SL}_2(\mathbb{R}))$ satisfies (9.13). By Proposition 7.8 $\pi_*(\psi)v$ is a smooth vector. Moreover, for $k \in K$ we then have

$$\begin{aligned}\pi_k\pi_*(\psi)v &= \int \psi(g)\pi_{kg}v\,\mathrm{d}m(g)\\ &= \int \psi(kgk^{-1})\pi_{kgk^{-1}}\pi_k v\,\mathrm{d}m(g) \qquad \text{(by (9.13))}\\ &= \int \psi(h)\pi_h\pi_k v\,\mathrm{d}m(h) = \pi_*(\psi)\pi_k v.\end{aligned}$$

Hence $\pi_k\pi_*(\psi) = \pi_*(\psi)\pi_k$. If now $v \in \mathcal{H}_\pi$ is in addition a K-eigenvector with K-weight $n \in \mathbb{Z}$ then

$$\pi_k\pi_*(\psi)v = \pi_*(\psi)\pi_k v = \pi_*(\psi)\chi_n(k)v = \chi_n(k)\pi_*(\psi)v$$

shows that $\pi_*(\psi)$ is also a K-eigenvector with K-weight n. □

Proof of Lemma 9.16. Let (ψ_ℓ) be an approximate identity in $C_c^\infty(\mathrm{SL}_2(\mathbb{R}))$ satisfying (9.13). Let $v \in \mathcal{H}_\pi$ and split v into a sum $v = \sum_{n\in\mathbb{Z}} v_n$ of K-eigenfunctions v_n of weight $n \in \mathbb{Z}$. Then

$$v = \lim_{N\to\infty}\lim_{\ell\to\infty}\sum_{n=-N}^{N} \pi_*(\psi_\ell)v_n,$$

where $\pi_*(\psi_\ell)v_n$ is smooth and has K-weight n by Lemma 9.17. Hence v is a limit of smooth K-finite vectors as claimed in the lemma.

Fix some $n \in \mathbb{Z}$ and suppose now that

$$\mathcal{V}_n = \{v \in \mathcal{H}_\pi \mid v \text{ has } K\text{-weight } n\}$$

is finite-dimensional. Fix a basis of $\mathcal{V}_n$. Using Lemma (9.17), we can find some $\psi \in C_c^\infty(G)$ satisfying (9.13) such that $\pi_*(\psi)v$ is close to v for each of the basis vectors v of $\mathcal{V}_n$. It follows that

$$\pi_*(\psi_n)|_{\mathcal{V}_n} : \mathcal{V}_n \longrightarrow \mathcal{V}_n$$

is as close to the identity on $\mathcal{V}_n$ as we desire. In particular, we may ensure that $\pi_*(\psi_n)(\mathcal{V}_n) = \mathcal{V}_n$, and the final claim of the lemma follows from Lemma 9.17. □

9.2.4 Irreducibility Proofs

Let us introduce three further types of irreducible unitary representations. We assume their existence for now and refer to Sections 9.3 and 9.5 for the detailed constructions.

- For $\xi \in \mathbb{R}$ there exists a unitary representation $\pi^{\xi,\mathrm{e}}$ with Casimir eigenvalue $\alpha_{\pi^{\xi,\mathrm{e}}} = -\xi^2$ and for which the dimension of the K-eigenspace for K-

weight n is 1 if n is even and 0 if n is odd. The representation $\pi^{\xi,\mathrm{e}}$ is called the *even principal series representation* (with frequency parameter ξ).

- For $\xi \in \mathbb{R}\smallsetminus\{0\}$ there exists a unitary representation $\pi^{\xi,\mathrm{o}}$ with Casimir eigenvalue $\alpha_{\pi^{\xi,\mathrm{o}}} = -\xi^2$ and for which the dimension of the K-eigenspace for K-weight n is 1 if n is odd and 0 if n is even. The representation $\pi^{\xi,\mathrm{o}}$ is called the *odd principal series representation* (with frequency parameter ξ).
- For $s \in (0,1)$ there exists a unitary representation γ^s with Casimir eigenvalue $\alpha_{\gamma^s} = s^2$ and for which the dimension of the K-eigenspace for K-weight n is 1 if n is even and 0 if n is odd. The representation γ^s is called the *complementary series representation* (with parameter s).

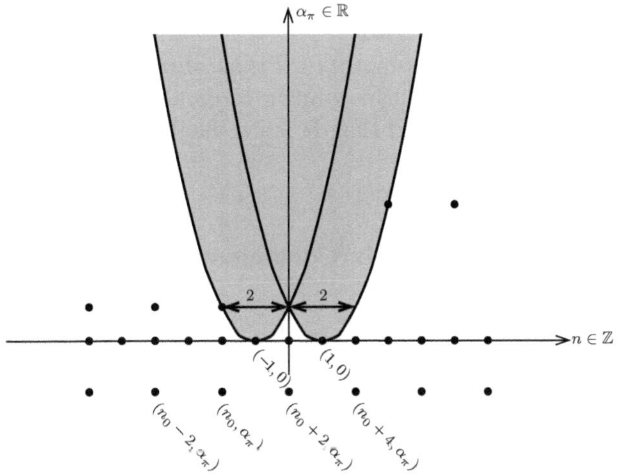

Fig. 9.1: The shaded region is the union of the region defined by the two inequalities $\alpha_\pi > (n+1)^2$ and $\alpha_\pi > (n-1)^2$. By (9.17), we know that this region cannot contain a pair (n,α_π) corresponding to a non-zero K-eigenvector for some representation $\pi \in \widehat{\mathrm{SL}_2(\mathbb{R})}$. In addition, we see how Proposition 9.12 and Corollary 9.13 create additional pairs from one pair unless we have $\alpha_\pi = (n \pm 1)^2$. We also note that the two parabolas have width 2 precisely at height $\alpha_\pi = 1$.

Proposition 9.18 (Irreducibility). *All of the above types of unitary representation are irreducible.*

Proof. For $\pi^{\xi,\mathrm{e}}$ with $\xi \in \mathbb{R}$ and γ^s with $s \in (0,1)$ we define $S = 2\mathbb{Z}$ while for $\pi^{\xi,\mathrm{o}}$ with $\xi \in \mathbb{R}\smallsetminus\{0\}$ we define $S = 1 + 2\mathbb{Z}$. By assumption the associated Hilbert space $\mathcal{H}$ contains a K-eigenspace (always one-dimensional) with K-weight n precisely if $n \in S$. For $n \in S$ let $F_n \in \mathcal{H}$ be a unit vector with K-weight n, so that $\{F_n \mid n \in S\}$ forms an orthonormal basis of $\mathcal{H}$. For the remainder of the proof we will simply write π for the unitary representation.

Now suppose $\mathcal{V} \subseteq \mathcal{H}$ is a closed invariant subspace. Restricting our representation to K, it follows that $\mathcal{V}$ contains an eigenvector for K and hence contains F_{n_0} for some $n_0 \in S$. By applying $\pi_\partial(\mathbf{r}^\pm)$ to F_{n_0} we obtain two further vectors with K-weight $n_0 \pm 2$ by Proposition 9.12. By Corollary 9.13 we have

$$\|\pi_\partial(\mathbf{r}^\pm)F_{n_0}\|^2 = \tfrac{1}{4}\big((n_0 \pm 1)^2 - \alpha_\pi\big)\|F_{n_0}\|^2. \tag{9.14}$$

Using our assumptions on the Casimir eigenvalue α_π of the representation and the parity of S, it follows that these two new vectors are non-zero.

In fact Figure 9.1 indicates when the raised or lowered vector vanishes geometrically. By (9.14), this happens only if (n_0, α_π) precisely belongs to one of the two parabolas. However, our assumption on the three types of unitary representation shows in all three cases that (n, α_π) is below the parabolas for all $n \in S$.

As the eigenspace for K-weight $n_0 \pm 2$ is assumed to be one-dimensional, we obtain $F_{n_0 \pm 2} \in \mathcal{V}$. Iterating this argument leads to $F_n \in \mathcal{V}$ for all $n \in S$, which in turn shows that $\mathcal{V} = \mathcal{H}$, and that $\mathcal{H}$ is irreducible. □

9.2.5 Determining the Matrix Coefficient

For the classification of the elements in $\widehat{\mathrm{SL}_2(\mathbb{R})}$ we need some more preparations.

Lemma 9.19 (Reducing to one variable). *Let π be a unitary representation of the group $\mathrm{SL}_2(\mathbb{R})$. Suppose that $v \in \mathcal{H}_\pi$ is a smooth eigenvector for K and Ω with K-weight $n \in \mathbb{Z}$ and Casimir eigenvalue $\alpha \in \mathbb{R}$. Then*

$$\varphi_v^\pi(k_\theta a_t k_\psi) = \mathrm{e}^{\mathrm{i}n(\theta+\psi)}\varphi_v^\pi(a_t)$$

for all $k_\theta a_t k_\psi \in \mathrm{SL}_2(\mathbb{R})$. Moreover, the function ϕ defined by

$$\phi(t) = \varphi_v^\pi(a_t)$$

for $t \in \mathbb{R}$ is real-valued, even, and smooth with derivatives given by

$$\phi^{(J)}(t) = \langle \pi_{a_t}\pi_\partial(\mathbf{a}^{\circ J})v, v\rangle \tag{9.15}$$

for $J \in \mathbb{N}_0$ and $t \in \mathbb{R}$.

Proof. The first part of the lemma follows simply because v is a K-eigenvector of weight n. Indeed, we have

$$\varphi_v^\pi(k_\theta a_t k_\psi) = \left\langle \pi_{a_t}\pi_{k_\psi}v, \pi_{k_{-\theta}}v\right\rangle = \langle \pi_{a_t}(\mathrm{e}^{\mathrm{i}n\psi}v), \mathrm{e}^{-\mathrm{i}n\theta}v\rangle = \mathrm{e}^{\mathrm{i}n(\theta+\psi)}\varphi_v^\pi(a_t)$$

for all $k_\theta a_t k_\psi \in \mathrm{SL}_2(\mathbb{R})$. We note that

$$k_{\frac{\pi}{2}} a_t k_{-\frac{\pi}{2}} = \begin{pmatrix} 0 & -1 \\ 1 & 0 \end{pmatrix} \begin{pmatrix} \mathrm{e}^t & 0 \\ 0 & \mathrm{e}^{-t} \end{pmatrix} \begin{pmatrix} 0 & 1 \\ -1 & 0 \end{pmatrix} = a_{-t},$$

for all $t \in \mathbb{R}$. For ϕ this implies

$$\phi(-t) = \varphi_v^\pi(a_{-t}) = \varphi_v^\pi\left(k_{\pi/2} a_t k_{-\pi/2}\right) = \varphi_v^\pi(a_t) = \phi(t)$$

for all $t \in \mathbb{R}$ by the above. In other words, ϕ is an even function on $\mathbb{R}$. Moreover,

$$\overline{\phi(t)} = \overline{\langle \pi_{a_t} v, v \rangle} = \langle v, \pi_{a_t} v \rangle = \langle \pi_{a_{-t}} v, v \rangle = \phi(-t) = \phi(t)$$

for all $t \in \mathbb{R}$, which also shows that ϕ only takes values in $\mathbb{R}$.

The formula for the derivatives of ϕ follows by induction using the definition of smoothness. Indeed, (9.15) holds by definition for $J = 0$ and if it holds for some $J \in \mathbb{N}_0$ then

$$\begin{aligned} \phi^{(J+1)}(t) &= \lim_{h \to 0} \frac{1}{h} \left(\phi^{(J)}(t+h) - \phi^{(J)}(t)\right) \\ &= \lim_{h \to 0} \frac{1}{h} \left\langle \pi_{a_{t+h}} \pi_\partial(\mathbf{a}^{\circ J}) v - \pi_{a_t} \pi_\partial(\mathbf{a}^{\circ J}) v, v \right\rangle \\ &= \lim_{h \to 0} \frac{1}{h} \langle \pi_{a_t} (\pi_{a_h} \pi_\partial(\mathbf{a}^{\circ J}) v - \pi_\partial(\mathbf{a}^{\circ J}) v), v \rangle \\ &= \left\langle \pi_{a_t} \pi_\partial(\mathbf{a}^{\circ(J+1)}) v, v \right\rangle. \end{aligned}$$

□

Proposition 9.20 (Determining the matrix coefficient). *Let π be a unitary representation of $\mathrm{SL}_2(\mathbb{R})$. Suppose that $v \in \mathcal{H}_\pi$ is a smooth eigenvector for K and Ω of weight $n \in \mathbb{Z}$ and Casimir eigenvalue $\alpha \in \mathbb{R}$. Then the matrix coefficient φ_v^π is uniquely determined by α, n, and $\|v\|$.*

Proof. By Lemma 9.19 it is sufficient to study the function $\phi(t) = \varphi_v^\pi(a_t)$ for $t \in \mathbb{R}$. The following two claims concerning ϕ will (together with analytic continuation) prove the proposition:

(A) ϕ is analytic on $\mathbb{R}$, and
(D) the derivatives $\phi^{(J)}(0)$ for $J \in \mathbb{N}_0$ are determined by α, n, and $\|v\|$.

To prove analyticity of the smooth function as in (A) we use (9.15) to bound the derivatives. For this, recall that $\mathbf{a} = \mathbf{r}^+ + \mathbf{r}^-$, so $\mathbf{a}^{\circ J}$ can be written in $\mathfrak{E}$ as a sum of 2^J non-commuting products $\mathbf{e}$ of $\mathbf{r}^+$ and $\mathbf{r}^-$ in any order. By Proposition 9.12 applying $\mathbf{r}^+$ or $\mathbf{r}^-$ to a K-eigenvector raises or lowers the K-weight by 2. Starting with v and iterating this j times for $j \leqslant J$ gives K-eigenvectors w of K-weight m satisfying

$$|m| \leqslant |n| + 2j \leqslant |n| + 2J.$$

If now $\mathbf{e}$ is a product of $j < J$ many $\mathbf{r}^+$ or $\mathbf{r}^-$ and $w = \pi_\partial(\mathbf{e})v$ has K-weight m, then we can use Corollary 9.13 to see that

$$\begin{aligned}\big\|\pi_\partial(\mathbf{r}^\pm)w\big\| &= \frac{1}{2}\sqrt{(m\pm 1)^2-\alpha}\|w\|\\ &\leqslant \frac{1}{2}\big(|m\pm 1|+|\alpha|\big)\|w\|\\ &\leqslant \frac{1}{2}\big(|n|+2J+1+|\alpha|\big)\|w\|\\ &\leqslant AJ\|w\|,\end{aligned}$$

where

$$A=\sup_{J\in\mathbb{N}}\frac{|n|+2J+1+|\alpha|}{2J}.$$

Iterating this J times gives

$$\|\pi_\partial(\mathbf{e})v\|\leqslant A^J J^J\|v\|$$

for any non-commuting product $\mathbf{e}$ of J many $\mathbf{r}^+$ or $\mathbf{r}^-$ in any order. Using the relation $\mathbf{a}^{\circ J}=(\mathbf{r}^++\mathbf{r}^-)^{\circ J}$ as indicated before, (9.15), and Cauchy–Schwarz we arrive at the estimate

$$\|\phi^{(J)}\|_\infty\leqslant(2A)^J J^J\|v\|^2 \tag{9.16}$$

for $k\in\mathbb{N}_0$.

Now recall Taylor's theorem in the form

$$\phi(t)=\sum_{j=0}^{J}\frac{\phi^{(j)}(t_0)}{j!}(t-t_0)^j+\mathrm{O}\Big(\frac{\|\phi^{(J+1)}\|_\infty}{(J+1)!}|t-t_0|^{J+1}\Big).$$

Together with (9.16) and Stirling's formula (see Section B.2) we see that the remainder term is bounded by

$$\begin{aligned}\frac{(2A)^J J^J\|v\|^2}{(J+1)!}|t-t_0|^{J+1} &\ll_v \frac{(2A)^J J^J}{\sqrt{J+1}\,\big(\frac{J+1}{\mathrm{e}}\big)^{J+1}}|t-t_0|^{J+1}\\ &\leqslant (2A)^J\mathrm{e}^{J+1}|t-t_0|^{J+1},\end{aligned}$$

which converges to 0 as $J\to\infty$ if $t\in\mathbb{R}$ is sufficiently close to $t_0\in\mathbb{R}$ This gives the claim (A).

For the proof of the claim (D) we set $t=0$ in (9.15) and fix $J\geqslant 0$. As before we expand $\mathbf{a}^{\circ k}$ into a sum of 2^J products $\mathbf{e}$ of J many $\mathbf{r}^+$ and $\mathbf{r}^-$ in any order. We show by induction on J that $\langle\pi_\partial(\mathbf{e})v,v\rangle_{\mathcal{H}_\pi}$ only depends on $\|v\|$, α, n, and $\mathbf{e}$. If $J=0$ (and so $\mathbf{e}=\mathbb{1}_{\mathfrak{E}}$) then $\langle\pi_\partial(\mathbb{1}_{\mathfrak{E}})v,v\rangle=\|v\|^2$. If $\mathbf{e}$ does not have the same number of $\mathbf{r}^+$ and $\mathbf{r}^-$ in its definition (as will always be the case when J is odd), then by Proposition 9.12 applied inductively $\pi_\partial(\mathbf{e})v$ and v have different K-weights, and so $\langle\pi_\partial(\mathbf{e})v,v\rangle=0$.

So we may assume $J>0$ and $\mathbf{e}=\mathbf{e}_1\circ\mathbf{r}^-\circ\mathbf{r}^+\circ\mathbf{e}_2$ or $\mathbf{e}=\mathbf{e}_1\circ\mathbf{r}^+\circ\mathbf{r}^-\circ\mathbf{e}_2$, where $\mathbf{e}_1$, $\mathbf{e}_2$ are again products of $\mathbf{r}^+$ and $\mathbf{r}^-$ in any order. By Proposition 9.12 the vector $\pi_\partial(\mathbf{e}_2)v$ is a K-eigenvector whose K-weight is determined by n and $\mathbf{e}_2$. The identity (9.11) allows us to express the products $\mathbf{r}^-\circ\mathbf{r}^+$

and $\mathbf{r}^+ \circ \mathbf{r}^-$ as linear combinations of Ω, $\mathbb{1}_{\mathfrak{C}}$, $\mathbf{k}$, and $\mathbf{k}^{\circ 2}$. Therefore $\pi_\partial(\mathbf{e}_2)v$ is an eigenvector for $\pi_\partial(\mathbf{r}^+ \circ \mathbf{r}^-)$ and $\pi_\partial(\mathbf{r}^- \circ \mathbf{r}^+)$ and the eigenvalue is determined by α, n, and $\mathbf{e}_2$. Taking the eigenvalue outside of the inner product leads us to study $\langle \pi_\partial(\mathbf{e}_1 \circ \mathbf{e}_2)v, v\rangle$, which by induction is already known to only depend on $\|v\|$, α, n, and $\mathbf{e}_1 \circ \mathbf{e}_2$. This concludes the proof of the induction, of claim (D), and hence of the lemma. □

The following will be our main tool for the classification of the elements of $\widehat{\mathrm{SL}_2(\mathbb{R})}$.

Theorem 9.21 (Isomorphisms). *Let π and τ be irreducible unitary representations of $\mathrm{SL}_2(\mathbb{R})$. Suppose that the Casimir eigenvalues $\alpha_\pi = \alpha_\tau$ agree, and that there exists some $n \in \mathbb{Z}$ such that both $\mathcal{H}_\pi$ and $\mathcal{H}_\tau$ contain a K-eigenvector of weight n. Then $\pi \cong \tau$.*

Proof. Let $v \in \mathcal{H}_\pi$ and $w \in \mathcal{H}_\tau$ be K-eigenvectors of weight n. Without loss of generality, we may assume that $\|v\| = \|w\| = 1$. By Proposition 9.20 this implies $\varphi_v^\pi = \varphi_w^\tau$. However, Proposition 1.65 now shows that

$$\mathcal{H}_\pi = \langle v\rangle_\pi \cong \langle w\rangle_\tau = \mathcal{H}_\tau$$

are isomorphic as unitary representations of $\mathrm{SL}_2(\mathbb{R})$. □

9.2.6 The Unitary Dual of $\mathrm{SL}_2(\mathbb{R})$

Using Theorem 9.21, we are now in a position to completely describe the unitary dual of $\mathrm{SL}_2(\mathbb{R})$. Since $\mathrm{SL}_2(\mathbb{R})$ has the non-trivial centre $\{\pm I\}$, the first distinction we can make between irreducible unitary representations of $\mathrm{SL}_2(\mathbb{R})$ is by use of the central character as in Corollary 1.32. Since the centre of $\mathrm{SL}_2(\mathbb{R})$ is given by $\{\pm I\}$, we say that $\pi \in \widehat{\mathrm{SL}_2(\mathbb{R})}$ is *even* if π_{-I} is I and *odd* if π_{-I} is $-I$. The second distinction is in terms of the 'infinitesimal character' obtained by applying Proposition 9.8 to the Casimir element Ω of Section 9.2.1, which gives the Casimir eigenvalue $\alpha_\pi \in \mathbb{R}$. The third and final distinction is in terms of which K-weights are present in the representation π.

The following result of Bargmann [4] contains all the possibilities of these three aspects. We depict $\widehat{\mathrm{SL}_2(\mathbb{R})}$ geometrically in Figure 9.2, where we draw even representations on the top half and odd ones on the bottom half. We also present $\widehat{\mathrm{SL}_2(\mathbb{R})}$ as a list in Table 9.1.

Theorem 9.22 (Unitary dual of $\mathrm{SL}_2(\mathbb{R})$). *Suppose that $\pi \in \widehat{\mathrm{SL}_2(\mathbb{R})}$ is even. Then one of the following four possibilities holds:*

- *$\alpha_\pi = 1$ and $\pi = \mathbb{1}$ is the trivial representation.*

Table 9.1: The different types of irreducible unitary representations and their main properties.

Notation	Name	Tempered?	Casimir eigenvalue	K-weights
$\delta^{n,\pm}$ for $n \geqslant 2$	discrete series representation	✓	$(n-1)^2$	$\pm(n+2\mathbb{N}_0)$
$\delta^{1,\pm}$	mock discrete series representation	✓	0	$\pm(1+2\mathbb{N}_0)$
$\pi^{\xi,\mathrm{e}}$ for $\xi \geqslant 0$	even principal series representation	✓	$-\xi^2$	$2\mathbb{Z}$
$\pi^{\xi,\mathrm{o}}$ for $\xi > 0$	odd principal series representation	✓	$-\xi^2$	$1+2\mathbb{Z}$
γ^s for $s \in (0,1)$	complementary series representation	✗	s^2	$2\mathbb{Z}$
$\mathbb{1}$	trivial representation	✗	1	$\{0\}$

- *$\alpha_\pi = (n-1)^2$ for some $n \in 2\mathbb{N}$, and $\pi = \delta^{n,\pm}$ is either the holomorphic or the anti-holomorphic discrete series representation with terminal weight $\pm n$ (see Section 8.4).*
- *$\alpha_\pi = -\xi^2$ for some $\xi \in [0,\infty)$ and $\pi = \pi^{\xi,\mathrm{e}}$ is the even principal series representation for the parameter ξ (see Section 9.3).*
- *$\alpha_\pi = s^2$ for some $s \in (0,1)$, and $\pi = \gamma^s$ is the complementary series representation for the parameter s (see Section 9.5).*

Suppose that $\pi \in \widehat{\mathrm{SL}_2(\mathbb{R})}$ is odd. Then one of the following three possibilities holds:

- *$\alpha_\pi = 0$ and $\pi = \delta^{1,\pm}$ is either the holomorphic or the anti-holomorphic mock discrete series representations (see Section 8.4).*
- *$\alpha_\pi = (n-1)^2$ for some $n \in (2\mathbb{N}+1)$, and $\pi = \delta^{n,\pm}$ is either the holomorphic or the anti-holomorphic discrete series representation with terminal weight $\pm n$ (see Section 8.4).*
- *$\alpha_\pi = -\xi^2$ for some $\xi \in (0,\infty)$ and $\pi = \pi^{\xi,\mathrm{o}}$ is the odd principal series representation for the parameter ξ (see Section 9.3).*

Proof of Theorem 9.22. In the following we let π be an irreducible unitary representation of $\mathrm{SL}_2(\mathbb{R})$. By Corollary 9.11, the closure of $\pi_\partial(\Omega)$ is multiplication by α_π for some $\alpha_\pi \in \mathbb{R}$. By Lemma 9.16, there also exists a smooth K-eigenvector $v \in \mathcal{H}_\pi$ with weight $n \in \mathbb{Z}$ and unit length $\|v\| = 1$. By Theorem 9.21, we have that (n, α_π) uniquely determines π up to isomorphism. Hence the question is really which $(n, \alpha_\pi) \in \mathbb{Z} \times \mathbb{R}$ are possible (in general, and within each irreducible representation). As already explained, Corollary 9.13 gives the constraint

$$\alpha_\pi \leqslant (n \pm 1)^2; \tag{9.17}$$

(see the discussion leading to (9.12), and Figure 9.1).

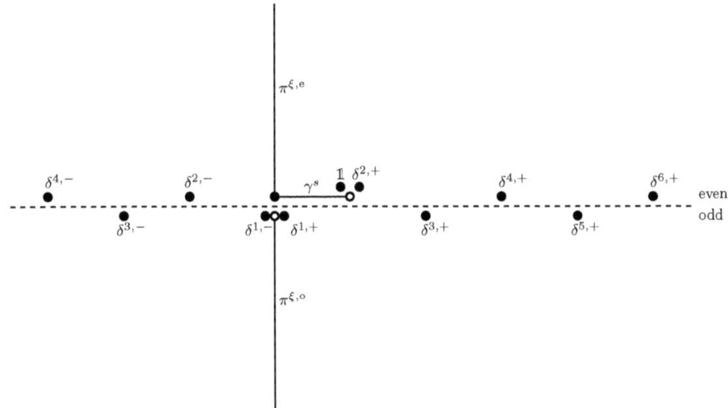

Fig. 9.2: The graphical representation of $\widehat{\mathrm{SL}_2(\mathbb{R})}$ as a subset of $\mathbb{C}$ (with the representations $\pi^{0,e}, \delta^{1,-}, \delta^{1,+}$ at the origin and both $\mathbb{1}$ and $\delta^{2,+}$ at $1 \in \mathbb{C}$) also has the property that α_π is the square of the position of π when drawn in $\mathbb{C}$ (except for the artificial small gap between the even and odd representations, and problems arising from $\{\pi^{0,e}, \delta^{1,+}, \delta^{1,-}\}$ and $\{\delta^{2,+}, \mathbb{1}\}$ which should be drawn at the same point).

We first go through the list of representations that we have already encountered. In addition to the three types of irreducible unitary representations listed in the beginning of Section 9.2.4, we have already seen the following irreducible representations.

- If $\pi = \mathbb{1}_G$ is the trivial representation on $\mathbb{C}$, then $v = 1 \in \mathbb{C}$ has K-weight $n = 0$ and Casimir eigenvalue 1, since $\pi_\partial(\Omega) = \pi_\partial(\mathbb{1}_{\mathfrak{C}}) = 1$. By the above, it follows that $\mathbb{1}_G$ is characterized by the pair $(n = 0, \alpha = 1)$.
- For the holomorphic discrete series representation $\delta^{n,+}$ or for the anti-holomorphic discrete series representation $\delta^{n,-}$ with $n \geqslant 2$ we have, by Theorem 8.23, that $\delta^{n,\pm}$ contains vectors with K-weights $\pm(n + 2\mathbb{N}_0)$ and only these. Moreover, Corollary 9.14 gives $\alpha_{\delta^{n,\pm}} = (n-1)^2$.
- This also holds similarly for the mock discrete series representations $\delta^{1,\pm}$ with K-weights $\pm(1 + 2\mathbb{N}_0)$ and $\alpha_{\delta^{1,\pm}} = 0$ (see Theorem 8.31 and Corollary 9.14).

It remains to show that the cases above give all possible irreducible unitary representations. So let π be an irreducible unitary representation with Casimir eigenvalue α_π and let $v_n \in \mathcal{H}_\pi$ be a smooth K-eigenvector of K-weight $n \in \mathbb{Z}$. We will now use Figure 9.1 to repeat and extend the argument that we used above in the proof of Proposition 9.18. By Proposition 9.12 and Corollary 9.13, the vectors $\pi_\partial(\mathbf{r}^\pm)v_n$ have K-weight $n \pm 2$ and satisfy (9.14). In particular, this implies that $\alpha_\pi \leqslant (n \pm 1)^2$, or equivalently that (n, α_π) does not belong to the 'forbidden' shaded region in Figure 9.1. Moreover, if (n, α_π) does not belong to either of the two parabolas defined by $\alpha = (n \pm 1)^2$, then we may replace n by $n \pm 2$ and iterate this argument to obtain further eigenvectors with different K-weights. If, however, $\alpha_\pi = (n+1)^2$ (or $\alpha_\pi = (n-1)^2$), then (9.14) shows that $\pi_\partial(\mathbf{r}^+)v_n = 0$ (or $\pi_\partial(\mathbf{r}^-)v_n = 0$, respectively).

This argument creates a chain of points (n, α_π) avoiding the forbidden region in Figure 9.1, possibly with end points belonging to either of the parabolas defined by $\alpha_\pi = (n \pm 1)^2$. There are a few possibilities for this chain of points, as follows.

- $\alpha_\pi < 0$, and the chain is bi-infinite.
- $\alpha_\pi = 0$, n is even, and the chain is bi-infinite.
- $\alpha_\pi = 0$, n is odd, and the chain is one of $\pm(1 + 2\mathbb{N}_0) \times \{0\}$.
- $\alpha_\pi \in (0, 1)$, n is even, and the chain is bi-infinite and jumps over the two shaded regions that have width less than two at the height α_π.
- However, $\alpha_\pi \in (0, 1]$ and n is odd is impossible, since the chain starting on either side would lead to the creation of one of the points $(\pm 1, \alpha_\pi)$ inside the forbidden region.
- $\alpha_\pi = 1$ and n even has three such chains, one starting at $(2, 1)$ going to the right, one starting at $(-2, 1)$ going to the left, and one consisting of $(0, 1)$ only.
- $\alpha_\pi > 1$ and $n > 0$ creates a half-infinite chain going to the right. It cannot go infinitely far to the left, as the forbidden region has width larger than 4 and the gaps in the chain have size 2. Hence the chain has to stop with a point on the right parabola. Replacing n by this minimal K-weight, we see that $\alpha_\pi = (n-1)^2 > 1$.
- $\alpha_\pi > 1$ and $n < 0$ gives rise to a half-infinite chain going to the left.

We leave it to the reader to match the cases above to the irreducible unitary representations discussed earlier and appearing in Table 9.1, which concludes the proof. □

Corollary 9.23. *Let π be a unitary representation of $\mathrm{SL}_2(\mathbb{R})$ and suppose that $v \in \mathcal{H}_\pi$ is a smooth eigenvector for K and Ω with K-weight $n \in \mathbb{Z}$ and Casimir eigenvalue $\alpha \in \mathbb{R}$. Then the restriction of π to the cyclic subspace $\langle v \rangle_\pi$ is irreducible, and its type is determined by (n, α) and Table 9.1 on page 426.*

PROOF. The last part of the proof of Theorem 9.22 shows that every pair (n, α_π) that does not lead via raising and lowering to a pair inside the forbidden region in Figure 9.1 is achieved by a unit vector $w \in \mathcal{H}_\rho$ for one of the irreducible representations ρ of $\mathrm{SL}_2(\mathbb{R})$. By Proposition 9.20 this shows that $\varphi_v^\pi = \varphi_w^\rho$, which implies $\langle v \rangle_\pi \cong \langle w \rangle_\rho = \mathcal{H}_\rho$ by Proposition 1.65. □

Using the description of $\widehat{\mathrm{SL}_2(\mathbb{R})}$ it is possible to derive similar descriptions of related groups.

Exercise 9.24. (a) Describe $\widehat{\mathrm{PSL}_2(\mathbb{R})}$.
(b) Describe $\widehat{\mathrm{GL}_2(\mathbb{R})^o}$.

For the following exercise, the reader may want to also use the concrete properties of the principal and complementary series representations in Sections 9.3 and 9.5.

Exercise 9.25. Let $G = \{g \in \mathrm{GL}_2(\mathbb{R}) \mid |\det g| = 1\}$, and note that $G = \mathrm{SL}_2(\mathbb{R}) \sqcup r\mathrm{SL}_2(\mathbb{R})$ where r is the diagonal matrix $r = \begin{pmatrix} 1 & 0 \\ 0 & -1 \end{pmatrix}$ with $r^2 = I$.
(a) Show that G has two one-dimensional representations.
(b) Describe how r interacts with K-weights.
(c) Use the (mock) discrete series representation to define for every $n \in \mathbb{N}$ an irreducible unitary representation $\delta^{n,G}$ whose restriction to $\mathrm{SL}_2(\mathbb{R})$ is equal to $\delta^{n,+} \oplus \delta^{n,-}$.
(d) Extend the principal series representation from $\mathrm{SL}_2(\mathbb{R})$ to G. Twist these by the non-trivial character to obtain a second non-isomorphic series of representations of G.
(e) Repeat (d) for the complementary series representation.
(f) Show that $\widehat{G}$ comprises precisely the representations found in (a), (c), (d), and (e).
(g) Describe $\widehat{\mathrm{GL}_2(\mathbb{R})}$.

Exercise 9.26. Let G denote a locally compact σ-compact metric group. Show that

$$\widehat{\mathrm{SL}_2(\mathbb{R}) \times G} \cong \widehat{\mathrm{SL}_2(\mathbb{R})} \times \widehat{G}$$

(in a natural sense, as in Proposition 3.17).

9.2.7 (Non-)Spherical Representations*

Definition 9.27. We say that a unitary representation π of $\mathrm{SL}_2(\mathbb{R})$ is *spherical* if $\mathcal{H}_\pi$ contains a non-trivial vector invariant under K (equivalently, of K-weight 0). Otherwise we say that π is *non-spherical.*

Essential Exercise 9.28. Let π be a unitary representation of $\mathrm{SL}_2(\mathbb{R})$. Show that π is non-spherical if

$$\int_K \varphi_v^\pi(k)\,\mathrm{d}m_K(k) = 0 \tag{9.18}$$

for all $v \in \mathcal{H}_\pi$.

Corollary 9.29 (Non-spherical representations are tempered). *Every non-spherical unitary representation of* $\mathrm{SL}_2(\mathbb{R})$ *is tempered, and so has decay exponent* $1 - \varepsilon$ *for all* $\varepsilon > 0$.

Proof. By Theorems 8.23 and 8.31 the discrete series representations and mock discrete series representations are tempered. By the discussions in the next section, the odd principal series representations $\pi^{\xi,\mathrm{o}}$ are also tempered for ξ in $\mathbb{R}\smallsetminus\{0\}$. Hence Theorem 9.22 (and Table 9.1), imply that every non-spherical irreducible unitary representation of $\mathrm{SL}_2(\mathbb{R})$ is tempered.

Suppose now that ρ is a non-spherical unitary representation of $\mathrm{SL}_2(\mathbb{R})$. By Exercise 9.28, being non-spherical is equivalent to the vanishing of the integral in (9.18). Using the definition of weak containment, it follows that every irreducible unitary representation of $\mathrm{SL}_2(\mathbb{R})$ that is weakly contained in ρ must be non-spherical also. By our discussion above, this shows that every irreducible unitary representation weakly contained in ρ is tempered. We now combine the definition of temperedness, the definition of weak containment, and Proposition 6.37. By the latter, we know that for any unit vector $v \in \mathcal{H}_\rho$ the matrix coefficient φ_v^ρ can be approximated in the compact-open topology by sums of the form $\sum_{j=1}^n \varphi_{v_j}^{\pi_j}$ for some irreducible unitary representations $\pi_j \prec \rho$ and vectors $v_j \in \mathcal{H}_{\pi_j}$ for $j = 1, \ldots, n$; since $\pi_j \prec \lambda_{\mathrm{SL}_2(\mathbb{R})}$ we can approximate $\varphi_{v_j}^{\pi_j}$ for $j = 1, \ldots, n$ by some sum of diagonal matrix coefficients for the regular representation. Putting these together, we obtain the same property for φ_v^ρ, which gives $\rho \prec \lambda_{\mathrm{SL}_2(\mathbb{R})}$. The last claim now follows from Theorem 8.42. □

9.2.8 A Differential Equation for Matrix Coefficients*

The following general lemma will also be useful. In fact to guess the asymptotic behaviour of the matrix coefficient ϕ, the reader may replace $\frac{\cosh(2t)}{\sinh(2t)}$ by 1 and $\frac{n^2}{\cosh^2 t}$ by 0, which simplifies the differential equation significantly.

Lemma 9.30 (The differential equation for the matrix coefficient). *Let π be a unitary representation of the group $\mathrm{SL}_2(\mathbb{R})$. Suppose that $v \in \mathcal{H}_\pi$ is a smooth eigenvector for K and Ω with K-weight $n \in \mathbb{Z}$ and Casimir eigenvalue $\alpha \in \mathbb{R}$. Then the smooth function $\phi\colon \mathbb{R} \to \mathbb{C}$ defined by*

$$\phi(t) = \varphi_v^\pi(a_t)$$

for $t \in \mathbb{R}$ satisfies the second order linear differential equation

$$\phi''(t) + 2\frac{\cosh(2t)}{\sinh(2t)}\phi'(t) + \left(1 - \alpha + \frac{n^2}{\cosh^2 t}\right)\phi(t) = 0$$

of degree two for all $t \in \mathbb{R} \smallsetminus \{0\}$.

Proof. To obtain the differential equation we will combine the information that v has K-weight $n \in \mathbb{Z}$, the assumption $\pi_\partial(\Omega)v = \alpha v$, and the formula

$$\Omega = \mathbb{1}_{\mathfrak{E}} + \mathbf{a}^{\circ 2} + 2\mathbf{e} \circ \mathbf{f} + 2\mathbf{f} \circ \mathbf{e}$$

in Lemma 9.2. For this, recall that v having K-weight n implies $\pi_\partial(\mathbf{k})v = inv$ and $\pi_\partial(\mathbf{k}^{\circ 2})v = -n^2 v$.

Also recall that

$$\phi'(t) = \left\langle \pi_{a_t} \pi_\partial(\mathbf{a})v, v \right\rangle$$

and

$$\phi''(t) = \left\langle \pi_{a_t} \pi_\partial(\mathbf{a}^{\circ 2})v, v \right\rangle$$

for all $t \in \mathbb{R}$ by (9.15).

We note that $\mathbf{a}^{\circ 2}$ is one of the terms in Ω, but that we do not yet have an interpretation of the term $2\mathbf{e} \circ \mathbf{f} + 2\mathbf{f} \circ \mathbf{e}$ in terms of ϕ. To obtain such an interpretation, we use the consequence of v having K-weight n mentioned above to express $-n^2\phi$ in three different ways. Indeed, we have

$$
\begin{aligned}
-n^2\phi(t) &= -n^2 \langle \pi_{a_t} v, v\rangle = \langle \pi_{a_t}(-n^2 v), v\rangle = \langle \pi_{a_t}\pi_\partial(\mathbf{k}^{\circ 2})v, v\rangle\,;\\
-n^2\phi(t) &= -\langle \pi_{a_t} inv, inv\rangle = -\langle \pi_{a_t}\pi_\partial(\mathbf{k})v, \pi_\partial(\mathbf{k})v\rangle = \langle \pi_\partial(\mathbf{k})\pi_{a_t}\pi_\partial(\mathbf{k})v, v\rangle\\
&= \Big\langle \pi_{a_t}\pi_\partial\big(\mathrm{Ad}_{a_{-t}}(\mathbf{k}) \circ \mathbf{k}\big)v, v\Big\rangle\,;\\
-n^2\phi(t) &= \langle \pi_{a_t} v, -n^2 v\rangle = \langle \pi_{a_t} v, \pi_\partial(\mathbf{k}^{\circ 2})v\rangle = \langle \pi_\partial(\mathbf{k}^{\circ 2})\pi_{a_t} v, v\rangle\\
&= \Big\langle \pi_{a_t}\pi_\partial\Big(\big(\mathrm{Ad}_{a_{-t}}(\mathbf{k})\big)^{\circ 2}\Big)v, v\Big\rangle
\end{aligned}
$$

for all $t \in \mathbb{R}$. We recall that $\mathbf{k} = -\mathbf{e} + \mathbf{f}$ and note that

$$
\mathrm{Ad}_{a_{-t}}(\mathbf{k}) = -\mathrm{e}^{-2t}\mathbf{e} + \mathrm{e}^{2t}\mathbf{f}
$$

for all $t \in \mathbb{R}$. We put these two formulas into the three expressions above, expand the resulting parentheses, and obtain from this that

$$
\begin{aligned}
-n^2\phi(t) &= \langle \pi_{a_t}\pi_\partial\big(\mathbf{e}^{\circ 2} - \mathbf{e} \circ \mathbf{f} - \mathbf{f} \circ \mathbf{e} + \mathbf{f}^{\circ 2}\big)v, v\rangle\\
&= \langle \pi_{a_t}\pi_\partial\big(\mathrm{e}^{-2t}\mathbf{e}^{\circ 2} - \mathrm{e}^{-2t}\mathbf{e} \circ \mathbf{f} - \mathrm{e}^{2t}\mathbf{f} \circ \mathbf{e} + \mathrm{e}^{2t}\mathbf{f}^{\circ 2}\big)v, v\rangle\\
&= \langle \pi_{a_t}\pi_\partial\big(\mathrm{e}^{-4t}\mathbf{e}^{\circ 2} - \mathbf{e} \circ \mathbf{f} - \mathbf{f} \circ \mathbf{e} + \mathrm{e}^{4t}\mathbf{f}^{\circ 2}\big)v, v\rangle
\end{aligned}
$$

for all $t \in \mathbb{R}$.

As our aim is to find a formula involving $2\mathbf{e}\circ\mathbf{f}+2\mathbf{f}\circ\mathbf{e}$, we wish to rid ourselves of all expressions involving the two terms $\mathbf{e}^{\circ 2}$ and $\mathbf{f}^{\circ 2}$. As we have three formulas for $-n^2\phi$, this is an exercise in linear algebra. In fact, multiplying the first line by 1, the second by $-\big(\mathrm{e}^{2t} + \mathrm{e}^{-2t}\big)$, the third by 1, and taking the sum gives

$$
-n^2\phi(t)\big(1 - \big(\mathrm{e}^{2t} + \mathrm{e}^{-2t}\big) + 1\big) = n^2\phi(t)\big(\mathrm{e}^{2t} - 2 + \mathrm{e}^{-2t}\big) = 4n^2 \sinh^2 t\, \phi(t)
$$

on the left-hand side.

On the right-hand side, we use the similarities between the three formulas and obtain the expression

$$
\langle \pi_{a_t}\pi_\partial(\mathbf{m}_t)v, v\rangle\,,
$$

where $\mathbf{m}_t$ is the element of $\mathfrak{E}_{\leqslant 2}$ defined by

$$
\begin{aligned}
\mathbf{m}_t = \big(\mathbf{e}^{\circ 2} &- \mathbf{e} \circ \mathbf{f} - \mathbf{f} \circ \mathbf{e} + \mathbf{f}^{\circ 2}\big)\\
&- \big(\mathrm{e}^{2t} + \mathrm{e}^{-2t}\big)\big(\mathrm{e}^{-2t}\mathbf{e}^{\circ 2} - \mathrm{e}^{-2t}\mathbf{e} \circ \mathbf{f} - \mathrm{e}^{2t}\mathbf{f} \circ \mathbf{e} + \mathrm{e}^{2t}\mathbf{f}^{\circ 2}\big)\\
&\qquad + \big(\mathrm{e}^{-4t}\mathbf{e}^{\circ 2} - \mathbf{e} \circ \mathbf{f} - \mathbf{f} \circ \mathbf{e} + \mathrm{e}^{4t}\mathbf{f}^{\circ 2}\big).
\end{aligned}
$$

Our choice of the three coefficients 1, $-\big(\mathrm{e}^{2t} + \mathrm{e}^{-2t}\big)$, and 1 was made so that the coefficient in front of $\mathbf{e}^{\circ 2}$ is

$$1-\left(\mathrm{e}^{2t}+\mathrm{e}^{-2t}\right)\mathrm{e}^{-2t}+\mathrm{e}^{-4t}=1-1-\mathrm{e}^{-4t}+\mathrm{e}^{-4t}=0$$

and the coefficient in front of $\mathbf{f}^{\circ 2}$ is

$$1-\left(\mathrm{e}^{2t}+\mathrm{e}^{-2t}\right)\mathrm{e}^{2t}+\mathrm{e}^{4t}=1-\mathrm{e}^{4t}-1+\mathrm{e}^{4t}=0$$

also. Therefore,

$$\begin{aligned}\mathbf{m}_t &= \left(-\mathbf{e}\circ\mathbf{f}-\mathbf{f}\circ\mathbf{e}\right)-\left(\mathrm{e}^{2t}+\mathrm{e}^{-2t}\right)\left(-\mathrm{e}^{-2t}\mathbf{e}\circ\mathbf{f}-\mathrm{e}^{2t}\mathbf{f}\circ\mathbf{e}\right)+\left(-\mathbf{e}\circ\mathbf{f}-\mathbf{f}\circ\mathbf{e}\right)\\ &= \left(-1+1+\mathrm{e}^{-4t}-1\right)\mathbf{e}\circ\mathbf{f}+\left(-1+\mathrm{e}^{4t}+1-1\right)\mathbf{f}\circ\mathbf{e}\\ &= \left(\mathrm{e}^{-4t}-1\right)\mathbf{e}\circ\mathbf{f}+\left(\mathrm{e}^{4t}-1\right)\mathbf{f}\circ\mathbf{e}.\end{aligned}$$

Using

$$\begin{aligned}\mathrm{e}^{4t}-1&=\mathrm{e}^{2t}\left(\mathrm{e}^{2t}-\mathrm{e}^{-2t}\right)=2\mathrm{e}^{2t}\sinh(2t),\\ \mathrm{e}^{-4t}-1&=\mathrm{e}^{-2t}\left(\mathrm{e}^{2t}-\mathrm{e}^{-2t}\right)=-2\mathrm{e}^{-2t}\sinh(2t),\end{aligned}$$

we also have

$$\mathbf{m}_t=-2\mathrm{e}^{-2t}\sinh(2t)\mathbf{e}\circ\mathbf{f}+2\mathrm{e}^{2t}\sinh(2t)\mathbf{f}\circ\mathbf{e}.$$

To summarize, we have shown that

$$\begin{aligned}4n^2\sinh^2 t\,\phi(t)&=\left\langle\pi_{a_t}\pi_\partial(\mathbf{m}_t)v,v\right\rangle\\ &=2\sinh(2t)\left\langle\pi_{a_t}\pi_\partial\left(-\mathrm{e}^{-2t}\mathbf{e}\circ\mathbf{f}+\mathrm{e}^{2t}\mathbf{f}\circ\mathbf{e}\right)v,v\right\rangle.\end{aligned}$$

Dividing by $2\sinh(2t)=4\sinh t\cosh t$, we obtain

$$n^2\frac{\sinh t}{\cosh t}\phi(t)=\left\langle\pi_{a_t}\pi_\partial\left(-\mathrm{e}^{-2t}\mathbf{e}\circ\mathbf{f}+\mathrm{e}^{2t}\mathbf{f}\circ\mathbf{e}\right)v,v\right\rangle.$$

Next we note that for $s_1,s_2\in\mathbb{R}$, we have

$$s_1\mathbf{e}\circ\mathbf{f}+s_2\mathbf{f}\circ\mathbf{e}=\frac{s_1+s_2}{2}(\mathbf{e}\circ\mathbf{f}+\mathbf{f}\circ\mathbf{e})+\frac{s_1-s_2}{2}\underbrace{[\mathbf{e},\mathbf{f}]}_{=\mathbf{a}},$$

which, with the choice $s_1=-\mathrm{e}^{-2t}$ and $s_2=\mathrm{e}^{2t}$, gives

$$\begin{aligned}n^2\frac{\sinh t}{\cosh t}\phi(t)&=\left\langle\pi_{a_t}\pi_\partial\left(\frac{\mathrm{e}^{2t}-\mathrm{e}^{-2t}}{2}(\mathbf{e}\circ\mathbf{f}+\mathbf{f}\circ\mathbf{e})-\frac{\mathrm{e}^{2t}+\mathrm{e}^{-2t}}{2}\mathbf{a}\right)v,v\right\rangle\\ &=\frac{\sinh(2t)}{2}\left\langle\pi_{a_t}\pi_\partial(2\mathbf{e}\circ\mathbf{f}+2\mathbf{f}\circ\mathbf{e})v,v\right\rangle-\cosh(2t)\phi'(t).\end{aligned}$$

Dividing by $\frac{\sinh(2t)}{2}=\sinh t\cosh t$, we also obtain

$$\left\langle\pi_{a_t}\pi_\partial(2\mathbf{e}\circ\mathbf{f}+2\mathbf{f}\circ\mathbf{e})v,v\right\rangle=\frac{n^2}{\cosh^2 t}\phi(t)+2\frac{\cosh(2t)}{\sinh(2t)}\phi'(t).$$

Using the relations $\pi_\partial(\Omega)v = \alpha v$ and $\Omega = \mathbb{1}_{\mathfrak{C}} + \mathbf{a}^{\circ 2} + 2\mathbf{e}\circ\mathbf{f} + 2\mathbf{f}\circ\mathbf{e}$, we finally arrive at the differential equation

$$\begin{aligned}\alpha\phi(t) &= \langle \pi_{a_t}(\alpha v), v\rangle \\ &= \langle \pi_{a_t}\pi_\partial(\Omega)v, v\rangle \\ &= \phi(t) + \phi''(t) + \langle \pi_{a_t}\pi_\partial(2\mathbf{e}\circ\mathbf{f} + 2\mathbf{f}\circ\mathbf{e})v, v\rangle \\ &= \phi''(t) + \left(1 + \frac{n^2}{\cosh^2 t}\right)\phi(t) + 2\frac{\cosh(2t)}{\sinh(2t)}\phi'(t),\end{aligned}$$

as in the lemma. □

9.3 The Principal Series Representations

We will now modify the representation π^0 from Section 8.5.2, which will give rise to the even and odd principal series representations appearing in Theorem 9.22. Along the way we will also explain the connection to Example 1.7.

Definition 9.31 (Principal series representation). For a given $\xi \in \mathbb{R}$, we define the character χ_ξ on

$$B = \{a_t u_x \mid t, x \in \mathbb{R}\}$$

by

$$\chi_\xi(a_t u_x) = \mathrm{e}^{\mathrm{i}\xi t}$$

for all $a_t u_x \in B$. The representation π^ξ of $G = \mathrm{SL}_2(\mathbb{R})$ is defined by

$$(\mathcal{H}_\xi, \pi^\xi) = \mathrm{Ind}_B^G(\mathbb{C}, \chi_\xi),$$

or, more concretely, by the left regular representation on the space $\mathcal{H}_\xi$ of those functions $f\colon G \to \mathbb{C}$ with the following properties:

(1) f is measurable,
(2) $f(gb) = \overline{\chi_\xi(b)}\Delta_B(b)^{\frac{1}{2}} f(g)$ for all $g \in G$ and $b \in B$, and
(3) $\|f|_K\|_{L^2(K)} < \infty$.

The *even principal series representation* $\pi^{\xi,\mathrm{e}} = \pi^{\xi,\mathrm{even}}$ (for frequency parameter ξ) is defined as the restriction of π^ξ to the subspace

$$\mathcal{H}_\xi^{\mathrm{even}} = \{f \in \mathcal{H}_\xi \mid f(-g) = f(g) \text{ for all } g \in G\}.$$

Similarly, the *odd principal series representation* $\pi^{\xi,\mathrm{o}} = \pi^{\xi,\mathrm{odd}}$ (for frequency parameter ξ) is defined as the restriction of π^ξ to the subspace

$$\mathcal{H}_\xi^{\mathrm{odd}} = \{f \in \mathcal{H}_\xi \mid f(-g) = -f(g) \text{ for all } g \in G\}.$$

Let us summarize the properties of the even and odd principal series representations that we will prove in this section.

Theorem 9.32 (Even and odd principal series representations). *The representations π^ξ, $\pi^{\xi,\mathrm{e}}$, and $\pi^{\xi,\mathrm{o}}$ are unitary representations with Casimir eigenvalue $-\xi^2$ for any $\xi \in \mathbb{R}$. The representation $\pi^{\xi,\mathrm{e}}$ is irreducible for any $\xi \in \mathbb{R}$ and $\pi^{\xi,\mathrm{o}}$ is irreducible for all $\xi \in \mathbb{R}\smallsetminus\{0\}$. Moreover, $\pi^{-\xi,\mathrm{e}}$ is isomorphic to $\pi^{\xi,\mathrm{e}}$, $\pi^{-\xi,\mathrm{o}}$ is isomorphic to $\pi^{\xi,\mathrm{o}}$ for all $\xi \in \mathbb{R}$, and $\pi^{0,\mathrm{o}}$ is isomorphic to the sum $\delta^{1,+} \oplus \delta^{1,-}$ of the holomorphic and anti-holomorphic mock discrete series representations. Finally, all of these representations are tempered with decay exponent $1-\varepsilon$ for all $\varepsilon > 0$ and are not discrete series representations.*

Proof of unitarity in Theorem 9.32. Recall from Example 1.7 that the group $\mathrm{SL}_2(\mathbb{R})$ acts on $\mathbb{S}^1 \subseteq \mathbb{R}^2$ via

$$\mathbb{S}^1 \ni v \longmapsto g\boldsymbol{\cdot}v = \frac{1}{\|gv\|} gv$$

for $g \in \mathrm{SL}_2(\mathbb{R})$ and $v \in \mathbb{S}^1$. Moreover, by Proposition 1.6 the formula

$$\pi_g^{\mathbb{S}^1,\xi}(f)(v) = \|g^{-1}v\|^{-1-\mathrm{i}\xi} f(g^{-1}\boldsymbol{\cdot}v)$$

for $g \in \mathrm{SL}_2(\mathbb{R})$, $f \in L^2_m(\mathbb{S}^1)$, and $v \in \mathbb{S}^1$ defines a unitary representation $\pi^{\mathbb{S}^1,\xi}$ of $\mathrm{SL}_2(\mathbb{R})$ on $L^2_m(\mathbb{S}^1)$.

We now show that π^ξ is $\pi^{\mathbb{S}^1,\xi}$ in disguise. In fact, we define for $f \in L^2_m(\mathbb{S}^1)$ the function

$$U(f)\colon \mathrm{SL}_2(\mathbb{R}) \ni g \longmapsto U(f)(g) = \|ge_1\|^{-1-\mathrm{i}\xi} f(g\boldsymbol{\cdot}e_1).$$

Then $\|U(f)\|_{L^2(K)} = \|f\|_{L^2_m(\mathbb{S}^1)}$ since the normalized Haar measure m_K is mapped under the action to the normalized length measure m on $\mathbb{S}^1$. Moreover, $g \in \mathrm{SL}_2(\mathbb{R})$ and $b = a_t u_x \in B$ imply $be_1 = \mathrm{e}^t e_1$ and so

$$\begin{aligned} U(f)(gb) &= \|gbe_1\|^{-1-\mathrm{i}\xi} f(gb\boldsymbol{\cdot}e_1) \\ &= \mathrm{e}^{-t-\mathrm{i}\xi t} \|ge_1\|^{-1-\mathrm{i}\xi} f(g\boldsymbol{\cdot}v) \\ &= \overline{\chi_\xi(b)} \Delta_B(b)^{\frac{1}{2}} U(f)(g) \end{aligned}$$

by (8.24), which shows that $U(f) \in \mathcal{H}_\xi$. We note that since $f \in L^2_m(\mathbb{S}^1)$ was arbitrary, this shows in particular that every $F \in L^2(K)$ has an extension to an element of $\mathcal{H}_\xi$. Moreover, by the Iwasawa decomposition and Definition 9.31(2) this extension is also uniquely determined.

Finally, we let $g_0 \in \mathrm{SL}_2(\mathbb{R})$ and calculate

$$U(f)(g_0^{-1}g) = \|g_0^{-1}ge_1\|^{-1-\mathrm{i}\xi} f(g_0^{-1}g\boldsymbol{\cdot}e_1)$$

and

$$\begin{aligned} U\big(\pi_{g_0}^{\mathbb{S}^1,\xi} f\big)(g) &= \|ge_1\|^{-1-\mathrm{i}\xi}\big(\pi_{g_0}^{\mathbb{S}^1,\xi} f\big)(g\boldsymbol{\cdot}e_1) \\ &= \underbrace{\|ge_1\|^{-1-\mathrm{i}\xi}\|g_0^{-1}(g\boldsymbol{\cdot}e_1)\|^{-1-\mathrm{i}\xi}}_{\|g_0^{-1}ge_1\|^{-1-\mathrm{i}\xi}} f(g_0^{-1}g\boldsymbol{\cdot}e_1). \end{aligned}$$

Together these show that $U\colon L^2_m(\mathbb{S}^1) \to \mathcal{H}_\xi$ is an intertwining isomorphism, and hence that π^ξ is a unitary representation.

Since $-I$ belongs to the centre of $\mathrm{SL}_2(\mathbb{R})$, the subspaces $\mathcal{H}_\xi^{\mathrm{even}}$, $\mathcal{H}_\xi^{\mathrm{odd}}$ of $\mathcal{H}_\xi$ are closed invariant subspaces. It follows that π^ξ, $\pi^{\xi,\mathrm{e}}$, $\pi^{\xi,\mathrm{o}}$ are well-defined unitary representations of $\mathrm{SL}_2(\mathbb{R})$. □

Exercise 9.33. As an alternative, use Corollary 8.37 to show that π^ξ defines a unitary representation of $\mathrm{SL}_2(\mathbb{R})$.

For the proof of irreducibility, we will use the following lemma.

Lemma 9.34 (Casimir eigenvalue for π^ξ). *Let $\xi \in \mathbb{R}$. Then the closure of $\pi^\xi_\partial(\Omega)$ is equal to multiplication by $\alpha_\xi = -\xi^2$. Moreover, for every $n \in \mathbb{Z}$ the extension of $\chi_{-n} \in L^2(K)$ to an element $F_{\xi,n} \in \mathcal{H}_\xi$ has K-weight n, and is given by*

$$F_{\xi,n}(k_\psi a_t u_x) = \mathrm{e}^{-\mathrm{i}n\psi-\mathrm{i}\xi t-t} \tag{9.19}$$

for all $k_\psi a_t u_x \in KAU = G$. These functions satisfy

$$\begin{aligned} \pi^\xi_\partial(\mathbf{r}^+)F_{\xi,n} &= \frac{n+1+\mathrm{i}\xi}{2}F_{\xi,n+2}, \\ \pi^\xi_\partial(\mathbf{r}^-)F_{\xi,n} &= \frac{-n+1+\mathrm{i}\xi}{2}F_{\xi,n-2}, \end{aligned}$$

and

$$\pi^\xi_\partial(\mathbf{a})F_{\xi,n} = \frac{n+1+\mathrm{i}\xi}{2}F_{\xi,n+2} + \frac{-n+1+\mathrm{i}\xi}{2}F_{\xi,n-2} \tag{9.20}$$

for all $n \in \mathbb{Z}$.

Proof. For any $n \in \mathbb{Z}$, we define $F_{\xi,n} \in \mathcal{H}_\xi$ by setting

$$F_{\xi,n}|_K = \chi_{-n} \in L^2(K)$$

to be the character defined by $-n$ and extending it by the defining properties of its elements to an element of $\mathcal{H}_\xi$. Using the formula $\Delta_B(a_t u_x) = \mathrm{e}^{-2t}$ for all $a_t u_x \in AN = B$ and the definition of $\mathcal{H}_\xi$, this gives (9.19).

To see that $F_{\xi,n}$ has K-weight n, we calculate

$$\pi^\xi_{k_\psi}(F_{\xi,n})(k_\theta) = F_{\xi,n}(k_\psi^{-1}k_\theta) = \mathrm{e}^{\mathrm{i}n\psi-\mathrm{i}n\theta} = \mathrm{e}^{\mathrm{i}n\psi}F_{\xi,n}(k_\theta)$$

for all $k_\psi, k_\theta \in K$. Since the characters χ_{-n} for $n \in \mathbb{Z}$ form an orthonormal basis of $L^2(K)$, it follows that the functions $F_{\xi,n}$ for $n \in \mathbb{Z}$ form an orthonormal basis of $\mathcal{H}_\xi$. We note that each $F_{\xi,n}$ is a smooth function on G, which implies, by dominated convergence, that it is also a smooth vector for π^ξ. Alternatively, the latter also follows from Lemma 9.16.

Next we wish to calculate $\pi^\xi_\partial(\mathbf{a})F_{\xi,n}$. For $t \in \mathbb{R}$ we have

$$\exp(t\mathbf{a}) = \begin{pmatrix} \mathrm{e}^t & \\ & \mathrm{e}^{-t} \end{pmatrix},$$

and

$$\big(\pi^\xi_{\exp(t\mathbf{a})} F_{\xi,n}\big)(k_\theta) = F_{\xi,n}\big(\exp(-t\mathbf{a})k_\theta\big) = F_{\xi,n}\left(\begin{pmatrix} \mathrm{e}^{-t} & \\ & \mathrm{e}^t \end{pmatrix} k_\theta\right).$$

In order to apply the definition of $F_{\xi,n}$ in (9.19), we need to write the argument in the form

$$\begin{pmatrix} \mathrm{e}^{-t} & \\ & \mathrm{e}^t \end{pmatrix} k_\theta = k_{\psi_0} a_{t_0} u_{x_0},$$

where in fact we are only interested in the angle parameter $\psi_0 = \psi_0(t,\theta)$ and the diagonal parameter $t_0 = t_0(t,\theta)$ considered as functions in t and θ. As in the proof of the estimate for the Harish-Chandra spherical function in Proposition 8.40, we obtain ψ_0 and t_0 by using polar coordinates in $\mathbb{R}^2$. Indeed,

$$k_{\psi_0} a_{t_0} u_{x_0} e_1 = \mathrm{e}^{t_0} \begin{pmatrix} \cos\psi_0 \\ \sin\psi_0 \end{pmatrix} \tag{9.21}$$

must equal

$$\begin{pmatrix} \mathrm{e}^{-t} & 0 \\ 0 & \mathrm{e}^t \end{pmatrix} k_\theta e_1 = \begin{pmatrix} \mathrm{e}^{-t} & 0 \\ 0 & \mathrm{e}^t \end{pmatrix} \begin{pmatrix} \cos\theta \\ \sin\theta \end{pmatrix} = \begin{pmatrix} \mathrm{e}^{-t}\cos\theta \\ \mathrm{e}^t \sin\theta \end{pmatrix}. \tag{9.22}$$

Therefore

$$\mathrm{e}^{2t_0} = \mathrm{e}^{-2t}\cos^2\theta + \mathrm{e}^{2t}\sin^2\theta.$$

We are interested in the partial derivative of $F_{\xi,n}$ with respect to t at $t = 0$. Hence we calculate from this that

$$\begin{aligned} 2\mathrm{e}^{2t_0}\frac{\partial}{\partial t}t_0 = \frac{\partial}{\partial t}(\mathrm{e}^{2t_0}) &= \frac{\partial}{\partial t}(\mathrm{e}^{-2t}\cos^2\theta + \mathrm{e}^{2t}\sin^2\theta) \\ &= -2\mathrm{e}^{-2t}\cos^2\theta + 2\mathrm{e}^{2t}\sin^2\theta. \end{aligned}$$

For $t = 0$, this gives, with $t_0(0,\theta) = 0$ for all $\theta \in \mathbb{R}$,

$$\frac{\partial}{\partial t}\Big|_{t=0}(t_0) = -\cos^2\theta + \sin^2\theta = -\cos(2\theta) = -\tfrac{1}{2}\big(\mathrm{e}^{2\theta\mathrm{i}} + \mathrm{e}^{-2\theta\mathrm{i}}\big). \tag{9.23}$$

For the angle $\psi_0 = \psi_0(\theta,t)$, we obtain from (9.21) and (9.22) that ψ_0 and θ correspond to directions in the same quadrant. Moreover,

$$\tan\psi_0 = \frac{\mathrm{e}^t\sin\theta}{\mathrm{e}^{-t}\cos\theta} = \mathrm{e}^{2t}\tan\theta;$$

$$(1+\tan^2\psi_0)\frac{\partial}{\partial t}\psi_0 = \frac{\partial}{\partial t}(\tan\psi_0) = \frac{\partial}{\partial t}\big(\mathrm{e}^{2t}\tan\theta\big) = 2\mathrm{e}^{2t}\tan\theta.$$

Setting $t=0$ and using in addition $\psi_0(0,\theta)=\theta$ for all $\theta\in\mathbb{R}$, we obtain

$$\frac{\partial}{\partial t}\Big|_{t=0}(\psi_0) = \frac{2\tan\theta}{1+\tan^2\theta} = 2\sin\theta\cos\theta = \sin 2\theta = \tfrac{1}{2\mathrm{i}}\big(\mathrm{e}^{2\theta\mathrm{i}} - \mathrm{e}^{-2\theta\mathrm{i}}\big). \tag{9.24}$$

Combining (9.23), (9.24), and using again $t_0(0,\theta)=0$ and $\psi_0(0,\theta)=\theta$ for all $\theta\in\mathbb{R}$, this gives

$$\begin{aligned}
\pi_\partial^\xi(\mathbf{a})F_{\xi,n}(k_\theta) &= \frac{\partial}{\partial t}\Big|_{t=0}\big(F_{\xi,n}(k_{\psi_0}a_{t_0}u_{x_0})\big) = \frac{\partial}{\partial t}\Big|_{t=0}\big(\mathrm{e}^{-\mathrm{i}n\psi_0-\mathrm{i}\xi t_0-t_0}\big)\\
&= \mathrm{e}^{-\mathrm{i}n\theta}\left(-\mathrm{i}n\Big(\frac{\partial}{\partial t}\Big|_{t=0}\psi_0\Big) - (\mathrm{i}\xi+1)\Big(\frac{\partial}{\partial t}\Big|_{t=0}t_0\Big)\right)\\
&= \mathrm{e}^{-\mathrm{i}n\theta}\left(-\mathrm{i}n\frac{1}{2\mathrm{i}}\big(\mathrm{e}^{2\theta\mathrm{i}}-\mathrm{e}^{-2\theta\mathrm{i}}\big) + (\mathrm{i}\xi+1)\frac{1}{2}\big(\mathrm{e}^{2\theta\mathrm{i}}+\mathrm{e}^{-2\theta\mathrm{i}}\big)\right)\\
&= \left(\frac{n+1+\mathrm{i}\xi}{2}\right)\mathrm{e}^{-\mathrm{i}(n+2)\theta} + \left(\frac{-n+1+\mathrm{i}\xi}{2}\right)\mathrm{e}^{-\mathrm{i}(n-2)\theta}\\
&= \left(\frac{n+1+\mathrm{i}\xi}{2}\right)F_{\xi,n+2}(k_\theta) + \left(\frac{-n+1+\mathrm{i}\xi}{2}\right)F_{\xi,n-2}(k_\theta)
\end{aligned}$$

for all $k_\theta\in K$. To summarize, we have shown (9.20).

Recalling that $\mathbf{a}=\mathbf{r}^+ + \mathbf{r}^-$ and that, by Proposition 9.12, $\pi_\partial(\mathbf{r}^\pm)F_{\xi,n}$ has weight $n\pm 2$, we obtain

$$\pi_\partial^\xi(\mathbf{r}^+)F_{\xi,n} = \frac{n+1+\mathrm{i}\xi}{2}F_{\xi,n+2}$$

and

$$\pi_\partial^\xi(\mathbf{r}^-)F_{\xi,n} = \frac{-n+1+\mathrm{i}\xi}{2}F_{\xi,n-2},$$

as claimed in the lemma.

Using the formula for Ω in (9.11) in terms of $\mathbf{r}^+$, $\mathbf{r}^-$, and $\mathbf{k}$, we obtain with $\pi_\partial^\xi(\mathbb{1}_{\mathfrak{C}}+\mathrm{i}\mathbf{k})F_{\xi,n} = (1-n)F_{\xi,n}$ that

$$\begin{aligned}
\pi_\partial^\xi(\Omega)F_{\xi,n} &= 4\pi_\partial^\xi(\mathbf{r}^+\circ\mathbf{r}^-)F_{\xi,n} + \pi_\partial^\xi\left((\mathbb{1}_{\mathfrak{C}}+\mathrm{i}\mathbf{k})^{\circ 2}\right)F_{\xi,n}\\
&= 2\pi_\partial^\xi(\mathbf{r}^+)(-n+1+\mathrm{i}\xi)F_{\xi,n-2} + (1-n)^2F_{\xi,n}\\
&= (n-1+\mathrm{i}\xi)(-n+1+\mathrm{i}\xi)F_{\xi,n} + (1-n)^2F_{\xi,n}\\
&= \big(-(n-1)^2-\xi^2+(1-n)^2\big)F_{\xi,n}\\
&= -\xi^2F_{\xi,n}
\end{aligned}$$

for all $n\in\mathbb{Z}$. Since the functions $F_{\xi,n}$ for $n\in\mathbb{Z}$ form an orthonormal basis of $\mathcal{H}_\xi$, the lemma follows. □

Proof of irreducibility and isomorphism claims in Theorem 9.32. Let $\xi \in \mathbb{R}$. By Lemma 9.34 the representations $\pi^{\xi,\mathrm{e}}$ and $\pi^{\xi,\mathrm{o}}$ have Casimir eigenvalue $-\xi^2$. By Proposition 9.18, we know that these are irreducible with the exception of $\pi^{0,\mathrm{o}}$. By Theorem 9.22 there is however only one even irreducible representation with Casimir eigenvalue $-\xi^2$, which gives $\pi^{\xi,\mathrm{e}} \cong \pi^{-\xi,\mathrm{e}}$. Similarly, for $\xi \in \mathbb{R}\smallsetminus\{0\}$ we have $\pi^{\xi,\mathrm{o}} \cong \pi^{-\xi,\mathrm{o}}$.

Let us now discuss $\pi^{0,\mathrm{o}}$ with Casimir eigenvalue 0. By Corollary 9.14 we also have $\alpha_{\delta^{1,+}} = \alpha_{\delta^{1,-}} = 0$. By the construction of $\pi^{0,\mathrm{o}}$, it contains all odd K-weights. Let $v_1 \in \mathcal{H}_{\pi^{0,\mathrm{o}}}$ be a unit vector with K-weight 1, and let $e_0 \in \mathcal{H}_{\delta^{1,+}}$ be the unit vector with K-weight 1 as in Lemma 8.25 (see also the paragraph after Theorem 8.31). By Proposition 9.20, we conclude that $\varphi_{v_1}^{\pi^{0,\mathrm{o}}} = \varphi_{e_0}^{\delta^{1,+}}$. However, Proposition 1.65 now implies that the cyclic representations $\langle v_1\rangle_{\pi^{0,\mathrm{o}}}$ and $\langle e_0\rangle_{\delta^{1,+}} = \mathcal{H}_{\delta^{1,+}}$ are isomorphic. That is, we have shown that $\delta^{1,+} < \pi^{0,\mathrm{o}}$ (up to isomorphisms). Using a unit vector $v_{-1} \in \mathcal{H}_{\pi^{0,\mathrm{o}}}$ of K-weight -1, we obtain, by the same argument again, that $\delta^{1,-} < \pi^{0,\mathrm{o}}$. Together we have shown that $\delta^{1,+}\oplus\delta^{1,-} < \pi^{0,\mathrm{o}}$. However, since in $\delta^{1,+}\oplus\delta^{1,-}$ and $\pi^{0,\mathrm{o}}$ each odd K-weight appears with multiplicity one, we must have equality. □

Proof of integrability and decay properties in Theorem 9.32. Let ξ be in $\mathbb{R}$ and n in $\mathbb{Z}$. By (9.19) we have $|F_{\xi,n}| = F_{0,0}$. For $m, n \in \mathbb{Z}$ this implies

$$\begin{aligned}\left|\left\langle \pi^{\xi}_g F_{\xi,m}, F_{\xi,n}\right\rangle\right| &= \left|\int_K F_{\xi,m}(g^{-1}k)\overline{F_{\xi,n}(k)}\,\mathrm{d}m_K(k)\right|\\ &\leqslant \int_K F_{0,0}(g^{-1}k)F_{0,0}(k)\,\mathrm{d}m_K(k) = \Xi(g)\end{aligned}$$

for $g \in \mathrm{SL}_2(\mathbb{R})$. By Proposition 8.40 and Theorem 8.42 we deduce that π^{ξ} is tempered with decay exponent $1-\varepsilon$ for any $\varepsilon > 0$ and $\xi \in \mathbb{R}$.

Also by Proposition 8.40 we have $\Xi \notin L^2(\mathrm{SL}_2(\mathbb{R}))$, which shows by Theorem 8.2 that $\pi^{0,\mathrm{e}}$ is not a discrete series representation. For the odd representation recall the isomorphism $\pi^{0,\mathrm{o}} \cong \delta^{1,+} \oplus \delta^{1,-}$ and that by Theorem 8.31 both $\delta^{1,+}$ and $\delta^{1,-}$ are also not discrete series representations. In the rest of the proof we verify by elementary analysis that $\pi^{\xi,\mathrm{e}}$ and $\pi^{\xi,\mathrm{o}}$ for $\xi \neq 0$ are not discrete series representations.

So suppose now that $\xi \in \mathbb{R}\smallsetminus\{0\}$ and $n \in \mathbb{Z}$. Let $\alpha = -\xi^2 < 0$. We recall that by Lemma 9.30 the function $\phi(t) = \varphi^{\pi^{\xi}}_{F_{\xi,n}}(a_t)$ satisfies for $t > 0$ the second order linear differential equation

$$\phi'' + f_1\phi' + f_0\phi = 0, \tag{9.25}$$

where $f_1(t) = 2 + \mathrm{O}(\mathrm{e}^{-2t})$ and $f_0(t) = 1 - \alpha + \mathrm{O}(\mathrm{e}^{-2t})$ for $t \geqslant 1$. Also recall that $\phi'(t) = \left\langle \pi^{\xi}_{a_t}\pi^{\xi}_{\partial}(\mathbf{a})F_{\xi,n}, F_{\xi,n}\right\rangle$ by (9.15). To see that $\pi^{\xi,\mathrm{e}}$ and $\pi^{\xi,\mathrm{o}}$ are not discrete series representations, fix $n \in \mathbb{Z}$ of the desired parity. We will show that $\varphi^{\pi^{\xi}}_{F_{\xi,n}}$ and $\varphi^{\pi^{\xi}}_{\pi_{\partial}(\mathbf{a})F_{\xi,n},F_{\xi,n}}$ are not both square integrable and apply Theorem 8.2. By using Lemma 9.19 and (8.11) it suffices to show that ϕ and ϕ' are not both square integrable on $[1,\infty)$ with respect to $\mathrm{e}^{2t}\,\mathrm{d}t$. We define $y(t) = \mathrm{e}^t\phi(t)$, so that

$$y'(t) = \mathrm{e}^t(\phi'(t) + \phi(t)),$$

and we wish to show that y and y' are not both square integrable on $[1, \infty)$ with respect to $\mathrm{d}t$. The differential equation (9.25) for y takes the form

$$\underbrace{y'' - 2y' + y}_{=\mathrm{e}^t\phi''} + f_1 \underbrace{(y' - y)}_{\mathrm{e}^t\phi'} + f_0 \underbrace{y}_{\mathrm{e}^t\phi} = 0,$$

or

$$y'' + F_1 y' + F_0 y = 0$$

for $F_1 = f_1 - 2 = \mathrm{O}(\mathrm{e}^{-2t})$ and $F_0 = 1 - f_1 + f_0 = -\alpha + \mathrm{O}(\mathrm{e}^{-2t})$. Finally, we define $z = (y')^2 - \alpha y^2$ so that

$$\begin{aligned} z' &= 2y'(y'' - \alpha y) \\ &= 2y'(-F_1 y' - F_0 y - \alpha y) \\ &= -2F_1(y')^2 - 2(F_0 + \alpha) y y' \\ &= \mathrm{O}(\mathrm{e}^{-2t}) z, \end{aligned} \tag{9.26}$$

where in the last step we applied the asymptotics for F_1 and F_0 and bounded both $(y')^2$ and yy' by $z = (y')^2 - \alpha y^2$. As π^ξ is tempered, we may use a multiple of Ξ to bound ϕ and ϕ'. By Proposition 8.40 we have $\Xi(a_t) \ll t\mathrm{e}^{-t}$ for $t \geqslant 1$, which gives $|y(t)| \ll t$, $|y'(t)| \ll t$, and $z(t) \ll t^2$ for $t \geqslant 1$. Therefore (9.26) implies that

$$z(t) - z(t_0) = \int_{t_0}^{t} z'(s)\, \mathrm{d}s = \mathrm{O}\left(\int_{t_0}^{t} \mathrm{e}^{-2s} s^2 \, \mathrm{d}s \right)$$

for $t \geqslant t_0 \geqslant 1$. However, this implies by the Cauchy criterion that $\lim_{t \to \infty} z(t)$ exists. In particular,

$$S(t_0) = \sup\{z(t) \mid t \geqslant t_0\} < \infty$$

for all $t_0 \geqslant 1$. Note that we have $S(t_0) \geqslant z(t_0) > 0$ as $z(t_0) = 0$ would give $\phi(t_0) = \phi'(t_0) = 0$ and contradict the uniqueness of the solution to (9.25) due to the Picard–Lindelöf theorem. Using (9.26) again, there exists a constant C so that

$$|z(t) - z(t_1)| \leqslant C \int_{t_1}^{t} \mathrm{e}^{-2s} \, \mathrm{d}s S(t_0) \leqslant \tfrac{1}{2} C \mathrm{e}^{-2t_1} S(t_0)$$

for all $t \geqslant t_1 \geqslant t_0 \geqslant 1$. Now choose $t_0 \geqslant 1$ so that $C\mathrm{e}^{-2t_0} \leqslant \frac{1}{2}$, choose $t_1 \geqslant t_0$ so that $z(t_1) > \frac{1}{2} S(t_0)$. Then

$$|z(t) - z(t_1)| \leqslant \tfrac{1}{4} S(t_0) \leqslant \tfrac{1}{2} z(t_1)$$

for $t > t_1$, which implies that $\lim_{t\to\infty} z(t) \geqslant \frac{1}{2}z(t_1)$. However, this also implies that

$$\int_1^\infty \left(|\alpha|y^2 + (y')^2\right) \mathrm{d}t = \int_1^\infty z \,\mathrm{d}t = \infty.$$

Hence y and y' cannot both be square-integrable on $[1,\infty)$. As discussed above, this shows that both $\pi^{\xi,\mathrm{e}}$ and $\pi^{\xi,\mathrm{o}}$ are not discrete series representations. □

Exercise 9.35. (a) Analyze the above argument to show that for any $\xi \in \mathbb{R}\smallsetminus\{0\}$ and $n \in \mathbb{Z}$ there exists a constant $C_{\xi,n}$ so that

$$\left|\varphi^{\pi^\xi}_{F_{\xi,n}}(g)\right| \ll C_{\xi,n}\|g\|^{-1}.$$

(b) Show that $C_{\xi,0} \to \infty$ as $\xi \to 0$, which makes the conclusion in part (a) less interesting.

9.4 Two Action-Associated Representations of $\mathrm{SL}_2(\mathbb{R})$*

We wish to show here how the principal series representations can naturally occur as components of other unitary representations. Since $\mathrm{SL}_2(\mathbb{R})$ acts both on the Euclidean plane $\mathbb{R}^2$ and on the hyperbolic plane $\mathbb{H}$, preserving area measure on the space in each case, this already gives rise to two natural unitary representations of $\mathrm{SL}_2(\mathbb{R})$. As we will see, the case of $\mathbb{R}^2$ will be relatively straightforward to analyze. On the other hand, understanding the case of $\mathbb{H}$ will require more work, but we will motivate the formulas arising.

9.4.1 The Action-Associated Representation on $\mathbb{R}^2$

Almost by definition, the group $\mathrm{SL}_2(\mathbb{R})$ acts continuously on $\mathbb{R}^2$, preserving the two-dimensional Lebesgue measure $m = m_{\mathbb{R}^2}$. Using Proposition 1.3, this gives rise to an action-associated representation $\pi^{\mathbb{R}^2}$ of $\mathrm{SL}_2(\mathbb{R})$ on $L^2_m(\mathbb{R}^2)$, where

$$\pi^{\mathbb{R}^2}_g(f)(x) = f(g^{-1}x)$$

for $g \in \mathrm{SL}_2(\mathbb{R})$, $f \in L^2_m(\mathbb{R}^2)$, and $x \in \mathbb{R}^2$.

Using polar coordinates

$$(r,\theta) \in (0,\infty) \times [0,2\pi)$$

for $\mathbb{R}^2\smallsetminus\{0\}$ with $\mathrm{d}m = r\,\mathrm{d}r\,\mathrm{d}\theta$, we make the following definition.

Definition 9.36 (Radial Mellin transform). For a function $f \in L^2_m(\mathbb{R}^2)$, an element $h \in \mathrm{SL}_2(\mathbb{R})$, and a frequency parameter $\xi \in \mathbb{R}$, we define the *radial Mellin transform* of f at (h,ξ) by

$$\widehat{f}^{\mathrm{rad}}(h,\xi) = \int_0^\infty f(rhe_1) r^{\mathrm{i}\xi}\,\mathrm{d}r, \tag{9.27}$$

where

$$e_1 = \begin{pmatrix}1\\0\end{pmatrix}$$

denotes the first basis vector of $\mathbb{R}^2$.

Just as in the case of the usual Fourier transform, the integral in (9.27) may not make sense as a Lebesgue integral. Thus we also need to discuss the meaning of this expression more carefully (which we will do in the proof of Proposition 9.40). The following lemma, together with the definition of the principal series representation in Definition 9.31 reveal why (9.27) is really the right definition.

Lemma 9.37 (Equivariance properties). *For $f \in C_c(\mathbb{R}^2)$ the radial Mellin transform $\widehat{f}^{\mathrm{rad}}(h,\xi)$ is well-defined and satisfies*

$$\begin{cases} \widehat{\pi_g^{\mathbb{R}^2}(f)}^{\mathrm{rad}}(h,\xi) = \widehat{f}^{\mathrm{rad}}(g^{-1}h,\xi) \\ \widehat{f}^{\mathrm{rad}}(hb,\xi) = \overline{\chi_\xi(b)}\Delta_B(b)^{\frac12}\widehat{f}^{\mathrm{rad}}(h,\xi) \end{cases}$$

for all $(h,\xi) \in \mathrm{SL}_2(\mathbb{R}) \times \mathbb{R}$, $g \in \mathrm{SL}_2(\mathbb{R})$, and $b \in B = AU$.

PROOF. It is clear that for $f \in C_c(\mathbb{R}^2)$, the domain of integration in 9.27 can be chosen to be a compact interval, which gives the first claim in the lemma.

Now fix some $g, h \in \mathrm{SL}_2(\mathbb{R})$ and $\xi \in \mathbb{R}$. Then

$$\begin{aligned} \widehat{\pi_g^{\mathbb{R}^2}(f)}^{\mathrm{rad}}(h,\xi) &= \int_0^\infty \big(\pi_g^{\mathbb{R}^2}(f)\big)(rhe_1) r^{\mathrm{i}\xi}\,\mathrm{d}r \\ &= \int_0^\infty f(g^{-1}rhe_1) r^{\mathrm{i}\xi}\,\mathrm{d}r \\ &= \widehat{f}^{\mathrm{rad}}(g^{-1}h,\xi), \end{aligned}$$

as claimed.

For the second claim, we calculate for $b = a_t u_x \in B = AU$ that

$$\begin{aligned} \widehat{f}^{\mathrm{rad}}(ha_t u_x,\xi) &= \int_0^\infty f\big(rh\underbrace{a_t u_x e_1}_{=\mathrm{e}^t e_1}\big) r^{\mathrm{i}\xi}\,\mathrm{d}r \\ &= \int_0^\infty f\big(\widetilde{r}he_1\big)\big(\widetilde{r}\mathrm{e}^{-t}\big)^{\mathrm{i}\xi}\mathrm{e}^{-t}\,\mathrm{d}\widetilde{r} \\ &= \mathrm{e}^{-\mathrm{i}\xi t}\mathrm{e}^{-t}\widehat{f}^{\mathrm{rad}}(h,\xi), \end{aligned}$$

where we used the substitution $\widetilde{r} = \mathrm{e}^t r$ with $\mathrm{d}\widetilde{r} = \mathrm{e}^t\,\mathrm{d}r$. The lemma follows by recalling that $\Delta_B(a_t u_x) = \mathrm{e}^{-2t}$ and $\chi_\xi(a_t u_x) = \mathrm{e}^{\mathrm{i}\xi t}$ for all $a_t u_x \in B$. □

In addition to the correct equivariance properties as shown above, the radial Mellin transform is also isometric in the following sense.

Lemma 9.38 (Isometry). *For $f \in C_c(\mathbb{R}^2)$ we have*

$$\|f\|_{L^2(\mathbb{R}^2)} = \left\| \widehat{f}^{\mathrm{rad}}|_{K\times\mathbb{R}} \right\|_{L^2(K\times\mathbb{R})},$$

where we equip $K \times \mathbb{R}$ with the Haar measure $\mathrm{d}m_K\,\mathrm{d}\xi = \frac{1}{2\pi}\,\mathrm{d}\theta\,\mathrm{d}\xi$.

PROOF. We first note that

$$\begin{aligned}\|f\|^2_{L^2(\mathbb{R}^2)} &= \int_0^{2\pi}\int_0^\infty \left|f\left(rk_\theta e_1\right)\right|^2 r\,\mathrm{d}r\,\mathrm{d}\theta\\ &= \int_0^{2\pi}\int_0^\infty \left|rf\left(rk_\theta e_1\right)\right|^2 \frac{\mathrm{d}r}{r}\,\mathrm{d}\theta.\end{aligned}$$

We now define, for $\theta \in [0, 2\pi)$, the function $F_\theta\colon \mathbb{R} \to \mathbb{C}$ by

$$F_\theta(s) = \mathrm{e}^s f\left(\mathrm{e}^s k_\theta e_1\right)$$

for all $s \in \mathbb{R}$, and note that F_θ corresponds, roughly speaking, to the restriction of f to the ray from 0 at angle θ to the positive x-axis. Using the fact that f has compact support in $\mathbb{R}^2$ and is bounded near 0, we see that $F_\theta \in L^1(\mathbb{R}) \cap L^2(\mathbb{R})$ for all $\theta \in \mathbb{R}$. Using the substitution $r = \mathrm{e}^s$ with $\frac{\mathrm{d}r}{r} = \mathrm{d}s$ we see that

$$\|f\|^2_{L^2(\mathbb{R}^2)} = \int_0^{2\pi} \underbrace{\int_{-\infty}^\infty |F_\theta(s)|^2\,\mathrm{d}s}_{=\|F_\theta\|^2_{L^2(\mathbb{R})}}\,\mathrm{d}\theta.$$

Next we use the fact that

$$\|F_\theta\|_{L^2(\mathbb{R})} = \|\widecheck{F_\theta}\|_{L^2(\mathbb{R})},$$

where $\widecheck{F_\theta}$ again denotes the Fourier back transform. Using the definitions and the substitution $r = \mathrm{e}^s$ with $\mathrm{d}r = \mathrm{e}^s\,\mathrm{d}s$ again, we obtain

$$\begin{aligned}\widecheck{F_\theta}(\zeta) &= \int_{-\infty}^\infty F_\theta(s)\mathrm{e}^{2\pi\mathrm{i}\zeta s}\,\mathrm{d}s\\ &= \int_{-\infty}^\infty \mathrm{e}^s f(\mathrm{e}^s k_\theta e_1)\mathrm{e}^{2\pi\mathrm{i}\zeta s}\,\mathrm{d}s\\ &= \int_0^\infty f(rk_\theta e_1)r^{2\pi\mathrm{i}\zeta}\,\mathrm{d}r = \widehat{f}^{\mathrm{rad}}(k_\theta, 2\pi\zeta)\end{aligned} \tag{9.28}$$

for all $\zeta \in \mathbb{R}$. Together, we obtain

$$\begin{aligned}\|f\|_{L^2(\mathbb{R}^2)}^2 &= \int_0^{2\pi} \|\widecheck{F_\theta}\|_{L^2(\mathbb{R})}^2 \,\mathrm{d}\theta \\ &= \int_0^{2\pi}\int_{-\infty}^{\infty} |\widehat{f}^{\mathrm{rad}}(k_\theta, 2\pi\zeta)|^2 \,\mathrm{d}\zeta\,\mathrm{d}\theta \\ &= \frac{1}{2\pi}\int_0^{2\pi}\int_{-\infty}^{\infty} |\widehat{f}^{\mathrm{rad}}(k_\theta, \xi)|^2 \,\mathrm{d}\xi\,\mathrm{d}\theta = \|\widehat{f}^{\mathrm{rad}}|_{K\times\mathbb{R}}\|_{L^2(K\times\mathbb{R})}^2\end{aligned}$$

by using the substitution $\xi = 2\pi\zeta$. □

Definition 9.31 and Lemmas 9.37 and 9.38 suggest the following definition.

Definition 9.39 (Integrals of principal series representations). For any σ-finite measure μ on $\mathbb{R}$, we define the space $\mathcal{H}_\mu$ of all functions

$$F\colon \mathrm{SL}_2(\mathbb{R}) \times \mathbb{R} \to \mathbb{C}$$

satisfying the following properties:

(1) F is measurable;
(2) $F(hb,\xi) = \overline{\chi_\xi(b)}\Delta_B(b)^{\frac{1}{2}}F(h,\xi)$ for all $h \in \mathrm{SL}_2(\mathbb{R})$, $b \in B$, $\xi \in \mathbb{R}$; and
(3) $\|f|_{K\times\mathbb{R}}\|_{L^2(K\times\mathbb{R}, m_K\times\mu)} < \infty$.

The unitary representation[†]

$$\pi^\mu = \int_{\mathbb{R}} \pi^\xi \,\mathrm{d}\mu(\xi)$$

is defined by the left regular representation on the first component; that is,

$$\pi_g^\mu(F)(h,\xi) = F(g^{-1}h,\xi)$$

for all $F \in \mathcal{H}_\mu$, $g, h \in \mathrm{SL}_2(\mathbb{R})$, and $\xi \in \mathbb{R}$. Moreover, we also define

$$\pi^{\mu,\mathrm{e}} = \int_{\mathbb{R}} \pi^{\xi,\mathrm{e}} \,\mathrm{d}\mu(\xi)$$

and

$$\pi^{\mu,\mathrm{o}} = \int_{\mathbb{R}} \pi^{\xi,\mathrm{o}} \,\mathrm{d}\mu(\xi)$$

to be the restrictions of π^μ to the subspaces

$$\mathcal{H}_\mu^{\mathrm{even}} = \{F \in \mathcal{H}_\mu \mid F(-g,x) = F(g,x) \text{ for all } g, x\}$$

and

$$\mathcal{H}_\mu^{\mathrm{odd}} = \{F \in \mathcal{H}_\mu \mid F(-g,x) = -F(g,x) \text{ for all } g, x\}$$

respectively.

[†] We did not discuss the integral of unitary representations, but have seen special cases before and believe that the notation is justified.

Proposition 9.40 (Spectral decomposition). *The action-associated representation of* $\mathrm{SL}_2(\mathbb{R})$ *on* $\mathbb{R}^2$ *is isomorphic to*

$$\pi^m = \pi^{m,\mathrm{e}} \oplus \pi^{m,\mathrm{o}} = \int_{\mathbb{R}} \pi^{\xi,\mathrm{e}}\,\mathrm{d}\xi \oplus \int_{\mathbb{R}} \pi^{\xi,\mathrm{o}}\,\mathrm{d}\xi,$$

where we use the Lebesgue measure $\mu = m$ *on* $\mathbb{R}$.

PROOF. By Lemmas 9.37 and 9.38, the radial Mellin transform

$$C_c(\mathbb{R}^2) \ni f \longmapsto \widehat{f}^{\mathrm{rad}} \in \mathcal{H}_m$$

is well-defined, intertwining, and an isometry. Hence it extends by the density of $C_c(\mathbb{R}^2)$ in $L^2(\mathbb{R}^2)$ to a well-defined, intertwining isometry

$$L^2(\mathbb{R}^2) \ni f \longmapsto \widehat{f}^{\mathrm{rad}} \in \mathcal{H}_m.$$

We keep referring to $\widehat{f}^{\mathrm{rad}}$ as the radial Mellin transform of $f \in L^2(\mathbb{R}^2)$.

It remains to show that this map is onto. For this, assume that $f_K \in C(K)$ and $f_{\mathbb{R}} \in C_c(\mathbb{R})$. We note that

$$f_K \otimes \widetilde{f_{\mathbb{R}}} \colon K \times \mathbb{R} \longrightarrow \mathbb{C}$$

can be extended using property (2) in Definition 9.39 to an element of $\mathcal{H}_m$. We claim that $f_K \otimes \widetilde{f_{\mathbb{R}}} = \widehat{f}^{\mathrm{rad}}$ for some $f \in C_c(\mathbb{R}^2)$. Also recall that the subspaces $C(K) \subseteq L^2(K)$ and $\widetilde{C_c(\mathbb{R})} \subseteq L^2(\mathbb{R}, m)$ are dense (for the latter, apply Theorem 2.17). Varying f_K and $f_{\mathbb{R}}$, we can then, for example, approximate any function of the form $\mathbb{1}_{B_K} \otimes \mathbb{1}_{B_{\mathbb{R}}}$ extended to an element of $\mathcal{H}_\mu$, where $B_K \subseteq K$ and $B_{\mathbb{R}} \subseteq \mathbb{R}$ are measurable with finite measures. For this reason, the claim implies that the image of the radial Mellin transform (extended to $L^2(\mathbb{R}^2)$) is indeed all of $\mathcal{H}_\mu$.

To prove the claim, we reuse the argument from the proof of Lemma 9.38. Let $f_K \in C(K)$ and $f_{\mathbb{R}} \in C_c(\mathbb{R})$ be as above. We define $f \in C_c(\mathbb{R}^2)$ using polar coordinates by

$$f(rk_\theta e_1) = \tfrac{1}{2\pi} f_K(\theta) r^{-1} f_{\mathbb{R}}\left(\tfrac{1}{2\pi}\log r\right).$$

For this f, the function F_θ for $\theta \in [0, 2\pi)$ appearing in the proof of Lemma 9.38 becomes

$$F_\theta(s) = \mathrm{e}^s f(\mathrm{e}^s k_\theta e_1) = \tfrac{1}{2\pi} f_K(\theta) f_{\mathbb{R}}\left(\tfrac{1}{2\pi}s\right)$$

for $s \in \mathbb{R}$. Hence by (9.28) and the substitution $\widetilde{s} = \frac{1}{2\pi}s$ we have

$$\begin{aligned}\widehat{f}^{\mathrm{rad}}(k_\theta, 2\pi\zeta) = \widetilde{F_\theta}(\zeta) &= \tfrac{1}{2\pi} f_K(\theta) \int_R f_{\mathbb{R}}\left(\tfrac{1}{2\pi}s\right) \mathrm{e}^{2\pi \mathrm{i} s\zeta}\,\mathrm{d}s \\ &= f_K(\theta) \int_{\mathbb{R}} f_{\mathbb{R}}(\widetilde{s}) \mathrm{e}^{2\pi\mathrm{i}\widetilde{s}2\pi\zeta}\,\mathrm{d}\widetilde{s} = f_K(\theta)\widetilde{f_{\mathbb{R}}}(2\pi\zeta)\end{aligned}$$

for all $\theta \in [0, 2\pi)$ and $\zeta \in \mathbb{R}$. Equivalently, we have $\widehat{f}^{\mathrm{rad}} = f_K \otimes \widecheck{f_{\mathbb{R}}}$ as claimed, which gives the proposition. □

Exercise 9.41. (a) Show that π^μ as in Definition 9.39 is indeed a unitary representation of $\mathrm{SL}_2(\mathbb{R})$ for any σ-finite measure μ on $\mathbb{R}$.
(b) Show that π^μ is tempered.

Exercise 9.42. Use Fourier inversion on $\mathbb{R}$ to prove a Fourier inversion formula that expresses $f \in C_c^\infty(\mathbb{R}^2)$ as an integral over values of $\widehat{f}^{\mathrm{rad}}$.

Exercise 9.43. Is the centralizer of the action-associated representation of $\mathrm{SL}_2(\mathbb{R})$ on $L^2(\mathbb{R})$ abelian? Prove your claim. Can you identify the centralizer?

9.4.2 Moving a Loudspeaker to Infinity

To better understand the formulas required for the hyperbolic Fourier transform we wish to discuss a physical interpretation of the Fourier transform on the two planes $\mathbb{R}^2$ and $\mathbb{H}$. As this is just meant as a motivation for the formal definitions coming later, we leave the details of these calculations as exercises.

To begin with, we imagine a loudspeaker L, which we assume will produce the desired sound for any given frequency and amplitude. We imagine the sound wave being represented by a $\mathbb{C}$-valued function f_L on the plane, where $|f_L|^2$ represents the energy of the wave, and the argument of f_L represents the phase shift of the wave. We also imagine that there is no energy loss in the passage of the wave through the medium. This physical interpretation suggests that $|f_L(P_0)|^2$ is inversely proportional to the length of the circle with centre L containing a point P, as illustrated in Figure 9.3.

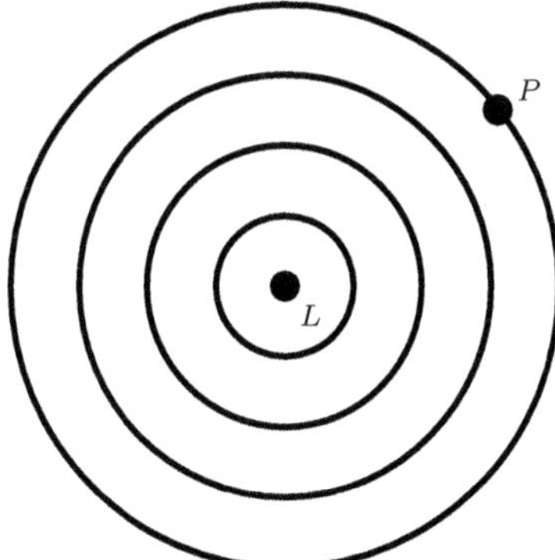

Fig. 9.3: The sound waves emanate from the loudspeaker L and decrease in loudness. The concentric circles indicate the phase of f_L.

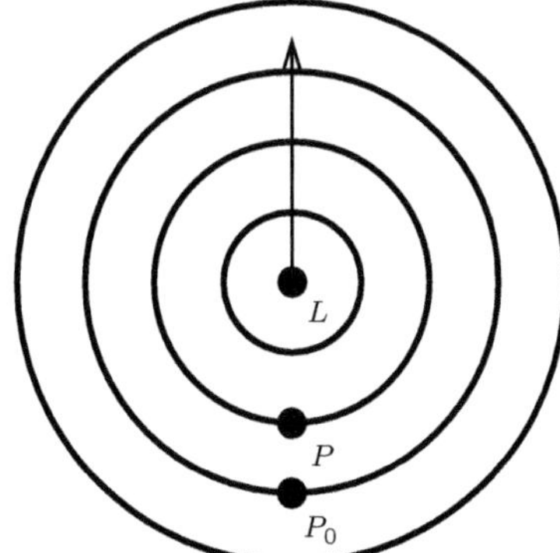

Fig. 9.4: We move the loudspeaker L upwards, and turn up the volume.

We now fix some origin P_0 in the plane, and move the loudspeaker L further away in some pre-determined direction, say upwards as in Figure 9.4. This of course means that we do not hear the sound much at P_0 if L is already far away. To get round this problem, we simultaneously turn up the volume at L so that $|f_L(P_0)|^2 = 1$. We now wish to move L to infinity and describe what will happen to f_L if we do so. However, to do this we have to distinguish between the cases of the Euclidean and hyperbolic planes.

Euclidean plane: In the Euclidean plane, the circle of radius r has circumference $2\pi r$. If P_0 belongs to a circle of radius r_0 (that is, if $r_0 = \|P_0 - L\|$) and P has distance $\|P - P_0\|$ to P_0, then P belongs to a circle of radius r with $\Delta r = r - r_0$ satisfying $|\Delta r| \leqslant \|P - P_0\|$ (by the triangle inequality). Hence

$$|f_L(P_0)|^2 \cdot 2\pi r_0 = |f_L(P)|^2 \cdot 2\pi r.$$

Letting L go to infinity, we have $r_0 \to \infty$ and $\frac{2\pi r}{2\pi r_0} = \frac{r_0+\Delta r}{r_0} \to 1$. Therefore the limiting sound distribution f will have the property that $|f|^2$ is constant and equal to 1. Moreover, the concentric circles degenerate to equidistant parallel lines, so that in the limit we may obtain in this way the function

$$f(x,y) = \mathrm{e}^{\mathrm{i}\xi y}$$

for $P = (x,y) \in \mathbb{R}^2$, where ξ represents the frequency of the wave. Allowing different frequencies and different directions along which L is moved, one obtains in this way any character $\chi_{(\xi_1,\xi_2)}$ for $(\xi_1,\xi_2) \in \mathbb{R}^2$, which we may think of as the elementary waves on $\mathbb{R}^2$.

This suggests the following interpretation for the Fourier transform of a function on $\mathbb{R}^2$. Given f, we first test the correlation of f against all elementary waves. Next we imagine infinitely many loudspeakers at infinity in all directions using various frequencies with well-chosen amplitudes. Fourier inversion now tells us that these then create the prescribed sound distribution f by superposition of the so-created (and correctly amplified) elementary waves.

Exercise 9.44. (a) For a given frequency $\xi \in \mathbb{R}$ and two points $P_0, L \in \mathbb{R}^2$, calculate the function f_L representing a wave of frequency ξ emanating from L with $f_L(P_0) = 1$.
(b) Calculate the limit f of f_L as $L = (0,y) \to \infty$.

Hyperbolic plane: To get some intuition for the hyperbolic Fourier transform, we repeat the above discussion on $\mathbb{H}$, which will lead to the functions that will take over the role of characters in the more formal discussions of the following sections.

We again imagine the loudspeaker L moving to infinity along the upward oriented geodesic and consider equidistant concentric circles with centre L, as in Figure 9.5. We note that in the limit we obtain circles (in the Euclidean sense, within $\mathbb{C} \supseteq \mathbb{D}$) touching the boundary. These are not hyperbolic geodesics; instead these curves are called *horocycles*.

To understand the limit function f of the sound distribution f_L for L going to the boundary, we need to calculate the circumference of a circle of radius R.

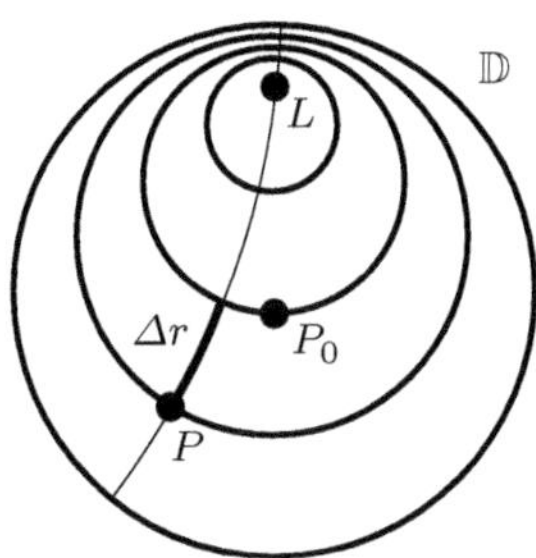

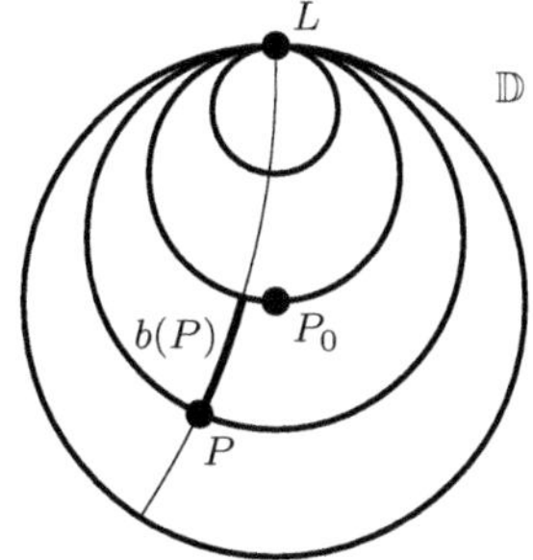

Fig. 9.5: A calculation reveals that any Möbius transformation $z \mapsto \frac{\alpha z+\beta}{\gamma z+\delta}$ on $\overline{\mathbb{C}}$ maps lines and circles to lines and circles. With this, it is straightforward to verify that concentric hyperbolic circles with centre L appear in the disk model of the hyperbolic plane as circles, with L appearing closer to the circle near the boundary. If L is moved to the boundary, these circles degenerate to circles touching the boundary.

To simplify matters, we let the centre be $0 \in \mathbb{D}$ as in Lemma 8.15. By (8.12) the Euclidean radius of this circle is given by $\rho = \tanh\left(\frac{R}{2}\right)$. Hence the circumference can be calculated using the path

$$[0, 2\pi] \ni \theta \longmapsto \rho \mathrm{e}^{\mathrm{i}\theta},$$

which, by definition of the Riemannian metric in (8.6), gives

$$\begin{aligned}\int_0^{2\pi} \frac{2}{(1-\rho^2)} \rho \,\mathrm{d}\theta = \frac{4\pi \tanh\left(\frac{R}{2}\right)}{\left(1-\tanh^2\left(\frac{R}{2}\right)\right)} &= 4\pi \frac{\frac{\sinh\left(\frac{R}{2}\right)}{\cosh\left(\frac{R}{2}\right)} \cdot \cosh^2\left(\frac{R}{2}\right)}{\left(\cosh^2\left(\frac{R}{2}\right) - \sinh^2\left(\frac{R}{2}\right)\right)} \\ &= 4\pi \sinh\left(\tfrac{R}{2}\right) \cosh\left(\tfrac{R}{2}\right) = 2\pi \sinh R.\end{aligned}$$

We let L go to infinity along the Northward geodesic, so that

$$r_0 = \mathsf{d}(P_0, L) \to \infty.$$

For a third point $P \in \mathrm{d}$, we let $r = \mathsf{d}(P, L)$. We also define the 'relative distance' from L compared to P_0 by setting it equal to

$$\Delta r = \mathsf{d}(P, L) - \mathsf{d}(P_0, L) = r - r_0,$$

see also Figure 9.5. Note that $\Delta r = r - r_0$ satisfies $|\Delta r| \leqslant \mathsf{d}(P, P_0)$. With the asymptotic $\sinh R \sim \mathrm{e}^R$ as $R \to \infty$ and

$$|f_L(P_0)|^2 2\pi \sinh r_0 = |f_L(P)|^2 2\pi \sinh r,$$

we obtain

$$\frac{|f_L(P)|^2}{|f_L(P_0)|^2} = \frac{\sinh r_0}{\sinh(r_0 + \Delta r)} \sim \mathrm{e}^{-\Delta r}$$

as $r_0 \to \infty$. Since we normalize the loudness of L to have $|f_L(P_0)|^2 = 1$, we expect that the limit sound wave satisfies

$$|f(P)| = \mathrm{e}^{-\frac{1}{2}b(P)}$$

for the limiting function b of Δr. Putting the phase with frequency ξ into the discussions, we expect that functions of the form†

$$f(P) = \mathrm{e}^{(-\frac{1}{2}+\frac{\mathrm{i}}{2}\xi)b(P)}$$

are the relevant elementary waves on the hyperbolic plane. This is indeed the case, so we will have to define the so-called Busemann function $b(P)$ more carefully (we also refer to Busemann's monograph [11]).

By varying both the frequency and the position of the loudspeakers on the boundary of the hyperbolic plane, we again expect that any sound distribution on the plane can be produced as a superposition of elementary waves.

Exercise 9.45. Repeat (a) and (b) from Exercise 9.44 for $\mathbb{H}$.

9.4.3 The Busemann Function

In the upper half-plane model $\mathbb{H}$, the desired function takes a particularly easy form. Indeed, if we move the loudspeaker L simply up to the point ∞ in $\partial\mathbb{H}$, then the concentric circles degenerate to horizontal lines near i, as in Figure 9.6.

Fig. 9.6: On the left we see that the concentric circles with centre $L = y\mathrm{i}$ for large y are almost horizontal Euclidean lines. On moving y to ∞, these become horizontal Euclidean lines or *horizontal horocycles* in the hyperbolic plane $\mathbb{H}$.

Definition 9.46 (Busemann function for $\infty \in \partial\mathbb{H}$). The *Busemann function* on $\mathbb{H}$ with respect to $\infty \in \partial\mathbb{H}$ (and origin $\mathrm{i} \in \mathbb{H}$) is defined for $z \in \mathbb{H}$ by

$$b_\infty^{\mathbb{H}}(z) = -\log \Im(z).$$

We note that the point $\infty \in \partial\mathbb{H}$ should be thought of as being the point 'at infinity' that is higher up than any $z \in \mathbb{H}$. Roughly speaking, $b_\infty^{\mathbb{H}}(z)$ is

† As earlier, we normalize the meaning of frequency in the following discussions to simplify some of the formulas arising.

comparing the distance of z and of i to ∞ (both of which are of course infinite). More precisely, $b_\infty^{\mathbb{H}}(z)$ measures the hyperbolic distance between the horizontal horocycle at $z = x + \mathrm{i}y \in \mathbb{H}$ and the horizontal horocycle at our designated origin i, given by

$$\mathsf{d}(\mathrm{i}y, \mathrm{i}) = \left| \int_1^y \frac{\mathrm{d}y}{y} \right| = |\log y|.$$

We should think of $b_\infty^{\mathbb{H}}(z)$ as an oriented relative distance, since $b_\infty^{\mathbb{H}}(z) > 0$ means that z is further from ∞ than i is, while $b_\infty^{\mathbb{H}}(z) < 0$ means that i is further away from ∞.

Using the discussion of Section 9.4.2 as in Figure 9.7, we are now led to the following definition.

Definition 9.47 (Hyperbolic wave). We define the *hyperbolic wave function* coming from ∞ with frequency $\xi \in \mathbb{R}$ (normalized for the origin $\mathrm{i} \in \mathbb{H}$) to be

$$\chi_{\infty,\xi}(z) = \mathrm{e}^{(-\frac{1}{2}+\frac{\mathrm{i}}{2}\xi)b_\infty^{\mathbb{H}}(z)} = \Im(z)^{\frac{1}{2}-\frac{\mathrm{i}}{2}\xi}$$

for all $z \in \mathbb{H}$.

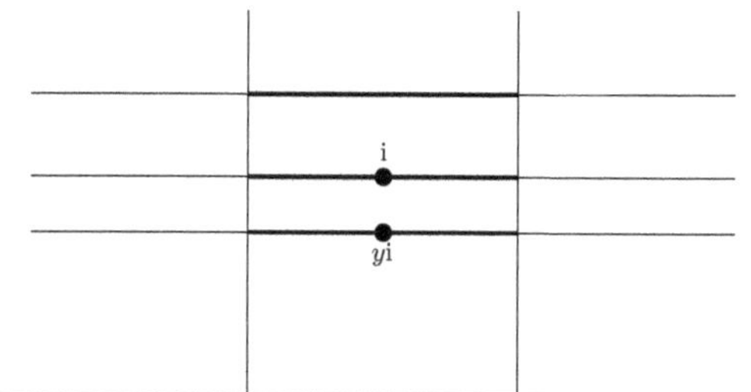

Fig. 9.7: We think of a hyperbolic wave coming from ∞ with horizontal horocycles being the wave fronts. Since an interval of Euclidean length 1 of the horizontal horocycle with vertical coordinate $y \in (0, \infty)$ has hyperbolic length $\frac{1}{y}$, the energy of the wave spreads over a larger region as it spreads down and so its intensity decreases, as in Definition 9.47.

We note that by definition of Möbius transformations in (8.3) and the description of $\varDelta_B$ in (8.24) we have

$$\begin{aligned}\chi_{\infty,\xi}(b^{-1}\bullet z) &= \Im\big(\mathrm{e}^{-2t}(z-x)\big)^{\frac{1}{2}-\frac{\mathrm{i}}{2}\xi} \\ &= \mathrm{e}^{-t(1-\mathrm{i}\xi)}\Im(z)^{\frac{1}{2}-\frac{\mathrm{i}}{2}\xi} = \varDelta_B(b)^{\frac{1}{2}}\chi_\xi(b)\chi_{\infty,\xi}(z)\end{aligned} \tag{9.29}$$

for all $b = u_x a_t \in B$ and $z \in \mathbb{H}$. These formulas suggest a possible link between the hyperbolic wave of frequency ξ and the principal series representation π^ξ corresponding to the frequency parameter ξ (see Section 9.3), which we will explain after defining the hyperbolic Fourier transform.

We also note that the identifications

$$B \ni b = u_x a_t \longmapsto bK \in \mathrm{SL}_2(\mathbb{R})/K \longmapsto z = b \cdot \mathrm{i} = x + \mathrm{e}^{2t}\mathrm{i} \in \mathbb{H}$$

are measure-preserving by our choice of the Haar measure m_B on B in Section 8.3.5 and the normalization $m_K(K) = 1$. We will also simply write m for the Haar measure $m = m_B \times m_K$ on $\mathrm{SL}_2(\mathbb{R}) = BK$, and will write $\int_G \cdot \,\mathrm{d}m$ for integration over $G = \mathrm{SL}_2(\mathbb{R})$.

9.4.4 The Fourier Transform on the Hyperbolic Plane

We recall from Section 8.3.1 that the action of $\mathrm{SL}_2(\mathbb{R})$ on $\mathbb{H}$ by Möbius transformations preserves the hyperbolic area measure defined by $\frac{\mathrm{d}x\,\mathrm{d}y}{y^2}$. By Proposition 1.3 this gives rise to an action-associated unitary representation of $\mathrm{SL}_2(\mathbb{R})$ on $L^2(\mathbb{H})$ defined by

$$\big(\pi_g^{\mathbb{H}}(f)\big)(z) = f(g^{-1} \cdot z)$$

for $g \in \mathrm{SL}_2(\mathbb{R})$, $f \in L^2(\mathbb{H})$, and $z \in \mathbb{H}$. We note that $\pi_{-I}^{\mathbb{H}} = I$ since

$$(-I) \cdot z = \frac{-1z + 0}{0z - 1} = z$$

for all $z \in \mathbb{H}$, so that $\pi^{\mathbb{H}}$ is an even representation. We also recall that we may use $\mathbb{H} \cong \mathrm{SL}_2(\mathbb{R})/K$ to identify $L^2(\mathbb{H})$ with the subspace of $L^2(\mathrm{SL}_2(\mathbb{R}))$ consisting of all right K-invariant functions. Under this identification, $\pi^{\mathbb{H}}$ becomes the restriction of the left regular representation $\lambda^{\mathrm{SL}_2(\mathbb{R})}$ to this subspace. In particular, $\pi^{\mathbb{H}}$ is tempered, and has uniform decay exponent $1 - \varepsilon$ for all $\varepsilon > 0$ by Theorem 8.32.

In analogy to the definition of the radial Mellin transform in Definition 9.36, and motivated by the discussions in Sections 9.4.2 and 9.4.3, we are led to the following definition.

Definition 9.48 (Hyperbolic Fourier transform). We define the hyperbolic Fourier transform of f at (h, ξ) for $f \in L^2_{\mathrm{vol}}(\mathbb{H})$, $h \in \mathrm{SL}_2(\mathbb{R})$, and a frequency parameter $\xi \in \mathbb{R}$, by

$$\widehat{f}^{\mathrm{hyp}}(h, \xi) = \int_{\mathbb{H}} f(h \cdot z)\overline{\chi_{\infty,\xi}(z)}\,\mathrm{dvol}(z) = \int_0^\infty \int_{-\infty}^\infty f(h \cdot (x + \mathrm{i}y))y^{\frac{1}{2} + \frac{\mathrm{i}}{2}\xi} \frac{\mathrm{d}x\,\mathrm{d}y}{y^2}.$$

As with the (radial) Fourier transform, this may not be a well-defined Lebesgue integral but, as we will see, can be defined for almost every (h, ξ) by an isometric extension of the transform on $C_c(\mathbb{H})$. We note that the measure-preserving substitution $w = h \cdot z$ in the definition implies that

$$\widehat{f}^{\mathrm{hyp}}(h,\xi) = \int_{\mathbb{H}} f(w)\overline{\chi_{\infty,\xi}(h^{-1}\cdot w)}\,\mathrm{dvol}(w). \tag{9.30}$$

Hence we will think of $\widehat{f}^{\mathrm{hyp}}(h,\xi)$ as the *correlation* of f with the hyperbolic wave function $\mathbb{H} \ni w \mapsto \chi_{\infty,\xi}(h^{-1}\cdot w)$ which we think of as 'coming from $h\cdot\infty$ normalized for the point $h\cdot\mathrm{i}$'.

Lemma 9.49 (Equivariance properties). *For $f \in C_c(\mathbb{H})$ the hyperbolic Fourier transform $\widehat{f}^{\mathrm{hyp}}$ is well-defined and satisfies*

$$\widehat{\pi_g^{\mathbb{H}} f}^{\mathrm{hyp}}(h,\xi) = \widehat{f}^{\mathrm{hyp}}(g^{-1}h,\xi)$$

and

$$\widehat{f}^{\mathrm{hyp}}(hb,\xi) = \overline{\chi_\xi(b)}\Delta_B(b)^{\frac{1}{2}}\widehat{f}^{\mathrm{hyp}}(h,\xi)$$

for all $g \in \mathrm{SL}_2(\mathbb{R})$, $(h,\xi) \in \mathrm{SL}_2(\mathbb{R}) \times \mathbb{R}$ and $b \in B = AU$.

In other words, the lemma says that for any $\xi \in \mathbb{R}$ the map

$$C_c(\mathbb{H}) \ni f \longmapsto \widehat{f}^{\mathrm{hyp}}(\cdot,\xi) \in \mathcal{H}_\xi^{\mathrm{even}}$$

intertwines the action-associated representation and the principal series representation $\pi^{\xi,\mathrm{e}}$.

Proof of Lemma 9.49. For $g,h \in \mathrm{SL}_2(\mathbb{R})$ and $\xi \in \mathbb{R}$ we have

$$\begin{aligned}\widehat{\pi_g^{\mathbb{H}}(f)}^{\mathrm{hyp}}(h,\xi) &= \int_{\mathbb{H}} (\pi_g^{\mathbb{H}} f)(h\cdot z)\overline{\chi_{\infty,\xi}(z)}\,\mathrm{dvol}(z)\\ &= \int_{\mathbb{H}} f(g^{-1}h\cdot z)\overline{\chi_{\infty,\xi}(z)}\,\mathrm{dvol}(z)\\ &= \widehat{f}^{\mathrm{hyp}}(g^{-1}h,\xi).\end{aligned}$$

Moreover, for $b = u_x a_t \in B$ we also have

$$\begin{aligned}\widehat{f}^{\mathrm{hyp}}(hb,\xi) &= \int_{\mathbb{H}} f\big(h\cdot(b\cdot z)\big)\overline{\chi_{\infty,\xi}(z)}\,\mathrm{dvol}(z)\\ &= \int_{\mathbb{H}} f(h\cdot w)\overline{\chi_{\infty,\xi}(b^{-1}\cdot w)}\,\mathrm{dvol}(w)\\ &= \Delta_B(b)^{\frac{1}{2}}\overline{\chi_\xi(b)}\int_{\mathbb{H}} f(h\cdot w)\overline{\chi_{\infty,\xi}(w)}\,\mathrm{dvol}(w)\end{aligned}$$

by using the measure-preserving substitution $w = b\cdot z$ and (9.29), which proves the lemma. □

We wish to explain Lemma 9.49 in another, more convenient, way using convolutions. For this, we let $g = kb \in KB = \mathrm{SL}_2(\mathbb{R})$ with $b = u_x a_t \in B$ and obtain

$$\chi_{\infty,\xi}(g^{-1}\raisebox{0.2ex}{\scriptsize$\bullet$}\mathrm{i}) = \chi_{\infty,\xi}(b^{-1}\raisebox{0.2ex}{\scriptsize$\bullet$}\mathrm{i}) = \mathrm{e}^{-t+\mathrm{i}t\xi} = \overline{F_\xi(b)} = \overline{F_\xi(g)}$$

where $F_\xi = F_{\xi,0} \in \mathcal{H}_\xi^{\mathrm{even}}$ is defined in (9.19). We also identify $\chi_{\infty,\xi}$ with the right K-invariant function

$$\chi_{\infty,\xi}\colon \mathrm{SL}_2(\mathbb{R}) \ni g \longmapsto \chi_{\infty,\xi}(g\raisebox{0.2ex}{\scriptsize$\bullet$}\mathrm{i}).$$

Recalling that $\mathrm{SL}_2(\mathbb{R})$ is unimodular, we can use the involution of Section 1.5.1 to put the above into the form

$$\chi_{\infty,\xi}^* = F_\xi. \tag{9.31}$$

The identification between functions on $\mathbb{H}$ with right $\mathrm{SO}_2(\mathbb{R})$-invariant functions on $\mathrm{SL}_2(\mathbb{R})$ allows us to use convolutions in $L^1(\mathrm{SL}_2(\mathbb{R}))$ as discussed in Section 1.5.1 for functions on $\mathbb{H}$. We will however also use convolutions of functions in $C_c(\mathrm{SL}_2(\mathbb{R}))$ and $C(\mathrm{SL}_2(\mathbb{R}))$, giving rise to functions in $C(\mathrm{SL}_2(\mathbb{R}))$ (see Exercise 1.49).

Lemma 9.50 (Convolution formula). *For a function $f \in C_c(\mathbb{H})$ we have*

$$\widehat{f}^{\mathrm{hyp}}(h,\xi) = f * F_\xi(h) = \int_G f(g\raisebox{0.2ex}{\scriptsize$\bullet$}\mathrm{i}) F_\xi(g^{-1}h)\,\mathrm{d}m_G(g)$$

for all $(h,\xi) \in \mathrm{SL}_2(\mathbb{R}) \times \mathbb{R}$.

We note that Lemma 9.50 implies both claims of Lemma 9.49. Indeed, we have $f * F_\xi \in \mathcal{H}_\xi$ since $F_\xi \in \mathcal{H}_\xi$ and $\mathcal{H}_\xi$ is defined by a formula using the right regular representation restricted to B, which commutes with the left convolution. Similarly, the equivariance under the action-associated representation (or, equivalently, under the left regular representation) also follows from the properties of convolutions.

Proof of Lemma 9.50. For a function $f \in C_c(\mathbb{H})$ and $(h,\xi) \in \mathrm{SL}_2(\mathbb{R}) \times \mathbb{R}$ we have

$$\begin{aligned}
\widehat{f}^{\mathrm{hyp}}(h,\xi) &= \int_{\mathbb{H}} f(w)\overline{\chi_{\infty,\xi}}(h^{-1}\raisebox{0.2ex}{\scriptsize$\bullet$}w)\,\mathrm{dvol}(w) && \text{(by (9.30))}\\
&= \int_G f(g\raisebox{0.2ex}{\scriptsize$\bullet$}\mathrm{i})\chi_{\infty,\xi}^*(g^{-1}h)\,\mathrm{d}m(g)\\
&= \int_G f(g\raisebox{0.2ex}{\scriptsize$\bullet$}\mathrm{i})F_\xi(g^{-1}h)\,\mathrm{d}m(g) = f * F_\xi(h), && \text{(by (9.31))}
\end{aligned}$$

where we also extended integration from $w = g\raisebox{0.2ex}{\scriptsize$\bullet$}\mathrm{i} \in \mathbb{H}$ to $g \in \mathrm{SL}_2(\mathbb{R})$ using $m_K(K) = 1$. □

Lemma 9.51 (Rapid decay of transform). *For any function $f \in C_c^\infty(\mathbb{H})$, we have*

$$\big\|\mathbb{R} \ni \xi \longmapsto \xi^\ell \widehat{f}^{\mathrm{hyp}}(I,\xi)\big\|_\infty < \infty$$

for any $\ell \in \mathbb{N}_0$.

PROOF. We first recall that for $F \in C_c^\infty(\mathbb{R})$ we have

$$\|\mathbb{R} \ni \xi \longmapsto \xi^\ell \check{F}(\xi)\|_\infty \ll_\ell \|F^{(\ell)}\|_1 \tag{9.32}$$

by partial integration and induction on ℓ (see, for example, [25, Prop. 9.43]). To apply this, we rewrite the definition of $\widehat{f}^{\text{hyp}}$ using the substitution $y = \mathrm{e}^{2t}$ with $\frac{\mathrm{d}y}{y} = 2\,\mathrm{d}t$, which gives

$$\begin{aligned}
\widehat{f}^{\text{hyp}}(I,\xi) &= \int_{\mathbb{H}} f(z)\overline{\chi_{\infty,\xi}(z)}\frac{\mathrm{d}x\,\mathrm{d}y}{y^2} \\
&= \int_0^\infty \int_{-\infty}^\infty f(x+\mathrm{i}y)\,\mathrm{d}x\, y^{-\frac{1}{2}} y^{\frac{\mathrm{i}}{2}\xi}\frac{\mathrm{d}y}{y} \\
&= \int_{-\infty}^\infty \underbrace{\int_{-\infty}^\infty f\bigl(x+\mathrm{i}\mathrm{e}^{2t}\bigr)\,\mathrm{d}x\, 2\mathrm{e}^{-t}}_{=F(t)}\, \mathrm{e}^{\mathrm{i}t\xi}\,\mathrm{d}t = \widehat{F}\bigl(\tfrac{1}{2\pi}\xi\bigr).
\end{aligned}$$

Differentiation under the integral sign shows that the function $F \in C_c(\mathbb{R})$ is indeed smooth, so that (9.32) proves the lemma. □

9.4.5 The Hyperbolic Fourier Inversion Formula

As explained at the end of Section 9.4.2, we expect to be able to write a given function on $\mathbb{H}$ as a superposition of elementary waves $z \mapsto \chi_{\infty,\xi}(k^{-1}\mathbin{\raisebox{0.2ex}{.}}z)$ of various frequencies ξ emanating from the boundary points $k\mathbin{.}\infty \in \partial\mathbb{H}$. For this the hyperbolic Fourier transform $\widehat{f}^{\text{hyp}}(k,\xi)$ for a pair $(k,\xi) \in K\times\mathbb{R}$ should be related to the desired volume at $k\mathbin{.}\infty \in \partial\mathbb{H}$ for frequency ξ. Assuming smoothness and compact support of the original function ensures that the desired integral representation converges.

Theorem 9.52 (Hyperbolic Fourier inversion). *Let $f \in C_c^\infty(\mathbb{H})$. Then*

$$f(z) = \frac{1}{16\pi}\int_{\mathbb{R}}\int_K \widehat{f}^{\text{hyp}}(k,\xi)\chi_{\infty,\xi}(k^{-1}\mathbin{.}z)\,\mathrm{d}m_K(k)\xi\tanh\frac{\pi\xi}{2}\,\mathrm{d}\xi$$

for all $z \in \mathbb{H}$.

The proof will rely on some elementary integral manipulations, Fourier inversion on $\mathbb{R}$ (applied in a surprising way), and a contour integration to determine the correct volume amplification factor $\xi\tanh\bigl(\frac{\pi\xi}{2}\bigr)$ for $\xi \in \mathbb{R}$. To reduce the complexity of the problem, we first consider a special class of functions.

Definition 9.53 (Spherical functions). A function $f\colon \mathbb{H} \to \mathbb{C}$ is called *spherical* if $f(k\mathbin{.}z) = f(z)$ for all $k \in K$ and $z \in \mathbb{H}$.

We note that due to the equivariance property in Lemma 9.49 the hyperbolic Fourier transform of a spherical function is again invariant under K. Because of this, for a spherical function f we will also use the simplified notation

$$\widehat{f}^{\text{hyp}}(\xi) = \int_{\mathbb{H}} f(z)\overline{\chi_{\infty,\xi}(z)}\,\mathrm{dvol}(z)$$

satisfying

$$\widehat{f}^{\text{hyp}}(k,\xi) = \int_{\mathbb{H}} f(k\cdot z)\overline{\chi_{\infty,\xi}(z)}\,\mathrm{dvol}(z) = \widehat{f}^{\text{hyp}}(\xi) \tag{9.33}$$

for all $(k,\xi) \in K\times\mathbb{R}$ by Definitions 9.48 and 9.53 above. Recall that Lemma 9.49 also shows that the function $\widehat{f}^{\text{hyp}}(\cdot,\xi)$ belongs to $\mathcal{H}_\xi^{\text{even}}$. With this, (9.33) becomes

$$\widehat{f}^{\text{hyp}}(\cdot,\xi) = \widehat{f}^{\text{hyp}}(\xi)F_\xi(\cdot). \tag{9.34}$$

The following lemma gives another connection between the hyperbolic Fourier transform and π^ξ, or more precisely its matrix coefficient $\phi_\xi = \varphi^{\pi^\xi}_{F_\xi}$.

Lemma 9.54 (Matrix coefficient giving symmetry). *Let $f \in C_c(\mathbb{H})$ be a spherical function. Then*

$$\widehat{f}^{\text{hyp}}(\xi) = \int_{\mathbb{H}} f\phi_\xi\,\mathrm{dvol}(z) \tag{9.35}$$

for all $\xi \in \mathbb{R}$. Moreover, we have $\phi_\xi = \phi_{-\xi}$ and

$$\widehat{f}^{\text{hyp}}(\xi) = \widehat{f}^{\text{hyp}}(-\xi)$$

for all $\xi \in \mathbb{R}$.

Proof. Using (9.33), the normalized Haar measure m_K on K, the convolution formula in Lemma 9.50, and Fubini's theorem we have

$$\begin{aligned}\widehat{f}^{\text{hyp}}(\xi) &= \int_K \widehat{f}^{\text{hyp}}(k,\xi)\,\mathrm{d}m_K(k)\\ &= \int_K\int_G f(g\cdot\mathrm{i})F_\xi(g^{-1}k)\,\mathrm{d}m_G(g)\,\mathrm{d}m_K(k)\\ &= \int_G f(g\cdot\mathrm{i})\underbrace{\int_K F_\xi(g^{-1}k)F_\xi(k)\,\mathrm{d}m_K(k)}_{=\langle\pi^\xi_g F_\xi,F_\xi\rangle=\phi_\xi(g)}\,\mathrm{d}m_G(g),\end{aligned}$$

which proves (9.35).

Finally, by Theorem 9.32, $\pi^{\xi,\mathrm{e}}$ and $\pi^{-\xi,\mathrm{e}}$ are unitarily isomorphic. Since the vector $F_{\pm\xi} \in \mathcal{H}^{\text{even}}_{\pm\xi}$ is, up to scalar multiples, the unique K-fixed vector and both have unit length, it follows that

$$\phi_{-\xi} = \varphi^{\pi^{-\xi,\mathrm{e}}}_{F_{-\xi}} = \varphi^{\pi^{\xi,\mathrm{e}}}_{F_\xi} = \phi_\xi.$$

Together with (9.35), this gives the lemma. □

We note that we are going to use the assumption that $f \in C_c^\infty(\mathbb{H})$ is spherical to reduce the number of free variables in the proof of the Fourier inversion formula. In fact, f is spherical if and only if it can be written as

$$f(z) = F_d\big(\mathsf{d}_{\mathrm{hyp}}(z, \mathrm{i})\big)$$

for $z \in \mathbb{H}$, where $F_d\colon [0,\infty) \to \mathbb{C}$ is a function, and $\mathsf{d}_{\mathrm{hyp}}$ denotes the hyperbolic metric on $\mathbb{H}$ (see Lemma 8.16). This follows from the transitivity of the action of K on every circle with centre i. In terms of the upcoming integral substitutions, it will be better to consider instead of $\mathsf{d}_{\mathrm{hyp}}(z,\mathrm{i})$ the closely related expression

$$r(z) = \cosh^2\big(\tfrac{1}{2}\mathsf{d}_{\mathrm{hyp}}(z, \mathrm{i})\big). \tag{9.36}$$

Lemma 9.55 (Inversion at i for spherical functions). *For a spherical function $f \in C_c^\infty(\mathbb{H})$ we have*

$$f(\mathrm{i}) = \frac{1}{16\pi^2}\int_{\mathbb{R}} \widehat{f}^{\mathrm{hyp}}(\xi)\xi \int_{\mathbb{R}} \frac{\sin \xi t}{\sinh t}\,\mathrm{d}t\,\mathrm{d}\xi.$$

PROOF. We will split the proof into smaller steps.

THE FUNCTION F. For the given spherical function $f \in C_c^\infty(\mathbb{H})$ and function $r\colon \mathbb{H} \to [1,\infty)$ defined by (9.36) there exists a continuous function F in $C_c([1,\infty))$ (and smooth on $(0,\infty)$) with $f(z) = F(r(z))$ for all $z \in \mathbb{H}$. In fact we can define F concretely by restricting f to $\{y\mathrm{i} \mid y > 0\}$ and setting $y = \mathrm{e}^{2t}$, so that $\mathsf{d}_{\mathrm{hyp}}(\mathrm{e}^{2t}\mathrm{i}, \mathrm{i}) = 2|t|$ and $r(\mathrm{e}^{2t}\mathrm{i}) = \cosh^2 t$ by (9.36). This leads to the definition

$$F(r) = f\left(\mathrm{i}\mathrm{e}^{2\,\mathrm{arcosh}\,\sqrt{r}}\right)$$

for $r \in [1,\infty)$. Equivalently, we have $F(\cosh^2 t) = f(\mathrm{i}\mathrm{e}^{2t})$ for $t \in \mathbb{R}$ and, since f is spherical, more generally, $F(r(z)) = f(z)$ for all $z \in \mathbb{H}$.

Using the identity $\cosh^2 t = \frac{1}{2}\cosh(2t) + \frac{1}{2}$ for $t \in \mathbb{R}$ and the formula for $\mathsf{d}_{\mathrm{hyp}}(\cdot,\cdot)$ in Lemma 8.16 we also have

$$\begin{aligned}
r(z) &= \frac{1}{2}\left(1 + \frac{\|z-\mathrm{i}\|^2}{2y}\right) + \frac{1}{2}\\
&= \frac{1}{2}\left(1 + \frac{x^2}{2y} + \frac{(y-1)^2}{2y}\right) + \frac{1}{2}\\
&= \frac{1}{2}\left(1 + \frac{x^2}{2y} + \frac{y+y^{-1}}{2} - 1\right) + \frac{1}{2} = \frac{1}{2} + \frac{y+y^{-1}}{4} + \frac{x^2}{4y}.
\end{aligned}$$

Below we will also use the coordinates $(x,t) \in \mathbb{R}^2$ for $z = x + \mathrm{i}\mathrm{e}^{2t} \in \mathbb{H}$, which gives

$$r(z) = \frac{1}{2} + \frac{\mathrm{e}^{2t} + \mathrm{e}^{-2t}}{4} + \frac{1}{4}x^2\mathrm{e}^{-2t} = \cosh^2 t + \frac{1}{4}x^2\mathrm{e}^{-2t}.$$

For our functions f and F, this gives

$$F\left(\cosh^2 t + \tfrac{1}{4}x^2 \mathrm{e}^{-2t}\right) = f(x + \mathrm{i}\mathrm{e}^{2t})$$

for $x, t \in \mathbb{R}$.

THE FUNCTIONS Φ AND Ψ. We now use this identity in the formula for the hyperbolic Fourier transform, and obtain

$$\begin{aligned}\widehat{f}^{\,\mathrm{hyp}}(\xi) &= \int_{\mathbb{H}} f(x+\mathrm{i}y) y^{\frac{1}{2}+\frac{1}{2}\mathrm{i}\xi} \frac{\mathrm{d}x\,\mathrm{d}y}{y^2} \\ &= \int_{\mathbb{R}}\int_{\mathbb{R}} F\left(\cosh^2 t + \tfrac{1}{4}x^2\mathrm{e}^{-2t}\right) \mathrm{e}^{t+\mathrm{i}\xi t}\mathrm{e}^{-4t}\,\mathrm{d}x\, 2\mathrm{e}^{2t}\,\mathrm{d}t \\ &= 2\int_{\mathbb{R}}\int_{\mathbb{R}} F\left(\cosh^2 t + \tfrac{1}{4}x^2\mathrm{e}^{-2t}\right) \mathrm{d}x\, \mathrm{e}^{-t}\mathrm{e}^{\mathrm{i}\xi t}\,\mathrm{d}t\end{aligned}$$

by using $y = \mathrm{e}^{2t}$ and $\mathrm{d}y = 2\mathrm{e}^{2t}\,\mathrm{d}t$. We now set $u = \frac{1}{2}\mathrm{e}^{-t}x$ with $2\,\mathrm{d}u = \mathrm{e}^{-t}\,\mathrm{d}x$ to arrive at

$$\widehat{f}^{\,\mathrm{hyp}}(\xi) = 4\int_{\mathbb{R}} \mathrm{e}^{\mathrm{i}\xi t}\int_{\mathbb{R}} F\left(\cosh^2 t + u^2\right)\mathrm{d}u\,\mathrm{d}t. \tag{9.37}$$

We use the inner integral to define the function

$$\Phi(s) = \int_{\mathbb{R}} F\left(s + u^2\right)\mathrm{d}u$$

for $s \in [1, \infty)$ and the composition

$$\Psi(t) = \Phi(\cosh^2 t) = \int_{\mathbb{R}} F\left(\cosh^2 t + u^2\right)\mathrm{d}u$$

for $t \in \mathbb{R}$. We note that $\Phi \in C_c([1,\infty))$ and $\Psi \in C_c(\mathbb{R})$.

APPLYING FOURIER INVERSION FOR Ψ. With the function Ψ to hand, we realize that $\widehat{f}^{\,\mathrm{hyp}}$ is essentially the Fourier transform of Ψ. More precisely, we may reformulate (9.37) as

$$\widehat{f}^{\,\mathrm{hyp}}(\xi) = 4\widecheck{\Psi}\left(\frac{1}{2\pi}\xi\right) \tag{9.38}$$

for all $\xi \in \mathbb{R}$.

By Lemma 9.51, we know that $\widehat{f}^{\,\mathrm{hyp}}$ decays rapidly. Hence we may apply Fourier inversion on $\mathbb{R}$, and the substitution $\xi = 2\pi\zeta$ to obtain from (9.38) that

$$\Psi(t) = \int_{\mathbb{R}} \widecheck{\Psi}(\zeta)\mathrm{e}^{-2\pi\mathrm{i}\zeta t}\,\mathrm{d}\zeta = \frac{1}{2\pi}\int_{\mathbb{R}} \widecheck{\Psi}\left(\frac{1}{2\pi}\xi\right)\mathrm{e}^{-\mathrm{i}\xi t}\,\mathrm{d}\xi = \frac{1}{8\pi}\int_{\mathbb{R}} \widehat{f}^{\,\mathrm{hyp}}(\xi)\mathrm{e}^{-\mathrm{i}\xi t}\,\mathrm{d}\xi$$

for all $t \in \mathbb{R}$. By Lemma 9.54 we have $\widehat{f}^{\,\mathrm{hyp}}(\xi) = \widehat{f}^{\,\mathrm{hyp}}(-\xi)$, which turns the above into

$$\Phi(\cosh^2 t) = \Psi(t) = \frac{1}{8\pi}\int_{\mathbb{R}} \widehat{f}^{\,\mathrm{hyp}}(\xi)\cos(\xi t)\,\mathrm{d}\xi$$

Using the rapid decay of $\widehat{f}^{\,\mathrm{hyp}}$ in Lemma 9.51 again, we may differentiate under the integral to obtain

$$\Psi'(t) = -\frac{1}{8\pi}\int_{\mathbb{R}} \widehat{f}^{\,\mathrm{hyp}}(\xi)\xi\sin(\xi t)\,\mathrm{d}\xi \tag{9.39}$$

for $t \in \mathbb{R}$. Using the chain rule for $\Psi(t) = \Phi(\cosh^2 t)$, we also have

$$\Psi'(t) = \Phi'(\cosh^2 t) 2\cosh t \sinh t$$

for $t \in \mathbb{R} \smallsetminus \{0\}$. We divide this by $2\sinh t$ to obtain, with (9.39),

$$\Phi'(\cosh^2 t)\cosh t = \frac{\Psi'(t)}{2\sinh t} = -\frac{1}{16\pi}\int_{\mathbb{R}} \widehat{f}^{\,\mathrm{hyp}}(\xi)\xi\frac{\sin\xi t}{\sinh t}\,\mathrm{d}\xi \tag{9.40}$$

for $t \in \mathbb{R} \smallsetminus \{0\}$.

Moreover, note that

$$\lim_{t\to 0}\frac{\sin\xi t}{\sinh t} = \xi$$

and

$$\left|\frac{\sin\xi t}{\sinh t}\right| = \left|\frac{\sin\xi t}{t}\right|\frac{t}{\sinh t} \leqslant |\xi|\frac{t}{\sinh t} \leqslant |\xi| \tag{9.41}$$

by the mean-value theorem applied to the function $\mathbb{R} \ni t \mapsto \sin\xi t$. Together with Lemma 9.51, this shows that Φ' extends continuously from $(1,\infty)$ to $[1,\infty)$. It follows from the mean value theorem applied to Φ that Φ also has a one-sided derivative as $s = 1$.

Interesting, but so what? The Fourier inversion formulas above allow us to obtain the value

$$\Phi(1) = \Psi(0) = \int_{\mathbb{R}} F(1+u^2)\,\mathrm{d}u$$

from the hyperbolic Fourier transform, but we wish instead to obtain the value $F(1) = f(\mathrm{i})$ of the integrand F in the definition of Φ. To obtain this, we rely on some more stunning but elementary integration trickery. In fact, we claim that one can recover F from Φ by the formula

$$F(s) = -\frac{1}{\pi}\int_{\mathbb{R}} \Phi'(s+v^2)\,\mathrm{d}v \tag{9.42}$$

for all $s \geqslant 1$.

To see this, note first that $f \in C_c^\infty(\mathbb{H})$ implies that $F|_{(1,\infty)}$ is smooth, that

$$\Phi'(s) = \int_{\mathbb{R}} F'(s+u^2)\,\mathrm{d}u$$

and

$$\Phi'(s+v^2) = \int_{\mathbb{R}} F'(s+u^2+v^2)\,\mathrm{d}u$$

for all $s>1$ and $v \in \mathbb{R}$. Integrating the latter over $v \in \mathbb{R}$, we obtain

$$\begin{aligned}\int_{\mathbb{R}} \Phi'(s+v^2)\,\mathrm{d}v &= \int_{\mathbb{R}^2} F'(s+u^2+v^2)\underbrace{\mathrm{d}u\,\mathrm{d}v}_{R\,\mathrm{d}R\,\mathrm{d}\theta}\\ &= 2\pi\int_0^\infty F'(s+R^2)R\,\mathrm{d}R\\ &= \pi\int_0^\infty F'(s+\rho)\,\mathrm{d}\rho,\end{aligned}$$

where we used polar coordinates (R,θ) for $(u,v) \in \mathbb{R}^2$ and the substitution $\rho = R^2$. For the latter integral we may now apply the fundamental theorem of calculus. Since $F \in C_c([1,\infty))$, this takes the form

$$\int_0^\infty F'(s+\rho)\,\mathrm{d}\rho = -F(s)$$

for all $s>1$, which proves the claim in (9.42) for $s>1$. Since F and Φ' both lie in $C_c([1,\infty))$, this extends by continuity to $s=1$.

Conclusion. We now set $s=1$ in (9.42), substitute $v=\sinh t$, and combine the resulting integral with (9.40), which leads to

$$\begin{aligned}F(1) &= -\frac{1}{\pi}\int_{\mathbb{R}} \Phi'(\underbrace{1+v^2}_{=\cosh^2 t})\,\mathrm{d}v\\ &= -\frac{1}{\pi}\int_{\mathbb{R}} \Phi'(\cosh^2 t)\cosh t\,\mathrm{d}t\\ &= \frac{1}{16\pi^2}\int_{\mathbb{R}}\left(\int_{\mathbb{R}} \widehat{f}^{\mathrm{hyp}}(\xi)\xi\frac{\sin\xi t}{\sinh t}\,\mathrm{d}\xi\right)\mathrm{d}t.\end{aligned}$$

Recall that by Lemma 9.49 we have

$$\left|\widehat{f}^{\mathrm{hyp}}(\xi)\right| \ll_f \frac{1}{(1+\xi^2)^2}.$$

Together with (9.41), we obtain

$$\left|\widehat{f}^{\mathrm{hyp}}(\xi)\xi\frac{\sin\xi t}{\sinh t}\right| \ll_f \frac{\xi^2}{(1+\xi^2)^2}\frac{t}{\sinh t},$$

which is easily seen to be integrable over $\mathbb{R}^2$. This implies that the integrand above lies in $L^1(\mathbb{R}^2)$. Hence we may apply Fubini's theorem and obtain, with

$$f(\mathrm{i}) = F(1) = \frac{1}{16\pi^2}\int_{\mathbb{R}} \widehat{f}^{\mathrm{hyp}}(\xi)\xi\int_{\mathbb{R}}\frac{\sin\xi t}{\sinh t}\,\mathrm{d}t\,\mathrm{d}\xi,$$

the lemma. □

Lemma 9.55 explains why the following result is of interest to us.

Lemma 9.56 (Volume factor). *For $\xi \in \mathbb{R}$ we have*

$$\int_{\mathbb{R}} \frac{\sin \xi t}{\sinh t}\,\mathrm{d}t = \pi \tanh\left(\frac{\pi\xi}{2}\right). \tag{9.43}$$

PROOF. To prove (9.43), we will apply the Cauchy integral formula to the meromorphic function f defined by

$$f(z) = \frac{\mathrm{e}^{\mathrm{i}\xi z}}{\sinh z}. \tag{9.44}$$

We start with some elementary observations about the function f. For the point $z = x + \mathrm{i}y$ with $x, y \in \mathbb{R}$ we may apply the reverse triangle inequality to see that

$$|\sinh z| = \left|\frac{\mathrm{e}^{x}\mathrm{e}^{\mathrm{i}y} - \mathrm{e}^{-x}\mathrm{e}^{-\mathrm{i}y}}{2}\right| \geqslant \frac{\mathrm{e}^{x} - \mathrm{e}^{-x}}{2} = \sinh x.$$

By symmetry, this gives

$$|\sinh z| \geqslant |\sinh x|, \tag{9.45}$$

which implies that $f(x + \mathrm{i}y)$ defined in (9.44) decays rapidly for $|x| \to \infty$ as long as $|y|$ is bounded. The estimate (9.45) also shows that $\sinh(x+\mathrm{i}y)$ can only vanish when $x = 0$. Since $\sinh(\mathrm{i}y) = \mathrm{i}\sin y$ for $y \in \mathbb{R}$, we see that f has poles precisely at the points in $\mathbb{Z}\pi\mathrm{i}$. At 0 the residue of f is given by 1. Finally, we have

$$\sinh(z + \pi\mathrm{i}) = \frac{\mathrm{e}^{z+\pi\mathrm{i}} - \mathrm{e}^{-z-\pi\mathrm{i}}}{2} = -\sinh z$$

for $z \in \mathbb{C}$, which implies that

$$f(z + \pi\mathrm{i}) = \frac{\mathrm{e}^{\mathrm{i}\xi(z+\pi\mathrm{i})}}{-\sinh z} = -\mathrm{e}^{-\xi\pi} f(z) \tag{9.46}$$

for $z \in \mathbb{C}\smallsetminus(\mathbb{Z}\pi\mathrm{i})$.

We now integrate f over the closed path

$$\gamma = \gamma_{\text{bottom}} \sqcup \gamma_{\text{right}} \sqcup \gamma_{\text{top}} \sqcup \gamma_{\text{top}} \sqcup \gamma_{\text{left}}$$

indicated in Figure 9.8.

The description of the poles of f given above now implies that

$$\oint_{\gamma} f(z)\,\mathrm{d}z = 2\pi\mathrm{i}$$

independent of $R > 1$ and $\varepsilon \in (0, 1)$. The decay properties of f discussed above also imply that

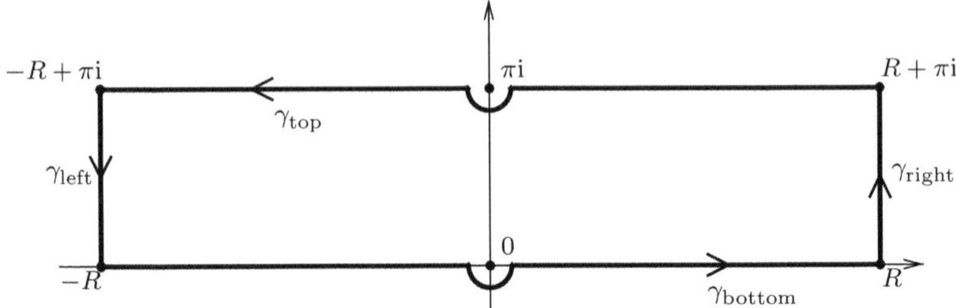

Fig. 9.8: The closed path $\gamma = \gamma_{R,\varepsilon}$ consists of four pieces. The first path γ_{bottom} goes from $-R$ to R but avoids the pole at 0 (but including it inside the contour) by following a semi-circle of radius ε around 0. The paths γ_{right}, γ_{top}, and γ_{left} go as indicated via $R+\pi\mathrm{i}$ and $-R+\pi\mathrm{i}$ back to $-R$, again avoiding the pole at $\pi\mathrm{i}$ (but leaving it outside the contour).

$$\lim_{R\to\infty} \oint_{\gamma_{\mathrm{right}}} f(z)\,\mathrm{d}z = \lim_{R\to\infty} \oint_{\gamma_{\mathrm{left}}} f(z)\,\mathrm{d}z = 0.$$

Moreover, we defined the paths γ_{bottom} and γ_{top} so that $\gamma_{\mathrm{bottom}} + \pi\mathrm{i}$ is equal to γ_{top} except for the orientation, which is reversed. Together with (9.46) this gives

$$\oint_{\gamma_{\mathrm{top}}} f(z)\,\mathrm{d}z = -\oint_{\gamma_{\mathrm{bottom}}} f(z+\pi\mathrm{i})\,\mathrm{d}z = \mathrm{e}^{-\xi\pi} \oint_{\gamma_{\mathrm{bottom}}} f(z)\,\mathrm{d}z$$

again independently of R and ε. Putting this together, we obtain

$$\begin{aligned} 2\pi\mathrm{i} &= \lim_{R\to\infty} \oint_{\gamma} f(z)\,\mathrm{d}z \\ &= \lim_{R\to\infty} \left(1+\mathrm{e}^{-\xi\pi}\right) \oint_{\gamma_{\mathrm{bottom}}} f(z)\,\mathrm{d}z \\ &= \left(1+\mathrm{e}^{-\xi\pi}\right)\left(\int_{-\infty}^{-\varepsilon} \frac{\mathrm{e}^{\mathrm{i}\xi t}}{\sinh t}\,\mathrm{d}t + \int_{\varepsilon}^{\infty} \frac{\mathrm{e}^{\mathrm{i}\xi t}}{\sinh t}\,\mathrm{d}t + \oint_{\gamma_\varepsilon} f(z)\,\mathrm{d}z\right), \end{aligned} \tag{9.47}$$

where $\gamma_\varepsilon\colon [\pi, 2\pi] \ni t \mapsto \varepsilon\mathrm{e}^{\mathrm{i}t}$ is the semi-circular path appearing in γ_{bottom}. To understand the asymptotics of

$$\oint_{\gamma_\varepsilon} f(z)\,\mathrm{d}z$$

as ε decreases to 0, we note that $f(z) = \frac{1}{z} + h(z)$ for a function h that is holomorphic at 0. By continuity of h, we have

$$\lim_{\varepsilon\searrow 0}\oint_{\gamma_\varepsilon} h(z)\,\mathrm{d}z = 0,$$

so we only have to calculate

$$\oint_{\gamma_\varepsilon}\frac{1}{z}\,\mathrm{d}z = \int_{\pi}^{2\pi}\frac{1}{\varepsilon\mathrm{e}^{\mathrm{i}t}}\varepsilon\mathrm{i}\mathrm{e}^{\mathrm{i}t}\,\mathrm{d}t = \pi\mathrm{i}.$$

We now take the imaginary part in (9.47) and let ε decrease to 0, which gives

$$2\pi = \left(1+\mathrm{e}^{-\xi\pi}\right)\left(\int_{-\infty}^{\infty}\frac{\sin\xi t}{\sinh t}\,\mathrm{d}t + \pi\right) = \left(1+\mathrm{e}^{-\xi\pi}\right)\int_{-\infty}^{\infty}\frac{\sin\xi t}{\sinh t}\,\mathrm{d}t+\pi+\mathrm{e}^{-\xi\pi}\pi.$$

Solving this equation for the integral gives

$$\int_{-\infty}^{\infty}\frac{\sin\xi t}{\sinh t}\,\mathrm{d}t = \frac{1-\mathrm{e}^{-\xi\pi}}{1+\mathrm{e}^{-\xi\pi}}\pi = \frac{\mathrm{e}^{\xi\pi/2}-\mathrm{e}^{-\xi\pi/2}}{\mathrm{e}^{\xi\pi/2}+\mathrm{e}^{-\xi\pi/2}}\pi,$$

and hence the lemma. □

Having obtained the hyperbolic Fourier inversion formula at i and for spherical functions in the last two lemmas, we are now in a position to prove the general case.

Proof of Theorem 9.52. Let $f \in C_c^\infty(\mathbb{H})$. Combining Lemmas 9.55 and 9.56, we see that if $f \in C_c^\infty(\mathbb{H})$ is spherical, then

$$f(\mathrm{i}) = \frac{1}{16\pi}\int_{\mathbb{R}}\widehat{f}^{\,\mathrm{hyp}}(\xi)\xi\tanh\left(\frac{\pi\xi}{2}\right)\mathrm{d}\xi.$$

To use this for a general $f \in C_c^\infty(\mathbb{H})$, we define

$$f_{\mathrm{sph}}(z) = \int_K f(k\raisebox{0.2ex}{\tiny$\bullet$}z)\,\mathrm{d}m_K(k).$$

Since $k\raisebox{0.2ex}{\tiny$\bullet$}\mathrm{i} = \mathrm{i}$ for all $k \in K$, and we may differentiate under the integral sign, it follows that $f_{\mathrm{sph}} \in C_c^\infty(\mathbb{H})$ has $f_{\mathrm{sph}}(\mathrm{i}) = f(\mathrm{i})$. Applying Lemma 9.56 to this spherical function, we obtain

$$f(\mathrm{i}) = f_{\mathrm{sph}}(\mathrm{i}) = \frac{1}{16\pi}\int_{\mathbb{R}}\widehat{f_{\mathrm{sph}}}^{\,\mathrm{hyp}}(\xi)\xi\tanh\left(\frac{\pi\xi}{2}\right)\mathrm{d}\xi. \tag{9.48}$$

Using Fubini's theorem and the definition of the hyperbolic Fourier transform, we also have

$$\begin{aligned}
\widehat{f}_{\mathrm{sph}}^{\mathrm{hyp}}(\xi) &= \int_{\mathbb{H}} f_{\mathrm{sph}}(z)\overline{\chi_{\infty,\xi}(z)}\,\mathrm{dvol}(z)\\
&= \int_K \int_{\mathbb{H}} f(k\boldsymbol{\cdot} z)\overline{\chi_{\infty,\xi}(z)}\,\mathrm{dvol}(z)\,\mathrm{d}m_K(k)\\
&= \int_K \widehat{f}^{\mathrm{hyp}}(k,\xi)\,\mathrm{d}m_K(k).
\end{aligned}$$

Putting this into (9.48), we obtain

$$f(\mathrm{i}) = \frac{1}{16\pi}\int_{\mathbb{R}}\int_K \widehat{f}^{\mathrm{hyp}}(k,\xi)\xi\tanh\left(\frac{\pi\xi}{2}\right)\mathrm{d}m_K(k)\,\mathrm{d}\xi.$$

Now let $h\in\mathrm{SL}_2(\mathbb{R})$ be arbitrary and define $\widetilde{f}=\pi_h^{\mathbb{H}}f$. Applying the previous formula to $\widetilde{f}$ and the equivariance claim in Lemma 9.49, we obtain

$$\begin{aligned}
f(h^{-1}\boldsymbol{\cdot}\mathrm{i}) = \bigl(\pi_h^{\mathbb{H}}f\bigr)(\mathrm{i}) &= \frac{1}{16\pi}\int_{\mathbb{R}}\int_K \widehat{\pi_h^{\mathbb{H}}f}^{\,\mathrm{hyp}}(k,\xi)\xi\tanh(\tfrac{\pi\xi}{2})\,\mathrm{d}m_K(k)\,\mathrm{d}\xi\\
&= \frac{1}{16\pi}\int_{\mathbb{R}}\int_K \widehat{f}^{\mathrm{hyp}}(h^{-1}k,\xi)\,\mathrm{d}m_K(k)\xi\tanh(\tfrac{\pi\xi}{2})\,\mathrm{d}\xi.
\end{aligned}$$

Next we note that

$$\begin{aligned}
\int_K \widehat{f}^{\mathrm{hyp}}(h^{-1}k,\xi)\,\mathrm{d}m_K(k) &= \Bigl\langle \pi_h^{\xi,\mathrm{e}}\widehat{f}^{\mathrm{hyp}}(\cdot,\xi),F_\xi\Bigr\rangle_{\mathcal{H}_\xi}\\
&= \Bigl\langle \widehat{f}^{\mathrm{hyp}}(\cdot,\xi),\pi_{h^{-1}}^{\xi,\mathrm{e}}F_\xi\Bigr\rangle_{\mathcal{H}_\xi}\\
&= \int_K \widehat{f}^{\mathrm{hyp}}(k,\xi)\overline{F_\xi(hk)}\,\mathrm{d}m_K(k)\\
&= \int_K \widehat{f}^{\mathrm{hyp}}(k,\xi)\chi_{\infty,\xi}(k^{-1}h^{-1}\boldsymbol{\cdot}\mathrm{i})\,\mathrm{d}m_K(k)
\end{aligned}$$

by (9.31). Combining this with the above, and setting $h^{-1}\boldsymbol{\cdot}\mathrm{i}=z$, we obtain

$$f(z) = \frac{1}{16\pi}\int_{\mathbb{R}}\int_K \widehat{f}^{\mathrm{hyp}}(k,\xi)\chi_{\infty,\xi}(k^{-1}\boldsymbol{\cdot}z)\,\mathrm{d}m_K(k)\xi\tanh(\tfrac{\pi\xi}{2})\,\mathrm{d}\xi,$$

which gives the theorem. □

9.4.6 The Hyperbolic Fourier Transform in the Disc Model

We recall from Section 8.3.3 that the action of $\mathrm{SL}_2(\mathbb{R})$ on $\mathbb{H}$ is conjugated to the action of $\mathrm{SU}_{1,1}(\mathbb{R})$ on $\mathbb{D}$ by the map

$$\Phi\colon \overline{\mathbb{C}} \ni w \longmapsto \begin{pmatrix} 1 & \mathrm{i} \\ \mathrm{i} & 1 \end{pmatrix} \bullet w = \frac{w+\mathrm{i}}{\mathrm{i}w+1} \in \overline{\mathbb{C}}$$

with $\Phi(\mathbb{D}) = \mathbb{H}$, $\Phi(0) = \mathrm{i}$ being our choice of origin in $\mathbb{H}$, and $\Phi(\mathrm{i}) = \infty$. Using this, we can move the Busemann function to $\mathbb{D}$.

Definition 9.57 (Busemann function for $\mathrm{i} \in \partial\mathbb{D}$). The Busemann function on $\mathbb{D}$ with respect to $\mathrm{i} \in \partial\mathbb{D}$ (and origin $0 \in \mathbb{D}$) is defined by

$$b_{\mathrm{i}}^{\mathbb{D}}(w) = b_{\infty}^{\mathbb{H}}(\Phi(w)) = -\log\left(\frac{1-|w|^2}{|w-\mathrm{i}|^2}\right).$$

We now verify that the two formulas in Definition 9.57 above are in fact equivalent. Indeed, for $w = x + \mathrm{i}y$ we have

$$\begin{aligned} \Im\Phi(w) &= \Im\frac{w+\mathrm{i}}{\mathrm{i}w+1} = \Im\frac{(x+\mathrm{i}y+\mathrm{i})(-\mathrm{i}x-y+1)}{(\mathrm{i}x-y+1)(-\mathrm{i}x-y+1)} \\ &= \Im\frac{-\mathrm{i}x^2-\mathrm{i}y^2+\mathrm{i}y-\mathrm{i}y+\mathrm{i}}{x^2+(y-1)^2} = \frac{1-x^2-y^2}{x^2+(y-1)^2} = \frac{1-|w|^2}{|w-\mathrm{i}|^2} \end{aligned}$$

as claimed.

Next we recall that

$$k_\theta = \begin{pmatrix} \mathrm{e}^{-\mathrm{i}\theta} & \\ & \mathrm{e}^{\mathrm{i}\theta} \end{pmatrix} \in K < \mathrm{SU}_{1,1}(\mathbb{R})$$

rotates $\mathbb{D}$ so that our origin $0 \in \mathbb{D}$ is fixed, and $k_\theta \bullet \mathrm{i} = \mathrm{e}^{-2\mathrm{i}\theta}\mathrm{i}$. This suggests the following more general definitions.

Definition 9.58 (Busemann functions and hyperbolic waves on $\mathbb{D}$). The *Busemann function on* $\mathbb{D}$ with respect to $p \in \partial\mathbb{D}$ (and origin $0 \in \mathbb{D}$) is defined by

$$b_p^{\mathbb{D}}(w) = -\log\left(\frac{1-|w|^2}{|w-p|^2}\right).$$

Moreover, the hyperbolic wave coming from p and with frequency $\xi \in \mathbb{R}$ (normalized for the origin $0 \in \mathbb{D}$) is defined by the function

$$\chi_{p,\xi}(w) = \mathrm{e}^{(-\frac{1}{2}+\frac{\mathrm{i}}{2}\xi)b_p^{\mathbb{D}}(w)} = \left(\frac{1-|w|^2}{|w-p|^2}\right)^{\frac{1}{2}-\frac{\mathrm{i}}{2}\xi}.$$

We imagine that the hyperbolic wave $\chi_{p,\xi}(w)$ is the sound produced by a loudspeaker at $p \in \partial\mathbb{D}$ using frequency $\xi \in \mathbb{R}$. The following reformulation of Theorem 9.52 establishes our goal to obtain any function $f \in C_c^\infty(\mathbb{D})$ as a superposition of such waves using loudspeakers at any point $p \in \partial\mathbb{D}$, and using all possible frequencies $\xi \in \mathbb{R}$.

Theorem 9.59 (Fourier inversion on $\mathbb{D}$). *Let $f \in C_c^\infty(\mathbb{D})$ and define the abbreviation*

$$\langle f, \chi_{p,\xi} \rangle = \int_{\mathrm{d}} f \overline{\chi_{p,\xi}} \, \mathrm{dvol}$$

for $p \in \partial\mathbb{D}$ and $\xi \in \mathbb{R}$. Then

$$f(z) = \frac{1}{16\pi} \int_{\mathbb{R}} \int_{\partial\mathrm{d}} \langle f, \chi_{p,\xi} \rangle \chi_{p,\xi}(z) \, \mathrm{d}p \, \xi \tanh\left(\tfrac{\pi\xi}{2}\right) \mathrm{d}\xi$$

for all $z \in \mathbb{D}$, where $\mathrm{d}p$ denotes the normalized Lebesgue measure on $\partial\mathbb{D}$.

9.4.7 The Unitary Isomorphism

We now show that the hyperbolic Fourier transform is in fact a unitary isomorphism between the action-associated representation $\pi^{\mathbb{H}}$ and an integral of all even principal series representations as in Definition 9.39.

Theorem 9.60 (Unitary isomorphism). *The hyperbolic Fourier transform satisfies the identity*

$$\|f\|^2_{L^2(\mathbb{H})} = \tfrac{1}{8\pi} \int_{[0,\infty)} \left\| \widehat{f}^{\mathrm{hyp}}(\cdot, \xi) \right\|^2_{\mathcal{H}_\xi} \xi \tanh\left(\tfrac{\pi\xi}{2}\right) \mathrm{d}\xi$$

for every $f \in C_c(\mathbb{H})$. Moreover, it extends to an intertwining unitary isomorphism between $\pi^{\mathbb{H}}$ and

$$\pi^{\mu,\mathrm{e}} = \int_{[0,\infty)} \pi^{\xi,\mathrm{e}} \, \mathrm{d}\mu(\xi),$$

where

$$\mathrm{d}\mu(\xi) = \tfrac{1}{8\pi} \xi \tanh\left(\tfrac{\pi\xi}{2}\right) \mathrm{d}\xi,$$

and $\mathrm{d}\xi$ denotes the Lebesgue measure on $[0,\infty)$.

The proof of the theorem above will be split into two parts. We start the proof of the isometric property on page 466, and surjectivity will be established on page 471. We begin with some preparatory material for the isometric property.

Lemma 9.61 (Convolution on $\mathbb{H}$). *Assume that $f \in C_c(\mathbb{H})$ and $\psi \in C(\mathbb{H})$. Then $f * \psi$ is again right K-invariant and so can be considered a function in $C(\mathbb{H})$.*

We refer to Figure 9.9 for the geometric meaning of $f * \psi$ for spherical ψ.

Proof of Lemma 9.61. By definition of convolution and the identification of right K-invariant functions on $\mathrm{SL}_2(\mathbb{R})$ and functions on $\mathbb{H}$, we have

$$f * \psi(g) = \int_G f(h \cdot \mathrm{i}) \psi(h^{-1} g \cdot \mathrm{i}) \, \mathrm{d}m(h)$$

for $g \in \mathrm{SL}_2(\mathbb{R})$. It follows that $f * \psi$ is also right K-invariant. This, together with Exercise 1.49, gives the lemma. We note that we may now also write

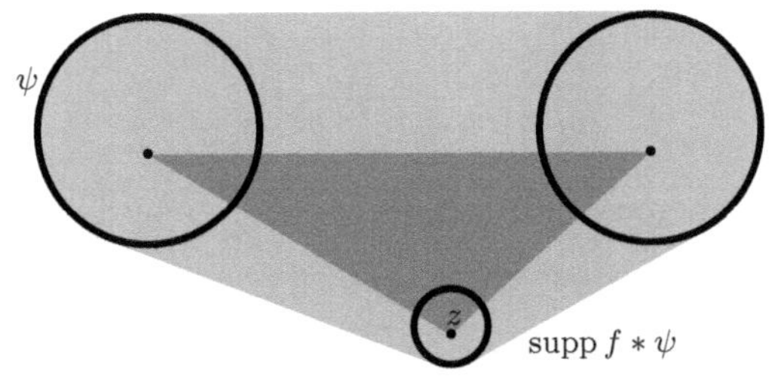

Fig. 9.9: Suppose ψ is the normalized characteristic function of a ball around the point $\mathrm{i} \in \mathbb{H}$ (or a continuous approximation of it). The geometric meaning of the value $f * \psi(z)$ is, in this case, that the ball is moved to $z \in \mathbb{H}$ and f is averaged over it. This creates a blurred-out version of f with support in a neighbourhood (drawn in light grey) of the original support (drawn in dark grey).

$$f * \psi(z) = \int_G f(h \cdot \mathrm{i}) \psi(h^{-1} \cdot z) \, \mathrm{d}m(h)$$

by using $z = g \cdot \mathrm{i} \in \mathbb{H}$ as the argument instead of $g \in \mathrm{SL}_2(\mathbb{R})$. □

Lemma 9.62 (Symmetry on $C_c(\mathbb{H})$). *Let $f \in C_c(\mathbb{H})$ and $\xi \in \mathbb{R}$. Then*

$$\int_K \left| \widehat{f}^{\,\mathrm{hyp}}(k, \xi) \right|^2 \mathrm{d}m_K(k) = \int_{\mathbb{H}} f * \phi_\xi \overline{f} \, \mathrm{dvol} = \int_K \left| \widehat{f}^{\,\mathrm{hyp}}(k, -\xi) \right|^2 \mathrm{d}m_K(k).$$

PROOF. By Lemma 9.50 we have the convolution formula $\widehat{f}^{\,\mathrm{hyp}}(\cdot, \xi) = f * F_\xi$. Combining this with Fubini's theorem, we obtain for $\int_K |\widehat{f}^{\,\mathrm{hyp}}(k, \xi)|^2 \, \mathrm{d}m_K(k)$ the formula

$$\begin{aligned}
&\int_K \int_G f(h_1 \cdot \mathrm{i}) F_\xi(h_1^{-1} k) \, \mathrm{d}m(h_1) \int_G \overline{f(h_2 \cdot \mathrm{i}) F_\xi(h_2^{-1} k)} \, \mathrm{d}m(h_2) \, \mathrm{d}m_K(k) \\
&\quad = \int_G \int_G f(h_1 \cdot \mathrm{i}) \underbrace{\int_K F_\xi(h_1^{-1} k) \overline{F_\xi(h_2^{-1} k)} \, \mathrm{d}m_K(k)}_{= \langle \pi^\xi_{h_1} F_\xi, \pi^\xi_{h_2} F_\xi \rangle_{\mathcal{H}_\xi}} \, \mathrm{d}m(h_1) \overline{f(h_2 \cdot \mathrm{i})} \, \mathrm{d}m(h_2) \\
&\quad = \int_G \int_G f(h_1 \cdot \mathrm{i}) \phi_\xi(h_2^{-1} h_1) \, \mathrm{d}m(h_1) \overline{f(h_2 \cdot \mathrm{i})} \, \mathrm{d}m(h_2) \\
&\quad = \int_G \int_G f(h_1 \cdot \mathrm{i}) \phi_\xi(h_1^{-1} h_2) \, \mathrm{d}m(h_1) \overline{f(h_2 \cdot \mathrm{i})} \, \mathrm{d}m(h_2) \\
&\quad = \int_G f * \phi_\xi \overline{f} \, \mathrm{d}m,
\end{aligned}$$

where we used the fact that

$$\phi_\xi(g) = \overline{\phi_\xi(g^{-1})} = \phi_\xi(g^{-1})$$

is real-valued (see Lemma 9.19). Since $\phi_\xi = \phi_{-\xi}$ by Lemma 9.54, the lemma follows. □

Proof of isometry formula in Theorem 9.60. Let $f \in C_c^\infty(\mathbb{H})$. Applying the hyperbolic Fourier inversion formula (Theorem 9.52) to f and Fubini's theorem, we see that

$$\begin{aligned}
\|f\|^2_{L^2(\mathbb{H})} &= \int_{\mathbb{H}} f(z)\overline{f(z)}\,\mathrm{dvol}(z) \\
&= \frac{1}{16\pi}\int_{\mathbb{H}}\int_{\mathbb{R}}\int_{K} \widehat{f}^{\mathrm{hyp}}(k,\xi)\chi_{\infty,\xi}(k^{-1}\bullet z)\xi\tanh(\tfrac{\pi\xi}{2})\,\mathrm{d}m_K(k)\,\mathrm{d}\xi\,\overline{f(z)}\,\mathrm{dvol}(z) \\
&= \frac{1}{16\pi}\int_{\mathbb{R}}\int_{K} \widehat{f}^{\mathrm{hyp}}(k,\xi)\underbrace{\int_{\mathbb{H}}\chi_{\infty,\xi}(k^{-1}\bullet z)\overline{f(z)}\,\mathrm{dvol}(z)}_{=\overline{\widehat{f}^{\mathrm{hyp}}(k,\xi)}}\mathrm{d}m_K(k)\xi\tanh(\tfrac{\pi\xi}{2})\mathrm{d}\xi \\
&= \frac{1}{16\pi}\int_{\mathbb{R}} \|\widehat{f}^{\mathrm{hyp}}(\cdot,\xi)\|^2_{\mathcal{H}_\xi}\xi\tanh(\tfrac{\pi\xi}{2})\,\mathrm{d}\xi.
\end{aligned}$$

Applying Lemma 9.62, we can also write this in the form

$$\|f\|^2_{L^2(\mathbb{H})} = \frac{1}{8\pi}\int_{[0,\infty)}\int_{K}\left|\widehat{f}^{\mathrm{hyp}}(k,\xi)\right|^2\,\mathrm{d}m_K(k)\xi\tanh(\tfrac{\pi\xi}{2})\,\mathrm{d}\xi.$$

It remains to show that this formula not only holds for all $f \in C_c^\infty(\mathbb{H})$ but also for $f \in C_c(\mathbb{H})$. For this we let (ψ_n) be an approximate identity in $C_c^\infty(\mathrm{SL}_2(\mathbb{R}))$ satisfying $\psi_n(kgk^{-1}) = \psi_n(g)$ for $k \in K$ and $g \in \mathrm{SL}_2(\mathbb{R})$ as in Lemma 9.17.

Now let $f \in C_c(\mathbb{H})$ and define

$$f_n = f * \psi_n \in C_c^\infty(G)$$

for $n \in \mathbb{N}$. Note that by continuity of f we have

$$\begin{aligned}
&f_n \to f \text{ as } n \to \infty; \\
&\operatorname{supp} f_n \subseteq (\operatorname{supp} f)B_1 \text{ for } n \in \mathbb{N}.
\end{aligned} \tag{9.49}$$

Moreover, much as in the proof of Lemma 9.17, it now follows that f_n is again right K-invariant, and so can be thought of as a function on $\mathbb{H}$.

For the smooth functions $f_n \in C_c^\infty(\mathbb{H})$, we already established the isometry formula. Moreover, (9.49) shows that $f_n \to f$ as $n \to \infty$ in $L^2(\mathbb{H})$ by dominated convergence. Together, these show that $\widehat{f_n}^{\mathrm{hyp}}|_{K\times\mathbb{R}}$, considered as an element of $\mathcal{H}_\mu^{\mathrm{even}} \cong L^2(K \times [0,\infty), m_K \times \mu)$ forms a Cauchy sequence, which will have an L^2 limit F with

$$\begin{aligned}\|F\|_{L^2(K\times\mathbb{R},m_K\times\mu)} &= \lim_{n\to\infty} \|\widehat{f_n}^{\mathrm{hyp}}|_{K\times[0,\infty)}\|_{L^2(K\times[0,\infty),m_K\times\mu)}\\ &= \lim_{n\to\infty} \|f_n\|_{L^2(\mathbb{H},\mathrm{vol})} = \|f\|_{L^2(\mathbb{H},\mathrm{vol})}.\end{aligned}$$

The convergence to F in $L^2(K\times\mathbb{R},m_K\times\mu)$ implies that along a subsequence $\widehat{f_n}^{\mathrm{hyp}}|_{K\times\mathbb{R}}$ converges almost surely to F. However, (9.49) and the definition of $\widehat{f}^{\mathrm{hyp}}$ in Definition 9.48 also imply that $\widehat{f_n}^{\mathrm{hyp}}$ converges to $\widehat{f}^{\mathrm{hyp}}$ as $n\to\infty$. Hence $F=\widehat{f}^{\mathrm{hyp}}$ and we obtain the isometry formula

$$\|\widehat{f}^{\mathrm{hyp}}|_{K\times[0,\infty)}\|_{L^2(K\times[0,\infty),m_K\times\mu)} = \|f\|_{L^2(\mathbb{H},\mathrm{vol})}.$$

□

It follows from the established isometry formula for the L^2-norms of the functions $f\in C_c(\mathbb{H})$ and $\widehat{f}^{\mathrm{hyp}}|_{K\times[0,\infty)}$ and Lemma 9.49 that the hyperbolic Fourier transform can be extended uniquely to an intertwining isometry from $L^2(\mathbb{H})$ into $\mathcal{H}_\mu^{\mathrm{even}}$. Moreover, the image $\mathcal{V}$ is then a closed π^μ-invariant subspace of $\mathcal{H}_\mu^{\mathrm{even}}$. This brings up the question of whether such subspaces can be classified. We answer this in a slightly more general case in the following proposition.

Proposition 9.63 (Invariant subspaces). *Let μ be a σ-finite measure on the half-line $[0,\infty)$ and define $\mathcal{H}_\mu^{\mathrm{even}}$ as in Definition 9.39. Then, for any closed $\pi^{\mu,\mathrm{e}}$-invariant subspace $\mathcal{V}\subseteq\mathcal{H}_\mu^{\mathrm{even}}$, there exists a measurable set $S_\mathcal{V}\subseteq[0,\infty)$, that may be thought of as the 'support' of the subspace, so that*

$$\mathcal{V}=\{F\in\mathcal{H}_\mu^{\mathrm{even}}\mid F(\cdot,\xi)=0 \text{ for } \mu\text{-almost every } \xi\in[0,\infty)\smallsetminus S_\mathcal{V}\}. \tag{9.50}$$

Exercise 9.64. Let μ be as above, and let $\mathcal{V}\subseteq\mathcal{H}_\mu^{\mathrm{even}}$ be a closed subspace. Show that a measurable subset $S_\mathcal{V}\subseteq[0,\infty)$ satisfying (9.50) is uniquely determined up to a null set by this property, assuming it exists.

Before starting the formal proof, we outline the structure of the argument. For simplicity, we write $\pi=\pi^{\mu,\mathrm{e}}$ and $\mathcal{H}_\pi=\mathcal{H}_\mu^{\mathrm{even}}$.

We will show, in turn, the following statements.

(a) The closure T_π of $\pi_\partial(\Omega)$ from Corollary 9.11 is given by

$$T_\pi = M_{-\mathrm{id}^2} \tag{9.51}$$

where $M_{-\mathrm{id}^2}$ is the multiplication operator defined by

$$M_{-\mathrm{id}^2}(F)(h,\xi) = -\xi^2F(h,\xi)$$

for all $(h,\xi)\in\mathrm{SL}_2(\mathbb{R})\times[0,\infty)$ and F in the domain

$$D_{M_{-\mathrm{id}^2}} = \left\{F\in\mathcal{H}_\pi \;\middle|\; \int_0^\infty\int_K |\xi^2F(k,\xi)|^2\,\mathrm{d}m_K(k)\,\mathrm{d}\mu(\xi)<\infty\right\}.$$

(b) We define

$$f_0\colon [0,\infty) \ni \xi \longmapsto (1+\xi^2)^{-1}$$

and obtain from the above that $M_{f_0} = (I - T_\pi)^{-1}$. We claim that this multiplication operator M_{f_0} commutes with the centralizer $C(\pi)$ of π.

(c) Using the functional calculus of M_{f_0}, we will prove that all multiplication operators M_f for $f \in L^\infty_\mu([0,\infty))$ commute with $C(\pi)$.

(d) Applying this to the orthogonal projection operator $P_{\mathcal{V}}$ of a π-invariant subspace $\mathcal{V} \subseteq \mathcal{H}_\pi$, we then obtain that $\mathcal{V}$ has $L^\infty_\mu([0,\infty))\mathcal{V} \subseteq \mathcal{V}$, which will allow us to conclude the proof.

We now discuss these four steps in detail.

Proof that $T_\pi = M_{\mathrm{id}^2}$ as claimed in (a). Let $T > 0$ and assume that f is a function in $L^2_\mu([0,\infty))$ with $f(\xi) = 0$ for μ-almost every $\xi > T$. For an integer $n \in 2\mathbb{Z}$ we now define

$$F_n(h,\xi) = f(\xi)F_{\xi,n}(h)$$

for $(h,\xi) \in \mathrm{SL}_2(\mathbb{R}) \times [0,\infty)$, where $F_{\xi,n} \in \mathcal{H}^{\mathrm{even}}_\xi$ is as defined in Lemma 9.34. By (9.19), we have that $F_{\xi,n}(h)$ depends smoothly on $h \in \mathrm{SL}_2(\mathbb{R})$ for any value of $\xi \in [0,\infty)$. Moreover, for a fixed n and for $\xi \in [0,T]$ the derivatives are uniformly bounded. In fact by (9.20) we have

$$\pi^\xi_\partial(\mathbf{a})F_n(h,\xi) = \frac{n+1+\mathrm{i}\xi}{2}F_{n+2}(h,\xi) + \frac{-n+1+\mathrm{i}\xi}{2}F_{n-2}(h,\xi). \tag{9.52}$$

As in the proof of Lemma 9.34, we can now use Proposition 9.12 to conclude that the first summand is equal to $\pi_\partial(\mathbf{r}^+)F_n$ and second is equal to $\pi_\partial(\mathbf{r}^-)F_n$. Using the fact that $\mathbf{k}, \mathbf{r}^+, \mathbf{r}^-$ span $\mathfrak{sl}_2(\mathbb{C})$, it follows that F_n is a smooth vector for π, and using the formula for Ω in (9.11), we also obtain

$$\pi_\partial(\Omega)F_n(h,\xi) = -\xi^2 F_n(h,\xi)$$

for $(h,\xi) \in \mathrm{SL}_2(\mathbb{R}) \times [0,\infty)$. Equivalently, the closed operator T_π satisfies (9.51) for the function F_n as above.

We now extend (9.51) to other functions $F \in \mathcal{H}_\pi$. To begin with, we may vary $n \in 2\mathbb{Z}$ over a finite set (using different functions $f_n \in L^2_\mu([0,\infty))$ with support in $[0,T]$). Note that $M_{-\mathrm{id}^2}$ is bounded on

$$\mathcal{H}_{\pi,\leqslant T} = \bigl\{F \in \mathcal{H}_\pi \mid F(\cdot,\xi) = 0 \text{ for } \mu\text{-almost every } \xi \in (T,\infty)\bigr\},$$

and these finite sums are precisely the K-finite vectors in $\mathcal{H}_{\pi,\leqslant T}$. Hence, it follows by continuity of $M_{-\mathrm{id}^2}$ on $\mathcal{H}_{\pi,\leqslant T}$ and closedness of T_π that (9.51) also holds for all $F \in \mathcal{H}_{\pi,\leqslant T}$.

Next let $F \in D_{M_{-\mathrm{id}^2}}$ and define

$$F_{\leqslant T} = \mathbb{1}_{[0,T]}F \in \mathcal{H}_{\pi,\leqslant T}$$

so that
$$T_\pi(F_{\leqslant T}) = M_{-\mathrm{id}^2}(F_{\leqslant T}).$$
For $T \to \infty$ we have $F_{\leqslant T} \to F$ and $M_{-\mathrm{id}^2}(F_{\leqslant T}) \to M_{-\mathrm{id}^2}(F)$ by dominated convergence. Since T_π is a closed operator, we see that (9.51) holds for all functions $F \in D_{M_{-\mathrm{id}^2}}$. Equivalently, we have $M_{-\mathrm{id}^2} \subseteq T_\pi$.

Suppose now that $F \in \mathcal{H}_\pi$ is a smooth vector. Note that multiplication with $\mathbb{1}_{[0,T]}$ is the intertwining orthogonal projection from $\mathcal{H}_\pi$ onto $\mathcal{H}_{\pi,\leqslant T}$. In particular $\mathbb{1}_{[0,T]} \cdot \pi_\partial(\mathbf{m})F = \pi_\partial(\mathbf{m})(\mathbb{1}_{[0,T]} \cdot F)$ for $\mathbf{m} \in \mathfrak{E}$. Using this for $\mathbf{m} = \Omega$ together with the above, we obtain $\mathbb{1}_{[0,T]} \cdot \pi_\partial(\Omega)F = M_{-\mathrm{id}^2}(F_{\leqslant T})$ for all $T \geqslant 0$. Letting $T \to \infty$ we obtain $\pi_\partial(\Omega) \subseteq M_{-\mathrm{id}^2}$.

To summarize, we have $\pi_\partial(\Omega) \subseteq M_{-\mathrm{id}^2} \subseteq T_\pi$. It is easy to see that $M_{-\mathrm{id}^2}$ is a closed operator. Taking the closure, we see that $T_\pi = M_{-\mathrm{id}^2}$ as claimed in (a). □

Proof that M_{f_0} commutes with $C(\pi)$ as claimed in (b). We define
$$f_0(\xi) = (1 + \xi^2)^{-1}$$
for $\xi \in [0, \infty)$, and first show that (a) implies that
$$M_{f_0} = (I - T_\pi)^{-1}.$$
Indeed,
$$I - T_\pi = M_{(1+\mathrm{id}^2)}$$
is injective on its domain, maps onto $\mathcal{H}_\pi$, and has M_{f_0} as its (bounded) inverse operator. We now show that M_{f_0} commutes with every intertwining bounded operator $B\colon \mathcal{H}_\pi \to \mathcal{H}_\pi$. For this, we first note that B maps every smooth vector v to a smooth vector Bv, and we have
$$\pi_\partial(\Omega)Bv = B\pi_\partial(\Omega)v.$$
If $v \in D_{T_\pi}$ there is a sequence of smooth vectors (v_n) with $v_n \to v$ and
$$\pi_\partial(\Omega)v_n \longmapsto T_\pi v$$
as $n \to \infty$. However, this implies $Bv_n \to Bv$ and $\pi_\partial(\Omega)Bv_n \to BT_\pi v$ as $n \to \infty$, and hence also $T_\pi B \supseteq BT_\pi$. Using $M_{f_0} = (I - T_\pi)^{-1}$, we now obtain the claim. Indeed, let $v \in \mathcal{H}_\pi$ and $(I - T_\pi)^{-1}v = w$ so that $v = (I - T_\pi)w$. Then we have $Bv = (I - T_\pi)Bw$, which implies that $(I - T_\pi)^{-1}Bv = Bw = B(I - T_\pi)^{-1}v$ as claimed in (b). □

Proof that M_f commutes with $C(\pi)$ as claimed in (c). Since
$$(I - T_\pi)^{-1} = M_{f_0}$$
is already realized as a multiplication operator and f_0 is injective, the following are now relatively easy claims to prove. The measurable functional calculus

for M_{f_0} (see [25, Sec. 12.6]) gives rise to other multiplication operators. By injectivity of f_0, every multiplication operator M_f for $f \in L^\infty_\mu([0,\infty))$ can be obtained from the measurable functional calculus of M_{f_0}. With the previous claim and the properties of the measurable functional calculus (see [25, Prop. 12.68]), this implies that every intertwining bounded operator $B\colon \mathcal{H}_\pi \to \mathcal{H}_\pi$ commutes with M_f for $f \in L^\infty_\mu([0,\infty))$. □

CONCLUSION OF THE PROOF OF PROPOSITION 9.63, AS OUTLINED IN (d). Now let $\mathcal{V} \subseteq \mathcal{H}_\pi$ be a closed π-invariant subspace, and let $B = P_\mathcal{V}$ be the orthogonal projection onto $\mathcal{V}$. By invariance of $\mathcal{V}$, the projection is intertwining and, by (c) above, we have $M_f\mathcal{V} \subseteq \mathcal{V}$ for all $f \in L^\infty_\mu([0,\infty))$.

Now let $F \in \mathcal{V}$ and define

$$S_F = \{\xi \in [0,\infty) \mid F(\cdot,\xi) \neq 0\}.$$

Using invariance of $\mathcal{V}$ under K, we can split F into a sum of K-eigenfunctions as

$$F = \sum_{n\in 2\mathbb{Z}} F_n$$

satisfying $F_n \in \mathcal{V}_n$ for all $n \in 2\mathbb{Z}$, and

$$S_F = \bigcup_{n\in 2\mathbb{Z}} S_{F_n}.$$

Since $\mathcal{H}_\xi$ contains (up to scalars) only one K-eigenfunction of weight $n \in 2\mathbb{Z}$ (namely $F_{\xi,n}$ in (9.19)) there exists some $f_n \in L^2_\mu([0,\infty))$ such that

$$F_n(h,\xi) = f_n(\xi)F_{\xi,n}(h)$$

for $(h,\xi) \in \mathrm{SL}_2(\mathbb{R}) \times [0,\infty)$. In particular,

$$S_{F_n} = \{\xi \in [0,\infty) \mid f_n(\xi) \neq 0\}.$$

Using $M_f\mathcal{V} \subseteq \mathcal{V}$ for all $f \in L^\infty_\mu([0,\infty))$, the fact that $F_n \in \mathcal{V}$, and dominated convergence, it follows that the function

$$\mathrm{SL}_2(\mathbb{R}) \times [0,\infty) \ni (h,\xi) \longmapsto f(\xi)F_{\xi,n}(h)$$

belongs to $\mathcal{V}$ for any $f \in L^2_\mu([0,\infty))$ with $\{\xi \mid f(\xi) \neq 0\} \subseteq S_{F_n}$. For $T > 0$ we define

$$S_{n,T} = S_{F_n} \cap [0,T].$$

By (9.52) and the argument directly following it, we see that $\pi_\partial(\mathbf{r}^\pm)$ are bounded operators on $\{F' \in \mathcal{H}_{\pi,\leqslant T} \mid F' \text{ has } K\text{-weight } n\}$. Since $\mathcal{V}$ is invariant under $\pi_\partial(\mathbf{r}^\pm)$ (where it is defined) it follows that the function

$$\mathrm{SL}_2(\mathbb{R}) \times [0,\infty) \ni (h,\xi) \longmapsto f(\xi)F_{\xi,m}(h)$$

belongs to $\mathcal{V}$ for any $f \in L^2_\mu([0,\infty))$ with $\{\xi \mid f(\xi) \neq 0\} \subseteq S_{F_n}$, where m, n lie in $2\mathbb{Z}$ are arbitrary. Varying $m, n \in 2\mathbb{Z}$ and the functions $f \in L^2_\mu([0,\infty))$, we can write any $F' \in \mathcal{H}_\pi$ with $\{\xi \mid F'(\cdot,\xi) \neq 0\} \subseteq S_F$ as a convergent sum of elements of $\mathcal{V}$ and obtain $F' \in \mathcal{V}$.

Since $\mathcal{V}$ is separable, we can find a dense set $\{F_{(k)} \mid k \in \mathbb{N}\}$ of vectors, apply the above argument to each $F_{(k)}$ and obtain the same statement for

$$S_{\mathcal{V}} = \bigcup_{k\in\mathbb{N}} S_{F_{(k)}} = \{\xi \in [0,\infty) \mid \text{there exists a } k \in \mathbb{N} \text{ with } F_{(k)}(\cdot,\xi) \neq 0\}.$$

Using the fact that the right-hand side of (9.50) defines a closed subspace of $\mathcal{H}_\mu$, this proves the proposition. □

Exercise 9.65. Let μ be a σ-finite measure on $[0,\infty)$ as in Proposition 9.63. Show that the centralizer of $\pi^{\mu,\mathrm{e}}$ is given by all multiplication operators M_f with $f \in L^\infty_\mu([0,\infty))$.

The following finishes our discussions of the hyperbolic Fourier transform.

Concluding the Proof of Theorem 9.60. Let μ be the measure on $[0,\infty)$ defined by $\frac{1}{8\pi}\xi \tanh \xi \,\mathrm{d}\xi$. By the first part of the proof on page 466, we know that

$$C_c(\mathbb{H}) \ni f \longmapsto \widehat{f}^{\mathrm{hyp}} \in \mathcal{H}^{\mathrm{even}}_\mu$$

is an intertwining isometry between the action-associated representation $\pi^{\mathbb{H}}$ and the integral $\pi^{\mu,\mathrm{e}}$ of the even principal series representation. Hence it can be extended to an intertwining isometry from $L^2(\mathbb{H})$ to a closed $\pi^{\mu,\mathrm{e}}$-invariant subspace $\mathcal{V} \subseteq \mathcal{H}^{\mathrm{even}}_\mu$. By Proposition 9.63, this subspace can be defined by a measurable subset $S_{\mathcal{V}} \subseteq [0,\infty)$ and the formula (9.50). We shall show that the set $S_{\mathcal{V}} = [0,\infty)$ (up to null sets) by finding a sequence (f_n) in $C_c(\mathbb{H})$ so that for every $\xi \in [0,\infty)$ there exists some $n \in \mathbb{N}$ with $\widehat{f_n}^{\mathrm{hyp}}(\cdot,\xi) \neq 0$.

In fact we let (f_n) be a sequence of spherical functions in $C_c(\mathbb{H})$ with

$$\int_{\mathbb{H}} f_n \,\mathrm{dvol} = 1$$

and $f_n \geqslant 0$ for all $n \in \mathbb{N}$, so that $\operatorname{supp} f_n$ is a shrinking neighbourhood of $\mathrm{i} \in \mathbb{H}$. For every $\xi \in [0,\infty)$ it now follows that

$$\widehat{f_n}^{\mathrm{hyp}}(k,\xi) = \int_{\mathbb{H}} f_n(z)\overline{\chi_{\infty,\xi}(z)} \,\mathrm{dvol}(z) \longrightarrow 1$$

as $n \to \infty$ and for all $k \in K$ by K-invariance of f_n and continuity of the function $z \mapsto \chi_{\infty,\xi}(z) = \Im(z)^{\frac{1}{2}-\frac{\mathrm{i}}{2}\xi}$.

Hence we have $S_{\mathcal{V}} = [0,\infty)$ (up to null sets), which gives $\mathcal{V} = \mathcal{H}^{\mathrm{even}}_\mu$ by (9.50), and Theorem 9.60 follows. □

9.5 The Complementary Series Representation

We recall from Section 9.3 that for $\xi \in \mathbb{R}$ the principal series representation $\pi^{\xi,\mathrm{e}}$ is constructed from the unitary character χ_ξ defined by

$$\chi_\xi(a_t u_x) = \mathrm{e}^{\mathrm{i}\xi t}$$

for $a_t u_x \in B$. Moreover, $\pi^{\xi,\mathrm{e}}$ then turned out to be an irreducible unitary representation with Casimir eigenvalue $\alpha_{\pi^{\xi,\mathrm{e}}} = -\xi^2$. According to Section 9.2.4 there should be another type γ^s of even irreducible unitary representation with Casimir eigenvalues $\alpha_{\gamma^s} = s^2$ for $s \in (0,1)$ that we have not yet seen. To construct γ^s we try to mimic the construction of $\pi^{\xi,\mathrm{e}}$ in Definition 9.31 while attempting to 'replace $\mathrm{i}\xi$ by s'.

Definition 9.66 (Complementary series representation). For $s \in (0,1)$ we define the non-unitary character $\chi_{(s)}$ on $B = \{a_t u_x \mid t, x \in \mathbb{R}\}$ by

$$\chi_{(s)}(a_t u_x) = \mathrm{e}^{st}$$

for $a_t u_x \in B$. The *complementary series representation* γ^s of $G = \mathrm{SL}_2(\mathbb{R})$ is initially defined as the left regular representation on the space $\mathcal{V}_s$ of those functions $f\colon G \to \mathbb{C}$ with the following properties:

(1) f is smooth,
(2) f is even (that is, $f(-g) = f(g)$ for all $g \in G$), and
(3) $f(gb) = \chi_{(s)}(b)^{-1} \Delta_B(b)^{\frac{1}{2}} f(g)$ for all $g \in G$ and $b \in B$.

The main difference between this and the construction of the principal series representation is, of course, that we are using here non-unitary characters, which means that the L^2-norm on K will not be preserved under the left regular representation (as was the case for the principal series representation). Instead, we will have to define a new norm and inner product on $\mathcal{V}_s$.

Theorem 9.67 (Complementary series representations). *Let $s \in (0,1)$. The regular representation on the completion $\mathcal{H}_{(s)}$ of $\mathcal{V}_s$ defines a non-tempered irreducible unitary representation γ^s with Casimir eigenvalue s^2, called the* complementary series representation.

9.5.1 The Space and its Inner Product

In the following, let $M = \{\pm I\}$ be the centre of $\mathrm{SL}_2(\mathbb{R})$. We will always identify functions on K/M with even functions on K.

Lemma 9.68 (Smooth functions on K/M). *Let $s \in (0,1)$. Then $\mathcal{V}_s$ is isomorphic to $C^\infty(K/M)$. In fact every element $f \in \mathcal{V}_s$ is uniquely determined by* *its restriction* $f|_K \in C^\infty(K/M)$ *and every even smooth function* f_K *on* K *can be extended via*

$$f(ka_tu_x) = f_K(k)\mathrm{e}^{-(s+1)t} \tag{9.53}$$

for $ka_tu_x \in KAU$ *to an element of* $\mathcal{V}_s$. *In particular, for every* $n \in 2\mathbb{Z}$ *the function* $F_{s,n}$ *defined by*

$$F_{s,n}(k_\psi a_t u_x) = \mathrm{e}^{-\mathrm{i}n\psi-(s+1)t} \tag{9.54}$$

for $k_\psi a_t u_x \in KAU$ *belongs to* $\mathcal{V}_s$ *and is a* K*-eigenfunction with* K*-weight* n.

PROOF. We note that Definition 9.66 implies that any function $f \in \mathcal{V}_s$ is uniquely determined by $f|_K$. Clearly the map

$$\Phi\colon K \times A \times U \ni (k_\psi, a_t, u_x) \longmapsto g = k_\psi a_t u_x \in \mathrm{SL}_2(\mathbb{R})$$

is smooth, and hence $f|_K \in C^\infty(K/M)$ for all $f \in \mathcal{V}_s$. The inverse of Φ is also smooth, since it maps $g \in \mathrm{SL}_2(\mathbb{R})$ first to the polar coordinates (ψ, r) of ge_1, and then to $k_\psi \in K$, $a_{\log r} \in A$, and $a_{\log r}^{-1}k_\psi^{-1}g = u_x \in U$. This shows that (9.53) defines a smooth function on G for any $f_K \in C^\infty(K/M)$. Moreover, $g = ka_tu_x$ is in KAU and $b = a_{t_0}u_{x_0}$ is in $B = AU$ and so

$$\begin{aligned} f(gb) &= f(ka_{t+t_0}u_{\mathrm{e}^{-2t_0}x+x_0}) \\ &= f_K(k)\mathrm{e}^{-(s+1)(t+t_0)} \\ &= f_K(k)\mathrm{e}^{-(s+1)t}\mathrm{e}^{-st_0}\mathrm{e}^{-t_0} \\ &= f(ka_tu_x)\chi_{(s)}(b)^{-1}\Delta_B(b)^{\frac{1}{2}}, \end{aligned}$$

which shows that $f \in \mathcal{V}_s$, as claimed in the lemma.

Applying the above to the character χ_{-n} on K for some $n \in 2\mathbb{Z}$ defines the function $F_{s,n} \in \mathcal{V}_s$ in (9.54). By definition,

$$\big(\gamma^s_{k_\theta}(F_{s,n})\big)(k_\psi) = F_{s,n}\big(k_\theta^{-1}k_\psi\big) = \mathrm{e}^{-\mathrm{i}n(\psi-\theta)} = \mathrm{e}^{\mathrm{i}n\theta}F_{s,n}(k_\psi)$$

for all $k_\theta, k_\psi \in K$. By the first part of the proof, this shows that $F_{s,n}$ has K-weight n for γ^s. □

Definition 9.69 (The inner product on $\mathcal{V}_s$). For $s \in (0,1)$ we define

$$\langle f_1, f_2\rangle_{\mathcal{V}_s} = \frac{1}{\pi^2}\int_0^\pi\int_0^\pi \frac{f_1(k_{\theta_1})\overline{f_2(k_{\theta_2})}}{|\sin(\theta_1-\theta_2)|^{1-s}}\,\mathrm{d}\theta_1\,\mathrm{d}\theta_2 \tag{9.55}$$

for $f_1, f_2 \in \mathcal{V}_s$.

For now, (9.55) simply falls from the sky. We will, however, give additional meaning to it after we have established the fundamental properties of this inner product.

Lemma 9.70 (The inner product on $\mathcal{V}_s$). *For any $s \in (0,1)$, the form* (9.55) *defines an inner product on $\mathcal{V}_s$.*

We note that the proof of this lemma will require both $s > 0$ and $1 - s > 0$.

PROOF OF LEMMA 9.70. We note that $\mathbb{R} \ni \theta \mapsto |\sin\theta|$ has period π. This gives us

$$\frac{1}{\pi^2}\int_0^\pi\int_0^\pi \frac{\mathrm{d}\theta_1\,\mathrm{d}\theta_2}{|\sin(\theta_1-\theta_2)|^{1-s}} = \frac{1}{\pi^2}\int\limits_{\mathbb{R}/\mathbb{Z}\pi}\int\limits_{\mathbb{R}/\mathbb{Z}\pi} \frac{\mathrm{d}\theta_1\,\mathrm{d}\theta_2}{|\sin(\underbrace{\theta_1-\theta_2}_{=\theta})|^{1-s}}$$
$$= \frac{1}{\pi}\int\limits_{\mathbb{R}/\mathbb{Z}\pi} \frac{\mathrm{d}\theta}{|\sin\theta|^{1-s}} < \infty$$

since $|\sin\theta| \asymp |\theta|$ as $\theta \to 0$ and $s > 0$ implies that $\int_0^1 \frac{\mathrm{d}\theta}{\theta^{1-s}} < \infty$. Therefore, the function $(\theta_1,\theta_2) \mapsto \frac{1}{|\sin(\theta_1-\theta_2)|^{1-s}}$ lies in $L^1([0,\pi]^2)$, and the integral in (9.55) converges for all $f_1, f_2 \in \mathcal{V}_s$.

Sesqui-linearity of $\langle\cdot,\cdot\rangle_{\mathcal{V}_s}$ follows directly from the definition in (9.55). Thus it remains to show that $\langle f, f\rangle_{\mathcal{V}_s} > 0$ for all $f \in \mathcal{V}_s \smallsetminus \{0\}$. The Fourier expansion of $f|_K$ allows us to write

$$f = \sum_{n\in 2\mathbb{Z}} c_n F_{s,n}$$

for some sequence of coefficients $(c_n) \in \ell^1(2\mathbb{Z})$. Using sesqui-linearity, we obtain

$$\langle f, f\rangle_{\mathcal{V}_s} = \sum_{m,n\in 2\mathbb{Z}} c_m\overline{c_n}\langle F_{s,m}, F_{s,n}\rangle_{\mathcal{V}_s}.$$

We claim that $m, n \in 2\mathbb{Z}$ with $m \neq n$ implies that $\langle F_{s,m}, F_{s,n}\rangle_{\mathcal{V}_s} = 0$ and that $\langle F_{s,n}, F_{s,n}\rangle_{\mathcal{V}_s} > 0$ for all $n \in 2\mathbb{Z}$. Together, these show that $\langle f, f\rangle_{\mathcal{V}_s} > 0$ for all $f \in \mathcal{V}_s \smallsetminus \{0\}$.

Suppose first that $m, n \in 2\mathbb{Z}$ with $m \neq n$. Then

$$\langle F_{s,m}, F_{s,n}\rangle_{\mathcal{V}_s} = \frac{1}{\pi^2}\int_0^\pi\int_0^\pi \frac{\mathrm{e}^{-\mathrm{i}m\theta_1+\mathrm{i}n\theta_2}}{|\sin(\theta_1-\theta_2)|^{1-s}}\,\mathrm{d}\theta_1\,\mathrm{d}\theta_2$$
$$= \frac{1}{\pi^2}\int\limits_{\mathbb{R}/\mathbb{Z}\pi}\int\limits_{\mathbb{R}/\mathbb{Z}\pi} \frac{\mathrm{e}^{\mathrm{i}(n(\theta_2'+\psi)-m(\theta_1'+\psi))}}{|\sin(\theta_1'-\theta_2')|^{1-s}}\,\mathrm{d}\theta_1'\,\mathrm{d}\theta_2'$$
$$= \mathrm{e}^{\mathrm{i}(n-m)\psi}\langle F_{s,m}, F_{s,n}\rangle_{\mathcal{V}_s}$$

by using the substitutions $\theta_1 = \theta_1' + \psi$ and $\theta_2 = \theta_2' + \psi$ for some $\psi \in \mathbb{R}$. This implies that $\langle F_{s,m}, F_{s,n}\rangle_{\mathcal{V}_s} = 0$ for $m \neq n$.

For $n \in 2\mathbb{Z}$, we define

$$I_n = \pi \langle F_{s,n}, F_{s,n} \rangle_{\mathcal{V}_s} = \frac{1}{\pi} \int\limits_{\mathbb{R}/\mathbb{Z}\pi} \int\limits_{\mathbb{R}/\mathbb{Z}\pi} \frac{\mathrm{e}^{\mathrm{i}n(\theta_2-\theta_1)}}{|\sin(\theta_1-\theta_2)|^{1-s}} \,\mathrm{d}\theta_1 \,\mathrm{d}\theta_2$$
$$= \int\limits_{\mathbb{R}/\mathbb{Z}\pi} \frac{\mathrm{e}^{\mathrm{i}n\theta}}{|\sin\theta|^{1-s}} \,\mathrm{d}\theta.$$

Note that

$$\overline{I_n} = \int_{\mathbb{R}/\mathbb{Z}\pi} \frac{\mathrm{e}^{-\mathrm{i}n\theta}}{|\sin\theta|^{1-s}} \,\mathrm{d}\theta = I_n$$

via the substitution $\theta' = -\theta$. Thus

$$I_n = \int_0^\pi \frac{\cos(n\theta)}{(\sin\theta)^{1-s}} \,\mathrm{d}\theta,$$

and so we wish to show that $I_n > 0$ for all $n \in 2\mathbb{Z}$. Since

$$I_0 = \int_0^\pi \frac{1}{(\sin\theta)^{1-s}} \,\mathrm{d}\theta > 0$$

and $I_{-n} = I_n$ for all $n \in 2\mathbb{Z}$, it remains to show that $I_n > 0$ for all $n \in 2\mathbb{N}$. For I_2 we have, using integration by parts,

$$\begin{aligned} I_2 &= \int_0^\pi \cos 2\theta (\sin\theta)^{s-1} \,\mathrm{d}\theta \\ &= \left[\frac{\sin 2\theta}{2} (\sin\theta)^{s-1}\right]_0^\pi - \int_0^\pi \frac{\sin 2\theta}{2} (s-1)(\sin\theta)^{s-2} \cos\theta \,\mathrm{d}\theta \\ &= \underbrace{[\cos\theta(\sin\theta)^s]_0^\pi}_{=0} + (1-s) \int_0^\pi \cos^2\theta (\sin\theta)^{s-1} \,\mathrm{d}\theta \\ &= (1-s) \int_0^\pi \frac{1+\cos 2\theta}{2} (\sin\theta)^{s-1} \,\mathrm{d}\theta \\ &= \frac{1-s}{2} (I_0 + I_2), \end{aligned}$$

where the boundary terms vanish since $s > 0$. We now solve this equation for I_2, to obtain

$$I_2 = \left(\frac{1-s}{1+s}\right) I_0 > 0$$

since $s < 1$.

For a general $n \in 2\mathbb{N}$, we again use integration by parts to see that

$$\begin{aligned} I_n &= \int_0^\pi \cos(n\theta)(\sin\theta)^{s-1}\,\mathrm{d}\theta \\ &= \underbrace{\left[\frac{\sin(n\theta)}{n}(\sin\theta)^{s-1}\right]_0^\pi}_{=0} - \int_0^\pi \frac{\sin n\theta}{n}(s-1)(\sin\theta)^{s-2}\cos\theta\,\mathrm{d}\theta \\ &= \frac{(1-s)}{n}\int_0^\pi \frac{\sin(n\theta)\cos\theta}{\sin\theta}(\sin\theta)^{s-1}\,\mathrm{d}\theta. \end{aligned}$$

We again wish to relate I_n to earlier values of the sequence, and hence use the fact that $n \in 2\mathbb{N}$ to calculate that

$$\begin{aligned} \frac{\sin(n\theta)\cos\theta}{\sin\theta} &= \frac{1}{2}\frac{(\mathrm{e}^{\mathrm{i}n\theta}-\mathrm{e}^{-\mathrm{i}n\theta})(\mathrm{e}^{\mathrm{i}\theta}+\mathrm{e}^{-\mathrm{i}\theta})}{(\mathrm{e}^{\mathrm{i}\theta}-\mathrm{e}^{-\mathrm{i}\theta})} \\ &= \frac{1}{2}\Big(\mathrm{e}^{\mathrm{i}(n-1)\theta}+\mathrm{e}^{\mathrm{i}(n-3)\theta}+\cdots+\mathrm{e}^{-\mathrm{i}(n-3)\theta}+\mathrm{e}^{-\mathrm{i}(n-1)\theta}\Big)\Big(\mathrm{e}^{\mathrm{i}\theta}+\mathrm{e}^{-\mathrm{i}\theta}\Big) \\ &= \frac{1}{2}\left(\mathrm{e}^{\mathrm{i}n\theta}+2\mathrm{e}^{\mathrm{i}(n-2)\theta}+\cdots+2\mathrm{e}^{-\mathrm{i}(n-2)\theta}+\mathrm{e}^{-\mathrm{i}n\theta}\right) \\ &= \cos(n\theta)+2\cos\left((n-2)\theta\right)+\cdots+2\cos(2\theta)+1. \end{aligned}$$

Putting this into the above formula for I_n gives

$$I_n = \frac{1-s}{n}\left(I_n+2I_{n-2}+\cdots+2I_2+I_0\right),$$

which may be solved for I_n to give the recursion formula

$$I_n = \frac{1-s}{n-1+s}\left(2I_{n-2}+\cdots+2I_2+I_0\right).$$

Using the fact that $s<1$, this implies once more that $I_n>0$ by induction on $n \in 2\mathbb{N}$. □

Definition 9.71 (The space $\mathcal{H}_{(s)}$). Let $s \in (0,1)$. We define $\mathcal{H}_{(s)}$ to be the completion of $\mathcal{V}_s$ with respect to the norm induced by $\langle\cdot,\cdot\rangle_{\mathcal{V}_s}$.

9.5.2 Unitarity of the Complementary Series

PROOF OF UNITARITY IN THEOREM 9.67. Recall from Example 1.7 that the group $\mathrm{SL}_2(\mathbb{R})$ acts on $v \in \mathbb{S}^1$ via $g\centerdot v = \frac{1}{\|gv\|}gv$ for $g \in \mathrm{SL}_2(\mathbb{R})$, and that the Radon–Nikodym derivative for the normalized length measure m on $\mathbb{S}^1$ satisfies

$$\frac{\mathrm{d}g_*^{-1}m}{\mathrm{d}m}(v) = \|gv\|^{-2}. \tag{9.56}$$

Also recall that we used this set-up earlier to discuss the principal series representation π^ξ on $\mathcal{H}_\xi \cong L^2(K) \cong L^2_m(\mathbb{S}^1)$.

For $u_1, u_2 \in \mathbb{R}^2 \smallsetminus \{(0,0)\}$ we define the function

$$\mathcal{D}(u_1, u_2) = |\det(u_1, u_2)|,$$

and note that $\mathcal{D}(r_1 u_1, r_2 u_2) = |r_1||r_2|\mathcal{D}(u_1, u_2)$ for all $r_1, r_2 \in \mathbb{R}^\times$. Moreover, we also have

$$\mathcal{D}(g^{-1}u_1, g^{-1}u_2) = |\det(g^{-1}(u_1, u_2))| = \mathcal{D}(u_1, u_2) \tag{9.57}$$

for $g \in \mathrm{SL}_2(\mathbb{R})$, by multiplicativity of the determinant. To understand the connection between $\mathcal{D}(\cdot,\cdot)$ and $\langle\cdot,\cdot\rangle_{\mathcal{V}_s}$, we let $\theta_1, \theta_2 \in [0, \pi)$, set

$$v_{\theta_j} = k_{\theta_j} e_1 = \begin{pmatrix} \cos\theta_j \\ \sin\theta_j \end{pmatrix}$$

for $j = 1, 2$ and calculate

$$\mathcal{D}(v_{\theta_1}, v_{\theta_2}) = \left| \det \begin{pmatrix} \cos\theta_1 & \cos\theta_2 \\ \sin\theta_1 & \sin\theta_2 \end{pmatrix} \right| = |\sin(\theta_1 - \theta_2)|. \tag{9.58}$$

Because of these formulas, it will be convenient to identify $k_\theta \in K$ with

$$v_\theta = k_\theta e_1 = \begin{pmatrix} \cos\theta \\ \sin\theta \end{pmatrix} \in \mathbb{S}^1$$

for $\theta \in [0, 2\pi)$. In this notation, and since $f \in \mathcal{V}_s$ is even, we may use (9.58) to express the norm in the form

$$\|f\|^2_{\mathcal{V}_s} = \int_{\mathbb{S}^1}\int_{\mathbb{S}^1} \frac{f(k_{\theta_1})\overline{f(k_{\theta_2})}}{\mathcal{D}(v_{\theta_1}, v_{\theta_2})^{1-s}} \, \mathrm{d}m(v_{\theta_1}) \, \mathrm{d}m(v_{\theta_2}),$$

where m denotes the normalized arc length measure on $\mathbb{S}^1$.

Fix some $f \in \mathcal{V}_s$ and $g \in \mathrm{SL}_2(\mathbb{R})$. Then, by definition, we have

$$\|\gamma^s_g f\|^2_{\mathcal{V}_s} = \int_{\mathbb{S}^1}\int_{\mathbb{S}^1} \frac{f(g^{-1}k_{\theta_1})\overline{f(g^{-1}k_{\theta_2})}}{\mathcal{D}(v_{\theta_1}, v_{\theta_2})^{1-s}} \, \mathrm{d}m(v_{\theta_1}) \, \mathrm{d}m(v_{\theta_2}).$$

Let us write $g^{-1}k_{\theta_j} = k_{\psi_j} a_{t_j} u_{x_j}$ for $j = 1, 2$ for the Iwasawa decomposition of these products, so that ψ_j, t_j, and x_j are functions of $v_{\theta_j} \in \mathbb{S}^1$ for $j = 1, 2$. Then

$$g^{-1}v_{\theta_j} = g^{-1}k_{\theta_j} e_1 = k_{\psi_j} a_{t_j} e_1 = \mathrm{e}^{t_j} v_{\psi_j},$$

in particular

$$\|g^{-1}v_{\theta_j}\| = \|a_{t_j} e_1\| = \mathrm{e}^{t_j}, \tag{9.59}$$

and

$$g^{-1}\bullet v_{\theta_j} = \|g^{-1}v_{\theta_j}\|^{-1}g^{-1}v_{\theta_j} = v_{\psi_j} \tag{9.60}$$

for $j = 1, 2$. Combining the definition of $\mathcal{V}_s$ with (9.59), we obtain

$$f(g^{-1}k_{\theta_j}) = f(k_{\psi_j}a_{t_j}u_{x_j}) = \mathrm{e}^{-t_j(s+1)}f(k_{\psi_j}) = \|g^{-1}v_{\theta_j}\|^{-(s+1)}f(k_{\psi_j})$$

for $f \in \mathcal{V}_s$. Using (9.57) and (9.60), we also have

$$\mathcal{D}(v_{\theta_1}, v_{\theta_2}) = \mathcal{D}(g^{-1}v_{\theta_1}, g^{-1}v_{\theta_2}) = \|g^{-1}v_{\theta_1}\|\|g^{-1}v_{\theta_2}\|\mathcal{D}(v_{\psi_1}, v_{\psi_2}).$$

For the norm of $\gamma_g^s f$, this leads to

$$\begin{aligned}\|\gamma_g^s f\|_{\mathcal{V}_s}^2 &= \iint\limits_{\mathbb{S}^1\mathbb{S}^1} \frac{\|g^{-1}v_{\theta_1}\|^{-(s+1)}f(k_{\psi_1})\|g^{-1}v_{\theta_2}\|^{-(s+1)}\overline{f(k_{\psi_2})}}{\|g^{-1}v_{\theta_1}\|^{1-s}\|g^{-1}v_{\theta_2}\|^{1-s}\mathcal{D}(v_{\psi_1}, v_{\psi_2})^{1-s}}\,\mathrm{d}m(v_{\theta_1})\mathrm{d}m(v_{\theta_2}) \\ &= \iint\limits_{\mathbb{S}^1\mathbb{S}^1} \frac{f(k_{\psi_1})\overline{f(k_{\psi_2})}}{\mathcal{D}(v_{\psi_1}, v_{\psi_2})^{1-s}}\|g^{-1}v_{\theta_1}\|^{-2}\|g^{-1}v_{\theta_2}\|^{-2}\,\mathrm{d}m(v_{\theta_1})\mathrm{d}m(v_{\theta_2}) \\ &= \iint\limits_{\mathbb{S}^1\mathbb{S}^1} \frac{f(k_{\psi_1})\overline{f(k_{\psi_2})}}{\mathcal{D}(v_{\psi_1}, v_{\psi_2})^{1-s}}\,\mathrm{d}g_*m(v_{\theta_1})\mathrm{d}g_*m(v_{\theta_2}). \qquad \text{(by (9.56))}\end{aligned}$$

However, by (9.60) this double integral now has the form

$$\begin{aligned}&\iint\limits_{\mathbb{S}^1\mathbb{S}^1} F\big(g^{-1}\bullet v_{\theta_1}, g^{-1}\bullet v_{\theta_2}\big)\,\mathrm{d}g_*m(v_{\theta_1})\,\mathrm{d}g_*m(v_{\theta_2}) \\ &\qquad\qquad = \iint\limits_{\mathbb{S}^1\mathbb{S}^1} F(v_{\theta_1}, v_{\theta_2})\,\mathrm{d}m(v_{\theta_1})\,\mathrm{d}m(v_{\theta_2}).\end{aligned}$$

Therefore the last expression for $\|\gamma_g^s f\|_{\mathcal{V}_s}^2$ turns into $\|f\|_{\mathcal{V}_s}^2$. This shows that γ_g^s is unitary on $\mathcal{V}_s$, and that it extends by continuity to a unitary operator on $\mathcal{H}_{(s)}$.

Finally, we verify the continuity of the complementary series representation γ^s. To see this, let $f_1, f_2 \in \mathcal{V}_s$ and note that

$$\langle \gamma_g^s f_1, f_2\rangle_{\mathcal{V}_s} = \iint\limits_{\mathbb{S}^1\mathbb{S}^1} \frac{f_1(g^{-1}k_{\theta_1})f_2(k_{\theta_2})}{\mathcal{D}(v_{\theta_1}, v_{\theta_2})^{1-s}}\,\mathrm{d}m(v_{\theta_1})\,\mathrm{d}m(v_{\theta_2})$$

depends continuously on $g \in \mathrm{SL}_2(\mathbb{R})$ by dominated convergence. This implies continuity of $g \mapsto \gamma_g^s f$ by simply expanding $\|\gamma_g^s f - \gamma_{g_0}^s f\|_{\mathcal{V}_s}^2$ into a sum of inner products. Hence Lemma 1.13 gives continuity of the unitary representation γ^s. □

9.5.3 The Casimir Eigenvalue for the Complementary Series

Lemma 9.72 (Casimir eigenvalue). *Let $s \in (0,1)$. For the complementary series representation, the Casimir element $\gamma^s_\partial(\Omega)$ is multiplication by s^2 on $\mathcal{V}_s$. Moreover,*

$$\gamma^s_\partial(\mathbf{r}^+)F_{s,n} = \frac{n+1+s}{2}F_{s,n+2} \quad \text{and}$$
$$\gamma^s_\partial(\mathbf{r}^-)F_{s,n} = \frac{-n+1+s}{2}F_{s,n-2}$$

for all $n \in 2\mathbb{Z}$.

PROOF. We reuse the calculation in the proof of Lemma 9.34. In fact we proved (9.20) by calculating a pointwise derivative, and this part of the argument would apply for any $\xi \in \mathbb{C}$. Using $\xi = -\mathrm{i}s$ replaces $\mathrm{i}\xi$ by s, and (9.20) takes the form

$$\gamma^s_\partial(\mathbf{a})F_{s,n} = \frac{n+1+s}{2}F_{s,n+2} + \frac{-n+1+s}{2}F_{s,n-2} \tag{9.61}$$

for $n \in 2\mathbb{Z}$ (see Exercise 9.73 below). We recall that $\mathbf{a} = \mathbf{r}^+ + \mathbf{r}^-$ and apply Proposition 9.12 for γ^s. From this the formulas for $\gamma^s_\partial(\mathbf{r}^+)F_{s,n}$ in the lemma follow.

We now apply Ω in the form (9.11) and obtain

$$\begin{aligned}\gamma^s_\partial(\Omega)F_{s,n} &= \gamma^s_\partial\big(4\mathbf{r}^+ \circ \mathbf{r}^- + (\mathbb{1}_{\mathfrak{C}} + \mathrm{i}\mathbf{k})^{\circ 2}\big)F_{s,n} \\ &= (-n+1+s)\gamma^s_\partial(2\mathbf{r}^+)F_{s,n-2} + \gamma^s_\partial(\mathbb{1}_{\mathfrak{C}} + \mathrm{i}\mathbf{k})^{\circ 2}F_{s,n} \\ &= (-n+1+s)(n-1+s)F_{s,n} + (1-n)^2F_{s,n} = s^2F_{s,n}\end{aligned}$$

for all $n \in 2\mathbb{Z}$. Using Fourier series for a smooth function on K/M, this extends to all smooth $f \in \mathcal{V}_s$. □

Exercise 9.73. Prove (9.61) as a partial derivative within $\mathcal{H}_{(s)}$.

We note that Lemma 9.72 establishes the remaining properties listed in the beginning of Section 9.2.4. Hence Proposition 9.18 now implies that the complementary series representation γ^s is irreducible for any $s \in (0,1)$.

9.5.4 Decay and Integrability Properties

The following shows, in particular, the remaining claim in Theorem 9.67 that the complementary series is not tempered. However, the precise information regarding the decay properties of the matrix coefficients of $F_{s,0} \in \mathcal{H}_{(s)}$ will be useful as well.

Lemma 9.74 (Matrix coefficient of $F_{s,0}$). *Let $s \in (0,1)$. The matrix coefficient $\phi_{(s)} = \varphi^{\gamma^s}_{F_{s,0}}$ is bi-K-invariant, satisfies the asymptotics*

$$\phi_{(s)}(g) \asymp_s \|g\|_{\mathrm{HS}}^{s-1}$$

for $g \in \mathrm{SL}_2(\mathbb{R})$, and belongs to $L^p(G)$ if and only if $p > \frac{2}{1-s}$. Moreover, $\phi_{(s)}$ converges for $s \nearrow 1$, uniformly on compact subsets of G, to the constant function 1.

PROOF. As $F_{s,0}$ has K-weight 0, it is clear that $\phi_{(s)}(g) = \langle \gamma^s_g F_{s,0}, F_{s,0}\rangle$ is bi-K-invariant. For that reason it suffices to consider $g = a_t$ for $t \geqslant 0$ in the proof of the asymptotics of $\phi_{(s)}$. We calculate

$$\begin{aligned}
\phi_{(s)}(a_t) &= \langle \gamma^s_{a_t} F_{s,0}, F_{s,0}\rangle \\
&= \frac{1}{\pi^2}\int_0^\pi\int_0^\pi \frac{F_{s,0}(a_t^{-1}k_{\theta_1})F_{s,0}(k_{\theta_2})}{|\sin(\theta_1-\theta_2)|^{1-s}}\,\mathrm{d}\theta_1\,\mathrm{d}\theta_2 \\
&= \frac{1}{\pi}\int_0^\pi F_{s,0}(a_t^{-1}k_{\theta_1}) \underbrace{\frac{1}{\pi}\int_0^\pi \frac{1}{|\sin(\theta_1-\theta_2)|^{1-s}}\,\mathrm{d}\theta_2}_{C_s}\,\mathrm{d}\theta_1,
\end{aligned} \tag{9.62}$$

where the inner integral contributes a constant C_s only depending on s (because we may use the substitution $\psi = \theta_1 - \theta_2$ for θ_2). As in the proof of Proposition 8.40, we apply the Iwasawa decomposition to $a_t^{-1}k_\theta = k_\psi a_{t_0} u_{x_0}$ for varying $k_\theta \in K$, which determines $k_\psi \in K$, $a_{t_0} \in A$, and $u_{x_0} \in U$. For t_0 this gives

$$\mathrm{e}^{2t_0} = \|k_\psi a_{t_0} u_{x_0} e_1\|^2 = \|a_t^{-1}k_\theta e_1\|^2 = \mathrm{e}^{-2t}\cos^2\theta + \mathrm{e}^{2t}\sin^2\theta.$$

Using the definition of $F_{s,0}$ in Lemma 9.68, we have

$$F_{s,0}(a_t^{-1}k_\theta) = \mathrm{e}^{-(s+1)t_0} = \left(\sqrt{\mathrm{e}^{-2t}\cos^2\theta + \mathrm{e}^{2t}\sin^2\theta}\right)^{-(s+1)} \tag{9.63}$$

Notice that this implies that $F_{s,0}(a_t^{-1}k_{\theta_1})$ is unchanged if we replace θ_1 by $\pi-\theta_1$. Hence we may also replace the outer normalized integral over $[0,\pi]$ in (9.62) by the normalized integral over $[0,\frac{\pi}{2}]$, which gives

$$\phi_{(s)}(a_t) = \frac{2C_s}{\pi}\int_0^{\frac{\pi}{2}} F_{s,0}(a_t^{-1}k_\theta)\,\mathrm{d}\theta. \tag{9.64}$$

To study the asymptotics of the matrix coefficient $\phi(s)$ of $F_{s,0}$, we first note that (9.63) implies that

$$F_{s,0}(a_t^{-1}k_\theta) \asymp \max\left(\mathrm{e}^{-t}|\cos\theta|, \mathrm{e}^{t}|\sin\theta|\right)^{-(s+1)}.$$

We now replace $F_{s,0}(a_t^{-1}k_{\theta_1})$ in the integral (9.64) by this maximum. The latter is given by $\mathrm{e}^t\sin\theta_1$ unless θ_1 is very close to 0—specifically, unless $\tan\theta_1 < \mathrm{e}^{-2t}$. Therefore

$$\begin{aligned}
\phi_{(s)}(a_t) &\asymp_s \int_0^{\arctan \mathrm{e}^{-2t}} \big(\mathrm{e}^{-t}\underbrace{\cos\theta}_{\asymp 1}\big)^{-(s+1)}\,\mathrm{d}\theta + \int_{\arctan \mathrm{e}^{-2t}}^{\frac{\pi}{2}} \big(\mathrm{e}^{t}\underbrace{\sin\theta}_{\asymp\theta}\big)^{-(s+1)}\,\mathrm{d}\theta \\
&\asymp_s \mathrm{e}^{(s+1)t}\arctan \mathrm{e}^{-2t} + \mathrm{e}^{-(s+1)t}\frac{1}{-s}\theta^{-s}\Big|_{\arctan \mathrm{e}^{-2t}}^{\frac{\pi}{2}} \\
&\asymp_s \mathrm{e}^{(s+1)t}\mathrm{e}^{-2t} + \frac{1}{s}\mathrm{e}^{-(s+1)t}\mathrm{e}^{2st} \\
&\asymp_s \mathrm{e}^{(s-1)t} + \frac{1}{s}\mathrm{e}^{(s-1)t} \asymp_s \mathrm{e}^{(s-1)t},
\end{aligned}$$

which gives the claimed asymptotic.

Now let $p > 0$. Using the asymptotics and the decomposition of the Haar measure in (8.11), we obtain

$$\int_{\mathrm{SL}_2(\mathbb{R})} \phi_{(s)}(g)^p\,\mathrm{d}m(g) \asymp \int_0^\infty \phi_{(s)}(a_t)^p \sinh 2t\,\mathrm{d}t.$$

Since we are only interested in whether this integral converges, we restrict the integral to $[1,\infty)$, use the estimate $\sinh 2t \asymp \mathrm{e}^{2t}$ for $t\in[1,\infty)$, and the asymptotics for $\phi_s(a_t)$ to see that

$$\int_1^\infty \phi_{(s)}(a_t)^p \sinh 2t\,\mathrm{d}t \asymp_s \int_1^\infty \mathrm{e}^{t(s-1)p}\mathrm{e}^{2t}\,\mathrm{d}t.$$

Notice that the exponent

$$t(s-1)p + 2t = (2-(1-s)p)t$$

of the integrand has a negative coefficient if and only if $p > \frac{2}{1-s}$, and that this characterizes finiteness of the integral.

It remains to prove the final claim in the lemma concerning the behaviour of $\phi_{(s)}$ as $s \nearrow 1$. For this, first note that C_s as in (9.62) depends continuously on $s\in(0,1)$, and that

$$C_s = \frac{1}{\pi}\int_0^\pi \frac{1}{|\sin\theta|^{1-s}}\,\mathrm{d}\theta \longrightarrow C_1 = \frac{1}{\pi}\int_0^\pi \mathrm{d}\theta = 1$$

for $s \nearrow 1$ by dominated convergence. Next note that $F_{s,0}(a_t^{-1}k_\theta)$ as in (9.63) also makes sense for $s \in [\frac{1}{2}, 1]$ and depends continuously on

$$(\theta, s, t) \in [0, \tfrac{\pi}{2}] \times [\tfrac{1}{2}, 1] \times \mathbb{R}.$$

On restricting t to a compact interval I, uniform continuity implies that the function $\phi_{(s)}(a_t)$ as defined in (9.64) makes sense and depends continuously on $(s, t) \in [\frac{1}{2}, 1] \times I$. Therefore by uniform continuity again and (9.63) we have that $\phi_{(s)}(a_t)$ converges to the function

$$I \ni t \longmapsto \phi_{(1)}(a_t) = \frac{2C_1}{\pi} \int_0^{\frac{\pi}{2}} \frac{1}{\mathrm{e}^{-2t}\cos^2\theta + \mathrm{e}^{2t}\sin^2\theta} \,\mathrm{d}\theta.$$

uniformly on I as $s \nearrow 1$. Now note that

$$\frac{\mathrm{d}}{\mathrm{d}\theta}\big(\arctan(\mathrm{e}^{2t}\tan\theta)\big) = \frac{1}{1 + \mathrm{e}^{4t}\tan^2\theta}\mathrm{e}^{2t}\frac{1}{\cos^2\theta} = \frac{1}{\mathrm{e}^{-2t}\cos^2\theta + \mathrm{e}^{2t}\sin^2\theta},$$

which together with $C_1 = 1$ gives

$$\phi_{(1)}(a_t) = \frac{2}{\pi}\lim_{b\nearrow\frac{\pi}{2}} \big[\arctan(\mathrm{e}^{2t}\tan\theta)\big]_0^b = 1$$

as required. □

9.5.5 A Sobolev Space of the Projective Line*

Recall that the real projective line $\mathbb{P}^1(\mathbb{R}) = \mathbb{R}^2 \smallsetminus \{0\}/\sim$ is defined as the quotient space of $\mathbb{R}^2 \smallsetminus \{0\}$ modulo the equivalence relation $u_1 \sim u_2$ if u_1 and u_2 are scalar multiples of each other. Note that $\mathbb{P}^1(\mathbb{R}) = \mathbb{S}^1/\sim$ can also be obtained from the circle by identifying opposite points. Moreover, we may also identify the equivalence class $[k_\theta e_1]_\sim \in \mathbb{P}^1(\mathbb{R})$ for some $k_\theta \in K$ with $k_\theta M \in K/M$. Consequently functions on $\mathbb{P}^1(\mathbb{R})$ correspond to even functions on K. In this sense, Lemma 9.68 shows that $\mathcal{V}_s$ can be identified with $C^\infty(\mathbb{P}^1(\mathbb{R}))$. For the completion $\mathcal{H}_{(s)}$ of $\mathcal{V}_s$ this leads to the following result.

Proposition 9.75 (A Sobolev space). *Let $s \in (0, 1)$. The norm on $\mathcal{V}_s$ is equivalent to the L^2-Sobolev norm with $-\frac{s}{2}$ derivatives. Hence $\mathcal{H}_{(s)}$ is the L^2-Sobolev space $\mathcal{W}^{-\frac{s}{2},2}(\mathbb{P}^1(\mathbb{R}))$ with $-\frac{s}{2}$ derivatives.*

PROOF. We recall that the L^2-Sobolev space $\mathcal{W}^{-\frac{s}{2},2}(\mathbb{T})$ with $-\frac{s}{2}$ derivatives is defined as the completion of $C^\infty(\mathbb{T})$ with respect to the norm defined by

$$\|f\|^2_{-\frac{s}{2},2} = \sum_{n\in\mathbb{Z}} |c_n|^2 |n|^{-s}$$

for $f = \sum_{n\in\mathbb{Z}} c_n \chi_n$. In the case of $\mathbb{P}^1(\mathbb{R}) = \mathbb{S}^1/\sim \cong K/M \cong \mathbb{T}/\langle \frac{1}{2} + \mathbb{Z}\rangle$, we simply restrict to even functions and $n \in 2\mathbb{Z}$. Due to the orthogonality relations satisfied by $F_{s,n}$ with respect to $\langle\cdot,\cdot\rangle_{\mathcal{V}_s}$ and $\langle\cdot,\cdot\rangle_{-\frac{s}{2},2} = \langle\cdot,\cdot\rangle_{\mathcal{W}^{-\frac{s}{2},2}}$ for $n \in 2\mathbb{Z}$, the proposition is equivalent to the statement that

$$|n|^{-s} \ll \|F_{s,n}\|^2_{\mathcal{V}_s} \ll \|F_{s,n}\|^2_{-\frac{s}{2},2} = |n|^{-s} \tag{9.65}$$

for all $n \in 2\mathbb{Z}$. For this, recall that $\|F_{s,-n}\|^2_{\mathcal{V}_s} = \|F_{s,n}\|^2_{\mathcal{V}_s}$ for all $n \in \mathbb{N}$. Hence to prove (9.65) it suffices to calculate the asymptotics of $\|F_{s,n}\|^2_{\mathcal{V}_s}$ as $n \to \infty$, which will follow by combining Corollary 9.13, Lemma 9.72, and Stirling's formula for the gamma function.

Let $n \in \mathbb{Z}$ and apply Lemma 9.72 to $F_{s,2n}$ to obtain

$$\gamma^s_\partial(\mathbf{r}^+)F_{s,2n} = \frac{2n+1+s}{2} F_{s,2n+2}. \tag{9.66}$$

On the other hand, Corollary 9.13 gives

$$\|\gamma^s_\partial(\mathbf{r}^+)F_{s,2n}\|^2 = \tfrac{1}{4}\big((2n+1)^2 - s^2\big)\|F_{s,2n}\|^2. \tag{9.67}$$

Together, we obtain the recursion formula

$$\begin{aligned}
\|F_{s,2n+2}\|^2 &= \frac{4}{(2n+1+s)^2}\|\gamma^s_\partial(\mathbf{r}^+)F_{s,2n}\|^2 && \text{(by (9.66))}\\
&= \frac{(2n+1)^2 - s^2}{(2n+1+s)^2}\|F_{s,2n}\|^2 && \text{(by (9.67))}\\
&= \frac{2n+1-s}{2n+1+s}\|F_{s,2n}\|^2 = \frac{n+\frac{1-s}{2}}{n+\frac{1+s}{2}}\|F_{s,2n}\|^2
\end{aligned}$$

for the norms.

Now recall the Gamma function

$$\Gamma(x) = \int_0^\infty t^{x-1}\mathrm{e}^{-t}\,\mathrm{d}t$$

for $x > 0$, and that integration by parts shows that $\Gamma(x+1) = x\Gamma(x)$. We define $c = c_s > 0$ by the formula

$$\|F_{s,2}\|^2 = \frac{\Gamma\big(1+\frac{1-s}{2}\big)}{\Gamma\big(1+\frac{1+s}{2}\big)}c,$$

and prove by induction on $n \in \mathbb{N}$ that

$$\|F_{s,2n}\|^2 = \frac{\Gamma\big(n+\frac{1-s}{2}\big)}{\Gamma\big(n+\frac{1+s}{2}\big)}c. \tag{9.68}$$

Indeed the definition of c is the start of the induction, and the recursion formula gives the inductive step

$$\begin{aligned}\|F_{s,2n+2}\|^2 = \frac{n+\frac{1-s}{2}}{n+\frac{1+s}{2}}\|F_{s,2n}\|^2 &= \frac{\left(n+\frac{1-s}{2}\right)\Gamma\left(n+\frac{1-s}{2}\right)}{\left(n+\frac{1+s}{2}\right)\Gamma\left(n+\frac{1+s}{2}\right)}c \\ &= \frac{\Gamma\left(n+1+\frac{1-s}{2}\right)}{\Gamma\left(n+1+\frac{1+s}{2}\right)}c\end{aligned}$$

for $n \in \mathbb{N}$.

Next we recall Stirling's formula for the gamma function, which states that

$$\Gamma(x) \sim \sqrt{\frac{2\pi}{x}}\left(\frac{x}{\mathrm{e}}\right)^x,$$

where as usual $\sim$ means that the ratio of the left-hand and right-hand side converges to 1 as $x \to \infty$ (see Section B.2). We will also write $\approx$ to mean that the ratio converges to a positive constant depending on $s \in (0,1)$. Using Stirling's formula on (9.68) gives

$$\begin{aligned}\|F_{s,2n}\|^2 = \frac{\Gamma\left(n+\frac{1-s}{2}\right)}{\Gamma\left(n+\frac{1+s}{2}\right)}c &\sim \underbrace{\sqrt{\frac{n+\frac{1+s}{2}}{n+\frac{1-s}{2}}}}_{\sim 1}\frac{\left(\frac{n+\frac{1-s}{2}}{\mathrm{e}}\right)^{n+\frac{1-s}{2}}}{\left(\frac{n+\frac{1+s}{2}}{\mathrm{e}}\right)^{n+\frac{1+s}{2}}}c \\ &\approx \underbrace{\left(\frac{n+\frac{1-s}{2}}{n+\frac{1+s}{2}}\right)^{n+\frac{1-s}{2}}}_{\approx 1}\left(n+\frac{1+s}{2}\right)^{\frac{1-s}{2}-\frac{1+s}{2}} \\ &\approx n^{-s}.\end{aligned}$$

Taking the square root gives the desired asymptotic in (9.65) for $\|F_{s,2n}\|_{\mathcal{V}_s}$ as $n \to \infty$. □

Exercise 9.76. Let $s \in (0,1)$. Show that γ^s has decay exponent $1-s$ while allowing a degree two Sobolev norm.

9.6 The Fell Topology on the dual of $\mathrm{SL}_2(\mathbb{R})$*

We conclude our discussion of unitary representations of $\mathrm{SL}_2(\mathbb{R})$ by characterizing the Fell topology on the unitary dual $\widehat{\mathrm{SL}_2(\mathbb{R})}$ of $\mathrm{SL}_2(\mathbb{R})$.

Proposition 9.77. *The Fell topology on $\widehat{\mathrm{SL}_2(\mathbb{R})}$ is characterized by the following properties:*

- *For every $\rho \in \widehat{\mathrm{SL}_2(\mathbb{R})}$ the singleton $\{\rho\}$ is closed.*

- *For a sequence in* $\{\delta^{n,\pm} \mid n \in \mathbb{N}\}$ *to converge it must be eventually constant.*
- *A sequence in* $\{\pi^{\xi,\mathrm{e}} \mid \xi \geqslant 0\}$ *converges if and only if the associated parameter* $\xi \in [0,\infty)$ *converges within* $[0,\infty)$.
- *A sequence in* $\{\pi^{\xi,\mathrm{o}} \mid \xi > 0\}$ *converges if and only if the associated parameter* $\xi \in (0,\infty)$ *converges within* $[0,\infty)$. *For* $\xi \searrow 0$ *the limit points are* $\delta^{1,+}$ *and* $\delta^{1,-}$.
- *A sequence in* $\{\gamma^s \mid s \in (0,1)\}$ *converges if and only if the associated parameter* $s \in (0,1)$ *converges within* $[0,1]$. *For* $s \searrow 0$ *the only limit is* $\pi^{0,\mathrm{e}}$. *For* $s \nearrow 1$ *the limit points are* $\mathbb{1}_{\mathrm{SL}_2(\mathbb{R})}$, $\delta^{2,+}$, *and* $\delta^{2,-}$.

In all cases the limits are the expected ones within the same parameter subset unless specifically said otherwise.

We summarize the properties in Proposition 9.77 in Figure 9.10 for the even part of the spectrum and in Figure 9.11 for the odd part of the spectrum.

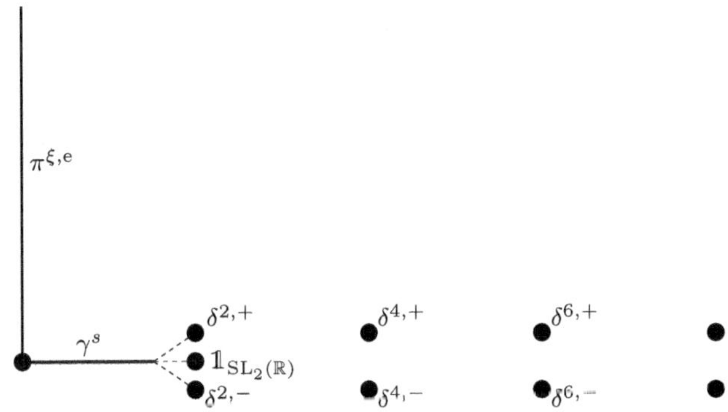

Fig. 9.10: The even part of $\widehat{\mathrm{SL}_2(\mathbb{R})}$ consists of an L shape (of infinite height and finite width) and countably many points that have no limits. The only failure of the Hausdorff property is in the bottom right tip of the L shape, where γ^s has three limits for $s \nearrow 1$.

The following lemma is an important first step in understanding the Fell topology on $\widehat{\mathrm{SL}_2(\mathbb{R})}$. For this result, recall that for any $\pi \in \widehat{\mathrm{SL}_2(\mathbb{R})}$ and $n \in \mathbb{Z}$ the subspace of vectors in $\mathcal{H}_\pi$ of K-weight n is at most one-dimensional. Hence if $\mathcal{H}_\pi$ contains non-zero vectors of K-weight and we choose a unit vector $v \in \mathcal{H}_\pi$ of K-weight n, then the diagonal matrix coefficient $\phi_n^\pi = \varphi_v^\pi$ only depends on π and n.

Lemma 9.78. *Let* $\pi \in \widehat{\mathrm{SL}_2(\mathbb{R})}$ *and let* (ρ_ℓ) *be a sequence in* $\widehat{\mathrm{SL}_2(\mathbb{R})}$. *Then* (ρ_ℓ) *converges to* π *as* $\ell \to \infty$ *in the Fell topology if and only if for one (or, equivalently, for all)* $n \in \mathbb{Z}$ *for which* $\mathcal{H}_\pi$ *contains non-trivial vectors of* K*-weight* n *we have that*

- $\mathcal{H}_{\rho_\ell}$ *contains a non-zero vector of* K*-weight* n *for all sufficiently large* ℓ, *and*

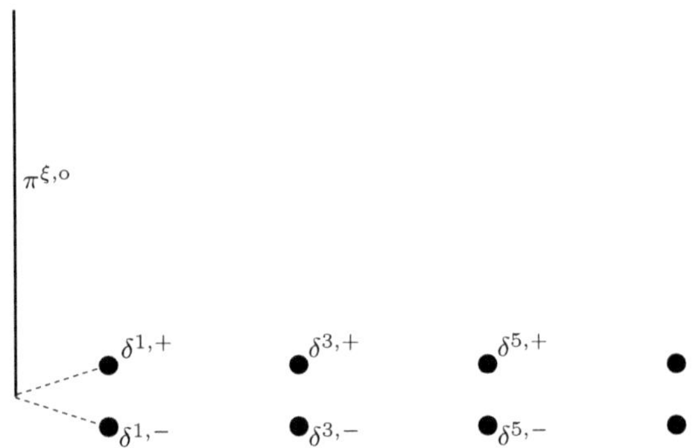

Fig. 9.11: The odd part of $\widehat{\mathrm{SL}_2(\mathbb{R})}$ consists of a line segment and countably many points that have no limits. The only failure of the Hausdorff property is at the bottom tip of the line segment where $\pi^{\xi,o}$ has two limits for $\xi \searrow 0$.

- *$\phi_n^{\rho_\ell} \to \phi_n^\pi$ in the compact-open topology as $\ell \to \infty$.*

If $\mathcal{H}_\pi$ contains non-zero vectors of K-weight n for all even $n \in 2\mathbb{Z}$ (or for all odd $n \in 2\mathbb{Z}+1$) and (ρ_ℓ) converges to π as $\ell \to \infty$, then π is the only limit of (ρ_ℓ) for $\ell \to \infty$.

Proof. Suppose first that (ρ_ℓ) converges to π as $\ell \to \infty$ and let $n \in \mathbb{Z}$ be such that $\mathcal{H}_\pi$ contains a unit vector v of K-weight n. Using Proposition 6.61 we see that there exists a sequence of unit vectors (w_ℓ) with $w_\ell \in \mathcal{H}_{\rho_\ell}$ so that $\varphi_{w_\ell}^{\rho_\ell}$ converges to φ_v^π in the compact-open topology as $\ell \to \infty$. We define $\widetilde{w}_\ell$ as the projection of w_ℓ to the subspace of K-eigenvectors with K-weight n. As in Lemma 8.43, we now obtain that $\phi_n^\pi = \varphi_v^\pi$ is still the limit of $\varphi_{\widetilde{\rho}_\ell}^{\rho_\ell}$. In particular $\|\widetilde{w}_\ell\|^2 \to \|v\|^2 = 1$ as $\ell \to \infty$ and we may normalize $\widetilde{w}_\ell$ without changing the limit. This gives the first implication.

The converse follows directly from Proposition 6.61 as follows. Since

$$p\colon \mathcal{E}^1(\mathrm{SL}_2(\mathbb{R})) \to \widehat{\mathrm{SL}_2(\mathbb{R})}$$

is continuous, we have $\rho_\ell = p(\phi_n^{\rho_\ell}) \to \pi = p(\phi_n^\pi)$ as $\ell \to \infty$.

Suppose now $\pi, \tau \in \widehat{\mathrm{SL}_2(\mathbb{R})}$, that $\mathcal{H}_\pi$ contains non-zero vectors of K-weight n for all $n \in 2\mathbb{Z}$, that the sequence (ρ_ℓ) in $\widehat{\mathrm{SL}_2(\mathbb{R})}$ converges both to π and to τ as $\ell \to \infty$. Applying the above to π we find some $n_0 \in 2\mathbb{Z}$ so that $\mathcal{H}_{\rho_\ell}$ contains non-zero vectors of K-weight n_0 for all sufficiently large ℓ. In particular, we obtain for those ℓ that ρ_ℓ is an even representation. Applying the above to τ it follows that τ is even and that there exists some $n \in 2\mathbb{Z}$ so that $\phi_n^{\rho_\ell} \to \phi_n^\tau$ as $\ell \to \infty$. However, as $\mathcal{H}_\pi$ also contains non-zero vectors of K-weight n we also have $\phi_n^{\rho_\ell} \to \phi_n^\pi$ as $\ell \to \infty$. Hence $\phi_n^\tau = \phi_n^\pi$ and so $\tau = \pi$ by Proposition 1.65. The argument for π odd is identical. □

Before starting the proof of Proposition 9.77, we note that most of the convergence claims could be understood by various instances of dominated convergence together with Lemma 9.78. However, we will give a different argument that handles all the cases at once.

PROOF OF PROPOSITION 9.77. We first show that a set of the form $\{\rho\}$ is closed for any $\rho \in \mathrm{SL}_2(\mathbb{R})$. Indeed, we may let $\rho_\ell = \rho$ for $\ell \in \mathbb{N}$ and then apply Lemma 9.78. If now π is a limit point, then there exists some $n \in \mathbb{Z}$ so that $\phi_n^\pi = \phi_n^\rho$. However, this shows that $\pi = \rho$ by Proposition 1.65.

Recall that the Fell topology on $\widehat{\mathrm{SL}_2(\mathbb{R})}$ is second countable by Lemma 6.44. Hence it suffices to study limits of sequences as in the proposition in order to characterize the Fell topology. Also note that the set of limits of a sequence (ρ_ℓ) is determined as the intersection of the set of limits of its subsequences if the sequence (ρ_ℓ) is split into finitely many subsequences. This observation allows us to restrict to sequences in $\{\delta^{n,\pm} \mid n \in \mathbb{N}\}$, in $\{\pi^{\xi,\mathrm{e}} \mid \xi \geqslant 0\}$, in $\{\pi^{\xi,\mathrm{o}} \mid \xi > 0\}$, or in $\{\gamma^s \mid s \in (0,1)\}$ as done in the proposition.

The case of a sequence (ρ_ℓ) in $\{\delta^{n,\pm} \mid n \in \mathbb{N}\}$ now follows quite directly. If (ρ_ℓ) is eventually constant, then the only limit is the obvious one (by the first part of the proof). If (ρ_ℓ) is not eventually constant and takes on only finitely many values, we can find two eventually constant subsequences with disjoint sets of limits. It remains to study the case of (ρ_ℓ) taking on infinitely many values. In this case we find an injective subsequence. We claim that this subsequence has no limits, which shows the same property for (ρ_ℓ). Suppose, for the purposes of the claim, that (ρ_ℓ) is injective. The properties of the (mock) discrete series representation in Theorem 8.23 (Theorem 8.31) combined with Lemma 9.78 contradict convergence: If the sequence had a limit, then there would exist a weight $n \in \mathbb{Z}$ so that $\mathcal{H}_{\rho_\ell}$ contains a unit vector of K-weight n for all sufficiently large ℓ. However, this only holds for finitely many (mock) discrete series representations.

It remains to understand limits of sequences with values in $S \subseteq \widehat{\mathrm{SL}_2(\mathbb{R})}$, where S consists either of the even or the odd principal series, or the complementary series. Fix some $n \in \mathbb{Z}$ and let $\rho \in S$. We suppose that $\mathcal{H}_\rho$ contains a unit vector v of K-weight n (that is, that n has the same parity as ρ). Then $\phi\colon \mathbb{R} \to \mathbb{C}$ defined by $\phi(t) = \langle \rho_{a_t} v, v\rangle$ for $t \in \mathbb{R}$ satisfies

$$\phi^{(j)}(t) = \big\langle \rho(a_t)\rho_\partial(\mathbf{a}^{\circ j})v, v\big\rangle$$

for $j \in \mathbb{N}_0$ and $t \in \mathbb{R}$ by (9.15). Moreover, the proof of claim (A) on page 423 (see (9.16) in particular) shows that we can bound $\|\phi^{(j)}\|_\infty$ in terms of a bound on n, α_ρ, and $j \in \mathbb{N}_0$ and that this bound implies that ϕ is analytic.

Suppose now that (ρ_ℓ) is a sequence in S and assume that $\alpha_{\rho_\ell} \to \alpha \in \mathbb{R}$ as $\ell \to \infty$. Let $v_\ell \in \mathcal{H}_{\rho_\ell}$ be a unit vector of K-weight $n \in \mathbb{Z}$ and let

$$\phi_\ell(t) = \langle \rho_\ell(a_t)v_\ell, v_\ell\rangle$$

for $t \in \mathbb{R}$. Then the above shows that $\|\phi_\ell^{(j)}\|_\infty \leqslant C_j$ with a uniform constant C_j for $j \in \mathbb{N}_0$ and $\ell \in \mathbb{N}$. However, this shows that $\phi_\ell^{(j)}$ is equicontinuous for $j \in \mathbb{N}_0$. Hence we can apply the Arzela–Ascoli theorem and find a subsequence along which the function and its derivatives converge with respect to the compact-open topology. Let ϕ denote the limit of this subsequence.

By the proof of claim (A) on page 423, ϕ is analytic. However, claim (D) shows even more: The derivatives $\phi_\ell^{(j)}(0)$ for $j \in \mathbb{N}_0$ are determined by α_{ρ_ℓ} and n by complicated but concrete recursive formulas. This implies that $\phi^{(j)}(0)$ for $j \in \mathbb{N}_0$ is similarly determined by α and n. As the limit of the converging subsequence of ϕ_ℓ does not depend on the subsequence, we see that ϕ_ℓ converges to ϕ.

In all cases ϕ is now the matrix coefficient of a K-weight n unit vector of an irreducible representation:

- Suppose $S = \{\pi^{\xi,\mathrm{e}} \mid \xi \geqslant 0\}$ and $\rho_\ell = \pi^{\xi_\ell,\mathrm{e}}$ for all $\ell \geqslant 1$. Then $n \in \mathbb{Z}$ is even, and $\alpha_\ell = -\xi_\ell^2 \to \alpha = -\xi^2$ as $\ell \to \infty$ for some $\xi \geqslant 0$. Note that $\mathcal{H}_\xi^{\mathrm{even}}$ also contains a unit vector v of K-weight n for which the analytic function $\mathbb{R} \ni t \mapsto \langle \pi_{a_t}^{\xi_\ell,\mathrm{e}} v, v\rangle$ has the same derivatives at $t = 0$ as ϕ. It follows that the expected limit $\pi^{\xi,\mathrm{e}}$ is the only limit of $(\pi^{\xi_\ell,\mathrm{e}})$.
- Suppose $S = \{\pi^{\xi,\mathrm{o}} \mid \xi > 0\}$ and $\rho_\ell = \pi^{\xi_\ell,\mathrm{o}}$ for all $\ell \geqslant 1$. Then $n \in \mathbb{Z}$ is odd, and $\alpha_\ell = -\xi_\ell^2 \to \alpha = -\xi^2$ as $\ell \to \infty$ for some $\xi \geqslant 0$. If $\xi > 0$, then just as in the even case the expected limit $\pi^{\xi,\mathrm{o}}$ is the only limit of $(\pi^{\xi_\ell,\mathrm{o}})$. If $\alpha = \xi = 0$, then the same arguments show that ϕ is the restriction of the matrix coefficient of a vector in $\mathcal{H}_{\delta^{1,+}}$ to A if $n > 0$ and otherwise of a vector in $\mathcal{H}_{\delta^{1,-}}$. Therefore the limits of $(\pi^{\xi_\ell,\mathrm{o}})$ for $\xi_\ell \searrow 0$ are $\delta^{1,+}$ and $\delta^{1,-}$.
- Suppose $S = \{\gamma^s \mid s \in (0,1)\}$ and $\rho_\ell = \gamma^{s_\ell}$ for all $\ell \geqslant 1$. Then $n \in \mathbb{Z}$ is even, and $\alpha_\ell = s_\ell^2 \to \alpha = s^2$ for some $s \in [0,1]$. If $s \in (0,1)$, then the expected limit γ^s is the only limit of (γ^{s_ℓ}). If $s = 0$, then ϕ is the restriction of the matrix coefficient of a vector in $\mathcal{H}_{\pi^{0,\mathrm{e}}}$. Hence (γ^{s_ℓ}) converges to $\pi^{0,\mathrm{e}}$ as $s_\ell \searrow 0$. If $s = 1$ and $n = 0$, then $\phi = 1$. If $s = 1$ and $n > 0$ (resp. $n < 0$) then ϕ is the restriction of the matrix coefficient of a vector in $\mathcal{H}_{\delta^{2,+}}$ (resp. in $\mathcal{H}_{\delta^{2,-}}$). Therefore the limits of γ^{s_ℓ} for $s_\ell \nearrow 1$ are $\mathbb{1}_{\mathrm{SL}_2(\mathbb{R})}$, $\delta^{2,+}$, and $\delta^{2,-}$.

Using subsequences as in the argument for the (mock) discrete series representation above, this argument handles all cases for which the Casimir eigenvalue is bounded.

It remains to show that the sequences $(\pi^{\xi_\ell,\mathrm{e}})$ or $(\pi^{\xi_\ell,\mathrm{o}})$ with $\xi_\ell \to \infty$ have no limit points. Hence we fix some $n \in \mathbb{Z}$ and consider

$$\phi_n^\xi(t) = \left\langle \pi_{a_t}^\xi F_{n,\xi}, F_{n,\xi} \right\rangle$$

for $t \in \mathbb{R}$. We claim that after some manipulations of the integral defining ϕ_n^ξ we are able to apply the Riemann–Lebesgue lemma to see that $\phi_n^\xi(t) \to 0$ as $\xi \to \infty$ for any $t \in \mathbb{R} \smallsetminus \{0\}$. By Lemma 9.78 this then implies that $\pi^{\xi,\mathrm{e}}$ and $\pi^{\xi,\mathrm{o}}$ have no limit points for $\xi \to \infty$. So let $t \neq 0$ and apply the definitions of $\langle\cdot,\cdot\rangle_{\mathcal{H}_\xi}$ and of $F_{n,\xi}$ as in (9.19) to obtain

$$\phi_n^\xi(t) = \frac{1}{2\pi}\int_0^{2\pi} F_{n,\xi}(a_t^{-1}k_\theta)\overline{F_{n,\xi}(k_\theta)}\,\mathrm{d}\theta = \frac{1}{\pi}\int_0^{\pi} \mathrm{e}^{-\mathrm{i}n\psi-\mathrm{i}\xi t_0-t_0}\mathrm{e}^{\mathrm{i}n\theta}\,\mathrm{d}\theta,$$

where we use the same notation as in the proof of Lemma 9.34 as follows. The functions $\psi\in[0,2\pi)$, $t_0,x_0\in\mathbb{R}$ are uniquely determined by

$$\begin{pmatrix}\mathrm{e}^{-t} & \\ & \mathrm{e}^{t}\end{pmatrix}k_\theta = k_\psi a_{t_0}u_{x_0},$$

which leads to

$$\mathrm{e}^{2t_0} = \mathrm{e}^{-2t}\cos^2\theta + \mathrm{e}^{2t}\sin^2\theta, \tag{9.69}$$

the fact that ψ and θ correspond to the directions in the same quadrant, and that $\tan\psi = \mathrm{e}^{2t}\tan\theta$.

However, unlike the situation in many of our discussions earlier, we are now interested in these expressions for a fixed $t\in\mathbb{R}\smallsetminus\{0\}$ and for $\xi\to\infty$. Note that we can use (9.69) to define $t_0 = t_0(\theta) = \frac{1}{2}\log\left(\mathrm{e}^{-2t}\cos^2\theta + \mathrm{e}^{2t}\sin^2\theta\right)$. For example, by taking the derivative, it is easy to see that on each of the intervals $[a,b]$ of the form $[0,\frac{\pi}{2}]$ and $[\frac{\pi}{2},\pi]$ the function $[a,b]\ni\theta\mapsto t_0(\theta)$ is a bijection with inverse $I\ni t_0\mapsto\theta(t_0)$ defined on a bounded interval $I\subseteq\mathbb{R}$. Considering only one of these intervals, we use a substitution to obtain

$$\int_a^b \mathrm{e}^{-\mathrm{i}n\psi-\mathrm{i}\xi t_0-t_0}\mathrm{e}^{\mathrm{i}n\theta}\,\mathrm{d}\theta = \int_I \underbrace{\mathrm{e}^{-\mathrm{i}n\psi(\theta(t_0))-t_0+\mathrm{i}n\theta(t_0)}\,\frac{\mathrm{d}\theta}{\mathrm{d}t_0}(t_0)}_{f(t_0)}\,\mathrm{e}^{-\mathrm{i}\xi t_0}\,\mathrm{d}t_0.$$

As I is bounded and

$$\int_I \frac{\mathrm{d}\theta}{\mathrm{d}t_0}\,\mathrm{d}t_0 = b-a,$$

we see that $f\in L^1(I)$. Hence we obtain from the Riemann–Lebesgue lemma that this integral converges to 0 as $\xi\to\infty$. It follows that $\phi_n^\xi(t)\to 0$ as $\xi\to 0$ for every $n\in\mathbb{Z}$ and $t\in\mathbb{R}\smallsetminus\{0\}$ as claimed earlier. □

9.7 Spectral Gap, Decay, and Integrability Exponents

Using the complete description of $\widehat{\mathrm{SL}_2(\mathbb{R})}$ and properties of the complementary series representation, we can upgrade the results concerning integrability and decay exponents from Section 8.6.

Definition 9.79 (Three spectral parameters). Let π be a unitary representation of $\mathrm{SL}_2(\mathbb{R})$. We define the *almost decay exponent of* π to be

$$\kappa_\pi = \sup\{\kappa \in [0,1] \mid \kappa \text{ is a decay exponent of } \pi\}$$

and the *almost integrability exponent of* π to be

$$p_\pi = \inf\{p \in [2,\infty] \mid p \text{ is an integrability exponent of } \pi\}.$$

Finally the *complementary series parameter* of π is defined by

$$s_\pi = \sup\{s \in [0,1) \mid s = 0 \text{ or } \gamma^s \prec \pi\}.$$

Theorem 9.80. *Let* π *be a unitary representation of* $\mathrm{SL}_2(\mathbb{R})$. *Then*

$$\kappa_\pi = \frac{2}{p_\pi} = 1 - s_\pi. \tag{9.70}$$

For $\kappa < \kappa_\pi$ *and any* $g \in G$ *we have*

$$\left|\langle \pi_g v, w\rangle\right| \ll_{\kappa_\pi,\kappa} \|v\|\|w\|\|g\|_{\mathrm{HS}}^{-\kappa} \tag{9.71}$$

for any K*-eigenvectors* $v, w \in \mathcal{H}_\pi$, *and*

$$\left|\langle \pi_g v, w\rangle\right| \ll_{\kappa_\pi,\kappa} \mathcal{S}(v)\,\mathcal{S}(w)\|g\|_{\mathrm{HS}}^{-\kappa} \tag{9.72}$$

for any C^1*-smooth vectors* $v, w \in \mathcal{H}_\pi$. *Moreover,* $\kappa_\pi > 0$, $p_\pi < \infty$, *and* $s_\pi < 1$ *are all equivalent to* π *having spectral gap.*

We note that, even for the proof of the first equality concerning κ_π and p_π, knowledge about the complementary series will be useful. We again write G for $\mathrm{SL}_2(\mathbb{R})$.

PROOF OF (9.71)–(9.72) AND THAT $\kappa_\pi = \frac{2}{p_\pi}$. Suppose that π is a unitary representation of $\mathrm{SL}_2(\mathbb{R})$. Comparing the definition of integrability and decay exponents in Definitions 7.25 and 8.44, we may assume that π has no fixed vectors.

By the definition of κ_π as a supremum and Lemma 8.46 any $p > \frac{2}{\kappa_\pi}$ is an integrability exponent. As $\kappa_\pi \leqslant 1$ and p_π is the infimum over all integrability exponents in $[2,\infty)$ we have $p_\pi \leqslant \frac{2}{\kappa_\pi}$ or, equivalently,

$$\kappa_\pi \leqslant \frac{2}{p_\pi}. \tag{9.73}$$

To prove the opposite inequality, suppose that $p \geqslant 2$ is such that π is p-integrable. This means that there exists a dense set of vectors $\mathcal{V} \subseteq \mathcal{H}_\pi$ so that $\varphi^\pi_{v,w} \in L^p(G)$ for all $v, w \in \mathcal{V}$. Let $\varepsilon > 0$, fix a positive $\widetilde{s} \in \left(\frac{2}{p} - \varepsilon, \frac{2}{p}\right)$, and note that $\widetilde{s} \in (0,1)$ since $p \geqslant p_\pi \geqslant 2$. We claim that this makes the inner tensor product $\pi \otimes \gamma^{\widetilde{s}}$ tempered.

Assuming the claim for now, we let $v, w \in \mathcal{H}_\pi$ be K-eigenvectors and let $F_{\widetilde{s},0}$ be the spherical function as in Lemma 9.68. Then both $v \otimes F_{\widetilde{s},0}$ and $w \otimes F_{\widetilde{s},0}$ are also K-eigenvectors. Therefore temperedness of $\pi \otimes \gamma^{\widetilde{s}}$ and Theorem 8.42(2), together with the estimate for the Harish-Chandra spherical function in Theorem 8.32, give

$$\begin{aligned}\left|\langle\pi_g v, w\rangle\right| \phi_{(\widetilde{s})}(g) &= \left|\langle(\pi\otimes\gamma^{\widetilde{s}})_g v\otimes F_{\widetilde{s},0}, w\otimes F_{\widetilde{s},0}\rangle\right| \\ &\ll_\varepsilon \|v\otimes F_{\widetilde{s},0}\|\|w\otimes F_{\widetilde{s},0}\|\|g\|_{\mathrm{HS}}^{-1+\varepsilon} \\ &\ll_{\varepsilon,\widetilde{s}} \|v\|\|w\|\|g\|_{\mathrm{HS}}^{-1+\varepsilon}\end{aligned}$$

for all $g\in G$, where $\phi_{(\widetilde{s})}$ is the matrix coefficient of $F_{\widetilde{s},0}$. Together with the lower bound for $\phi_{(\widetilde{s})}$ in Lemma 9.74, we obtain after dividing by $\phi_{(\widetilde{s})}$ the estimate

$$\left|\langle\pi_g v, w\rangle\right| \ll_{\varepsilon,\widetilde{s}} \|v\|\|w\|\|g\|_{\mathrm{HS}}^{-\widetilde{s}+\varepsilon}.$$

Recalling the assumption $\widetilde{s}\in\left(\frac{2}{p}-\varepsilon,\frac{2}{p}\right)$ we also obtain

$$\left|\langle\pi_g v, w\rangle\right| \ll_{p,\varepsilon} \|v\|\|w\|\|g\|_{\mathrm{HS}}^{-\frac{2}{p}+2\varepsilon}$$

for all K-eigenvectors $v,w\in\mathcal{H}_\pi$. By Proposition 7.29 this upgrades automatically to all C^1-smooth vectors $v,w\in\mathcal{H}_\pi$ if we replace the norm of v,w by the degree-one Sobolev norms of v,w (and multiply the implicit constant by an absolute constant). Hence $\kappa=\frac{2}{p}-2\varepsilon$ is a decay exponent for π satisfying (9.71) and (9.72). Recalling that $\varepsilon>0$ and $p\geqslant 2$ with π being p-integrable were arbitrary, we see that the claim implies that $\kappa_\pi\geqslant\frac{2}{p_\pi}$. Together with (9.73), this gives the desired equality.

Turning to the claim that $\pi\otimes\gamma^{\widetilde{s}}$ is tempered, notice first that the linear hull $\langle\gamma^{\widetilde{s}}(G)F_{\widetilde{s},0}\rangle$ of the G-orbit of $F_{\widetilde{s},0}$ is dense in $\mathcal{H}_{(\widetilde{s})}$ by irreducibility of the complementary series representation. Therefore $\langle\mathcal{V}\rangle\otimes_{\mathrm{la}}\langle\gamma^{\widetilde{s}}(G)F_{\widetilde{s},0}\rangle$ is dense in $\mathcal{H}_\pi\otimes\mathcal{H}_{(\widetilde{s})}$ and, by sesqui-linearity of matrix coefficients, it suffices to consider the matrix coefficient ϕ of $v\otimes\gamma^{\widetilde{s}}_{g_1}F_{\widetilde{s},0}$ and $w\otimes\gamma^{\widetilde{s}}_{g_2}F_{\widetilde{s},0}$. This gives

$$\phi(g)=\langle\pi_g v,w\rangle\langle\gamma^{\widetilde{s}}_{gg_1}F_{\widetilde{s},0},\gamma^{\widetilde{s}}_{g_2}F_{\widetilde{s},0}\rangle=\varphi^\pi_{v,w}(g)\lambda_{g_2}\rho_{g_1}\phi_{(\widetilde{s})}(g).$$

By assumption, $\varphi^\pi_{v,w}\in L^p(G)$ and, by Lemma 9.74, $\phi_{(\widetilde{s})}\in L^{\frac{2}{1-\frac{2}{p}}}$ since $\widetilde{s}<\frac{2}{p}$ implies $\frac{2}{1-\widetilde{s}}<\frac{2}{1-\frac{2}{p}}$. Using (B.1), this implies that ϕ belongs to $L^q(G)$ for

$$q=\frac{p\frac{2}{1-\frac{2}{p}}}{p+\frac{2}{1-\frac{2}{p}}}=\frac{2p}{p(1-\frac{2}{p})+2}=2,$$

which proves the claim. □

Corollary 9.81. *Let $s\in(0,1)$. Then the complementary series γ^s has almost decay exponent*

$$\kappa_{\gamma^s}=1-s.$$

Proof. Applying the already established first part of Theorem 9.80 to γ^s and the vector $F_{s,0} \in \mathcal{H}_s$, we obtain the upper bound in

$$\|g\|_{\mathrm{HS}}^{s-1} \ll_s |\phi_{(s)}(g)| \ll_{s,\kappa} \|g\|_{\mathrm{HS}}^{-\kappa} \tag{9.74}$$

for all $g \in G$ and any $\kappa < \kappa_{\gamma^s}$. The lower bound in (9.74) comes from Lemma 9.74. Together we obtain $s-1 \leqslant -\kappa$ or, equivalently, $\kappa \leqslant 1-s$ by letting $g \to \infty$. Since $\kappa < \kappa_{\gamma^s}$ was arbitrary, this gives $\kappa_{\gamma^s} \leqslant 1-s < 1$.

By Lemma 9.74, $\phi_{(s)} = \varphi^{\gamma^s}_{F_{s,0}}$ belongs to $L^p(G)$ precisely when $p > \frac{2}{1-s}$. By Exercise 8.45 and irreducibility of $\gamma^{(s)}$, this implies that γ^s is p-integrable for all $p > \frac{2}{1-s}$ and so

$$p_{\gamma^s} \leqslant \frac{2}{1-s}.$$

Together with the first part of Theorem 9.80 again, we obtain from this

$$\kappa_{\gamma^s} = \frac{2}{p_{\gamma^s}} \geqslant 1-s,$$

and hence the corollary. □

Concluding the proof of Theorem 9.80. We start by proving the inequality $\kappa_\pi \leqslant 1 - s_\pi$. For this we suppose that $\gamma^s \prec \pi$ for some $s \in (0,1)$ and $\kappa < \kappa_\pi$. By the first part of the theorem, we know that κ satisfies the decay estimate (9.71) for all K-eigenvectors $v, w \in \mathcal{H}_\pi$. Using Lemma 8.43 just as in the proof of (1) $\Longrightarrow$ (2) in Theorem 8.42 on page 378, it follows that κ is also a decay exponent for γ^s. By Corollary 9.81 this implies $\kappa \leqslant \kappa_{\gamma^s} = 1-s$. As $\kappa < \kappa_\pi$ and $s \in (0,1)$ with $\gamma^s \prec \pi$ were arbitrary, we obtain

$$\kappa_\pi \leqslant 1 - s_\pi. \tag{9.75}$$

For this, we also note that (9.75) holds trivially if there is no complementary series γ^s weakly contained in π, since in this case $s_\pi = 0$.

Note that if $s_\pi = 1$ then (9.75) shows that $\kappa_\pi = 0$ and so there is equality in (9.75). If $s_\pi = 0$ (that is, if no complementary series are weakly contained in π) then π is tempered and so $\kappa_\pi = 1$. This follows, for example, from the argument in Corollary 9.29 but also from the discussion below. So we now suppose that $s_\pi \in [0,1)$ and claim that there exists a countable set $S \subseteq (0, s_\pi]$ (with $S = \emptyset$ if $s_\pi = 0$) so that

$$\pi \prec \lambda \oplus \bigoplus_{s \in S} \gamma^s. \tag{9.76}$$

To prove the claim let $v \in \mathcal{H}_\pi$ be a unit vector, $Q \subseteq G$ a compact set, and $\varepsilon > 0$. Applying Proposition 6.37 we can find finitely many irreducible representations $\pi_j \prec \pi$ and vectors $v_j \in \mathcal{H}_{\pi_j}$ for $j = 1, \ldots, n$ so that $\sum_{j=1}^n \|v_j\|^2 = 1$ and φ^π_v is equal to $\sum_{j=1}^n \varphi^{\pi_j}_{v_j}$ on Q up to $\mathrm{O}(\varepsilon)$. In Theorem 9.22 we found all irreducible representations of $G = \mathrm{SL}_2(\mathbb{R})$, and by Table 9.1 (see Theorem 8.23, Theorem 8.31, and Theorem 9.32) we have $\pi_j \prec \lambda$ or $\pi_j = \gamma^s$ for some $s \in (0, s_\pi]$. In the former case we may apply the definition of weak containment $\pi_j \prec \lambda$ and replace $\varphi_{v_j}^{\pi_j}$ by a sum of matrix coefficients for the regular representation. In other words, we may assume instead that $\pi_j = \lambda$ or $\pi_j = \gamma^s$ for some $s \in (0, s_\pi]$.

We now vary v within a dense countable subset of the unit sphere in $\mathcal{H}_\pi$, set $Q = B_n^{\|\cdot\|_{\mathrm{HS}}}$ and $\varepsilon = \frac{1}{n}$ for $n \in \mathbb{N}$. This way we obtain a subset $S \subseteq (0, s_\pi]$ that is at most countable so that (9.76) holds by the definition of weak containment in Definition 6.1.

Let $\kappa \in (0, 1 - s_\pi)$ so that $\kappa < 1 - s$ for all $s \in S$. By Corollary 9.81 this shows that κ is a decay exponent for γ^s for all $s \in S$. In fact, we have

$$|\langle \gamma^s v, w \rangle| \ll \|v\| \|w\| \|g\|_{\mathrm{HS}}^{-\kappa} \tag{9.77}$$

for all $s \in S$ and K-eigenfunctions $v, w \in \mathcal{H}_{(s)}$. As $\kappa < 1 - s_\pi \leqslant 1$ the estimate (9.77) holds similarly for λ. This implies the same estimate for the representation $\lambda \oplus \bigoplus_{s \in S} \gamma^s$ (see Exercise 9.82).

Using (9.76), applying Lemma 8.43, and using the argument for (1) $\Longrightarrow$ (2) in the proof of Theorem 8.42 on page 378, we obtain

$$\left|\left\langle \pi_g v, w \right\rangle\right| \ll \|v\| \|w\| \|g\|_{\mathrm{HS}}^{-\kappa}$$

for any K-eigenvectors $v, w \in \mathcal{H}_\pi$ (without changing the implicit constant), which once more implies (9.72) for C^1-smooth vectors. We therefore see that any $\kappa < 1 - s_\pi$ is a decay exponent, which gives $\kappa_\pi \geqslant 1 - s_\pi$ for the supremum, and hence equality in (9.75).

This completes the proof of (9.70) except for a tiny detail that we have intentionally kept hidden until now. To prove that $\kappa \in (0, 1 - s_\pi)$ is a decay exponent for π, we used the estimate (9.77) but did not discuss the dependency of the implicit multiplicative constant on s in S. (Note that Corollary 9.81 makes no claim concerning this.) Assuming that we can choose the implicit constant so that it does not depend on $s \in S$ but only on κ and s_π, the above argument applies as explained.

To see that (9.77) holds with an implicit constant that only depends on κ and s_π we will review the proof of Corollary 9.81. So let $s \in S$. By Lemma 9.74 we have that $|\phi(s)|$ belongs to $L^p(G)$ if and only if $p > \frac{2}{1-s}$, which implies that γ^s is p-integrable for $p > \frac{2}{1-s}$ by Exercise 8.45. Note that $\kappa < 1 - s_\pi$ and $s \leqslant s_\pi$ imply $\frac{2}{1-s} \leqslant \frac{2}{1-s_\pi} < \frac{2}{\kappa}$, which allows us to fix some $p \in \left(\frac{2}{1-s_\pi}, \frac{2}{\kappa}\right)$ so that γ^s is p-integrable for all $s \in S$. Since $p < \frac{2}{\kappa}$ we may also fix some $\widetilde{s}$ in $\left(\kappa, \frac{2}{p}\right)$. By the first part of the proof of Theorem 9.80 we have that $\gamma^s \otimes \gamma^{\widetilde{s}}$ is tempered. which gives

$$\left|\left\langle \gamma_g^s v, w \right\rangle \phi_{(\widetilde{s})}(g)\right| \ll_\varepsilon \|v\| \|w\| \|g\|_{\mathrm{HS}}^{-1+\varepsilon}$$

for any K-eigenvectors $v, w \in \mathcal{H}_s$. Dividing by $|\phi_{(\widetilde{s})}(g)| \asymp_{\widetilde{s}} \|g\|_{\mathrm{HS}}^{\widetilde{s}-1}$ we obtain

$$|\langle \gamma_g^s v, w\rangle| \ll_{\widetilde{s},\varepsilon} \|v\|\|w\|\|g\|_{\mathrm{HS}}^{-\widetilde{s}+\varepsilon}.$$

As $\widetilde{s} > \kappa$ we may set $\varepsilon = \widetilde{s} - \kappa$ and obtain (9.77) with an implicit constant that only depends on κ and s_π.

Finally we note that $\kappa_\pi > 0$ (and so, equivalently, $p_\pi < \infty$ or $s_\pi < 1$) implies that π has spectral gap by Proposition 7.27. Assume for the converse that $\kappa_\pi = 0$ and so, equivalently, that $s_\pi = 1$. However, this means by definition that there exists a sequence $s_n \nearrow 1$ with $\gamma^{s_n} \prec \pi$. By the definition of weak containment and Lemma 9.74 this shows that $\mathbb{1}_G \prec \pi$. Using (for example) Proposition 6.25 and the condition ($\prec_{\mathrm{op}}$) in Theorem 6.31, this shows that π cannot have spectral gap. □

Exercise 9.82. Let $S \subseteq (0,1)$ be countable, $s_\pi = \sup S$, and $\kappa \in (0, 1-s_\pi)$. Prove that (9.77) also holds for $\lambda \oplus \bigoplus_{s\in S} \gamma^s$ and K-eigenfunctions in its associated Hilbert space.

Exercise 9.83. Suppose the unitary representation π of $\mathrm{SL}_2(\mathbb{R})$ is a countable direct sum of irreducible representations. Define

$$s_\pi^\oplus = \sup\{s \in [0,1) \mid s = 0 \text{ or } \gamma^s \text{ is one of the summands of } \pi\}.$$

Show that in this case Theorem 9.80 also holds for $s_\pi^\oplus$ instead of s_π.

9.8 Compact Quotients of $\mathrm{SL}_2(\mathbb{R})$

We will study in this section the action-associated representation on compact quotients $X = \Gamma\backslash\mathrm{SL}_2(\mathbb{R})$ by uniform lattices.

9.8.1 Effective Decay of Matrix Coefficients

Corollary 9.84. *Let $\Gamma < \mathrm{SL}_2(\mathbb{R})$ be a uniform lattice and $X = \Gamma\backslash\mathrm{SL}_2(\mathbb{R})$. Then the action-associated representation π^X of $\mathrm{SL}_2(\mathbb{R})$ has effective decay of matrix coefficients.*

PROOF. By Proposition 6.29 the action-associated representation π^X has spectral gap. By Theorem 9.80 this implies that π^X also has effective decay of matrix coefficients. □

We note that the above result also holds more generally for finite volume quotients $X = \Gamma\backslash\mathrm{SL}_2(\mathbb{R})$ by any lattice $\Gamma < \mathrm{SL}_2(\mathbb{R})$. Proving this requires a different argument, for example using an analysis of the so-called Eisenstein series and the cuspidal spectrum. We will not pursue this further here.

9.8.2 Complete Decomposability

The following result is special to compact quotients.

Theorem 9.85. *Let $\Gamma < \mathrm{SL}_2(\mathbb{R})$ be a uniform lattice and $X = \Gamma\backslash\mathrm{SL}_2(\mathbb{R})$. Then the action-associated representation π^X is a countable direct sum of irreducible unitary representations.*

PROOF. For the proof it will be useful to first refine part of Proposition 6.29. Indeed we claim that for $\psi \in C_c^\infty(G)$ the convolution operator $\pi_*(\psi)$ maps $L^2(X)$ into $C^\infty(X)$.

To see the claim, let $\psi \in C_c^\infty(X)$ and $h = \exp(t\mathbf{m})$ for some $t \in \mathbb{R}$ and element $\mathbf{m} \in \mathfrak{sl}_2(\mathbb{R})$. For $f \in L^2(X)$ we than have

$$\pi_*(\psi)f(xh) = \int_G \psi(g)f(xhg)\,\mathrm{d}m_G(g) = \int_G \psi(h^{-1}g)f(xg)\,\mathrm{d}m_G(g).$$

Using this together with (6.16) we have for $t \in [-1,1]\smallsetminus\{0\}$ that

$$\begin{aligned}\left\|\frac{1}{t}\left((\pi_*(\psi)f)\left(\cdot\exp(t\mathbf{m})\right) - (\pi_*(\psi)f)\,(\cdot)\right) - \pi_*\left(\lambda_\partial(\mathbf{m})\psi\right)f\right\|_\infty \\ = \left\|\pi_*\left(\frac{1}{t}\left(\lambda_{\exp(t\mathbf{m})}\psi - \psi\right) - \lambda_\partial(\mathbf{m})\psi\right)f\right\|_\infty \\ \ll \left\|\frac{1}{t}\left(\lambda_{\exp(t\mathbf{m})}\psi - \psi\right) - \lambda_\partial(\mathbf{m})\psi\right\|_\infty \|f\|_2\end{aligned}$$

with the implicit constant depending on $\operatorname{supp}\psi$, $\mathbf{m} \in \mathfrak{sl}_2(\mathbb{R})$, and X only. However, as $\psi \in C_c^\infty(G)$ the final supremum norm converges to 0 as $t \to 0$. This implies that the derivative of $\pi_*(\psi)f$ in the direction of $\mathbf{m}$ is equal to $\pi_*\left(\lambda_\partial(\mathbf{m})\psi\right)f$. Iterating this statement shows that $\pi_*(\psi)f \in C^\infty(X)$ as claimed.

We show next that $\pi_\partial(\Omega)$ commutes with convolution operators. To see this, we let μ be a compactly supported complex measure on $\mathrm{SL}_2(\mathbb{R})$ and let f_1, f_2 be functions in $C^\infty(X)$. Then

$$\pi_*(\mu)f_1(x) = \int_G f_1(xg)\,\mathrm{d}\mu(g)$$

for $x \in X$. Differentiating under the integral sign shows $\pi_*(\mu)f_1 \in C^\infty(X)$. Moreover,

$$\begin{aligned}\langle \pi_\partial(\Omega)\pi_*(\mu)f_1, f_2\rangle &= \langle \pi_*(\mu)f_1, \pi_\partial(\Omega)f_2\rangle \\ &= \int_G \langle \pi_g f_1, \pi_\partial(\Omega)f_2\rangle \,\mathrm{d}\mu(g) \\ &= \int_G \langle \pi_g \pi_\partial(\Omega)f_1, f_2\rangle \,\mathrm{d}\mu(g) \\ &= \langle \pi_*(\mu)\pi_g(\Omega)f_1, f_2\rangle .\end{aligned}$$

As this holds for all $f_2 \in C^\infty(X)$, we obtain

$$\pi_\partial(\Omega)\pi_*(\mu)f_1 = \pi_*(\mu)\pi_\partial(\Omega)f_1. \tag{9.78}$$

We now show that any non-trivial invariant subspace $\mathcal{V} \subseteq L^2(X)$ contains an irreducible closed subspace. For this let $f_0 \in \mathcal{V}$ be a unit vector and choose from a suitable approximate identity (ψ_n) an element $\psi \in C_c^\infty(G)$ with the properties that $\psi \geqslant 0$, $\psi^* = \psi$, $\|\psi\|_1 = 1$ and $\|\pi_*(\psi)f_0 - f_0\| < 1$. This shows that $\pi_*(\psi)|_\mathcal{V} \neq 0$. By Proposition 6.29 and invariance of $\mathcal{V}$, we have that $\pi_*(\psi)|_\mathcal{V}\colon \mathcal{V} \to \mathcal{V}$ is a compact self-adjoint operator. Let $\mu \neq 0$ be an eigenvalue of $\pi_*(\psi)|_\mathcal{V}$ and $\mathcal{V}_\mu$ its associated finite-dimensional eigenspace. By the above claim we have that $\mathcal{V}_\mu = \pi_*(\psi)\mathcal{V}_\mu \subseteq C^\infty(X)$. In particular, $\pi_\partial(\Omega)$ is defined on $\mathcal{V}_\mu$. For $f \in \mathcal{V}_\mu$ we obtain from (9.78) that

$$\pi_*(\psi)\pi_\partial(\Omega)f = \pi_\partial(\Omega)\pi_*(\psi)f = \pi_\partial(\Omega)\mu f = \mu\pi_\partial(\Omega)f,$$

which shows that $\pi_\partial(\Omega)\mathcal{V}_\mu \subseteq \mathcal{V}_\mu$. Therefore the restriction of $\pi_\partial(\Omega)$ to the finite-dimensional subspace $\mathcal{V}_\mu$ has a smooth eigenfunction $g \in \mathcal{V}_\mu \subseteq \mathcal{V}$.

Next we decompose $g = \sum_{n\in\mathbb{Z}} g_n$ into a sum of K-eigenfunctions and choose $n \in \mathbb{Z}$ so that $g_n \neq 0$. Note that

$$F = g_n = \left(\pi^X|_K\right)_* (\overline{\chi_n})\, g \in \mathcal{V}$$

as $\mathcal{V}$ is invariant. Using (9.78) again, we see that F is also an eigenfunction for $\pi_\partial(\Omega)$.

To summarize, we have shown that any closed invariant subspace $\mathcal{V}$ contains a non-zero K-eigenvector with K-weight $n \in \mathbb{Z}$ so that F is also an eigenvector for $\pi_\partial^X(\Omega)$ and eigenvalue $\lambda \in \mathbb{R}$. By Corollary 9.23 this implies that the restriction of π^X to the cyclic subspace $\langle F\rangle_{\pi^X} \subseteq \mathcal{V}$ is irreducible.

The theorem now follows from a simple application of Zorn's lemma. Let

$$\mathcal{C} = \{S \mid S \text{ is a set of pairwise orthogonal irreducible subspaces of } L^2(X)\}$$

ordered by inclusion. It is straightforward to see that any linearly ordered chain in $\mathcal{C}$ has an upper bound, namely the union of the chain. Hence there exists a maximal element in $\mathcal{C}$. That is, there exists a maximal set S_0 of pairwise orthogonal subspaces in $L^2(X)$. As $L^2(X)$ is separable, S_0 is at most countable. Let $\mathcal{W}$ be the direct sum of the subspaces in S_0 and let $\mathcal{V} = \mathcal{W}^\perp$. If $\mathcal{V} \neq 0$ we can apply the above argument to find an irreducible subspace of $\mathcal{V}$, which contradicts maximality of S_0. Hence $\mathcal{V} = 0$ and $\mathcal{W} = L^2(X)$ is a direct sum of the irreducible subspaces contained in S. Note that each irreducible subspace satisfies that the space of K-invariant vectors is at most one-dimensional. As the space $L^2(X)^K = L^2(X/K) = L^2(\Gamma\backslash\mathbb{H})$ is infinite-dimensional, it follows that $|S_0| = \infty$. □

9.8.3 The First Non-trivial Eigenvalue

We continue our excursion into hyperbolic surfaces by establishing a link between effective decay of matrix coefficients and the first non-trivial eigenvalue for the Laplace–Beltrami operator on the surface. Following the choices made in Section 8.3 concerning $\mathbb{H}$ we define the Laplace–Beltrami operator Δ_{hyp} on $\mathbb{H}$ by

$$\Delta_{\mathrm{hyp}} f(z) = y^2 \left(\partial_x^2 f(z) + \partial_y^2 f(z)\right)$$

for $f \in C^\infty(\mathbb{H})$ and $z \in \mathbb{H}$. It can be verified directly that the action of an element $g \in \mathrm{SL}_2(\mathbb{R})$ by Möbius transformations on $\mathbb{H}$ satisfies

$$(\Delta_{\mathrm{hyp}} f) \circ g = \Delta_{\mathrm{hyp}}(f \circ g)$$

for $f \in C^\infty(\mathbb{H})$ (see Exercise 9.88). This also shows that Δ_{hyp} descends to a well-defined operator on $C^\infty(\Gamma\backslash\mathbb{H})$ for any discrete subgroup $\Gamma < \mathrm{SL}_2(\mathbb{R})$ (which we will also obtain in Lemma 9.87 by a different argument). To avoid technicalities concerning cone points of $\Gamma\backslash\mathbb{H}$ we will assume in the following that $\Gamma/\{\pm I\}$ is torsion-free, which implies in particular that no $\gamma \in \Gamma$ other than $\pm I$ has a fixed point in $\mathbb{H}$. Indeed, if $\gamma \cdot z = z$ for some $z \in \mathbb{H}$ we find some $g \in \mathrm{SL}_2(\mathbb{R})$ with $z = g \cdot \mathrm{i}$. This gives $g^{-1}\gamma g \cdot \mathrm{i} = \mathrm{i}$ and hence $g^{-1}\gamma g$ generates a discrete subgroup of $\mathrm{Stab}_{\mathrm{SL}_2(\mathbb{R})}(\mathrm{i}) = \mathrm{SO}_2(\mathbb{R})$, and so must be a torsion element.

For a compact surface $M = \Gamma\backslash\mathbb{H}$ defined by a uniform lattice $\Gamma < \mathrm{SL}_2(\mathbb{R})$ with no non-central torsion elements the Laplace–Beltrami operator Δ_{hyp} has a satisfying spectral theory. Indeed, there exists a sequence of eigenvalues

$$\lambda_0 = 0 < \lambda_1 \leqslant \lambda_2 \leqslant \cdots \tag{9.79}$$

with $\lambda_n \to \infty$ for $n \to \infty$, and a sequence of eigenfunctions

$$f_0 = \mathbb{1}_M, f_1, f_2, \cdots \in C^\infty(M)$$

so that

$$\Delta_{\mathrm{hyp}} f_n = -\lambda_n f_n$$

for all $n \in \mathbb{N}_0$.

The first non-trivial eigenvalue $\lambda_1 > 0$ measures in a sense the amount of connectivity of the surface M. For us it is of interest because of the following result.

Corollary 9.86 (First eigenvalue and almost decay exponent). *Let Γ be a torsion-free uniform lattice in $\mathrm{SL}_2(\mathbb{R})$, let $X = \Gamma\backslash\mathrm{SL}_2(\mathbb{R})$, and let*

$$M = \Gamma\backslash\mathbb{H} \cong X/K.$$

Then the action-associated representation π^X has almost decay exponent

$$\kappa_{\pi^X} = \begin{cases} 1 & \text{if } \lambda_1 \geqslant \frac{1}{4}, \text{ and} \\ 1-\sqrt{1-4\lambda_1} & \text{if } \lambda_1 < \frac{1}{4}. \end{cases}$$

For the proof we first need to establish a link between Δ_{hyp} and the Casimir operator

$$\Omega = \mathbb{1}_{\mathfrak{E}} + \mathbf{a}^{\circ 2} + \mathbf{d}^{\circ 2} - \mathbf{k}^{\circ 2}$$

considered so often in this chapter.

Lemma 9.87. *Let Γ, X, and M be as in Corollary 9.86. Let $f \in C^\infty(M)$, which we may identify with a smooth K-invariant function on X. Then*

$$\pi_\partial^X(\Omega)f = f + 4\Delta_{hyp}f.$$

PROOF. We identify $f \in C^\infty(M)$ with the smooth functions $\mathbb{H} \ni z \mapsto f(\Gamma z)$ and $\mathrm{SL}_2(\mathbb{R}) \ni g \mapsto f(\Gamma g{\cdot}\mathrm{i})$. In order to prove the lemma, we have to first calculate $\pi_\partial^X(\mathbf{a})f$ and $\pi_\partial^X(\mathbf{d})f$ as functions of $g \in \mathrm{SL}_2(\mathbb{R})$.

By definition we have

$$\pi_\partial^X(\mathbf{a})f(g) = \partial_t|_{t=0} f(g a_t{\cdot}\mathrm{i}) \tag{9.80}$$

for all $g \in \mathrm{SL}_2(\mathbb{R})$ as $a_t = \exp(t\mathbf{a})$. To calculate (9.80) we use the chain rule for differentiation, while always expressing total derivatives using the standard basis of $\mathbb{R}^2$. Hence the total derivative of f at $z = g{\cdot}\mathrm{i} = x + \mathrm{i}y \in \mathbb{H}$ is simply

$$\left(\partial_x f(z), \partial_y f(z)\right).$$

Next we write $g = u_x \begin{pmatrix} y^{\frac{1}{2}} & \\ & y^{-\frac{1}{2}} \end{pmatrix} k_\theta$ for some $k_\theta \in K$ and apply (8.5) for the Möbius transformation corresponding to u_x, $\begin{pmatrix} y^{\frac{1}{2}} & \\ & y^{-\frac{1}{2}} \end{pmatrix}$, and k_θ respectively, to see that their derivatives are the matrices representing multiplication by the complex numbers 1, y, and $\frac{1}{(\sin\theta\mathrm{i}+\cos\theta)^2} = \mathrm{e}^{-2\theta\mathrm{i}}$ respectively. Finally we note that the total derivative of $a_t{\cdot}\mathrm{i} = \mathrm{e}^{2t}\mathrm{i}$ at $t = 0$ is simply $2\mathrm{i}$, which we identify with $2e_2$. Putting these together, we obtain

$$\begin{aligned} \left(\pi_\partial^X(\mathbf{a})f\right)(g) &= \left(\partial_x f(z), \partial_y f(z)\right) \cdot y \cdot \begin{pmatrix} \cos 2\theta & \sin 2\theta \\ -\sin 2\theta & \cos 2\theta \end{pmatrix} 2e_2 \\ &= 2y\left(\sin(2\theta)\partial_x f(z) + \cos(2\theta)\partial_y f(z)\right) \end{aligned}$$

where $g = u_x \begin{pmatrix} y^{\frac{1}{2}} & \\ & y^{-\frac{1}{2}} \end{pmatrix} k_\theta$ and $z = g\cdot\mathrm{i}$.

For $\pi_\partial^X(\mathbf{d})(f)$ we only have to change the last step of the calculation. Indeed a simple calculation reveals that

$$\frac{\mathrm{d}}{\mathrm{d}t}\Big|_{t=0}(\exp(t\mathbf{d})\cdot\mathrm{i}) = \frac{\mathrm{d}}{\mathrm{d}t}\Big|_{t=0}\left(\begin{pmatrix}\cosh t & \sinh t\\ \sinh t & \cosh t\end{pmatrix}\cdot\mathrm{i}\right) = 2,$$

and so we simply have to replace $2e_2$ as above by $2e_1$. This gives

$$\pi_\partial^X(\mathbf{d})f(g) = 2y\left(\cos(2\theta)\partial_x f(z) - \sin(2\theta)\partial_y f(z)\right).$$

Fix some $z_0 \in \mathbb{H}$. As $\Gamma < \mathrm{SL}_2(\mathbb{R})$ is discrete and has no non-central torsion elements, it follows that there exists some $r > 0$ so that

$$\Psi\colon B_r^{\mathbb{H}}(z_0) \ni z \longmapsto \Gamma z \in M$$

is injective. We let $O = B_r^{\mathbb{H}}(z_0)$ and choose some $F \in C_c^\infty(O)$. Using the chart map Ψ we may also consider F as a function on M. We now calculate

$$\begin{aligned}\left\langle \pi_\partial^X(\Omega)f, F\right\rangle &= \left\langle \pi_\partial^X\left(\mathbb{1}_{\mathfrak{E}} + \mathbf{a}^{\circ 2} + \mathbf{d}^{\circ 2} - \mathbf{k}^{\circ 2}\right)f, F\right\rangle \\ &= \langle f, F\rangle - \left\langle \pi_\partial^X(\mathbf{a})f, \pi_\partial^X(\mathbf{a})F\right\rangle - \left\langle \pi_\partial^X(\mathbf{d})f, \pi_\partial^X(\mathbf{d})F\right\rangle.\end{aligned}$$

Using our preparations above for f and F we have that $\pi_\partial^X(\mathbf{a})f\overline{\pi_\partial^X(\mathbf{a})F}$ is equal to

$$2y\Big(\sin(2\theta)\partial_x f(z) + \cos(2\theta)\partial_y f(z)\Big)2y\Big(\sin(2\theta)\partial_x\overline{F(z)} + \cos(2\theta)\partial_y\overline{F(z)}\Big).$$

Next we use the fact that $m = \frac{1}{\pi}\,\mathrm{d}\theta\,\mathrm{dvol} = \frac{1}{\pi}\,\mathrm{d}\theta\frac{1}{y^2}\,\mathrm{d}x\,\mathrm{d}y$. Integrating over θ, we see that $\left\langle \pi_\partial^X(\mathbf{a})f, \pi_\partial^X(\mathbf{a})F\right\rangle$ is equal to

$$2\int_O\Big(\partial_x f(x+\mathrm{i}y)\partial_x\overline{F(x+\mathrm{i}y)} + \partial_y f(x+\mathrm{i}y)\overline{\partial_y F(x+\mathrm{i}y)}\Big)\mathrm{d}x\,\mathrm{d}y.$$

Finally, we use the fact that $F \in C_c(O)$ and apply integration by parts along x and along y separately, which leads to

$$\begin{aligned}-2\int_O\Big(\partial_x^2 f(x+\mathrm{i}y)\overline{F(x+\mathrm{i}y)} + \partial_y^2 f(x+\mathrm{i}y)\,\overline{F(x+\mathrm{i}y)}\Big)\mathrm{d}x\,\mathrm{d}y \\ = -2\int_O(\Delta_{\mathrm{hyp}}f)\,\overline{F}\,\mathrm{dvol}(z).\end{aligned}$$

The expression $\left\langle \pi_\partial^X(\mathbf{d})f, \pi_\partial^X(\mathbf{d})F\right\rangle$ gives the same result, which shows that

$$\left\langle \pi_\partial^X(\Omega)f, F\right\rangle = \left\langle f + 4\Delta_{\mathrm{hyp}}f, F\right\rangle.$$

As $F \in C_c^\infty(O)$ was arbitrary, we see that $\pi_\partial^X(\Omega)f$ is equal to $f + 4\Delta_{\mathrm{hyp}}f$ on the image of O. Varying $z_0 \in \mathbb{H}$ proves the lemma. □

PROOF OF COROLLARY 9.86. By Theorem 9.85 we have

$$L^2(X) \cong \bigoplus_{j=0}^{\infty} \mathcal{V}_j \tag{9.81}$$

for countably many irreducible subspaces $\mathcal{V}_j < L^2(X)$. We may assume that we have $\mathcal{V}_0 = \mathbb{C}\mathbb{1}$, which is the only trivial representation (by ergodicity of the action of $\mathrm{SL}_2(\mathbb{R})$ on X). For $j \geqslant 1$ and a non-spherical $\mathcal{V}_j$ we apply Corollary 9.29 to see that the restriction of π^X to $\mathcal{V}_j$ is tempered and has decay exponent $1-\varepsilon$ for all $\varepsilon > 0$.

Suppose now $j \geqslant 1$ and $\mathcal{V}_j$ is spherical, and let $f \in \mathcal{V}_j$ be a non-zero K-invariant function. Let α_j be the eigenvalue of $\pi_\partial^X(\Omega)|_{\mathcal{V}_j}$ so $\pi_\partial^X(\Omega)f = \alpha_j f$. Combining Lemma 9.16 with the first part of the proof of Theorem 9.85, we see that $f \in C^\infty(M)$ for $M = \Gamma\backslash\mathbb{H}$. By Lemma 9.87 this shows that f is also an eigenfunction of Δ_{hyp}; that is, $\Delta_{\mathrm{hyp}}(f) = -\lambda_n f$ for some $n \geqslant 1$ and, moreover,

$$\alpha_j = 1 - 4\lambda_n \leqslant 1 - 4\lambda_1. \tag{9.82}$$

If now $\lambda_1 \geqslant \frac{1}{4}$, then $\alpha_j \leqslant 0$ for all $j \in \mathbb{N}$. However, this implies by Theorem 9.32 that $\mathcal{V}_j$ is isomorphic to a principal series representation and is tempered with almost decay exponent 1. Hence in this case all direct summands of (9.81) with $j \geqslant 1$ are tempered.

Suppose now that $\lambda_1 < \frac{1}{4}$. If $\mathcal{V}_j$ is isomorphic to the complementary series representation γ^{s_j}, then (9.82) shows that

$$\alpha_j = s_j^2 \leqslant 1 - 4\lambda_1.$$

Hence the complementary series parameter satisfies

$$s_{\pi^X} \leqslant \sqrt{1 - 4\lambda_1}.$$

On the other hand the eigenfunction for the first non-trivial eigenvalue generates an irreducible representation by Corollary 9.23, which must be a complementary series representation for parameter $\sqrt{1-4\lambda_1}$. Therefore

$$s_{\pi^X} = \sqrt{1 - 4\lambda_1}.$$

Theorem 9.80 and Exercise 9.83 show that π^X has almost decay exponent

$$\kappa_{\pi^X} = 1 - \sqrt{1 - 4\lambda_1}.$$

□

Exercise 9.88. Prove that $(\Delta_{\mathrm{hyp}} f) \circ g = \Delta_{\mathrm{hyp}}(f \circ g)$ for $f \in C^\infty(\mathbb{H})$ and $g \in \mathrm{SL}_2(\mathbb{R})$.

9.8.4 Effective Equidistribution of Horocycle Orbits

Furstenberg [33] showed in 1972 that 'the horocycle flow is uniquely ergodic' on compact quotients: For any uniform lattice $\Gamma < \mathrm{SL}_2(\mathbb{R})$, continuous function ϕ on $X = \Gamma\backslash\mathrm{SL}_2(\mathbb{R})$, and point $x \in X$ we have

$$\frac{1}{T}\int_0^T \phi(xu_s)\,\mathrm{d}s \longrightarrow \frac{1}{m_X(X)}\int_X \phi\,\mathrm{d}m_X$$

as $T \to \infty$ (uniformly in x). Our final goal is to obtain an error estimate for this convergence using an argument due to Burger [10] from 1990.

Theorem 9.89 (Effective equidistribution of horocycle orbits). *Let Γ be a uniform lattice in $\mathrm{SL}_2(\mathbb{R})$ and let $X = \Gamma\backslash\mathrm{SL}_2(\mathbb{R})$. There exists a $\kappa > 0$ so that for all $\phi \in C^4(X)$, $x \in X$, and $S > 0$ we have*

$$\left|\frac{1}{S}\int_0^S \phi(xu_s)\,\mathrm{d}s - \frac{1}{m_X(X)}\int_X \phi\,\mathrm{d}m_X\right| \ll_{X,\kappa} S^{-\kappa}\,\mathcal{S}_4(\phi).$$

In fact κ only depends on the summands of $L^2(X)$ with respect to the action-associated unitary representation $\pi = \pi^X$: If the complementary series representations do not appear, then any $\kappa < \frac{1}{2}$ would work. If a complementary series representation appears in $L^2(X)$, then $\kappa = \frac{1-s_\pi}{2}$ would work.

As we will see, we can associate to our problem a concrete second-order ordinary differential equation with constant coefficients—assuming that ϕ is an eigenfunction of the Casimir operator. We then solve this ordinary differential equation explicitly and obtain explicit decay properties. Finally we will need a (weak version of the) Weyl law to generalize the estimates to an arbitrary function.

Expanding Horocycle Orbits

For the proof of Theorem 9.89 a switch in our point of view will be useful. Instead of studying longer and longer pieces of the horocycle orbit of a given starting point x, we instead wish to expand a fixed piece $\{x_0u_s \mid s \in [0,1]\}$ by a diagonal element. Using compactness of X and uniformity in x_0, we will convert the resulting statement back into the statement needed for the theorem.

More precisely, let $X = \Gamma\backslash\mathrm{SL}_2(\mathbb{R})$ be as in Theorem 9.89 and $\phi \in C^\infty(X)$. Also fix some $x_0 \in X$ and define

$$\Phi(t) = \int_0^1 \phi(x_0 u_s a_t^{-1})\,\mathrm{d}s \tag{9.83}$$

for $t \in \mathbb{R}$. Using $a_t u_s a_t^{-1} = u_{e^{2t}s}$ for $s \in \mathbb{R}$ we also have

$$\Phi(t) = \int_0^1 \phi(x_0 a_t^{-1} a_t u_s a_t^{-1})\,\mathrm{d}s = \int_0^1 \phi(x_0 a_t^{-1} u_{\mathrm{e}^{2t}s})\,\mathrm{d}s,$$

so Φ is the average of ϕ along a piece of the U-orbit of $x_0 a_t^{-1}$ of length $S = \mathrm{e}^{2t}$. Our goal is to show that

$$\Phi(t) \longrightarrow \frac{1}{m_X(X)} \int_X \phi\,\mathrm{d}m_X$$

as $t \to \infty$ with a precise and uniform estimate on the speed of convergence. To prove this we will first assume that ϕ is an eigenfunction for the Casimir operator.

If $\phi = c\mathbb{1}_X$ is a constant then $\Phi(t) = c = \frac{1}{m_X(X)} \int \phi\,\mathrm{d}m_X$ for all $t \in \mathbb{R}$ and there is nothing to prove. In all other cases we wish to show that $\Phi(t)$ decays for $t \to \infty$ with a certain speed, and that the speed depends on the irreducible representations generated by ϕ.

A Second Order Differential Equation

We suppose now that $\phi \in C^\infty(X)$ is an eigenfunction for the Casimir operator $\pi_\partial(\Omega)$ with eigenvalue α (for example, because ϕ belongs to an irreducible summand of $L^2(X)$). We claim that this implies that Φ as defined by (9.83) satisfies the second-order differential equation

$$\Phi''(t) + 2\Phi'(t) + (1-\alpha)\Phi(t) = \mathrm{e}^{-2t}\Psi(t) \tag{9.84}$$

where $\Psi \in C^\infty(\mathbb{R})$ satisfies

$$\|\Psi\|_\infty \leqslant \|\phi\|_{C^1(X)}. \tag{9.85}$$

For this we note that

$$\Omega = \mathbb{1}_{\mathfrak{E}} + \mathbf{a}^{\circ 2} + 2\mathbf{e}\circ\mathbf{f} + 2\mathbf{f}\circ\mathbf{e} = \mathbb{1}_{\mathfrak{E}} + \mathbf{a}^{\circ 2} - 2\mathbf{a} + 4\mathbf{e}\circ\mathbf{f}$$

by the definition of Ω in Lemma 9.2 and since $[\mathbf{e},\mathbf{f}] = \mathbf{a}$. Next we apply the definition of $\pi_\partial(\mathbf{a})$ and note that

$$\big(\pi_\partial(\mathbf{a})\phi\big)(x) = \lim_{h\to 0} \frac{1}{h}\big(\phi(x a_h) - \phi(x)\big)$$

converges uniformly for all $x \in X$. Setting x to be $x_0 u_s a_t^{-1}$ and integrating over $s \in [0,1]$ gives, by our definition of Φ in (9.83),

$$\int_0^1 \big(\pi_\partial(\mathbf{a})\phi\big)(x_0 u_s a_t^{-1})\,\mathrm{d}s = \lim_{h\to 0}\frac{1}{h}\big(\Phi(t-h) - \Phi(t)\big) = -\Phi'(t).$$

Applying this argument a second time gives

$$\Phi''(t) = \int_0^1 \big(\pi_\partial(\mathbf{a}^{\circ 2})\phi\big)(x_0 u_s a_t^{-1})\,\mathrm{d}s.$$

For the term $\mathbf{e} \circ \mathbf{f}$ in Ω we let $\psi = \pi_\partial(\mathbf{f})\phi$. Similarly to the above we then have

$$\big(\pi_\partial(\mathbf{e})\psi\big)(x) = \lim_{h \to 0} \frac{1}{h}\big(\psi(xu_h) - \psi(x)\big)$$

uniformly for all $x \in X$. We again set $x = x_0 u_s a_t^{-1} = x_0 a_t^{-1} u_{\mathrm{e}^{2t}s}$ and integrate over $s \in [0,1]$. This gives

$$\begin{aligned}\int_0^1 \big(\pi_\partial(\mathbf{e} \circ \mathbf{f})\phi\big)(xa_t^{-1}u_{\mathrm{e}^{2t}s})\,\mathrm{d}s &= \mathrm{e}^{-2t}\int_0^{\mathrm{e}^{2t}} \big(\pi_\partial(\mathbf{e})\psi\big)(xa_t^{-1}u_r)\,\mathrm{d}r \\ &= \mathrm{e}^{-2t}\big(\psi(xa_t^{-1}u_{\mathrm{e}^{2t}}) - \psi(xa_t^{-1})\big)\end{aligned}$$

by the substitution $r = \mathrm{e}^{2t}s$ and the fundamental theorem of calculus.

We define

$$\Psi(t) = -4\big(\psi(xa_t^{-1}u_{\mathrm{e}^{2t}}) - \psi(xa_t^{-1})\big) = -4\big(\psi(xu_1 a_t^{-1}) - \psi(xa_t^{-1})\big).$$

Together we obtain

$$\alpha\Phi = \int_0^1 \big(\pi_\partial(\mathbb{1}_{\mathfrak{E}} + \mathbf{a}^{\circ 2} - 2\mathbf{a} + 4\mathbf{e} \circ \mathbf{f})\phi\big)(xu_s a_t^{-1})\,\mathrm{d}s = \Phi + \Phi'' + 2\Phi' - \mathrm{e}^{-2t}\Psi$$

which proves (9.84) and (9.85).

Next we consider the various possible values for α and prove the desired estimates. For this note that the ordinary differential equation (9.84) has characteristic polynomial $T^2 + 2T + (1-\alpha)$, with zeros $-1 \pm \sqrt{\alpha}$.

Vanishing Casimir Eigenvalue

We suppose first that ϕ generates a sum of representations that are isomorphic to $\pi^{0,\mathrm{e}}$, $\delta^{1,+}$, or $\delta^{1,-}$. We can handle these simultaneously as in these cases the Casimir eigenvalue is $\alpha = 0$ and (9.84) becomes

$$\Phi''(t) + 2\Phi'(t) + \Phi(t) = \mathrm{e}^{-2t}\Psi(t) \tag{9.86}$$

for $t \in \mathbb{R}$. We define $y(t) = \mathrm{e}^t\Phi(t)$ for $t \in \mathbb{R}$ and substitute $\Phi(t) = \mathrm{e}^{-t}y(t)$ into (9.86) to obtain the simpler differential equation

$$y''(t) = \mathrm{e}^{-t}\Psi(t) \tag{9.87}$$

for $t \in \mathbb{R}$. Now let

$$F(t) = \int_0^t \mathrm{e}^{-r}\Psi(r)\,\mathrm{d}r$$

for $t \in \mathbb{R}$. The general solution to (9.87) is then given by

$$y(t) = c_0 + c_1 t + \int_0^t F(r)\,\mathrm{d}r$$

for $t \in \mathbb{R}$ and some constants $c_0, c_1 \in \mathbb{C}$, which gives

$$\Phi(t) = \mathrm{e}^{-t}\Big(c_0 + c_1 t + \int_0^t F(r)\,\mathrm{d}r\Big)$$

for $t \in \mathbb{R}$. Setting $t = 0$ in Φ and Φ' gives

$$|c_0| = |\Phi(0)| = \Big|\int_0^1 \phi(x_0 u_s)\,\mathrm{d}s\Big| \leqslant \|\phi\|_\infty$$

and

$$|-c_0 + c_1| = |\Phi'(0)| = \Big|\int_0^1 \pi_\partial(\mathbf{a})\phi(x_0 u_s)\,\mathrm{d}s\Big| \leqslant \|\phi\|_{C^1(X)}$$

as $F(0) = 0$. Moreover, by definition and (9.85) we also have

$$F(t) \ll \|\phi\|_{C^1(X)}$$

for $t \geqslant 0$. Together these estimates lead to

$$|\Phi(t)\| \ll \|\phi\|_{C^1(X)}(1+t)\mathrm{e}^{-t}$$

for $t \geqslant 0$.

In the remaining cases the characteristic polynomial does not have a multiple root and the solutions take a different form as in the following exercise.

Essential Exercise 9.90. (a) Let $s \neq 0$ be a complex number and let $\alpha = s^2$. Show that the general solution of the ordinary differential equation

$$y''(t) - \alpha y(t) = f(t) \tag{9.88}$$

for $f \in C^1(\mathbb{R})$ is given by

$$y(t) = \mathrm{e}^{st}\left(c_1 + \frac{1}{2s}\int_0^t \mathrm{e}^{-sr} f(r)\,\mathrm{d}r\right) + \mathrm{e}^{-st}\left(c_2 - \frac{1}{2s}\int_0^t \mathrm{e}^{sr} f(r)\,\mathrm{d}r\right).$$

(b) Use the substitution $y(t) = \mathrm{e}^t\Phi(t)$ and $f(t) = \mathrm{e}^{-t}\Psi(t)$ to relate solutions of (9.88) to solutions of (9.84).

The Complementary Series

Suppose that ϕ has Casimir eigenvalue $\alpha = s_0^2$ for some $s_0 \in (0,1)$ corresponding to the complementary series representation γ^{s_0}. Using Exercise 9.90 we obtain that

$$\begin{aligned}\Phi(t) = \mathrm{e}^{(-1+s_0)t}\Big(c_1 + \frac{1}{2s_0}\int_0^t \mathrm{e}^{(-1-s_0)r}\Psi(r)\,\mathrm{d}r\Big)\\ + \mathrm{e}^{(-1-s_0)t}\Big(c_2 - \frac{1}{2s_0}\int_0^t \mathrm{e}^{(-1+s_0)}\Psi(r)\,\mathrm{d}r\Big) \qquad (9.89)\end{aligned}$$

for all $t \in \mathbb{R}$ and some fixed constants $c_1, c_2 \in \mathbb{C}$. We calculate the constants by setting $t = 0$ in Φ and Φ', which gives

$$\Phi(0) = c_1 + c_2 = \int_0^1 \phi(x_0 u_s)\,\mathrm{d}s$$

$$\Phi'(0) = (-1+s_0)c_1 + (-1-s_0)c_2 = -\int_0^1 \pi_\partial(\mathbf{a})\phi(x_0 u_s)\,\mathrm{d}s$$

since the derivatives of the integrals cancel each other. Solving this pair of equations for c_1 and c_2 gives the estimate

$$|c_1|, |c_2| \ll \frac{1}{s_0}\|\phi\|_{C^1(X)}.$$

As only finitely many complementary series representations appear in $L^2(X)$ by (9.79) on page 497, we have $\frac{1}{s_0} \ll_X 1$ and $\frac{1}{1-s_0} \ll_X 1$. Combined with (9.85) this allows us to uniformly bound the integrals appearing in (9.89) and obtain

$$\begin{aligned}\|\Phi(t)\| \ll_X \|\phi\|_{C^1(X)}\Big(\mathrm{e}^{(-1+s_0)t} + \mathrm{e}^{(-1+s_0)t}\int_0^t \mathrm{e}^{(-1-s_0)r}\,\mathrm{d}r\\ + \mathrm{e}^{(-1-s_0)t}\int_0^t \mathrm{e}^{(-1+s_0)r}\,\mathrm{d}r\Big) \ll_X \|\phi\|_{C^1(X)}\mathrm{e}^{(-1+s_0)t}\end{aligned}$$

for all $t \geqslant 0$.

The Principal Series Representations with Non-zero Parameter

Suppose that ϕ has the Casimir eigenvalue $\alpha = -\xi^2$ for some $\xi > 0$ corresponding to even or odd principal series representations. Using Exercise 9.90 with $s = \mathrm{i}\xi$ we obtain that

$$\Phi(t) = \mathrm{e}^{(-1+\mathrm{i}\xi)t}\left(c_1 + \frac{1}{2\mathrm{i}\xi}\int_0^t \mathrm{e}^{(-1-\mathrm{i}\xi)r}\Psi(r)\,\mathrm{d}r\right) \\ + \mathrm{e}^{(-1-\mathrm{i}\xi)t}\left(c_1 - \frac{1}{2\mathrm{i}\xi}\int_0^t \mathrm{e}^{(-1+\mathrm{i}\xi)r}\Psi(r)\,\mathrm{d}r\right)$$

for all $t \in \mathbb{R}$ and for some fixed $c_1, c_2 \in \mathbb{C}$. We calculate the constants by setting $t = 0$ in Φ and Φ' as above. This gives the estimates

$$|c_1|, |c_2| \ll \frac{1}{\xi}\|\phi\|_{C^1(X)}.$$

Combined with (9.85) we obtain the estimate

$$|\Phi(t)| \ll_X \frac{1}{\xi}\|\phi\|_{C^1(X)}\mathrm{e}^{-t}$$

for all $t \geqslant 0$ in this case.

The Discrete Series Representation

Suppose ϕ has the Casimir eigenvalue $(n-1)^2$ for an integer $n \geqslant 2$ and set k to be $n - 1 \geqslant 1$. Once more this gives

$$\Phi(t) = \mathrm{e}^{(-1+k)t}\left(c_1 + \frac{1}{2k}\int_0^t \mathrm{e}^{(-1-k)r}\Psi(r)\,\mathrm{d}r\right) \\ + \mathrm{e}^{(-1-k)t}\left(c_2 - \frac{1}{2k}\int_0^t \mathrm{e}^{(-1+k)r}\Psi(r)\,\mathrm{d}r\right)$$

for all $t \in \mathbb{R}$ and some $c_1, c_2 \in \mathbb{C}$ satisfying

$$|c_1|, |c_2| \ll \frac{1}{k}\|\phi\|_{C^1(X)}.$$

However, this time some exponential functions have non-negative coefficients which require more care.

Let us start with the second summand in our formula for $\Phi(t)$. Using (9.85) we have

$$\begin{aligned} \mathrm{e}^{(-1-k)t}\left|c_2 - \frac{1}{2k}\int_0^t \mathrm{e}^{(-1+k)r}\Psi(r)\,\mathrm{d}r\right| \\ &\ll \frac{1}{k}\|\phi\|_{C^1(X)}\mathrm{e}^{(-1-k)t}\left(1 + \int_0^t \mathrm{e}^{(-1+k)r}\,\mathrm{d}r\right) \\ &\ll \frac{1}{k}\|\phi\|_{C^1(X)}\mathrm{e}^{-2t}(1+t) \end{aligned}$$

for $t \geqslant 0$, where the factor $(1+t)$ is only needed in case $k=1$.

For the first summand in $\Phi(t)$ we note that the integral converges as $t \to \infty$. Thus we may define

$$\widetilde{c_1} = c_1 + \frac{1}{2k}\int_0^\infty \mathrm{e}^{(-1-k)r}\psi(r)\,\mathrm{d}r,$$

which gives

$$\begin{aligned}\Phi(t) &= \mathrm{e}^{(-1+k)t}\Big(\widetilde{c_1} - \frac{1}{2k}\int_t^\infty \mathrm{e}^{(-1-k)r}\Psi(r)\,\mathrm{d}r\Big) + \mathrm{O}\Big(\frac{1}{k}\|\phi\|_{C^1(X)}\mathrm{e}^{-2t}(1+t)\Big)\\ &= \widetilde{c_1}\mathrm{e}^{(-1+k)t} + \mathrm{O}\Big(\frac{1}{k}\|\phi\|_{C^1(X)}\mathrm{e}^{-2t}(1+t)\Big)\end{aligned}$$

for $t \geqslant 0$. By definition Φ is bounded for $t \geqslant 0$, so if $k>1$ this implies that $\widetilde{c_1}=0$. The case $k=1$ corresponds to $\delta_2^{\pm}$ and Casimir eigenvalue 1 (which agrees with the Casimir eigenvalue of the trivial representation). In this case

$$\lim_{t\to\infty}\Phi(t) = \widetilde{c_1}.$$

By the (ineffective) unique ergodicity result of Furstenberg [33] (see also [24, Th. 11.27]) we have

$$\lim_{t\to\infty}\Phi(t) = \frac{1}{m_X(X)}\int_X \phi\,\mathrm{d}m_X.$$

Assuming that ϕ generates a sum of discrete series representations (that is, $\int_X \phi\,\mathrm{d}m_X = 0$), we conclude that $\widetilde{c_1}=0$. Hence we obtain in the case where ϕ generates a sum of discrete series representations isomorphic to $\delta^{n,+}$ or $\delta^{n,-}$ the estimate

$$|\Phi(t)| \ll \frac{1}{n}\|\phi\|_{C^1(X)}\mathrm{e}^{-2t}(1+t)$$

for $t \geqslant 0$.

A Brief Summary

Let $x \in X$ and $S \geqslant 1$. We define $t = \frac{1}{2}\log S$ and apply the discussion above to $x_0 = xa_t$. In this way we obtain uniform statements about the average along the orbit xu_s for $s \in [0,S]$ as follows:

- If $\phi = c$ is a constant then the average is equal to $c = \frac{1}{m_X(X)}\int_X \phi\,\mathrm{d}m_X$.
- If ϕ generates a sum of representations isomorphic to γ^{s_0} for some fixed s_0 lying in $(0,1)$ then the average is bounded by $\ll_X S^{-\frac{1}{2}(1-s_0)}\|\phi\|_{C^1(X)}$.
- If ϕ generates a sum of representations isomorphic to $\pi^{0,\mathrm{e}}$, $\delta^{1,+}$, or $\delta^{1,-}$ then the average is bounded by $\ll_X (1+\log S)S^{-\frac{1}{2}}\|\phi\|_{C^1(X)}$.

- If ϕ generates a sum of representations isomorphic to $\pi^{\xi,\mathrm{e}}$ or $\pi^{\xi,\mathrm{o}}$ for $\xi > 0$ then the average is bounded by $\ll_X S^{-\frac{1}{2}}\frac{1}{\xi}\|\phi\|_{C^1(X)}$.
- If ϕ generates a sum of discrete series representations $\delta^{n,+}$ or $\delta^{n,-}$ for some fixed $n \geqslant 2$ then the average is bounded by $\ll_X (1+\log S)S^{-1}\frac{1}{n}\|\phi\|_{C^1(X)}$.

By considering the worst case this explains the description of $\kappa \in (0, \frac{1}{2})$ in Theorem 9.89. For those $\kappa > 0$ and a Casimir eigenfunction ϕ we therefore have

$$\left|\frac{1}{S}\int_0^S \phi(xu_s)\,\mathrm{d}s - \frac{1}{m_X(X)}\int_X \phi\,\mathrm{d}m_X\right| \ll_{X,\kappa} S^{-\kappa}C(\phi),$$

where $C(\phi)$ is equal to $\|\phi\|_{C^1(X)}$ potentially multiplied by $\frac{1}{\xi}$ in the principal series case or $\frac{1}{n}$ in the discrete series case.

Using Uniform Convergence of a Convenient Splitting

We claim that any function $\phi \in C^\infty(X)$ can be split into a countable sum

$$\phi = \sum_{n=0}^{\infty} \phi_n \tag{9.90}$$

where $\phi_0 = \frac{1}{m_X(X)}\int_X \phi\,\mathrm{d}m_X$, each ϕ_n for $n \geqslant 1$ satisfies the decay estimate for long horocycle orbits by one of the cases considered above, and

$$\sum_{n=1}^{\infty} C(\phi_n) \ll_X \mathcal{S}_4(\phi). \tag{9.91}$$

Using linearity we obtain

$$\begin{aligned}\left|\frac{1}{S}\int_0^S \phi(xu_s)\,\mathrm{d}s - \frac{1}{m_X(X)}\int_X \phi\,\mathrm{d}m_X\right| &= \left|\sum_{n=1}^{\infty}\int_X \phi_n(xu_s)\,\mathrm{d}s\right| \\ &\ll_{X,\kappa} S^{-\kappa}\sum_{n=1}^{\infty} C(\phi_n) \ll_X S^{-\kappa}\,\mathcal{S}_4(\phi)\end{aligned}$$

from the claim. Theorem 9.89 follows by the density of $C^\infty(X) \subseteq C^4(X)$.

To obtain the splitting (9.90) and the bound (9.91) we will combine Theorem 9.85, the Sobolev embedding theorem, and the Weyl law. First recall that the Sobolev embedding theorem on X with $1+\frac{\dim X}{2} = 2 < 3$ gives the estimate

$$\|\phi_n\|_{C^1(X)} \ll_X \mathcal{S}_3(\phi) \tag{9.92}$$

for any $\phi \in C^\infty(X)$ (see [25, Sec. 5.2] for an introduction and Evans [27] for more details).

By applying Theorem 9.85 we can decompose $L^2(X)$ into a countable direct sum of irreducible representations. In applying this to $\phi \in C^\infty(X)$ we collect some of the terms to obtain the splitting

$$\phi = \int_X \phi \, \mathrm{d}m_X + \phi_{\mathrm{cs}} + \phi_{\mathrm{ps,e}} + \phi_{\mathrm{ps,o}} + \phi_{\mathrm{mds}} + \phi_{\mathrm{ds}}.$$

Here the summands satisfy the following properties:

- ϕ_{cs} is a finite sum of Casimir eigenfunctions with eigenvalues in $(0,1)$, each generating a complementary series representation;
- $\phi_{\mathrm{ps,e}}$ and $\phi_{\mathrm{ps,o}}$ are countable sums of Casimir eigenfunctions in $(-\infty,0]$ generating even, respectively odd, principal series representations;
- ϕ_{mds} is a Casimir eigenfunction for eigenvalue 0 generating a finite sum of mock discrete series representations; and
- $\phi_{\mathrm{ds}} = \sum_{n\geqslant 1} \phi_{\mathrm{ds},n}$ is countable sum of Casimir eigenfunctions $\phi_{\mathrm{ds},n}$ for eigenvalues $(n-1)^2$ for $n \geqslant 2$ corresponding to discrete series representations.

We note that $\phi \in C^\infty(X)$ implies that ϕ is also a smooth vector for π^X. As the splitting above corresponds to invariant subspaces it follows that the summands are also smooth vectors for π^X. The Sobolev embedding theorem in turn now shows that the summands also belong to $C^\infty(X)$.

Our discussions combined with (9.92) apply directly to ϕ_{mds}. Moreover, as the number of eigenfunctions in ϕ_{cs} depends only on X (the number of non-tempered eigenfunctions for Δ_{hyp} on $M = \Gamma\backslash\mathbb{H}^2$ with eigenvalue in $(0,\frac{1}{4})$; see (9.79)), the same holds for ϕ_{cs}. However, the functions $\phi_{\mathrm{ps,e}}, \phi_{\mathrm{ps,o}}$ and ϕ_{ds} are countable sums and so require more care.

We start with

$$\phi_{\mathrm{ds}} = \sum_{n\geqslant 1} \phi_{\mathrm{ds},n}. \tag{9.93}$$

Here we have

$$\sum_{n\geqslant 2} C(\phi_{\mathrm{ds},n}) = \sum_{n\geqslant 2} \frac{1}{n} \|\phi_{\mathrm{ds},n}\|_{C^1(X)} \ll \sum_{n\geqslant 2} \frac{1}{n} \mathcal{S}_3(\phi_{\mathrm{ds},n})$$
$$\ll \sqrt{\sum_{n\geqslant 2} \mathcal{S}_3(\phi_{\mathrm{ds},n})^2} \ll \mathcal{S}_3(\phi_{\mathrm{ds}}) \leqslant \mathcal{S}_3(\phi),$$

where we used (in turn) the definition of $C(\phi_{\mathrm{ds},n})$, the Sobolev embedding (9.92), the Cauchy–Schwarz inequality in $\ell^2(\mathbb{N})$, and the fact that the splitting (9.93) corresponds to a sum of invariant subspaces. This proves (9.90)–(9.91) for ϕ_{ds}.

We split the function $\phi_{\mathrm{ps,e}} = \sum_{n=1}^\infty \phi_n$ into summands generating even principal series representations with Casimir eigenvalues $\alpha_n = 1 - 4\lambda_n$ corresponding to non-decreasing Laplace–Beltrami eigenvalues λ_n on $M = \Gamma\backslash\mathbb{H}^2$. We are interested in the growth of the α_n (respectively, the λ_n) so that we may argue along the same lines as was done for the discrete series representation above. Weyl's law (ignoring the finitely many eigenvalues in $(0,\frac{1}{4})$; see Zworski [94] for an accessible source) states in this setting that

$$\lim_{T\to\infty}\frac{|\{n \mid \lambda_n \leqslant T\}|}{T} = c_0$$

for some positive constant c_0 depending on the total volume of M. In particular

$$|\{n \mid \lambda_n \leqslant T\}| < cT$$

for some constant c and all $T \geqslant 1$. Applying this for $T = \frac{1}{c}n$ gives

$$\lambda_n > \frac{1}{c}n \tag{9.94}$$

for all $n \geqslant 1$. We also note that

$$\begin{aligned}
|\alpha_n|\,\mathcal{S}_3(\phi_n)^2 &= \sum_{\mathbf{e}} |\alpha_n|\,\langle \pi_\partial(\mathbf{e})\phi_n, \pi_\partial(\mathbf{e})\phi_n\rangle \\
&= -\sum_{\mathbf{e}} \big\langle \pi_\partial\big((\mathbb{1}_{\mathfrak{E}} + \mathbf{a}^{\circ 2} + \mathbf{d}^{\circ 2} - \mathbf{k}^{\circ 2})\mathbf{e}\big)\phi_n, \pi_\partial(\mathbf{e})\phi_n\big\rangle \\
&= -\,\mathcal{S}_3(\phi)^2 + \sum_{\mathbf{e}} \big(\|\pi_\partial(\mathbf{a}\circ\mathbf{e})\phi_n\|^2 + \|\pi_\partial(\mathbf{d}\circ\mathbf{c})\phi_n\|^2 - \|\pi_\partial(\mathbf{k}\circ\mathbf{e})\phi_n\|^2\big) \\
&\ll \mathcal{S}_4(\phi)^2,
\end{aligned}$$

where $\mathbf{e}$ varies in the sums over the elements of $\mathfrak{E}$ appearing in the definition of $\mathcal{S}_3$. Using the fact that $|\alpha_n| = \xi_n^2 = 4\lambda_n - 1$ has linear growth by (9.94), it follows that

$$\begin{aligned}
\sum_{n=1}^{\infty} C(\phi_n) = \sum_{n=1}^{\infty}\frac{1}{\xi_n}\|\phi_n\|_{C^1(X)} &\ll \sum_{n=1}^{\infty}\frac{1}{\xi_n}\,\mathcal{S}_3(\phi_n) \leqslant \sum_{n=1}^{\infty}\frac{1}{|\alpha_n|}\,\mathcal{S}_4(\phi_n) \\
&\ll \sqrt{\sum_{n=1}^{\infty}\mathcal{S}_4(\phi_n)^2} \ll \mathcal{S}_4(\phi_{\mathrm{ps,e}}) \leqslant \mathcal{S}_4(\phi).
\end{aligned}$$

Once again we deduce (9.90)–(9.91).

If $-I \in \Gamma$ then π is an even representation and we do not have to consider $\phi_{\mathrm{ps,o}}$. If not, then the treatment above for $\phi_{\mathrm{ps,e}}$ is readily adapted to $\phi_{\mathrm{ps,o}}$ (assuming an analogue to the growth estimate (9.94) is known, see Exercise 9.93).

A Weak Version of Weyl's Law

As we only need a weak version of the Weyl law we indicate through a series of exercises how to obtain this in an elementary way. We suppose for simplicity that Γ has no torsion elements other than possibly $-I$. In fact any lattice Γ

has a finite index subgroup $\Gamma_0 < \Gamma$ satisfying this, and it suffices to prove our estimate for Γ_0.

As Γ is assumed to be a uniform lattice in $\mathrm{SL}_2(\mathbb{R})$ there exists a uniform $\varepsilon > 0$ such that the canonical projection

$$O = B_\varepsilon^{\mathbb{H}}(I) \ni z \longmapsto \Gamma g z \in M = \Gamma\backslash\mathbb{H}$$

is an isometry for every fixed $g \in \mathrm{PSL}_2(\mathbb{R})$.

Using compactness once more we may suppose that $\Gamma g_1 O, \dots, \Gamma g_J O$ forms an open cover of M and that $\chi_1, \dots, \chi_J$ is a smooth partition of unity on M with $\operatorname{supp}\chi_j \subseteq \Gamma g_j O$ for $j = 1, \dots, J$.

Exercise 9.91. Use Fourier series on $L^2(\mathbb{T}^2)$ to show that there exists a constant c_0 so that any subspace $V_T \subseteq C^\infty(\mathbb{T}^2)$ with

$$\mathcal{S}_1^{\mathbb{T}^2}(\phi) = \sqrt{\|\phi\|_2^2 + \sum_{k=1}^{2} \|\partial_k \phi\|_2^2} \leqslant \sqrt{1+T}\|\phi\|_2$$

for all $\phi \in V_T$ has dimension at most $c_0 T$.

Suppose now that $(\psi_n)_{n\in\mathbb{N}_0}$ is a sequence of eigenfunctions for Δ_{hyp} with eigenvalues $-\lambda_n$ satisfying $\lambda_0 = 0 < \lambda_1 \leqslant \lambda_2 \leqslant \cdots$. We are interested in obtaining an upper bound of the form

$$N(T) = |\{n \mid \lambda_n \leqslant T\}| < cT \tag{9.95}$$

for some c (depending on X) and all $T \geqslant 1$. We define V_T to be the span of all ψ_n with $\lambda_n \leqslant T$. With this (9.95) is a version of the dimension estimates appearing in Exercise 9.91.

Exercise 9.92 (Even case). (a) Show that any $\phi \in V_T$ satisfies $\mathcal{S}_1(\phi) \leqslant \sqrt{1+4T}\|\phi\|_2$ (if we use the basis $\mathbf{a}, \mathbf{d}, \mathbf{k}$ of $\mathfrak{sl}_2(\mathbb{R})$ to define $\mathcal{S}_1$).
(b) Show that any $\phi \in V_T$ and $j \in \{1, \dots, J\}$ satisfy $\mathcal{S}_1(\chi_j \phi) \ll \sqrt{1+4T}\|\phi\|_2$, where the implicit constant depends on our choice of the partition of unity.
(c) Show that for any $T \geqslant 1$ there exists some $j(T) \in \{1, \dots, J\}$ and a subspace $V_T^* \leqslant V_T$ of dimension at least $N(T)/J$ so that $\phi \in V_T^*$ implies $\|\chi_{j(T)}\phi\|_2 \geqslant \frac{1}{J}\|\phi\|_2$.
(d) Now use Exercise 9.91 for $\chi_{j(T)} V_T^*$ to prove (9.95).

Exercise 9.93 (Odd case). Prove that the number of direct summands of $L^2(X)$ isomorphic to an odd principal series representation with Casimir eigenvalue α satisfying $|\alpha| \leqslant T$ grows at most linearly with T.

9.9 Further Topics

Many of our discussions were directed towards $\mathrm{SL}_2(\mathbb{R})$ and related groups. From this it should be clear that we have barely scratched the surface of the theory of unitary representations of Lie groups, and that there are many ways one could go from here. We finish this chapter by highlighting a few of these further directions.

- Naturally one could also be interested in studying $\widehat{G}$ for other simple Lie groups, and one of the next candidates for this might be $\mathrm{SL}_2(\mathbb{C})$. As it turns out, the unitary dual of $\mathrm{SL}_2(\mathbb{C})$ is somewhat less complicated than that of $\mathrm{SL}_2(\mathbb{R})$, since $\mathrm{SL}_2(\mathbb{C})$ has no discrete series representations.
- For number theory the restriction to real and complex Lie groups is unnatural, and one should also study the unitary duals of the form $\mathbb{G}(\mathbb{Q}_p)$ for simple algebraic groups $\mathbb{G}$ over $\mathbb{Q}_p$.
- We briefly touched on the study of $L^2(\Gamma\backslash\mathrm{SL}_2(\mathbb{R}))$ for (cocompact) lattices $\Gamma < \mathrm{SL}_2(\mathbb{R})$. Here any appearance of the principal or complementary series representations corresponds to an eigenfunction of Δ_{hyp} on the space $M = \Gamma\backslash\mathrm{SL}_2(\mathbb{R})/K$. We did not discuss another extraordinary connection to number theory: Any appearance of the discrete series representation in $L^2(\Gamma\backslash\mathrm{SL}_2(\mathbb{R}))$ corresponds to modular forms for Γ.
- A surprising uniform spectral gap property was conjectured by Selberg in 1965 [86]. According to this the action-associated representation on the space $L^2_0(\Gamma_N\backslash\mathrm{SL}_2(\mathbb{R}))$ is tempered for all $N \in \mathbb{N}$, where

$$\Gamma_N = \{\gamma \in \mathrm{SL}_2(\mathbb{Z}) \mid \gamma \equiv I \pmod{N}\}$$

  is the principal congruence subgroup of level N. This amounts to saying that the first non-trivial eigenvalue λ_1 of Δ_{hyp} on $\Gamma_N\backslash\mathbb{H}$ is always at least $\frac{1}{4}$; Selberg already proved that $\lambda_1 \geqslant \frac{3}{16}$ for all $N \in \mathbb{N}$. As this is written, the best results in the direction of Selberg's conjecture include work of Kim and Sarnak [51, App. 2] giving the bound $\frac{1}{4} - (\frac{7}{64})^2$, and numerical verification of the conjecture for $N < 857$ by Booker and Strömbergsson [8].
- Using this and works of Jacquet and Langlands [45] and of Burger and Sarnak [9] many further cases of the so-called property (τ) could be established. The complete solution was obtained by Clozel in 2003 [16]: There is uniformity of spectral gap for all congruence quotients defined by simple $\mathbb{Q}$-groups.

As mentioned in the Preface on page v, the latter topic was our original goal for this volume. Even though we did not achieve this ambitious goal, we hope that the text will help students and researchers to establish sound base camps for their own journeys into various mathematical mountain ranges.

Appendix A
Linear Algebra

A.1 Generating the Special Linear Group

Let $\mathbb{K}$ be a field, $n \geqslant 2$ an integer, and consider the group $\mathrm{SL}_n(\mathbb{K})$. For indices $j,k \in \{1,\dots,n\}$ with $j \neq k$ we also define the elementary unipotent subgroup $U_{j,k} = \{u_{j,k}(s) \mid s \in \mathbb{K}\}$ where $u_{j,k}(s)$ is the matrix with 1s on the diagonal, the entry $s \in \mathbb{K}$ in the jth row and kth column, and 0s in all other positions. We claim that these elementary unipotent subgroups generate, meaning that

$$\mathrm{SL}_n(\mathbb{K}) = \langle u_{j,k}(s) \mid j \neq k \text{ and } s \in \mathbb{K} \rangle.$$

To see this, we modify the Gauss elimination algorithm.[†] Recall that multiplication of an element $g \in \mathrm{SL}_n(\mathbb{K})$ by $u_{j,k}(s)$ on the left corresponds to the row operation of adding the multiple of the kth row of g by s to the jth row of g. If $g_{j,1} = 0$ for all $j > 1$, then we first multiply by $u_{2,1}(1)$ to achieve $g_{2,1} \neq 0$. So assume that $g_{j,1} \neq 0$ for some $j > 1$. Then we may multiply by $u_{1,j}(s)$ for an appropriate choice of $s \in \mathbb{K}$ to achieve $g_{1,1} = 1$. Using $u_{j,1}(s)$ for $j > 1$ and appropriate $s \in \mathbb{K}$ will then achieve a new matrix with $g_{j,1} = \delta_{j,1}$ for $j \in \{1,\dots,n\}$.

Repeating this for the second column and also multiplying by appropriate $u_{j,k}(s)$ with $j < k$ and $s \in \mathbb{K}$ on the right (corresponding to row operations), we eventually arrive at a diagonal matrix for which the first $n-1$ diagonal entries are 1. Since by assumption g has determinant 1 and all the elementary unipotent matrices also have determinant 1, we have thus obtained the identity matrix.

Equivalently, we have elementary unipotent matrices $u_1,\dots,u_p$ and $v_1,\dots,v_q$ with

$$u_1 u_2 \cdots u_p g v_1 \cdots v_q = I,$$

which shows that $g = (u_1 \cdots u_p)^{-1}(v_1 \cdots v_q)^{-1}$ and so $\mathrm{SL}_n(\mathbb{K})$ is generated by the elementary unipotent matrices as claimed.

[†] We will modify the initial group element $g \in \mathrm{SL}_n(\mathbb{K})$ by multiplication with elementary unipotent matrices, but will call the resulting matrix g at each stage of the argument.

M. Einsiedler and T. Ward, *Unitary Representations and Unitary Duals*,
Graduate Texts in Mathematics 308, https://doi.org/10.1007/978-3-032-03899-9

Exercise A.1 (Commutator of special linear group). Show that

$$U_{j,k} \subseteq [\mathrm{SL}_n(\mathbb{K}), \mathrm{SL}_n(\mathbb{K})]$$

for $j \neq k$, and conclude that $[\mathrm{SL}_n(\mathbb{K}), \mathrm{SL}_n(\mathbb{K})] = \mathrm{SL}_n(\mathbb{K})$.

A.2 The Cartan Decomposition

We briefly recall the so-called polar or Cartan decomposition in $\mathrm{GL}_d(\mathbb{R})$: Every element $g \in \mathrm{GL}_d(\mathbb{R})$ can be written in the form

$$g = kak'$$

with $k, k' \in \mathrm{O}_d(\mathbb{R})$ and a positive diagonal matrix a whose eigenvalues are arranged in decreasing size. Moreover, if $g \in \mathrm{SL}_d(\mathbb{R})$ then it is possible to achieve $k, k' \in \mathrm{SO}_d(\mathbb{R})$.

To obtain this decomposition, we only have to apply basic linear algebra. Indeed, notice that gg^{t} is positive and symmetric. Hence gg^{t} has an orthonormal basis of eigenvectors. Ordering them by size of the respective eigenvalues, and switching the sign of the last eigenvector if necessary, these eigenvectors together form the columns of a matrix $k \in \mathrm{SO}_d(\mathbb{R})$ with the property that $D = k^{\mathrm{t}}gg^{\mathrm{t}}k$ is diagonal, with positive eigenvalues arranged in decreasing size. We define the diagonal matrix a to be the positive square root of $D = a^2$, and define k' to be $a^{-1}k^{\mathrm{t}}g$. Then $g = kak'$ and

$$k'(k')^{\mathrm{t}} = a^{-1} \underbrace{k^{\mathrm{t}}gg^{\mathrm{t}}k}_{=D} a^{-1} = a^{-1}Da^{-1} = I$$

shows that $k' \in \mathrm{O}_d(\mathbb{R})$ as claimed.

If $g \in \mathrm{SL}_d(\mathbb{R})$ then $\det k = 1$ and $\det a > 0$ forces $\det k' = 1$, and so we also have $k' \in \mathrm{SO}_d(\mathbb{R})$ as claimed.

Appendix B
Analysis

B.1 Integrability Exponents of Products

Lemma B.1. *Let (X,μ) be a σ-finite measure space, and assume that p_1 and p_2 lie in $(0,\infty)$. Then*

$$L^{p_1}(X)L^{p_2}(X) \subseteq L^{\frac{p_1p_2}{p_1+p_2}}(X). \tag{B.1}$$

More generally,

$$L^{\frac{1}{s_1}}(X)\cdots L^{\frac{1}{s_m}}(X) \subseteq L^{\frac{1}{s_1+\cdots+s_m}}(X) \tag{B.2}$$

for all $m \in \mathbb{N}$ and $s_1,\dots,s_m \in (0,\infty)$

PROOF. Let $f_j \in L^{p_j}(X)$ for $j \in \{1,2\}$. Then $|f_1|^{\frac{p_1p_2}{p_1+p_2}} \in L^{\frac{p_1+p_2}{p_2}}(X)$, since

$$\int \left(|f_1|^{\frac{p_1p_2}{p_1+p_2}}\right)^{\frac{p_1+p_2}{p_2}} \mathrm{d}\mu = \int |f_1|^{p_1}\,\mathrm{d}\mu < \infty,$$

and, similarly, $|f_2|^{\frac{p_1p_2}{p_1+p_2}} \in L^{\frac{p_1+p_2}{p_1}}(X)$. Using $\frac{p_1+p_2}{p_2} \geqslant 1$, $\frac{p_1+p_2}{p_1} \geqslant 1$,

$$\frac{1}{\frac{p_1+p_2}{p_2}} + \frac{1}{\frac{p_1+p_2}{p_1}} = 1,$$

and Hölder's inequality, we deduce that

$$|f_1f_2|^{\frac{p_1p_2}{p_1+p_2}} = |f_1|^{\frac{p_1p_2}{p_1+p_2}}|f_2|^{\frac{p_1p_2}{p_1+p_2}} \in L^1(X),$$

which proves (B.1). To iterate this, it is convenient to note that (B.1) is equivalent to

$$L^{\frac{1}{s_1}}(X)L^{\frac{1}{s_2}}(X) \subseteq L^{\frac{1}{s_1+s_2}}(X)$$

for all $s_1,s_2 \in (0,\infty)$. This gives (B.2) by induction. □

M. Einsiedler and T. Ward, *Unitary Representations and Unitary Duals*,
Graduate Texts in Mathematics 308, https://doi.org/10.1007/978-3-032-03899-9

B.2 Stirling's Formula

Exercise B.2. Use integration by parts to find a recurrence relation for $\int_0^\infty t^n \mathrm{e}^{-t}\,\mathrm{d}t$ as a function of $n \in \mathbb{N}_0$, and use this to deduce that $\int_0^\infty t^n \mathrm{e}^{-t}\,\mathrm{d}t = n!$.

The gamma function is defined[(23)] by permitting $n \in \mathbb{N}_0$ to be a real (or complex) variable in Exercise B.2. More precisely, we define the gamma function by

$$\Gamma(x) = \int_0^\infty t^{x-1}\mathrm{e}^{-t}\,\mathrm{d}t$$

for $x > 0$. Just as integration by parts gives Exercise B.2, we have

$$\Gamma(x+1) = \int_0^\infty t^x \mathrm{e}^{-t}\,\mathrm{d}t = -t^x\mathrm{e}^{-t}\Big|_0^\infty + x\int_0^\infty t^{x-1}\mathrm{e}^{-t}\,\mathrm{d}t = x\Gamma(x)$$

for all $x > 0$.

Stirling's formula for the gamma function states that

$$\Gamma(x) \sim \sqrt{\frac{2\pi}{x}}\left(\frac{x}{\mathrm{e}}\right)^x, \tag{B.3}$$

where $\sim$ means that the ratio of the left-hand and right-hand side converges to 1 as $x \to \infty$.

Exercise B.3. Prove the asymptotic Stirling formula (B.3) by the following steps.
(a) Use the substitution $t = xu$ to obtain

$$\Gamma(x+1) = \int_0^\infty \mathrm{e}^{x\log t - t}\,\mathrm{d}t = \mathrm{e}^{x\log x}x\int_0^\infty \mathrm{e}^{(x(\log u - u)}\,\mathrm{d}u. \tag{B.4}$$

(b) Now use *Laplace's method* for the integral term in (B.4) as follows. Calculate that $\log u - u$ has its maximum at $u_0 = 1$, and apply Taylor approximation at u_0 to see that

$$\log u - u = -1 - \tfrac{1}{2}(u-1)^2 + \mathrm{O}\big((u-1)^3\big)$$

for u near u_0. Show that the integral over the regions $(0, 1-\delta)$ and $(1+\delta, \infty)$ are of smaller order, and apply another convenient substitution to study

$$\int_{1-\delta}^{1+\delta} \mathrm{e}^{x(\log u - u)}\,\mathrm{d}u.$$

(c) Combine the above and use the asymptotic $\left(1+\frac{1}{x}\right)^x \to \mathrm{e}$ as $x \to \infty$ to obtain the desired asymptotic for $\Gamma(x)$.

Appendix C
Topological Groups

C.1 Metrizability

Recall that a collection $\mathcal{U}$ of neighbourhoods of a point x in a topological space X is a *neighbourhood basis* if for any neighbourhood $V \ni x$ there is some neighbourhood $U \in \mathcal{U}$ with $U \subseteq V$. A topological space is *first-countable* if every point has a countable neighbourhood basis. We note that a metric space (X, d) is automatically first-countable, since the open neighbourhoods $B_{1/n}(x)$ for $n \in \mathbb{N}$ form a countable neighbourhood basis at $x \in X$.

Also recall that a topological group is a group G together with a Hausdorff topology with respect to which the maps $g \mapsto g^{-1}$ and $(g_2, g_2) \mapsto g_1 g_2$ are continuous. This means that

- if U is a neighbourhood of a product $g_1 g_2 \in G$, then there are neighbourhoods $U_1 \ni g_1$ and $U_2 \ni g_2$ with $U_1 U_2 \subseteq U$;
- if U is a neighbourhood of $g \in G$, then there is a neighbourhood V of g^{-1} with $V^{-1} \subseteq U$.

Lemma C.1 (Birkhoff–Kakutani [6, 46]). *The following properties of a topological group G are equivalent.*

(1) *G has a left-invariant metric, that is a metric d_ℓ giving the topology of G which additionally has $\mathsf{d}_\ell(gh_1, gh_2) = \mathsf{d}_\ell(h_1, h_2)$ for all $g, h_1, h_2 \in G$.*
(2) *G is metrizable.*
(3) *G is first-countable.*
(4) *The identity $e \in G$ has a countable basis of open neighbourhoods.*

PROOF.[(24)] It is clear that (1) $\Longrightarrow$ (2) $\Longrightarrow$ (3) $\Longrightarrow$ (4).

So let us assume (4), and let $\mathcal{V} = \{V_1, V_2, \ldots\}$ be a countable neighbourhood basis at the identity e consisting of open sets. Without loss of generality, we may assume that

$$V_1 \supseteq V_2 \supseteq \cdots \supseteq \{e\},$$

M. Einsiedler and T. Ward, *Unitary Representations and Unitary Duals*,
Graduate Texts in Mathematics 308, https://doi.org/10.1007/978-3-032-03899-9

and, since G is Hausdorff, we have $\bigcap_{n\geqslant 1} V_n = \{e\}$. We wish to construct sets that mimic the behaviour of nested metric open sets. To that end, we use the continuity properties of the two group operations to inductively construct from the sequence of sets (V_n) another nested sequence of open neighbourhoods of the identity

$$U_1 = G \supseteq U_{2^{-1}} \supseteq U_{2^{-2}} \supseteq \cdots \supseteq \{e\}$$

with the property that $U_{2^{-n}}^{-1} = U_{2^{-n}}$ (each set is symmetric), $U_{2^{-n}} \subseteq V_n$, and $U_{2^{-(n+1)}}U_{2^{-(n+1)}} \subseteq U_{2^{-n}}$ for each $n \geqslant 1$. We note that $U_{2^{-1}}U_{2^{-1}}$ is automatically contained in $U_1 = G$. It follows that

$$\bigcap_{n\geqslant 0} U_{2^{-n}} \subseteq \bigcap_{n\geqslant 1} V_n = \{e\}. \tag{C.1}$$

Moreover, for any dyadic rational $a \in \mathbb{Z}[\frac{1}{2}] \cap (0,1]$ we define

$$U_a = U_{2^{-n_1}} \cdots U_{2^{-n_r}} \tag{C.2}$$

where

$$a = 2^{-n_1} + \cdots + 2^{-n_r} \tag{C.3}$$

is the binary expansion of a arranged in the natural order; that is, with

$$1 \leqslant n_1 < \cdots < n_r.$$

Using these neighbourhoods, we define the function

$$\begin{aligned} f\colon G &\longrightarrow [0,1] \\ g &\longmapsto \inf\{a \in \mathbb{Z}[\tfrac{1}{2}] \cap (0,1] \mid g \in U_a\}. \end{aligned}$$

As we will see, f captures all the essential information about the sets U_a for a lying in $\mathbb{Z}[\frac{1}{2}] \cap (0,1]$. For instance, we have

$$g \in U_a \Longrightarrow f(g) \leqslant a \tag{C.4}$$

for all $g \in G$ and $a \in \mathbb{Z}[\frac{1}{2}] \cap (0,1]$, by definition.

Using this function, we also define

$$\begin{aligned} \mathsf{d}_\ell\colon G \times G &\longrightarrow [0,1] \\ (g_1, g_2) &\longmapsto \sup_{k\in G} |f(kg_1) - f(kg_2)|. \end{aligned}$$

Our aim is to show that d_ℓ is a left-invariant metric inducing the topology of G, as claimed in (1).

The following properties are easy to verify from the definition:

(i) (Semi-positivity) $\mathsf{d}_\ell(g_1, g_2) \geqslant 0$ for all $g_1, g_2 \in G$.
(ii) (Symmetry) $\mathsf{d}_\ell(g_1, g_2) = \mathsf{d}_\ell(g_2, g_1)$ for all $g_1, g_2 \in G$.
(iii) (Triangle inequality) $\mathsf{d}_\ell(g_1, g_3) \leqslant \mathsf{d}_\ell(g_1, g_2) + \mathsf{d}_\ell(g_2, g_3)$ for $g_1, g_2, g_3 \in G$.

(iv) (Left-invariance) $\mathsf{d}_\ell(hg_1, hg_2) = \mathsf{d}_\ell(g_1, g_2)$ for all $g_1, g_2, h \in G$.

In other words, d_ℓ defines a left-invariant pseudo-metric. We still have to show that d_ℓ is a metric, and that it induces the original topology of G. For this we have to establish further properties of the neighbourhoods defined in (C.2) and of the function f. In fact, we claim that $a = k2^{-n}$ with $k, n \in \mathbb{N}$ and $k < 2^n$ implies

$$U_a U_{2^{-n}} \subseteq U_{a+2^{-n}}. \tag{C.5}$$

Furthermore, we claim that the neighbourhoods are nested in the sense that

$$a < b \Longrightarrow U_a \subseteq U_b \tag{C.6}$$

for all $a, b \in \mathbb{Z}[\frac{1}{2}] \cap (0, 1]$. We note that this implies a partial converse to (C.4). Indeed, by combining (C.6) with the definition of f we have that

$$f(g) < b \Longrightarrow g \in U_b \tag{C.7}$$

for all $g \in G$ and $b \in \mathbb{Z}[\frac{1}{2}] \cap (0, 1]$.

For the proof of (C.5), we note that if there is no carry in the binary addition of $a = k2^{-n}$ and 2^{-n} then this is just the definition in (C.2), but if there is a carry one uses the defining properties of $U_{2^{-j}}$ for $j \leqslant n$ (possibly more than once) to obtain (C.5).

To see (C.6), we suppose that a is given by (C.3). If $b < a+2^{-n_r}$, this follows since by definition U_b is the product of U_a and a set containing e. If $b = a+2^{-n_r}$, this follows from (C.5). If $b > a + 2^{-n_r}$, the conclusion follows from the latter case and induction on n_r.

Our next aim is to show that d_ℓ is indeed a metric, meaning that it also satisfies the following property:

(v) (Definiteness) If $\mathsf{d}_\ell(g_1, g_2) = 0$ for some $g_1, g_2 \in G$, then $g_1 = g_2$.

So assume that some $g_1, g_2 \in G$ satisfy $\mathsf{d}_\ell(g_1, g_2) = 0$. By left-invariant, this is equivalent to $\mathsf{d}_\ell(e, g_1^{-1}g_2) = 0$, which by definition of d_ℓ implies that

$$|f(e) - f(g_1^{-1}g_2)| = 0.$$

As U_a for $a \in \mathbb{Z}[\frac{1}{2}] \cap (0, 1]$ is a neighbourhood of e, we see that $f(e) = 0$ and so also $f(g) = 0$ for $g = g_1^{-1}g_2$. By (C.7), this shows that $g \in U_{2^{-n}}$ for all $n \in \mathbb{N}$, which by (C.1) implies $g = e$ and so $g_1 = g_2$ as required.

As d_ℓ is a left-invariant metric, the metric topology induced by d_ℓ has the property that U is a neighbourhood of e if and only if gU is a neighbourhood of g. As G is assumed to be a topological group, this also holds for the original topology. Hence it suffices to compare the neighbourhoods of e with respect to the metric and the original topology.

Let us write $B_r(e)$ for the metric ball of radius $r > 0$ around e with respect to d_ℓ. For $n \in \mathbb{N}$ and $g \in B_{2^{-n}}(e)$ we have $f(g) \leqslant \mathsf{d}_\ell(g, e) < 2^{-n}$, which, together with (C.7), implies $g \in U_{2^{-n}}$. Using the fact that $\{U_{2^{-n}} \mid n \in \mathbb{N}\}$ is a neighbourhood basis in the original topology, this implies the following property:

(vi) (First comparison) Any neighbourhood of e in the original topology is also a metric neighbourhood with respect to d_ℓ.

So it remains to prove the converse to (vi):

(vii) (Second comparison) Any metric neighbourhood of e is also a neighbourhood in the original topology of G.

We note that (vii) follows from $U_{2^{-n-1}} \subseteq B_{2^{-n+1}}(e)$ for $n \in \mathbb{N}$. To see the latter, let $g \in U_{2^{-n-1}}$ and $k \in G$ be arbitrary. If

$$f(k) \in [j2^{-(n+1)}, (j+1)2^{-(n+1)})$$

for some $j \in \mathbb{N}_0$, then $k \in U_{(j+1)2^{-(n+1)}}$ by (C.7), and so $kg \in U_{(j+2)2^{-(n+1)}}$ by (C.5), which implies that

$$f(kg) < (j+2)2^{-(n+1)} \leqslant f(h) + 2^{-n}.$$

Using $U^{-1} = U$ and the fact that $h \in G$ and $g \in U$ were arbitrary, the inequality $f(k) < f(kg) + 2^{-n}$ follows from the former. Together, we obtain

$$\mathsf{d}_\ell(g, e) = \sup_{k \in G} |f(kg) - f(k)| \leqslant 2^{-n}.$$

As $g \in U_{2^{-n-1}}$ was arbitrary, we obtain $U_{2^{-n-1}} \subseteq B_{2^{-n+1}}(e)$, which implies (vii) as $n \in \mathbb{N}$ was arbitrary. □

If d_ℓ is a left-invariant metric then $\mathsf{d}_{\mathrm{r}}(x, y) = \mathsf{d}_\ell(x^{-1}, y^{-1})$ is a right-invariant metric defining the same topology. A bi-invariant metric d_{bi} with

$$\mathsf{d}_{\mathrm{bi}}(xgy, xhy) = \mathsf{d}_{\mathrm{bi}}(g, h)$$

for all $x, y, g, h \in G$ giving the topology of G only exists in special cases (see Exercise C.2). A striking result of Milnor [71] is that a connected Lie group admits a bi-invariant metric if and only if it is isomorphic to $K \times \mathbb{R}^n$ for some compact Lie group K.

Exercise C.2. (a) Show that a metrizable compact group has a bi-invariant metric that gives its topology.
(b) Show that $G = \mathrm{SL}_d(\mathbb{R})$ cannot be equipped with a bi-invariant metric that gives its topology for $d \geqslant 2$.

C.2 Quotient Groups

As in the main text, we will for brevity write equality to denote some canonical isomorphisms. The following result contains an analogue of the open mapping theorem.

Proposition C.3 (Quotient group and open maps). *Let G be a locally compact, σ-compact metric group and let $H \subseteq G$ be a closed normal subgroup of G.*

Then G/H is again a locally compact σ-compact metric group. The canonical projection map

$$p\colon G \ni g \longmapsto gH \in G/H$$

is continuous, open, and a subset of G/H is compact if and only if it is the image under p of a compact subset in G. Moreover, if $\theta\colon G \to G'$ is a continuous surjective homomorphism between locally compact, σ-compact metric groups G and G', then θ is open and induces an isomorphism $\overline{\theta}$ between the topological groups $G/\ker\theta$ and G'.

PROOF. Recall that G/H is a group since H is normal and that a set $O \subseteq G/H$ is open with respect to the quotient topology if and only if $p^{-1}(O) \subseteq G$ is open. In particular, $p\colon G \to G/H$ is continuous by definition of the quotient topology. If $U \subseteq G$ is open, then

$$p^{-1}\big(p(U)\big) = UH = \bigcup_{h\in H}(Uh)$$

is open in G. By definition of the quotient topology, this shows that $p(U)$ is open in G/H. That is, $p\colon G \to G/H$ is continuous and open.

Since G is second countable and p is continuous and open, it follows that G/H is also second countable. In particular, every $gH \in G/H$ has a countable basis of neighbourhoods, and it suffices to study converging sequences to describe the topology of G/H.

Suppose now that $g_nH \to gH$ as $n \to \infty$. We claim that we can lift this convergence to G, that is to show that there exists some (h_n) in H with $g_nh_n \to g$ as $n \to \infty$. Indeed, if this is not the case then there is a neighbourhood U of g for which $(g_nH) \cap U = \emptyset$ for infinitely many n, and this would imply that g_nH lies outside the neighbourhood $p(U)$ of gH infinitely often, which is a contradiction.

Suppose now that $g_nH \to gH$ and $g_nH \to g'H$ as $n \to \infty$. By the above, there exist sequences (h_n) and (h'_n) in H with $g_nh_n \to g$ and $g_nh'_n \to g'$ as $n \to \infty$, which implies that $h_n^{-1}h'_n = (g_nh_n)^{-1}(g_nh'_n) \to g^{-1}g'$ as $n \to \infty$, so $g^{-1}g' \in H$ and $gH = g'H$ since H is closed. Therefore G/H is Hausdorff. As we have already obtained second countability, we also have that G/H is metrizable by Lemma C.1. Since p is open and continuous we now also see that G/H is locally compact and σ-compact.

To see continuity of the group operation, we suppose that $g_nH \to gH$ and $g'_nH \to g'H$ as $n \to \infty$. By the lifting claim above, there exist sequences (h_n) and (h'_n) in H with $g_nh_n \to g$ and $g'_nh'_n \to g'$ as $n \to \infty$. This implies that $g_nh_ng'_nh'_n \to gg'$ and $(g_nh_n)^{-1} \to g^{-1}$ as $n \to \infty$ and hence $(g_nH)(g'_nH) \to gg'H$ and $(g_nH)^{-1} \to (gH)^{-1}$ as $n \to \infty$, by applying the continuous projection map, as required.

We show next that a compact subset in G/H is precisely the image $p(K)$ of some compact set $K \subseteq G$. By continuity of p the image $p(K)$ of a compact set $K \subseteq G$ is compact. Suppose that $K' \subseteq G/H$ is compact. For every element $gH \in K'$ we choose an open neighbourhood U_g of g with compact closure. Hence we obtain the open cover $\{p(U_g) \mid gH \in K'\}$ of K', and can choose a finite subcover $K' \subseteq p(U_{g_1} \cup \cdots \cup U_{g_n})$. Defining

$$K = \overline{U_{g_1} \cup \cdots \cup U_{g_n}} \cap p^{-1}(K'),$$

we obtain the claim.

Suppose now that $\theta\colon G \to G'$ is a surjective continuous homomorphism as in the last part of the proposition. Let U be an open neighbourhood of $0 \in G$, and let $V = V^{-1}$ be a compact neighbourhood of 0 with $VV \subseteq U$. By σ-compactness of G there exists a sequence (g_n) in G with $G = \bigcup_{n=1}^{\infty} g_n V$. By surjectivity of θ, this implies that $G' = \bigcup_{n=1}^{\infty} \theta(g_n)\theta(V)$, and by the properties of the Haar measure we have $m_{G'}(\theta(V)) > 0$. For the function $f = \mathbb{1}_{\theta(V)}$ we have

$$\begin{aligned} f * f(h) &= \int_{G'} \mathbb{1}_{\theta(V)}(k) \underbrace{\mathbb{1}_{\theta(V)}(k^{-1}h)}_{=\mathbb{1}_{\theta(V)}(h^{-1}k)} \,\mathrm{d}m_{G'}(k) \\ &= \langle \mathbb{1}_{\theta(V)}, \lambda_h \mathbb{1}_{\theta(V)} \rangle \\ &= m_{G'}\bigl(\theta(V) \cap (h\theta(V))\bigr) \end{aligned}$$

since $\theta(V) = \theta(V)^{-1}$. Using continuity of the regular representation and Cauchy–Schwarz, this implies that $f * f \in C_c(G')$ with $f * f(0) > 0$. Therefore

$$(f * f)^{-1}\bigl((0, \infty)\bigr) \subseteq \theta(V)\theta(V)^{-1} = \theta(VV) \subseteq \theta(U)$$

is a neighbourhood of $0 \in G'$. As U was an arbitrary neighbourhood of $0 \in G$ and θ is a homomorphism, this shows that θ is an open map.

We define $\overline{\theta}\colon G/\ker\theta \to G'$ by $\overline{\theta}(g\ker\theta) = \theta(g)$. If $O \subseteq G'$ is open, then $\theta^{-1}(O) \subseteq G$ is open by continuity of θ, and so $\overline{\theta}^{-1}(O)$ is open in $G/\ker\theta$ by definition of the quotient topology. If $U \subseteq G/\ker\theta$ is open, then

$$V = p^{-1}(U) = \{g \in G \mid g\ker\theta \in U\}$$

is open, and so $\overline{\theta}(U) = \theta(V)$ is open in G' by the above. This concludes the proof. □

C.3 Locally Homogeneous Spaces

Some of the discussions in this volume concern the action of a group G on quotients $\Gamma\backslash G$ by a discrete subgroup Γ. We only briefly discuss this set-up, and refer to [24, Ch. 9, 11] and [26, Ch. 1] for more details.

In the following, we assume that G is a locally compact, σ-compact, metric group and that $\Gamma < G$ is a discrete subgroup. By the Birkhoff–Kakutani lemma (Lemma C.1) we may assume that the metric d_G on G is left-invariant. This allows us to define the metric d_X on the quotient space

$$X = \Gamma\backslash G = \{\Gamma g \mid g \in G\}$$

by

$$\mathsf{d}_X(\Gamma g_1, \Gamma g_2) = \inf_{\gamma_1,\gamma_2\in\Gamma} \mathsf{d}_G(\gamma_1 g_1, \gamma_2 g_2) = \inf_{\gamma\in\Gamma} \mathsf{d}_G(g_1, \gamma g_2) \tag{C.8}$$

for all $\Gamma g_1, \Gamma g_2 \in X$. We note that the equality of the two infima is due to the left invariance of d_G (specifically, under translation by elements of Γ).

Now fix some $x = \Gamma h \in X$. Then the map

$$G \ni g \longmapsto hg \longmapsto \Gamma hg = xg$$

is Lipschitz with Lipschitz constant 1. Indeed, for $g_1, g_2 \in G$ we have

$$\mathsf{d}_X(xg_1, xg_2) = \inf_{\gamma\in\Gamma} \mathsf{d}_G(hg_1, \gamma hg_2) \leqslant \mathsf{d}_G(hg_1, hg_2) = \mathsf{d}_G(g_1, g_2)$$

by definition of d_X and the left-invariance of d_G by h. Restricting the map above to a suitable ball $B_r^G(e)$ around the identity $e \in G$, we even obtain an isometry. To see this, let

$$r = \tfrac{1}{4} \inf_{\gamma\in\Gamma\smallsetminus\{e\}} \mathsf{d}_G(e, h^{-1}\gamma h) \tag{C.9}$$

and note that the discreteness of Γ (and hence of $h^{-1}\Gamma h$) implies that $r > 0$. If now $g_1, g_2 \in B_r^G(e)$ and $\gamma \in \Gamma$ satisfy

$$\mathsf{d}_G(hg_1, \gamma hg_2) \leqslant \mathsf{d}_G(g_1, g_2) < 2r,$$

then

$$\begin{aligned}
\mathsf{d}_G(e, h^{-1}\gamma h) &\leqslant \mathsf{d}_G(e, g_1) + \mathsf{d}_G(g_1, h^{-1}\gamma h) \\
&< r + \mathsf{d}_G(h^{-1}\gamma^{-1}hg_1, e) \\
&< r + \mathsf{d}_G(h^{-1}\gamma^{-1}hg_1, g_2) + \mathsf{d}_G(g_2, e) \\
&< 2r + \mathsf{d}_G(hg_1, \gamma hg_2) < 4r.
\end{aligned}$$

However, this implies $\gamma = e$ by the choice of r, and hence that

$$B_r^G(e) \ni g \longmapsto xg \in B_r^X(x) \tag{C.10}$$

is an isometry. In this sense X is locally isometric to the group G, which is the reason we refer to X as a *locally homogeneous space*.

We note that if X is compact, then one can find (for example, by varying h in (C.9) within a compact set $C \subseteq G$ with $X = \Gamma C$) a uniform $r > 0$ with the property that (C.10) is an isometry for every $x \in X$. In this case Γ is called a *uniform lattice*.

Recall that we assumed that G is σ-compact. Hence we can find a sequence (h_n) of elements of G and a sequence (r_n) of positive real numbers so that

$$B_{r_n}^G(e) \ni g \longmapsto \Gamma h_n g \in B_{r_n}^X(\Gamma h_n)$$

is isometric for all $n \in \mathbb{N}$ and

$$X = \bigcup_{n=1}^{\infty} B_{r_n}^X(\Gamma h_n).$$

This allows the definition of the measurable set

$$F = \bigsqcup_{n=1}^{\infty} \left(h_n B_{r_n}^G(e) \smallsetminus \bigcup_{k<n} \bigcup_{\gamma \in \Gamma} \gamma h_k B_{r_k}^G(e) \right)$$

which is a *fundamental domain* in the sense that

$$F \ni g \longmapsto \Gamma g \in X \tag{C.11}$$

is a bijection.

Assuming that G is unimodular, we can define a measure m_X, which we will also call the Haar measure, as the push-forward of m_G restricted to F with respect to the map $F \ni g \mapsto \Gamma g \in X$. Using left invariance of m_G, it can be shown that m_X does not depend on the choice of the fundamental domain F. Using this and right invariance of m_G, it can be shown that m_X is invariant under the natural action

$$G \times X \ni (g, \Gamma h) \longmapsto \Gamma h g^{-1} \in X$$

of G on X.

Exercise C.4. (a) Show that (C.9) does indeed define a metric inducing the quotient topology on $X = \Gamma \backslash G$.
(b) Show that for a compact quotient $X = \Gamma \backslash G$ one can find a uniform $r > 0$ so that (C.10) is an isometry for all $x \in X$.
(c) Show that (C.11) is indeed a bijection.
(d) Show that the measure m_G is well-defined and G-invariant.

Hints for Selected Problems

Exercise 1.8 (p. 11): Choose, for example, a non-zero function $f \in C_c(\mathbb{R})$ so that

$$\mathrm{supp}(\lambda_g f) \cap \mathrm{supp}(f) = \emptyset.$$

Alternatively, construct a non-zero function $f \in C_c(\mathbb{R})$ so that so that $\lambda_g f$ is 'almost' equal to $-f$.

Exercise 1.9 (p. 11): As μ and ν define the same measure class, $\{x \in X \mid \frac{\mathrm{d}\nu}{\mathrm{d}\mu}(x) = 0\}$ is a μ-null set. For a measurable non-negative function f consider the identity

$$\int f \,\mathrm{d}\mu = \int f \Big(\frac{\mathrm{d}\nu}{\mathrm{d}\mu}\Big)^{-1} \frac{\mathrm{d}\nu}{\mathrm{d}\mu} \,\mathrm{d}\mu = \int f \Big(\frac{\mathrm{d}\nu}{\mathrm{d}\mu}\Big)^{-1} \mathrm{d}\nu.$$

Exercise 1.10 (p. 11): For $g_1, g_2 \in \mathrm{SL}_2(\mathbb{R})$ and $v \in \mathbb{S}^1$ we have $g_2 v = \|g_2 v\| g_2 \bullet v$ by our definitions, and so $\|g_1 g_2 v\| = \|g_1(g_2 \bullet v)\| \|g_2 v\|$. Now raise this identity to the power -2.

Exercise 1.11 (p. 11): Let $F\colon X \to [0, \infty)$ be measurable and let $J_g(x) = |\det D_x(g \bullet x)|$ for $x \in X$ denote the Jacobian of $X \ni x \mapsto g \bullet x \in X$. Then by the substitution rule for Lebesgue measure we have

$$\begin{aligned}\int_X F(y) \,\mathrm{d}m_X(y) &= \int_X F(g^{-1} \bullet x) J_{g^{-1}}(x) \,\mathrm{d}m_X(x) \\ &= \int_X F(g^{-1} \bullet x) J_{g^{-1}}(g g^{-1} \bullet x) \,\mathrm{d}m_X(x) = \int_X F(y) J_{g^{-1}}(g \bullet y) \,\mathrm{d}g_*^{-1} m_X(y).\end{aligned}$$

Conclude that $J(g, x) = \frac{\mathrm{d}m_X}{\mathrm{d}g_*^{-1} m_X}(x) = J_{g^{-1}}(g \bullet x)$ for almost every $x \in X$.

Exercise 1.12 (p. 11): Show first that $c(e, x) = 1$ for all $x \in X$ and then use the cocycle equation for $c(g, g^{-1} \bullet x) c(g^{-1}, x)$.

Exercise 1.14 (p. 12): It is clearly sufficient to show that weak continuity implies continuity. For this let $u \in \mathcal{H}_\pi$, $g, g_0 \in G$, expand $\|\pi_g u - \pi_{g_0} u\|^2$ in terms of inner products, and use weak continuity for $g \to g_0$.

Exercise 1.18 (p. 16): For the modular character, you should obtain $\Delta\left(\begin{pmatrix} a & b \\ & 1 \end{pmatrix}\right) = |a|^{-2}$ for (a), and

M. Einsiedler and T. Ward, *Unitary Representations and Unitary Duals*, Graduate Texts in Mathematics 308, https://doi.org/10.1007/978-3-032-03899-9

$$\Delta\left(\begin{pmatrix} a_1 & x & z \\ & a_2 & y \\ & & a_3 \end{pmatrix}\right) = \frac{a_2}{a_1}\frac{a_3}{a_2}\frac{a_3}{a_1} = \frac{a_3^2}{a_1^2}$$

for (b).

Exercise 1.19 (p. 16): For (a) use $m(Bg) = \Delta(g)m(B)$ for $B = G$. For (b), assuming finite left Haar measure, let $U = U^{-1}$ be a compact neighbourhood of the identity e of G. Find a maximal collection of disjoint left translates $g_1U, g_2U, \dots$ and show that this collection must be finite. Now show that $G = \bigcup g_i U^2$.

Exercise 1.23 (p. 19): If $v \in \mathcal{V}$, $w \in \mathcal{V}^\perp$, and $g \in G$, then $\langle \pi_g w, v\rangle = \langle w, \pi_{g^{-1}} v\rangle = 0$, which implies that $\pi_g w \in \mathcal{V}^\perp$. Hence $\pi_g(v+w) = \pi_g v + \pi_g w$, $\pi_g v \in \mathcal{V}$, and $\pi_g w \in \mathcal{V}^\perp$ for $v \in \mathcal{V}$ and $w \in \mathcal{V}^\perp$.

Exercise 1.24 (p. 19): To show continuity for $S = \mathbb{N}$, let

$$D = \{(w_n) \mid w_n = 0 \text{ for all but finitely many } n\}$$

and use Lemma 1.13.

Exercise 1.27 (p. 21): Use induction on $\dim \mathcal{H}$ and Exercise 1.23.

Exercise 1.30 (p. 23): Apply Exercise 1.23 to see that $\mathcal{H}_\pi$ has no non-trivial proper π-invariant subspaces.

Exercise 1.31 (p. 23): Clearly $\{0\}$, $\mathcal{H}_\pi \times \{0\}$, $\{0\} \times \mathcal{H}_\rho$, ad $\mathcal{H}_\pi \oplus \mathcal{H}_\rho$ are closed invariant subspaces. Show that if there is another closed invariant subspace, then it must be the graph of a bounded intertwining isomorphism between $\mathcal{H}_\pi$ and $\mathcal{H}_\rho$ (see also Corollary 1.38).

Exercise 1.35 (p. 24): The centre of $\mathrm{SL}_2(\mathbb{R})$ is given by $\{\pm I\}$. To see that $\mathrm{SL}_2(\mathbb{R})$ has no characters, see Exercise A.1 in Appendix A. For (b) show that $\pi^{\mathbb{S}^1,\xi}$ splits into two closed invariant subspaces corresponding to the two central characters.

Exercise 1.39 (p. 26): After replacing $\mathcal{H}_\pi$ by the complement of the kernel of T one can assume that T is injective. Consider again the unitary representation $\pi \oplus \rho$ and its restriction to the invariant closed subspace $\mathrm{Graph}(T)$. Apply Lemma 1.21 to the restrictions to $\mathrm{Graph}(T)$ of the intertwining projection operators onto $\mathcal{H}_\pi$ resp. $\mathcal{H}_\rho$.

Exercise 1.42 (p. 28): Consider the map $\mathbb{C}[G] \ni r \mapsto \pi_*(r)v_0$ and apply Lemma 1.21 on the orthogonal complement of the kernel.

Exercise 1.43 (p. 29): For a given $v \in \mathcal{H}_\pi$ and $t \in \{0, \dots, n-1\}$ consider the sum

$$v_t = \frac{1}{n}\sum_{j=0}^{n-1} \mathrm{e}^{-2\pi \mathrm{i} tj/n} \pi_k^j v.$$

Show that $v_t \in \mathcal{V}_t = \{w \in \mathcal{H}_\pi \mid \pi_k w = \mathrm{e}^{2\pi \mathrm{i} t/n} w\}$ and $v = \sum_{t=0}^{n-1} v_t$.

Exercise 1.45 (p. 32): For (a), the two characters on $\mathrm{S}_3/\mathrm{A}_3$ give rise to two old representations, and any of the two non-trivial characters on A_3 give rise to the same new representation of S_3. For (b) define $P_{\mathrm{s}}v = \frac{1}{6}\sum_{\tau\in\mathrm{S}_3}\rho(\tau)v$, $P_{\mathrm{ss}}v = \frac{1}{6}\sum_{\tau\in\mathrm{S}_3}\mathrm{sign}(\tau)\rho(\tau)v$, and $P_{cs}v = v - \frac{1}{3}(v + \rho(\sigma)v + \rho(\sigma^2)v)$ for all $v \in V$. Finally, show that a non-zero cycle-symmetric vector $v \in V_{cs}$ generates a representation that is either the irreducible new representation of S_3 or the square of the irreducible new representation.

Exercise 1.47 (p. 35): If $u \in L^1(G)$ is a unit, consider $u*\psi_n$ for an approximate identity (ψ_n) as in Proposition 1.46.

Exercise 1.48 (p. 35): Apply Proposition 1.46 to $\psi_n' = \left(\int_{B_n} \psi_n \,\mathrm{d}m\right)^{-1} \psi_n|_{B_n}$ and estimate $\|\psi_n - \psi_n'\|_1$.

Exercise 1.50 (p. 38): Due to the involution it is enough to show that μ in $M(G)$ and f in $L^1(G)$ implies $\mathrm{d}(\mu * \nu_f) = \mathrm{d}\nu_{f'}$ for some f' in $L^1(G)$. Now use the substitution $h = g_1 g_2$ for g_2 to see that

$$\int F \,\mathrm{d}\mu * \nu_f = \iint F(g_1 g_2) f(g_2) \,\mathrm{d}\mu(g_1) \,\mathrm{d}m(g_2) = \int F(h) \int f(g_1^{-1} h) \,\mathrm{d}\mu(g_1) \,\mathrm{d}m(h)$$

and set $f'(h) = \displaystyle\int f(g_1^{-1} h) \,\mathrm{d}\mu(g_1)$.

Exercise 1.54 (p. 41): Apply the definition to the identity $\langle B \circ \pi_*(\nu) u, w \rangle = \langle \pi_*(\nu) u, B^* w \rangle$ for $\nu \in M(G)$, $u \in \mathcal{H}_\pi$, and $w \in \mathcal{H}_\rho$.

Exercise 1.56 (p. 41): Let $a, w \in \mathcal{H}_\pi$ and calculate $\langle \pi_{g_0} \pi_*(f) u, w \rangle = \langle \pi_*(f) u, \pi_{g_0^{-1}} w \rangle$ using the definition of the convolution operator.

Exercise 1.57 (p. 41): We refer to the discussion in [25, Sec. 3.5.3].

Exercise 1.58 (p. 42): Assume first $f \in \mathscr{L}^\infty(X)$ is real-valued and non-negative, and that ν is a probability measure on G. Apply Jensen's inequality to obtain

$$\left(\int_G f(g^{-1} \bullet x) \,\mathrm{d}\nu(g) \right)^2 \leqslant \int_G f(g^{-1} \bullet x)^2 \,\mathrm{d}\nu(g)$$

for every $x \in X$. Now integrate with respect to μ to obtain

$$\int_X \left(\int_G f(g^{-1} \bullet x) \,\mathrm{d}\nu(g) \right)^2 \mathrm{d}\mu(x) \leqslant \|f\|_2^2$$

by Fubini's theorem. Use this to show that for any $f \in L^2_\mu(X)$ and $\nu \in M(G)$, the map

$$x \longmapsto \int_G f(g^{-1} \bullet x) \,\mathrm{d}\nu(g)$$

defines an $L^2_\mu(X)$ function satisfying the defining property of $\pi_*(\nu) f$. See also [25, Exercise 3.86(b) & Lem. 3.75].

Exercise 1.59 (p. 42): For $u \in \mathcal{H}_\pi \smallsetminus \{0\}$, we have

$$\begin{aligned} |\langle \pi_*(\nu) v - \nu(G) w, u \rangle| &= \left| \int \langle \pi_g v - w, u \rangle \,\mathrm{d}\nu(g) \right| \\ &\leqslant \int_{\mathrm{supp}(\nu)} |\langle \pi_g v - w, u \rangle| |f_\nu(g)| \,\mathrm{d}\mu(g) \\ &\leqslant \int_{\mathrm{supp}(\nu)} \|\pi_g v - w\| \|u\| |f_\nu(g)| \,\mathrm{d}\mu(g) \leqslant \varepsilon \|\nu\| \|u\|, \end{aligned}$$

where $d\nu = f_\nu \, d\mu$ as in Section 1.5.2. Now set $u = \pi_*(\nu)v - \nu(G)w$ to derive the conclusion.

Exercise 1.60 (p. 42): For (a), note that

$$\left|\langle \pi_*(f)v, v\rangle\right| = \left|\int_G f(g)\langle \pi_g v, v\rangle \, dm(g)\right| \leqslant \int_G |f(g)| \|v\|^2 \, dm(g) = \|v\|^2,$$

and analyze the case of equality. For (b), assume first that $\mathcal{H}_\pi$ has no invariant vectors, let $v \in \mathcal{H}_\pi$, and apply the spectral theorem for the self-adjoint operator $\pi(f)$ to find a measure μ_v on $[-1, 1]$ with

$$\|\pi_*(f)^n v\| = \langle \pi_*(f)^{2n} v, v\rangle = \int t^{2n} \, d\mu_v$$

for all $n \in \mathbb{N}$. Show that $\mu_v(\{-1, 1\}) = 0$.

Exercise 1.63 (p. 43): Note that $\mathcal{H}_\pi = \mathcal{H}_\pi^G \oplus (\mathcal{H}_\pi^G)^\perp$ and that cyclicity of π implies that $\pi|_{\mathcal{H}_\pi^G}$ is also cyclic.

Exercise 1.67 (p. 45): Show that the map has trivial kernel.

Exercise 1.68 (p. 45): Show that the graph of U is the cyclic subspace generated by the element $(v, a) \in \mathcal{H}_\pi \oplus \mathcal{H}_\rho$. Use the assumption $\varphi^\pi_{v,w} = \varphi^\rho_{a,b}$ to see that this cyclic subspace is not equal to $\mathcal{H}_\pi \oplus \mathcal{H}_\rho$, and apply Corollary 1.38.

Exercise 1.69 (p. 45): Define $B\colon \mathcal{H}_\pi \to L^2(G)$ by $B(v) = \varphi^\pi_{v,w}$ for $v \in \mathcal{H}_\pi$ and some fixed $w \in \mathcal{H}_\pi$ of norm one. Use the equivariance of matrix coefficients and Schur's lemma (Theorem 1.29) to see that B is an isometric embedding, so that $\pi < \rho$. Finally, use Section 1.2 to see that $\pi < \lambda$.

Exercise 1.75 (p. 51): By Proposition 1.72, ϕ is a matrix coefficient $\phi = \varphi_v$ for some vector $v \in \mathcal{H}_\pi$ and a unitary representation π. Calculate $\langle \pi_*(f_1)v, \pi_*(f_2)v\rangle$ for $f_1, f_2 \in L^1(G)$ and set $f_1 = f_2$.

Exercise 1.80 (p. 60): If $\|\phi_0\| = \|\varphi^\pi_v\| = \|v\|^2 \in (0, 1)$ and $\|\varphi^\rho_0\| = \|a\|^2 \leqslant 1$, then the proof of (1.37) fails since $\langle a', a\rangle = \|v\|^2 + \mathrm{O}(\varepsilon)$ might not be sufficient for (1.37). The case of $\phi_0 = 0$ is even worse, as we cannot even start the proof. In Chapter 2 we will see how different the neighbourhoods of $\phi_0 = 0$ are, even when $G = \mathbb{R}$.

Exercise 1.82 (p. 61): Use Corollary 1.81 to derive a contradiction from the assumption that there are $g_1, g_2 \in G$ with $g_1 g_2 \neq g_2 g_1$.

Exercise 1.84 (p. 62): Restricting to $G \times \mathcal{E}^1(G)$, we even have continuity by Proposition 1.79, which together with $(g, 0) \mapsto 0$ for $g \in G$ gives measurability.

Exercise 1.85 (p. 62): Let $K \subseteq G$ be compact and recall that $C(K)$ is separable with respect to the supremum norm. Use this to find for $\varepsilon > 0$ a sequence (ϕ_j) in $\mathcal{E}^1(G)$ with the property that

$$\mathcal{E}^1(G) = \bigcup_{j=1}^\infty \underbrace{\{\phi \in \mathcal{E}^1(G) \mid \|\phi - \phi_j\|_{K,\infty} < \varepsilon\}}_{=B_j}$$

and $n \in \mathbb{N}$ with

$$\mu\left(\mathcal{E}^1(G) \smallsetminus \bigcup_{j=1}^n B_j\right) < \varepsilon.$$

Now restrict the integral in Corollary 1.83 to $\bigcup_{j=1}^n B_j$ and, for $j = 1, \ldots, n$, approximate this integral over

$$\widetilde{B}_j = B_j \smallsetminus \bigcup_{k=1}^{j-1} B_k$$

by $\mu(\widetilde{B}_j)\phi_j$.

Exercise 1.86 (p. 62): Let $f \in L^1(G)$ with $f = f^*$, $\int f \,\mathrm{d}m = 1$, and $f > 0$. By Exercise 1.60 we have $\rho_*(f)^n v \to P_{\mathcal{H}_\rho^G}(v)$ as $n \to \infty$ for any unitary representation ρ and $v \in \mathcal{H}_\rho$. For $\rho \neq \mathbb{1}$ this shows that $\rho_*(f)^n v \to 0$ as $n \to \infty$ for all $v \in \mathcal{H}_\rho$ or, equivalently, $\int f^{*n}\phi \,\mathrm{d}m \to 0$ as $n \to \infty$ for all $\phi \in \mathcal{E}^1(G) \smallsetminus \{\mathbb{1}\}$. Now we use Fubini's theorem and dominated convergence to get

$$\begin{aligned}\langle \pi_*(f)^n v, \pi_*(f)^n v\rangle &= \int f^{*2n}\phi_0 \,\mathrm{d}m = \int_{\mathcal{E}^1(G)} \int f^{*2n}\phi \,\mathrm{d}m \,\mathrm{d}\mu(\phi)\\ &= \mu(\{\mathbb{1}\}) + \underbrace{\int_{\mathcal{E}^1(G)\smallsetminus\{\mathbb{1}\}} \int f^{*2n}\phi \,\mathrm{d}m \,\mathrm{d}\mu(\phi)}_{\longrightarrow 0 \text{ as } n \longrightarrow \infty}.\end{aligned}$$

Exercise 1.88 (p. 64): Apply Theorem 1.87 to two copies of $\mathrm{SL}_2(\mathbb{R})$ inside $\mathrm{SL}_3(\mathbb{R})$ containing the matrices

$$\begin{pmatrix} 1 & 1 & \\ & 1 & \\ & & 1 \end{pmatrix}, \begin{pmatrix} 1 & & \\ & 1 & 1 \\ & & 1 \end{pmatrix}$$

respectively.

Exercise 1.89 (p. 64): Use induction on the dimension of $\mathcal{H}_\pi$. In the induction step, pick an eigenvector $v \in \mathcal{H}_\pi$ for $a = \begin{pmatrix} 2 & \\ & \frac{1}{2} \end{pmatrix} \in \mathrm{SL}_2(\mathbb{R})$. Now apply the argument for the first case in Theorem 1.87 to show that v is fixed by $\mathrm{SL}_2(\mathbb{R})$, and apply the induction hypothesis to $\langle v\rangle^\perp$.

Exercise 1.92 (p. 65): Show that if $\pi\left(\begin{pmatrix} 1 & 1 \\ 0 & 1 \end{pmatrix}\right) = I$ is trivial, then $\pi\left(\begin{pmatrix} 1 & 0 \\ 1 & 1 \end{pmatrix}\right) = I$ is also, and that this then implies that π is trivial. Now use the subgroups

$$\left\{\begin{pmatrix} 1 & k \\ 0 & 1 \end{pmatrix} \,\middle|\, k \in \mathbb{F}_p\right\} \lhd \left\{\begin{pmatrix} a & k \\ 0 & a^{-1} \end{pmatrix} \,\middle|\, a \in \mathbb{F}_p^\times, k \in \mathbb{F}_p\right\}$$

in the same way as in the discussion in Section 1.4.4, and note that $\frac{p-1}{2} = |(\mathbb{F}_p^\times)^2|$.

Exercise 2.2 (p. 67): For (a) use the translation invariance of m to show that $\int \chi \,\mathrm{d}m = 0$ for $\chi \neq \mathbb{1}_G$, and deduce that the characters form an orthonormal set. For completeness of this orthonormal set use the Stone–Weierstrass theorem (see [25, Sec. 2.3.2]). Finally, use the assumption that G is compact and metrizable to see that $\widehat{G}$ is countable. For (b) show that the terms of an approximate identity (ψ_n) can be approximated by finite linear combinations of characters (using the Stone–Weierstrass theorem) and apply Proposition 1.52 (see also [25, Sec. 3.5]).

Exercise 2.3 (p. 68): For (a) there is no continuity requirement, so the generator of $\mathbb{Z}$ can be sent to any element of $\mathbb{S}^1 \cong \mathbb{T}$. Part (b) can be deduced from (c), by composing with the canonical map $\mathbb{R} \to \mathbb{T}$. For (c), let $\chi\colon \mathbb{R} \to \mathbb{S}^1$ be a continuous character and consider, for example,

$$\mathbb{R} \ni x \longmapsto \int_x^{x+\delta} \chi \,\mathrm{d}m = \chi(x) \int_0^\delta \chi \,\mathrm{d}m$$

for a suitable $\delta > 0$ to show that χ is differentiable and satisfies $\chi' = a\chi$ for some $a \in \mathbb{C}$.

Exercise 2.6 (p. 72): For the topology on $\widehat{\mathbb{R}} \cong \mathbb{R}$, show that

$$B_{\varepsilon/2\pi N}(0) \subseteq \left\{t \in \mathbb{R} \mid |\mathrm{e}^{2\pi\mathrm{i}xt} - 1| < \varepsilon \text{ for all } x \in [-N, N]\right\}$$

and

$$\left\{t \in \mathbb{R} \mid |\mathrm{e}^{2\pi\mathrm{i}xt} - 1| < \tfrac{\varepsilon}{2} \text{ for all } x \in [0, 1]\right\} \subseteq B_{\varepsilon}(0)$$

for all $\varepsilon \in (0, 1]$ and $N > 0$. For the isomorphism $\widehat{\mathbb{R}^d} \cong \mathbb{R}^d$ use the definition $\chi_t(g) = \mathrm{e}^{2\pi\mathrm{i}g\cdot t}$ for $g, t \in \mathbb{R}^d$ and the inner product $g \cdot t = \sum_{j=1}^{d} g_j t_j$. Given an unknown $\chi \in \widehat{\mathbb{R}^d}$ consider the restriction to the jth coordinate axis to find $t_j \in \mathbb{R}$ for $j = 1, \dots, d$.

Exercise 2.7 (p. 72): If $a \in \mathcal{A}$ has $\|a\| < 1$ and $\chi(a) = 1$ then $b = \sum_{n=1}^{\infty} a^n$ converges and has $a + ab = b$, which implies $\chi(a) + \chi(a)\chi(b) = \chi(b)$ or equivalently $1 + \chi(b) = \chi(b)$.

Exercise 2.8 (p. 72): For $t \in \mathbb{R}$ define $\chi_t \colon \mathbb{R} \ni x \mapsto \mathrm{e}^{2\pi\mathrm{i}xt}$. Then

$$\mathcal{E}^1(\mathbb{R}) \ni \chi_t \to 0 \in \mathcal{E}^{\leqslant 1}(\mathbb{R})$$

as $t \to \infty$ in the weak* topology but not in the compact-open topology.

Exercise 2.11 (p. 76): Show that $\mu_v = \sum_{n\in\mathbb{N}} |v_n|^2 \delta_{t_n}$.

Exercise 2.14 (p. 79): Use the spectral theorem to show that the unitary representation χ_{t_0} associated to $t_0 \in \widehat{G}$ is contained in π if and only if there exists a spectral measure μ_v for some $v \in \mathcal{H}_\pi$ with $\mu_v(\{t_0\}) > 0$.

Exercise 2.16 (p. 79): Use, for example, the trivial representation on $\ell^2(\mathbb{N})$.

Exercise 2.25 (p. 87): Use the two equivariance properties in the Plancherel formula (Theorem 2.17) to pass from λ_{g_n} to M_{g_n} and then from M_{g_n} to $\lambda_{\imath(g_n)}$.

Exercise 2.28 (p. 88): If G is discrete, then the compact-open topology on $\widehat{G}$ is equal to the product topology inherited from $(\mathbb{S}^1)^G$.

Exercise 2.35 (p. 94): Generalize the proof of Corollary 2.32.

Exercise 2.37 (p. 94): Let $H = \overline{\langle A \rangle}$ be the closure of the subgroup generated by A. Show that $H^\perp = A^\perp$ and apply Proposition 2.33(3).

Exercise 2.38 (p. 95): Redefine H_n as a subgroup of $\mathbb{Q}$, so that $\imath_n$ becomes the inclusion map $H_n \hookrightarrow H_{n+1}$ for every $n \in \mathbb{N}$.

Exercise 2.49 (p. 105): Notice that $\mathbb{Z}[\frac{1}{p} \mid p \in S]$ is a dense subset of $\mathbb{R}$ since $S \neq \emptyset$. Use this to construct an injective map from $\widehat{\mathbb{R}}$ to $\widehat{\mathbb{Z}[\frac{1}{p} \mid p \in S]}$ by restricting characters on $\mathbb{R}$ to the dense subgroup.

Exercise 2.54 (p. 109): If $\langle v \rangle_\pi \perp \langle w \rangle_\pi$ then $\mu_{v,w} = 0$ satisfies the characterizing property (2.22). Conversely, if $\mu_{v,w} = 0$, then $\langle \pi_g v, w \rangle = 0$ for all $g \in G$ by (2.22).

Exercise 2.56 (p. 110): Show that equality holds if and only if v and w are linearly dependent.

Exercise 2.59 (p. 114): If $\mathcal{V}$ is invariant, use Exercise 2.54 to show that $\pi_{\mathrm{FC}}(F)\mathcal{V} \perp \mathcal{V}^\perp$. The converse follows from Corollary 1.53 and Proposition 2.58.

Exercise 2.60 (p. 114): To see this we calculate

$$\int \langle g,t\rangle\langle g_0,t\rangle \,\mathrm{d}\mu_{v,w}(t) = \int \langle gg_0,t\rangle \,\mathrm{d}\mu_{v,w}(t) = \langle \pi_g \pi_{g_0} v, w\rangle = \int \langle g,t\rangle \,\mathrm{d}\mu_{\pi_{g_0} v,w}(t)$$

for $g, g_0 \in G$ to obtain $\mathrm{d}\mu_{\pi_{g_0} v,w}(t) = \langle g_0,t\rangle \,\mathrm{d}\mu_{v,w}(t)$. This gives

$$\begin{aligned}\int \langle g_0,t\rangle F \,\mathrm{d}\mu_{v,w}(t) &= \int F \,\mathrm{d}\mu_{\pi_{g_0} v,w}(t) = \langle \pi_{\mathrm{FC}}(F)\pi_{g_0} v, w\rangle \\ &= \langle \pi_{g_0}\pi_{\mathrm{FC}}(F)v, w\rangle = \int \langle g_0,t\rangle \,\mathrm{d}\mu_{\pi_{\mathrm{FC}}(F)v,w}\end{aligned}$$

for all $g_0 \in G$, and this implies the second claim in the exercise. Alternatively, apply Proposition 2.53(5)–(6) and Proposition 2.58(6).

Exercise 2.63 (p. 116): For $v \in \mathcal{H}_\pi$ and $B = \bigsqcup_{n=1}^\infty B_n$, we have

$$\begin{aligned}\left\| \Pi_B v - \sum_{n=1}^N \Pi_{B_n} v \right\|^2 &= \left\| \Pi_B v - \Pi_{\bigsqcup_{n=1}^N B_n} v \right\|^2 \\ &= \left\| \pi_{\mathrm{FC}}\big(\mathbb{1}_{\bigsqcup_{n=N+1}^\infty}\big) v \right\|^2 = \mu_v\Big(\bigsqcup_{n=N+1}^\infty B_n \Big) \longrightarrow 0\end{aligned}$$

as $N \to \infty$, since μ_v is a finite measure on $\widehat{G}$. For the counter-example, any unitary representation with $\Pi_{B_n} \neq 0$ for all $n \in \mathbb{N}$ (as is the case, for example, for the regular representation on G and a countable partition $\{B_1, B_2, \dots\}$ decomposing $\widehat{G}$ into subsets with positive measure) will do.

Exercise 2.66 (p. 118): For one direction rephrase Proposition 2.58(1) using $\mu_{\max}$. For the second, note that the operator norm of M_F on $L^2_{\mu_{\max}}(\widehat{G})$ is equal to the essential supremum norm of F.

Exercise 2.68 (p. 118): Apply the spectral theorem twice and define a multiplication operator using the Radon–Nikodym derivative.

Exercise 2.73 (p. 124): To see that every 'strong measurable' set is also 'weak measurable' show that $B_M = \{T \in \mathrm{B}(\mathcal{V}_\infty) \mid \|T\| \leqslant M\}$ is both strong and weak measurable, that the strong operator topology restricted to B_M is separable and metrizable, that $T \mapsto \|Tv\|$ is weak measurable on B_M, and that the strong neighbourhood $\{T \in B_M \mid \|Tv - w\| \leqslant \varepsilon\}$ is weak measurable for any $v, w \in \mathcal{V}_\infty$ and $\varepsilon > 0$.

Exercise 3.2 (p. 133): For (b), suppose that there exists a π-invariant closed real subspace $\mathcal{V}_\mathbb{R}$ such that $\mathcal{H}_\pi = \mathcal{V}_\mathbb{R} \oplus \mathrm{i}\mathcal{V}_\mathbb{R}$, and define $U\colon \mathcal{H}_\pi \to \mathcal{H}'_\pi$ by

$$U(v_1 + \mathrm{i}v_2) = (v_1 - \mathrm{i}v_2)' = v_1' + \mathrm{i}v_2'.$$

For (a) apply (b) with $\mathcal{V}_\mathbb{R} = L^2_\mathbb{R}(G)$.

Exercise 3.3 (p. 133): If $\mathcal{H}_\pi = \bigoplus_{n\in\mathbb{N}} L^2_{\mu_n}(\widehat{G})$, then

$$\mathcal{H}_{\overline{\pi}} = \mathcal{H}'_\pi \cong \bigoplus_{n\in\mathbb{N}} L^2_{\mu_n^*}(\widehat{G})$$

where μ_n^* is the image of μ_n under the map $\widehat{G} \ni t \mapsto -t \in \widehat{G}$. Simplifying the notation, suppose $\mathcal{H}_\pi = L^2_\mu(\widehat{G})$ is defined by the multiplication representation and a finite measure μ

on $\widehat{G}$. Then $L^2_\mu(\widehat{G}) \ni f \mapsto f^* \in L^2_{\mu^*}(\widehat{G})$ defined by $f^*(t) = \overline{f(-t)}$ is conjugate-linear, an isometry, and gives rise to the intertwining unitary isomorphism.

Exercise 3.5 (p. 136): If $\mathcal{V} \otimes' \mathcal{W}$ together with the map $(v,w) \mapsto v \otimes' w$ is another Hilbert space with the same properties, use the universal property of $\mathcal{V} \otimes_{\mathrm{la}} \mathcal{W}$ to show that the subspace spanned by $\{v \otimes' w \mid v \in \mathcal{V}, w \in \mathcal{W}\}$ is isometrically isomorphic to $\mathcal{V} \otimes_{\mathrm{la}} \mathcal{W}$ equipped with $\langle \cdot, \cdot \rangle_\otimes$. Extend this isomorphism to $\mathcal{V} \otimes \mathcal{W}$.

Exercise 3.8 (p. 138): Use continuity of $\otimes$ from Proposition 3.4.

Exercise 3.10 (p. 139): First note that $\|A \otimes B\| = \|(A \otimes I)(I \otimes B)\| \leqslant \|A\|\|B\|$ by the argument in the proof, and then calculate $\sup\{\|(A \otimes B)(v \otimes w)\| \mid \|v\| \leqslant 1, \|w\| \leqslant 1\}$. Next note that the last formula in Corollary 3.9 follows from (3.4) and density of the linear span of the vectors of the form $v \otimes w$ for $v \in \mathcal{V}$ and $w \in \mathcal{W}$. For $v_1, v_2 \in \mathcal{V}$ and $w_1, w_2 \in \mathcal{W}$ we also have

$$\begin{aligned}\langle (A \otimes B)^* v_1 \otimes w_1, v_2 \otimes w_2 \rangle &= \langle v_1 \otimes w_1, (A \otimes B) v_2 \otimes w_2 \rangle \\ &= \langle v_1, A v_2 \rangle \langle w_1, B w_2 \rangle = \langle A^* v_1, v_2 \rangle \langle B^* w_1, w_2 \rangle \\ &= \langle (A^* \otimes B^*) v_1 \otimes w_1, v_2 \otimes w_2 \rangle,\end{aligned}$$

which implies that $(A \otimes B)^* = A^* \otimes B^*$. The remaining properties follow from this.

Exercise 3.12 (p. 140): Given a unit vector $w \in \mathcal{H}_\rho$, recall that the map $v \mapsto v \otimes w$ from $\mathcal{H}_\pi$ to $\mathcal{H}_\pi \otimes \mathcal{H}_\rho$ is intertwining for π and for the restriction of $\pi \otimes \rho$ to G. Conclude from this that $\langle v_0 \otimes w_0 \rangle_{\pi \otimes \rho} \supseteq \mathcal{H}_\pi \otimes w_0$ and repeat for a fixed $v \in \mathcal{H}_\pi$ the argument using the map $w \mapsto v \otimes w$ from $\mathcal{H}_\rho$ to $\mathcal{H}_\pi \otimes \mathcal{H}_\rho$.

Exercise 3.13 (p. 140): Use the two constructions introduced in this chapter and Corollary 1.81 to verify that the assumptions of the Stone–Weierstrass theorem are satisfied for the closure of the linear hull of all matrix coefficients of all irreducible representations of G.

Exercise 3.14 (p. 140): Show that the spectral measure for $v \otimes w$ is the product $\mu_v \times \mu_w$ of the spectral measures of v and w. For the inner tensor product, take the push-forward under $\widehat{G} \times \widehat{G} \ni (t_1, t_2) \mapsto t_1 + t_2$, obtaining $\mu_v * \mu_w$.

Exercise 3.24 (p. 149): The sign representation $(a,x) \mapsto (-1)^a$ for $(a,x) \in G$ and the trivial representation are the only one-dimensional representations. All other irreducible unitary representations are two-dimensional, defined by $\pi^{(n)}_{(0,x)}(v_1, v_2) = (\chi_n(x) v_1, \overline{\chi_n(x)} v_2)$ and $\pi^{(n)}_{(1,0)}(v_1, v_2) = (v_2, v_1)$ for some $n \in \mathbb{N}$, for $x \in \mathbb{T}$ and $(v_1, v_2) \in \mathbb{C}^2$.

Exercise 3.35 (p. 157): By our standing assumption G is metrizable, which implies that $L^2(G)$ is separable. Now apply the Peter–Weyl theorem (Theorem 3.34).

Exercise 3.37 (p. 157): Combine Lemma 3.22 with the Peter–Weyl theorem (Theorem 3.34).

Exercise 3.38 (p. 157): Counting dimensions of representations shows that $|G| = \sum_{[\pi] \in \widehat{G}} d_\pi^2$.

Exercise 3.49 (p. 164): Use Schur orthogonality and the equivariance properties from the proof of Theorem 3.34 to describe $G \ni g \mapsto d_\pi \pi_{m,n} * \pi_{k,\ell}(g) = d_\pi \langle \varphi_{\pi_g w_k, w_\ell}, \varphi_{w_n, w_m} \rangle$.

Exercise 3.52 (p. 165): In fact we have $|G^\sharp| = \dim L^2(G^\sharp) = |\widehat{G}|$.

Exercise 4.1 (p. 169): For (a), note that

$$t \longmapsto g_t \int_0^\delta g_s \, \mathrm{d}s = \int_t^{t+\delta} g_s \, \mathrm{d}s$$

is differentiable by the fundamental theorem of calculus. For sufficiently small $\delta > 0$, this gives the result. For (b), prove that if $t \mapsto g_t$ has derivative $\mathbf{m}$ at 0, then $g_t' = \mathbf{m} g_t = g_t \mathbf{m}$ for all $t \in \mathbb{R}$. Then assume that $t \mapsto g_t$ and $t \mapsto h_t$ have the same derivative at 0, and calculate the derivative of $t \mapsto g_t^{-1} h_t = g_{-t} h_t$ for all $t \in \mathbb{R}$.

Exercise 4.5 (p. 176): For $\mathbf{c} \in \mathfrak{g}$, apply Exercise 4.1 for the homomorphism

$$\mathbb{R} \ni t \longmapsto \rho(\exp(t\mathbf{c})),$$

which shows that ρ has directional derivatives satisfying (4.7). To see the differentiability of ρ, let $\mathbf{c}_1, \dots, \mathbf{c}_d$ be a basis of $\mathfrak{g}$ so that

$$\mathbb{R}^d \ni (t_1, \dots, t_d) \longmapsto \exp(t_1 \mathbf{c}_1) \cdots \exp(t_d \mathbf{c}_d)$$

is a smooth coordinate system of G near 0, and apply ρ.

Exercise 4.7 (p. 181): Only the step with the subtitle 'Assuming irreducibility' needs to be generalized.

Exercise 4.12 (p. 183): Analyze the weights to show that $\pi_m \otimes \pi_n$ is isomorphic to the sum of $\pi_{m+n}, \pi_{m+n-2}, \dots, \pi_{|m-n|}$.

Exercise 4.18 (p. 191): Let A denote the area measure on $\mathbb{S}^2 \cong \mathrm{SU}_2(\mathbb{R})/T$. Show that the measure μ defined by

$$\int_{\mathrm{SU}_2(\mathbb{R})} f \, \mathrm{d}\mu = \int_{\mathrm{SU}_2(\mathbb{R})/T} \int_T f(gt) \, \mathrm{d}m_T(t) \, \mathrm{d}A(gT)$$

for $f \in C(\mathrm{SU}_2(\mathbb{R}))$ is a Haar measure on $\mathrm{SU}_2(\mathbb{R})$. Alternatively, use (4.24) and argue as on pages 192–194.

Exercise 4.20 (p. 194): Since $\mathrm{SO}_3(\mathbb{R}) \cong \mathrm{SU}_2(\mathbb{R})/C$ we can identify $\mathrm{SO}_3(\mathbb{R})^\sharp$ with $[0,2]$ and the characters are given by the characters $\chi_0, \chi_2, \chi_4, \dots$ of $\mathrm{SU}_2(\mathbb{R})$ corresponding to even highest weights.

Exercise 5.3 (p. 198): If v is an eigenvector for t, then

$$\pi_a \pi_g v = \pi_g \pi_{g^{-1}ag} v = \pi_g \langle g^{-1}ag, t\rangle v = \langle a, g{\scriptstyle\bullet}t\rangle \pi_g v$$

for $a \in A$ and $g \in G$.

Exercise 5.9 (p. 202): It is tempting to think that this is tautological, but in fact it is an elementary consequence of properties of Haar measure. In fact for a Borel subset $B \subseteq G$ satisfying $m(B) > 0$ and $m(B \triangle gB) = 0$ for all $g \in G$, define μ by $\mathrm{d}\mu = \mathbb{1}_B \, \mathrm{d}m$ and show that μ is a Haar measure to conclude that $m(G \smallsetminus B) = 0$.

Exercise 5.12 (p. 203): Apply the definition of ergodicity to the sets $F^{-1}([\alpha, \beta))$ for rational $\alpha < \beta$ and take the intersection of all such sets of full measure.

Exercise 5.13 (p. 204): Show that $F = \frac{\mathrm{d}\mu_1}{\mathrm{d}\mu_2}$ is S-invariant, and use this function to define an intertwining isometric multiplication operator from $L^2(\widehat{H}, \mu_1)$ into $L^2(\widehat{H}, \mu_2)$.

Exercise 5.14 (p. 205): For (a), define $\mathcal{H}_\pi = \mathbb{C}^{K_d}$ and

$$\big(\pi_a v\big)(k) = \langle a, kt_0\rangle v(k),$$
$$\big(\pi_{k_0} v\big)(k) = v(k_0^{-1}k)$$

for $a \in A$, $v \in \mathcal{H}_\pi$, and $k, k_0 \in K_d$. For (c), prove that the new representations are isomorphic if and only if $K_d t_0 = K_d t_1$.

Exercise 5.17 (p. 209): Let $\{B_1, B_2, \dots\}$ be a countable partition of $r\mathbb{S}^1$ with $\mu(B_n) > 0$ for all $n \in \mathbb{N}$. Define $v = \big(\mathbb{1}_{B_n}\big)_n \in \mathcal{H}^\infty_{\pi^r}$, and show that v is a cyclic vector for $\mathcal{H}^\infty_{\pi^r}$ (by once again using the functional calculus for $\pi|_A$).

Exercise 5.18 (p. 209): For (a) the new representations are their own contragredient representations, which follows from Exercise 3.3 since $\mu_r^* = \mu_r$ for all $r > 0$.

Exercise 5.19 (p. 209): Mimic the discussion in Section 1.4.4 using the normal subgroup $\mathrm{SO}_2(\mathbb{R}) \lhd \mathrm{O}_2(\mathbb{R})$.

Exercise 5.21 (p. 209): Let π be a new irreducible of $G = \mathrm{O}_2(\mathbb{R}) \ltimes \mathbb{R}^2$ and let $A = \mathbb{R}^2$. Apply the arguments in Section 5.2.2 to find $r > 0$ and a unit vector $v_0 \in \mathcal{H}_\pi$ that is fixed under the action of $\mathrm{SO}_2(\mathbb{R})$ and let

$$g_0 = \begin{pmatrix} 1 & 0 & 0 \\ 0 & -1 & 0 \\ & & 1 \end{pmatrix} \in G.$$

Show that both $v_+ = v_0 + \pi_{g_0} v_0$ and $v_- = v_0 - \pi_{g_0} v_0$ are fixed under the action of $\mathrm{SO}_2(\mathbb{R})$. Show also that the cyclic subspaces $\langle v_+\rangle_{\pi|_A}$ and $\langle v_-\rangle_{\pi|_A}$ are invariant under G, and conclude the proof.

Exercise 5.24 (p. 216): For (a), use Fourier series. For (b), use the fact that any orbit is dense and the frame to prove injectivity. Also apply Lusin's theorem to find a T-invariant F_σ-set in $\mathbb{T}^2$ of full measure on which Φ_B is bi-measurable and restrict Φ_B to this set. For (c), assume that $\mu = \mu_{T,B_1} = \mu_{T,B_2}$ and define $\rho = \big(\Phi_{B_1}^{-1} \times \Phi_{B_2}^{-1}\big)_* \mu$. Show that ρ is an ergodic probability measure for the rotation by (α, α) on $\mathbb{T}^2 \times \mathbb{T}^2$, and hence the Haar measure on a coset of $\Delta = \{(x, x) \mid x \in \mathbb{T}^2\}$. Now use the frame again to see that ρ is the Haar measure on Δ itself, and conclude that $m_{\mathbb{T}^2}(B_1 \triangle B_2) = 0$.

Exercise 5.25 (p. 218): If $\alpha_2 - \alpha_1 = m\xi$ use $U\colon L^2(\mathbb{T}) \to L^2(\mathbb{T})$ defined by

$$U(f)(x) = \mathrm{e}^{2\pi i m x} f(x)$$

for the intertwining map. For the converse suppose that $U\colon L^2(\mathbb{T}) \to L^2(\mathbb{T})$ satisfies

$$U \circ (\chi^{\alpha_1} \otimes \pi^\mu)_g = (\chi^{\alpha_2} \otimes \pi^\mu)_g \circ U$$

for all $g \in G$. Restricting to A gives $U(f)(x) = \mathrm{e}^{2\pi i k(x)} f(x)$ for $f \in L^2(\mathbb{T})$, $x \in \mathbb{T}$, and a measurable function $k\colon \mathbb{T} \to \mathbb{T}$ by Proposition2.72. Restrict to S to obtain a transformation law for the function k. Pushing the Lebesgue measure on $\mathbb{T}$ to the graph of k gives a probability measure μ on $\mathbb{T}^2$ with translation invariance. As μ gives full measure to a graph, the translation invariance gives a constraint on α_1 and α_2.

Exercise 5.29 (p. 222): Apply Fubini's theorem in the case of a translate by some $a_0 \in A$.

Exercise 5.30 (p. 222): Note that the integral in (5.6) may not converge. To get round this problem, first define the operator on bounded measurable functions with compact support using Fubini's theorem, and show that it extends to a unitary isomorphism.

Exercise 5.50 (p. 236): You may use $G = \mathrm{SO}_2(\mathbb{R}) \ltimes \mathbb{R}^2$ as in Section 5.2.2 and a character on the normal abelian subgroup $\mathbb{R}^2$.

Exercise 5.60 (p. 244): Define the subgroups

$$S = \left\{ \begin{pmatrix} s & & 0 \\ & s^{-1} & 0 \\ & & 1 \end{pmatrix} \,\middle|\, s > 0 \right\}$$

and

$$A = \left\{ h_a = \begin{pmatrix} 1 & & a_1 \\ & 1 & a_2 \\ & & 1 \end{pmatrix} \,\middle|\, a = \begin{pmatrix} a_1 \\ a_2 \end{pmatrix} \in \mathbb{R}^2 \right\}.$$

Notice that the S-orbits on $\widehat{A}$ are free except for the orbit of $0 \in \widehat{A}$ (see Hints Figure 1) and that the hyperbolas can be parametrized by their intersection points with the lines $y = \pm x$.

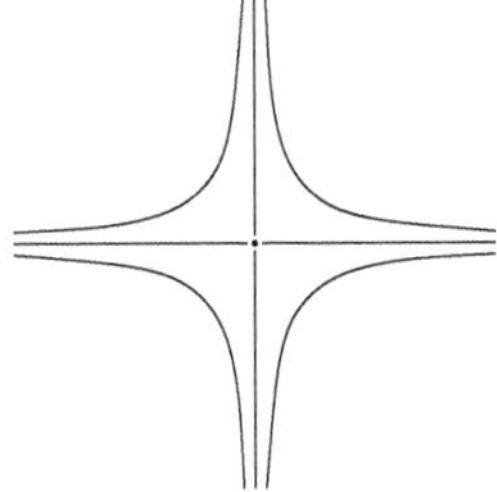

Hints Fig. 1: The S-orbits on $\widehat{A}$ are 0, the coordinate rays, and hyperbolas.

Exercise 6.3 (p. 258): Apply Theorem 2.17 and Proposition 2.27.

Exercise 6.4 (p. 258): For a given compact $K \subseteq G$, $\varepsilon > 0$, and $v \in \mathcal{H}_\pi$ use $\pi \prec \rho$ to find $n \geqslant 1$ and $w_1, \dots, w_n \in \mathcal{H}_\rho$ such that $\|\varphi_v^\pi - \sum_{j=1}^n \varphi_{w_j}^\rho\|_{K,\infty} < \varepsilon$. Then apply $\rho \prec \gamma$ to find for every w_j some $n_j \geqslant 1$ and $w'_{j,1}, \dots, w'_{j,n_j} \in \mathcal{H}_\gamma$ with $\|\varphi_{w_j}^\rho - \sum_{k=1}^{n_j} \varphi_{w'_{j,k}}^\gamma\|_{K,\infty} < \frac{\varepsilon}{n}$. Summing over $j = 1, \dots, n$ and combining it with the above gives the conclusion.

Exercise 6.5 (p. 258): Multiply the defining equation (6.1) for $\pi \prec \rho$ by χ.

Exercise 6.10 (p. 262): Note that $\|\pi_g w - \chi(g) w\|^2 = 2\|w\|^2 - 2\Re \overline{\chi(g)} \varphi_w^\pi$ for $w \in \mathcal{H}_\pi$ and $g \in G$, and apply Proposition 6.8. See also the proof of Corollary 6.15.

Exercise 6.12 (p. 263): For (b) use Exercise 6.5, and for (a) use

$$\|\pi_+(g_a h_b) v_\varepsilon - v_\varepsilon\| \leqslant \|\pi_+(g_a h_b) v_\varepsilon - \pi_+(g_b) v_\varepsilon\| + \|\pi_+(g_b) v_\varepsilon - v_\varepsilon\|.$$

Exercise 6.24 (p. 266): Let $Q \subseteq G$ be a finite set as in Lemma 6.23 and let $\Gamma = \langle Q \rangle < G$ be the subgroup generated by Q. Now define and study a unitary representation of G on $\ell^2(G/\Gamma)$ and the special vector $v_0 = \mathbb{1}_{I\Gamma}$.

Exercise 6.27 (p. 272): The left-hand side involves the integral of $|\pi_*(f)v|^2$, v, and $\pi_*(f)v$ with respect to μ_n.

Exercise 6.28 (p. 272): Apply the mean ergodic theorem in (6.13) and use the Borel–Cantelli lemma (see [24, Th. A.9]).

Exercise 6.33 (p. 281): Suppose first that there exists a compact set $K \subseteq G$ with $\nu(B) = 0$ for all measurable sets $B \subseteq G \smallsetminus K$, and apply the argument in the first part of the proof of Theorem 6.31.

Exercise 6.38 (p. 286): One direction follows from transitivity in Exercise 6.4. For the converse, let $v \in \mathcal{H}_\rho$ and apply Proposition 6.37 to find $\pi^{(1)}, \dots, \pi^{(n)} \in \widehat{G}$ satisfying the weak containment $\pi^{(j)} \prec \rho$ for $j = 1, \dots, n$ and $w_j \in \mathcal{H}_{\pi^{(j)}}$ for $j = 1, \dots, n$ with $\sum_{j=1}^n \|w_j\|^2 = 1$, so that φ_v^ρ is approximated by $\sum_{j=1}^n \varphi_{w_j}^{\pi^{(j)}}$. Now apply the assumption $\pi^{(j)} \prec \gamma$, which holds for $j = 1, \dots, n$, to approximate the latter sum by a matrix coefficient associated to γ.

Exercise 6.48 (p. 291): Use Corollary 6.45 together with Corollary 6.13.

Exercise 6.51 (p. 294): Use the transitivity in Exercise 6.4 and Proposition 6.37.

Exercise 6.52 (p. 294): Argue as in the proof of Corollary 6.39, replacing (for example) the use of Corollary 6.34 by Definition 6.50.

Exercise 6.58 (p. 296): By our standing assumptions, $G = \bigcup_{\ell=1}^\infty Q_\ell$ is a countable union of compact subsets and $C(Q_\ell)$ is separable. Now write $CO(\phi, Q, \varepsilon)$ as a union of sets of the form $CO(\phi_{\ell,m}; Q_\ell, \frac{1}{n})$ with $Q \subseteq Q_\ell$ and $(m,n) \in \mathbb{N}^2$, and apply Lemma 6.56.

Exercise 6.65 (p. 300): Let $\overline{\pi}$ be the contragredient representation of π, show that

$$\mathbb{1} < \pi \otimes \overline{\pi} \prec \rho \otimes \overline{\pi},$$

and recall that the latter is a unitary representation on $\mathcal{H}_\rho \otimes \mathcal{H}'_\pi \cong \operatorname{End}(\mathcal{H}_\pi, \mathcal{H}_\rho)$.

Exercise 6.68 (p. 303): The description of $\widehat{G_d}$ involves the quotient of $\mathbb{R}^d \smallsetminus \{0\}$ by K_d and d discrete points corresponding to $\widehat{K_d}$.

Exercise 6.69 (p. 305): Show that the irreducible representations corresponding to the orbits $A \bullet (r,r)$ depend continuously on $r \in (0, \infty)$, and converge for $r \to 0$ to any old representation in $\widehat{A}$, the irreducible representations corresponding to the orbit $A \bullet (1,0)$, and also to the irreducible representation corresponding to the orbit $A \bullet (0,1)$. The latter two also converge to all old representations. These statements apply similarly to the other four quadrants, and together with the Fell topology on $\widehat{A}$ describe the Fell topology on $\widehat{G}$ completely.

Exercise 7.5 (p. 310): Show first that

$$\pi_{\exp(t\mathbf{a})} v - v = \int_0^t \pi_{\exp(s\mathbf{a})} v_{\mathbf{a}} \, \mathrm{d}s$$

for $t \in \mathbb{R}$ and apply continuity of the representation.

Exercise 7.10 (p. 313): Start by expressing $\mathbf{a}$ as a sum $\mathbf{a} = \sum_{j=1}^d s_j \mathbf{b}_j$ in terms of the orthonormal basis used to define $\mathcal{S}$. Apply (7.1) to get

$$\|\pi_{\exp \mathbf{a}} v - v\| \leqslant \|\pi_\partial(\mathbf{a}) v\| \leqslant \sum_{j=1}^d |s_j| \|\pi_\partial(\mathbf{b}_j) v\|$$

and apply Cauchy–Schwarz to the last sum.

Exercise 7.14 (p. 314): Apply continuity of B in the definition of $\pi_\partial(a)v$.

Exercise 7.16 (p. 316): Note that

$$v(x - se_j) - v(x) = -\int_0^s \partial_j v(x - re_j)\,\mathrm{d}r$$

for all $v \in C^\infty(\mathbb{R}^d)$, $x \in \mathbb{R}^d$, $s \in \mathbb{R}$, and $j \in \{1, \ldots, d\}$. Now translate this to the weak integral statement

$$\lambda_{se_j} v - v = \int_0^s \lambda_{re_j}(-\partial_j v)\,\mathrm{d}r$$

in L^2, and apply Lemma 7.4. Note that the minus sign comes from the definition of the left regular representation.

Exercise 7.17 (p. 316): For (a), we first note that if π is an old representation defined by a character, then every vector in $\mathcal{H}_\pi = \mathbb{C}$ is smooth. So assume that $\pi = \pi^r$ for some $r > 0$. Since the spectral measures for $\pi|_{\mathbb{R}^2}$ are compactly supported, and $\pi|_{\mathbb{R}^2}$ is given by a multiplication representation, it follows from dominated convergence that every $v \in \mathcal{H}_\pi$ is smooth for $\pi|_{\mathbb{R}^2}$. The restriction $\pi_{\mathrm{SO}_2(\mathbb{R})}$ is isomorphic to the regular representation. Example 7.3 shows that v is smooth for $\pi_{\mathrm{SO}_2(\mathbb{R})}$ if and only if its Fourier coefficients have super-polynomial decay, which in turn is equivalent to smoothness in the usual sense. For (b) we again note that $\pi|_S$ is isomorphic to the regular representation of S, so Example 7.15 shows that a vector $v \in \mathcal{H}_+$ is smooth for $\pi|_S$ if and only if $v \in C^\infty\big((0,\infty)\big)$ so that $(0,\infty) \ni t \mapsto t^m v^{(m)}(t)$ (the derivatives of v for $\pi|_S$) belongs to $L^2_{\mu_+}\big((0,\infty)\big)$ for all $m \in \mathbb{N}_0$. Smoothness of $v \in \mathcal{H}_\pi$ under $\pi|_A$, on the other hand, is, by Example 7.15, equivalent to $t \mapsto t^m v(t)$ belonging to $L^2_{\mu_+}\big((0,\infty)\big)$ for all $m \in \mathbb{N}_0$. Putting these together, we obtain that v is smooth if and only if $v \in C^\infty\big((0,\infty)\big)$ and for every $m, n \in \mathbb{N}_0$ the function $(0,\infty) \ni t \mapsto t^{m+n} v^{(m)}(t)$ belongs to $L^2_{\mu_+}\big((0,\infty)\big)$. For (c) one can again apply Example 7.15 separately for the x-direction and the y-direction and obtain that a vector $v \in L^2(\mathbb{R})$ is smooth for the representation π_ξ if and only if v is a Schwartz function (that is, if $t \mapsto t^m v^{(n)}(t)$ lies in $L^2(\mathbb{R})$ for all $m, n \in \mathbb{N}_0$).

Exercise 7.22 (p. 318): Suppose that $v, w \in D_{\pi_\partial(a)}$. Then

$$\begin{aligned}\langle v, \pi_\partial(a)w\rangle &= \lim_{t\to 0}\left\langle v, \frac{1}{t}\left(\pi_{\exp(ta)}w - w\right)\right\rangle \\ &= \lim_{t\to 0}\frac{1}{t}\left\langle \pi_{\exp(-ta)}v - v, w\right\rangle = -\left\langle \pi_\partial(a)v, w\right\rangle,\end{aligned}$$

which shows $-\pi_\partial(a) \subseteq \pi_\partial(a)^*$. For the converse, suppose that $v \in D_{\pi_\partial(a)^*}$ and link this to Lemma 7.4.

Exercise 7.34 (p. 332): The subgroup in $\widetilde{\mathrm{SL}_3(\mathbb{R})}$ corresponding to the subgroup $\mathrm{ASL}_2(\mathbb{R})$ of $\mathrm{SL}_3(\mathbb{R})$ is the semi-direct product of a double cover of $\mathrm{SL}_2(\mathbb{R})$ and $\mathbb{R}^2$, but the action on $\widehat{\mathbb{R}^2}$ remains unchanged. Hence the argument for Theorem 7.30 only needs to be changed a little.

Exercise 7.35 (p. 333): For (a) use Fubini in the form

$$\begin{aligned}\|D_t f\|_2^2 &= \int_X \left|\frac{1}{t}\int_0^t f(g_s{\boldsymbol\cdot}x)\,\mathrm{d}s - \int_X f\,\mathrm{d}\mu\right|^2 \mathrm{d}\mu \\ &= \frac{1}{t^2}\int_0^t\int_0^t \left(\langle \pi_{s_1}f, \pi_{s_2}f\rangle_{L^2_\mu(X)} - \left|\int_X f\,\mathrm{d}\mu\right|^2\right)\mathrm{d}s_1\,\mathrm{d}s_2\end{aligned}$$

and apply the assumption (7.18).

Exercise 7.37 (p. 334): Define $n = \lfloor t^{\frac{1}{\alpha}} \rfloor \geqslant N = \lfloor T^{\frac{1}{\alpha}} \rfloor$ so that by the definition of X_T we have

$$\left| \frac{1}{n^\alpha} \int_0^{n^\alpha} f(g_s \bullet x) \, \mathrm{d}s - \int_\alpha f \, \mathrm{d}\mu \right| \leqslant n^{-\delta\alpha} S(f) \ll t^{-\delta} S(f).$$

Estimate $|t - n^\alpha|$ and use the bound $\|f\|_\infty \leqslant S(f)$ to conclude the argument.

Exercise 8.12 (p. 347): Assume that $h \in H \smallsetminus K$, and apply the Cartan (or KAK, or singular value) decomposition.

Exercise 8.14 (p. 350): Here are two ways to prove this: Show that

$$m_{\mathrm{SL}_d(\mathbb{R})}(B) = m_{\mathbb{R}^{d^2}}\big(\{sg \mid s \in [0,1], g \in B\}\big)$$

for measurable $B \subseteq \mathrm{SL}_d(\mathbb{R})$ defines the claimed bi-invariant Haar measure. Alternatively, show that the elementary unipotent subgroups belong to $[\mathrm{SL}_d(\mathbb{R}), \mathrm{SL}_d(\mathbb{R})]$ and generate $\mathrm{SL}_d(\mathbb{R})$, which together imply that $\Delta(\mathrm{SL}_d(\mathbb{R})) = 1$.

Exercise 8.17 (p. 357): Extract the definition of $c(\cdot, \cdot)$ from our definition of π^n. Note that $|c(\cdot, \cdot)|$ must satisfy the assumption (1.6) as π is unitary and c must satisfy the cocycle equation (1.5) as we already know (8.16).

Exercise 8.26 (p. 363): Modify the proof of Lemma 8.19.

Exercise 8.45 (p. 380): Let $\mathcal{V} = \langle \pi(G) v_0 \rangle_{\mathrm{la}}$ be the linear hull of $\pi(G) v_0$. Now study the matrix coefficient for $v, w \in \mathcal{V}$, use sesqui-linearity of the inner product, Lemma 1.15, and the assumption on v_0 to show that $\varphi^\pi_{v,w} \in L^p(G)$.

Exercise 8.47 (p. 381): For (a) suppose that $\mathcal{H}_\pi$ has no invariant vectors. Let $\mathcal{V} \subseteq \mathcal{H}_\pi$ be a dense subset as in the definition of the integrability exponent, such that $\varphi^\pi_{v,w} \in L^p(G)$ for $v, w \in \mathcal{V}$ and $p = 2d > p_\pi$. Use (B.2) (from Lemma B.1 on page 515) for $m = d$ and $s_j = \frac{1}{p}$ for $j = 1, \ldots, d$ to see that the d-fold tensor product $\rho = \otimes^d \pi$ (defined inductively, using Section 3.2 as the inductive step) is 2-integrable. This implies that ρ is tempered by Theorem 8.5. For (b) combine (a) with Theorem 8.42(2), Proposition 8.40, and Proposition 7.29.

Exercise 8.48 (p. 382): Show that for two measurable fundamental domains $E, F \subseteq G$ we have $E = \bigsqcup_{\gamma \in \Gamma} E \cap \gamma F$ and $F = \bigsqcup_{\gamma \in \Gamma} \gamma^{-1}(E \cap \gamma F)$. Now use the fact that m is bi-invariant. For the last part use a finite G-invariant measure to define a right G-invariant measure on G.

Exercise 8.50 (p. 382): Note that β acts on $z \in \mathbb{H}$ via $\beta \bullet z = \frac{z}{2z+1}$, which may appear to be more complicated than the action of α. To understand the action of β better, conjugate β by $\sigma = \begin{pmatrix} 0 & -1 \\ 1 & 0 \end{pmatrix}$ and verify that $\sigma \bullet F = F$.

Exercise 8.51 (p. 383): This assertion is simply (8.33) and (8.34) in the case $\gamma = \alpha^{\pm 1}$ or $\gamma = \beta^{\pm 1}$ has length one. If we assume the claim for a word γ of a given length then (8.33) and (8.34) again show the same claim after increasing the length of γ (on the left) by one. It follows that the image of F under the Möbius action of γ determines the left-most group element $\alpha^{\pm 1}$ or $\beta^{\pm}$ in the reduced word. By induction, it follows that $\gamma \bullet F$ determines the powers $m_1, n_1, \ldots, m_k, n_k$ of the generators.

Exercise 8.52 (p. 383): The hard part is to show that

$$D = \bigcup_{\gamma \in \Gamma} \gamma \cdot \overline{F} = \mathbb{H}.$$

For this show that there exists some uniform $\varepsilon > 0$ so that for all $z \in D$ we have $B_\varepsilon^{\mathbb{H}}(z) \subseteq D$. This unusual 'uniform openness' implies that the set D is both open and closed, which by connectedness of $\mathbb{H}$ implies that $D = \mathbb{H}$. Note that it is sufficient to consider $z \in F$ (since Γ acts by isometries on $\mathbb{H}$). Split F into subregions as in Hints Figure 2 and adjust your argument: For the middle part only finitely many neighbouring copies of F are important. For the top part the neighbours $\alpha^n F$ with $n \in \mathbb{Z}$ are needed.

Exercise 8.53 (p. 383): Estimate the area of each region of Hints Figure 2.

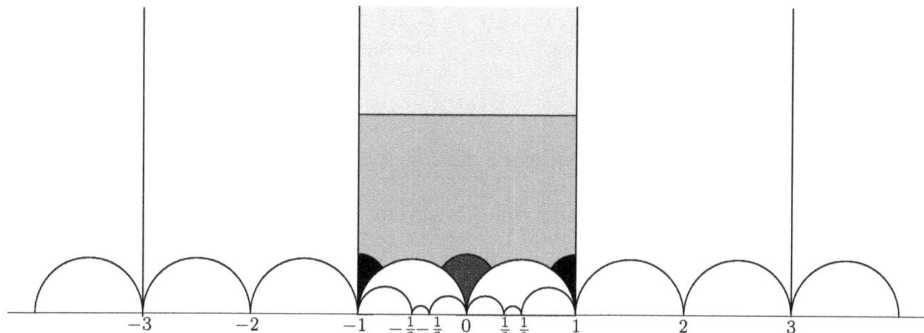

Hints Fig. 2: The subgroup Γ is a lattice.

Exercise 8.55 (p. 384): Fix some $\psi \in C_c^\infty(\mathrm{SL}_2(\mathbb{R}))$ with $\psi \geqslant 0$ and $\int \psi \,\mathrm{d}m = 1$ and define $f_j = \pi_*^{X_n}(\psi)\mathbb{1}_{K_j}$ for $j = 1, 2$. Estimate $\mathcal{S}(f_j)$ and apply the uniform decay in Exercise 8.47 to $\widetilde{f_j} = f_j - \frac{\int f_j \,\mathrm{d}m_X}{v_n}\mathbb{1}_{X_n}$. This leads to the estimate $|\langle \pi_g \widetilde{f_1}, \widetilde{f_2}\rangle \leqslant c_1 \|g\|_{\mathrm{HS}}^{-\frac{1}{2d}} v_n$ for a constant c_1 depending on d and ψ. This in turn forces $\langle \pi_g f_1, f_2\rangle \neq 0$ for $\|g\|_{\mathrm{HS}}$ sufficiently large and leads to the desired existence.

Exercise 8.56 (p. 384): Start with $K_1 g^{-1} \cap K_2 \neq \emptyset$ and use the definition in (8.35).

Exercise 8.57 (p. 384): Recall that Λ is a free group. Show that γ as in Exercise 8.56 is a non-trivial element of length at least $n - 3\ell \geqslant \frac{1}{4}n - 3$. As $\|\gamma\|_{\mathrm{HS}} \ll \|g\|_{\mathrm{HS}}$ is uniformly bounded and Γ is discrete, this shows that n must be bounded.

Exercise 8.60 (p. 389): Consider $S = \{g \in \mathrm{SL}_2(\mathbb{R}) \mid \|g\|_{\mathrm{HS}} \leqslant R\}$ and $f_\varepsilon = \frac{1}{\sqrt{m(B)}}\mathbb{1}_{\Gamma B_\varepsilon}$, where B_ε is the ε-neighbourhood of $I \in \mathrm{SL}_2(\mathbb{R})$. Apply Corollary 8.59, take the inner product with f_ε, and relate $\langle \pi_*^X(\mathbb{1}_S) f_\varepsilon, f_\varepsilon\rangle$ to the count of lattice elements in certain balls.

Exercise 8.62 (p. 391): Verify that π is obtained from π^1 by composition with the automorphism obtained from $W = \begin{pmatrix} 0 & -1 \\ 1 & 0 \end{pmatrix} \in \mathrm{SL}_2(\mathbb{R})$ (see page 392) and agrees with π^1 on the centre of H.

Exercise 8.63 (p. 392): Note that

$$U_{g_1 g_2} \pi_h U_{g_1 g_2}^{-1} = \pi_{\Phi_{g_1 g_2}(h)} = \pi_{\Phi_{g_1}(\Phi_{g_2}(h))} = U_{g_1} \pi_{\Phi_{g_2}(h)} U_{g_1}^{-1} = U_{g_1} U_{g_2} \pi_h U_{g_2}^{-1} U_{g_1}^{-1}$$

for $h \in H$. Now apply Schur's lemma (Theorem 1.29).

Exercise 8.68 (p. 397): Set $z = a + ib$ with $b > 0$ and $u = x + iy$. Show that

$$\left|e^{\pi i z u^2}\right| \leqslant e^{-b(x^2-y^2)-2axy}$$

and consider the piecewise linear closed path from $-s$ to s, then to $s - \frac{t}{2}$, to $-s - \frac{t}{2}$, and back to $-s$.

Exercise 9.3 (p. 409): Part (a) follows from Theorem 4.6. For (b), apply Theorem 4.11 to find $\mathbb{C}\Omega$ as a one-dimensional invariant complement to $\mathfrak{E}_{\leqslant 1}$ and the representation found in (a). For (c) and (d), one again has to analyze dimensions and weights.

Exercise 9.5 (p. 410): Let G be a closed linear group with Lie algebra $\mathfrak{g} = \operatorname{Lie} G$, and recall the definition of the exponential map $\exp\colon \mathfrak{g} \to G$ from Section 4.1.1. Fix $\mathbf{a}, \mathbf{b} \in \mathfrak{g}$ and a function $\varphi \in C^\infty(G)$. Now consider, for a fixed g, the smooth function

$$\Phi\colon \mathbb{R}^2 \ni (t_1, t_2) \longmapsto \varphi\bigl(\exp(-t_2\mathbf{b})\exp(-t_1\mathbf{a})g\bigr)$$

and calculate $\partial_{t_1}\partial_{t_2}\Phi|_{t_1=t_2=0}$ in two ways. First note that

$$(\partial_{t_2}\Phi)(t_1, 0) = \lambda_\partial(\mathbf{b})\varphi(\exp(-t_1\mathbf{a})g)$$

for all t_1, which gives $\partial_{t_1}\partial_{t_2}\Phi(0,0) = \lambda_\partial(\mathbf{a})\lambda_\partial(\mathbf{b})\varphi(g)$. Next recall that

$$\partial_{t_1}\partial_{t_2}\Phi(0,0) = \partial_{t_2}\partial_{t_1}\Phi,$$

write Φ in the form

$$\Phi(t_1, t_2) = \varphi\bigl(\exp(-t_1(\mathbf{a}))\exp(-\operatorname{Ad}_{\exp(t_1\mathbf{a})} t_2\mathbf{b})g\bigr),$$

and apply the chain rule to obtain

$$\partial_{t_1}\Phi(0, t_2) = \lambda_\partial(\mathbf{a})\varphi\bigl(\exp(-t_2\mathbf{b})g\bigr) + \partial_{t_1}|_{t_1=0}\varphi\bigl(\exp(-\operatorname{Ad}_{\exp(t_1\mathbf{a})}(t_2\mathbf{b}))g\bigr).$$

Moreover,

$$\operatorname{Ad}_{\exp(t_1\mathbf{a})}(t_2\mathbf{b}) = t_2\mathbf{b} + t_1t_2[\mathbf{a},\mathbf{b}] + \mathrm{O}(t_1^2t_2)$$

and

$$\varphi\bigl(\exp(\operatorname{Ad}_{\exp(t_1\mathbf{a})}(-t_2\mathbf{b}))g\bigr) = \varphi(\exp(-t_2\mathbf{b})) + t_1t_2\lambda_\partial([\mathbf{a},\mathbf{b}])\varphi(\exp(-t_2\mathbf{b})g) + \mathrm{O}_\varphi(t_1^2t_2)$$

for all sufficiently small $t_1, t_2 \in \mathbb{R}$ imply that

$$\partial_{t_1}|_{t_1=0}\varphi\bigl(\exp(-\operatorname{Ad}_{\exp(t_1\mathbf{a})}(t_2\mathbf{b})g)\bigr) = t_2\lambda_\partial([\mathbf{a},\mathbf{b}])\varphi(\exp(-t_2\mathbf{b})g).$$

Taking the derivative ∂_{t_2}, one obtains $\partial_{t_1}\partial_{t_2}\Phi(0,0) = \lambda_\partial(\mathbf{b})\lambda_\partial(\mathbf{a})\varphi(g) + \lambda_\partial([\mathbf{a},\mathbf{b}])\varphi(g)$.

Exercise 9.7 (p. 411): Apply the assumption and

$$\operatorname{Ad}_{\exp\mathbf{a}}|_{\mathfrak{E}_{\leqslant d}}(\Omega) = \exp\bigl(\operatorname{ad}_{\mathbf{a}}|_{\mathfrak{E}_{\leqslant d}}\bigr)\Omega = \Omega|_{\mathfrak{E}_{\leqslant d}}$$

to see that $\operatorname{Ad}_g(\Omega) = \Omega$ for all $g \in \exp(\mathfrak{g})$. Finally, note that $\exp(\mathfrak{g})$ generates an open subgroup of G.

Exercise 9.24 (p. 428): For (a) note that a representation π of $\mathrm{SL}_2(\mathbb{R})$ induces a representation of $\mathrm{PSL}_2(\mathbb{R}) = \mathrm{SL}_s(\mathbb{R})/\{\pm I\}$ if and only if π is even. For (b), note that

$$\mathrm{GL}_2(\mathbb{R})^o \cong \mathrm{SL}_2(\mathbb{R}) \times \mathbb{R}$$

and apply Lemma 1.28 and Corollary 1.32 to see that $\widehat{\mathrm{GL}_2(\mathbb{R})^o} \cong \widehat{\mathrm{SL}_2(\mathbb{R})} \times \mathbb{R}$ (in a natural sense).

Exercise 9.25 (p. 429): For (a), note that every character on $\mathrm{SL}_2(\mathbb{R})$ is trivial by Exercise 1.35. As $G/\mathrm{SL}_2(\mathbb{R}) \cong \mathbb{Z}/2\mathbb{Z}$, there exist two characters on G, namely $\mathbb{1}_G$ and det. For (b), note that $rkr^{-1} = k^{-1}$ for $k \in K = \mathrm{SO}_2(\mathbb{R})$. Use this to show that for any representation r maps a vector with K-weight $n \in \mathbb{Z}$ to a vector with K-weight $-n$. For (c), given $n \in \mathbb{N}$ define $\mathcal{H}_{\delta^{n,G}} = \mathcal{H}_{\delta^{n,+}} \oplus \mathcal{H}_{\delta^{n,+}}$, $\delta_g^{n,G} = \delta_g^{n,+} \oplus \delta_{rgr^{-1}}^{n,+}$ for $g \in \mathrm{SL}_2(\mathbb{R})$, and $\delta_r^{n,G}(v_+, v_-) = (v_-, v_+)$ for $(v_+, v_-) \in \mathcal{H}_{\delta^{n,G}}$. Now verify that this defines an irreducible unitary representation of G. For (d), recall that the principal series representation π^ξ can be obtained from the measure class preserving action of $\mathrm{SL}_2(\mathbb{R})$ on $\mathbb{S}^1$. This action extends to G, which allows us to extend π^ξ to all of G. Restricting to the even and odd subspaces again defines $\pi^{\xi,\mathrm{e}}$ and $\pi^{\xi,\mathrm{o}}$ as unitary representations of G. Twisting with det defines further unitary representations. Except for the case of $\pi^{0,\mathrm{o}}$, these representations are different (by applying Schur's lemma for the restriction to $\mathrm{SL}_2(\mathbb{R})$). For (e), in order to avoid having to study the complementary series γ^s in detail, argue as follows. Note that $\mathrm{SL}_2(\mathbb{R}) \ni g \mapsto \gamma^s_{rgr^{-1}}$ defines a unitary representation of $\mathrm{SL}_2(\mathbb{R})$ on $\mathcal{H}_{(s)}$ that is isomorphic to γ^s. Hence there exists a unitary isomorphism $U \colon \mathcal{H}_{(s)} \to \mathcal{H}_{(s)}$ so that $\gamma^s_{rgr^{-1}} = U\gamma^s_g U^{-1}$. By Schur's lemma we have $U^2 = \alpha I$ for some $\alpha \in \mathbb{S}^1$. Dividing U by a square root of α, we may assume that $U^2 = I$, which allows an extension of γ^s_r to G by setting $\gamma^s_r = U$. Twisting by det defines another class of unitary representations of G. For (f), first show that $\mathrm{Ad}_r(\Omega) = \Omega$. For a given $\pi \in \widehat{G}$ apply Schur's lemma and Corollary 9.23 to relate $\pi|_{\mathrm{SL}_2(\mathbb{R})}$ to one of the irreducible unitary representations of $\mathrm{SL}_2(\mathbb{R})$. Now match π to one of the representations discussed in (c) or (e). For (g) use $\mathrm{GL}_2(\mathbb{R}) \cong G \times \mathbb{R}$ and Example 1.34.

Exercise 9.26 (p. 429): Let π be an irreducible unitary representation of $\mathrm{SL}_2(\mathbb{R}) \times G$ and let Ω be the Casimir element in the first factor $\mathrm{SL}_2(\mathbb{R})$. Schur's lemma in the form of Proposition 9.8 extended to allow $\mathrm{SL}_2(\mathbb{R}) \times G$ shows that $\pi_\partial(\Omega) = \alpha_\Omega I$ for some $\alpha_\Omega \in \mathbb{R}$. Now fix a basis of the space of vectors of K-weight n for a suitable $n \in \mathbb{Z}$ and apply Corollary 9.23. Show that the cyclic spaces spanned by these basis vectors with respect to $\mathrm{SL}_2(\mathbb{R})$ are orthogonal, and apply Proposition 3.17.

Exercise 9.28 (p. 429): Note that

$$\int_K \varphi_v^\pi(k) \, \mathrm{d}m_K(k) = \langle (\pi|_K)_*(\mathbb{1}_K)v, v\rangle$$

and that $(\pi|_K)_*(\mathbb{1}_K)$ is the projection to the subspace of vectors with K-weight 0.

Exercise 9.35 (p. 440): In the last part of the proof of Theorem 9.32 we improved the bound $|z(t)| \ll t^2$ for $t \geqslant 1$ to $|z(t)| \ll 1$ for $t \geqslant 1$, which translates to the claim in (a). For (b), show by dominated convergence that $\varphi^{\pi^\xi}_{F_{\xi,0}}$ converges for $\xi \to 0$ to $\Xi = \varphi^{\pi^0}_{F_{0,0}}$.

Exercise 9.41 (p. 445): For (b), use Theorem 8.42(2) for π^ξ and integrate the estimate to obtain the same for π^μ.

Exercise 9.42 (p. 445): Show that F_θ appearing in the proofs of Lemma 9.38 and Proposition 9.40 is a Schwarz function (even though it may not be compactly supported).

Exercise 9.43 (p. 445): The centralizer is not abelian. Indeed, because of the existence of the isomorphism between $\pi^{\xi,\mathrm{e}}$ and $\pi^{-\xi,\mathrm{e}}$ (and similarly for the odd principal series representations), we can also rephrase Proposition 9.40 by saying that $\pi^{\mathbb{R}^2}$ is isomorphic to

$$\int_0^\infty \underbrace{\pi^{\xi,\mathrm{e}} \oplus \pi^{\xi,\mathrm{e}}}_{\pi^{\xi,\mathrm{e}}\otimes\mathbb{C}^2} \oplus \underbrace{\pi^{\xi,\mathrm{o}} \oplus \pi^{\xi,\mathrm{o}}}_{\pi^{\xi,\mathrm{o}}\otimes\mathbb{C}^2} \,\mathrm{d}\xi.$$

Now use operators defined using elements of $\mathrm{Mat}_{2,2}\big(\mathscr{L}^\infty([0,\infty))\big) \times \mathrm{Mat}_{2,2}\big(\mathscr{L}^\infty([0,\infty))\big)$ (as in Section 2.8).

Exercise 9.73 (p. 479): By Lemma 9.16, $F_{s,n} \in \mathcal{H}_{(s)}$ is smooth. Let $f \in \mathcal{V}_s$ and apply dominated convergence to the inner product $\langle \frac{1}{t}\big(\gamma^{(s)}_{\exp(t\mathbf{a})} F_{s,n} - F_{s,n}\big), f\rangle_{\mathcal{V}_s}$ as $t \to 0$ to conclude that the pointwise derivative in (9.61) is equal to the derivative in the sense of Definition 7.1.

Exercise 9.76 (p. 484): Let $m, n \in 2\mathbb{Z}$ and consider $F_{s,m}, F_{s,n} \in \mathcal{V}_s$ as in Lemma 9.68. Using $|F_{s,m}| = |F_{s,n}| = F_{s,0}$ and Lemma 9.74 we may estimate

$$|\langle \gamma^s_g F_{s,m}, F_{s,n}\rangle| \leqslant \langle \gamma^s_g F_{s,0}, F_{s,0}\rangle = \phi_{(s)}(g) \asymp_s \|g\|_{\mathrm{HS}}^{\frac{s-1}{2}}.$$

Notice that this does not match the assumption of Proposition 7.29 as $F_{s,m}$ and $F_{s,n}$ are not normalized. Instead we have

$$|\langle \gamma^s_g F_{s,m}, F_{s,n}\rangle| \ll_s |m|^{\frac{s}{2}} |n|^{\frac{s}{2}} \|g\|_{\mathrm{HS}}^{\frac{s-1}{2}} \|F_{s,m}\| \|F_{s,n}\|$$

by (9.65). Now generalize the argument for Proposition 7.29.

Exercise 9.82 (p. 494): Combine the estimate in (9.77) for the summands of $\lambda \oplus \bigoplus_{s\in S} \gamma^s$ with Cauchy–Schwarz in $\mathbb{C}\times\mathbb{C}^{|S|}$. The uniformity of the implicit constant in (9.77) is discussed in the proof of Theorem 9.80.

Exercise 9.83 (p. 494): Note that the first part of the proof does not involve s_π. For the second part (starting on page 492) show that the claim (9.76) holds for the set

$$S = \{s \in (0,1) \mid \gamma^s \text{ is one of the direct summands of } \pi\}.$$

Exercise 9.91 (p. 511): Show that the projection from V_T to the linear hull of the characters corresponding to elements of $\mathbb{Z}^2 \cap B_{c\sqrt{T}}(0)$ must be injective. The dimension of the latter space grows at most linearly. See [25, Prop. 6.65 & Lem. 6.68] for more details.

Exercise 9.92 (p. 511): For (a) use Lemma 9.87. For (b) simply apply the product rule and estimate. For (c) consider the linear maps $L_j\colon V_T \ni \phi \to \chi_j\phi$ for $j = 1,\ldots,J$. Apply the singular value decomposition to L_j and the L^2 norms in domain and range: There exists a subspace $V_{T,j} \subseteq V_T$ so that $\phi \in V_{T,j}$ implies that $\|\chi_j\phi\|_{L^2} \geqslant \frac{1}{J}\|\phi\|_{L^2}$ and $\phi \in V_{T,j}^\perp$ implies that $\|\chi_j\phi\|_{L^2} < \frac{1}{J}\|\phi\|_{L^2}$. If $\dim V_{T,j} \geqslant \frac{N(T)}{J}$ we may set $j(T) = j$ and $V_T^* = V_{T,j}$. So suppose that $\dim V_{T,j} < \frac{N(T)}{J}$ (or equivalently $\dim V_{T,j}^\perp > N(T)\big(1 - \frac{1}{J}\big)$) for $j = 1,\ldots,J$. By taking intersections we obtain

$$\dim \bigcap_{j=1}^{J} V_{T,j}^\perp > 0$$

and that ϕ belonging to this intersection satisfies $\|\chi_j\phi\| < \frac{1}{J}\|\phi\|$ for all $j \in \{1,\ldots,J\}$. This contradicts the assumption that $\chi_1 + \cdots + \chi_J = 1$. For (d) note that a switch from a smooth measure with bounded density to Lebesgue measure only affects multiplicative constants.

Exercise 9.93 (p. 511): Note that any odd principal series representation contains a one-dimensional subspace of vectors with K-weight 1. Hence it suffices to count the number of Casimir eigenfunctions with bounded eigenvalue inside the eigenspace for K with K-weight 1.

Argue as in Exercise 9.92 (using a smooth partition of unity consisting of K-invariant functions) to localize the count to functions supported in a fixed compact subset of $\mathrm{SL}_2(\mathbb{R})$. Use the Iwasawa decomposition and K-weight 1 to relate this once again to Exercise 9.91.

Exercise A.1 (p. 514): Calculate the commutator $[a, u_{j,k}(s)]$ of a diagonal matrix a and $u_{j,k}(s)$ for $j \neq k$ and $s \in \mathbb{K}$.

Exercise C.2 (p. 520): For (a), start with a metric d on G and define

$$\mathsf{d}_{\mathrm{new}}(g_1, g_2) = \max_{x,y \in G} \mathsf{d}(xg_1y, xg_2y).$$

For (b), notice that such a metric would be invariant under conjugation, and consider conjugation of elementary unipotent matrices by diagonal matrices.

Notes

(1)(p. v) The concept of property (τ) goes back to the work of Selberg [86], Kazhdan [50], Burger and Sarnak [9], and Clozel [16]; we refer to overviews by Lubotzky [63] and Sarnak [81] for more on this. It is a weaker version of property (T) which we discuss in Chapter 6, and has found diverse applications in number theory, geometry, topology, dynamical systems, and computer science.

(2)(p. 22) This was used by Schur [83] to prove orthogonality relations and develop the representation theory of finite groups. There are important generalizations to Lie groups, Lie algebras, and to non-simple modules.

(3)(p. 45) This correspondence between representations of operator algebras and states (distinguished linear functionals on the algebra) is achieved by an explicit construction of the representation from the function. It plays a central role in the Gelfand–Naimark theorem characterizing C^*-algebras. The construction was first shown by Gelfand and Naimark [34], and further developed by Segal [84].

(4)(p. 105) These compact groups are known as *solenoids* as they have a copy of the group $\mathbb{R}$ 'wrapped' inside them; this terminology seems to have been used first by van Dantzig [21, p. 75].

(5)(p. 129) This was shown by Wiener in the case $G = \mathbb{R}$ as part of a study of Tauberian theorems [93].

(6)(p. 130) This perspective, in which properties of locally compact fields are placed at the heart of algebraic number theory, is eloquently developed in the monograph by Weil [90].

(7)(p. 144) This important result appears implicitly in the analysis of almost periodic functions on groups in work of Bochner and von Neumann [7, Th. 39] and explicitly in work of Hurevitsch [43, Th. 1].

(8)(p. 156) This fundamental result in harmonic analysis, due to Peter and Weyl [77] also may be used to show that any compact group is a projective limit of Lie groups.

(9)(p. 158) If the compact group G is semi-simple, then classical work of Cartan [13, Ex. 6] shows that $G^\sharp$ can be realized by identifying certain boundary points of a compact convex polyhedron in the Lie algebra of a maximal torus in G.

(10)(p. 182) We refer to Knapp [54, Prop. 7.15] for this argument, which goes back to work of Hurwitz [44] in the case of SL_n and to Weyl [92] in the general semi-simple case, and is therefore sometimes called Weyl's unitarity trick.

(11)(p. 213) This is part of a number of different formulations of results in quantum mechanics concerning commutation relations between position and momentum operators. These results may be expressed in representation theory as a classification of unitary representations of the Heisenberg group; we refer to Stone [88] and von Neumann [74] for the original formulations.

M. Einsiedler and T. Ward, *Unitary Representations and Unitary Duals*, Graduate Texts in Mathematics 308, https://doi.org/10.1007/978-3-032-03899-9

(12)(p. 219) Results of this form were found by Segal [85] and Mautner [69, 70]; we refer to surveys by Gross [40] and Mackey [68] and their references for an account of the history.
(13)(p. 229) There are many results of this form under various hypotheses on the groups. Banach [3, Th. 4, Ch. 1] showed that a Borel homomorphism between Polish groups is continuous, and similar results are used by Mackey. The statement is true for G and U locally compact without any additional countability hypotheses, as shown by Kleppner [53].
(14)(p. 232) Results of this form with various hypotheses appear in several places including the work of Mackey [67] and Kallman [47].
(15)(p. 256) Type I has a formal definition using the connection to C^* algebras, and was shown to be equivalent to the Fell topology being T_0 by work of Glimm [36]. We refer to Kirillov [52, Sec. 8.4] for the tame and wild terminology and for further details.
(16)(p. 321) The Casimir element (or invariant, or operator) is a distinguished element of the center of the universal enveloping algebra of any Lie algebra. The name comes from work on rigid body dynamics by Casimir [14], where the natural example is the squared angular momentum operator (a Casimir element associated to $\mathrm{SO}_3(\mathbb{R})$). The more general development in Lie theory goes back to work of Casimir and van der Waerden [15].
(17)(p. 335) This characterization was shown by Bargmann [4] in a concrete setting and by Godement [37, 38] abstractly.
(18)(p. 381) These groups were first systematically studied by Poincaré [78], who gave them the name to reflect the motivation from earlier work of Fuchs [31]. We refer to the monograph of Katok [48] for an introduction to the theory of Fuchsian groups in general.
(19)(p. 382) This was presented by Sanov [80] as an example of a free group defined by matrices with integer entries (earlier examples had relied on transcendental entries to preclude relations between entries). He also showed that the elements of this group are precisely those integer matrices of the form

$$\begin{pmatrix} 1+4m & 2k \\ 2f & 1+4h \end{pmatrix}$$

with determinant one.
(20)(p. 390) The Baker–Campbell–Hausdorff formula expresses $\log(\exp(\mathbf{a})\exp(\mathbf{b}))$ for $\mathbf{a}, \mathbf{b}$ in the Lie algebra of a Lie group as a series involving repeated commutators of $\mathbf{a}$ and $\mathbf{b}$. We refer to Knapp [54] for the general formulation and proof, and a paper of Achilles and Bonfiglioli [1] for the rather complex history of the result.
(21)(p. 398) The more general construction of the metaplectic group Mp_{2n} is a double cover of the symplectic group Sp_{2n}, and this was used by Weil [91] to interpret theta functions in terms of representation theory. In the case $n = 1$ the symplectic group Sp_{2n} coincides with $\mathrm{SL}_2(\mathbb{R})$, which is the language we have chosen to use here. The metaplectic group Mp_{2n} is not a 'matrix group', meaning that it has no faithful finite-dimensional representations.
(22)(p. 403) For more on the decomposition of representations of $\mathrm{SL}_2(\mathbb{R})$, we refer to Lang [61].
(23)(p. 516) In fact this choice of how to extend the factorial function is more natural than at first appears. The Bohr–Mollerup theorem states that if $f\colon (0,\infty) \to (0,\infty)$ has $f(1) = 1$, satisfies $f(x+1) = xf(x)$, and has the property that $x \mapsto \log f(x)$ is convex, then $f(x) = \Gamma(x)$ for all $x > 0$. We refer to the monograph of Artin [2] for an accessible account, and for more on Stirling's formula.
(24)(p. 517) The proof of Lemma C.1 given here is taken from the monograph of Montgomery and Zippin [72, Sec. 1.22] and Tao's blog [89].

References

1. R. Achilles and A. Bonfiglioli, 'The early proofs of the theorem of Campbell, Baker, Hausdorff, and Dynkin', *Arch. Hist. Exact Sci.* **66** (2012), no. 3, 295–358. https://doi.org/10.1007/s00407-012-0095-8.
2. E. Artin, *The Gamma Function*, in *Athena Series: Selected Topics in Mathematics* (Holt, Rinehart and Winston, New York–Toronto–London, 1964). Translated by Michael Butler.
3. S. Banach, *Théorie des opérations linéaires*, in *Monogr. Mat., Warszawa* **1** (PWN - Panstwowe Wydawnictwo Naukowe, Warszawa, 1932) (French). https://eudml.org/doc/271931.
4. V. Bargmann, 'Irreducible unitary representations of the Lorentz group', *Ann. of Math. (2)* **48** (1947), 568–640. https://doi.org/10.2307/1969129.
5. B. Bekka, P. de la Harpe, and A. Valette, *Kazhdan's property (T)*, in *New Mathematical Monographs* **11** (Cambridge University Press, Cambridge, 2008). https://doi.org/10.1017/CBO9780511542749.
6. G. Birkhoff, 'A note on topological groups', *Compositio Math.* **3** (1936), 427–430. http://www.numdam.org/article/CM_1936__3__427_0.pdf.
7. S. Bochner and J. von Neumann, 'Almost periodic functions in groups. II', *Trans. Amer. Math. Soc.* **37** (1935), no. 1, 21–50. https://doi.org/10.2307/1989694.
8. A. R. Booker and A. Strömbergsson, 'Numerical computations with the trace formula and the Selberg eigenvalue conjecture', *J. Reine Angew. Math.* **607** (2007), 113–161. https://doi.org/10.1515/CRELLE.2007.047.
9. M. Burger and P. Sarnak, 'Ramanujan duals. II', *Invent. Math.* **106** (1991), no. 1, 1–11. https://doi.org/10.1007/BF01243900.
10. M. Burger, 'Horocycle flow on geometrically finite surfaces', *Duke Math. J.* **61** (1990), no. 3, 779–803. https://doi.org/10.1215/S0012-7094-90-06129-0.
11. H. Busemann, *The geometry of geodesics* (Academic Press Inc., New York, N. Y., 1955).
12. J. Cannon, W. Floyd, R. Kenyon, and W. Parry, 'Hyperbolic geometry', in *Flavors of geometry*, in *Math. Sci. Res. Inst. Publ.* **31**, pp. 59–115 (Cambridge Univ. Press, Cambridge, 1997).
13. E. Cartan, 'La géométrie des groupes simples', *Ann. Mat. Pura Appl.* **4** (1927), no. 1, 209–256. https://doi.org/10.1007/BF02409989.
14. H. Casimir, *Rotation of a rigid body in quantum mechanics* (Ph.D. thesis, Leiden University, 1931). https://ilorentz.org/history/proefschriften.
15. H. Casimir and B. L. van der Waerden, 'Algebraischer Beweis der vollständigen Reduzibilität der Darstellungen halbeinfacher Liescher Gruppen', *Math. Ann.* **111** (1935), no. 1, 1–12. https://doi.org/10.1007/BF01472196.
16. L. Clozel, 'Démonstration de la conjecture τ', *Invent. Math.* **151** (2003), no. 2, 297–328. https://doi.org/10.1007/s00222-002-0253-8.

M. Einsiedler and T. Ward, *Unitary Representations and Unitary Duals*, Graduate Texts in Mathematics 308, https://doi.org/10.1007/978-3-032-03899-9

17. I. P. Cornfeld, S. V. Fomin, and Y. G. Sinaĭ, *Ergodic theory*, in *Grundlehren der Mathematischen Wissenschaften* **245** (Springer-Verlag, New York, 1982). https://doi.org/10.1007/978-1-4615-6927-5. Translated from the Russian by A. B. Sosinskiĭ.
18. M. Cowling, 'The Kunze–Stein phenomenon', *Ann. Math. (2)* **107** (1978), no. 2, 209–234. https://doi.org/10.2307/1971142.
19. M. Cowling, U. Haagerup, and R. Howe, 'Almost L^2 matrix coefficients', *J. Reine Angew. Math.* **387** (1988), 97–110. https://doi.org/10.1515/crll.1988.387.97.
20. C. W. Curtis and I. Reiner, *Representation theory of finite groups and associative algebras*, in *Pure and Applied Mathematics, Vol. XI* (Interscience Publishers, a division of John Wiley & Sons, New York-London, 1962). https://www.ams.org/books/chel/356/.
21. D. v. Dantzig, 'Topologisch-algebraische verkenning', in *Zeven voordrachten over topologie*, in *Centrumreeks, no. 1. Math. Centrum Amsterdam*, pp. 56–79 (J. Noorduijn en Zoon, Gorinchem, 1950).
22. M. Einsiedler, E. Lindenstrauss, and T. Ward, *Entropy in ergodic theory and topological dynamics* (to appear). https://tbward0.wixsite.com/books/entropy.
23. M. Einsiedler, G. Margulis, and A. Venkatesh, 'Effective equidistribution for closed orbits of semisimple groups on homogeneous spaces', *Invent. Math.* **177** (2009), no. 1, 137–212. https://doi.org/10.1007/s00222-009-0177-7.
24. M. Einsiedler and T. Ward, *Ergodic theory with a view towards number theory*, in *Graduate Texts in Mathematics* **259** (Springer-Verlag London Ltd., London, 2011). https://doi.org/10.1007/978-0-85729-021-2.
25. M. Einsiedler and T. Ward, *Functional analysis, spectral theory, and applications*, in *Graduate Texts in Mathematics* **276** (Springer-Verlag London Ltd., London, 2017). http://dx.doi.org/10.1007/978-3-319-58540-6.
26. M. Einsiedler and T. Ward, *Homogeneous dynamics and applications* (to appear). https://tbward0.wixsite.com/books/homogeneous.
27. L. C. Evans, *Partial differential equations*, in *Graduate Studies in Mathematics* **19** (American Mathematical Society, Providence, RI, second ed., 2010). https://doi.org/10.1090/gsm/019.
28. P. Eymard, 'L'algèbre de Fourier d'un groupe localement compact', *Bull. Soc. Math. France* **92** (1964), 181–236. https://doi.org/10.24033/bsmf.1607.
29. J. M. G. Fell, 'Weak containment and induced representations of groups', *Canad. J. Math.* **14** (1962), 237–268. https://doi.org/10.4153/CJM-1962-016-6.
30. G. B. Folland, *A course in abstract harmonic analysis*, in *Studies in Advanced Mathematics* (CRC Press, Boca Raton, FL, 1995). https://doi.org/10.1201/b19172.
31. L. Fuchs, 'Über eine Klasse von Functionen mehrerer Variablen, welche durch Umkehrung der Integrale von Lösungen der linearen Differentialgleichungen mit rationalen Coefficienten entstehen', *J. Reine Angew. Math.* **89** (1880), 151–169. https://doi.org/10.1515/crll.1880.89.151.
32. W. Fulton and J. Harris, *Representation theory*, in *Graduate Texts in Mathematics* **129** (Springer-Verlag, New York, 1991). https://doi.org/10.1007/978-1-4612-0979-9.
33. H. Furstenberg, 'The unique ergodicity of the horocycle flow', in *Recent advances in topological dynamics (Proc. Conf. Topological Dynamics, Yale Univ., New Haven, Conn., 1972; in honor of Gustav Arnold Hedlund)*, in *Lecture Notes in Math.* **Vol. 318**, pp. 95–115 (Springer, Berlin-New York, 1973).
34. I. Gelfand and M. Naimark, 'On the imbedding of normed rings into the ring of operators in Hilbert space', *Rec. Math. [Mat. Sbornik] N.S.* **12(54)** (1943), 197–213. http://mi.mathnet.ru/eng/msb6155.
35. E. Glasner and B. Weiss, 'Kazhdan's property T and the geometry of the collection of invariant measures', *Geom. Funct. Anal.* **7** (1997), no. 5, 917–935. https://doi.org/10.1007/s000390050030.
36. J. Glimm, 'Type I C^*-algebras', *Ann. of Math. (2)* **73** (1961), 572–612. https://doi.org/10.2307/1970319.

37. R. Godement, 'Sur les relations d'orthogonalité de V. Bargmann. I. Résultats préliminaires; II. Démonstration générale', *C. R. Acad. Sci. Paris* **225** (1947), 521–523. https://gallica.bnf.fr/ark:/12148/bpt6k3177x/f521.
38. R. Godement, 'Sur les relations d'orthogonalité de V. Bargmann. II. Démonstration générale', *C. R. Acad. Sci. Paris* **225** (1947), 657–659. https://gallica.bnf.fr/ark:/12148/bpt6k3177x/f657.
39. A. Gorodnik and A. Nevo, *The ergodic theory of lattice subgroups*, in *Annals of Mathematics Studies* **172** (Princeton University Press, Princeton, NJ, 2010). https://press.princeton.edu/titles/9106.html.
40. K. I. Gross, 'On the evolution of noncommutative harmonic analysis', *Amer. Math. Monthly* **85** (1978), no. 7, 525–548. https://doi.org/10.2307/2320861.
41. Harish-Chandra, 'Spherical functions on a semisimple Lie group. I', *Amer. J. Math.* **80** (1958), 241–310. https://doi.org/10.2307/2372786.
42. T. W. Hungerford, *Algebra*, in *Graduate Texts in Mathematics* **73** (Springer-Verlag, New York-Berlin, 1980). http://dx.doi.org/10.1007/978-1-4612-6101-8. Reprint of the 1974 original.
43. A. Hurevitsch, 'Unitary representation in Hilbert space of a compact topological group', *Rec. Math. [Mat. Sbornik] N. S.* **13(55)** (1943), 79–86. http://mi.mathnet.ru/eng/msb6173.
44. A. Hurwitz, 'Über die Erzeugung der Invarianten durch Integration', *Nachr. Ges. Wiss. Göttingen, Math.-Phys. Kl.* **1897** (1897), 71–90. http://eudml.org/doc/58378.
45. H. Jacquet and R. P. Langlands, *Automorphic forms on* GL(2), in *Lecture Notes in Mathematics, Vol. 114* (Springer-Verlag, Berlin-New York, 1970). https://doi.org/10.1007/BFb0058988.
46. S. Kakutani, 'Über die Metrisation der topologischen Gruppen', *Proc. Imp. Acad.* **12** (1936), no. 4, 82–84. https://doi.org/10.3792/pia/1195580206.
47. R. R. Kallman, 'Certain quotient spaces are countably separated. III', *J. Funct. Anal.* **22** (1976), 225–241. https://doi.org/10.1016/0022-1236(76)90010-0.
48. S. Katok, *Fuchsian groups*, in *Chicago Lectures in Mathematics* (University of Chicago Press, Chicago, IL, 1992). https://press.uchicago.edu/ucp/books/book/chicago/F/bo3621076.html.
49. Y. Katznelson, *An introduction to harmonic analysis*, in *Cambridge Mathematical Library* (Cambridge University Press, Cambridge, third ed., 2004). https://doi.org/10.1017/CBO9781139165372.
50. D. A. Každan, 'On the connection of the dual space of a group with the structure of its closed subgroups', *Funkcional. Anal. i Priložen.* **1** (1967), 71–74. https://doi.org/10.1007/BF01075866.
51. H. H. Kim, 'Functoriality for the exterior square of GL_4 and the symmetric fourth of GL_2', *J. Amer. Math. Soc.* **16** (2003), no. 1, 139–183. http://dx.doi.org/10.1090/S0894-0347-02-00410-1. With Appendix 1 by D. Ramakrishnan and Appendix 2 by Kim and P. Sarnak.
52. A. A. Kirillov, *Elements of the theory of representations*, in *Grundlehren der Mathematischen Wissenschaften, Band 220* (Springer-Verlag, Berlin-New York, 1976). https://doi.org/10.1007/978-3-642-66243-0. Translated from the Russian by Edwin Hewitt.
53. A. Kleppner, 'Measurable homomorphisms of locally compact groups', *Proc. Am. Math. Soc.* **106** (1989), no. 2, 391–395. https://doi.org/10.2307/2048818.
54. A. W. Knapp, *Lie groups beyond an introduction*, in *Progress in Mathematics* **140** (Birkhäuser Boston, Inc., Boston, MA, 1996). http://dx.doi.org/10.1007/978-1-4757-2453-0.
55. A. W. Knapp, *Representation theory of semisimple groups*, in *Princeton Landmarks in Mathematics* (Princeton University Press, Princeton, NJ, 2001). https://doi.org/10.1515/9781400883974. An overview based on examples, Reprint of the 1986 original.
56. B. O. Koopman, 'Hamiltonian systems and transformations in Hilbert space', *Proc. Natl. Acad. Sci. USA* **17** (1931), 315–318. https://doi.org/10.1073/pnas.17.5.315.

57. E. Kowalski, *An introduction to the representation theory of groups*, in *Graduate Studies in Mathematics* **155** (American Mathematical Society, Providence, RI, 2014). https://bookstore.ams.org/gsm-155/.
58. U. Krengel, 'On the speed of convergence in the ergodic theorem', *Monatsh. Math.* **86** (1978/79), no. 1, 3–6. https://doi.org/10.1007/BF01300052.
59. R. A. Kunze, 'Book Review: The theory of unitary group representations', *Bull. Amer. Math. Soc.* **84** (1978), no. 1, 73–75. https://doi.org/10.1090/S0002-9904-1978-14410-3.
60. R. A. Kunze and E. M. Stein, 'Uniformly bounded representations and harmonic analysis of the 2×2 real unimodular group', *Amer. J. Math.* **82** (1960), 1–62. https://doi.org/10.2307/2372876.
61. S. Lang, $SL_2(\mathbf{R})$, in *Graduate Texts in Mathematics* **105** (Springer-Verlag, New York, 1985). https://doi.org/10.1007/978-1-4612-5142-2. Reprint of the 1975 edition.
62. E. Lindenstrauss, 'Invariant measures and arithmetic quantum unique ergodicity', *Ann. of Math. (2)* **163** (2006), no. 1, 165–219. http://dx.doi.org/10.4007/annals.2006.163.165.
63. A. Lubotzky, *Discrete groups, expanding graphs and invariant measures*, in *Modern Birkhäuser Classics* (Birkhäuser Verlag, Basel, 2010). https://doi.org/10.1007/978-3-0346-0332-4. With an appendix by Jonathan D. Rogawski, Reprint of the 1994 edition.
64. G. W. Mackey, 'Induced representations of locally compact groups. I', *Ann. of Math. (2)* **55** (1952), 101–139. https://doi.org/10.2307/1969423.
65. G. W. Mackey, 'Induced representations of locally compact groups. II. The Frobenius reciprocity theorem', *Ann. of Math. (2)* **58** (1953), 193–221. https://doi.org/10.2307/1969786.
66. G. W. Mackey, 'Induced representations of locally compact groups and applications', in *Functional Analysis and Related Fields (Proc. Conf. for M. Stone, Univ. Chicago, Chicago, Ill., 1968)* (Springer, New York, 1970), 132–166. https://doi.org/10.1007/978-3-642-48272-4_6.
67. G. W. Mackey, *The theory of unitary group representations* (University of Chicago Press, Chicago, Ill.-London, 1976). Based on notes by James M. G. Fell and David B. Lowdenslager of lectures given at the University of Chicago, Chicago, Ill., 1955, Chicago Lectures in Mathematics.
68. G. W. Mackey, 'Harmonic analysis as the exploitation of symmetry—a historical survey', *Bull. Amer. Math. Soc. (N.S.)* **3** (1980), no. 1, part 1, 543–698. https://doi.org/10.1090/S0273-0979-1980-14783-7.
69. F. I. Mautner, 'Unitary representations of locally compact groups. I', *Ann. of Math. (2)* **51** (1950), 1–25. https://doi.org/10.2307/1969494.
70. F. I. Mautner, 'Unitary representations of locally compact groups. II', *Ann. of Math. (2)* **52** (1950), 528–556. https://doi.org/10.2307/1969431.
71. J. Milnor, 'Curvatures of left invariant metrics on Lie groups', *Advances in Math.* **21** (1976), no. 3, 293–329. https://doi.org/10.1016/S0001-8708(76)80002-3.
72. D. Montgomery and L. Zippin, *Topological transformation groups* (Interscience Publishers, New York-London, 1955).
73. J. v. Neumann, 'Allgemeine Eigenwerttheorie Hermitescher Funktionaloperatoren.', *Math. Ann.* **102** (1930), 49–131. https://doi.org/10.1007/BF01782338.
74. J. v. Neumann, 'Über einen Satz von Herrn M. H. Stone', *Ann. of Math. (2)* **33** (1932), no. 3, 567–573. https://doi.org/10.2307/1968535.
75. J. v. Neumann, 'Zur Operatorenmethode in der klassischen Mechanik', *Ann. of Math. (2)* **33** (1932), 587–642. https://doi.org/10.2307/1968537.
76. H. Oh, 'Uniform pointwise bounds for matrix coefficients of unitary representations and applications to Kazhdan constants', *Duke Math. J.* **113** (2002), no. 1, 133–192. https://doi.org/10.1215/S0012-7094-02-11314-3.
77. F. Peter and H. Weyl, 'Die Vollständigkeit der primitiven Darstellungen einer geschlossenen kontinuierlichen Gruppe', *Math. Ann.* **97** (1927), no. 1, 737–755. http://dx.doi.org/10.1007/BF01447892.

78. H. Poincaré, 'Théorie des groupes fuchsiens', *Acta Math.* **1** (1882), no. 1, 1–76. https://doi.org/10.1007/BF02391835.
79. Z. Rudnick and P. Sarnak, 'The behaviour of eigenstates of arithmetic hyperbolic manifolds', *Comm. Math. Phys.* **161** (1994), no. 1, 195–213. https://doi.org/10.1007/BF02099418.
80. I. N. Sanov, 'A property of a representation of a free group', *Doklady Akad. Nauk SSSR (N. S.)* **57** (1947), 657–659.
81. P. Sarnak, *Some applications of modular forms*, in *Cambridge Tracts in Mathematics* **99** (Cambridge University Press, Cambridge, 1990). https://doi.org/10.1017/CBO9780511895593.
82. P. Sarnak, 'Recent progress on the quantum unique ergodicity conjecture', *Bull. Amer. Math. Soc. (N.S.)* **48** (2011), no. 2, 211–228. http://dx.doi.org/10.1090/S0273-0979-2011-01323-4.
83. I. Schur, 'Neue Begründung der Theorie der Gruppencharaktere', *Sitzungsberichte der Akad. der Wiss. zu Berlin 1905. Physikalisch-Math. Kl.* (1905), 406–432.
84. I. E. Segal, 'Irreducible representations of operator algebras', *Bull. Amer. Math. Soc.* **53** (1947), 73–88. https://doi.org/10.1090/S0002-9904-1947-08742-5.
85. I. E. Segal, 'An extension of Plancherel's formula to separable unimodular groups', *Ann. of Math. (2)* **52** (1950), 272–292. https://doi.org/10.2307/1969470.
86. A. Selberg, 'On the estimation of Fourier coefficients of modular forms', in *Proc. Sympos. Pure Math., Vol. VIII*, pp. 1–15 (Amer. Math. Soc., Providence, R.I., 1965). http://dx.doi.org/10.1090/pspum/008.
87. K. Soundararajan, 'Quantum unique ergodicity for $\mathrm{SL}_2(\mathbb{Z})\backslash\mathbb{H}$', *Ann. of Math. (2)* **172** (2010), no. 2, 1529–1538. https://doi.org/http://doi.org/10.4007/annals.2010.172.1529.
88. M. H. Stone, 'On one-parameter unitary groups in Hilbert space', *Ann. of Math. (2)* **33** (1932), no. 3, 643–648. https://doi.org/10.2307/1968538.
89. T. Tao, *The Birkhoff–Kakutani theorem* (http://terrytao.wordpress.com/). Accessed:28 April 2013.
90. A. Weil, *Basic number theory*, in *Die Grundlehren der mathematischen Wissenschaften, Band 144* (Springer-Verlag New York, Inc., New York, 1967). https://doi.org/10.1007/978-3-642-61945-8.
91. A. Weil, 'Sur certains groupes d'opérateurs unitaires', *Acta Math.* **111** (1964), 143–211. https://doi.org/10.1007/BF02391012.
92. H. Weyl, 'Theorie der Darstellung kontinuierlicher halbeinfacher Gruppen durch lineare Transformationen. I', *Math. Z.* **23** (1925), 271–309. http://dx.doi.org/10.1007/BF01506234.
93. N. Wiener, 'Tauberian theorems', *Ann. of Math. (2)* **33** (1932), no. 1, 1–100. https://doi.org/10.2307/1968102.
94. M. Zworski, *Semiclassical analysis*, in *Graduate Studies in Mathematics* **138** (American Mathematical Society, Providence, RI, 2012). https://doi.org/10.1090/gsm/138.

Author Index

Index of Notation

General Index

GPSR Compliance

The European Union's (EU) General Product Safety Regulation (GPSR) is a set of rules that requires consumer products to be safe and our obligations to ensure this.

If you have any concerns about our products, you can contact us on ProductSafety@springernature.com

In case Publisher is established outside the EU, the EU authorized representative is:

Springer Nature Customer Service Center GmbH
Europaplatz 3
69115 Heidelberg, Germany

Batch number: 10370734

Printed by Printforce, the Netherlands